In Memory of

Jeanne Sharkey Anderson '43

Dedicated by the

Carleton Alumni Association

Terrestrial Global Productivity

This is a volume in the

PHYSIOLOGICAL ECOLOGY series
Edited by Harold A. Mooney

A complete list of books in this series appears at the end of the volume.

Terrestrial Global Productivity

Edited by

Jacques Roy
Centre d' Ecologie, Fonctionnelle et Evolutive
Centre National de la Recherche Scientifique
Montpellier, France

Bernard Saugier
Ecologie, Systématique et Evolution
Université Paris-Sud
Paris, France

Harold A. Mooney
Department of Biological Sciences
Stanford University
Stanford, California

A Harcourt Science and Technology Company

San Diego San Francisco New York Boston London Sydney Tokyo

This book is printed on acid-free paper. ♾

Academic Press
A Harcourt Science and Technology Company
525 B Street, Suite 1900, San Diego, California 92101-4495, USA
http://www.academicpress.com

Academic Press
Harcourt Place, 32 Jamestown Road, London NW1 7BY, UK
http://www.academicpress.com

Library of Congress Catalog Card Number: 00-111104

International Standard Book Number: 0-12-505290-1

PRINTED IN THE UNITED STATES OF AMERICA
01 02 03 04 05 06 GP 9 8 7 6 5 4 3 2 1

Contents

*Also contributing are the following members of the Global Primary Production Team: Alberte Bondeau, Michael Coughenour, Gérard Dedieu, Tagir Gilmanov, Stith T. Gower, Kathy Hibbard, David W. Kicklighter, William J. Parton, Wilfred M. Post, David Price, and Larry Tieszen.

Contributors

Numbers in parentheses indicate the pages on which the authors' contributions begin.

Göran I. Ågren (83), Department of Ecology and Environmental Research, Swedish University of Agricultural Sciences, S-750 07 Uppsala, Sweden

Jeffrey S. Amthor (9, 33), Oak Ridge National Laboratory, Oak Ridge, Tennessee 37831

Dennis D. Baldocchi (9, 33), Ecosystems Science Division, Department of Environmental Science, University of California, Berkeley, Berkeley, California 94780

Paul Bolstad (245), Department of Forest Resources, University of Minnesota, St. Paul, Minnesota 55108

Philippe Ciais (449), Laboratoire de Modélisation du Climate et de l'Environnement, Gif sur Yvette, France

Wolfgang Cramer (429), Potsdam Institute for Climate Impact Research, D-144 73 Potsdam, Germany

Bert G. Drake (123), Smithsonian Environmental Research Center, Edgewater, Maryland 21037

James R. Ehleringer (345), Department of Biology, University of Utah, Salt Lake City, Utah 84112

Pierre Freidlingstein (449), Goddard Institute for Space Studies, New York, New York

Andrew Friend (449), Institute of Terrestrial Ecology, Edinburgh Research Station, Penicuik Midlothian EH26-0QB, United Kingdom

Jan Goudriaan (301), Department of Theoretical Production Ecology, Agricultural University, 6708 PD Wageningen, The Netherlands

John Grace (401), Institute of Ecology and Resource Management, University of Edinburgh, Edinburgh EH9 3JU, United Kingdom

J. J. Rob Groot (301), Research Institute for Agribiology and Soil Fertility (AB-DLO), 670 AA Wageningen, The Netherlands

Joël Guiot (479), CNRS-IMEP, Faculte de St-Jerôme, F-13397 Marseille, France

David O. Hall† (363), Division of Life Sciences, King's College of London, London W8 7AH, United Kingdom

†Deceased.

Niro Higuchi (401), Instituto Nacional de Pesquisas de Amazonia, CEP 69, 0111-970 Manaus Amazonas, Brazil

Richard A. Houghton (499), Woods Hole Research Center, Woods Hole, Massachusetts 02543

Jo I. House (363), Division of Life Sciences, King's College of London, London W8 7AH, United Kingdom

Bruce Hungate (123), Department of Biological Sciences, Northern Arizona University, Flagstaff, Arizona 86011

Robert B. Jackson (61), Department of Botany and Nicholas School of the Environment, Duke University, Durham, North Carolina 27708

Paul G. Jarvis (211), Institute of Ecology and Resource Management, University of Edinburgh, Edinburgh EH9 3JU, United Kingdom

Richard Joffre (83), Centre d' Ecologie Fonctionnelle et Evolutive, Centre National de la Recherche Scientifique, F-34293 Montpellier Cedex 5, France

Dominique Jolly (479), Global Systems Group, School of Ecology, Lund University, S-223 62 Lund, Sweden

Sven Jonasson (189), Department of Plant Ecology, University of Copenhagen, DK-1353 Copenhagen, Denmark

Fouzia Laarif (479), Dynamic Palaeoclimatology, Lund University, S-22100 Lund, Sweden

Martin J. Lechowicz (61), Department of Biology, McGill University, Montreal, Quebec, Canada H3A 1B1

Susan E. Lee (521), Department of Animal and Plant Sciences, University of Sheffield, Sheffield S20 2TN, United Kingdom

Xia Li (61), Department of Biological Sciences, Stanford University, Stanford, California 94305

Mark R. Lomas (521), Department of Animal and Plant Sciences, University of Sheffield, Sheffield, S20 2TN, United Kingdom

Denis Loustau (123), INRA Unité de Recherches Forestières, F-33611 Gazinet, France

Yadvinder Malhi (401), Institute of Ecology and Resource Management, University of Edinburg, Edinburgh EH9 3JU, United Kingdom

Sam J. McNaughton (101), Biological Research Laboratories, Syracuse University, Syracuse, New York 13244

Patrick Meir (401), Institute of Ecology and Resource Management, University of Edinburgh, Edinburgh EH9 3JU, United Kingdom

Harold A. Mooney (61, 543), Department of Biological Sciences, Stanford University, Stanford, California 94305

Richard J. Olson (429), Oak Ridge National Laboratory, Oak Ridge, Tennessee 37831

Changhui Peng (479), Natural Resources Canada, Canadian Forest Service, Northern Forestry Center, Edmonton, Alberta, Canada T6H 3S5

Colin J. Prentice (479), Global Systems Group, School of Ecology, Lund University, S-223 62 Lund, Sweden

Stephen D. Prince (429), Department of Geography, University of Maryland, College Park, Maryland 20742

Serge Rambal (315), Centre d' Ecologie Fonctionnelle et Evolutive, Centre National de la Recherche Scientifique, F-34293 Montpellier Cedex 5, France

Peter B. Reich (245), Department of Forest Resources, University of Minnesota, St. Paul, Minnesota 55108

Jacques Roy (1, 169, 543), Centre d'Ecologie Fonctionnelle et Evolutive and GDR 1936 DIV-ECO, Centre National de la Recherche Scientifique, F-34293 Montpellier Cedex 5, France

Osvaldo E. Sala (285), Cátedra de Ecología, Falcultad de Agronomía, Universidad de Buenos Aires and IFEVA, Consejo Nacional de Investigaciones Cientifícas y Técnicas, Buenos Aires C1417DSE, Argentina

Bernard Saugier (1, 211, 543) Laboratoire d'Ecologie Vegetale, Batiment 362, Universite de Paris-Sud, 91-405 Orsay Cedex 7, France

David S. Schimel (449), National Center for Atmospheric Research, Boulder, Colorado

E.-Detlef Schulze (211), Max-Planck-Institut für Biogeochemie, D-07 701 Jena, Germany

Jonathan M. O. Scurlock (429), Oak Ridge National Laboratory, Oak Ridge, Tennessee 37831

Gaius R. Shaver (189), The Ecosystems Center, Marine Biological Laboratory, Woods Hole, Massachusetts 02453

Ben Smith (479), Global Systems Group, School of Ecology, Lund University, 223 62 Lund, Sweden

Peter W. J. Uithol (301), Research Institute for Agribiology and Soil Fertility (AB-DLO), 670 AA Wageningen, The Netherlands

F. Ian Woodward (521), Department of Animal and Plant Sciences, University of Sheffield, Sheffield S202TN, United Kingdom

Preface

The Earth climate system and the functioning of the biosphere interact strongly. A change in one affects the other and feedback mechanisms occur. These interactions are now influenced by human activities at an unprecedented scale, both in their rate and in their geographical extent. Global change is now a fact: atmospheric CO_2 concentration has increased by nearly 30% since 1750, the Earth's surface has warmed by 0.6°C during the twentieth century, and there is an increasing pressure for converting natural vegetation to agricultural land and pastures. This trend is not likely to reverse in the next decades, but we have large uncertainties in predicting the magnitudes of these changes. Productivity of terrestrial ecosystems, a major integrative process between living organisms and the atmosphere, is strongly affected by climate and land use and feeds back on the climate mainly through the CO_2 and H_2O cycles.

In the early 1970s the International Biological Programme (IBP) provided a wealth of information on the biological basis of productivity in different environments. However, global changes were not an issue at that time, remote-sensing research was not developed enough to extrapolate local measurements to larger areas, and global productivity models were in their infancy. A lot has happened since the inception of the IBP and it is now time to assess our current state of knowledge and research and to identify gaps in knowledge and critical issues that need to be addressed. In this volume we update and synthesize our current knowledge on global primary productivity, its current rate at the planetary level, and how it varies among biomes, and, where possible, make predictions of future productive capacities according to scenarios of human impacts.

This volume is divided into three sections. First we assess the component processes of primary productivity. Then we examine our current knowledge of the productive capacities of the major biomes of the Earth. Finally we look at global productivity from different perspectives as well as consider how productivity has varied in the past and what we might expect in the future.

We thank O. W. Archibald, professor at the University of Saskatchewan, for kindly supplying us with maps of the world biomes used at the beginning of Chapters 9–17, redrawn from Archibald (1995).

We thank the Conseil de la Région Languedoc-Roussillon, the Centre d'Ecologie Fonctionnelle et Evolutive (CNRS), and the U.S. National Aeronautics and Space Administration for support of this endeavor. We are particularly indebted to Diane Wickland for her considerable help and understanding.

Frode Eckardt

This volume was inspired by the work of Frode Eckardt, who performed a comparable synthesis in 1968. Two of us (Mooney and Saugier) had the privilege of working with Professor Eckardt and observed and participated in his pioneering work and innovative approaches.

Saugier met Frode Eckardt in Paris in 1965 or 1966. Eckardt was looking for a young physicist to work in the new research team he had in Montpellier, in what is now the Centre d'Ecologie Fonctionnelle et Evolutive (CNRS). The job was to measure CO_2 and water vapor fluxes in the air above plant canopies. He was enthusiastic about this venture; Saugier recognized this as the chance of a lifetime. Eckardt's idea was to study plant productivity utilizing various approaches: controlled-environment chambers and micrometeorological techniques at the canopy level in the field, and leaf cuvettes in the field and in the lab to study the response of photosynthesis to environmental factors. He also designed methods to describe canopy architecture (not fully used at the time because of the lack of desk computers) and had André Berger working on plant–water relationships. At that time de Wit had just published his 1965 paper on a model of photosynthesis for plant canopies and Eckardt asked Robert Sauvezon to develop a similar model. Another member of the team, Georges Heim, was in charge of measuring biomass and its energy content. Maurice Méthy dealt with optical properties of leaves and plant canopies. By the spring of 1967 the techniques were in hand. What vegetation to study? Heim introduced the sunflower model and had a crop growing in a plot next to the laboratory. Eckardt thought it was a good idea to test all of our methods on a simple model such as sunflowers before attacking the complexities of mediterranean ecosystems. By late spring we had all of the instruments installed and started to record various parameters. In those days, data were recorded on 12-channel paper recorders. As the study progressed Eckardt was constantly asking: "How are you doing? What did you learn? Have you computed your fluxes?" Those were good times for research. Eventually, this work was published in *Oecologia Plantarum* (Eckardt *et al.*, 1971).

After working on sunflowers, Eckardt went on to study more natural ecosystems: a mature evergreen forest (Eckardt *et al.*, 1975) and a saltmarsh ecosystem (Eckardt, 1972). For the forest he managed to get a fairly complete picture of the carbon balance of the system and its components.

Eckardt often seemed more interested in the methods and in the measurements than in the data. He often had clever ideas and wanted them to be realized right away, which did not stop him from revising a nascent idea devised only a few days earlier. He was dynamic and liked big science and good scientists. He spent 1 year at the California Institute of Technology, working with Frits Went, and he found the work exciting and very different from his own.

Eckardt was an influential member of the French IBP and worked briefly for UNESCO. He organized two UNESCO meetings, one in 1962 in Montpellier on the methods of plant ecophysiology (Eckardt, 1965), and the other in Copenhagen on net primary productivity (Eckardt, 1968). These meetings, and the books that resulted from them, were landmarks in the development of plant ecophysiology.

Mooney took a sabbatical leave in Eckardt's laboratory in 1967. It was his first opportunity to see how integrated plant ecology research could become—the process was unprecedented for the times, with physicists, engineers, and biologists working together as a team. Their program was comprehensive and pathbreaking. Utilizing a battery of approaches, some of which would be considered innovative even today, they were doing pioneering work on scaling from leaf-level to whole-ecosystem measurements. There is no question that Frode Eckardt was one of the founders of modern plant eco-physiology.

We remember Eckardt as a lively person (his laughing was famous), very dedicated to science, with a deep intuition and a great enthusiasm. It was a privilege to have been associated, if even briefly, with this remarkable man and the approaches that he pioneered.

B. SAUGIER
J. ROY
H. A. MOONEY

References

Archibald, O. W. (1995). "Ecology of World Vegetation." Chapman and Hall, Toronto.

de Wit, C. T. (1965). "Photosynthesis of Leaf Canopies." Agricultural Research Report 663, pp. 1–57. Wageningen, The Netherlands.

Eckardt, F. E., ed. (1965). "Methodology of Plant Eco-physiology: Proceedings of the Montpellier Symposium." UNESCO, Paris.

Eckardt, F. E., ed. (1968). "Functioning of Terrestrial Ecosystems at the Primary Production Level." UNESCO, Paris.

Eckardt, F. E. (1972). Dynamique de l'écosystème, stratégie des végétaux et échanges gazeux: Cas des enganes à *Salicornia fruticosa*. *Oecol. Planta.* **7,** 333–345.

Eckardt, F. E., Heim, G., Methy, M., Saugier, B., and Sauvezon, R. (1971). Fonctionnement d'un écosystème au niveau de la production primaire. Mesures effectuées dans une culture d'*Helianthus annuus*. *Oecol. Plant.* **6,** 51–100.

Eckardt, F. E., Heim, G., Methy, M., and Sauvezon, R. (1975). Interception de l'énergie rayonnante, échanges gazeux et croissance dans une forêt méditerranéenne à feuillage persistant (Quercetum ilicis). *Photosynthetica* **9,** 145–156.

1

Terrestrial Primary Productivity: Definitions and Milestones

Jacques Roy and Bernard Saugier

Primary production is a complex set of processes in which chemical or solar energy is converted to produce biomass. By far the main primary producers are the green plants, which convert solar energy, carbon dioxide, and water to glucose and eventually to plant tissue. Primary production is the main force driving the functioning of ecosystems, providing resources for a diversity of consumers, as well as participating in the regulation of the global climate through the carbon and water cycles.

Primary productivity is the rate at which energy is converted into biomass. Units of energy or biomass (per unit of area and per unit of time) can be used as measures of productivity, although the latter is more common and will be used in this book.

Estimations of primary productivity can be obtained through different methods. The oldest one is by destructive measurements of the plant biomass, aboveground and belowground. Frequent samplings are necessary to avoid an underestimation due to the loss of plant material through herbivory or senescence (Long *et al.*, 1989). A major difficulty lies in the recovery of fine roots and in the separation of live and dead roots: too often approximate root:shoot coefficients are used instead of direct measurements of belowground productivity (Nadelhoffer and Raich, 1992; Chapter 4, this volume). This method measures the actual accumulation of biomass after some of the products of photosynthesis are expended for the plant's own maintenance through respiration. This accumulation of biomass per unit of time and per unit of ground surface [g dry matter (DM) m^{-2} yr^{-1}] is called net primary productivity (NPP; the symbol for use in equations is P_n). Foresters also use Mg ha^{-1} yr^{-1} (*i.e.*, t ha^{-1}; 1 Mg ha^{-1} = 100 g m^{-2}). For conversion into carbon, 1 g DM contains on average 0.5 g C for woody tissues and slightly less for herbaceous plants.

More recent methods involve the measurement of carbon dioxide fluxes at the vegetation–atmosphere interface. Enclosures connected to infrared gas analyzers or micrometeorological profiles above the vegetation (CO_2, wind speed, air temperature, and humidity) have been used since the end of the 1960s (*e.g.*, Eckardt *et al.*, 1971), mainly on low-stature vegetation. For forests, enclosures are not practical and the profile method does not work well due to the high surface rugosity. The eddy correlation method is used instead. It requires the measurement at a high frequency (typically 10 times per second) of CO_2 concentration as well as the vertical wind speed. This approach was used in the 1950s to record sensible heat flux (Swinbank, 1951) but was not applied to CO_2 flux until the 1980s due to the lack of a fast-response CO_2 sensor. Both enclosures and micrometeorological methods measure net ecosystem productivity (NEP; symbol for use in equations is P_e) because they include not only the CO_2 fluxes resulting from plant activity, but also the CO_2 fluxes from heterotrophic respiration (mainly soil decomposers) (see Chapters 2 and 3, this volume, for a discussion of these measurements).

Figure 1-1 shows the instantaneous CO_2 fluxes that, once integrated over time, are the components of NEP. A_g is the gross canopy assimilation (net assimilation of the foliage + dark respiration). R_a is the autotrophic respiration, i.e., the sum of the respiration rates (day and night) of the various plant organs, including roots. It is rarely measured because root respiration in particular is quite difficult to separate from soil respiration. A_n is the net vegetation photosynthesis ($A_n = A_g - R_a$). The time-integrated value of A_n is NPP. The CO_2 flux density (F) measured by micrometeorologists above plant canopies is the difference between the canopy assimilation and the respiration of autotrophic and heterotrophic organisms (mainly soil decomposers). The time-integrated value of F is NEP. By convention we choose here F positive when the ecosystem is gaining carbon, which is the opposite of the convention used by micrometeorologists. A_g will be positive or null, and we will take all respiration terms positive, so we may write

$$F = A_g - R_a - R_h.$$

Soil respiration R_s is easily measured and is equal to the sum of root respiration R_r and of heterotrophic respiration R_h. Thus

$$F + R_s = A_g - R_a + R_r.$$

$F + R_s$ may thus be used to compute A_g if some estimate of aboveground respiration R_{ag} ($= R_a - R_r$) is possible.

The available direct measurements of NPP have some limitations for the estimation of the global terrestrial productivity. Their reliability is variable,

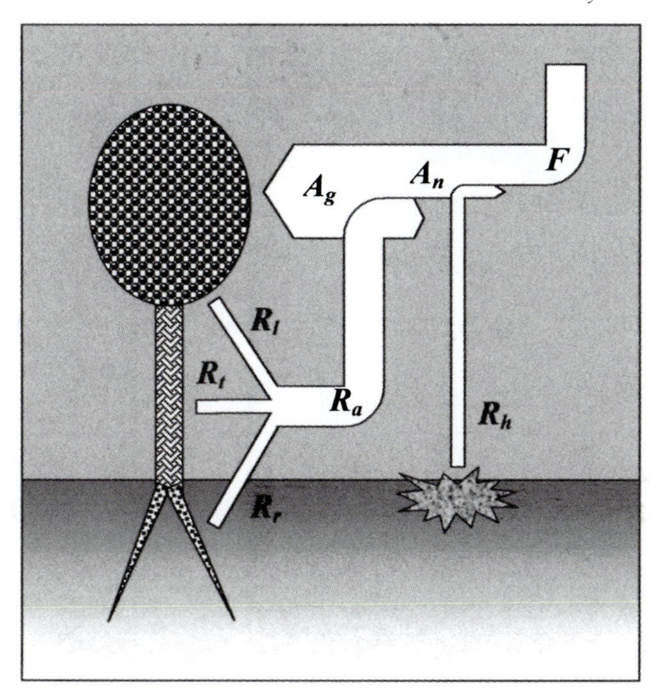

Figure 1-1 Carbon fluxes between ecosystems and the atmosphere. Absorption of CO_2 by photosynthesis (A_g) is more or less compensated for by autotrophic respiration (R_a) from the leaves (R_l), trunks (R_t), and roots (R_r) and by heterotrophic respiration (R_h). The time-integrated value of A_n is the net primary productivity, and the time-integrated value of F is the net ecosystem productivity. With a mature vegetation, $R_h \sim A_n$ and $R_h + R_a \sim A_g$, thus $F \sim O$.

they are limited in number, and they are not evenly distributed among the various types of ecosystems (see Chapter 18, this volume). The estimation of global terrestrial productivity depends also on how well a set of direct measurements is representative of a given biome and takes into account land use changes (see Chapter 21, this volume). There are also uncertainties as to the area to attribute to each biome (see Chapter 23, this volume). To overcome these difficulties, several types of models have been developed. There are either physiological models simulating ecosystem fluxes from environmental variables, remote-sensing-based models interpreting the light spectrum reflected by the land surfaces, or inverse models deducing fluxes from time and space variations in atmospheric CO_2 and ^{13}C data (see Chapter 19, this volume). Process models with physiological functions offer the opportunity to simulate past or future changes in NPP according to environmental changes (see Chapters 20 and 22, this volume).

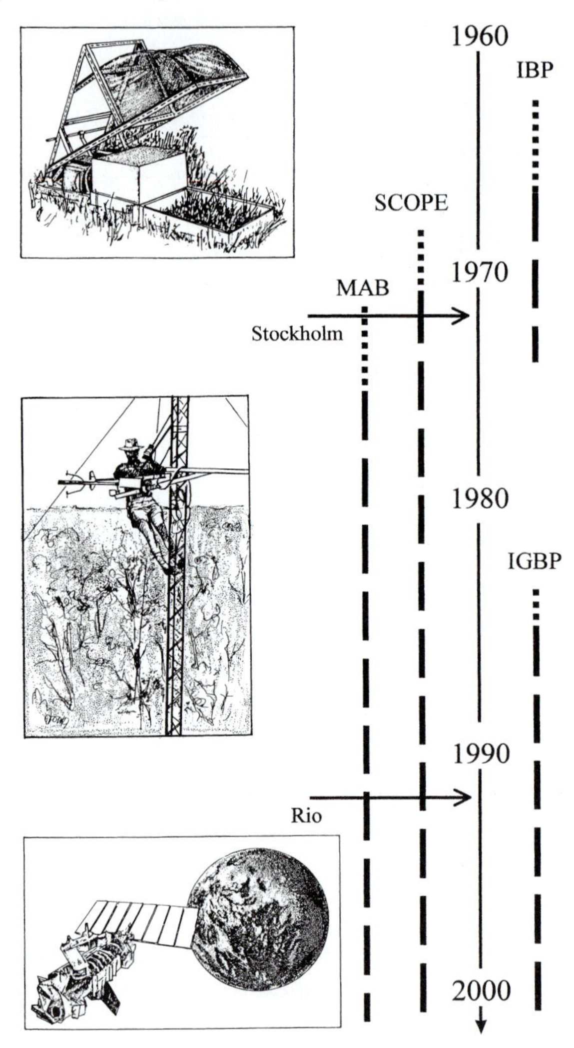

Figure 1-2 Milestones in the study of terrestrial ecosystems (from left to right): the change in scale of measurement techniques; international conferences raising concern for the environment; scientific international organizations shaping environmental research; and international research programs concerned with terrestrial ecosystems. Influential books published over these decades include those by Carson (1962), Rodin and Bazilevich (1967), Eckardt (1968), Sestak *et al.* (1971), Lieth and Whittaker (1975), Lovelock (1979), Ehrlich and Ehrlich (1981), Bolin and Cook (1983), Pearcy *et al.* (1989), Houghton *et al.* (1990), Schulze and Mooney (1993), UNEP (1995), and Walker *et al.* (1999) Drawings by R. Ferris.

The milestones in the investigation of terrestrial productivity correspond to increased capabilities to integrate measurements over larger scales, as indicated in Fig. 1-2. This development is largely the outcome of the efforts of international scientific organizations to stimulate research in this field [e.g., the Scientific Committee on Problems of the Environment (SCOPE) and Man and the Biosphere (MAB)]. Two international programs, the International Biological Program (IBP) and subsequently the International Geosphere–Biosphere Program (IGBP), have been, and are, instrumental in the coordination and direction of research. Two conferences played an influential role in putting environmental problems into the political and public arena: Stockhom in 1972 on humans and the environment and Rio in 1992 on the environment and development. Numerous publications attest to the progress made during the past decade. Some books remain as landmarks because they shed new light on environmental or scientific issues (Carson, 1962; Lovelock, 1979; Ehrlich and Ehrlich, 1981; Schulze and Mooney, 1993) had a large impact on the development of techniques (Eckardt, 1965; Sestak *et al.,* 1971; Pearcy *et al.,* 1989), or synthesized our knowledge (Rodin and Bazilevich, 1967; Lieth and Whittaker, 1975; Bolin and Cook, 1983; Houghton *et al.,* 1990; UNEP, 1995; Walker and Steffen, 1999). Triggered by the realization of the global impacts of human action, the second half of the twentieth century has been characterized by the "globalization of ecological thought" as developed by Mooney (1998). However, a lot of effort remains to be expended to improve our knowledge at all scales, as well as to influence policy based on this knowledge.

References

Bolin, B., and Cook, R. B., eds. (1983). "The Major and Biochemical Cycles and Their Interactions." John Wiley & Sons, Chichester.

Carson, R. L. (1962). "Silent Spring." Houghton Mifflin, Boston.

Eckardt, F. E., ed. (1965). "Methodology of Plant Ecophysiology: Proceedings of the Montpellier Symposium." UNESCO, Paris.

Eckardt, F. E., Heim, G., Méthy, M., Saugier, B., and Sauvezon, R. (1971). "Fonctionnement d'un Écosystème au Niveau de la Production Primaire. Mesures Effectuées dans une Culture d'*Helianthus annuus.*" *Oecol. Planta.* **6,** 51–100.

Ehrlich, P. R., and Ehrlich, A. H. (1981). "Extinction. The Causes and Consequences of the Disappearance of Species." Random House, New York.

Houghton, J. T., Jenkins, G. J., and Ephraums, J. J., eds. (1990). "Climate Change: The IPCC Scientific Assessment." Cambridge Univ. Press, Cambridge.

Lieth, H., and Whittaker, R. H., eds. (1975). "Primary Productivity of the Biosphere." Springer-Verlag, Berlin.

Long, S. P., Garcia Moya, E., Imbamba, S. K., Kamnalrut, A., Piedade, M. T. F., Scurlock, J. M. O., Shen, Y. K., and Hall, D. O. (1989). Primary productivity of natural grass ecosystems of the tropics: A reappraisal. *Plant Soil* **115,** 155–166.

Lovelock, J. E. (1979). "Gaia: A New Look at Life on Earth." Oxford Univ. Press, Oxford.

Mooney, H. A. (1998). "The Globalization of Ecological Thought." Ecology Institute, Oldendorf/Luhe.

Pearcy, R. W., Ehleringer, J., Mooney, H. A., and Rundel, P. W. (1989). "Plant Physiological Ecology. Field Methods and Instrumentation." Chapman and Hall, London.

Nadelhoffer, K. J., and Raich, J. W. (1992). Fine root production estimates and belowground carbon allocation in forest ecosystems. *Ecology* **73,** 1139–1147.

Rodin, L. E., and Bazilevich, N. I. (1967). "Production and Mineral Cycling in Terrestrial Vegetation." Olover and Boyd, Edingburgh.

Schulze, E.-D., and Mooney, H. A., eds. (1993). "Biodiversity and Ecosystem Function." Springer-Verlag, Berlin.

Sestak, Z., Catsky, J., and Jarvis, P. G., eds. (1971). "Plant Photosynthetic Production. Manual of Methods." Dr. W. Junk, The Hague.

Swinbank, W. C. (1951). The measurement of vertical transfer of heat and water vapor by eddies in the lower atmosphere. *J. Meteorol.* **8,** 135–145.

UNEP (1995). "Global Biodiversity Assessment." Cambridge Univ. Press, Cambridge.

Walker, B., and Steffen, W., eds. (1999). "Global Change and Terrestrial Ecosystems." Cambridge Univ. Press, Cambridge.

I

Component Processes

2

Canopy Photosynthesis: History, Measurements, and Models

Dennis D. Baldocchi and Jeffrey S. Amthor

I. Introduction

A plant canopy consists of an assemblage of plants with leaves that possess a particular spatial distribution and assortment of angle orientations (de Wit, 1965; Monsi *et al.*, 1973). How a collection of leaves intercepts sunlight and uses light energy to assimilate CO_2 is the basis of canopy photosynthesis. The major factors affecting canopy photosynthesis, through light interception, include the angular relationship between leaves and Earth–sun geometry and the leaves' vertical and horizontal positions. Other factors affecting canopy photosynthesis include environmental conditions (temperature, wind speed, humidity, and CO_2 concentration), the life history of leaves, the availability of soil moisture and nutrients, stomatal conductance, and specific photosynthetic pathways (Schulze, 1986; Gutschick, 1991; Stenberg *et al.*, 1995).

The complexity of canopy photosynthesis can be illustrated with a closed canopy on a fair day during the growing season. Inside the canopy some leaves are fully sunlit, others are exposed to frequent sunflecks, and the remainder exist in deep, but occasionally punctuated, shade (Chazdon, 1988; Oker-Blom *et al.*, 1991). Many sunlit leaves of C_3 species will experience saturated rates of carboxylation if CO_2 supply is limited (Farquhar and von Caemmerer, 1982). Furthermore, sunlit leaves are often warmer than shaded leaves. This situation promotes dark respiration (Amthor, 1994), decreases the solubility of CO_2 relative to O_2 (Farquhar *et al.*, 1980), and decreases the CO_2 specificity factor of ribulose, bisphosphate, and carboxylase/oxygenase (rubisco) (Harley and Tenhunen, 1991).

The photosynthetic response of leaves to sunflecks does not have the same relationship observed during steady light conditions. As the light exposure

Terrestrial Global Productivity

of a leaf transcends from a dark to bright condition, a dynamic response, known as induction, occurs if the previous dark exposure period was prolonged and it deactivated rubisco. The consequence of this enzymatic deactivation is a delay in the attainment of the next steady-state level of photosynthesis and a reduction of its magnitude. The duration of this delayed response can exceed 20 min, but this time response diminishes if the leaf is exposed to repeated light flecks (Chazdon, 1988; Pearcy, 1990). Postillumination carbon fixation is another important dynamic response experienced by leaves in fluctuating light. When a leaf's exposure to the sun is eclipsed, rates of carbon fixation can be sustained for a spell, as accumulated pools of photosynthetic metabolites are consumed. In fluctuating light environments, the occurrence of postillumination photosynthesis enhances assimilation rates in comparison to rates that would otherwise occur under steady conditions with the same mean level of light exposure (Pearcy, 1990).

The photosynthetic capacity of leaves deep in a canopy adapts to shade (Stenberg *et al.*, 1995). In the case of C_3 plants, less rubisco is allocated to shaded leaves than to sunlit ones (Field, 1991; Hollinger, 1996). This effect causes their photosynthetic rates to saturate at lower light levels (Bjorkman, 1981). It also causes their photosynthetic capacity to be lower than that of leaves at the top of the canopy.

The light climate of needles in conifer stands is even more complex, because the geometric structure of shoots causes needles to cast penumbral and umbral shade on other needles (Oker-Blom *et al.*, 1991). This shoot structure causes their photosynthetic rate, under direct light, to be less than would occur if the needles were displayed in a planar pattern (Oker-Blom *et al.*, 1991; Stenberg *et al.*, 1995). On the other hand, the structure of shoots enhances their ability to harvest diffuse radiation and it allows conifer stands to maintain more leaves compared to angiosperms.

The goal of this chapter is to discuss several aspects of canopy photosynthesis. To accomplish this goal, we will extract information from the disciplines of plant biochemistry, ecophysiology, radiative transfer theory, micrometeorology, and biogeochemistry. First, we give a general history of measuring and modeling canopy photosynthesis. Next, we discuss micrometeorological and ecophysiological concepts that are being adopted to evaluate canopy photosynthesis. Third, we describe some general attributes of canopy photosynthesis that have been derived from field studies. Finally, we close this chapter with a few comments on some avenues of research that we think are worth exploring.

II. History

A. Measurements

The earliest measurements of whole-plant CO_2 exchange occurred during the first half of the twentieth century (Boysen-Jensen, 1918; Henrici, 1921;

Lundegardh, 1922). Pioneering investigators evaluated canopy CO_2 exchange by encasing plants in translucent chambers. Dr. Eckardt, whom the symposium honors, also made early contributions toward chamber design and the measurement of canopy photosynthesis (Eckardt, 1968).

Over the years, closed, semiclosed, and open chamber systems have been used (Jarvis and Catsky, 1971; Field *et al.*, 1989). Closed chamber systems measure photosynthesis as a function of the time rate of change of [CO_2]. This method is subject to bias error if CO_2 in the chamber deviates from the background concentration, because photosynthesis is a function of [CO_2]. One way to circumvent this artifact is to use a semiclosed chamber. This method operates on the null-balance principle. CO_2 in the chamber is kept relatively constant by delivering regulated amounts of CO_2 to compensate for its drawdown as a leaf photosynthesizes, or buildup as it respires. Open systems evaluate carbon assimilation fluxes by measuring concentration differences between air flowing into and exiting the chamber and the volumetric flow rate through the chamber. This method enables a plant in a chamber to be relatively coupled to its external environment.

Overheating, humidification, disturbance of turbulent mixing, and imperfect light transmission are bias errors introduced by chamber methods (Musgrave and Moss, 1961; Tranquillini, 1964; Billings *et al.*, 1966; Denmead *et al.* 1993). When electrical power is ample, air conditioning and dehumidification systems can be employed to minimize these problems (Tibbits and Langhans, 1993).

The accuracy of canopy photosynthesis measurements is also a function of the physicochemical techniques used to measure [CO_2] (see Sestak *et al.* 1971; Jarvis and Catsky, 1971). The simplest analytical method involves colorimetric detection of CO_2 by a dry absorbent (typical sensitivity is 50 μl liter^{-1}). Wet analytical methods are more sensitive (0.5 to 1 μl liter^{-1}) and accurate (±2–7%). Typically, air samples are bubbled through a dilute solution of alkali hydroxide (KOH, NaOH). The amount of CO_2 absorbed by the solution is detected by electrical conductivity or titration measurements. Errors associated with chemical titration and conductivity analytical methods include sensitivity to temperature and difficulty maintaining the dilution of the alkali solution low enough to absorb all of the CO_2 (Sestak *et al.*, 1971). Wet analytical methods are also labor and time intensive, taking 5–10 min to make each measurement.

The invention of the nondispersive infrared absorption spectrometer (or IRGA), in the 1950s, led to a renaissance of photosynthesis measurements. Infrared absorption spectrometers are sensitive (0.5 μl liter^{-1}), accurate, and can be operated continuously and automatically. Musgrave and Moss (1961), Tranquillini (1964), and Billings *et al.* (1966) were among the first workers to employ IRGAs to measure whole-plant photosynthesis with chambers. At the same time, micrometeorologists started using IRGAs to measure CO_2 fluxes across the canopy–atmosphere interface. Inoue (1957), Lemon

(1960), and Monteith and Szeicz (1960) conducted some of the earliest micrometeorological studies on canopy CO_2 exchange.

In comparison to chambers, micrometeorological methods have several appealing features. They are *in situ,* they do not disturb the vegetation, and they sample a relatively large area. They can also be applied over forests, which are cumbersome to enshroud. Nevertheless, micrometeorological methods have drawbacks, especially the application of flux-gradient theory over forests. Besides the need to erect tall towers, eddy exchange coefficients do not conform to similarity theory in the roughness sublayer (Raupach and Legg, 1984). It is also difficult to measure vertical concentration gradients that resolve physiologically meaningful fluxes over well-mixed forests.

Advances in the development of fast-responding IRGAs, sonic anemometers, and digital software are making some criticisms of micrometeorological methods *passe.* Today, the eddy covariance method is being employed to measure canopy-scale CO_2 fluxes directly and for extended periods (Wofsy *et al.,* 1993; Greco and Baldocchi, 1996). A major unresolved problem with this technology involves evaluating CO_2 fluxes at night, when the atmosphere is stable and turbulent transfer is intermittent.

At this writing, only a handful of comparison studies between chambers and micrometeorological methods exist (Held *et al.,* 1990; Denmead *et al.,* 1993; Dugas *et al.,* 1997). On one hand, the imperfect transmission and diffusion of light through chamber walls can cause assimilation rates of enclosed trees to outperform trees outside a chamber (Denmead *et al.,* 1993). On the other hand, warmer soil within a chamber system can cause it to measure smaller rates of CO_2 uptake compared to a micrometeorological method (Dugas *et al.,* 1997).

In practice, chambers or micrometeorological methods rarely measure canopy photosynthesis, specifically. Rather, leaf-scale measures of CO_2 exchange (F_l) include both photosynthesis and dark respiration (R_d). Similarly, measurements of CO_2 exchange (F_c) made at the biosphere–atmosphere interface include contributions from photosynthesis, plant respiration, and heterotrophic respiration. Although gross canopy photosynthesis is difficult to attain, an estimate of net canopy photosynthesis (P_n) can be acquired. This involves simultaneous measurements of CO_2 exchange at the plant–atmosphere interface and over the soil, plus a measure of the storage of CO_2 in the air between the soil and flux measurement height (Baldocchi *et al.,* 1987; Ruimy *et al.,* 1995). Groups led by Monteith (Monteith *et al.,* 1964; Biscoe *et al.,* 1975), Baumgartner (1969), Denmead (1976), and Rosenberg (Brown and Rosenberg, 1971; Verma and Rosenberg, 1976) pioneered the approach of measuring P_n by this difference approach.

B. Models

Saeki (1960), de Wit (1965), and Duncan *et al.* (1967) are credited with some of the earliest models of canopy photosynthesis. Their pioneering efforts stemmed from the development of algorithms to compute leaf photosynthesis and the transfer of solar radiation through plant canopies (Monsi and Saeki, 1953; de Wit, 1965).

The earliest canopy photosynthesis models assumed solar radiation was the only independent variable and that the canopy was a turbid medium. Later models considered the distinct geometric configuration of the canopy, as radiative transfer models became more sophisticated, or focused on microenvironmental variables that controlled mass and heat transfer and stomatal conductance (Cowan, 1968; Waggoner *et al.*, 1969). Over the years, canopy photosynthesis models have been adapted that consider the distinct geometry of row crops (Fukai and Loomis, 1976), orchards (Cohen and Fuchs, 1987), and conifer trees (Wang and Jarvis, 1990) and the clumping of foliage within the volume of a canopy (Gutschick, 1991; Baldocchi and Harley, 1995).

III. Current Theoretical Concepts

A. Leaf Photosynthesis and Stomatal Conductance

Most terrestrial plants accomplish photosynthesis using the C_3 biochemical pathway. Alternative means for performing photosynthesis involve the C_4 and CAM pathways. The C_4 photosynthesis pathway is more efficient than the C_3 pathway because leaf anatomy and metabolism concentrate CO_2 within chloroplasts, a mechanism that inhibits photorespiration (Ehleringer and Monson, 1993). Ecologically, C_4 photosynthesis tends to be associated with tropical and subtropical grassland ecosystems. Spatially, the C_4 pathway may account for 10–25% of global photosynthesis (Lloyd and Farquhar, 1996).

Modern physiological CO_2 exchange models link calculations of photosynthesis, respiration, stomatal conductance, and transpiration to one another. At present, Farquhar's photosynthesis model (Farquhar *et al.*, 1980) is used in many models for C_3 leaves (Collatz *et al.*, 1991; Nikolov *et al.*, 1995; Su *et al.*, 1996). For C_4 leaves, the physiological model of Collatz *et al.* (1992) is a reliable candidate.

Farquhar's photosynthesis model is gaining wide acceptance because it is based on sound biochemical principles. This feature allows it to predict effects of light, temperature, CO_2, and rubisco amount on photosynthesis. Moreover, a huge database of model parameters is available (Wullschleger, 1993), allowing the Farquhar model to be used widely.

Farquhar *et al.* (1980) evaluate leaf CO_2 assimilation A_l as the balance of photosynthesis, photorespiration, and respiration:

$$A_l = V_c - 0.5V_o - R_d, \tag{1}$$

where V_c is the rate of carboxylation, V_o is the rate of oxygenation, and R_d is the rate of dark respiration. The coeffcient, 0.5, implies that two oxygenations lead to one photorespiratory decarboxylation. The term, $V_c - 0.5V_o$, is evaluated as

$$V_c - 0.5V_o = \min[W_c, W_j](1 - \Gamma_*/C_i), \tag{2}$$

where W_c is the rate of carboxylation when ribulose bisphosphate (RuBP) is saturated, W_j is the carboxylation rate when RuBP regeneration is limited by electron transport, Γ_* is the CO_2 compensation point in the absence of dark respiration, and C_i is the intercellular CO_2 concentration (Farquhar and von Caemmerer, 1982).

The value of W_c is calculated from

$$W_c = \frac{V_{c\,max}(C_i - \Gamma)}{C_i + K_C(1 + [O_2]/K_O)}, \tag{3}$$

where $V_{c\,max}$ is the maximum carboxylation rate when RuBP is saturated and K_O and K_C are the Michaelis–Menten coefficients for O_2 and CO_2. Here W_j is defined as

$$W_j = \frac{J(C_i - \Gamma)}{4C_i + 8\Gamma}, \tag{4}$$

where J is the potential rate of electron transport; J is evaluated as a function of incident photosynthetic photon flux density (I), the quantum yield (α), and the maximum rate of electron transport (J_{max}):

$$J = \frac{\alpha I}{\sqrt{1 + \alpha^2 I^2 / J_{max}}}. \tag{5}$$

The evaluation of some photosynthetic model parameters merits further comment. Several photosynthetic model parameters (e.g., $V_{c\,max}$) depend on leaf nitrogen because rubisco is a nitrogen-rich compound (Field, 1991; Wullschleger, 1993). Furthermore, maximum rates of carboxylation ($V_{c\,max}$) are strongly correlated with the maximum rate of electron transport (J_{max}) (Wullschleger, 1993) and rates of dark respiration are constrained by rates of photosynthesis (Gifford, 1994). Model coefficients describing kinetic reactions and photosynthetic capacity have strong nonlinear dependencies on temperature (Thornley and Johnson, 1990; Harley and Tenhunen, 1991). Therefore, the application of photosynthesis models, in the field, requires an evaluation of the leaf energy balance.

The supply of CO_2 to the intercellular spaces is regulated by diffusion through the leaf boundary layer and stomata. Contemporary models evaluate stomatal conductance (g_s), for well-watered plants, as an empirical function of leaf photosynthesis, humidity, and [CO2] (Collatz *et al.*, 1991). Typically, the proportionality constant between stomatal conductance and photosynthesis is a constrained parameter. Limited soil moisture availability and limited hydraulic conductivity are among the factors causing the stomatal conductance factor to deviate below its cardinal value (Sala and Tenhunen, 1996; Falge *et al.*, 1996; Ryan and Yoder, 1997).

B. Scaling or Integrating Carbon Dioxide Fluxes from Leaf to Canopy Dimensions

By definition, canopy photosynthesis (A_c) is equal to the integrated sum of photosynthesis by leaves throughout the canopy volume. Three classes of models dominate the field of canopy photosynthesis (see Jarvis, 1993). Two of them treat the canopy as a layer of vegetation overlying the soil, and are denoted as "big-leaf" models. The third model class divides a canopy into multiple layers, shells, or cubes and simulates the impact of spatial gradients of microclimatic variables on the system of equations defining leaf photosynthesis.

The simplest big-leaf photosynthesis model is a scaling model. One version stems from Monteith (1972), and evaluates canopy photosynthesis as a function of the canopy's photosynthetic efficiency (ϵ), the canopy's light absorption efficiency (f), and the amount of incident solar radiation (S_0):

$$A_c = \epsilon f S_0. \tag{6}$$

Another version of the big-leaf scaling model considers a dependence on physiological quantities, such as rubisco kinetic parameters (Amthor *et al.* 1994; Lloyd *et al.*, 1995; Aber *et al.*, 1996). The appeal of big-leaf scaling models is their dependence on a limited number of variables that have a linear dependence on one another. They can also be evaluated using data observed from satellites to yield estimates of global net primary productivity (Ruimy *et al.*, 1994). In practice the parameters of big-leaf scaling models do not relate to measurable physiological or physical quantities, so they must be tuned (Lloyd *et al.*, 1995).

Integrating leaf photosynthetic rates with respect to leaf area index (L) derives the second type of a big-leaf model. Examples of this approach have been used by Saeki (1960), Thornley and Johnson (1990), and Sellers *et al.* (1996). Derivation of A_c depend on the adoption of a functional form for A_l that can be integrated. A common approach is to model A_l as a function of sunlight, using the rectangular hyperbola formula, and assume that maximum photosynthesis, A_m, is constant with depth (Thornley and Johnson, 1990). This approach yields

$$A_c = \frac{A_m}{k} \ln\left[\frac{\alpha k I_0 + A_m(1 - m)}{\alpha k I_0 \exp(-kL) + A_m(1 - m)}\right], \tag{7}$$

where I_0 is the incident photon flux density, α is the photosynthetic efficiency, m is the leaf transmission coefficient, and k is the radiation extinction coefficient.

When one assumes that A_m varies with depth in the canopy at the same rate that the mean light environment is attenuated, Eq. (7) simplifies to

$$A_c = A_l(I_0)\frac{(1 - \exp(-kL))}{k}. \tag{8}$$

The advantage of an integrated, big-leaf photosynthesis model is its ability to be parameterized in terms of leaf-level photosynthetic measurements (Sellers *et al.*, 1996). On the other hand, compromising assumptions must be made sometimes to allow a system of equations to be integrated analytically. For example, a 5–10% overestimate of canopy photosynthesis can occur when leaf nitrogen is not distributed optimally throughout the canopy (Hollinger, 1996; de Pury and Farquhar, 1997). Big-leaf integration models also rely on flux-gradient, K-theory, which is violated within plant canopies because countergradient transfer occurs within the volume of vegetation (Raupach, 1988).

Multilayer or shell models can deal with the geometric properties of open vegetation (Wang and Jarvis, 1990; Myneni, 1991) and are able to consider the impacts of nonlinear physiological and physical processes on canopy photosynthesis (Jarvis, 1993; Stenberg *et al.*, 1995). Multilayer models can also be coupled with micrometeorological theories, such as higher order closure (Su *et al.*, 1996) and Lagrangian (Baldocchi and Harley, 1995) models, that evaluate turbulent diffusion and accommodate countergradient transfer. The current state-of-the-art computing of canopy photosynthesis couples modules associated with (1) turbulent diffusion, (2) radiative transfer, (3) C_3 (or C_4) leaf photosynthesis, (4) stomatal conductance, (5) leaf energy balance, and (6) soil and stem respiration. A canopy micrometeorology model is needed to assess the light, temperature, humidity, CO_2, and wind environment within and above vegetation, which drive physiological functions. Examples of coupled, multilayer canopy photosynthesis models include those of Norman (1979), Norman and Polley (1989), Wang and Jarvis (1990), Gutschick (1991), Baldocchi and Harley (1995), Leuning *et al.* (1995), Su *et al.* (1996), and Williams *et al.* (1996).

Special considerations are needed to integrate photosynthesis rates of conifer needles to the canopy scale. These issues involve the effects of shoot structure, penumbra, and age stratification. Because these topics are beyond the scope of this chapter, we refer the reader to Jarvis (1993) and Stenberg *et al.* (1995).

C. Model Validation

Although numerous models on canopy photosynthesis exist, relatively few have been tested with field data. One constraint has been the lack of field data. Consequently, the earliest tests of canopy photosynthesis models were performed against fewer than 30 data points (Sinclair *et al.*, 1976; Norman and Polley, 1989; Baldocchi, 1993). In retrospect, these tests yielded inconclusive results because atmospheric variability caused fluxes to possess rather large sampling errors for statistical confidence tests. Nor did small data sets encompass a wide enough range of conditions to put a model through its paces. Within the past 5 years, data sets with over 200 data points have become available (Kim and Verma, 1991; Amthor *et al.*, 1994; Baldocchi and Harley, 1995), and now, with seasonal and annual scale studies underway (Goulden *et al.*, 1996; Greco and Baldocchi, 1996; Valentini *et al.*, 1996), data sets one to two orders of magnitude larger are available for model testing.

Independent tests show that big-leaf models can yield accurate (within $\pm 20\%$) assessments of canopy CO_2 exchange rates (Sinclair *et al.*, 1976; Amthor *et al.*, 1994; Lloyd *et al.*, 1995; Aber *et al.*, 1996; Frolking *et al.*, 1996; dePury and Farquhar, 1997). Favorable model results stem from several facts. First, many accurate big-leaf models are an artifact of tuning (e.g., Lloyd *et al.*, 1995). Second, canopy photosynthesis is relatively insensitive to gradients in CO_2 concentration and turbulent mixing (Baldocchi, 1993). Third, marked improvements in the performance of big-leaf models are obtained by simply partitioning the canopy into two light classes, the sunlit and the shaded fractions (Sinclair *et al.*, 1976; dePury and Farquhar, 1997).

Tests of multilayer canopy photosynthesis models have yielded favorable results, too (Baldocchi and Harley, 1995; Williams *et al.*, 1996; Baldocchi and Meyers, 1998; dePury and Farquhar, 1997). They are able to mimic the mean diurnal patterns and magnitudes of CO_2 exchange over a diverse range of vegetation classes (crops and broadleaved and conifer forests) during the growing season, when soil moisture is ample. To get the favorable results over forests, however, the models must consider the effect of clumped foliage on radiative transfer through forest stands and the effects of age and nutrition on reducing the ability of trees to conduct water (see Ryan and Yoder, 1997).

A present challenge is to apply canopy photosynthesis models to the task of computing seasonal patterns of canopy CO_2 exchange. Provided enough information on soil moisture and phenology is available, initial studies are yielding encouraging results (Aber *et al.*, 1996; Frolking *et al.*, 1996).

IV. Processes: Response of Canopy Photosynthesis to External Forcings

A. Sunlight

Although leaf photosynthesis is a hyperbolic function of sunlight, many micrometeorological studies shows that canopy CO_2 exchange rates (F_c) are

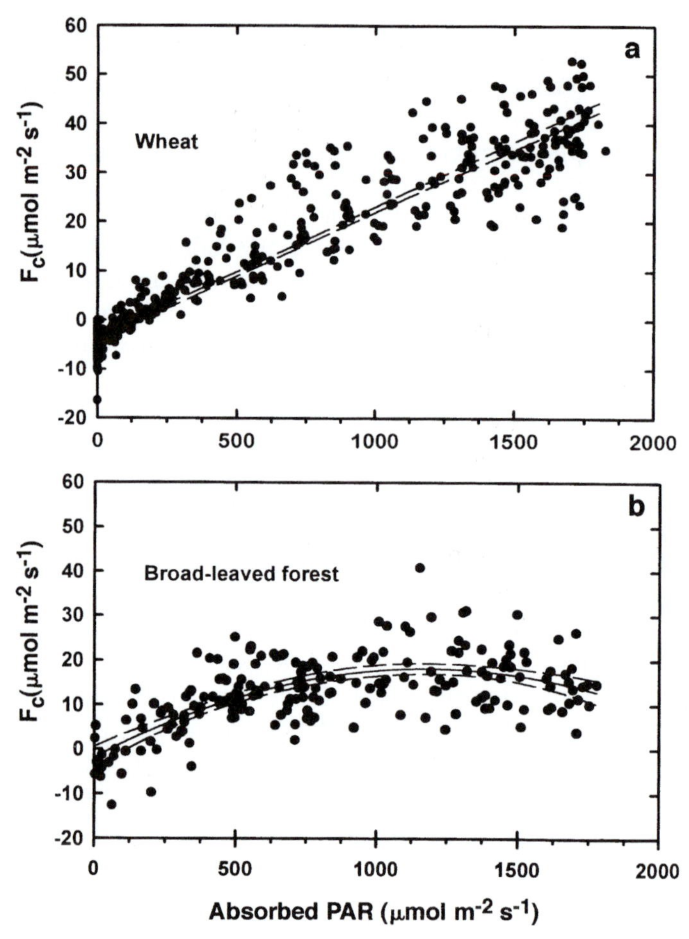

Figure 2-1 The relationship between canopy scale measurements of CO_2 flux density and available photosynthetic photon flux (Q_a). Carbon fluxes were measured with the eddy covariance method. (a) Wheat: variations in Q_a accounted for 91% of the variance in F_c (after Baldocchi, 1994). (b) Temperate broadleaved forest: variations in Q_a accounted for 42% of the variance in F_c (after Baldocchi, 1997).

a quasilinear function of absorbed sunlight (Baldocchi, 1994; Ruimy *et al.*, 1995; Rochette *et al.*, 1996). This quasilinear response tends to be associated with closed, well-watered crop canopies (Fig. 2-1a) or with data averaged over the course of a day (Leuning *et al.*, 1995). Forests and sparse vegetation, on the other hand, experience a markedly nonlinear response between F_c and absorbed sunlight (Fig. 2-1b) (Hollinger *et al.*, 1994; Fan *et al.*, 1995; Valentini *et al.*, 1996).

The linear and nonlinear dependencies of canopy photosynthesis on absorbed sunlight can be explained, in part, by an interaction between leaf area index (L) and carboxylation capacity, $V_{c\ max}$. Theoretically, canopy photosynthesis (P_c) of vegetation with a high carboxylation capacity ($V_{c\ max}$ = 100 μmol m^{-2} s^{-1}) becomes saturated at moderate photon flux densities, when a canopy is sparse ($L = 1$). In contrast, the theoretical response of P_c to light is quasilinear when the canopy is closed ($L = 5$) (Fig. 2-2). Model calculations also reveal that the photosynthesis–light response of a closed crop canopy ($L = 5$) changes from a saturated response to a more linear one as $V_{c\ max}$ increases from 25 to 100 μmol m^{-2} s^{-1} (Fig. 2-3).

The slope of the relationship between canopy CO_2 exchange rates and available sunlight (Q_a) is affected by whether the sky is clear or cloudy. Often, F_c can double when sky conditions change from clear to cloudy and Q_p remains the same (Hollinger *et al.*, 1994; Fan *et al.*, 1995; Rochette *et al.*, 1996). Photosynthetic rates can be lower on clear days at a given Q_p because sunlit leaves can be light saturated. They also experience a higher heat load, compared to leaves exposed to diffuse radiation, which enhances respiration of the former (Baldocchi and Harley, 1995; Rochette *et al.*, 1996).

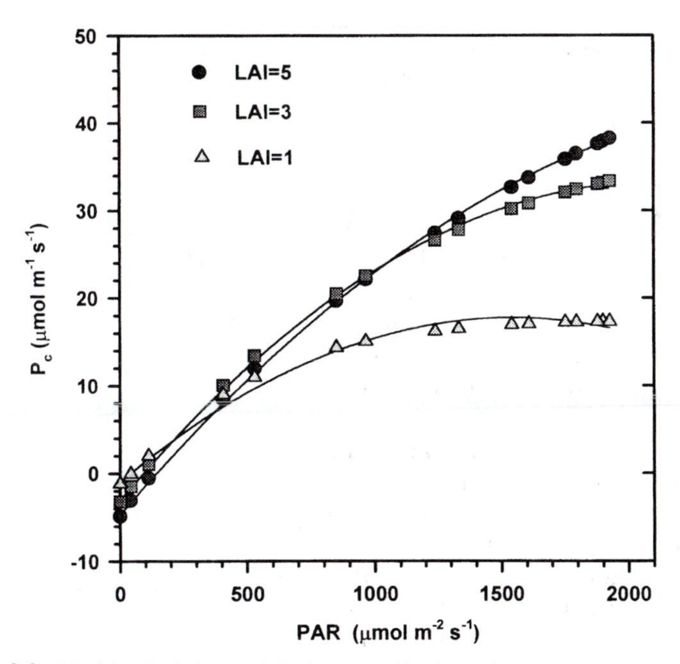

Figure 2-2 Model calculations on the impact of leaf area index (LAI) on canopy photosynthesis of a highly productive canopy ($V_{c\ max}$ = 100 μmol m^{-2} s^{-1}). Calculations were generated by a coupled canopy photosynthesis/micrometeorology model (CANVEG) (Baldocchi and Meyers, 1998).

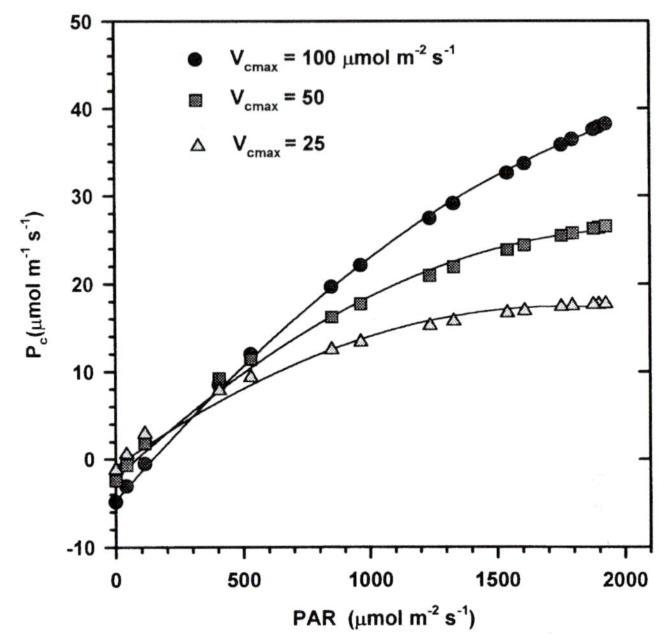

Figure 2-3 Model calculations on the impact of maximum carboxylation velocity on canopy photosynthesis of a closed canopy (LAI = 5). Calculations were generated by a coupled canopy photosynthesis/micrometeorology model (CANVEG) (Baldocchi and Meyers, 1998).

B. Leaf Architecture

Theoretical calculations predict that photosynthetic rates of canopies with erect leaves, and high leaf area indices, are less inclined to light-saturate. Consequently, canopies with erect leaves can achieve photosynthetic rates that are 70–100% greater than those whose leaves are arrayed horizontally (de Wit, 1965; Duncan *et al.*, 1967; Monsi *et al.*, 1973; Wang *et al.*, 1992). Conversely, field data from subalpine communities show that canopy photosynthesis was greatest over tall herb and dwarf shrubs that concentrated flat leaves in the upper layers (Tappeiner and Cernusca, 1996).

The spatial pattern of plant stands and leaves also affects canopy photosynthesis. Crowns that shade 100% of the ground attain canopy photosynthetic rates that are almost double those that shade only 25% of the ground (Wang *et al.*, 1992). Clumping of leaves within a crown enhances the probability of beam penetration through canopies and increases rates of canopy photosynthesis as compared to a canopy with leaves that have a random spatial distribution and spherical angle distribution (Gutschick, 1991; Baldocchi and Harley, 1995).

C. Wind

Some studies report a positive correlation between CO_2 fluxes and wind speed, whereas others indicate no significant relationship between these two variables. Lemon (1960) and Uchijima (1976), for example, conclude that photosynthesis is limited on sunny days when wind speeds are low because a lack of turbulence limits the CO_2 supply to the crop. Yet, this concept is suspect because those fluxes were derived from the aerodynamic method. Data from a method independent of wind speed (Brown and Rosenberg, 1971) and theoretical calculations (Baldocchi, 1993) suggest that turbulent mixing supplies adequate amounts of CO_2 to a crop during the day. If wind does affect canopy photosynthesis, it may stem from the impact that movement of the canopy (*honami*) has on the distribution of light through the canopy (Baldocchi *et al.*, 1981) or its leaf energy balance and respiration.

D. Temperature

The response of canopy CO_2 exchange rates to temperature is parabolic. The temperature optimum of canopy CO_2 exchange rates of many crops and forests growing in temperate continental climates, under full sunlight, is on the order of 20–30°C (Jeffers and Shibles, 1969; Baldocchi, 1997; Price and Black, 1990). The temperature optimum, however, is very plastic and can vary with species, ecotype, site, and time of year (Stenberg *et al.*, 1995). Hollinger *et al.* (1994), for example, reported a 5°C shift in the temperature optimum of a *Nothofagus* (beech) forest between spring and summer.

In general, leaf photosynthesis decreases markedly at leaf temperatures exceeding 37°C. This diminution occurs from a decrease in membrane stability, a decrease in the relative solubility of CO_2 as compared to O_2, a decrease in the specificity factor of rubisco, an exponential increase in dark respiration rates, and an accumulation of carbohydrates (Harley and Tenhunen, 1991). Only temperatures exceeding 40–50°C cause damage to photosynthetic machinery (Bjorkman, 1981).

The zero crossing for canopy CO_2 exchange occurs in the range between 30 and 35°C (Jeffers and Shibles, 1969; Baldocchi, 1997). This threshold is too low to be the sole artifact of photosynthetic kinetics and solubility (Bjorkman, 1981). Elevated soil respiration, which increases exponentially with soil temperature, normally explains the premature decrease of canopy photosynthesis with temperature.

Freezing is another temperature-related phenomenon that affects canopy photosynthesis. Several field studies on evergreen conifers (Tranquillini, 1964; Jarvis *et al.*, 1976; Stenberg *et al.*, 1995) and corn (McGuinn and King, 1990) report that freezing reduces subsequent photosynthetic capacity appreciably. Low soil temperatures can also reduce photosynthesis through effects on water balance and stomatal conductance (Stenberg *et al.*, 1995).

E. Vapor Pressure Deficits

High vapor pressure deficits (D) can limit CO_2 uptake rates over a variety of forests (Jarvis *et al.*, 1976; Price and Black, 1990; Fan *et al.*, 1995), crops (Pettigrew *et al.*, 1990), and savanna shrublands (Verhoef *et al.*, 1996). On the other hand, the influence of D and temperature on CO_2 exchange of temperate broadleaved forests is difficult to distinguish because the two variables are correlated (Verma *et al.*, 1986; Hollinger *et al.*, 1994). Within aerodynamically smooth vegetation, the air remains humid near actively transpiring leaves. Consequently, the atmosphere's vapor pressure deficit (vpd) will have a limited effect on canopy conductance and photosynthesis (Grantz and Meinzer, 1991).

F. Leaf Nitrogen

In the past decade, many investigators have reported that photosynthetic capacity varies with depth in a canopy (Field, 1991; Hollinger, 1996). Theoreticians suggest that plants either distribute leaf nitrogen optimally through the canopy (Leuning *et al.*, 1995; Sands, 1995; Hollinger, 1996; dePury and Farquhar, 1997) or they coordinate the vertical distribution of N to maintain a balance between W_c and W_j (Chen *et al.*, 1993). In either case, the photosynthetic rates of plant stands that distribute leaf nitrogen with depth exceed those of canopies that distribute N uniformly.

G. Water Relations

Soil moisture deficits impact the magnitude and the diurnal course of canopy-scale CO_2 exchange rates (Biscoe *et al.*, 1975; Kim and Verma, 1991; Olioso *et al.*, 1996; Verhoef *et al.*, 1996; Baldocchi, 1997). When plants are exposed to cool and humid air and adequate soil moisture, the diurnal pattern of canopy CO_2 exchange rates is single peaked and the maximum occurs near midday. For plants suffering from modest soil moisture deficits, peak rates of CO_2 uptake occur in the morning. Double-peaked patterns of daily photosynthesis tend to occur when the air is hot ($T_{air} > 30°C$) and dry (vpd > 3 kPa) or when leaf temperature exceeds the optimum for photosynthesis (Olioso *et al.*, 1996; Verhoef *et al.*, 1996). These conditions cause midday stomatal closure, promote dark and photorespiration, and suppress photosynthesis (Schulze, 1986).

Typically, reductions in canopy photosynthesis during periods of soil moisture defecits are associated with stomatal closure. However, it must be remembered that drought and high-temperature stress often co-occur. Hence, enhanced respiration, during these periods, will also limit canopy CO_2 uptake rates (Baldocchi, 1997).

H. Season

Photosynthetic rates of plant systems vary over the course of the growing season as photosynthetic capacity, the availability of solar radiation and soil

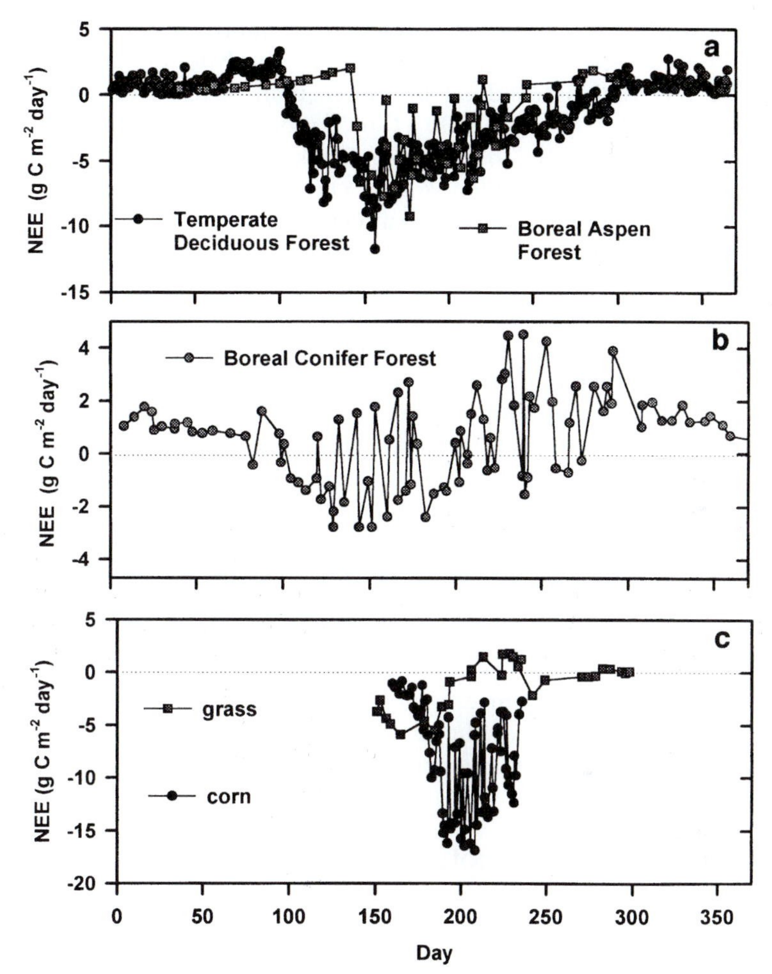

Figure 2-4 Seasonal variations in the daily sum of the net atmosphere/ecosystem CO_2 exchange (NEE). Negative values denote a net loss of carbon from the atmosphere (but a gain by the biosphere). (a) A temperate (Greco and Baldocchi, 1996) and boreal, broadleaved, deciduous, forest (Black *et al.* 1996). (b) Boreal conifer forest (Grelle, 1997). (c) Temperate grass (Kim and Verma, 1991) and corn crop (Desjardins, 1985).

moisture, and air and soil temperature vary (Thomas and Hill, 1949; Tranquillini, 1964; Monteith *et al.*, 1964). Leaf form (needle, broadleaf), leaf habit (evergreen, deciduous), and latitude also affect the seasonal pattern of CO_2 exchange. Figure 2-4a shows that broadleaved forest canopies lose carbon when the canopy is dormant. During springtime leaf expansion the direction and magnitude of carbon fluxes switch rapidly as forests change from losing 1–3 g C m^{-2} d^{-1} to gaining 5–10 g C m^{-2} d^{-1}. A marked influence in latitude is noted by the observation that the onset of carbon uptake by a boreal aspen forest lags behind a southerly temperate forest by a month. Because of this latitudinal difference, temperate broadleaved forests have the potential to gain 100–200 g C m^{-2} more per year compared to boreal forests (Greco and Baldocchi, 1996; Black *et al.*, 1996) and more northern temperate forests (Goulden *et al.*, 1996). Year-to-year variations in the length of the growing season can also affect the net carbon uptake of broadleaved and conifer forests by this magnitude (Goulden *et al.*, 1996; Frolking *et al.*, 1996).

Although conifers also lose carbon during the winter, they can achieve carbon gain in the spring before deciduous forests are able to do so (Fig. 2-4b). On the other hand, net carbon uptake by conifer forests is very sensitive to soil moisture and temperature. Adverse summer heat and dry spells or days with low light levels can cause them to lose carbon during the heart of the growing season (Baldocchi *et al.*, 1997).

The onset of photosynthesis in temperate grasslands and crops is much delayed behind the dates experienced in forests, because annual crops and grasses must germinate from seeds. Native grasslands are more apt to experience soil moisture deficits than are crops (which are either irrigated or grow in more humid regions). Hence, grasslands experience many days with a net transfer of carbon to the atmosphere. Among other observations of agricultural crops, several investigators have observed that canopy photosynthesis by cereal decreases after anthesis (Thomas and Hill, 1949; Biscoe *et al.*, 1975; Baldocchi, 1994) because respiration is stimulated.

V. Canopy Photosynthesis in the Future

Several key global and regional environmental change factors have the potential to alter photosynthesis significantly by terrestrial ecosystems. Perhaps the most frequently mentioned factor is increasing atmospheric CO_2 concentration. In C_3 leaves, future elevated CO_2 will stimulate photosynthesis and at the same time inhibit photorespiration. The overall effect is likely to be an increase in CO_2 uptake by canopies (Amthor, 1995). In addition, water-use efficiency, light-use efficiency, and leaf N-use efficiency will increase with an increase in ambient CO_2. In the long run, photosynthesis may acclimate to long-term elevated CO_2, reducing the positive response to increasing CO_2 (e.g., Amthor, 1995; Lloyd and Farquhar, 1996). We expect relatively small effects of increasing CO_2 on C_4 canopy photosynthesis, although the increase in water-use efficiency resulting from reduced stomatal conductance may enhance canopy development and indirectly increase canopy photosynthesis.

A second important environmental change is the land-surface warming expected to accompany further increases in atmospheric CO_2. In the short term, warming can alter several of the component processes of leaf and canopy photosynthesis. We expect, however, that various acclimation and adaptation processes might compensate moderate warming, over many years.

Increasing atmospheric N deposition is a third environmental change that might affect future photosynthesis. In ecosystems that are limited by N availability, increasing N deposition has the potential to enhance N uptake and therefore photosynthesis (through increased leaf N levels). On the other hand, increased N deposition, associated with increased acidity, might be stressful for plants, resulting in reduced photosynthesis and growth.

Changes in tropospheric pollutant levels may also affect future photosynthesis in several ecosystems. For example, an increase in levels of the regional air pollutant O_3 might damage leaves and reduce canopy photosynthesis.

VI. Research Directions

Many research questions associated with time and space remain unanswered. With regard to time, we do not know how dynamic effects associated with photosynthesis and sunfleck integrate to the canopy scale. At longer time scales, rates of leaf photosynthesis in old trees are lower than in younger trees (Ryan and Yoder, 1997). How this effect sums to the canopy scale and over the course of a growing season remains an unknown.

As we move up in spatial scale and apply canopy photosynthesis to exercises on landscape and global ecology, we will need to confirm our attempts to extrapolate canopy-level concepts to the scale of landscape, regions, and continents. Estimates of canopy photosynthesis on hilly terrain remain untested, too. Networks of long-term eddy covariance measurements are being implemented through the EUROFLUX and AmeriFlux programs to address some of these problems.

With regard to measuring canopy photosynthesis and respiration, there are bias errors associated with the episodic and intermittent nature of nighttime eddy fluxes between the biosphere and atmosphere. More studies are needed on how to measure and evaluate these fluxes.

Acknowledgments

DDB would like to thank Dr. Norman Rosenberg and Dr. Shashi Verma, for being mentors and for providing the opportunity and resources to initiate a study of canopy photosynthesis, and Dr. John Norman for teaching much about canopy photosynthesis modeling. We also thank Dr. Verma and Dr. R. Valigura for providing internal reviews of this work.

Both DDB and JSA are supported by a research grant from the U.S. Department of Energy (DOE), Terrestrial Carbon Processes Program. Work by JSA, at Oak Ridge National Laboratory, is conducted under the auspices of DOE subordinate to Contract No. DE-ACOS-96OR22462, with Lockheed Martin Energy Research Corp.

References

Aber, J. D., Reich, P. B., and Goulden, M. L. (1996). Extrapolating leaf CO_2 exchange to the canopy, a generalized model of forest photosynthesis compared with measurements by eddy correlation. *Oecologia* **106,** 257–265.

Amthor, J. S. (1994). Plant respiratory responses to the environment and their carbon balance. *In* "Plant–Environmental Interactions" (R. E. Wilkenson, ed.), pp. 501–554. Marcel Dekker, New York.

Amthor, J. S. (1995). Terrestrial higher-plant response to increasing atmospheric [CO_2] in relation to the global carbon cycle. *Global Change Biol.* **1,** 243–274.

Amthor, J. S., Goulden, M. L., Munger, J. W., and Wofsy, S. C. (1994). Testing a mechanistic model of forest-canopy mass and energy exchange using eddy correlation: Carbon dioxide and ozone uptake by a mixed oak-maple stand. *Aus. J. Plant Physiol.* **21,** 623–651.

Baldocchi, D. D. (1993). Scaling water and carbon exchange from leaves to a canopy: Rules and tools. *In* "Scaling Physiological Processes: Leaf to Globe" (J. Ehleringer and C. B. Field, eds.), pp. 77–114. Academic Press, San Diego.

Baldocchi, D. D. (1994). A comparative study of mass and energy exchange rates over a closed C_3 (wheat) and an open C_4 (corn) crop. II. CO_2 exchange and water use efficiency. *Agric. For. Meteorol.* **67,** 291–321.

Baldocchi, D. D. (1997). Measuring and modeling carbon dioxide and water vapor exchange over a temperate broad-leaved forest during the 1995 summer drought. *Plant Cell Environ.* **20,** 1098–1107.

Baldocchi, D. D., and Harley, P. C. (1995). Scaling carbon dioxide and water vapor exchange from leaf to canopy in a deciduous forest: model testing and application. *Plant Cell Environ.* **18,** 1157–1173.

Baldocchi, D. D., and Meyers, T. P. (1998). On using eco-physiological, micrometeorological and biogeochemical theory to evaluate carbon dioxide, water vapor and gaseous deposition fluxes over vegetation. *Agric. For. Meteorol.* **90,** 1–18.

Baldocchi, D. D., Verma, S. B., and Rosenberg, N. J. (1981). Environmental effects on the CO_2 flux and CO_2-water flux ratio of alfalfa. *Agric. Meteorol.* **24,** 175–184.

Baldocchi, D. D., Verma, S. B., and Anderson, D. E. (1987). Canopy photosynthesis and water use efficiency in a deciduous forest. *J. Appl. Ecol.* **24,** 251–260.

Baldocchi, D. D., Vogel, C. A., and Hall, B. (1997). Seasonal variation of carbon dioxide exchange rates above and below a boreal jackpine forest. *Agric. For. Meteorol.* **83,** 147–170.

Baumgartner, A. (1969). Meteorological approach to the exchange of CO_2 between atmosphere and vegetation, particularly forests stands. *Photosynthetica* **3,** 127–149.

Billings, W. D., Clebsch, E. E. C., and Mooney, H. A. (1966). Photosynthesis and respiration rates of Rocky Mountain alpine plants under field conditions. *Am. Midland Natural.* **75,** 34–44.

Biscoe, P. V., Scott, R. K., and Monteith, J. L. (1975). Barley and its environment. III. Carbon budget of the stand. *J. Appl. Ecol.* **12,** 269–293.

Bjorkman, O. (1981). The response of photosynthesis to temperature. *In* "Plants and Their Atmosperic Environment" (J. Grace, E. D. Ford, and P. G. Jarvis, eds.), pp. 273–302. Blackwell Scientific Publ., Oxford, UK.

Black, T. A., den Hartog, G., Neumann, H., Blanken, P., Yang, P., Nesic, Z., Chen, S., Russel, C.,

Voroney, P., and Stabeler, R. (1996). Annual cycles of CO_2 and water vapor fluxes above and within a Boreal aspen stand. *Global Change Biol.* **2,** 219–230.
Boysen-Jensen, P. (1918). Studies on the production of matter in light and shadow plants. *Bot. Tidskr.* **36,** 219–262.
Brown, K. W., and Rosenberg, N. J. (1971). Energy and CO2 balance of an irrigated sugar beet field in the Great Plains. *Agron. J.* **63,** 207–213.
Chazdon, R. L. (1988). Sunflecks and their importance to forest understory plants. *Adv. Ecol. Res.* **18,** 1–63.
Chen, J. L., Reynolds, J. F., Harley, P. C., and Tenhunen, J. D. (1993). Coordiation theory of leaf nitrogen distribution in a canopy. *Oecologia* **93,** 63–69.
Cohen, S., and Fuchs, M. (1987). The distribution of leaf area, radiation, photosynthesis and transpiration in a Shamouti orange hedgerow orchard. 1. Leaf area and radiation. *Agric. For. Meteorol.* **40,** 123–144.
Collatz, G. J., Ball, J. T., Grivet, C., and Berry, J. A. (1991). Regulation of stomatal conductance and transpiration, a physiological model of canopy processes. *Agric. For. Meteorol.* **54,** 107–136.
Collatz, G. J., Ribas-Carbo, M., and Berry, J. A. (1992). Coupled photosynthesis–stomatal model for leaves of C_4 plants. *Austr. J. Plant Physiol.* **19,** 519–538.
Cowan, I. R. (1968). Mass, heat and momentum exchange between stands of plants and their atmospheric environment. *Q. J. R. Meteorol. Soc.* **94,** 523–544.
Denmead, O. T. (1976). Temperate cereals. *In* "Vegetation and the Atmosphere, II" (J. L. Monteith, ed.), pp. 1–31. Academic Press, London.
Denmead, O. T., Dunin, F. X., Wong, S. C., and Greenwood, E. A. N. (1993). Measuring water use efficiency of Eucalypt trees with chambers and micrometeorological techniques. *J. Hydrol.* **150,** 649–664.
dePury, D., and Farquhar, G. D. (1997). Simple scaling of photosynthesis from leaves to canopies without the errors of big-leaf models. *Plant Cell Environ.* **20,** 537–557.
de Wit, C. T. (1965). "Photosynthesis of Leaf Canopies." Agricultural Research Report 663, pp. 1–57. Wageningen, The Netherlands.
Dugas, W. A., Reicosky, D. C., and Kiniry, J. R. (1997). Chamber and micrometeorological measurements of CO_2 and H_2O fluxes for three C_4 grasses. *Agric. For. Meteorol.* **83,** 113–134.
Duncan, W. G., Loomis, R. S., Williams, W. A., and Hanau, R. (1967). A model for simulating photosynthesis in plant communities. *Hilgardia* **38,** 181–205.
Eckardt, F. E. (1968). Techniques de mesure de la photosynthèse sur le terrain basées sur l'emploi d'enceintes climatisées. *In* "Functioning of Terrestrial Ecosystems at the Primary Production Level" (F. E. Eckardt, ed.), pp. 289–319. UNESCO, Paris.
Ehleringer, J. R., and Monson, R. K. (1993). Evolutionary and Ecological Aspects of Photosynthetic Pathway Variation. *Annu. Rev. Ecol. Syst.* **24,** 411–439.
Falge, E., Graber, W., Siegwolf, R., and Tenhunen, J. D. (1996). A model of the gas exchange response of *Picea abies* to habitat condition. *Trees* **10,** 277–287.
Fan, S. M., Goulden, M. L., Munger, J. W. *et al.* (1995). Environmental controls on the photosynthesis and respiration of a boreal lichen woodland, a growing season of whole-ecosystem exchange measurements by eddy correlation. *Oecologia* **102,** 443–452.
Farquhar, G. D., and von Caemmerer, S. (1982). Modeling photosynthetic response to environmental conditions. *In* "Encyclopedia of Plant Physiology 12B" (O. L. Lange, P. S. Nobel, C. B. Osmond, and H. Ziegler, eds.), pp. 549–587. Springer-Verlag, Berlin.
Farquhar, G. D., von Caemmerer, S., and Berry, J. A. (1980). A biochemical model of photosynthetic CO_2 assimilation in leaves of C_3 species. *Planta* **149,** 78–90.
Field, C. B. (1991). Ecological scaling of carbon gain to stress and resource availability. *In* "Integrated Responses of Plants to Stress" (H. A. Mooney, W. E. Winner, and E. J. Pell, eds.), pp. 1–32. Academic Press, San Diego.

Field, C. B., Ball, J. T., and Berry, J. E. (1989). Photosynthesis: Principles and field techniques. *In* "Plant Physiological Ecology: Field Methods and Instrumentation" (R. W. Pearcy, J . Ehleringer, H. A. Mooney, and P. W. Rundel, eds.), pp. 208–253. Chapman-Hall, London.

Frolking, S., Goulden, M. L., Wofsy, S. C., Fan, S. M., Sutton, D. J., Munger, J. W., Bazzaz, A. M., Daube, B. C., Crill, P. M., Aber, J. D., Band, L. E., Wang, X., Savage, K., Moore, T., and Harriss, R. C. (1996). Modelling temporal variability in the carbon balance of a spruce/moss boreal forest. *Global Change Biol.* **2,** 343–366.

Fukai, S., and Loomis, R. S. (1976). Leaf display and light environments in row-planted cotton communities. *Agric. Meteorol.* **17,** 353–379.

Gifford, R. M. (1994). The global carbon cycle: A viewpoint on the missing sink. *Aust. J. Plant Physiol.* **21,** 1–15.

Goulden, M. L., Munger, J. W., Fan, S. M., Daube, B. C., and Wofsy, S. C. (1996). Exchange of carbon dioxide by a deciduous forest: response to interannual climate variability. *Science* **271,** 1576–1578.

Grantz, D. A., and Meinzer, F. C. (1991). Regulation of transpiration in field-grown sugarcane: evaluation of the stomatal response to humidity in the Bowen ratio technique. *Agric. For. Meteorol.* **53,** 169–183.

Greco, S., and Baldocchi, D. D. (1996). Seasonal variations of CO_2 and water vapor exchange rates over a temperate deciduous forest. *Global Change Biol.* **2,** 183–198.

Grelle, A. (1997). Long term water and carbon dioxide fluxes from a boreal forest. Methods and applications. Silvestra 28 Acta Universtatis, Agriculturae Sueciae. Swedish Univ. Agricultural Sciences, Uppsala.

Gutschick, V. P. (1991). Joining leaf photosynthesis models and canopy photon-transport models. *In* "Photon–Vegetation Interaction: Application in Remote Sensing and Plant Physiology" (R. Myneni and J. Ross, eds.), pp. 501–535. Springer-Verlag, Berlin.

Harley, P. C., and Tenhunen, J. D. (1991). Modeling the photosynthetic response of C_3 leaves to environmental factors. *In* "Modeling Photosynthesis—From Biochemistry to Canopy," pp. 17–39. Am. Soc. Agronomy, Madison, Wisconsin.

Held, A. A., Steduto, P., Orgas, F., Matista, A., and Hsiao, T. C. (1990). Bowen ratio/energy balance technique for estimating crop net CO_2 assimilation and comparison with a canopy chamber. *Theor. Appl. Climatol.* **42,** 203–213.

Henrici, M. (1921). *Verh. Naturforsch. Ges. (Basel)* **32,** 107–168.

Hollinger, D. (1996). Optimality and nitrogen allocation in a tree canopy. *Tree Physiol.* **16,** 627–634.

Hollinger, D. Y., Kelliher, F. M., Byers, J. N., Hunt, J. E., McSeveny, T. M., and Weir, P. L. (1994). Carbon dioxide exchange between an undisturbed old-growth temperate forest and the atmosphere. *Ecology* **75,** 134–150.

Inoue, I. (1957). An aerodynamic measurement of photosynthesis over a paddy field. *Proc. 7th Jp. Natl. Congr. Appl. Mech.*, pp. 211–214.

Jarvis, P. G. (1993). Prospects for bottom-up models. *In* "Scaling Physiological Processes: Leaf to Globe" (J. Ehleringer and C. B. Field, eds.), pp. 115–126. Academic Press, San Diego.

Jarvis, P. G., and Catsky, J. (1971). General principles of gasometric methods and the main aspects of installation design. *In* "Plant Photosynthetic Production: Manual of Methods" (Z. Sestak, J. Catsky, and P. G. Jarvis, eds.), pp. 49–110. Dr. W. Junk, The Hague.

Jarvis, P. G., James, G. B., and Landsberg, J. J. (1976). Coniferous forest. *In* "Vegetation and the Atmosphere, Vol. 2" (J. L. Monteith, ed.), pp. 171–240. Academic Press, London.

Jeffers, D. L., and Shibles, R. M. (1969). Some effects of leaf area, solar radiation, air temperature and variety on net photosynthesis in field grown soybeans. *Crop Sci.* **13,** 82–84.

Kim, J., and Verma, S. B. (1991). Modeling canopy photosynthesis: scaling up from a leaf to canopy in a temperate grassland ecosystem. *Agric. For. Meteorol.* **57,** 187–208.

Lemon, E. R. (1960). Photosynthesis under field conditions. II. An aerodynamic method for determining the turbulent carbon dioxide exchange between the atmosphere and a corn field. *Agron. J.* **52,** 697–703.

Leuning, R., Kelliher, F. M., dePury, D., and Schulze, E. D. (1995). Leaf nitrogen, photosynthesis, conductance and transpiration: scaling from leaves to canopies. *Plant Cell Environ.* **18,** 1183–1200.

Lloyd, J. J., and Farquhar, G. D. (1996). The CO_2 dependence of photosynthesis, plant growth responses to elevated atmospheric CO_2 concentrations and their interaction with soil nutrient status. I. General principles and forest ecosystems. *Oecologia* **10,** 4–32.

Lloyd, J., Grace, J., Miranda, A. C., *et al.* (1995). A simple calibrated model of Amazon rainforest productivity based on leaf biochemical properties. *Plant Cell Environ.* **18,** 1129–1145.

Lundegardh, H. (1922). Neue Apparate zur Analyse de Kohlensauregehalts der Luft. *Biochem Z.* **131,** 109–115.

McGuinn, S. M., and King, K. M. (1990). Simultaneous measurements of heat, water vapor and CO_2 fluxes above alfalfa and maize. *Agric. For. Meteorol.* **49,** 331–349.

Monsi, M., and Saeki, T. (1953). Uber de Lichtfaktor in den Pflanzengesellschaften und seine Bedeutung fur die Stoffenproduktion. *Jpn. J. Bot.* **14,** 22–52.

Monsi, M., Uchijima, Z., and Oikawa, T. (1973). Structure of foliage canopies and photosynthesis. *Annu. Rev. Ecol. Syst.* **4,** 301–327.

Monteith, J. L. (1972). Solar radiation and productivity of terrestrial ecosystems. *J. Appl. Ecol.* **9,** 747–766.

Monteith, J. L., and Szeicz, G. (1960). The CO_2 flux over a field of sugar beets. *Q. J. R. Meteorol. Soc.* **86,** 205–214.

Monteith, J. L., Szeicz, G., and Yabuki, K. (1964). Crop photosynthesis and the flux of carbon dioxide below the canopy. *J. Appl. Ecol.* **1,** 321–337.

Musgrave, R. B., and Moss, D. N. (1961). Photosynthesis under field conditions. I. A portable, closed system for determining the rate of photosynthesis and respiration in corn. *Crop Sci.* **1,** 37–41.

Myneni, R. B. (1991). Modeling radiative transfer and photosynthesis in three-dimensional vegetation canopies. *Agric. For. Meteorol.* **55,** 323–344.

Nikolov, N. T., Massman, W. J., and Schoettle, A. W. (1995). Coupling biochemical and biophysical processes at the leaf level: an equilibrium photosynthesis model for leaves of C_3 plants. *Ecol. Model.* **80,** 205–235.

Norman, J. M. (1979). Modeling the complete crop canopy. *In* "Modification of the Aerial Environment of Crops" (B. Barfield and J. Gerber, eds.), pp. 249–280. American Society of Agricultural Engineers, St. Joseph, Missouri.

Norman, J. M., and Polley, W. (1989). Canopy photosynthesis. *In* "Photosynthesis," pp. 227–241. Alan Liss, New York.

Oker-Blom, P., Lappi, J., and Smolander, H. (1991). Radiation regime and photosynthesis of coniferous stands. *In* "Photon-Vegetation Interactions" (R. B. Myneni and J. Ross, eds.), pp. 469–499. Springer-Verlag, Berlin.

Olioso, A., Carlson, T. N., and Brisson, N. (1996). Simulation of diurnal transpiration and photosynthesis of a water stressed soybean crop. *Agric. For. Meteorol.* **81,** 41–59.

Pearcy, R. W. (1990). Sunflecks and photosynthesis in plant canopies. *Annu. Rev. Plant Physiol. Mol. Biol.* **41,** 421–453.

Pettigrew, W. T., Hesketh, J. D., Peters, D. B., and Woolley, J. T. (1990). A vapor pressure deficit effect on crop canopy photosynthesis. *Photosynthet. Res.* **24,** 27–34.

Price, D. T., and Black, T. A. (1990). Effects of short-term variation in weather on diurnal canopy CO_2 flux and evapotranspiration of a juvenile Douglas fir stand. *Agric. For. Meteorol.* **50,** 139–158.

Raupach, M. R. (1988). Canopy transport processes. *In* "Flow and Transport in the Natural Environment" (W. L. Steffen and O. T. Denmead, eds.). Springer-Verlag, Berlin.

Raupach, M. R., and Legg, B. J. (1984). The uses and limitations of flux-gradient relationships in micrometeorology. *Agric. Water Manage.* **8,** 119–131.

Rochette, P., Desjardins, R. L., Pattey, E., and Lessard, R. (1996). Instantaneous measurement of radiation and water use efficiencies of a maize crop. *Agron. J.* **88,** 627–635.

Ruimy, A., Saugier, B., and Dedieu, G. (1994). Methodology for the estimation of terrestrial net primary production from remotely sensed data. *J. Geophys. Res.* **99,** 5263–5283.

Ruimy, A., Jarvis, P. G., Baldocchi, D. D., and Saugier, B. (1995). CO_2 fluxes over plant canopies: A literature review. *Adv. Ecol. Res.* **26,** 1–68.

Ryan, M. G., and Yoder, B. J. (1997). Hydraulic limits to tree height and tree growth. *Bioscience* **47,** 235–242.

Saeki, T. (1960). Interrelationships between leaf amount, light distribution and total photosynthesis in a plant community. *Bot. Mag. (Tokyo)* **73,** 55–63.

Sala, A., and Tenhunen, J. (1996). Simulations of canopy net photosynthesis and transpiration of *Quercus ilex* L. under the influence of seasonal drought. *Agric. For. Meteorol.* **78,** 203–222.

Sands, P. (1995). Modelling canopy production. 1. Optimal distribution of photosynthesis resources. *Aust. J. Plant Physiol.* **22,** 593–601.

Schulze, E. D. (1986). Carbon dioxide and water vapor exchange in response to drought in the atmosphere and in the soil. *Annu. Rev. Plant. Physiol.* **37,** 247–274.

Sellers, P. J., Randall, D. A., Collatz, G. J., Berry, J. A., Field, C. B., Dazlich, D. A., Zhang, C., Collelo, G. D., and Bounoua, L. (1996). A revised land surface parameterization (SIB2) for atmospheric GCMs. Part 1. Model formulation. *J. Climate* **9,** 676–705.

Sestak, Z., Catsky, J., and Jarvis, P. G. (1971). "Plant Photosynthetic Production: Manual of Methods." Dr. W. Junk, The Hague.

Sinclair, T. R., Murphy, C. E., and Knoerr, K. R. (1976). Development and evaluation of simplified models for simulating canopy photosynthesis and transpiration. *J. Appl. Ecol.* **13,** 813–829.

Stenberg, P., deLucia, E. H., Schoettle, A. W., and Smolander, H. (1995). Photosynthetic light capture and processing from cell to canopy. *In* "Resource Physiology of Conifers" (W. Smith and T. Hinckley, eds.), pp. 3–38. Academic Press, San Diego.

Su, H. B., Paw, K. T., and Shaw, R. H. (1996). Development of a coupled leaf and canopy model for the simulation of plant-atmosphere interaction. *J. Appl. Meteorol.* **35,** 733–748.

Tappeiner, U., and Cernusca, A. (1996). Microclimate and fluxes of water vapor, sensibleheat and carbon dioxide in structurally differing subalpine plant communities in the central Caucasus. *Plant Cell Environ.* **19,** 40–417.

Thomas, M. D., and Hill, G. R. (1949). Photosynthesis under field conditions. *In* "Photosynthesis in Plants" (J. Franck and W. E. Loomis, eds.), pp. 19–52. Iowa State Univ. Press., Ames, Iowa.

Thornley, J. H. M., and Johnson, I. R. (1990). "Plant and Crop Modeling" Oxford Univ. Press, Oxford, UK.

Tibbits, T. W., and Langhans, R. W. (1993). Controlled-environment studies. *In* "Photosynthesis and Production in a Changing Environment" (D. O. Hall *et al.*, eds.), pp. 65–78. Chapman-Hall, London.

Tranquillini, W. (1964). Photosynthetic and dry matter production of trees at high altitudes. *In* "The Formation of Wood in Forest Trees," pp. 505–518.

Uchijima, Z. (1976). "Maize Vegetation and the Atmosphere, vol. 2." (J. L. Monteith, ed.). Academic Press, London.

Valentini, R., de Angelis, P., Matteucci, G., Monaco, R., Dore, S., and Scarascia-Mugnozza, G. E. (1996). Seasonal net carbon dioxide exchange of a beech forest with the atmosphere. *Global Change Biol.* **2,** 199–208.

Verhoef, A., Allen, S. J., deBruin, H. A. R., Jacobs, C. M. J., and Heusinkveld, B. G. (1996). Fluxes of carbon dioxide and water vapor from a Sahelian savanna. *Agric. For. Meteorol.* **80,** 231–248.

Verma, S. B., and Rosenberg, N. J. (1976). Carbon dioxide concentration and flux in a large agricultural region of the Great Plains of North America. *J. Geophys. Res.* **81,** 399–405.

Verma, S. B., Baldocchi, D. D., Anderson, D. E., Matt, D. R., and Clement, R. E. (1986). Eddy fluxes of CO_2, water vapor, and sensible heat over a deciduous forest. *Bound. Layer Meteorol.* **36,** 71–91.

Waggoner, P. E., Furnival, G. M., and Reifsnyder, W. E. (1969). Simulation of the microclimate in a forest. *For. Sci.* **15,** 37–45.

Wang, Y. P., and Jarvis, P. G. (1990). Description and validation of an array model-MAESTRO. *Agric. For. Meteorol.* **51,** 257–280.

Wang, Y. P., McMurtrie, R. E., and Landsberg, J. J. (1992). Modelling canopy photosynthetic productivity. *In* "Crop Photosynthesis and Temporal Determinations" (N. R. Baker and H. Thomas, eds.), pp. 43–67. Elsevier, Amsterdam, The Netherlands.

Williams, M., Rastetter, E. B., Fernades, D. N., Goulden, M. L., Wofsy, S. C., Shaver, G. R., Meillo, J. M., Munger, J. W., Fan, S. M., and Nadelhoffer, K. J. (1996). Modelling the soil-plant-atmosphere continuum in a *Quercus–Acer* stand at Harvard Forest: The regulation of stomatal conductance by light, nitrogen and soil/plant hydraulic properties. *Plant Cell Environ.* **19,** 911–927.

Wofsy, S. C., Gouldin, M. L., Munger, J. W., Fan, S. M., Bakwin, P. S., Daube, B. C., Bassow, S. L., and Bazzaz, F. A. (1993). Net exchange of CO_2 in a mid-latitude forest. *Science.* **260,** 1314–1317.

Wullschleger, S. D. (1993). Biochemical limitations to carbon assimilation in C_3 plants. A retrospective analysis of A/C_i curves from 109 species. *J. Exp. Bot.* **44,** 907–920.

3

Terrestrial Higher Plant Respiration and Net Primary Production[1]

Jeffrey S. Amthor and Dennis D. Baldocchi

The survival of green plants is based on their ability to maintain a positive balance between gain of energy through photosynthesis and loss due to respiration, death of tissues and grazing.

F. E. Eckardt (1975), "Photosynthesis and Productivity in Different Environments"

Neither the rate nor the extent of production need bear a close relation to photosynthetic rate, or be determined by it. . . . The processes that follow photosynthesis, such as respiration and translocation, or other limitations on the capacity of plants to grow and utilize photosynthate, can be major determinants of productivity.

L. T. Evans (1975), "Photosynthesis and Productivity in Different Environments"

I. Plant Respiration in Relation to Terrestrial Ecosystem Net Primary Production

A large fraction of the C assimilated in higher plant photosynthesis is released back to the atmosphere as CO_2 during subsequent plant respiration. Thus, plant respiration is a large negative component of the C budget of plants and ecosystems. It contributes to the control of ecosystem net primary production (NPP) because NPP is gross primary production (GPP) minus plant respiration (hereafter simply *respiration*). The relationship between ecosystem NPP and GPP is therefore dictated by respiration.

[1]This chapter was authored by a U.S. Government contractor under Contract No. DE-AC05-96OR22464. Accordingly, the U.S. Government retains a nonexclusive, royalty-free license to publish or reproduce the published form of this contribution, or allow others to do so, for U.S. Government purposes.

In spite of the fact that respiration leads to a large loss of C from plants, its true importance is related to the functions it performs. That is to say, respiration is more than a loss of CO_2. The CO_2 released is a necessary by-product of biochemical reactions that support nearly all the growth, transport, and maintenance processes occurring in plants (except those intimately associated with photosynthesis). Most importantly, plant growth and health are impossible without respiration, so at least much of the CO_2 released in respiration is essential (Beevers, 1961, 1970). In short, respiration is the metabolic bridge from photosynthesis to growth; it consumes photosynthate to generate usable forms of energy that then drive plant growth and maintenance processes. It also supplies many of the C-skeleton building blocks needed for biosynthesis.

When considering respiration in the context of NPP, we find it appropriate to relate amounts of CO_2 released in respiration to amounts assimilated in gross photosynthesis (i.e., GPP). We prefer the use of a ratio. This ratio is the amount of CO_2 released in respiration divided by the amount assimilated in photosynthesis. This is denoted R_a/P_g, where R_a and P_g are plant respiration and photosynthesis integrated over time periods ranging from "growing seasons" to years. Annual totals are most meaningful with respect to NPP. This ratio reflects the C costs of respiratory processes relative to C gains in photosynthesis. The question arises: "How conservative is the ratio R_a/P_g within and among terrestrial plants and ecosystems?"

II. Regulation of Respiration Rate

Because respiration is a large part of a plant's C budget, it is relevant to ask "What controls respiration rate?" because those controls may be important regulators of NPP. Biochemically, respiration is presumably regulated in large part by needs for—and uses and turnover of—its products (Beevers, 1961, 1970; Dry *et al.*, 1987). Notable among those products are (1) a range of C-skeleton intermediates generated all along the respiratory pathways, (2) ATP, and (3) the reductants NADH and NADPH. It is thought that respiratory products accumulate when they are not used by other processes, and in so doing they retard respiration through negative feedback mechanisms. Also, without the use (i.e., turnover) of ATP, NADH, and NADPH, the required respiratory substrates ADP, NAD^+, and $NADP^+$ are unavailable for further respiration. When respiratory products are used rapidly, however, negative feedbacks on respiration are released and associated respiratory substrates are regenerated, with the result that respiration rate increases. Widespread circumstantial evidence in favor of this scheme of respiratory control is found in the strong positive relationship between respiration rate and growth rate; young, rapidly growing organs have fast respiration compared to mature and more slowly growing organs, presumably because rapidly growing organs use respiratory products rapidly. And, biochemical uncouplers—which break the links between respiration rate and the use or turnover of respiratory products—generally stimulate respiration rate. The uncoupled respiration rate is a measure of respiratory *capacity,* which is set by the amount of respiratory machinery (enzymes, transporters) present. Active meristematic tissue may respire at the rate limited by respiratory capacity, whereas in more mature tissue, the uncoupled rate of respiration may be double the normal rate (Beevers, 1961). Interspecific variation in response to uncouplers may occur, however. For example, *Alocasia odora* leaf respiration increased 60–80% in the presence of an uncoupler, but *Spinacia oleracea* leaf respiration was unaffected (Noguchi and Terashima, 1997). In this case, nighttime *S. oleracea* leaf respiration was apparently regulated by carbohydrate availability (Noguchi and Terashima, 1997), carbohydrates being a main respiratory substrate.

These biochemical controls of respiration presumably occur at all levels of plant organization, from the single mitochondrion or parcel of cytoplasm to the global biosphere. Hence, the regulation of respiration is inherently scale independent and the same metabolic principles used to predict and explain respiration over the short term in cells, organs, and single plants can be used to make predictions of (and understand) respiration at larger spatial and longer temporal scales. It is, therefore, in theory possible to calculate global respiration from knowledge of the global extent of the processes consuming respiratory products. Because the rates of processes consuming respiratory products at the large spatial and long temporal scales are imprecisely known, however, present estimates of annual global terrestrial higher plant respiration are crude (and see Sprugel *et al.,* 1995). Moreover, ecosystem respiration (R_e) includes metabolism by heterotrophs, which may not be so easily generalizable.

We emphasize that direct evidence that respiration is fully coupled to useful processes in nature is lacking. We suppose that evolution favors tightly regulated respiration, but it is possible that some degree of "idling respiration" (*sensu* Beevers, 1970) is also generally present—and perhaps indirectly serves essential functions. Thus, we expect that respiration is generally coupled to the processes that use respiratory products, but we must allow for the possibility that "inefficiencies" could also be a normal aspect of respiration. To the extent that respiration is controlled by the rate of processes using respiratory products, temperature will affect respiration to the same degree that it affects the processes using respiratory products. Importantly, various stresses might reduce the coupling of respiration to "normal" growth and maintenance processes.

III. The Fundamental (Semi)Mechanistic Model of Plant Respiration

Early numerical models of respiration were empirical. They related respiration rate to easily measured plant properties such as dry mass or surface area. Although useful for summarizing data, they often lacked explanatory and predictive power. Also, relationships between respiration rate and surface area or dry mass fail to account for much of the observed variability in respiration rate (Ryan, 1990; Ryan *et al.*, 1994a; Sprugel *et al.*, 1995).

A significant step forward was to set respiration rate at the uncoupled rate in meristematic tissue, and to slow it with tissue aging (de Wit *et al.*, 1970). It had long been known that respiration rate per unit dry mass declines with increasing age in organs and whole plants; the research of Kidd *et al.* (1921) was especially significant (see also, e.g., Inamdar *et al.*, 1925; Price, 1960).

Limitations of these models in explaining and predicting respiration rate were realized 30 years ago by R. S. Loomis. He suggested to C. T. de Wit that one could, in theory, sum up all the nonphotosynthetic biochemical reactions shown in wall charts of metabolic pathways in proportion to their occurrence and rate in plants. The total CO_2 released per unit of dry mass per unit of time calculated in this way would be a mechanistic estimate of respiration rate. Although Loomis was skeptical that such a project could be carried out successfully, de Wit encouraged F. W. T. Penning de Vries to give it a try (Amthor, 2000). Penning de Vries met with considerable success. The fruits of his labors included an elegant mechanistic model of the respiratory costs (i.e., CO_2 released) of growth of tissues of a specified composition from specified substrates (see, e.g., de Wit *et al.*, 1970, 1978; Penning de Vries, 1972, 1974, 1975b; Penning de Vries *et al.*, 1974, 1983, 1989; Penning de Vries and van Laar, 1975). He also produced a refined, semimechanistic model of plant maintenance costs or maintenance respiration (Penning de Vries 1975a; Penning de Vries *et al.*, 1983). While this model development was occurring, novel approaches to measurement and analysis were used to obtain experimental estimates of growth and maintenance respiration in plants (e.g., McCree 1970; Thornley, 1970; Hesketh *et al.*, 1971).

We suggest that the simplest (clearest) way simultaneously to understand, quantify, and predict the amount of respiration—and its relationships to GPP and NPP—is to describe it with a semimechanistic model, based on the preceding work, as follows:

$$R_a = r_B B + r_I I + r_N N_a + r_T T + r_M S,$$

where B is the rate of biosynthesis of new tissue; r_B is the respiratory cost of (i.e., CO_2 released during) biosynthesis; I is the rate of active ion uptake by roots; r_I is the respiratory cost of active ion uptake; N_a is the rate of N assimilation (excluding N assimilated directly by photosynthetic metabolism);

r_N is the respiratory cost of nonphotosynthetic N assimilation; T is the rate of translocation of carbohydrates, amino acids, and other compounds; r_T is the respiratory cost of translocation (mainly phloem loading?); S is the amount of structural tissue (total mass less temporary nonstructural storage compounds); and r_M is the cost of maintaining existing structure. We believe this accounts for most of the quantitatively important aspects of respiration. The coefficients r_B, r_I, r_N, and r_T are ratios, but r_M is a rate, because B, I, N_a, and T are rates, whereas S is a state. Estimation of the coefficients r from biochemical principles was an outcome of Penning de Vries's work. Overviews of values of the coefficients are given in Thornley and Johnson (1990) and Amthor (1994a). The following may be true in general for whole plants: $r_M\, S = r_B\, B > r_N\, N_a \approx r_T\, T > r_I I$.

A simplified model is also often used. It divides respiration into the two components, growth and maintenance, as follows (Thornley, 1970):

$$R_a = g\,B + m\,S,$$

where g is the growth respiration coefficient, which includes costs of biosynthesis, ion uptake, N assimilation, and the translocation of substrates used in growth; and m is the maintenance respiration coefficient, which includes costs of maintenance and the translocation of substrates used in maintenance. Over an annual cycle, the total of B is essentially the same as NPP. The term gB is growth respiration rate (R_g) and the term mS is maintenance respiration rate (R_m). We note that NPP must be less than GPP/(1 + g) because some maintenance respiration always occurs.

A. Growth Respiration

Growth is the conversion of temporary pools of substrates such as carbohydrates and amides into new structures (including enzymes) and "permanent" storage such as starch in seeds. Growth includes monomer synthesis, polymerization of monomers into polymers such as cellulose and proteins, organization of polymers into organelles and cells, and "tool maintenance" or the turnover of molecules catalyzing growth (Penning de Vries *et al.*, 1974). (Development of the secondary and tertiary structure of polymers and the organization of polymers into organelles and cells apparently uses little energy.) Nutrient uptake and assimilation, and translocation of substances used in growth, are also growth processes in the two-component scheme of respiration. Growth respiration R_g is the CO_2 generated by growth processes, but some of that CO_2 arises outside the respiratory pathways per se (Penning de Vries *et al.*, 1974, 1989). The respiratory pathways per se that contribute to R_g are the same as the respiratory pathways contributing to R_m.

In addition to detailed reaction-by-reaction analyses based on Penning de Vries *et al.* (1974, 1983, 1989), shortcut methods of estimating g (and r_B)

from plant elemental composition and/or energy content are available (McDermitt and Loomis, 1981; Vertregt and Penning de Vries, 1987; Williams *et al.*, 1987). Comparisons of various methods of estimating g (and r_B) are in Williams *et al.* (1987), Lafitte and Loomis (1988), and Wullschleger *et al.* (1997).

A key concept is that R_g is proportional to NPP—or growth—through the ratio g. Values of g depend on the compositions of substrates used for growth and the tissue grown. Values are relatively small for high-carbohydrate tissue, and large for high-fat, high-lignin, and high-protein tissue (Amthor, 2000). At the whole-plant and ecosystem scales, typical values of g may be 0.25–0.35 mol CO_2 mol^{-1} C. Any factor that stimulates NPP without greatly altering plant chemical composition will probably stimulate R_g proportionally. Thus, rapid growth requires rapid growth respiration! Knowledge of the ratio R_g/NPP (= g) is central to understanding respiration–NPP connections. To the extent that g is a constant, R_g will be affected by temperature (and other environmental factors) in the same way that NPP is (de Wit *et al.*, 1970). If an environmental change reduces NPP by 30%, for example, we expect that R_g will also be reduced about 30%.

According to the model, respiration per unit dry mass and R_g/R_a increase with specific growth rate (Fig. 3-1). The release of metabolic heat also increases with growth rate, and from the perspective of plant metabolic energy balance, the ratio metabolic heat release/CO_2 release is indicative of specific growth rate for a constant g and m (Fig. 3-1).

B. Maintenance Respiration

Maintenance respiration includes the energy and C used for turnover of labile cellular constituents, active intracellular transport to counteract membrane leaks and ion imbalances, and repair and acclimation processes that result from a stressful or variable environment (see Penning de Vries, 1975a). It is difficult to measure higher plant R_m, and a wide range of estimates have been derived (Amthor, 1989; Ryan, 1990; Sprugel, 1990; Ryan *et al.*, 1996). In particular, the coefficient m can be very small in woody tissue. Moreover, a distinction should be made between metabolically inactive heartwood and more active sapwood in trees (Ryan, 1990; Sprugel, 1990; Ryan *et al.*, 1994a).

Because maintenance presumably involves mainly protein turnover and other processes that are likely to be related to protein content, R_m may be better related to plant N content (N) than to structural dry mass. This leads to the following modified model (Barnes and Hole, 1978):

$$R_a = gB + m_N N,$$

where m_N is maintenance respiration rate per unit N, and $m_N N = mS = R_m$. Variation in m_N can be smaller than variation in m (Jones *et al.*, 1978; Ryan,

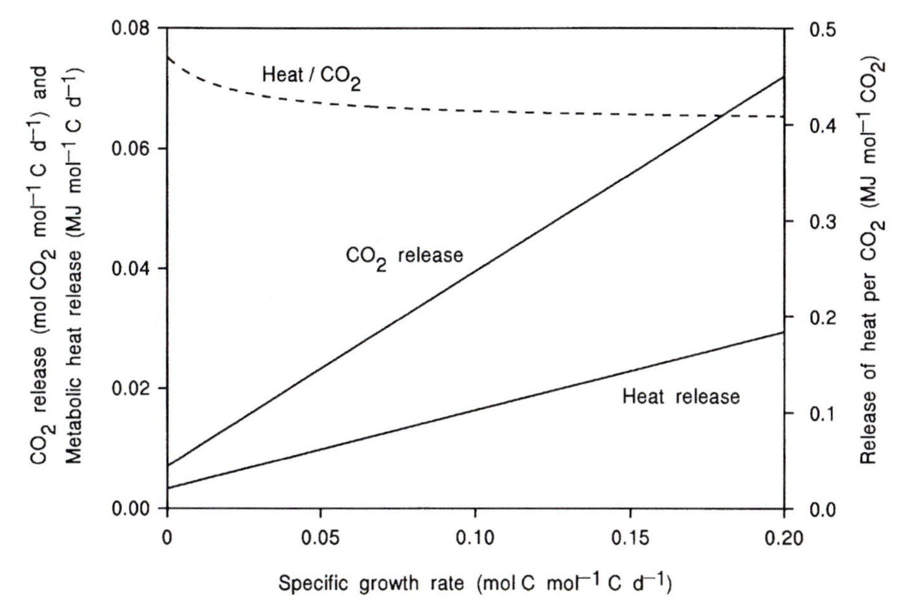

Figure 3-1 Relationships between specific growth rate and release of CO_2 and metabolic heat in plant metabolism (solid lines, left axis) and the ratio heat release/CO_2 release (dashed line, right axis). The case shown is for tissue with a (1) maintenance respiration coefficient *m* of 0.007 mol CO_2 mol^{-1} C d^{-1}, (2) growth respiration coefficient *g* of 0.325 mol CO_2 mol^{-1} C, (3) energy content of the substrates of growth of 470 kJ mol^{-1} C, and (4) energy content of new structural matter of 492 kJ mol^{-1} C. That is, 470 kJ is released as heat per mol CO_2 released in maintenance processes, whereas ~402 kJ is released as heat per mol CO_2 released in growth processes. In this example, growth conserves energy relative to C. [We note that Figs. 4.A2 and 4.A3 in Amthor (1994c) are in error with respect to metabolic heat release per CO_2 release as a function of specific growth rate; due to a typographical error in the computer program generating those figures, the heat release rate slows too much per unit increase in specific growth rate.]

1991, 1995; Collier and Grodzinski, 1996; Ryan *et al.*, 1996), indicating that indeed *N* may be a better predictor of R_m than is *S*. Although estimates of m_N vary, many are in the range 2–5 μmol CO_2 mol^{-1} N s^{-1} in the temperature range 10–20°C (Table 3-1). In some instances, it may be that only a part of R_m is related to tissue N content (Li and Jones, 1992). In any case, it seems that R_m is related not only to plant size, but also plant composition. There is a corollary: increased standing stock increases whole-ecosystem plant R_m to the extent that plant N per unit ground area increases.

The value of R_m increases about exponentially with temperature in the shortterm, but the fraction of annual daytime net canopy CO_2 assimilation used for stem maintenance increased linearly, over the range 5–13%, with site temperature in four conifer ecosystems (Ryan *et al.*, 1995).

Table 3-1 Selected Estimates of the Maintenance Respiration Coefficient m_N in Higher Plants[a]

Species	Plant part	Growth conditions	Temperature (°C)	m_N (μmol CO_2 mol^{-1} N s^{-1})	Notes	Ref.[b]
General	Leaf	Chamber	15–30	0.5–0.9	Theoretical	Penning de Vries (1975a)
Abies amabilis	Stem	Field	20	4.2		Ryan (1991)
Abies balsamea	Stem	Field	15	7–14		Lavigne *et al.* (1996)
Abies lasiocarpa	Leaf	Field	10	1.5		Ryan (1995)
Alnus crispa	Leaf	Field	10	3.0		Ryan (1995)
Betula papyrifera	Leaf	Field	10	3.1		Ryan (1995)
Cucumis sativus	Shoot	Chamber	26	3.3		Szaniawski (1985)
	Root	Chamber	26	21		Szaniawski (1985)
Diplacus aurantiacus	Leaf	Field	20	3.5–6.6		Merino *et al.* (1982)
Glycine max	Leaf	Greenhouse	23	22	Ambient CO_2	Thomas and Griffin (1994)
Gossypium hirsutum	Leaf	Chamber	29	14	Ambient CO_2	Thomas *et al.* (1993)
Helianthus annuus	Shoot	Chamber	20	2.1		Szaniawski and Kielkiewicz (1982)
	Root	Chamber	20	4.7		Szaniawski and Kielkiewicz (1982)
Heteromeles arbutifolia	Leaf	Field	20	6.6		Merino *et al.* (1982)
Larix laricina	Leaf	Field	10	2.2		Ryan (1995)
Lepechinia calycina	Leaf	Field	20	5.5		Merino *et al.* (1982)
Liriodendron tulipifera	Leaf	Field chamber	Ambient	21	Ambient CO_2	Wullschleger *et al.* (1992)
Lolium perenne	Shoot	Field	15	1.8–3.7		Jones *et al.* (1978)
Medicago sativa	Plant	Chamber	18–26	2.3	Nitrate grown, no osmotic stress	Shone and Gale (1983)
Oryza sativa	Shoot	Chamber	31	1.6		Baker *et al.* (1992)
Pennisetum clandestinum	Leaf lamina	Chamber	20	2.3		Murtagh *et al.* (1987)
Picea engelmannii	Leaf	Field	10	2.0		Ryan (1995)
Picea mariana	Leaf	Field	10	2.1		Ryan (1995)
	Stem	Field	15	2.8–2.9[c]		Lavigne and Ryan (1997)
Pinus banksiana	Leaf	Field	10	2.0		Ryan (1995)
	Stem	Field	15	2.3–3.5[c]	Young	Lavigne and Ryan (1997)
	Stem	Field	15	1.7–2.2[c]	Old	Lavigne and Ryan (1997)
Pinus contortia	Stem	Field	20	1.2		Ryan (1991)
	Leaf	Field	10	2.8		Ryan (1995)
Pinus ponderosa	Stem	Field	10–25	3.9		Carey *et al.* (1996)

Pinus radiata	Leaf	Field	15	1.7		Ryan *et al.* (1996)
	Fine root	Field	15	11		Ryan *et al.* (1996)
Populus balsamifera	Leaf	Field	10	2.9		Ryan (1995)
Populus tremuloides	Leaf	Field	10	2.1–2.4		Ryan (1995)
	Stem	Field	15	0.6–0.7[c]		Lavigne and Ryan (1997)
Quercus alba	Stem	Field chamber	Ambient	0.4	Ambient CO_2	Wullschleger *et al.* (1995, 1997)
Quercus rubra	Leaf	Field chamber	Ambient	6.1	Spring/seedlings, ambient O_3	Wullschleger *et al.* (1996)
	Leaf	Field chamber	Ambient	4.4	Fall/seedlings, ambient O_3	Wullschleger *et al.* (1996)
	Leaf	Field chamber	Ambient	2.9	Spring/mature trees, ambient O_3	Wullschleger *et al.* (1996)
	Leaf	Field chamber	Ambient	4.6	Fall/mature trees, ambient O_3	Wullschleger *et al.* (1996)
Salix plantifolia	Leaf	Field	10	2.1		Ryan (1995)
Secale cereale	Leaf	Field	20	3.5[c]		McCullough and Hunt (1993)
	Stem	Field	20	4.3[c]		McCullough and Hunt (1993)
Shepherdia canadensis	Leaf	Field	10	2.2		Ryan (1995)
Solanum tuberosum	Root	Hydroponic	18	3.1		Bouma *et al.* (1996)
Sorghum bicolor	Plant	Chamber	30	3.3		McCree (1983)
	Plant	Chamber	30	2.7	At anthesis	Stahl and McCree (1988)
Trifolium repens	Plant	Chamber	25	3.8–4.4		McCree (1982)
× *Triticosecale*	Leaf lamina	Field	20	5.7[c]		McCullough and Hunt (1993)
	Stem	Field	20	2.6[c]		McCullough and Hunt (1993)
Triticum aestivum	Leaf lamina	Field	20	6.4–12		Winzeler *et al.* (1988)
	Leaf lamina	Field	20	7.4[c]		McCullough and Hunt (1993)
	Stem	Field	20	6.1[c]		McCullough and Hunt (1993)
Vicia faba	Leaf	Chamber	20	2.4		Irving and Silsbury (1988)

[a] Different methods and assumptions are used in different studies. We think the values given may be representative of many plants, but emphasize that there is considerable uncertainty in estimates of m_N.

[b] Many of these references do not include a direct estimate of m_N; we calculated it from data given therein. For O_2 uptake measurements, a respiratory quotient of unity was assumed (i.e., mol O_2 uptake = mol CO_2 release).

[c] In some cases, R_m may be unrelated to N content of leaves and stems both within and among species (e.g., Byrd *et al.*, 1992; McCullough and Hunt, 1993), including woody stems (Lavigne and Ryan, 1997). Also, fruit R_m may be poorly related to fruit N content (e.g., Hole and Barnes, 1980) because of marked changes in "physiological activity" of N with fruit development.

C. Relative Magnitudes of Growth and Maintenance Respiration

In crops, seasonal R_m/R_a may be ~50% (Amthor, 1989). In temperate grasslands, annual R_m may account for 75–84% of R_a, and in temperate forests, it may be 78–88% of annual R_a (Ryan, 1991). When considering individual organs, estimates of R_m/R_a in a 20-year-old *Pinus radiata* plantation are 95% in foliage, 96% in branches, 52% in stem plus bark, 62% in coarse roots, and 76% in fine roots (Ryan *et al.*, 1996, for "control" trees). Other estimates of the fraction of tree stem respiration that is due to maintenance include 40–65% in cool-temperate evergreen forests (Ryan, 1990), 54 and 82% in a tropical rainforest (Ryan *et al.*, 1994a), 86–88% in cool-temperate pines (Lavigne, 1996), 56–65% in a temperate deciduous forest (Edwards and Hanson, 1996), and 49–74% in boreal forests (Ryan *et al.*, 1998). Thus, R_m is at least as important as R_g is to a plant's annual C balance.

IV. Respiration following and during Photosynthesis

In the immediate term, light (or perhaps photosynthesis) may inhibit leaf respiration (we are not concerned in this chapter with *photo*respiration, a component of photosynthesis). Moderate light level may slow respiration 15–80% (e.g., Sharp *et al.*, 1984; Brooks and Farquhar, 1985; Kirschbaum and Farquhar, 1987; Villar *et al.*, 1995; Atkin *et al.*, 1997). On the other hand, photosynthesis apparently requires concomitant respiration, at least respiratory electron transport processes (reviewed by Krömer, 1995). Knowledge of any effects of light on leaf respiration is needed to estimate daily (24 h) leaf respiration amount because leaf respiration is measured in the dark. Presently, it is impossible to state *a priori* the quantitative effects of light on leaf respiration in any particular situation. Nonetheless, when constructing daily respiratory budgets, it is imperative to state what assumptions are made about leaf respiration in the light.

In the medium term, respiration is often most rapid following a period of rapid photosynthesis. According to Weintraub (1944), these observations date to at least the 1876 and 1881 publications of J. Borodin, the 1893 publication of F. Aereboe, and the 1905 publication of G. L. C. Matthaei. Since then, these observations are common (e.g., Sale, 1974).

Why should this be the case? When photosynthesis is rapid, translocation of photosynthate will tend to be rapid, and this may entail extra respiratory metabolism for phloem loading in leaves as well as active processes associated with transport and compartmentation of recent photosynthate in sinks. Increased levels of nonstructural carbohydrates may in turn contribute to more rapid growth processes (and R_g) throughout a plant (see references in Amthor, 1997). And, perhaps idling respiration responds positively to elevated levels of carbohydrates resulting from rapid photosynthesis. After all, carbohydrates are a key substrate of respiration.

In the long term, photosynthesis supplies the substrates of respiration and growth, and annual R_a plus growth is limited by annual P_g. Growth and respiration are required for the construction and maintenance of photosynthetic organs, however, so photosynthesis is impossible without previous respiration. Thus, photosynthesis and respiration are codependent across a range of time scales.

V. Respiration in Leaves, Stems, and Roots

Even though typical leaf respiration rates are slow compared with photosynthetic capacity, leaf respiration can be a significant fraction of the respiratory budget and therefore plant C balance. For example, leaves may account for 40–60% of tropical forest R_a (reviewed in Allen and Lemon, 1976). In three tropical forest stands, Yoda (1983) estimated that leaf respiration was 51–56% of R_a, with stems and branches contributing 32–38% of R_a, and roots the remaining 9–15%. In a warm-temperate evergreen oak forest, leaf respiration was estimated at 50–60% of R_a (Yoda, 1978). In a temperate deciduous forest, leaf respiration (including forest floor herbs) was put at 28% of R_a, with roots contributing 26% of R_a, and branch–bole–stump respiration responsible for the other 46% (Edwards *et al.*, 1981). Annual respiration by 20-year-old *Pinus radiata* trees was reported to be 28–39% from foliage, 11–12% from branches, 24–31% from stems, 7–11% from coarse roots, and 10–20% from fine roots (Ryan *et al.*, 1996, for all treatments). In nonwoody biomes such as grasslands, we expect that leaf respiration is a large fraction of R_a, although root–crown respiration can equal shoot respiration in grasslands (based on estimates of Andrews *et al.*, 1974).

The great unknown with respect to both R_a and NPP involves roots. The partitioning of respiration between roots and shoots depends in part on the root/shoot ratio and the N content of root and shoot, the amount of growth in roots and shoots, and the extent of other processes using respiratory products such as translocation, nutrient uptake from the soil, and nutrient assimilation in roots versus shoots. Quantifying these processes *in situ* is problematic. Difficulties with access to roots and measurements of their activities are obvious. Nonetheless, insights have been gained. For example, fine root respiration rate per unit dry mass can exceed that of other tissues, and may be well coupled to N content (e.g., Ryan *et al.*, 1996). On the other hand, coarse root respiration per unit dry mass is often slow. The maintenance coefficient m is thus very different for fine and coarse roots, and coefficients derived from measurements of fine roots should not be used to estimate coarse root R_m. Soil temperature may affect forest fine root turnover rate, or longevity, and this may be the result of temperature effects on root R_m (Hendrick and Pregitzer, 1993). Thus, soil temperature might regulate the relationship between root production rate and root standing stock through its influence on root R_m.

A perennial problem in constructing respiratory budgets is deciding what to do about mycorrhizal respiration. Are mycorrhizae part of the "effective" roots, or simply soil heterotrophs? It is perhaps most appropriate to consider root and mycorrhizal respiration together, because the activities of mycorrhizae are presumably linked to root activity (and vice versa?). In root exudates, carbon that is respired by microorganisms associated with roots might also be considered part of overall "root" respiration, but such definitions are tricky.

VI. Respiration in Comparison to Photosynthesis at the Ecosystem Scale

Unfortunately, it is impossible to directly measure whole-plant or plant-community respiration during the course of a 24-h day (or season or year) because of simultaneous daytime photosynthetic CO_2 uptake. Also, separation of *in situ* root respiration from the CO_2 released by other soil organisms is difficult (Hanson *et al.*, 2000). This means that all estimates of daily, seasonal, and annual R_a are equivocal to some degree. In spite of these difficulties, potentially useful estimates of R_a and the ratio R_a/P_g—in nature—exist (Table 3-2). These estimates are, however, just that: estimates. They all include errors in measurements and assumptions. They are primarily useful as educated guesses placing respiration into a semiquantitative framework for evaluating the efficiency with which plants use photosynthate in growth and storage. Presentation of even two digits in Table 3-2 may imply greater precision than there is in reality. In fact, estimates of R_a are sometimes made without underlying measurements of CO_2 (or O_2) exchange. For example, *Fagus sylvatica* root respiration estimates by Möller *et al.* (1954) were simply set to 20% of stem plus branch respiration estimates, whereas whole-tree respiration estimates by Grier and Logan (1977) were made without any respiration measurements at the study site! Nonetheless, educated guesses summarized in Table 3-2 place considerable significance on R_a in the C balance of terrestrial ecosystems and encourage improved understanding of its regulation.

Temperate forests are relatively well studied, and our view is that the most reliable studies give R_a/P_g in the range ~0.50–0.60 in those forests. Boreal forest R_a/P_g may be larger (Table 3-2). The ratio may also be larger in trop-

Table 3-2 Estimates of the Fundamental Ratio R_a/P_g for Terrestrial Ecosystems[a]

Ecosystem	R_a/P_g	Ref.
Crop		
Alfalfa	0.35–0.49	Thomas and Hill (1949)
Maize, rice, and wheat	~0.3–0.6	Amthor (1989)
Grassland		
Shortgrass prairie	0.34	Andrew *et al.*, (1974)
	0.51	Detling (1979)
Tall grass prairie		
No grazing	0.61	Risser *et al.*, (1981)
Seasonal grazing	0.65	Risser *et al.*, (1981)
Year-round grazing	0.62	Risser *et al.*, (1981)
Forest		
Tropical moist		
Ivory Coast	0.75	Müller and Nielsen (1965)
Puerto Rico	0.88	Derived from Odum (1970)
Southern Thailand	0.66	Kira (1975)
Temperate		
Warm evergreen	0.72	Kira (1975)
Warm evergreen "oak"	0.66	Kira and Yabuki (1978)
Abies sachalinensis	0.53	Kira (1975)
Castanopis cuspidata	0.575	Kira (1975)
Chamaecyparis obtusa plantation	0.62	Hagihara and Hozumi (1991)
Cryptomeria japonico plantation	0.71	Kira (1975), mean of five estimate
Fagus crenata		
Secondary forest	0.44	Kira (1975)
Plantation	0.56	Kira (1975)
Fagus sylvatica		
8 years old	0.46	Möller *et al.* (1954)
25 years old	0.39	Möller *et al.* (1954)
46 years old	0.43	Möller *et al.* (1954)
85 years old	0.47	Möller *et al.* (1954)
Fraxinus excelsior plantation	0.37	Kira (1975)
Liriodendron tulipifera	0.66	Harris *et al.* (1975)
Picea abies plantation	0.32	Kira (1975)
Pinus densiflora plantation	0.71	Kira (1975)
Pinus ponderosa	0.55	Law *et al.* (1999)
Pinus taeda plantation	0.58	Kinerson (1975)
Pinus spp.	0.39–0.71[b]	Ryan *et al.* (1944b)
Pseudotsuga–Tsuga	0.93	Grier and Logan (1977)
Quercus–Acer	0.44–0.55	Amthor (2000)
Quercus–Acer	0.54	M. L. Goulden (personal communication, 1997)
Quercus–Pinus	0.55	Whittaker and Woodwell (1969)
Quercus spp.	0.61	Satchell (1973), as cited by Edwards *et al.* (1981)
Quercus–Carpinus	0.38	Medwecka-Kornas *et al.* (1974), as cited by Edwards *et al.* (1981)

(*continues*)

Table 3-2 *(Continued)*

Ecosystem	R_a/P_g	Ref.
Forest *(continued)*		
Subalpine		
Coniferous	0.72	Kitazawa (1977), as cited by Edwards *et al.* (1981)
Abies	0.675	Kira (1975)
Abies veitchii	0.61	Kira (1975), mean of three estimates
Boreal		
Picea mariana	0.69	M. L. Goulden (personal communication, 1997)
Picea mariana	0.72–0.77	Ryan *et al.* (1998)
Pinus banksiana	0.68	Baldocchi *et al.* (1997)
Pinus banksiana	0.69–0.74	Ryan *et al.* (1998)
Populus tremuloides	0.55[c]	Black *et al.* (1996)
Populus tremuloides	0.64–0.67	Ryan *et al.* (1998)
Temperate coastal salt marsh		
Spartina	0.77	Teal (1962)
Spartina–Distichlis	0.69	Woodwell *et al.* (1979)
Arctic tundra	0.50	Reichle (1975)

[a]Both R_a and P_g must have the same units, e.g., mol C m^{-2} [ground] $year^{-1}$. As implied by the sample units just given, estimates of R_a and P_g should be for an entire year, except for crops, where they apply to the "growing season." Nonetheless, some of the values in this table represent data from periods less than a full year. To our knowledge, these estimates of R_a/P_g are based on the assumption that leaf respiration occurs at about the same rate in the light as in the dark, at a given temperature. Data from controlled-environment chambers are included in this summary.

[b]Range of values for seven young (16–40 years olds) *Pinus* stands. Ryan *et al.* (1994b) gave daily (24-h) stem, branch, and root respiration, but foliage respiration was for nights only. The measure of photosynthesis presented was daytime canopy net CO_2 assimilation. Here, to obtain R_a, we doubled the nighttime foliage respiration amounts. To obtain P_g, we added nighttime foliage respiration amount to daytime canopy net CO_2 assimilation. Both our transformations are based on the assumption that daytime foliage respiration was similar to nighttime foliage respiration.

[c]Assuming belowground NPP is 35% of NPP.

ical forests, though "modern" measurements there are limited. Many estimates for crops indicate a relatively low R_a/P_g, often in the range 0.35–0.50 (see Amthor, 1989). We believe that crop R_a/P_g is generally small, and we attribute this to the selection of plants that efficiently convert photosynthate into storage compounds in seeds and tubers. Other ecosystems are poorly represented, but ratios in the range 0.50–0.70 are typical. In a speculative vein, we note that R_m may be positively related to R_g and growth rate (Amthor, 1989; Lavigne and Ryan, 1997), which might contribute to a relatively conservative ratio R_a/NPP and therefore a conservative R_a/P_g.

An obvious requirement for accurately assessing R_a/P_g is accurate measurement of R_a (and P_g). Nighttime eddy covariance measurements of ecosystem CO_2 exchange provide estimates of R_e when daytime values of R_e can be derived from the nighttime measurements (Goulden *et al.*, 1996). If, in addition, CO_2 efflux from ecosystem heterotrophs (R_h) can be estimated, R_a would be given by difference (i.e., $R_a = R_e - R_h$). Unfortunately, nighttime measurements of R_e by eddy covariance contain uncertainty. In eight forest stands, eddy covariance estimates of R_e were compared to estimates of R_e derived from chamber measurements of soil surface, bole, and leaf respiration (Goulden *et al.*, 1996; Lavigne *et al.*, 1998; Law *et al.*, 1998). Although the estimates of R_e made by the two methods were correlated, it seems that the eddy covariance approach often underestimated R_e by 15–40% in those forests. This implies that direct measurements of R_e using the eddy covariance technique are presently unreliable, at least in those forests. And, nighttime eddy covariance measurements are variable (noisy). There are also, however, uncertainties in estimates of R_e derived from chamber measurements. In addition, eddy-covariance-measurement "footprints" in forests are relatively large and dynamic (Baldocchi, 1997), and may not be adequately sampled with a small number of chamber locations. In any case, R_a cannot presently be determined from R_e in forests because of difficulties in determining both R_e and R_h.

The R_a/P_g values in Table 3-2 may be contrasted with typical leaf-level estimates of respiration and photosynthetic capacity, which typically indicate that high-light photosynthesis assimilates CO_2 at rates 40–100 times faster than CO_2 is released in nighttime leaf respiration (e.g., Pearcy and Sims, 1994). Thus, comparisons between leaf respiration rate and photosynthetic capacity greatly underestimate the significance of respiration to a plant's C balance. Obviously, it is 24-h (or seasonal or annual) totals of plant-community respiration and photosynthesis that are the quantities relevant to NPP (Edwards *et al.*, 1981).

VII. Optimum Leaf Area Index: Does It Exist?

It is sometimes stated that an optimum leaf area index (L) for NPP exists (e.g., Larcher, 1995, p. 149). This notion comes from a simple conceptual model that assumes (1) P_g per unit ground area increases asymptotically with increasing L and (2) R_a increases about linearly with increasing L. The result is that NPP increases, passes through a maximum, and then decreases with increasing L. The L giving maximum NPP is called the "optimum" L. An underlying idea is that shaded leaves at the canopy bottom are "parasitic."

Photosynthesis is indeed expected to approach a maximum with increasing L as light interception becomes complete. The precise form of this relationship depends on things such as the angles of leaves, sun elevation, and the ratio of direct beam to diffuse radiation (see Chapter 2, this volume). But, canopy and whole-plant respiration is not expected to increase *linearly* with *L*. Instead, respiration by shaded leaves within a canopy acclimates to low light so incremental increases in *L* do not result in proportional increases in leaf respiration. Part of the acclimation to shade is a reduction in N per unit leaf area, arising because shade leaves are generally thin (Ellsworth and Reich, 1993; Pearcy and Sims, 1994), and this can contribute to lower R_m per leaf area. The smaller N per unit leaf area and thinner leaves in the shade also reduce leaf growth respiration per unit leaf area in the shade. Moreover, slow photosynthesis in shade leaves—due to low light—limits the amount of translocation from shade leaves, and this would limit translocation respiration proportionally. All the metabolic controls on respiration in sun versus shade leaves are as yet, however, incompletely understood (Noguchi *et al.*, 1996). Our view is that respiration generally proceeds as rapidly as is required to meet metabolic demands—limited perhaps by substrate availability—and shade leaves have limited metabolic demands. The overall result is commonly observed: leaf respiration per unit leaf area and per unit leaf mass declines with depth in a canopy (e.g., Nishioka *et al.*, 1978; Yoda, 1983).

Asymptotic increases in photosynthesis and respiration with increasing *L* result in an asymptotic increase in NPP with *L*. This is most easily studied in crops, for which *L* can be experimentally controlled and precisely measured (e.g., Fig. 3-2). In other ecosystems, *L*–NPP relationships are not so easily studied. Nonetheless, some insight can be gained into the possibility of an optimum *L* in other ecosystems with available observations. For example, no obvious optimum *L* for aboveground NPP (ANPP) exists in forests, according to a pooling of data (Fig. 3-3). The considerable scatter in Fig. 3-3 is expected given the diversity in forest types, climates, and other factors included in the data set. In any case, it is clear that ANPP does not generally decline with increase in *L* at high *L*. For example, with $L > 12$, all ANPP estimates are large (Fig. 3-3). Broadleaf forests in Fig. 3-3 (●) generally have *L* in the range 4–8, with no obvious relationship to ANPP. We also examined the relationship between leaf area duration (LAD) and ANPP using data represented in Fig. 3-3, but there was no indication of an "optimum" LAD for ANPP (not shown). An obvious difficulty in assessing NPP–*L* relationships from a literature survey is the significant uncertainty in many estimates of *L*.

Whether an optimum *L* could exist is an open question. Available data indicate that canopies do not grow enough leaves to surpass an optimum *L*. They also indicate that any optimum *L* would greatly exceed the *L* needed for nearly complete solar radiation capture by a canopy. In short, the notion of an optimum *L* for NPP has little, if any, application to nature. That is, respiration is not a slave to *L;* rather, respiration responds to the environment

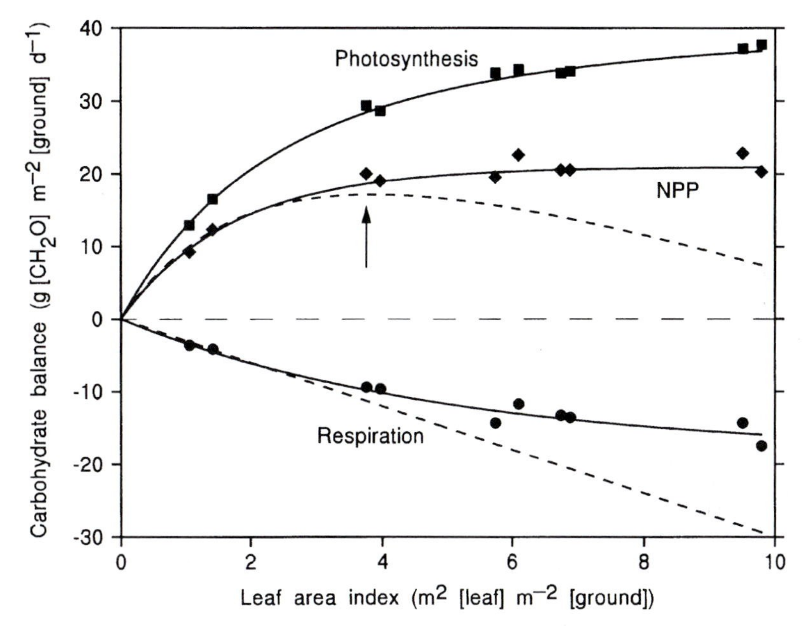

Figure 3-2 Daily plant respiration, NPP, and photosynthesis in a rice crop in the field as a function of leaf area index (L) for variety Peta, 4 weeks before flowering (Cock and Yoshida, 1973). Photosynthesis is given by the sum of measured respiration and NPP (in units of mass of carbohydrate). ■, Photosynthesis; ●, respiration; ♦, NPP. Solid lines are drawn to approximate the data. Also shown is an unverified, but often suggested, linear increase in respiration with increasing L and the resulting NPP (both shown with dashed lines). The arrow marks the optimum L for the case of the fictional linear R_a–L relationship. The dashed lines may be compared to, e.g., the simplistic model results shown in Zelitch (1971, Figs. 9.8–9.10).

(including within-canopy light gradients) and the metabolic needs of the canopy and whole plant.

VIII. Big Trees and Declining Forest Net Primary Production

Forest NPP may decline with increasing tree age (see Ryan *et al.*, 1997). A common but unverified assumption is that the biomass accumulated in old, large trees leads directly to large amounts of respiration, in particular R_m. This presumably increases R_a/P_g because canopy closure occurs early in stand development with little prospect for increasing GPP with further age increases (Ryan and Waring, 1992). This assumption is similar to that underlying the optimum-L notion, and indicts respiration as a drain on NPP. Additional considerations, however, indicate that rapid respiration is un-

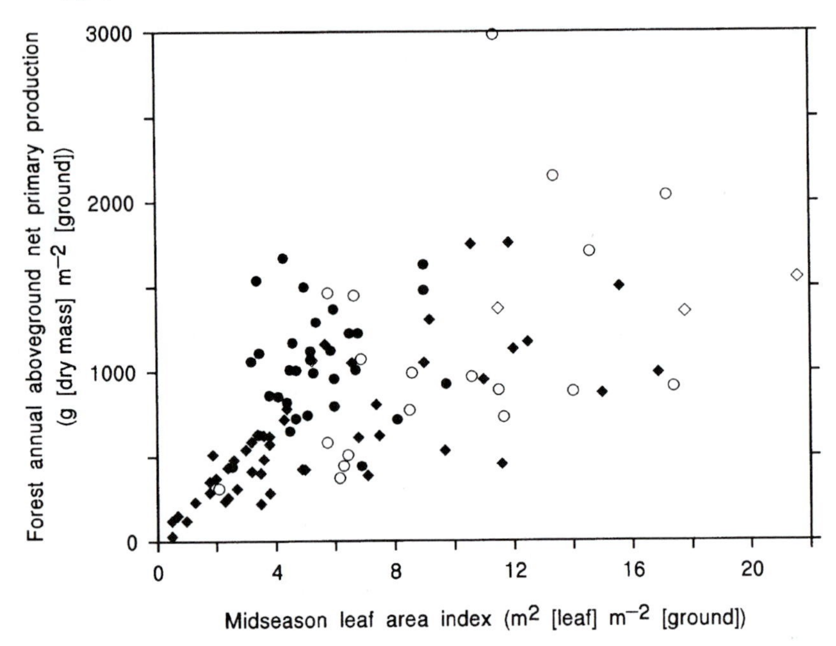

Figure 3-3 Aboveground NPP as a function of maximum or summer leaf area index (L) as reported for 83 forests and tree plantations in the International Biological Programme woodlands data set (DeAngelis *et al.*, 1981), 15 diverse conifer stands in Oregon (Gholz, 1982; Runyon *et al.*, 1994), and four boreal pine stands (Vogel, 1997). Some of the conifer values of L are for total leaf surface area, and some are for half-total surface area. ●, Broadleaf forests; ○, broadleaf plantations; ◆, conifer forests; ◇, conifer plantations.

likely to be the (main) cause of relatively low NPP in old stands, in spite of large biomass (Ryan and Waring, 1992; Ryan *et al.*, 1994b, 1997; Yoder *et al.*, 1994; Ryan and Yoder, 1997). Note also that once canopy closure occurs during early stand development (for closed-canopy forests), whole-forest leaf and fine-root masses may remain more or less constant over time. So, therefore, should their R_m. And, to the extent that NPP declines with age, we expect that R_g declines.

If it is not fast respiration, then what causes low NPP in old stands? Declining GPP with stand age increase, along with a relatively conservative ratio R_a/P_g, would result in declining NPP (Möller *et al.*, 1954). Reduced photosynthetic activity in old, large trees could be caused by low vascular-system hydraulic conductance, which in turn causes reduced stomatal conductance and CO_2 uptake (see Yoder *et al.*, 1994; Ryan *et al.*, 1997). Slow photosynthesis in old trees might also be caused, in part, by limited N availability to leaves, or other factors (see, e.g., Landsberg and Gower, 1997).

IX. Respiratory Responses to Environmental Change: The Future

Several regional and global environmental-change factors might affect R_a and the ratio R_a/P_g in terrestrial ecosystems. Global warming is a present concern, and we assume that warming affects respiration to the extent that it affects processes using respiratory products. Growth respiration is expected to respond to temperature in parallel with the temperature response of NPP, and R_m is presumably affected by temperature to the degree that maintenance processes are affected. Short-term responses of respiration to temperature are positive and strong, but long-term responses may be more moderate. Two processes seem important [see Precht *et al.* (1973) and references in Amthor (1994b)]: (1) acclimation to temporary, e.g., weekly and seasonal, changes in temperature and (2) adaptation to the prevailing climate. Acclimation and adaptation of R_a to future warming (if any) may be incomplete, but our expectation is that they will diminish effects of long-term warming on respiration. If there is a change in R_a/P_g with warming, it might be a slight increase, and driven by an increase in the ratio R_m/R_a. The extent of any increase in R_a/P_g or R_m/R_a with warming may depend on ecosystem type and present climate. In winter-deciduous ecosystems, warming may increase growing season length, which may also influence R_a/P_g and R_m/R_a.

The threat of global warming comes from ongoing atmospheric changes, most notably increasing CO_2. Increasing CO_2 itself is likely to affect most plant processes, including respiration (reviewed in Amthor, 1997). Our general expectations with respect to respiration are (1) R_g will increase about in proportion to increasing NPP; (2) the growth coefficient g could be affected by changes in plant composition, most notably a slight decrease associated with reduced N (or protein) concentration; (3) R_m might increase with increase in plant size, except that the ratio N/S may decline, resulting in a decline in the maintenance coefficient m but perhaps not m_N; and (4) to the extent that g or m values are lower in the future due to plant compositional changes, the ratio R_a/P_g may decline slightly. In addition to these indirect effects of CO_2 on respiration (*sensu* Amthor, 1991), there are several reports of a direct (short-term) inhibition of leaf and root respiration by CO_2 in the dark (e.g., Amthor *et al.*, 1992; reviewed in Amthor, 1997). There are no reports that CO_2 directly inhibits woody-stem respiration. There are also many reports of a lack of an effect of nighttime CO_2 on leaf respiration (Amthor, 1997), and our research with eight species using six methods to measure gas exchange, including measurements in the field, indicates that leaf respiration (CO_2 efflux and O_2 uptake) is unaffected by CO_2 in the dark (G. W. Koch and J. S. Amthor, unpublished data, 1992–1997). And, another report indicates that root respiration in the tree species

Citrus volkameriana is largely insensitive to CO_2 (Bouma *et al.*, 1997), albeit this contrasts previous reports for roots of the tree species *Pseudotsuga menziesii* and *Pinus* sp. (see Amthor, 1997). On balance, we think that direct effects of CO_2 on future respiration will be at most small. Nonetheless, a direct effect of increasing CO_2 on leaf and root respiration is an open issue.

Other environmental changes might also affect respiration. For example, increasing N deposition in temperate ecosystems might stimulate GPP, NPP, and R_a (Melillo *et al.*, 1996), but perhaps not R_a/P_g. Also, increases in regional tropospheric O_3 levels might inhibit photosynthesis and subsequent growth and respiration. Increased O_3 pollution might also increase the ratios R_m/R_a and R_a/P_g (Amthor, 1994b).

X. Summary

Plant respiration is the metabolic link between GPP and NPP. It is also a large component of a plant's C budget; perhaps typically 50–70% of C assimilated in GPP is released back to the atmosphere as CO_2 during subsequent plant respiration. Because great uncertainty remains concerning *in situ* measurements of R_a (and P_g), it is hard to quantify more precisely the role of R_a in C cycles of various ecosystems. We judge the available data to be too imprecise to assess properly whether R_a/P_g is at the present time conservative within or among ecosystems. Moreover, environmental change such as warming and increasing CO_2 concentration may affect R_a and P_g differently, so the ratio R_a/P_g may change in the future. In any case, future studies of the relationship between R_a and NPP or GPP will be more enlightening than simple measurements of respiration rate.

Acknowledgments

We thank Mike Ryan, Paul Hanson, Stan Wullschleger, Marnie Goodbody, and Beverly Law for reviewing this chapter, and Mike Goulden and Beverly Law for sharing unpublished data. Preparation of this condensed chapter was supported through the U.S. Department of Energy's (DOE) Office of Biological and Environmental Research (OBER) Terrestrial Carbon Processes program and the DOE/NSF/NASA/USDA/EPA Interagency Program on Terrestrial Ecology and Global Change (TECO) by the U.S. DOE's OBER under Contract No. DE-AC05-96OR22464 with Lockheed Martin Energy Research Corporation, Publication No. 5021, Environmental Sciences Division, Oak Ridge National Laboratory.

References

Allen, L. H., Jr., and Lemon, E. R. (1976). Carbon dioxide exchange and turbulence in a Costa Rican tropical rain forest. *In* "Vegetation and the Atmosphere, Vol. 2, Case Studies" (J. L. Monteith, ed.), pp. 265–308. Academic Press, London.

Amthor, J. S. (1989). "Respiration and Crop Productivity." Springer-Verlag, New York.

Amthor, J. S. (1991). Respiration in a future, higher-CO_2 world. *Plant Cell Environ.* **14,** 13–20.

Amthor, J. S. (1994a). Respiration and carbon assimilate use. *In* "Physiology and Determination of Crop Yield" (K. J. Boote *et al.*, eds.), pp. 221–250. American Society of Agronomy, Madison, Wisconsin.

Amthor, J. S. (1994b). Plant respiratory responses to the environment and their effects on the carbon balance. *In* "Plant–Environment Interactions" (R. E. Wilkinson, ed.), pp. 501–554. Marcel Dekker, New York.

Amthor, J. S. (1994c). Higher plant respiration and its relationships to photosynthesis. *In* "Ecophysiology of Photosynthesis" (E.-D. Schulze and M. M. Caldwell, eds.), pp. 71–101. Springer-Verlag, Berlin.

Amthor, J. S. (1997). Plant respiratory responses to elevated carbon dioxide partial pressure. *In* "Advances in Carbon Dioxide Effects Research" (L. H. Allen, Jr. *et al.*, eds.), pp. 35–77. American Society of Agronomy, Madison, Wisconsin.

Amthor, J. S. (2000). The McCree–de Wit–Penning de Vries–Thornley respiration paradigms: 30 years later. *Ann. Bot.* **86,** 1–20.

Amthor, J. S., Koch, G. W., and Bloom, A. J. (1992). CO_2 inhibits respiration in leaves of *Rumex crispus* L. *Plant Physiol.* **98,** 757–760.

Andrews, R., Coleman, D. C., Ellis, J. E., and Singh, J. S. (1974). Energy flow relationships in a shortgrass prairie ecosystem. *In* "Proceedings of the First International Congress of Ecology," pp. 22–28. Pudoc, Wageningen, The Netherlands.

Atkin, O. K., Westbeek, M. H. M., Cambridge, M. L., Lambers, H., and Pons, T. L. (1997). Leaf respiration in light and darkness. A comparison of slow- and fast-growing *Poa* species. *Plant Physiol.* **113,** 961–965.

Baker, J. T., Laugel, F., Boote, K. J., and Allen, L. H., Jr. (1992). Effects of daytime carbon dioxide concentration on dark respiration in rice. *Plant Cell Environ.* **15,** 231–239.

Baldocchi, D. D. (1997). Flux footprints within and over forest canopies. *Boundary-Layer Meteorol.* **85,** 273–292.

Baldocchi, D. D., Vogel, C. A., and Hall, B. (1997). Seasonal variation of carbon dioxide exchange rates above and below a boreal jack pine forest. *Agric. For. Meteorol.* **83,** 147–170.

Barnes, A., and Hole, C. C., (1978). A theoretical basis of growth and maintenance respiration. *Ann. Bot.* **42,** 1217–1221.

Beevers, H. (1961). "Respiratory Metabolism in Plants." Row, Peterson & Company, Evanston, Illinois.

Beevers, H. (1970). Respiration in plants and its regulation. *In* "Prediction and Measurement of Photosynthetic Productivity" (I. Setlik, ed.), pp. 209–214. Pudoc, Wageningen, The Netherlands.

Black, T. A., den Hartog, G., Neumann, H. H., Blanken, P. D., Yang, P. C., Russell, C., Nesic, Z., Lee, X., Chen, S. G., Staebler, R., and Novak, M. D. (1996). Annual cycles of water vapour and carbon dioxide fluxes in and above a boreal aspen forest. *Global Change Biol.* **2,** 219–229.

Bouma, T. J., Broekhuysen, A. G. M., and Veen, B. W. (1996). Analysis of root respiration of *Solanum tuberosum* as related to growth, ion uptake and maintenance of biomass. *Plant Physiol. Biochem.* **34,** 795–806.

Bouma, T. J., Nielsen, K. L., Eissenstat, D. M., and Lynch, J. P. (1997). Estimating respiration of roots in soil: Interactions with soil CO_2, soil temperature and soil water content. *Plant Soil* **195,** 221–232.

Brooks, A., and Farquhar, G. D. (1985). Effect of temperature on the CO_2/O_2 specificity of ribulose-1,5-bisphosphate carboxylase/oxygenase and the rate of respiration in the light. *Planta* **165,** 397–406.

Byrd, G. T., Sage, R. F., and Brown, R. H. (1992). A comparison of dark respiration between C_3 and C_4 plants. *Plant Physiol.* **100,** 191–198.

Carey, E., V., DeLucia, E. H., and Ball, J. T. (1996). Stem maintenance and construction respiration in *Pinus ponderosa* grown in different concentrations of atmospheric CO_2. *Tree Physiol.* **16,** 125–130.

Cock, J. H., and Yoshida, S. (1973). Photosynthesis, crop growth, and respiration of a tall and short rice varieties. *Soil Sci. Plant Nutr.* **19,** 53–59.

Collier, D. E., and Grodzinski, B. (1996). Growth and maintenance respiration of leaflet, stipule, tendril, rachis, and petiole tissues that make up the compound leaf of pea (*Pisum sativum*). *Can. J. Bot.* **74,** 1331–1337.

DeAngelis, D. L., Gardner, R. H., and Shugart, H. H. (1981). Productivity of forest ecosystems studied during the IBP: The woodlands data set. *In* "Dynamic Properties of Forest Ecosystems" (D. E. Reichle, ed.), pp. 567–672. Cambridge Univ. Press, Cambridge.

Detling, J. K. (1979). Processes controlling blue grama production on the shortgrass prairie. *In* "Perspectives in Grassland Ecology" (N. R. French, ed.), pp. 25–42. Springer-Verlag, New York.

de Wit, C. T., Brouwer, R., and Penning de Vries, F. W. T. (1970). The simulation of photosynthetic systems. *In* "Prediction and Measurement of Photosynthetic Productivity" (I. Setlik, ed.), pp. 47–70. Pudoc, Wageningen, The Netherlands.

de Wit, C. T., *et al.* (1978). "Simulation of Assimilation, Respiration and Transpiration of Crops." Pudoc, Wageningen, The Netherlands.

Dry, I. B., Bryce, J. H., and Wiskich, J. T. (1987). Regulation of mitochondrial respiration. *In* "The Biochemistry of Plants, Vol. 11, Biochemistry of Metabolism" (D. D. Davies, ed.), pp. 213–252. Academic Press, San Diego.

Edwards, N. T., and Hanson, J. J. (1996). Stem respiration in a closed-canopy upland oak forest. *Tree Physiol.* **16,** 433–439.

Edwards, N. T., Shugart, H. H., Jr., McLaughlin, S. B., Harris, W. F., and Reichle, D. E. (1981). Carbon metabolism in terrestrial ecosystems. *In* "Dynamic Properties of Forest Ecosystems" (D. E. Reichle, ed.), pp. 499–536. Cambridge Univ. Press, Cambridge.

Ellsworth, D. S., and Reich, P. B. (1993). Canopy structure and vertical patterns of photosynthesis and related leaf traits in a deciduous forest. *Oecologia* **96,** 169–178.

Gholz, H. L. (1982). Environmental limits on aboveground net primary production, leaf area, and biomass in vegetation zones of the Pacific northwest. *Ecology* **63,** 469–481.

Goulden, M. L., Munger, J. W., Fan, S.-M., Daube, B. C., and Wofsy, S. C. (1996). Measurements of carbon sequestration by long-term eddy covariance methods and a critical evaluation of accuracy. *Global Change Biol.* **2,** 169–182.

Grier, C. G., and Logan, R. S. (1977). Old-growth *Pseudotsuga menziesii* communities of a western Oregon watershed: Biomass distribution and production budgets. *Ecol. Monogr.* **47,** 373–400.

Hagihara, A., and Hozumi, K. (1991). Respiration. *In* "Physiology of Trees" (A. S. Raghavendra, ed.), pp. 87–110. Wiley, New York.

Hanson, P. J., Edwards, N. T., Garten, C. T., Jr., and Andrews, J. A. (2000). Separating root and soil microbial contributions to soil respiration: A review of methods and results. *Biogeochemistry* **48,** 115–146.

Harris, W. F., Sollins, P., Edwards, N. T., Dinger, B. E., and Shugart, H. H. (1975). Analysis of carbon flow and productivity in a temperate deciduous forest ecosystem. *In* "Productivity of World Ecosystems" (D. E. Reichle, J. F. Franklin, and D. W. Goodall, eds.), pp. 116–122. National Academy of Sciences, Washington, D.C.

Hendrick, R. L., and Pregitzer, K. S. (1993). Patterns of fine root mortality in two sugar maple forests. *Nature (London)* **361,** 59–61.

Hesketh, J. D., Baker, D. N., and Duncan, W. G. (1971). Simulation of growth and yield in cotton: Respiration and the carbon balance. *Crop Sci.* **11,** 394–398.

Hole, C. C., and Barnes, A. (1980). Maintenance and growth components of carbon dioxide efflux from growing pea fruits. *Ann. Bot.* **45,** 295–307.

Inamdar, R. S., Singh, S. B., and Pande, T. D. (1925). The growth of the cotton plant in India. I. The relative growth-rates during successive periods of growth and the relation between growth-rate and respiratory index throughout the life-cycle. *Ann. Bot.* **39,** 281–311.

Irving, D. E., and Silsbury, J. H. (1988). The respiration of mature field bean (*Vicia faba* L.) leaves during prolonged darkness. *Ann. Bot.* **62,** 473–479.

Jones, M. B., Leafe, E. L., Stiles, W., and Collett, B. (1978). Pattern of respiration of a perennial ryegrass crop in the field. *Ann. Bot.* **42,** 693–703.

Kidd, F., West, C., and Briggs, G. E. (1921). A quantitative analysis of the growth of *Helianthus annuus.* Part I. The respiration of the plant and its parts throughout the life cycle. *Proc. R. Soc. Lond., Ser.B* **B92,** 368–384.

Kinerson, R. S. (1975). Relationships between plant surface area and respiration in loblolly pine. *J. Appl. Ecol.* 12, 965–971.

Kira, T. (1975). Primary production of forests. *In* "Photosynthesis and Productivity in Different Environments" (J. P. Cooper, ed.), pp. 5–40. Cambridge Univ. Press, Cambridge.

Kira, T., and Yabuki, K. (1978). Primary production rates in the Minamata Forest. *In* "Biological Production in a Warm-Temperate Evergreen Oak Forest of Japan (JIBP Synthesis Vol. 18)" (T. Kira, Y. Ono, and T. Hosokawa, eds.), pp. 131–138. Univ. Tokyo Press, Tokyo.

Kirschbaum, M. U. F., and Farquhar, G. D. (1987). Investigation of the CO_2 dependence of quantum yield and respiration in *Eucalyptus pauciflora. Plant Physiol.* **83,** 1032–1036.

Krömer, S. (1995). Respiration during photosynthesis. *Annu. Rev. Plant Physiol. Plant Mol. Biol.* **46,** 45–70.

Lafitte, H. R., and Loomis, R. S. (1988). Calculation of growth yield, growth respiration and heat content of grain sorghum from elemental and proximal analyses. *Ann. Bot.* **62,** 353–361.

Landsberg, J. J., and Gower, S. T. (1997). "Applications of Physiological Ecology to Forest Management." Academic Press, San Diego.

Larcher, W. (1995). "Physiological Plant Ecology, 3rd Ed." Springer-Verlag, New York.

Lavigne, M. B. (1996). Comparing stem respiration and growth of jack pine provenances from northern and southern locations. *Tree Physiol.* **16,** 847–852.

Lavigne, M. B., and Ryan, M. G. (1997). Growth and maintenance respiration rates of aspen, black spruce and jack pine stems at northern and southern BOREAS sites. *Tree Physiol.* **17,** 543–551.

Lavigne, M. B., Franklin, S. E., and Hunt, E. R., Jr. (1996). Estimating stem maintenance respiration rates of dissimilar balsam fir stands. *Tree Physiol.* **16,** 687–695.

Lavigne, M. B., Ryan, M. G., Anderson, D. E., Baldocchi, D. D., Crill, P. M., Fitzjarrald, D. R., Goulden, M. L., Gower, S. T., Massheder, J. M., McCaughey, J. H., Rayment, M., and Striegl, R. G. (1998). Comparing nocturnal eddy covariance measurements to estimates of ecosystem respiration made by scaling chamber measurements. *J. Geophys. Res.* **102,** 28,977–28,985.

Law, B. E., Ryan, M. G., and Anthoni, P. M. (1999). Seasonal and annual respiration of a ponderosa pine ecosystem. *Global Change Biol.* **5,** 169–182.

Li, M., and Jones, M. B. (1992). Effect of nitrogen status on the maintenance respiration of C_3 and C_4 *Cyperus* species. *In* "Molecular, Biochemical and Physiological Aspects of Plant Respiration" (H. Lambers and L. H. W. van der Plas, eds.), pp. 509–514. SPB Academic Publishing, The Hague, The Netherlands.

McCree, K. J. (1970). An equation for the rate of respiration of white clover plants grown under controlled conditions. *In* "Prediction and Measurement of Photosynthetic Productivity" (I. Setlik, ed.), pp. 221–229. Pudoc, Wageningen, The Netherlands.

McCree, K. J. (1982). Maintenance requirements of white clover at high and low growth rates. *Crop Sci.* **22,** 345–351.

McCree, K. J. (1983). Carbon balance as a function of plant size in sorghum plants. *Crop Sci.* **23,** 1173–1177.

McCullough, D. E., and Hunt, L. A. (1993). Mature tissue and crop canopy respiratory characteristics of rye, triticale and wheat. *Ann. Bot.* **72,** 269–282.

McDermitt, D. K., and Loomis, R. S. (1981). Elemental composition of biomass and its relation to energy content, growth efficiency, and growth yield. *Ann. Bot.* **48,** 275–290.

Melillo, J. M., Prentice, I. C., Farquhar, G. D., Schulze, E. D., and Sala, O. E. (1996). Terrestrial biotic responses to environmental change and feedbacks to climate. *In* "Climate Change 1995—The Science of Climate Change" (J. T. Houghton, L. G. Meira Filho, B. A. Callander, N. Harris, A. Katenberg, and K. Maskell, eds.), pp. 445–481. Cambridge Univ. Press, Cambridge.

Merino, J., Field, C., and Mooney, H. A. (1982). Construction and maintenance costs of mediterranean-climate evergreen and deciduous leaves. I. Growth and CO_2 exchange analysis. *Oecologia* **53,** 208–213.

Möller, C. M., Müller, D., and Nielsen, J. (1954). Graphic presentation of dry matter production of European beech. *Det Forstlige Forsøgsvæsen i Danmark* **21,** 327–335.

Müller, D., and Nielsen, J. (1965). Production brute, pertes par respiration et production nette dans la forêt ombrophile tropicale. *Forst. Forsøgsv. Danm.* **29,** 69–160.

Murtagh, G. J., Halligan, E. A., and Greer, D. H. (1987). Components of growth and dark respiration of kikuyu (*Pennisetum clandestinum* Chiov.) at various temperatures. *Ann. Bot.* **59,** 149–157.

Murty, D., McMurtrie, R. E., and Ryan, M. G. (1996). Declining forest productivity in aging forest stands: A modeling analysis of alternative hypotheses. *Tree Physiol.* **16,** 187–200.

Nishioka, M., Hozumi, K., Kirita, H., and Nagano, M. (1978). Estimation of canopy photosynthesis and respiration. *In* "Biological Production in a Warm-Temperate Evergreen Oak Forest of Japan (JIBP Synthesis Vol. 18)" (T. Kira, Y. Ono, and T. Hosokawa, Eds.), pp. 99–111. Univ. Tokyo Press, Tokyo.

Noguchi, K., and Terashima, I. (1997). Different regulation of leaf respiration between *Spinacia oleracea,* a sun species, and *Alocasia odora,* a shade species. *Physiol. Plant.* **101,** 1–7.

Noguchi, K., Sonoike, K., and Terashima, I. (1996). Acclimation of respiratory properties of leaves of *Spinacia oleracea* L., a sun species, and of *Alocasia macrorrhiza* (L.) G. Don., a shade species, to changes in growth irradiance. *Plant Cell Physiol.* **37,** 377–384.

Odum, H. T. (1970). An emerging view of the ecological system at El Verde. *In* "A Tropical Rain Forest" (H. T. Odum and R. F. Pigeon, eds.), pp. I-191–I-289. U.S. Atomic Energy Commission, Washington, D.C.

Pearcy, R. W., and Sims, D. A. (1994). Photosynthetic acclimation to changing light environments: Scaling from the leaf to the whole plant. *In* "Exploitation of Environmental Heterogeneity by Plants" (M. M. Caldwell and R. W. Pearcy, eds.), pp. 145–174. Academic Press, San Diego.

Penning de Vries, F. W. T. (1972). Respiration and growth. *In* "Crop Processes in Controlled Environments" (A. R. Rees, K. E. Cockshull, D. W. Hand, and R. G. Hurd, Eds.), pp. 327–348. Academic Press, London.

Penning de Vries, F. W. T. (1974). Substrate utilization and respiration in relation to growth and maintenance in higher plants. *Neth. J. Agric. Sci.* **22,** 40–44.

Penning de Vries, F. W. T. (1975a). The cost of maintenance processes in plant cells. *Ann. Bot.* **39,** 77–92.

Penning de Vries, F. W. T. (1975b). Use of assimilates in higher plants. *In* "Photosynthesis and Productivity in Different Environments" (J. P. Cooper, ed.), pp. 459–480. Cambridge Univ. Press, Cambridge.

Penning de Vries, F. W. T., and van Laar, H. H. (1975). Substrate utilization in germinating seeds. *In* "Environmental Effects on Crop Physiology" (J. J. Landsberg and C. V. Cutting, eds.), pp. 217–228. Academic Press, London.

Penning de Vries, F. W. T., Brunsting, A. H. M., and van Laar, H. H. (1974). Products, requirements and efficiency of biosynthesis: A quantitative approach. *J. Theor. Biol.* **45,** 339–377.

Penning de Vries, F. W. T., van Laar, H. H., and Chardon, M. C. M. (1983). Bioenergetics of growth of seeds, fruits, and storage organs. *In* "Potential Productivity of Field Crops Under Different Environments," pp. 37–59. International Rice Research Institute, Los Baños, Laguna, Philippines.

Penning de Vries, F. W. T., Jansen, D. M., ten Berge, H. F. M., and Bakema, A. (1989). "Simulation of Ecophysiological Processes of Growth in Several Annual Crops." Pudoc, Wageningen, The Netherlands.

Precht, H., Christophersen, J., Hensel, H., and Larcher, W. (1973). "Temperature and Life." Springer-Verlag, New York.

Price, C. A. (1960). Respiration and development of vegetative plant organs and tissues. *In* "Handbuch der Pflanzenphysiologie, Band XII, Pflanzenatmung Einschliesslich Gärunge und Säurestoffwechsel, Teil 2" (W. Ruhland, ed.), pp. 493–520. Springer-Verlag, Berlin.

Reichle, D. E. (1975). Advances in ecosystem analysis. *BioScience* **25,** 257–264.

Risser, P. G., Birney, E. C., Blocker, H. D., May, S. W., Parton, W. J., and Wiens, J. A. (1981). "The True Prairie Ecosystem." Hutchinson Ross, Stroudsburg, Pennsylvania.

Runyon, J., Waring, R. H., Goward, S. N., and Welles, J. M. (1994). Environmental limits on net primary production and light-use efficiency across the Oregon transect. *Ecol. Applic.* **4,** 226–237.

Ryan, M. G. (1990). Growth and maintenance respiration in stems of *Pinus contorta* and *Picea engelmannii. Can. J. For. Res.* **20,** 48–57.

Ryan, M. G. (1991). Effects of climate change on plant respiration. *Ecol. Applic.* **1,** 157–167.

Ryan, M. G. (1995). Foliar maintenance respiration of subalpine and boreal trees and shrubs in relation to nitrogen content. *Plant Cell Environ.* **18,** 765–772.

Ryan, M. G., and Waring, R. H. (1992). Maintenance respiration and stand development in a subalpine lodgepole pine forest. *Ecology* **73,** 2100–2108.

Ryan, M. G., and Yoder, B. J. (1997). Hydraulic limits to tree height and tree growth. *BioScience* **47,** 235–242.

Ryan, M. G., Hubbard, R. M., Clark, D. A., and Sanford, R. L., Jr. (1994a). Woody-tissue respiration for *Simarouba amara* and *Minquartia guianensis,* two tropical wet forest trees with different growth habits. *Oecologia* **100,** 213–220.

Ryan, M. G., Linder, S., Vose, J. M., and Hubbard, R. M. (1994b). Dark respiration of pines. *Ecol. Bull.* **43,** 50–63.

Ryan, M. G., Gower, S. T., Hubbard, R. M., Waring, R. H., Gholz, H. L., Cropper, W. P., Jr, and Running, S. W. (1995). Woody tissue maintenance respiration of four conifers in contrasting climates. *Oecologia* **101,** 133–140.

Ryan, M. G., Hubbard, R. M., Pongracic, S., Raison, R. J., and McMurtrie, R. E. (1996). Foliage, fine-root, woody-tissue and stand respiration in *Pinus radiata* in relation to nitrogen status. *Tree Physiol.* **16,** 333–343.

Ryan, M. G., Binkley, D., and Fownes, J. H. (1997). Age-related decline in forest productivity: Pattern and process. *Adv. Ecol. Res.* **27,** 213–262.

Ryan, M. G., Lavigne, M. B., and Gower, S. T. (1998). Annual carbon cost of autotrophic respiration in boreal forest ecosystems in relation to species and climate. *J. Geophys. Res.* **102,** 28,871–28,883.

Sale, P. J. M. (1974). Productivity of vegetable crops in a region of high solar input. III. Carbon balance of potato crops. *Aust. J. Plant Physiol.* **1,** 283–296.

Sharp, R. E., Matthews, M. A., and Boyer, J. S. (1984). Kok effect and the quantum yield of photosynthesis. Light partially inhibits dark respiration. *Plant Physiol.* **75,** 95–101.

Shone, M. G. T., and Gale, J. (1983). Effect of sodium chloride stress and nitrogen source on respiration, growth and photosynthesis in lucerne (*Medicago sativa* L.). *J. Exp. Bot.* **34,** 1117–1125.

Sprugel, D. G. (1990). Components of woody-tissue respiration in young *Abies amabilis* (Dougl.) Forbes trees. *Trees* **4,** 88–98.

Sprugel, D. G., Ryan, M. G., Brooks, R. J., Vogt, K. A., and Martin, T. A. (1995). Respiration from the organ level to the stand. *In* "Resource Physiology of Conifers" (W. K. Smith and T. M. Hinckley, eds.), pp. 255–299. Academic Press, San Diego.

Stahl, R. S., and McCree, K. J. (1988). Ontogenetic changes in the respiration coefficients of grain sorghum. *Crop Sci.* **28,** 111–113.

Szaniawski, R. K. (1985). Homeostasis in cucumber plants during low temperature stress. *Physiol. Plant.* **64,** 212–216

Szaniawski, R. K., and Kielkiewicz, M. (1982). Maintenance and growth respiration in shoots and roots of sunflower plants grown at different root temperatures. *Physiol. Plant.* **54,** 500–505.

Teal, J. M. (1962). Energy flow in the salt marsh ecosystem of Georgia. *Ecology* **43,** 614–624.

Thomas, R. B., and Griffin, K. L. (1994). Direct and indirect effects of atmospheric carbon dioxide enrichment on leaf respiration of *Glycine max* (L.) Merr. *Plant Physiol.* **104,** 355–361.

Thomas, M. D., and Hill, G. R. (1949). Photosynthesis under field conditions. *In* "Photosynthesis in Plants" (J. Franck and W. E. Loomis, eds.), pp. 19–52. Iowa State College Press, Ames, Iowa.

Thomas, R. B., Reid, C. D., Ybema, R., and Strain, B. R. (1993). Growth and maintenance components of leaf respiration of cotton grown in elevated carbon dioxide partial pressure. *Plant Cell Environ.* **16,** 539–546.

Thornley, J. H. M. (1970). Respiration, growth and maintenance in plants. *Nature (London)* **227,** 304–305.

Thornley, J. H. M., and Johnson, I. R. (1990). "Plant and Crop Modelling: A Mathematical Approach to Plant and Crop Physiology." Oxford Univ. Press, Oxford.

Vertregt, N., and Penning de Vries, F. W. T. (1987). A rapid method for determining the efficiency of biosynthesis of plant biomass. *J. Theor. Biol.* **128,** 109–119.

Villar, R., Held, A. A., and Merino, J. (1995). Dark leaf respiration in light and darkness of an evergreen and a deciduous plant species. *Plant Physiol.* **107,** 421–427.

Vogel, J. (1997). "Carbon and Nitrogen Dynamics for Boreal Jack Pine Stands with Different Understory Vegetation." M.Sc. thesis, Univ. Wisconsin, Madison, Wisconsin.

Weintraub, R. L. (1944). Radiation and plant respiration. *Bot. Rev.* **10,** 383–459.

Whittaker, R. H., and Woodwell, G. M. (1969). Structure, production and diversity of the oak–pine forest at Brookhaven, New York. *J. Ecol.* **57,** 155–174.

Williams, K. Percival, F., Merino, J., and Mooney, H. A. (1987). Estimation of tissue construction cost from heat of combustion and organic nitrogen content. *Plant Cell Environ.* **10,** 725–734.

Winzeler, M., McCullough, D. E., and Hunt, L. A. (1988) Genotypic differences in dark respiration of mature leaves in winter wheat (*Triticum aestivum* L.). *Can J. Plant Sci.* **68,** 669–675.

Woodwell, G. M., Houghton, R. A., Hall, C. A. S., Whitney, D. E., Moll, R. A., and Juers, D. W. (1979). The flax pond ecosystem study: The annual metabolism and nutrient budgets of a salt marsh. *In* "Ecological Processes in Coastal Environments" (R. L. Jefferies and A. J. Davy, eds.), pp. 491–511. Blackwell, Oxford.

Wullschleger, S. D., Norby, R. J., and Gunderson, C. A. (1992). Growth and maintenance respiration in leaves of *Liriodendron tulipifera* L. exposed to long-term carbon dioxide enrichment in the field. *New Phytol.* **121,** 515–523.

Wullschleger, S. D., Norby, R. J., and Hanson, P. J. (1995). Growth and maintenance respiration in stems of *Quercus alba* after four years of CO_2 enrichment. *Physiol. Plant.* **93,** 47–54.

Wullschleger, S. D., Hanson, P. J., and Edwards, G. S. (1996). Growth and maintenance respiration in leaves of northern red oak seedlings and mature trees after 3 years of ozone exposure. *Plant Cell Environ.* **19,** 577–584.

Wullschleger, S. D., Norby, R. J., Love, J. C., and Runck, C. (1997). Energetic costs of tissue construction in yellow-poplar and white oak trees exposed to long-term CO_2 enrichment. *Ann. Bot.* **80,** 289–297.

Yoda, K. (1978). Estimation of community respiration. *In* "Biological Production in a Warm-Temperate Evergreen Oak Forest of Japan (JIBP Synthesis Vol. 18)" (T. Kira, Y. Ono, and T. Hosokawa, eds.), pp. 112–131. Univ. Tokyo Press, Tokyo.
Yoda, K. (1983). Community respiration in a lowland rain forest in Pasoh, peninsular Malaysia. *Jpn. J. Ecol.* **33,** 183–197.
Yoder, B. J., Ryan, M. G., Waring, R. H., Schoettle, A. W., and Kaufmann, M. R. (1994). Evidence of reduced photosynthetic rates in old trees. *For. Sci.* **40,** 513–527.
Zelitch, I. (1971). "Photosynthesis, Photorespiration, and Plant Productivity." Academic Press, New York.

4

Phenology, Growth, and Allocation in Global Terrestrial Productivity

Robert B. Jackson, Martin J. Lechowicz, Xia Li, and Harold A. Mooney

I. Introduction

A symposium held in Montpellier, France, in 1965 on the functioning of terrestrial ecosystems and the link to primary production (Eckardt, 1968) was an important milestone in our understanding of the controls and distribution of the Earth's primary productivity. At this symposium Professor M. Monsi presented his classic paper on mathematical models of plant communities (Monsi, 1968). This extraordinarily rich paper contained new and fundamental approaches to understanding the relationships between canopy distribution in time and space and stand productivity. Among the many issues discussed was the relationship between allocation and productivity. Monsi showed, for example, how production was dependent on the proportion of dry matter allocated to photosynthetic tissue, as well as the proportion of time this tissue was active on the plant (its phenology). Because of the analogy of "compound interest," the gain in dry matter is higher for photosynthetic tissue than for nonphotosynthetic tissue when there are no environmental limitations.

The ideas we discuss in this chapter on the importance of phenology, growth, and allocation for terrestrial primary productivity build on the foundation laid by Monsi (1968). The topic is far too broad to be reviewed comprehensively here; consequently, we provide background and examples for a subset of issues relevant to global change. We highlight trees and forest ecosystems because these systems contain the predominant fraction of global terrestrial carbon stores in plants. We end by highlighting several promising areas of research: (1) extension of the present theory of the timing of allocation to include the phenology of belowground organs and reproduction, (2) techniques for improving estimates of belowground productivity and patterns of global allocation above- and belowground, and (3) the use of remote sensing combined with modeling for estimating global phenology and productivity.

II. Phenology

A. The History of Phenological Studies

Phenology is the study of the seasonal timing of biological events (Lieth, 1974, 1994), typically those that are readily observable such as the timing of bud burst in spring, the ripening of fruit, or the coloring of leaves in fall. Less observable but equally important is belowground phenology, including the seasonal timing of fine root growth. Intimate knowledge of phenology was essential in preindustrial societies because phenological events were used to decide when to plant crops or gather natural products. The aphorism that "seed should be sown when oak leaves are as big as a sow's ear" springs from folk knowledge accumulated over millennia, knowledge held by people throughout north temperate zones where oaks predominate. The species of oak change from place to place, but the wood anatomy that characterizes oaks ensures that in a local forest they will be among the last trees to produce leaves (Lechowicz, 1984, 1995). The leafing of oak is a reliable natural indicator that the danger of late frost is past, and it is not surprising that this phenological event became a benchmark in characterization of the seasons.

Beginning in about the seventeenth century, folk knowledge of phenological events began to be transformed by the ongoing scientific revolution in western societies. There was a great vogue for publishing "calendars of nature" (Stillingfleet, 1791; Aikin, 1799; Howitt, 1835), essentially collations of phenological sequences drawn from folk knowledge. These were superseded as formal records of the phenology of agricultural and natural systems began to be kept, sometimes at the behest of governments (Clarke, 1936) or as a pastime of the aristocracy (Jenyns, 1846; Molesworth, and Ormerod, 1880; Sparks and Carey, 1995). There is an extraordinary number of these historical records of phenology, although many are published obscurely if published at all (Balatova-Tulackova, 1970). These records, which are assuming greater importance as we struggle to assess the possible impacts of climate change on natural ecosystems, are beginning to be collected and made more available in electronic databases (Lieth, 1994; Lechowicz and Koike, 1995; see also links to the International Phenology Network, http://mmf.math.rsu.ru/phenology/links.en.html; contact, listserv@nic.surfnet.nl). In the past few decades (see Schnelle, 1955; Lieth, 1974, 1994), phenology has also emerged as a scientific discipline dedicated to modeling and predicting the seasonal timing of biological events, not simply observing and recording them. It is this science that concerns us here, with an emphasis on woody species in natural ecosystems.

Most of the science has focused on the prediction of the seasonal timing of bud burst for leaves and flowers. The large part of the work is phenomenological and not well rooted in an understanding of the mechanisms underlying a phenological response. Emphasis is often on the prediction of the timing of an event in a given species using readily available data (daily temperature or hours of daylight). The most notable exception to this phenomenological approach is the detailed work on *Pinus sylvestris* by Sarvas (1974) and others who have studied the biochemistry and physiology of phenological responses such as dormancy and flowering (Luomajoki, 1986; Champagnat, 1989; Hänninen *et al.*, 1985; Cannell, 1990). Such detailed work is important for parameterizing future generations of phenological models.

B. Phenology and Phenomenological Models

Three elements are woven together in phenomenological models: a period of chilling in fall and winter, a period of warming in late winter and spring, and photoperiodic signals that may affect responses to temperature (Hunter and Lechowicz, 1992). Differences in the detail and form of the inputs define three types of phenomenological models: warming models, sequential models, and parallel models. The simplest and oldest phenological models use only the daily progression of temperature to predict the timing of events such as bud burst. An event is predicted to occur when a measure of cumulative warming (measured as degree-days) reaches a species- and event-specific threshold; tallying degree-days requires specifying a day of the year when the tally will begin and a base temperature above which temperature increments will be accumulated. Hence, despite the apparent simplicity of their dependence on only a single datum (daily mean temperature), these warming models in fact require the estimation of three parameters: a starting date, a base temperature, and a temperature sum.

A second class of phenomenological models, referred to as sequential (Hänninen, 1987), focuses on what sets the starting date when a plant begins to respond to and accumulate warming temperatures toward its threshold for a particular phenological event. Rather than assume a fixed starting date (e.g., January 1), these sequential models mandate that a chilling requirement be met before the plant responds to warming. By analogy to the accumulation of warmth in late winter and spring, the models tally chilling degree-days through the fall and winter. They also require specification of a day on which the chilling will begin to be tallied, but this is often readily tied to such recognizable events as leaf fall in deciduous species or the day of the first killing frost.

The third model class, parallel (Hänninen, 1987), is so named because these models allow an interaction between chilling and warming (usually more warming is required if less chilling has occurred). A variant in this class of models brings photoperiod into consideration, with the warming requirement reduced as photoperiod lengthens. The net effect of these parallel models is to reduce the likelihood that a period of warmth in midwinter, such as a February thaw, will trigger bud burst when the risk of a return to freezing conditions is still great.

With their focus on the timing of an event in relation to environmental conditions, present phenological models fail to account for the possible dependence of a phenological response on endogenous factors. We might expect coordination of different phenological responses in plants, not simply because they are responding to similar environmental cues but also because they depend on common internal resources. For example, both leafing and flowering of trees draw on stores of carbohydrates, proteins, and minerals (Kozlowski, 1992). There is of necessity some coordination in the timing of these demands on stored resources. Some species produce flowers and leaves simultaneously, whereas others separate these two competing resource demands to different parts of the season (Lechowicz, 1995). Similar differences arise in the timing of root and shoot growth; some species synchronize allocation to above- and belowground growth and others separate them (Lyr and Hoffmann, 1967; Riedacker, 1976; Harris *et al.*, 1995). In this view it is clear that phenology is really a question of the timing of allocation in plants, not simply the prediction of responses to exogenous environmental cues. Phenological responses to exogenous cues are likely to have evolved as a way to coordinate resource allocation for competing plant functions in a seasonal climate. Our understanding of the mechanisms underlying a phenological response cannot be sought solely in the biochemistry of chilling factors or phytochrome-mediated responses to photoperiodic cues. We also need to consider the costs imposed by a phenological event and the return in acquired resources that follows from alternative timing of the event—the legacy of Monsi's (1968) pioneering insight.

C. Toward a Theory of Functional Types in Phenology

Kikuzawa (1991, 1995; Kikuzawa *et al.*, 1996) developed a theory of phenology that begins to move beyond phenomenological models toward an analysis of the costs and benefits that accrue from alternative strategies of phenology and allocation. He focuses on the timing of allocation of resources to leaves, recognizing two alternative strategies in leaf phenology: one favored in high-light environments that are not limited by nutrient or water resources and the other in environments that are more resource limited. He shows that resource-rich environments favor a steady production of short-lived leaves through the growing season and their array in multilayer canopies with vertically oriented shoots (Kikuzawa *et al.*, 1996). Resource-poor situations conversely favor the production of a single cohort of long-lived leaves each season and their array in a monolayer canopy with more horizontally oriented shoots. These analyses are an important first step toward a general theory that would place the many observations of phenological responses to exogenous factors (e.g., temperature, photoperiod) in the context of alternative plant production strategies. This is the fundamental context of phenology that Monsi (1968) recognized when he identified the consequences of the timing of allocation to different plant functions.

There are some important challenges to be met if we are to come to a comprehensive theory of phenology relevant both to functional ecology and to ecosystem ecology. From the viewpoint of functional ecology, we must extend the rudimentary theory to include allocation to belowground tissues and to reproduction. Kikuzawa's analysis (Kikuzawa, 1991, 1995) emphasizes light resources, which have different temporal patterns of availability compared to water and nutrient resources. Consequently the cost–benefit analysis of allocations to above- versus belowground organs involved in resource acquisition is likely to differ and lead to a greater diversity of phenological strategies at the whole-plant level. The same argument applies to reproduction, whereby the seasonality of pollinator availability and suitable conditions for seedling establishment will influence the timing of allocation to reproduction within and among years. If we secure a more comprehensive theory of the timing of allocation to different vegetative organs and to reproduction, we will also serve the needs of ecosystem ecologists interested in global change. For example, our relative ignorance of the contribution of belowground biomass to net primary productivity is a significant problem in the analysis of ecosystem responses to global change. If we have a sound theoretical basis to link the wealth of observational data on the timing of aboveground production to belowground production, then we can refine our estimates of seasonal production in diverse ecosystems.

D. Phenology and Climate Change

It is clear that the timing of phenological events can directly and significantly affect plant growth and reproduction. Tree breeders recognize that genotypes of deciduous trees with earlier leafing and/or later leaf senescence (longer periods of annual production) have higher rates of growth (e.g., Wang and Tigerstedt, 1993 for *Betula;* Farmer, 1996 for *Populus*). The loss of flowers in a late spring frost and the failure to mature fruit before an early fall frost are problems familiar to growers of fruit and nut trees (Alston and Tobutt, 1989; Germain, 1990).

As climate at a location changes, the timing of a particular phenological response may shift, as may the costs and benefits associated with the phenological response. Photoperiodic cues on ecological time scales and in a specific location are stable, whereas temperature regimes are not. Consequently, trees under climate change could end up responding to environmental cues that are poorly synchronized with local seasonal cycles (i.e., climate change could decouple existing photoperiod/climate relationships). On the other hand, though the phenology of some trees is apparently influenced by photoperiod (Nizinski and Saugier, 1988; Heide, 1993), in general it appears that tree phenology responds predominantly to thermal cues rather than to photoperiodic ones (Hunter and Lechowicz, 1992; Kramer, 1995a).

There is a fairly large, but somewhat inconsistent, literature on the effects of phenology when trees are moved hundreds of kilometers within or close to the limits of their present geographic range. Most commonly, trees moved to higher latitudes respond with an earlier phenology than do the local stock—they leaf out and flower earlier (Kramer, 1995a; Deans and Harvey, 1995; Farmer, 1996). In general, displacements of a few hundred kilometers seem to be without penalty and often lead to growth advantages (Farmer, 1996), at least in fairly short provenance trials. In the longer term, climatic extremes that occur only very infrequently may pose problems for the displaced stock compared to the local provenance.

Excellent examples of the importance of phenology for ecosystem productivity come from recent eddy covariance and satellite studies in forests. Goulden *et al.* (1996) examined changes in productivity at the Harvard Forest in response to interannual climate variability. In 2 years of the study, leaves emerged 6–10 days earlier than in other years, leading to a cumulative increase in net ecosystem exchange of 20–40%. The timing of the end of the growing season was also important. Canopy senescence occurred 5–10 days later in 1992 and 1993 than in the other years of the study, increasing annual gross production by 50 g C m^{-2} (Goulden *et al.*, 1996). Such a lengthening of the growing season may be likely during the next few centuries, with large potential consequences for ecosystem productivity. In fact, recent atmospheric and satellite analyses show up to a 12-day advancement in the initiation of CO_2 uptake from 45°N to 70°N latitudes in the past few decades (Keeling *et al.*, 1996; Myneni *et al.*, 1997).

Storage poses a significant complication in viewing phenology as an expression of the timing of allocation in that the production and utilization of stored resources are uncoupled. In general there is a regular annual cycle of movement of photosynthate into and out of storage pools. In both coniferous and hardwood trees, stored carbohydrate reserves are highest in winter, drop to a low midway through the growing season, and then gradually recover to winter levels (Wargo, 1971; Essiamah and Eschrich, 1985; Egger *et al.*, 1996; Hansen *et al.*, 1996). There can also be short-term cycling into and out of storage reserves within the growing season that is associated with rhythmic shoot growth (Alaoui-Sosse *et al.*, 1994). In many trees there are also interannual cycles of storage buildup, transformation (e.g., the conversion of starch to soluble sugars in winter), and depletion associated with intervals between reproductive events (Silvertown, 1980; Monselise and Goldschmidt, 1982; Kozlowski, 1992).

Phenological studies traditionally emphasize the timing of leaf emergence in spring, but in fact this only marks the onset of the annual production cycle in deciduous trees. In temperate forests the timing of leaf-out typically spans 3–6 weeks (Lechowicz, 1984) and it is this "green wave" that most satellite indices track (Schwartz, 1994). To put this green wave in the context of seasonal patterns of actual production, it helps to consider the different ways that trees organize the actual timing of shoot extension growth through the warm season, a phenological character that applies equally to evergreen and deciduous species. There are fewer data of this sort available, but we can draw on a careful report by Anic (1964) to illustrate the diversity of shoot growth phenology among European forest trees. Between 1948 and 1960, Anic regularly observed thousands of trees growing in the vicinity of Zagreb and recorded their shoot extension every few days throughout the growing season. He monitored 9 gymnosperm and 30 angiosperm tree species that occurred naturally. The local forests, which are described by phytosociologists as a Querceto roboris–Carpinetum betuli type, occur on brown forest soils in a region with a continental climate showing strong seasonality. The mean annual temperature is 10.6°C with an annual mean maximum temperature of 34.9°C in summer and a minimum of −15.7°C in winter. Mean annual precipitation is 871 mm and is evenly distributed throughout the year. The forests are part of the Balkan refugium from which trees recolonized northern Europe after the Quaternary glaciations (Bennett *et al.,* 1991; Willis, 1994). Anic's (1964) observations of the seasonality of shoot growth are therefore relevant to a broader geographic range in Europe than might initially be apparent.

The annual cycle of shoot growth of trees in this region may be characterized primarily by (1) the time at which the shoot growth begins and (2) how long it proceeds (Fig. 4-1). The onset of shoot growth spans about 4–5 weeks in spring, beginning in early April for some species but as late as early May for others. There is a much greater difference among species in their duration of shoot growth, which ranges from 44 to 127 days. Such interspecific variation in the duration of shoot growth is correlated with annual height growth (Fig. 4-1), which is a good indicator of tree productivity. These different seasonal patterns of shoot growth could indicate important differences among species in their responses to climate change. We might assume, for example, that the productivity of species with longer periods of shoot extension would be subject not only to adverse climatic events such as late spring frosts, but also summer droughts. With a few exceptions, however,

Forest saplings near Zagreb, 1948-1960

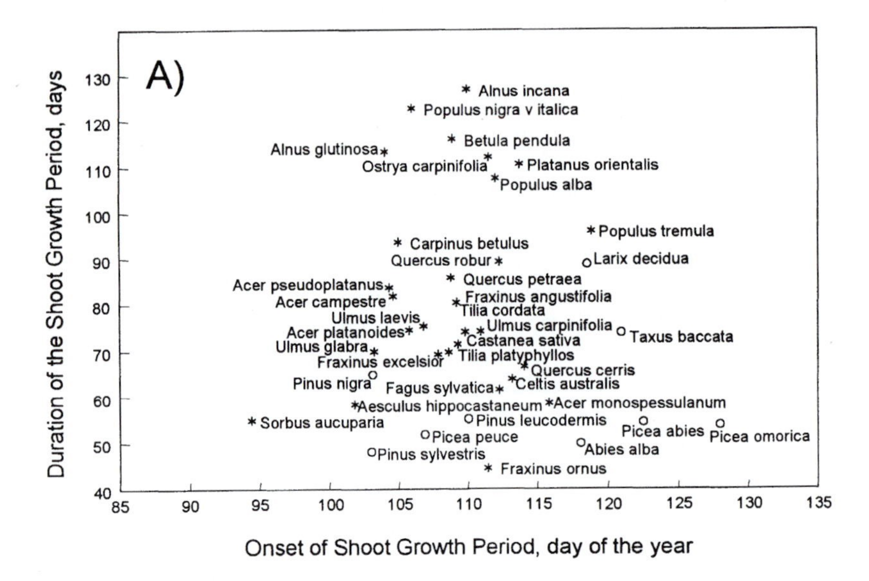

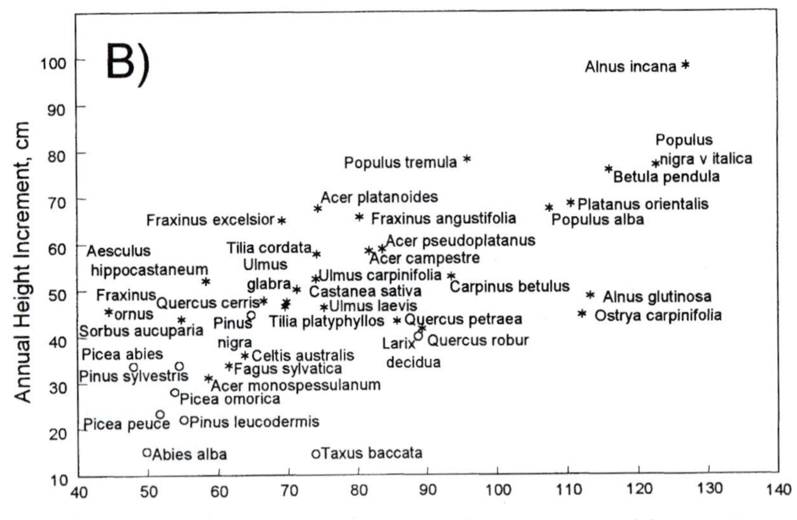

Figure 4-1 The seasonality of shoot growth in 30 angiosperm and 9 gymnosperm tree species growing near Zagreb in former Yugoslavia (data from Anic, 1964). (A) The relationship between the timing of the onset of shoot growth and its duration; (B) the correlation between annual height growth and the duration of shoot growth. Open symbols are needle leaf gymnosperms; asterisks are broadleaf angiosperms. The values are the means of observations made between 1948 and 1960.

that may be an unwarranted assumption because most species achieve the majority of their annual height increase in an initial flush of growth.

The diversity of phenological responses and associated variation in growth among this broad sample of European trees is an indication that the forests we have now are not necessarily the forests of the future. If climate change alters seasonality sufficiently, the competitive interactions among these species will likely shift and lead to forests of different composition and different predominant phenology. In the previous interglacial period *Picea omorica* was dominant throughout Europe, but in the current interglacial it is restricted to a small part of former Yugoslavia (Bennett *et al.*, 1991). This species has the latest onset and one of the shortest durations of shoot growth of any of the European forest trees (Fig. 4-1). *Picea abies,* the dominant European spruce today, is similar in phenology to *P. omorica.* At the ecosystem scale, should we take the shifting dominance in the European spruces as an indication that the part of the annual cycle in which the forest functions as a carbon sink might be altered under climate change? We do not know, but the possibility that forest composition may differ in a future climate, with effects on the seasonality of production, is an important unknown. This possibility also illustrates the potential longer term importance of biodiversity to ecosystem functioning: species that are a minor component of present systems may become more important under climate change, and their elimination now might limit ecosystem functioning in the future.

III. Growth and Allocation

A. Growth Models

Ecosystem productivity depends not only on the timing of allocation to different aspects of plant growth, but also the nature of the growth functions that determine future rates of production. A critical issue is balancing allocation to belowground tissues involved in soil resource uptake with allocation to shoots involved in the capture of carbon and energy. For forest trees, growth models fall into two broad categories, those emphasizing the development of tree form and those emphasizing the process of growth. Models of tree form define rules for branching architecture in considerable detail, but devote less attention to productivity processes that determine how much material is available for growth. Conversely, there are a number of detailed models of the physiological processes underlying growth that adopt simplified tree forms (e.g., root/stem/leaf without explicit architectural detail). Models such as LIGNUM (Pertunnen *et al.*, 1996) and AMAPpara (Forcaud *et al.*, 1998) set out to combine models of form and process to more realistically mimic tree growth.

This is not the place for a full review of whole-tree growth models, but it is useful to consider how the available models treat allocation processes. If we consider phenology as essentially the timing of resource allocation, then the way such models deal with allocation is important (Nikinmaa, 1992; Cannell and Dewar, 1994). Approaches to modeling the allocation of photosynthate to competing functions or structures in whole-plant growth models basically fall into two classes. One group uses the observed forms of trees to dictate allocational rules that lead to similar tree form; the other attempts to identify allocational rules with reference to actual physiological processes. In general, there are far too few data on the mechanisms governing the partitioning of biomass, and no compelling reason at this time to favor one modeling strategy over another.

The models that depend on observed form to guide allocation are easier to parameterize and were generally developed earlier. The simplest approach (partitioning rules) is to assign fixed partitioning coefficients that define the allocation of photosynthate. The effect of this fixed partitioning is a changing ratio in the accumulated investment in different structures and functions over time. A second approach (allometric rules) posits fixed ratios between certain structural parts of the plant, with photosynthate partitioned to maintain those ratios. A variant of this approach (functional rules) posits the maintenance of some functional balance in the plant that is maintained through appropriate allocation of photosynthate. The former variant emphasizes form, and the latter, function, but both act to keep a fixed set of proportions among plant parts and activities. All these approaches are essentially descriptive and do not allow much dynamic response to changing environmental conditions. The proportionate investments among competing structures are set by empirical measurements of existing plants rather than by general physiological processes.

Two other approaches to modeling allocation build on processes that might affect changes in allocation during plant development or across environments. The first of these (priority rules) assigns priorities to each structure or function through a series of equations describing allocation when photosynthate is insufficient to meet all demands. The steeper the allocation curve for a particular structure or function, the more likely that demand will be met even when photosynthate is in short supply. On the other hand, this approach allows some photosynthate to be allocated to most structures or functions in all but the most adverse situations. Competing demands are met in parallel, not sequentially. The other more mechanistic approach (transport rules) determines the resistances to movement of photosynthate between sources and sinks and assumes that these resistances determine partitioning to competing structures or functions. These two approaches are really variants on a common theme. Both try to define the basis for allocation by measurements of allocational process, rather than by simple measurements of the plant.

B. The Coupling of Root and Shoot Growth

Had we a more complete understanding of allocation, we would be better able to predict the general patterns of coordination between root and shoot growth. Despite the difficulties of studying belowground systems, the importance of root phenology and its coupling to shoot phenology have been recognized for centuries (see Lyr and Hoffmann, 1967). Belowground phenology is less likely to be useful as an indicator of global change than aboveground phenology because monitoring roots is more difficult and because there are fewer discrete events in root systems, such as flowering. But there can be no doubt that the effects of global change on belowground phenology and allocation will have important consequences for global productivity. Global warming may extend the length of the belowground growing season, and phenological differences among species may lead to important changes in species composition in response to global change.

The initiation of root growth in the spring generally occurs before the initiation of shoot growth, a fact observed by Theophrastus more than 2000 years ago (Lyr and Hoffmann, 1967). Roots of *Acer saccharinum* resume growth at a cooler temperature than do shoots (5° and 10°C, respectively), but root and shoot growth are not independent; an auxin signal from shoot buds of *Acer* appears necessary for the initiation of spring root growth (Richardson, 1958). In addition to beginning earlier, root growth also tends to extend longer into the fall and winter than does shoot growth, even after leaf abscission in deciduous forests (Hendrick and Pregitzer, 1996). In general, root growth slows or ceases during winter, but it may continue at soil temperatures very near freezing (Bhar *et al.*, 1970). It may also continue in deep soil layers when roots of the same individual plant are in frozen soil layers near the surface (Fernandez and Caldwell, 1975). Soil warming can extend the active period of root growth (Lyr and Hoffmann, 1967) and roots have been shown to grow 2–4 months longer in unfrozen soil covered in snow than in frozen soil (Kolesnikov, 1971). Clearly global warming has the potential to alter belowground productivity through an extended growing season.

In addition to generalizations about the initiation and timing of root and shoot growth, there are strong species effects on the coupling of above- and belowground phenology. The timing of root and shoot growth in seedlings of *Robinia pseudoacacia* was relatively synchronous, whereas that of *Pinus sylvestris* was almost completely segregated (Lyr and Hoffmann, 1967). Other trees studied in the experiment showed intermediate coupling. For cold-desert shrubs in the field, more than half of all root growth of *Artemisia tri-*

dentata occurred before any shoot growth was visible, and root growth was minimal during fruit development (Fernandez and Caldwell, 1975). For the co-occurring shrub *Atriplex confertifolia,* there was much greater overlap in the timing of root and shoot growth and almost all root growth in deeper layers occurred during fruit development. Although such phenological differences are important for niche partitioning, it not clear to what degree such differences are genetic compared to environmental. The greater the role of genetics in controlling phenology, the greater the potential for large changes in community composition with global warming and other aspects of environmental change.

IV. Future Directions

Estimating patterns of global phenology and productivity have been priorities since the efforts of the International Biological Programme. Refining current estimates is especially important for identifying the missing carbon sink and for establishing an accurate baseline for measuring the effects of global change. Two areas of pressing importance are (1) the use of remote sensing coupled with global models to refine estimates of phenology and aboveground net primary production (NPP), and (2) improving estimates of carbon turnover in roots and the soil.

A. Remote Sensing and Modeling of Global Phenology and Productivity

Remote sensing provides a powerful new tool for addressing patterns of global phenology, particularly when combined with global models. Satellite data from the Advanced Very High Resolution Radiometer (AVHRR) and normalized difference vegetation indices (NDVIs) have been used successfully to examine regional and global vegetation types and phenology for more than a decade (Tucker, 1986; Townshend and Justice, 1986; Moulin *et al.*, 1997). NDVIs are based on the normalized ratio of the visible and near-infrared (IR) spectral bands, because photosynthesizing vegetation reflects proportionally more radiation in the near-IR range than at visible wavelengths. Higher NDVI values indicate greater photosynthetically active leaf area (Hobbs, 1990). A framework for using the seasonality of NDVI to estimate phenology and productivity was developed by Running *et al.* (1995) (Fig. 4-2). These satellite data essentially integrate the tree-to-tree and place-to-place variation in observational records of phenological events.

Although satellite data can be used successfully for determining regional phenology, they are more useful in regions with pronounced phenological changes (e.g., temperate deciduous forests) than in regions without (e.g., tropical evergreen forest). Li *et al.* (unpublished observations) investigated

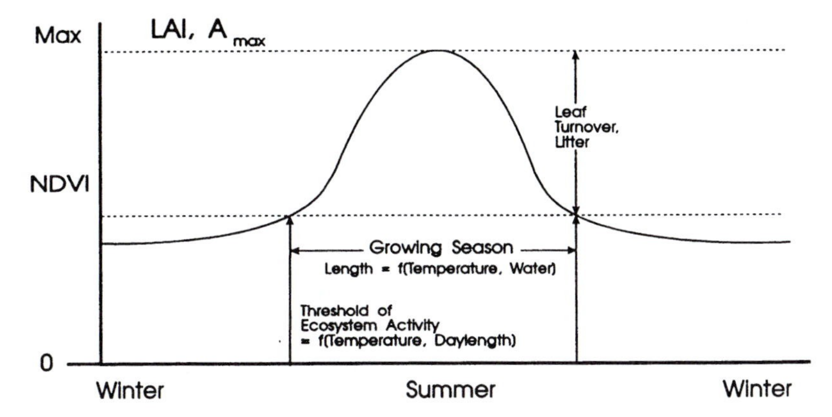

Figure 4-2 NVDI seasonality analysis. The potential use of satellite-derived NDVI data for determining phenology and primary productivity. The idealized curve shows how the amount of productivity (ΔNDVI), the length of the growing season, and the timing of leaf onset and offset are all potentially available from satellite data (from Running *et al.*, 1995). LAI, Leaf area index.

the use of changes in the NDVI [ΔNDVI = NDVI($t-1$) − NDVI(t), where t is a monthly time period] as an indicator or correlate of litterfall. They found that the metric worked well for systems with clear on/off leaf cycles (Fig. 4-3), but was not universally good at predicting litterfall (Fig. 4-4). The NDVI typically has a relatively poor temporal resolution for most grasslands and evergreen systems (though the relationship for the Spanish coniferous forest in Fig. 4-4 is relatively good, analyses for three other coniferous forests showed poor correlations).

White *et al.* (1997) developed a method of using satellite data to determine phenological development at the continental scale. The raw NDVI is used as an input to create a smoothed index that defines the beginning and end of the green period (Fig. 4-5). Based on this method they mapped the onset and offset of greenness for the vegetation of the United States (excluding shrubland and evergreen forests). From their analyses they noted that during a 3-year period from 1990 to 1992 the continentally averaged length of the growing season varied by 17 days, an amount that could have a large effect on regional productivity.

Remotely sensed data are especially useful for clarifying global patterns of phenology and productivity when combined with ecosystem and global modeling (Running, 1990). Phenology is becoming increasingly integrated into such models. Jorg and Heimann (1996) were among the first to address the role of phenology in regulating vegetative growth in a global ecosystem model. Their model simulated two phenological sequences, bud burst and

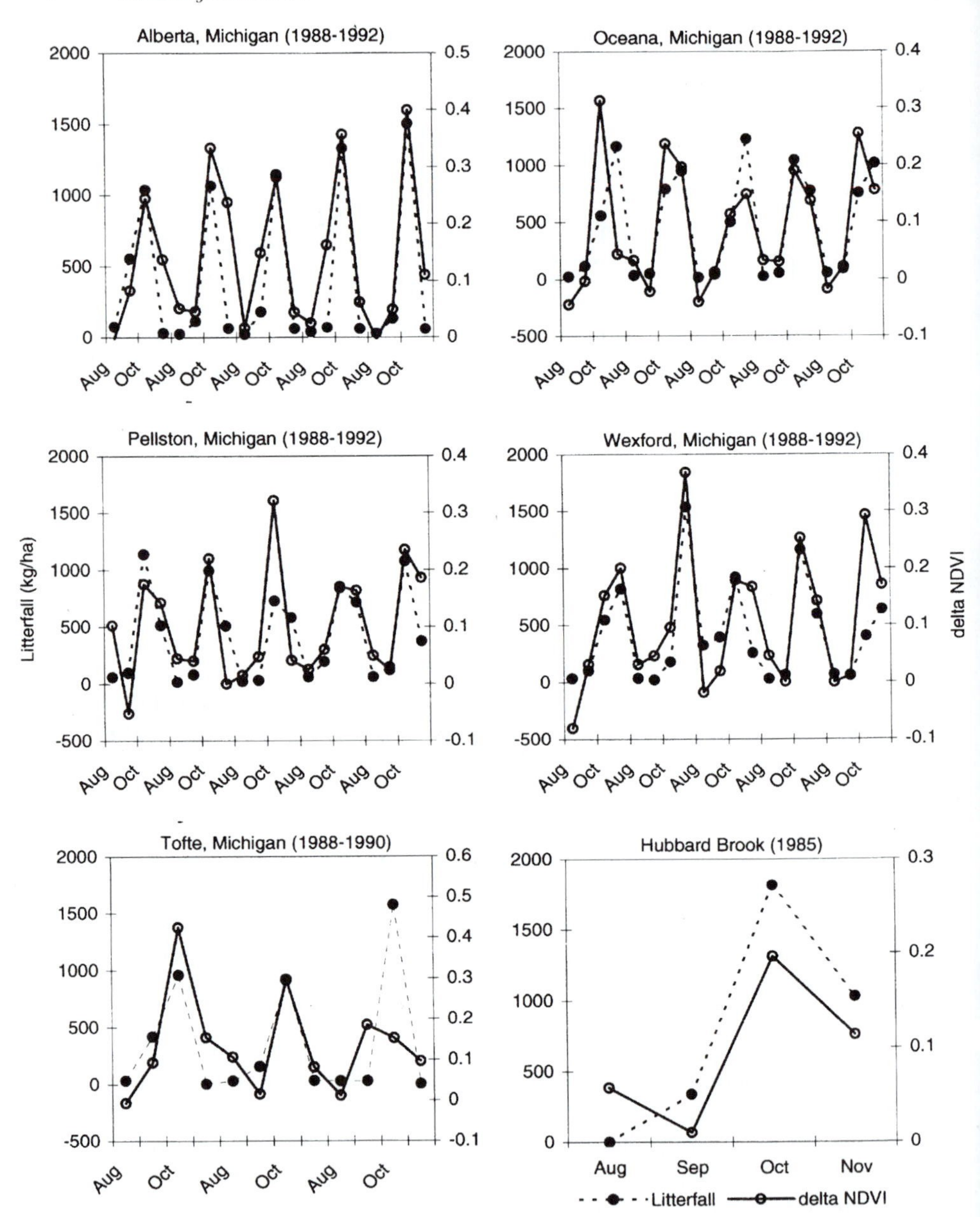

Figure 4-3 Correlations between ΔNDVI and litterfall for six deciduous forests sites (Li *et al.*, unpublished data). ●---●, litterfall; ○—○, ΔNDVI. ΔNDVI = NDVI($t-1$) − NDVI(t), where t is a monthly time period. Each year has four measurement dates of NDVI (Aug 15, Sept 15, Oct 15, and Nov 15) and the ΔNDVI value for each August measurement is approximately zero.

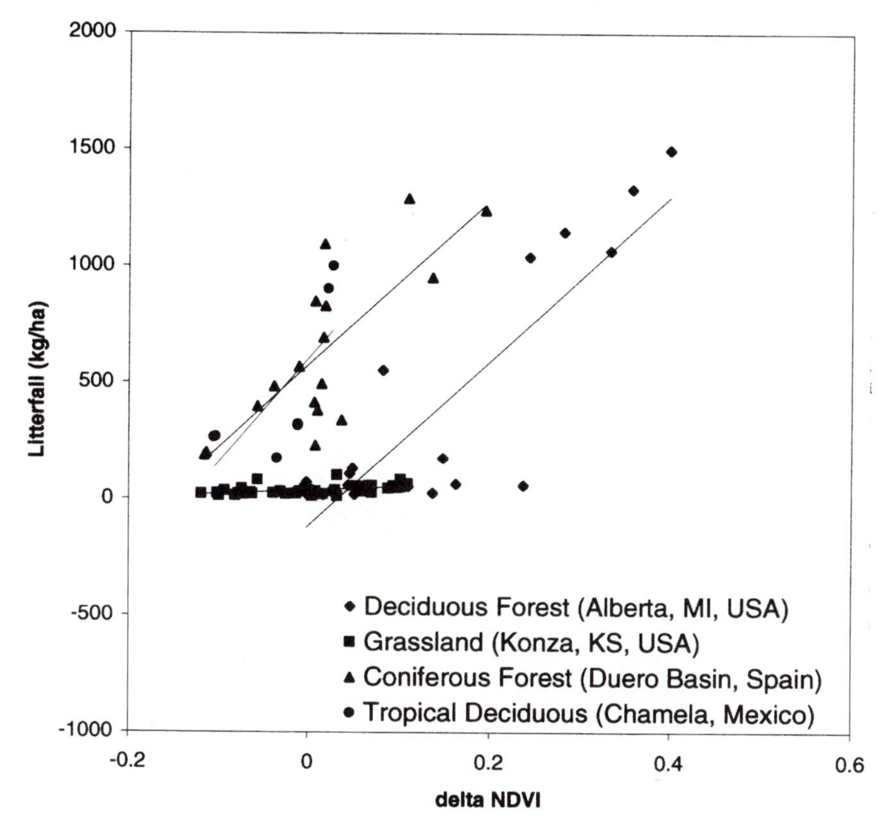

Figure 4-4 Relationships between litterfall and ΔNDVI for a hardwood forest and a temperate grassland in the United States, a coniferous forest in Spain, and a tropical deciduous forest in Mexico (Li *et al.*, unpublished data). $\Delta NDVI = NDVI(t-1) - NDVI(t)$, where t is a monthly time period.

leaf fall, based on a set of climate-dependent, biome-specific rules. Bud burst was predicted after the growing-degree-day requirements were satisfied. Litterfall was predicted based on the temperature below 5°C (for temperate deciduous ecosystems), the ratio of actual to potential evapotranspiration for tropical deciduous ecosystems, or a constant litterfall production for evergreen ecosystems. Because of almost universal limitations in knowledge of the timing of belowground allocation, their analysis did not explicitly consider root turnover or biomass; this limited the success of the model in systems with relatively large belowground allocation, such as grasslands (Jorg and Heimann, 1996). Burton and Cumming (1995) simulated plant response to global change using a forest patch model that integrates species-specific phenology and site-specific frost events. The simulations showed

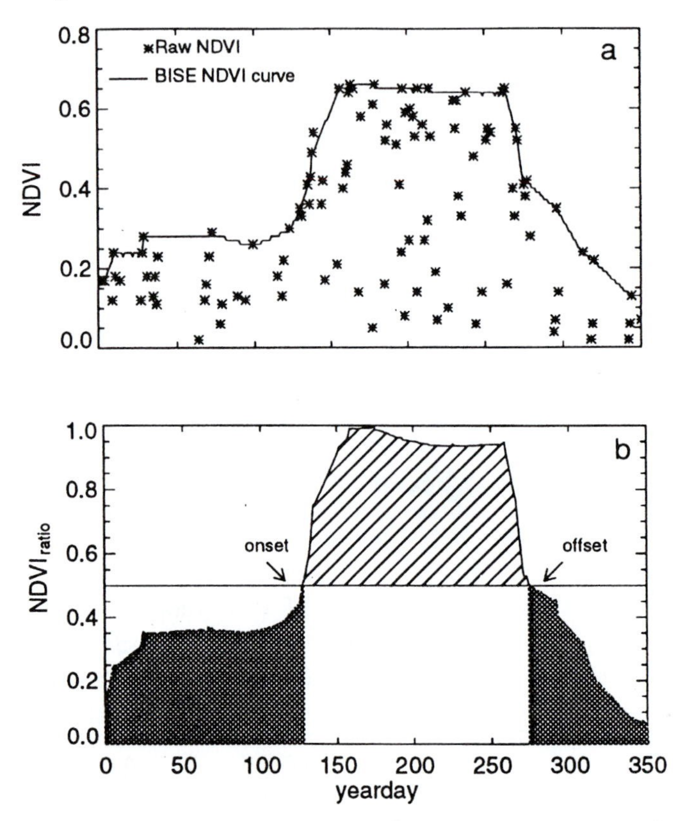

Figure 4-5 Daily AVHRR observations of a 1-km^2 pixel in a deciduous, broadleaf forest in the northeast United States (a) and 400 NDVI curves for the same site averaged and transformed to generate the best fit to the NDVI data (b). (a) The solid line represents the Best Index Slope Extraction (BISE) of the phenological time course. (b) The striped area corresponds to the growing season, the dark area corresponds to the nongrowing season, and the points where the curve crosses NDVI = 0.5 represent leaf onset and offset (from White *et al.*, 1997).

major shifts in equilibrium forest composition and productivity. Kramer (1995b, 1996) carried out a comprehensive analysis of the potential impact of altered phenology on the productivity of European trees under climate change. He compared estimates from a number of different models of tree growth and forest productivity involving contrasting assumptions about allocation and controls on net primary productivity. Phenological effects were important in all instances, about the same order of magnitude as changes in other key parameters influencing net production. By using remote-sensing data to establish regional estimates of production and litterfall, and to parameterize global models, significant progress should be forthcoming for global estimates of aboveground phenology and productivity. The challenge is to generate climate-driven phenological models that are applicable at the global scale.

B. Belowground Productivity and Its Distribution in the Soil

A second area where progress is needed is in refining estimates of belowground productivity and allocation to above- and belowground structures (Casper and Jackson, 1997; Jackson, 1998). Despite the difficulty in estimating belowground production, several relatively new techniques offer promise for improving our understanding of belowground phenology and production. Minirhizotrons have made the *in situ* study of root growth and demography feasible for field studies, highlighting the role of fine roots in ecosystem functioning (Taylor, 1987; Hendrick and Pregitzer, 1993). Although the technique is not new, the use of increasingly small rhizotron tubes, the ability to place these tubes in almost any field setting, and improvements in video technology have made the technique more widespread and much more fruitful. A second promising approach is the use of stable isotopes (e.g., ^{13}C, ^{2}H, and ^{18}O) for partitioning ecosystem carbon fluxes. In the next decade great strides will likely be made in using these isotopes to separate root respiration from microbial decomposition, to partition soil fluxes from canopy fluxes, and to estimate the coupling of CO_2 and H_2O fluxes in plant canopies (Broadmeadow and Griffiths, 1993).

A database of climate, soil, and root attributes was constructed to examine patterns of root biomass, length, and surface area with depth in the soil and their relationship to environmental variables globally (Jackson *et al.*, 1996, 1997; Canadell *et al.*, 1996). Based on results from hundreds of field studies, the data were used to examine patterns of global root biomass and annual belowground NPP. Total root biomass is approximately 290×10^{15} g globally (or approximately 140×10^{15} g C) (Table 4-1). Global fine root biomass is approximately 80×10^{15} g, equivalent in size to 5% of the atmospheric carbon pool, and the live fine root fraction is approximately half the total fine root pool (Table 4-1). If one assumes that fine roots turn over on average once a year, then 20×10^{15} g C of net primary production cycles through fine roots annually, one-third of total net primary production for plants globally (Jackson *et al.*, 1997).

Such an analysis necessarily glosses over many important phenomena, particularly spatial and temporal variation in ecosystems. Temporal variation is in part a synonym for phenology, and, as discussed above, such phenological data are especially lacking for root systems. Minirhizotrons, stable isotopes, and other techniques should go a long way to filling in such gaps and improving global estimates of belowground primary productivity.

Table 4-1 Global Classification Scheme[a]

Biome	Land area (10^6 km^2)	Total root biomass (10^9 Mg)	Total fine root biomass (10^9 Mg)	Live fine root biomass (10^9 Mg)
Tropical rainforest	17.0	83	9.7	5.7
Tropical seasonal forest	7.5	31	4.3	2.1
Temperate evergreen forest	5.0	22	4.1	2.5
Temperate deciduous forest	7.0	29	5.6	3.1
Boreal forest	12.0	35	7.2	2.8
Woodland and Shrubland	8.5	41	4.4	2.4
Savanna	15.0	21	14.9	7.7
Temperate grassland	9.0	14	13.6	8.5
Tundra/alpine	8.0	10	7.7	2.7
Desert	18.0	6.6	4.9	2.3
Cultivated	14.0	2.1	2.1	1.1
Totals	**121**	**292**	**78.2**	**40.8**

[a]Scheme of Whittaker (1975); estimates are based on data in Jackson *et al.* (1996, 1997).

V. Summary and Conclusions

We have emphasized that phenology is best viewed as the timing of resource allocation to different plant functions (above- and belowground growth, reproduction, etc.) that in turn set the long-term trajectory of plant growth and productivity. Various models of phenology, allocation, and growth are available, but there is as yet no comprehensive theory to integrate such models. Our empirical and theoretical analysis of belowground growth is especially deficient given the importance of belowground biomass and turnover for primary production. Both the literature on tree breeding and ecosystem models incorporating measures of aboveground phenological events show that the phenology of vegetation is important to ecosystem net primary production. In addition, satellite data using the greening of the landscape as an indication of production are increasingly important in global models of the carbon cycle. It would be especially useful if we could relate these remotely sensed data of aboveground events to seasonal patterns of belowground production. Our ability to use such global satellite data and to adjust the regional signals for vegetation heterogeneity should advance our understanding of the roles of phenology, growth, and allocation in global plant productivity.

Acknowledgments

This research was supported by grants from the National Science Foundation (DEB 97–33333), NIGEC/DOE (Tul-038/95–98), and the Andrew W. Mellon Foundation to RBJ; the Natural Sciences and Engineering Research Council of Canada to MJL; and NASA-EOS (NAS 5-31726) to HAM. We thank T. Hinckley, W. Hoffmann, and S. Thomas for helpful comments on the manuscript.

References

Aikin, A. (1799). "The Natural History of the Year," 2nd Ed. J. Johnson, London.

Alaoui-Sosse, B., Parmentier, C., Dizengremel, P., and Barnola, P. (1994). Rhythmic growth and carbon allocation in *Quercus robur.* 1. Starch and sucrose. *Plant Physiol. Biochem.* **32,** 331–339.

Alston, F. H., and Tobutt, K. R. (1989). Breeding and selection for reliable cropping in apples and pears. *In* "Proceedings of the 47th University of Nottingham Easter School in Agricultural Science," pp. 329–339. Sutton Bonington, England.

Anic, M. (1964). Il dinamismo di crescita longitudinale di alcune specie arboree durante il periodo vegetativo. *Annal. Accad. Ital. Sci. Forestal.* **30,** 317–354.

Balatova-Tulackova, E. (1970). Bibliographie der phanospektrum—Diagramme von pflanzengesellschaften. *Excerpt. Botan. Sec. B Bd.* **10** 243–257.

Bennett, K. D., Tzedakis, P. C., and Willis, K. J. (1991). Quaternary refugia of north European trees. *J. Biogeogr.* **18,** 103–115.

Bhar, D. S., Mason, G. F., and Hilton, R. J. (1970). *In situ* observations of plum root growth. *J. Am. Soc. Hort. Sci.* **95,** 237–239.

Broadmeadow, M. S. J., and Griffiths, H. (1993). Carbon isotope discrimination and the coupling of CO_2 fluxes within forest canopies. *In* "Stable Isotopes and Plant Carbon–Water Relations" (J. R. Ehleringer, A. E. Hall, and G. D. Farquhar, eds.), pp. 102–129. Academic Press, San Diego.

Burton, P. J., and Cumming, S. G. (1995). Potential effects of climate change on some western Canadian forests, based on phenological enhancements to a patch model of forest succession. *Water Air Soil Pollut.* **84,** 401–414.

Cannell, M. G. R. (1990). Modelling the phenology of trees. *Silva Carelica.* **15,** 11–27.

Cannell, M. G. R., and Dewar, R. C. (1994). Carbon allocation in trees: A review of concepts for modelling. *Adv. Ecol. Res.* **25,** 59–104.

Canadell, J., Jackson, R. B., Ehleringer, J. R., Mooney, H. A., Sala, O. E., and Schulze, E.-D. (1996). Maximum rooting depth of vegetation types at the global scale. *Oecologia* **108,** 583–595.

Casper, B. B., and Jackson, R. B. (1997). Plant competition underground. *Annual Review of Ecol. Systemat.* **28**, 545–570.

Champagnat, P. (1989). Rest and activity in vegetative buds of trees. *Ann. Sci. For.* **46** (suppl.) 9s–26s.

Clarke, J. E. (1936). The history of British phenology. *Q. J. Roy. Meteorol. Soc.* **62,** 19–23

Deans, J. D., and Harvey, F. J. (1995). Phenologies of sixteen European provenances of sessile oak growing in Scotland. *Forestry* **68,** 265–273.

Eckardt, F. E. (ed.) (1968). "Functioning of Terrestrial Ecosystems at the Primary Production Level." UNESCO, Paris.

Egger, B., Einig, W., Schlereth, A., Wallenda, T., Magel, E., Loewe, A., and Hampp, R. (1996). Carbohydrate metabolism in one- and two-year old spruce needles, and stem carbohydrates from three months before to three months after bud break. *Physiol. Planta.* **96,** 91–100.

Essiamah, S., and Eschrich, W. 1985. Changes of starch content in the storage tissues of deciduous trees during winter and spring. *Int. Assoc. Wood Ana. Bull.* **6,** 97–106.

Farmer, R. E., Jr. (1996). The genecology of *Populus*. *In* "Biology of *Populus* and its Implications for Management and Conservation" (R. F. Stettler, H. D. Bradshaw, Jr., P. E. Heilman, and T. M. Hinckley, eds.), pp. 33–55.

Fernandez , O. A., and Caldwell, M. M. (1975). Phenology and dynamics of root growth of three cool semi-desert shrubs under field conditions. *J. Ecol.* **63,** 703–714.

Forcaud, T., Blaise, F., Bartholemy, D., Houllier, F., and deReffye, P. (1998). A physiological approach to tree growth modelling in the software AMAPpara. *In* "Connection Between Silviculture and Wood Quality through Modelling." Proc. IUFRO Workshop, Kruger National Park, South Africa. In press.

Germain, E. (1990). Inheritance of late leafing and lateral bud fruitfulness in walnut (*Juglans regia* L.), phenotypic correlations among some traits of the trees. *Acta Hort.* **284,** 125–135.

Goulden, M. L., Munger, J. W., Fan, S.-M., Daube, B. C., and Wofsy, S. C. (1996). Exchange of carbon dioxide by a deciduous forest: Response to interannual climate variability. *Science* **271,** 1576–1578.

Hänninen, H. (1987). Effects of temperature on dormancy release in woody plants: Implications of prevailing models. *Silva Fenn.* **21,** 279–299.

Hänninen, H., Kanninen, M., and Smolander, H. (1985). The annual cycle of forest trees: The Sarvas approach revisited. *In* "Crop Physiology of Forest Trees" (P. M. A. Tigerstedt, P. Puttonen, and V. Koski, eds.). Helsinki University Press, Helsinki.

Hansen, J., Vogg, G., and Beck, E. (1996). Assimilation, allocation and utilization of carbon by 3-year-old Scots pine (*Pinus sylvestris* L.) trees during winter and early spring. *Trees* **11,** 83–90.

Harris, J. R., Bassuk, N. L., Zobel, R. W., and Whitlow, T. H. (1995). Root and shoot growth periodicity of green ash, scarlet oak, turkish hazelnut, and tree lilac. *J. Am. Soc. Hort. Sci.* **120,** 211–216.

Heide, O. M. (1993). Daylength and thermal time responses of budburst during dormancy release in some northern deciduous trees. *Physiol. Planta.* **88,** 531–540.

Hendrick, R. L., and Pregitzer, K. S. (1993). Patterns of fine root mortality in two sugar maple forests. *Nature (London)* **361,** 59–61.

Hendrick, R. L., and Pregitzer, K. S. (1996). Temporal and depth-related patterns of fine root dynamics in northern hardwood forests. *J. Ecol.* **84,** 167–176.

Hobbs, R. J. (1990). Remote sensing of spatial and temporal dynamics of vegetation. *In* "Remote Sensing of Biosphere Functioning" (R. J. Hobbs, and H. A. Mooney, eds.), pp. 203–219. Springer Verlag, New York.

Howitt, William (1835). "The Book of the Seasons, or The Calendar of Nature," 3rd Ed. Bentley, London.

Hunter, A. F., and Lechowicz, M. J. (1992). Predicting the timing of budburst in temperate trees. *J. Appl. Ecol.* **29,** 597–604.

Jackson, R. B. (1999) The importance of root distributions for hydrology, biogeochemistry, and ecosystem functioning. *In* "Integrating Hydrology, Ecosystem Dynamics, and Biogeochemistry in Complex Landscapes" (J. Tenhunen and P. Kabat, eds., Dahlem Conf.), pp. 219–240. John Wiley and Sons, Chichester.

Jackson, R. B., Canadell, J., Ehleringer, J. R., Mooney, H. A., Sala, O. E., and Schulze, E.-D. (1996). A global analysis of root distributions for terrestrial biomes. *Oecologia* **108,** 389–411.

Jackson, R. B., Mooney, H. A., and Schulze, E.-D. (1997). A global budget for fine root biomass, surface area, and nutrient contents. *Proc. Natl. Acad. Sci. U.S.A.* **94,** 7362–7366.

Jenyns, L. (1846). "Observations in Natural History." John van Voorst, London.

Jorg, K., and Heimann, M. (1996). A prognostic phenology scheme for global terrestrial carbon cycle models. *Climate Res.* **6,** 1–19.

Keeling, C. D., Chin, J. F. S., and Whorf, T. P. (1996). Increased activity of northern vegetation inferred from atmospheric CO_2 measurements. *Nature (London)* **382,** 146–149.

Kikuzawa, K. (1991). A cost-benefit analysis of leaf habit and leaf longevity of trees and their geographical pattern. *Am. Naturalist* **138,** 1250–1263.

Kikuzawa, K. (1995). Leaf phenology as an optimal strategy for carbon gain in plants. *Can. J. Bot.* **73,** 158–163.

Kikuzawa, K., Koyama, H., Umeki, K., and Lechowicz, M. J. (1996). Some evidence for an adaptive linkage between leaf phenology and shoot architecture in sapling trees. *Funct. Ecol.* **10,** 252–257.

Kolesnikov, V. (1971). "The Root System of Fruit Plants." Izdatelstvo Mir, Moscow.

Kozlowski, T. T. (1992). Carbohydrate sources and sinks in woody plants. *Bot. Rev.* **58,** 107–222.

Kramer, K. (1995a). Phenotypic plasticity of the phenology of seven European tree species in relation to climatic warming. *Plant Cell Environ.* **18,** 93–104.

Kramer, K. (1995b). Modelling comparison to evaluate the importance of phenology for the effects of climate change on growth of temperate-zone deciduous trees. *Climate Res.* **5,** 119–130.

Kramer, K. (1996). Phenology and growth of European trees in relation to climate change. Thesis, Landbouw Universiteit, Wageningen. CIP-Data Koninklijke Bibliotheek, Den Haag, Netherlands.

Lechowicz, M. J. (1984). Why do temperate deciduous trees leaf out at different times? Adaptation and the ecology of forest communities. *Am. Natural.* **124,** 821–842.

Lechowicz, M. J. (1995). Seasonality of flowering and fruiting in temperate forest trees. *Can. J. Bot.* **73,** 175–182.

Lechowicz, M. J., and Koike, T. (1995). Phenology and seasonality of woody plants: An unappreciated element in global change research? *Can. J. Bot.* **73,** 147–148.

Lieth, H., ed. (1974). "Phenology and Seasonality Modelling." Springer-Verlag, New York.

Lieth, H. (1994). Aims and methods in phenological modelling. *G. Bot. Ital.* **128,** 151–182.

Luomajoki, A. (1986). Timing of microsporogenesis in trees with a reference to climatic adaptation: A review. *Acta For. Fenn.* **196,** 1–33.

Lyr, H., and Hoffmann, G. (1967). Growth rates and growth periodicity of tree roots. *Int. Rev. For. Res.* **2,** 181–235.

Molesworth, C., and Ormerod, E. A. (1880). "Cobham Journals: Abstracts and Summaries of Meteorological and Phenological Observations in 1825 to 1850." E. Stanford, London.

Monselise, S. P., and Goldschmidt, E. E. (1982). Alternate bearing in fruit trees. *Hort. Rev.* **4,** 128–173.

Monsi, M. (1968). Mathematical models of plant communities. *In* "Functioning of Terrestrial Ecosystems at the Primary Production Level" (F. Eckardt, ed.), pp. 131–149. UNESCO, Paris.

Moulin, S., Kergoat, L., Viovy, N., and Dedieu, G. (1997). Global scale assessment of vegetation phenology using NOAA/AVHRR satellite measurements. *J. Climate* **10,** 1154–1170.

Myneni, R. B., Keeling, C. D., Tucker, C. J., Asrar, G., and Nemani, R. R. (1997). Increased plant growth in the northern high latitudes from 1981 to 1991. *Nature (London)* **386,** 698–702.

Nikinmaa, E. (1992). The allocation of carbohydrates. *In* "Simulation of Forest Development" (H. Salminen and T. Katerman, eds.), pp. 21–34. Finnish Forestry Research Institute Paper 407, Helsinki.

Nizinski, J. J., and Saugier, B. (1988). A model of leaf budding and development for a mature *Quercus* forest. *J. Appl. Ecol.* **25,** 643- 652.

Perttunen, J., Sievanen, R., Nikinmaa, E., Saarenmaa, H., and Vakeva, J. (1996). LIGNUM: A tree model based on simple structural units. *Ann. Bot.* **77,** 87–98.

Richardson, S. D (1958). Bud dormancy and root development in *Acer saccharinum*. *In* "The Physiology of Forest Trees" (K. V. Thimann, ed.), pp. 409–425. Ronald Press, New York.

Riedacker, A. (1976). Growth and regeneration rhythms of ligneous species. *Ann. Sci. For. (Paris)* **33,** 109–138.

Running, S. W. (1990). Estimating terrestrial primary productivity by combining remote sensing and ecosystem simulation. *In* "Remote Sensing of Biosphere Functioning" (R. J. Hobbs and H. A. Mooney, eds.), pp. 65–86. Springer Verlag, New York.

Running, S. W., Loveland, T. R., Pierce, L. L., Nemani, R. R., and Hunt, E. R., Jr. (1995). A remote sensing based vegetation classification logic for global land cover analysis. *Remote Sens. Environ.* **51,** 39–48.

Sarvas, R. (1974). Investigations on the annual cycle of development of forest trees. II. Autumn dormancy and winter dormancy. *Comm. Instit. Forestal. Fenn.* **84.**

Schnelle, F. S. (1955). "Probleme der Bioklimatologie. Band 3. Planzen-Phanologie." Akademische Verlagsgesellschaft Geest und Portig K-G, Leipzig.

Schwartz, M. D. (1994). Monitoring global change with phenology: The case of the spring green wave. *Int. J. Biometeorol.* **38,** 18–22.

Silvertown, J. W. (1980). The evolutionary ecology of mast seeding in trees. *Biol. J. Linn. Soc.* **14,** 235–250.

Sparks, T. H., and Carey, P. D. (1995). The response of species to climate over two centuries: An analysis of the Marsham phenological record, 1736–1947. *J. Ecol.* **83,** 321–329.

Stillingfleet, B. (1791). The calendar of flora, Swedish and English. *In* "Miscellaneous Tracts Relating to Natural History, Husbandry, and Physick," pp. 223–337. J. Dodsley, London.

Taylor, H. M. (1987). "Minirhizotron Observation Tubes: Methods and Applications for Measuring Rhizosphere Dynamics." American Society of Agronomy, Madison, Wisconsin.

Townshend, J. R. G., and Justice, C. O. (1986). Analysis of the dynamics of African vegetation using the normalized difference vegetation index. *Int. J. Remote Sens.* **7,** 1435–1445.

Tucker, C. J. (1986). Maximum normalized difference vegetation index images for sub-Saharan Africa for 1983–1985. *Int. J. Remote Sens.* **7,** 1383–1384.

Wang, T., and Tigerstedt, P. M. A. (1993). Variation of growth rhythm among families and correlation between growth rhythm and growth rate in *Betula pendula* Roth. *Scand. J. For. Res.* **8,** 489–497.

Wargo, P. M. (1971). Seasonal changes in carbohydrate levels in roots of sugar maple. USDA Forest Service Research Paper NE-213, Northeastern Forest Experiment Station.

White, M. E., Thornton, P. E., and Running, S. W. (1997). A continental phenology model for monitoring vegetation responses to interannual climatic variability. *Global Biogeochem. Cycles* **11,** 217–234.

Whittaker, R. H. (1975). "Communities and Ecosystems." Macmillan, London.

Willis, K. J. (1994). The vegetational history of the Balkans. *Q. Sci. Rev.* **13,** 769–788.

5

From Plant to Soil: Litter Production and Decomposition

Richard Joffre and Göran I. Ågren

I. Introduction

In a world in which nitrogen is a limiting factor for net primary production in almost all terrestrial ecosystems (Vitousek and Howarth, 1991)—and with 30–40 times more nitrogen found in soil organic matter than in land plants (Schlesinger, 1997)—it is clear that even small changes in soil organic matter turnover may have large effects on ecosystem productivity. With a changing climate and atmospheric CO_2 concentration, the circulation of matter between plants and soils will change as a result of both direct impacts on processes (e.g., plant growth is temperature dependent) and indirect changes (e.g., plant growth rate could affect allocation of carbon to leaves, wood, and roots, as well as litter chemistry, which in turn may control decomposition rates). We discuss in this chapter some problems in our understanding of what determines litter production and decomposition rates of litters and soil organic matter.

Discussions of climate change effects on ecosystems can be found in Koch and Mooney (1996) and Walker and Steffen (1996). What we add here is a further illumination of what we know and where knowledge is uncertain and where that uncertainty has a potential to be important for our ability to predict future changes in status and function of ecosystems. One aspect that we emphasize is the need to identify whether changes will be of a quantitative or qualitative nature (Gorissen, 1996). Quantitative changes are those whereby the basic structure of the ecosystem remains unchanged but the rates are changed. A typical example would be an increased rate of decomposition under a higher temperature, but without effect on products of decomposition. We argue that many of the observed effects in climate change experiments belong to this category. Qualitative changes are more dramatic because they involve significant structural changes. We are not aware of any experiments showing such changes, but that might be because there are no experiments of long enough duration for such effects to manifest. Replacement of plant functional groups, because of differential responses to climate variables, and soil community changes, because of changes in litter quality, are possible candidates.

II. Litter Inputs: Quantity and Quality

In the long term, litter input rates are determined by plant production, and hence by the average climate; in the short term, the variability of climate plays an essential role. Although both the interannual and the intraannual variabilities in litter input can be high (e.g., Bray and Gorham, 1964; Gower and Son, 1992; Miller *et al.*, 1996), very little is quantitatively known about the causes. Estimates of aboveground biomass and leaf litter production are more abundant for the different biomes (see Chapter 9–17, this volume), compared to data concerning root biomass and its turnover (Jackson *et al.*, 1996, 1997; Cairns *et al.*, 1997; Vogt *et al.*, 1998).

Studies of climate change effects on plant growth have primarily focused on CO_2 effects (Saxe *et al.*, 1998), and some reviews (Curtis and Wang, 1998) indicate that plant biomass might increase about 30% under doubled CO_2. The consequences for litter production depend also on how longevity of plant tissue is affected. There are few studies of this, but leaf longevity has been shown to be correlated with other leaf traits (Reich *et al.*, 1997). Nitrogen availability affects longevity of fine roots (Majdi and Kangas, 1997) and other factors are likely to do the same. We leave the discussion of changes in aboveground production to others (see Chapter 7, this volume).

A. Belowground Production

There are two major sources of organic matter production belowground—roots and rhizodeposition, defined here as C loss in the soil from roots during the lifetime of the plant (Whipps and Lynch, 1985, 1986). Root production can be viewed as a matter of allocation between above- and belowground production, whereas exudates and rhizodeposition normally are considered separately. Luxmoore (1981) hypothesizes that an increase of belowground carbon translocation due to elevated atmospheric CO_2 concentration may result in the extension of (1) the root system and/or (2) the rhizospheric activity. These modifications would also impact nutrient mobilization and soil organic matter turnover and storage (Curtis *et al.*, 1994; Norby, 1994).

1. Allocation Root:Shoot Changes in the allocation between root and shoot will affect the composition of the litter because in general these different sources will have different decomposition patterns. Rogers *et al.* (1994) reviewed observed effects on allocation in CO_2 experiments. In general, an increase in root:shoot ratio was observed. Saxe *et al.* (1998) concluded in a review on trees and forest functioning that mostly no effects are observed and that both increases and decreases in root:shoot ratios occur. A problem with many of the studies of root:shoot partitioning under elevated CO_2 is that direct effects are not separate from indirect effects through, e.g., changes in plant nutrient status. In a study by Pettersson *et al.* (1993), one of the very few studies with a strict control over both plant nutrition and CO_2, there was no effect of CO_2 on allocation that could not be attributed to changes in plant nutrient status. An examination of the studies cited by Rogers *et al.* (1994) shows that in only 6 (13%) out of 45 studies are changes in plant nutrient status given. Comparing the root:shoot ratios in these six studies at the same nutrient concentration reveals that there is no effect of CO_2. Thus, the major effect of the increased CO_2 in these experiments has been a decrease in plant nitrogen concentration, which can be taken as the driving force behind the changed allocation. Climatic change can, therefore, as discussed above, be expected to change allocation patterns through influences on nutrient availability. van Nordwiijk *et al.* (1998) discuss this question further and suggest that changes in patterns of water availability might alter root turnover.

2. Exudates–Rhizodeposition–Mycorrhizas Roots act as continuous sources of organic compounds (Rovira, 1956, 1969; Martin, 1975; Barber and Martin, 1976; Warembourg and Billès, 1978; Rovira *et al.,* 1979; Lambers, 1987). Various substances released by roots can be classified according to mode of release. Lynch and Whipps (1990) propose the following four groups: (1) water-soluble exudates (sugars, amino acids, organic acids, hormones, and vitamins), which leak from the root without involvement of metabolic energy; (2) secretions (polymeric carbohydrates and enzymes), which depend on metabolic processes for their release; (3) lysates, released when cells autolyse; and (4) gases. From a soil perspective (Cardon, 1996), the release of organic materials from roots, even though it represents a small proportion of the total rhizodeposition, is extremely important because these compounds influence nutrient availability through the regulation of the microbial biomass community and activity (Helal and Sauerbeck, 1984; van Veen *et al.,* 1989, 1991; Griffiths and Robinson, 1992; van de Geijn and van Veen, 1993).

Estimations of the amount of carbon lost as exudates and secretions vary considerably between investigators. Methodological problems in measurements do not always allow separate estimations of carbon losses in natural ecosystems [see Meharg (1994) for a comprehensive review of labeling techniques]. In controlled experiments involving annual crops during a few weeks, between 30 and 60% of net fixed carbon is transferred to roots (see Lynch and Whipps, 1990). The percentage of C transferred to roots lost as exudates (*sensu lato,* including respiration) in the soil ranges from 11 to 70% in 10 of the 11 summarized studies. In one of the cases (Johnen and Sauerbeck, 1977), the experiment was conducted during 73 days (tomato) and 153 days (wheat), and the proportion of C lost as respiration increased, respectively, to 78 and 80% of the C transferred to root. In general, the fraction of net fixed C transferred to root is higher for perennial plants, resulting in higher total rhizodeposition (Grayston *et al.,* 1996). The total root-derived carbon (corresponding to root respiration and microorganism respiration resulting from the mineralization of cells sloughed as the roots grow through the soil plus exudates) increased with the age of tree seedlings, ranging from 5% of net C uptake at 3 months to 21% at 18 months for chestnut trees (Rouhier *et al.,* 1994).

Rogers *et al.* (1994) hypothesized that types and amounts of organic and inorganic chemicals released from roots of plants under high CO_2 conditions will change. Few studies of rhizodeposition exist, and direct evidence of quantitative changes has still to be shown (Stulen and den Hertog, 1993; Hodge, 1996). An increase of belowground translocation of C is generally obtained for plants grown under elevated CO_2 levels (Norby *et al.,* 1987; Billès *et al.,* 1993). But, in Norby's study on *Pinus echinata* seedlings, exudation per unit root mass was not changed and the apparent effect on exudation was attributed to a greater fine root mass in elevated CO_2. Similarly, in a long-term experiment (18 months) with chestnut trees grown under 350 and 700 ppm of atmospheric CO_2, Rouhier *et al.* (1994) showed that the higher total root-derived carbon for elevated CO_2 was a consequence of higher net C uptake and not because of a change in allocation.

Interactions between mineral nitrogen content and elevated CO_2 levels could also affect quality (e.g., C:N ratio) of root-derived compounds (Lekkerkek *et al.,* 1990), modifying the dynamics of soil organic matter (van Veen *et al.,* 1991). Liljeroth *et al.* (1990) showed that during the growth of wheat in a nitrogen-fertilized soil, the decomposition of native soil organic matter was retarded compared with a soil to which no fertilizer nitrogen was applied. This result led them to hypothesize that in case of richer soils, the root-derived exudates have a high energy value and, in consequence, are more easily decomposable compared to native soil organic matter. As stated by van der Linden *et al.* (1987), van de Geijn *et al.* (1993), and Gorissen (1996), the incorporation of plant organic compounds into soil organic matter may strongly depend on the quality of the input (e.g., C:N ratio), be it litter- or root-derived material. Reviewing the effects on mycorrhizal growth of growing plants in elevated CO_2 (Lewis *et al.,* 1994; Monz *et al.,* 1994; Morgan *et al.,* 1994; O'Neill, 1994; Seegmüller and Rennenberg, 1994; Ineichen *et al.,* 1995; Jongen *et al.,* 1996; Lewis and Strain, 1996; Markkola *et al.,* 1996; Rouhier and Read, 1998; Walker *et al.,* 1998) shows that a general trend seems to be no or a positive response of mycorrhizal colonization to elevated CO_2. Lewis *et al.* (1994) found that although elevated CO_2 may significantly increase root carbohydrate levels, the increases may not affect the percentage of fine roots that are mycorrhizal. Nevertheless, Ineichen *et al.* (1995) showed that the mycorrhizal fungus *Pisolithus tinctorius,* which depended entirely on the *Pinus sylvestris* assimilates, grew much faster at increased CO_2. As pointed out with respect to the effects on allocation in CO_2 experiments, and in agreement with the results of Klironomos *et al.* (1996), it seems that interactions between nutrient conditions and CO_2 atmospheric conditions greatly affect the responses of microbial biomass and mycorrhiza and may lead to far less predictable feedback patterns than previously thought. Staddon and Fitter (1998), reviewing the effect of elevated atmospheric CO_2 on arbuscular mycorrhizas, emphasized that the observed effects are indirect and are a result of faster plant growth at higher CO_2 concentrations.

B. Litter Quality

Several different measures have been used to express quality of litter [see Heal *et al.* (1997) for an historical overview]. The most commonly used measurements are concentrations of different elements (mostly N), the ratio between elements (C:N, C:P, etc.), or the concentrations of certain chemical compounds (lignin of phenolics) or some combination thereof [e.g., the lignocellulose index (LCI) = lignin/(lignin + cellulose)]. Several of these measures express basically the same thing, e.g., C:N and %N are, because of the constancy of carbon in plant material, almost identical. Other measures suffer by only accounting for a small fraction of the total plant material, e.g., N constitutes about only 1% of the total litter mass. The use of measures of quality that are not stable during the whole decomposition process also seems less satisfactory; N concentration normally increases during decomposition and is accompanied by decreasing decomposition rates, yet high N concentrations are initially taken as an indicator of easily decomposing material. Two new ways that take into account the complete composition of the litter and that apply to all stages of the decomposition are the continuous quality approach [see Eq. (1)] and near-infrared reflectance spectroscopy (NIRS) (Joffre *et al.,* 1992; Gillon *et al.,* 1993, 1999). Ågren and Bosatta (1996a,b) calculate litter quality q from litter chemistry as

$$q = 1.25c_{as} + 0.65c_{ai} + c_{ex}, \tag{1}$$

where c_{as}, c_{ai}, and c_{ex} correspond to relative amounts of the carbon fractions in the litter ($c_{as} + c_{ai} + c_{ex} = 1$). These fractions are, respectively, acid soluble (c_{as}), acid insoluble (also often denoted lignin) (c_{ai}), and water soluble and nonpolar soluble (c_{ex}).

Using data from Aber *et al.* (1984, 1990), Berg *et al.* (1987), Taylor *et al.* (1989), and Wessén and Berg (1986), we have calculated the correlations between some of these measures of quality for 28 different litter types (Table 5-1). The only measures that are strongly correlated are those between q and

Table 5-1 Correlation (r) between Measures of Initial Quality[a]

Component	q_0[b]	N (%)	LCI[c]
C:N	0.24 ns	—	−0.11 ns
Lignin:N	0.09ns	—	0.03 ns
Cellulose concentration	0.84***	−0.003 ns	
Lignin concentration	−0.80***	−0.28 ns	
Soluble concentration	−0.31 ns	0.22 ns	
LCI	−0.94***	−0.26 ns	

[a]Based on 28 different litter types [Aber *et al* (1984): *Acer rubrum* wood, *Pinus strobus* wood, *Acer saccharum* leaves, *Populus grandidentata* leaves, *Quercus alba* leaves, *Pinus strobus* needles, *Tsuga canadensis* needles, *Quercus borealis* leaves, *Tsuga canadensis* bark, *Acer rubrum* bark, *Pinus strobus* roots, *Acer saccharum* roots; Aber *et al.* (1990): *Betula papyrifera* foliage, two types of *Acer rubrum* foliage, two types of *Quercus borealis* foliage, *Pinus resinosa* foliage; Berg *et al.* (1987): *Trifolium pratense* roots; Taylor *et al.* (1989): *Populus tremuloides* leaves, *Populus balsamifera* leaves, *Pseudotsuga menziesii* needles, *Picea glaua* needles, *Cornus alba* leaves, *Rosa woodsii* and *Rosa acicularis* leaves, *Heracleum lanatum* leaves, *Bouteloua gracilis, Stipa comata,* and *Agropyron* spp. leaves; Wessén and Berg (1986): *Hordeum distichum* straw].

[b]Initial litter quality calculated fom Eq. (1); ns, nonsignificant; ***$p < 0.001$.

[c]Lignocellulose index.

other measures of quality based on the same chemical fractionation, e.g., LCI. This is trivial and uninteresting, but the lack of correlation between other measures is a concern. The success of both C:N and LCI to explain initial decomposition rates can therefore only be fortuitous; both should not be appropriate. The form in which carbon occurs in plant cells, the concentration of other nutrients, and the composition of the various secondary compounds also seem to play major roles in the regulation of the decomposition process. For example, Berendse *et al.* (1987) constructed a decomposition model that distinguished the importance of carbohydrates masked by lignin and of free carbohydrates relative to nitrogen and lignin. Currently, the importance of the different chemical compounds in the regulation of decomposition is still controversial. One of the probable reasons is that these studies are based on a rather rough understanding of the biochemical composition of litter and particularly of the carbon fractions measured by standard chemical methods (Ryan *et al.*, 1990). For instance, lignin, which is recognized as being one of the main factors contributing to the resistance of litters to decompose, is the plant constituent with the most complex structure and is most difficult to measure. According to Paustian *et al.* (1997), litter quality is also related to the structural complexity of the constituents, in terms of the size of the molecules and the diversity of their chemical bonds.

Near-infrared spectral signatures of organic materials in the region of 1100–2500 nm are mainly due to the vibrations of molecules possessing functional groups of atoms, including —CH, —OH, and —NH. During the decomposition process, spectra of litter reveal the existence of chemical bonds, some of which vary directly and proportionately as a function of the stage of litter decomposition (Gillon *et al.*, 1993). As a result, using a long-term microcosm experiment involving 12 different species, accurate predictive calibration equations were obtained between the spectral data and the stage of decay (Gillon *et al.*, 1993). Using a similar approach conducted over a wider range of species, Gillon *et al.* (1999) confirmed this first result and showed that the litter mass remaining as well as the decay rate constant loss can be accurately predicted by near-infrared spectra (Gillon *et al.*, 1999) (Fig. 5-1). Moreover, the initial spectral characteristics of the litters could be correlated with their decay rate and a litter decomposability index (LDI). It should therefore be possible, at each stage of decomposition or at each time (t), to correlate the spectral characteristics of the decomposed litters with an index of decomposability relevant to this stage. This possibility of direct-

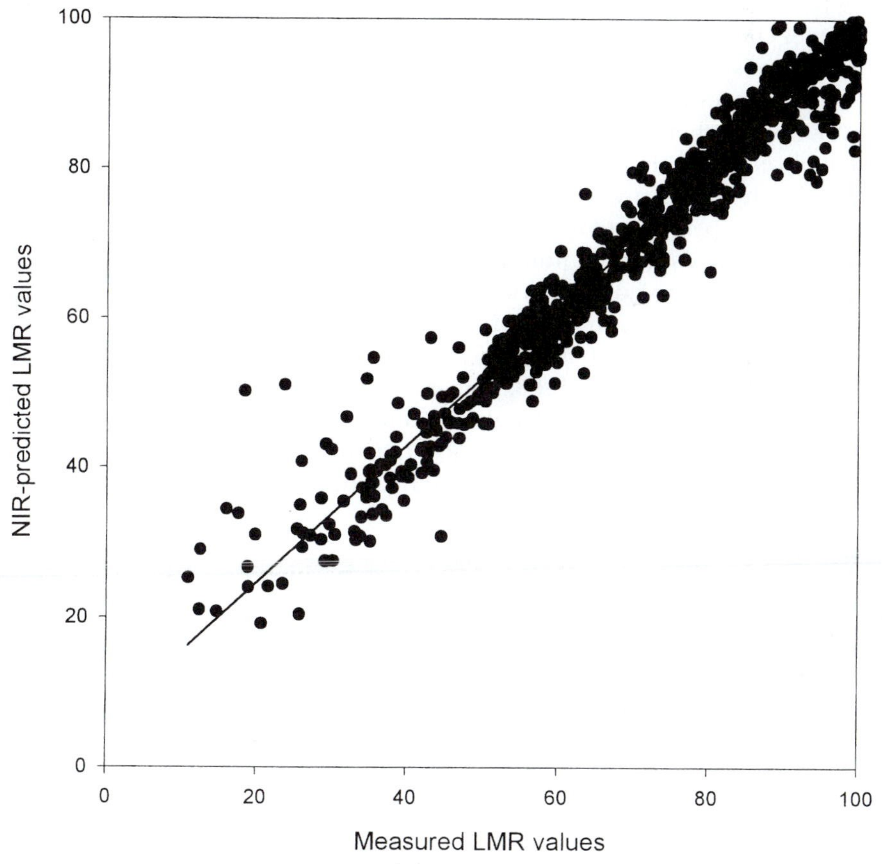

Figure 5-1 Near-infrared (NIR)-predicted litter mass remaining (LMR) (% initial organic matter) and measured LMR values. The data sets represent 34 litter types decomposing in microcosm during 8 weeks (858 data points). Data from Gillon *et al.* (1999).

ly measuring the change in global chemical composition provides new perspectives in the "litter quality" determination. We have shown that NIRS can be used to estimate q (Joffre *et al.*, 2000). With the simplicity and power of NIRS, this might be the future way to measure organic matter quality.

III. From Plant Organic Matter to Soil Organic Matter

A. Processes in Litter Decomposition

Three component processes are involved in the decomposition of any plant residue (Swift *et al.*, 1979). The initial phase has an important influence on subsequent processes, and the mass loss of litter during this first decomposition phase has been attributed largely to leaching (Berg, 1984; Ibrahima *et al.*, 1995). Leaching changes the chemical composition of litter by lowering the sugar concentration and increasing the fiber concentration. These changes could therefore modify litter quality and transform litter into a material that is less readily biodegradable (Parsons *et al.*, 1990; Gillon *et al.*, 1994). Changes in litter chemistry due to elevated CO_2 could affect leaching, although there is currently little evidence for this. The material is next catabolized through the action of the decomposer community. This second phase of the decay process corresponds to chemical changes such as mineralization (liberation of inorganic elements) and, in parallel, the synthesis of decomposer tissues and humus (Heal *et al.*, 1997). A third process, due to the action of macrofauna and occurring from the first moment of litterfall, is comminution, through which there is a physical reduction in particle size, increasing the surface area for microbial activity, and an incorporation of litter into the soil (Lavelle, 1988; Tian *et al.*, 1995; Wardle and Lavelle, 1997). Litter decomposition is regulated by numerous factors, of which the most important are probably (1) the physicochemical environment under which decay takes place and (2) the resource quality, acting through the decomposer organisms.

B. Changes in Quality and Decomposition Processes

Are differences in litter properties caused by the CO_2 environment in which the plants have grown, such that each decomposition process is qualitatively different, or do the differences fit into already existing frameworks for describing decomposition, although with quantitative differences? Norby and Cotrufo (1998) conclude that the changes in chemical compositions of litters as a result of a rising level of atmospheric CO_2 are not likely to alter decomposition rates significantly. Some studies indicate a qualitative change (e.g., Coûteaux *et al.*, 1991). These studies are, however, normally concerned only with changes in mass loss and cannot therefore distinguish between factors that only change rates and those that change patterns of decomposition. Moreover, results of Franck *et al.* (1997) show that the effects of elevated CO_2 on decomposition and N release from litter are highly species specific.

van Ginkel *et al.* (1996) studied, during 2 years, the rate of decomposition of grass roots of plants grown under two different CO_2 concentrations and two different nutrient supplies. They also monitored changes in the chemical composition. We can therefore use the model of Ågren and Bosatta (1996a,b) to calculate litter quality from litter chemistry, Eq. (1), and its effects on decomposition rates. The fraction of a litter remaining, *g*, after a time *t* of incubation, is then given by

$$g(t) = [q(t) / q(0)]^{\frac{1-e_0}{\eta_{11}e_0}} = \left[1 + f_c\eta_{11}\beta u_0 q(0)^{\beta} t\right]^{-\frac{1-e_0}{\eta_{11}e_0\beta}} \tag{2}$$

where $q(t)$ is the quality of the litter at time t, and f_c, η_{11}, β, and e_0 are parameters with fixed values. Only $u_0 = 0.371$, which reflects climatic conditions, has been adjusted to give the best possible fit between prediction and measurements. The value of u_0 agrees well with what Ågren and Bosatta (1996b) predict for temperatures such as the ones used in this experiment. As can be seen in Fig. 5-2, decomposition rates of plants grown under ambient or elevated CO_2 or high or low nutrient supply can be described with the same model and with the same parameters (f_c, η_{11}, β, e_0, and u_0). The treatment has changed only the chemistry of the roots in such a way that initial quality and, hence, rates are changed (Table 5-2).

The temperature dependence of the decomposition rate (u_0) has been extensively studied and is often assumed to be exponential, although this can be questioned (Kirschbaum, 1995; Kätterer *et al.*, 1998). Functions other than the exponential (e.g., a quadratic) might therefore better describe the temperature dependency of the decomposition rate. This can be particularly important at temperatures close to 0°C, where rates are low and small temperature changes will have large relative effects. Water availability is less studied, but Seyferth and Persson (2000) found that carbon mineral-

Table 5-2 Decompostion of Grass Roots[a]

Treatment	Initial c_{ex} (%)	Initial c_{as} (%)	Initial c_{ai} (%)	$q(0)$	$q(1)$	$q(2)$
Ambient CO_2, low N	31.7	59.6	8.6	1.125	0.912	0.862
High CO_2, low N	33.5	56.9	9.7	1.100	0.921	0.875
Ambient CO_2, high N	28.7	63.1	8.2	1.136	0.907	0.857
High CO_2, high N	31.5	59.2	9.2	1.055	0.937	0.896

[a]Data show initial chemical compositions (c) and quality (q) initially and after 1 and 2 years. Data from van Ginkel *et al.* (1996).

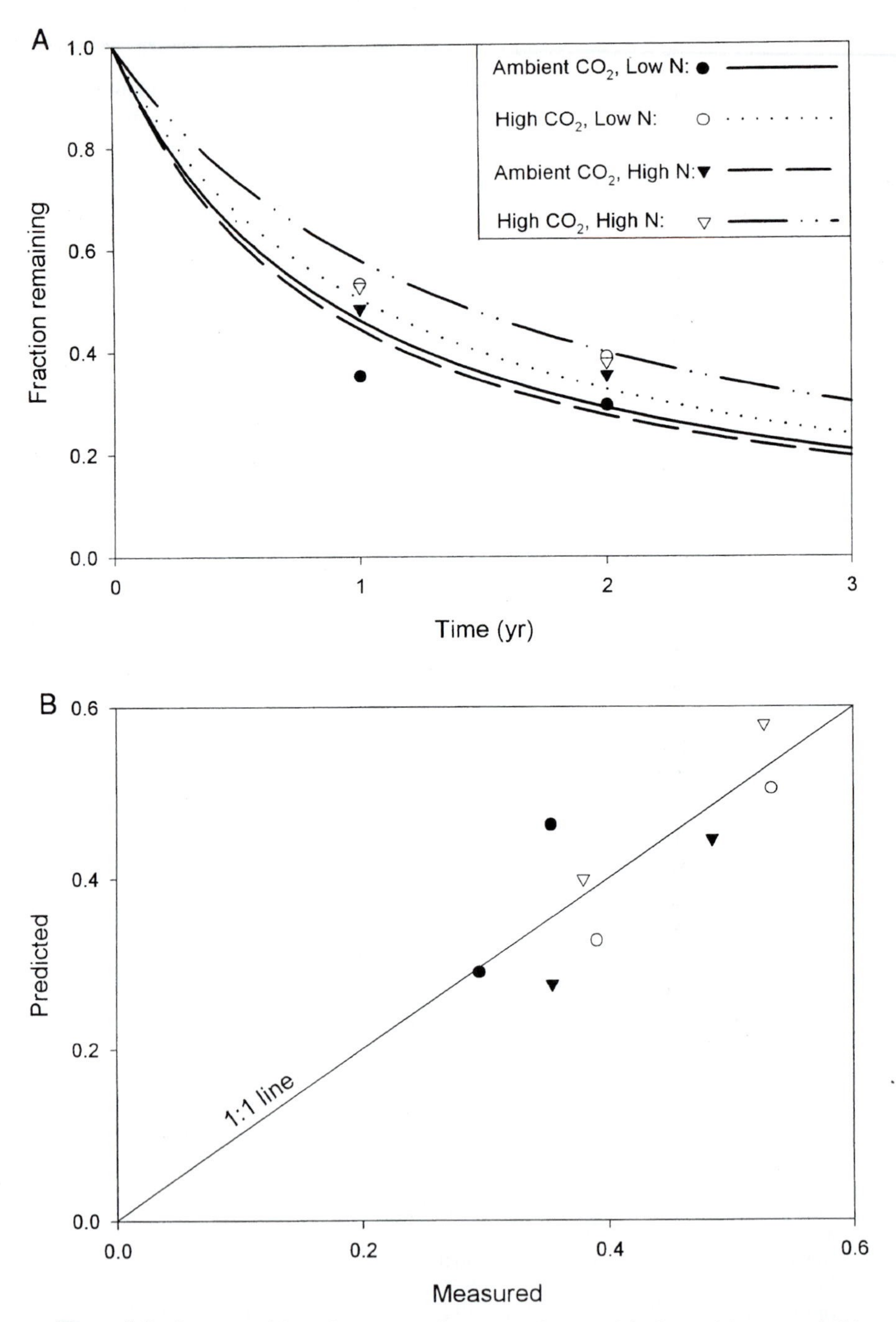

Figure 5-2 Decomposition of grass roots from experiments with elevated (open symbols) and ambient CO_2 (solid symbols) and high (△, ▲) and low (○, ●) nutrition. (A) Measured (symbols) and predicted (lines) remaining mass as a function of time. (B) Predicted versus measured remaining mass. The line is the 1:1 line. Data from van Ginkel *et al.* (1996).

ization rates could be linearly related to log(−soil water potential). Interactions between temperature and water are even less studied but could be critical under extreme conditions.

C. Importance of Biological Interactions between Fauna and the Microbial Community

Characteristics of ecosystems determine the types of biological interactions within the soil community: in temperate forests, a close relationship between humus type and animal activity has been emphasized (Petersen and Luxton, 1982). Results obtained by David *et al.* (1993) and Schaeffer and Schauermann (1990) in beech forests show that the composition and abundance of the soil macrofauna vary greatly according to type of humus: mesotrophic mull, moder, or dysmoder. In the mesotrophic mull, the saprophagous fauna showed the highest diversity with species of Diplopoda, Isopoda, Gastropoda, and Lumbricidae. The macrofauna was also significantly more abundant (in individual number as well as in biomass) in mull than in the other types of humus. Mycophagous mesofauna groups dominated in mor and moder types and the microbial community was mainly composed of fungi. In contrast, the action of earthworms in mull humus promotes the formation of soil aggregates where bacterial communities are very active. As stated by Schaeffer and Schauermann (1990), the comminuting effect of the saprophagous macrofauna in the mull leads to translocation of organic material onto and into the mineral soil, whereas comminution of litter in the moder system is confined to the organic topsoil. Long-term experiments conducted on the organic layers of a *Pinus sylvestris* forest demonstrated that soil fauna species diversity and succession, microhabitat specification, and degradation stages of organic matter are strongly interrelated (Berg *et al.*, 1998). Describing in a more general way the interactions between the microflora and the soil fauna, Wardle and Lavelle (1997) detailed three different scales of resolution: the "microfood webs," involving soil nematodes, protozoa, and their predators; the "litter-transforming systems," involving soil meso- and macrofauna; and the "ecosystem engineers," involving larger organisms such as termites, ants, and endogeic and anecic earthworms. The composition of the decomposer community is, therefore, strongly connected to the decomposition processes, a problem also analyzed theoretically by Zheng *et al.* (1997). However, the question that we cannot answer at the moment is that of whether a change in litter input following climate change can be strong enough to induce an ecologically significant change in the decomposer community (Swift *et al.*, 1998). Wardle *et al.* (1998) noted that changes in net primary production could influence components of the decomposer food web, but because their regulation involved a complex series of interactions influenced by top-down and bottom-up forces, the direction of changes could not be easily established.

IV. Research Needs

It should be clear from our discussion here that there is a need for better understanding in several areas. The following key areas seem to be where we need to improve our knowledge:

1. We still do not understand allocation of C and other elements in plants. In particular, we need a better quantification of the belowground inputs. New techniques are needed that can bring down the variability in the estimates of inputs (Hodge *et al.*, 1996).
2. The feedback between plants and soils must be taken into account, and the two entities should not be studied in isolation from each other. Nevertheless, the interactions are species dependent (Hungate *et al.*, 1996; Berntson and Bazzaz, 1996), indicating that the response at the ecosystem level might not be easily predictable (Franck *et al.*, 1997).
3. Direct effects of interactions between macrofauna and microbial communities remain as one of a major unknown key issues. However, interactions are not symmetrical, and, as stated by Wardle and Giller (1996), the disproportionate effects of some components of soil biota on other soil components were appreciated by soil biologists well before the keystone concept gained recognition by ecologists. Periods of activity of different groups are different, and their regulation is at different time scales (bacteria vs. earthworms) (Lavelle, 1997). This may require the use of a hierarchy of embedded models to better predict activities.

Because processes occur in different layers of the soil and depend on different groups of biological actors, the vertical distribution of soil organic matter should be considered when we test the sensitivity of different processes to regulating factors.

V. Concluding Remarks

It is clear that a changing climate will change plant production both quantitatively and qualitatively. Most of these changes can be incorporated into existing frameworks for analyzing soil organic matter turnover, because the major effects seem to be on the quantities of litter entering the decomposition system. Observations of changes that would lead to qualitatively new structures in the decomposer community have so far been few, but this could be a result of observations extending over too narrow a range. The differential response of plant functional types to climate change indicates that large changes in plant species composition might occur in certain ecosystems with subsequent consequences for litter quantities and qualities.

References

Aber. J. D., McClaugherty, C. A., and Melillo, J. M. (1984). Litter decomposition in Wisconsin forests—Mass loss, organic-chemical constituents and nitrogen. Research Report R3284, pp. 1–72. Univ. of Wisconsin—Madison, College of Agriculture and Life Sciences.

Aber, J. D., Melillo, J. M., and McClaugherty, C. A. (1990). Predicting long-term patterns of mass loss, nitrogen dynamics, and soil organic matter formation from initial fine litter chemistry in temperate forest ecosystems. *Can. J. Bot.* **68,** 2201–2208.

Ågren, G. I., and Bosatta, E. (1996a). "Theoretical Ecosystem Ecology—Understanding Element Cycles." Cambridge Univ. Press, Cambridge, UK.

Ågren, G. I., and Bosatta, E. (1996b). Quality: A bridge between theory and experiment in soil organic matter studies. *Oikos* **76,** 522–528.

Ågren, G. I., Shaver, G. R., and Rastetter, E. B. (1999). Nutrients: Dynamics and limitations. *In* "Carbon Dioxide and Environmental Stress" (Y. Lou and H. A. Mooney, eds.). Academic Press, San Diego (in press).

Barber, D. A., and Martin, J. K. (1976). The release of organic substances by cereal roots into soil. *New Phytol.* **76,** 69–80.

Berendse, F., Berg, B., and Bosatta, E. (1987). The effect of lignin and nitrogen on the decomposition of litter in nutrient-poor ecosystems: A theoretical approach. *Can. J. Bot.* **65,** 1116–1120.

Berg, B. (1984). Decomposition of root litter and some factors regulating the process: Long-term root litter decomposition in a Scots pine forest. *Soil Biol. Biochem.* **16,** 609–617.

Berg, B., Müller, M., and Wessén, B. (1987). Decomposition of red clover (*Trifolium pratense*) roots. *Soil Biol. Biochem.* **19,** 589–593.

Berg, M. P., Kniese, J. P., Bedaux, J. J. M., and Verhoef, H. A. (1998). Dynamics and stratification of functional groups of micro- and mesoarthropods in the organic layer of a Scots pine forest. *Biol. Fert. Soils* **26,** 268–284.

Berntson, G. M., and Bazzaz, F. A. (1996). Belowground positive and negative feedbacks on CO_2 growth enhancement. *Plant Soil* **187,** 119–131.

Billès, G., Rouhier, H., and Bottner, P. (1993). Modifications of the carbon and nitrogen allocations in the plant (*Triticum aestivum* L.) soil system in response to increased atmospheric CO_2 concentration. *Plant Soil* **157,** 215–225.

Bray, J. R., and Gorham, E. (1964). Litter production in forests of the world. *Adv. Ecol. Res.* **2,** 101–157.

Cairns, M. A., Brown, S., Helmer, E. H., and Baumgardner, G. A. (1997). Root biomass allocation in the world's upland forests. *Oecologia* **111,** 1–11.

Cardon, Z. G. (1996). Influence of rhizodeposition under elevated CO_2 on plant nutrition and soil organic matter. *Plant Soil* **187,** 277–288.

Coûteaux, M. M., Mousseau, M., Célérier, M. L., and Bottner, P. (1991). Increased atmospheric CO_2 and litter quality: Decomposition of sweet chestnut leaf litter with animal food webs of different complexities. *Oikos* **61,** 54–64.

Curtis, P. S., O'Neill, E. G., Teeri, J. A., Zak, D. R., and Pregitzer, K. S. (1994). Belowground responses to rising atmospheric CO_2: Implications for plants, soil biota and ecosystem processes. *Plant Soil* **165,** 1–6.

Curtis, P. S., and Wang, X. (1998). A meta-analysis of elevated CO_2 effects on woody plant mass, form, and physiology. *Oecologia* **113,** 299–313.

David, J. F., Ponge, J. F., and Delecour, F. (1993). The saprophagous macrofauna of different types of humus in beech forests of Ardenne (Belgium). *Pedobiologia* **37,** 49–56.

Franck, V. M., Hungate, B. A., Chapin, F. S. III, and Field, C. B. (1997). Decomposition of litter produced under elevated CO_2: Dependence on plant species and nutrient supply. *Biogeochemistry* **36,** 223–237.

Gillon, D., Joffre, R., and Dardenne, P. (1993). Predicting the stage of decay of decomposing leaves by near infrared reflectance spectroscopy. *Can. J. For. Res.* **23,** 2552–2559.

Gillon, D., Joffre, R., and Ibrahima, A. (1994). Initial litter properties and decay rate: A microcosm experiment on Mediterranean species. *Can. J. Bot.* **72,** 946–954.

Gillon, D., Joffre, R., and Ibrahima, A. (1999). Can litter decomposability be predicted by near infrared reflectance spectroscopy? *Ecology* **80,** 175–186.

Gorissen, A. (1996). Elevated CO_2 evokes quantitative and qualitative changes in carbon dynamics in a plant/soil system: Mechanisms and implications. *Plant Soil* **187,** 289–298.

Gower, S. T., and Son, Y. (1992). Differences in soil and leaf litterfall nitrogen dynamics for five forest plantations. *Soil Sci. Soc. Am. J.* **56,** 1959–1966.

Grayston, S. J., Vaughan, D., and Jones, D. (1996). Rhizosphere carbon flow in trees, in comparison with annual plants: The importance of root exudation and its impact on microbial activity and nutrient availability. *Appl. Soil Ecol.* **5,** 29–56.

Griffiths, B., and Robinson, D. (1992). Root-induced nitrogen mineralisation: A nitrogen balance model. *Plant Soil* **139,** 253–263.

Heal, O. W., Anderson, J. M., and Swift, M. J. (1997). Plant litter quality and decomposition: An historical overview. *In* "Driven by Nature. Plant Litter Quality and Decomposition" (G. Cadish and K. E. Giller, eds.), pp. 3–30. CAB International, Wallingford, UK.

Helal, H. M., and Sauerbeck, D. R. (1984). Influence of plant roots on C and P metabolism in soil. *Plant Soil* **76,** 175–182.

Hodge, A. (1996). Impact of elevated CO_2 on mycorrhizal associations and implications for plant growth. *Biol. Fert. Soils* **23,** 388–398.

Hodge, A., Grayston, S. J., and Ord, B. G. (1996). A novel method for characterisation and quantification of plant root exudates. *Plant Soil* **184,** 97–104.

Hungate, B. A., Canadell, J. G., and Chapin III, F. S. (1996). Plant species mediate changes in soil microbial N in response to elevated CO_2. *Ecology* **77,** 2505–2515.

Ibrahima, A., Joffre, R., and Gillon, D. (1995). Changes in litter during the initial leaching phase: An experiment on the leaf litter of Mediterranean species. *Soil Biol. Biochem.* **27,** 931–939.

Ineichen, K., Wiemken, V., and Wiemken, A. (1995). Shoots, roots and ectomycorrhiza formation of pine seedlings at elevated atmospheric carbon dioxide. *Plant Cell. Environ.* **18,** 703–707.

Jackson, R. B., Canadell, J., Ehleringer, J. R., Mooney, H. A., Sala, O. E., and Schulze, E. D. (1996). A global analysis of root distributions for terrestrial biomes. *Oecologia* **108,** 389–411.

Jackson, R. B., Mooney, H. A., and Schulze, E. D. (1997). A global budget for fine root biomass, surface area, and nutrient contents. *Proc. Natl. Acad. Sci. U.S.A.* **94,** 7362–7366.

Joffre, R., Gillon, D., Dardenne, P., Agneessens, R., and Biston, R. (1992). The use of near-infrared reflectance spectroscopy in litter decomposition studies. *Ann. Sci. For.* **49,** 481–488.

Joffre, R., Ågren, G. I., Gillon, D., and Bosatta, E. (2000). In progress.

Johnen, B. G., and Sauerbeck, D. R. (1977). A tracer technique for measuring growth, mass and microbial breakdown of plants roots during vegetation. *In* "Soil Organisms as Components of Ecosystems" (U. Lohm and T. Persson, eds.), pp. 366–373. Ecological Bulletin 25, Stockholm.

Jongen, M., Fay, P., and Jones, M. B. (1996). Effects of elevated carbon dioxide and arbuscular mycorrhizal infection on *Trifolium repens. New Phytol.* **132,** 413–423.

Kätterer, T., Reichstein, M., Andrén, O., and Lomander, A. (1998). Temperature dependence of organic matter decomposition: A critical review using literature data analyzed with different models. *Biol. Fert. Soils* **27,** 258–262.

Kirschbaum, M. U. F. (1995). The temperature dependence of soil organic matter decomposition and the effect of global warming on soil organic C storage. *Soil Biol. Biochem.* **27,** 753–760.

Klironomos, J. N., Rillig, M. C., and Allen, M. F. (1996). Below-ground microbial and microfaunal responses to *Artemisia tridentata* grown under elevated atmospheric CO_2. *Funct. Ecol.* **10,** 527–534.

Koch, G. W., and Mooney, H. A. (eds.) (1996). "Carbon Dioxide and Terrestrial Ecosystems." Academic Press, San Diego.

Lambers, H. (1987). Growth, respiration, exudation and symbiotic association: The fate of carbon translocated to the roots. *In* "Root Development and Function" (P. J. Gregory, J. V. Lake, and A. Rose, eds.), pp. 125–145. Cambridge Univ. Press, Cambridge, UK.

Lavelle, P. (1988). Earthworm activities and the soil system. *Biol. Fert. Soils* **6,** 237–251.

Lavelle, P. (1997). Faunal activities and soil processes: Adaptive strategies that determine ecosystem function. *Adv. Ecol. Res.* **27,** 93–132.

Lekkerkek, L. J. A., van de Geijn, S. C., and van Veen, J. A. (1990). Effects of elevated atmospheric CO_2 levels on the carbon economy of a soil planted with wheat. *In* "Soils and the Greenhouse Effect" (A. F. Bouwman, ed.), pp. 423–429. John Wiley & Sons, Chichester.

Lewis, J. D., and Strain, B. R. (1996). The role of mycorrhizas in the response of *Pinus taeda* seedlings to elevated CO_2. *New Phytol.* **133,** 431–443.

Lewis, J. D., Thomas, R. B., and Strain, B. R. (1994). Effect on elevated CO_2 on mycorrhizal colonization on loblolly pine (*Pinus taeda* L.) seedlings. *Plant Soil* **165,** 81–88.

Liljeroth, E., van Veen, J. A., and Miller, H. J. (1990). Assimilate translocation to the rhizosphere of two wheat lines and subsequent utilisation by rhizosphere microorganisms at two soil nitrogen concentrations. *Soil Biol. Biochem.* **22,** 1015–1021.

Luxmoore, R. J. (1981). CO_2 and phytomass. *BioScience* **31,** 626.

Lynch, J. M., and Whipps, J. M. (1990). Substrate flow in the rhizosphere. *Plant Soil* **129,** 1–10.

Majdi, H., and Kangas, P. (1997). Demography of fine roots in response to nutrient application in a Norway spruce stand in southwestern Sweden. *Ecoscience* **4,** 199–205.

Markkola, A. M., Ohtonen, A., Ahonen-Jonnarth, U., and Ohtonen, R. (1996). Scots pine responses to CO_2 enrichment. I. Ectomycorrhizal fungi and soil fauna. *Environ. Pollut.* **94,** 309–316.

Martin, J. K. (1975). ^{14}C-Labelled material leached from the rhizosphere of plants, supplied continuously with $^{14}CO_2$. *Soil Biol. Biochem.* **7,** 395–399.

Meharg, A. A. (1994). A critical review of labelling techniques used to quantify rhizosphere carbon flow. *Plant Soil* **166,** 55–62.

Miller, J. D., Cooper, J. M., and Miller, H. G. (1996). Amounts and nutrient weights in litterfall, and their annual cycles, from a series of fertilizer experiments on pole-stage Sitka spruce. *Forestry* **69,** 289–302.

Monz, C. A., Hunt, H. W., Reeves, F. B., and Elliott, E. T. (1994). The response of mycorrhizal colonization to elevated CO_2 and climate change in *Pascopyrum smithii* and *Bouteloua gracilis*. *Plant Soil* **165,** 75–80.

Morgan, J. A., Knight, W. G., Dudley, L. M., and Hunt, H. W. (1994). Enhanced root system C-sink activity, water relations and aspects of nutrient acquisition in mycotrophic *Bouteloua gracilis* subjected to CO_2 enrichment. *Plant Soil* **165,** 139–146.

Norby, R. J. (1994). Issues and perspectives for investigating root responses to elevated atmospheric carbon dioxide. *Plant Soil* **165,** 9–20.

Norby, R. J., O'Neill, E. G., Gregory, H. W., and Luxmoore, R. J. (1987). Carbon allocation, root exudation and mycorrhizal colonization of *Pinus echinata* seedlings grown under CO_2 enrichment. *Tree Physiol.* **3,** 203–210.

Norby, R. J., and Cotrufo, M. F. (1998). A question of litter quality. *Nature (London)* **396,** 17–18.

O'Neill, E. G. (1994). Responses of soil biota to elevated atmospheric carbon dioxide. *Plant Soil* **165,** 55–65.

Parsons, W. F. J., Taylor, B. R., and Parkinson, D. (1990). Decomposition of aspen (*Populus tremuloides*) leaf litter modified by leaching. *Can. J. For. Res.* **20,** 943–951.

Paustian, K., Ågren, G. I., and Bosatta, E. (1997). Modelling litter quality effects on decomposition and soil organic matter dynamics. *In* "Driven by Nature: Plant Litter Quality and Decomposition" (G. Cadisch and K. E. Giller, eds.), pp. 313–335. CAB International, Wallingford, UK.

Petersen, H., and Luxton, M. (1982). A comparative analysis of soil fauna populations and their rôle in decomposition processes. *Oikos* **39,** 287–388.

Pettersson, R., McDonald, A. J. S., and Stadenberg, I. (1993). Response of small birch plants (*Betula pendula* Roth.) to elevated CO_2 and nitrogen supply. *Plant Cell Environ.* **16,** 1115–1121.

Poorter, H. (1993). Interspecific variation in the growth response of plants to an elevated ambient CO_2 concentration. *Vegetatio* **104/105,** 77–97.

Reich, P. B., Walters, M. B., and Ellsworth, D. S. (1997). From tropics to tundra: Global convergence in plant functioning. *Proc. Natl. Acad. Sci. U.S.A.* **94,** 13730–13734.

Robinson, C. H., Michelsen, A., Lee, J. A., Whitehead, S. J., Callaghan, T. V., Press, M. C., and Jonasson, S. (1997). Elevated atmospheric CO_2 affects decomposition of *Festuca vivipara* (L.) Sm. litter and roots in experiments simulating environmental change in two contrasting arctic ecosystems. *Global Change Biol.* **3,** 37–49.

Rogers, H. H., Runion, G. B., and Krupa, S. V. (1994). Plant responses to atmospheric CO_2 enrichment with emphasis on roots and the rhizosphere. *Environ. Pollut.* **83,** 155–189.

Rouhier, H., Billès, G., El Kohen, A., Mousseau, M., and Bottner, P. (1994). Effects of elevated CO_2 on carbon and nitrogen distribution within a tree (*Castanea sativa* Mill)–soil system. *Plant Soil* **162,** 281–292.

Rouhier, H., and Read, D. J. (1998). The role of mycorrhiza in determining the response of *Plantago lanceolata* to CO_2 enrichment. *New Phytol.* **139,** 367–373.

Rovira, A. D. (1956). Plant root excretions in relation to the rhizosphere effect. 1. The nature of root exudates from oats and peas. *Plant Soil* **7,** 178–194.

Rovira, A. D. (1969). Plant root exudates. *Bot. Rev.* **35,** 35–59.

Rovira, A. D., Foster, J. K., and Martin, J. K. (1979). Origin, nature and nomenclature of the organic materials in the rhizosphere. *In* "The Soil–Root Interface" (J. L. Harley and R. S. Russell, eds.), pp. 1–4. Academic Press, London.

Ryan, M. G., Mellilo, J. M., and Ricca, A. (1990). A comparison of methods for determining proximate carbon fractions of forest litter. *Can. J. For. Res.* **20,** 166–171.

Saxe, H., Ellsworth, D. S., and Heath, J. (1998). Tree and forest functioning in an enriched CO_2 atmosphere. *New Phytol.* **139,** 395–436.

Schaeffer, M., and Schauermann, J. (1990). The soil fauna of beech forests: Comparison between a mull and a moder soil. *Pedobiologia* **34,** 299–314.

Schlesinger, W. H. (1997). "Biogeochemistry—An Analysis of Global Change." Academic Press, San Diego.

Seegmüller, S., and Rennenberg, H. (1994). Interactive effects of mycorrhization and elevated carbon dioxide on growth of young pedunculate oak (*Quercus robur* L.) trees. *Plant Soil* **167,** 325–329.

Seyferth, U., and Persson, T. (2000). Effects of soil temperature and moisture on carbon mineralisation in a coniferous forest soil. *Soil Biol. Biochem.* Submitted.

Staddon, P. L., and Fitter, A. H. (1998). Does elevated atmospheric carbon dioxide affect arbuscular mycorrhizas? *Trends Ecol. Evol.* **13,** 455–458.

Stulen, I., and den Hertog, J. (1993). Root growth and functioning under atmospheric CO_2 enrichment. *Vegetatio* **104/105,** 99–115.

Swift, M. J., Heal, O. W., and Anderson, J. M. (1979). "Decomposition in Terrestrial Ecosystems." Blackwell Scientific Publ., Oxford, UK.

Swift, M. J., Andrén, O., Brussard, L., Briones, M., Coûteaux, M.-M., Ekschmitt, K., Kjoller, A., Loiseau, A., and Smith, P. (1998). Global change, soil biodiversity, and nitrogen cycling in terrestrial ecosystems: three case studies. *Global Change Biol.* **4,** 729–743.

Taylor, B. R., Parkinson, D., and Parsons, W. F. J. (1989). Nitrogen and lignin content as predictors of litter decay rates: A microcosm test. *Ecology* **70,** 97–104.

Tian, G., Brussaard, L., and Kang, B. T. (1995). Breakdown of plant residues with contrasting chemical compositions: Effect of earthworms and millipedes. *Soil Biol. Biochem.* **27,** 277–280.

van de Geijn, S. C., and van Veen, J. A. (1993). Implications of increased carbon dioxide levels for carbon input and turnover in soils. *Vegetatio* **104/105,** 283–295.

van der Linden, A. M. A., van Veen, J. A., and Frissel, M. J. (1987). Modelling soil organic matter levels after long-term applications of crop residues, farmyards and green manures. *Plant Soil* **101,** 21–28.

van Ginkel, J. H., Gorissen, A., and van Veen, J. A. (1996). Long-term decomposition of grass roots as affected by elevated atmospheric carbon dioxide. *J. Environ. Qual.* **25,** 1122–1128.

van Noordwijk, M., Martikainen, P., Bottner, P., Cuevas, E., Rouland, C., and Dhillion, S. S. (1998). Global change and root function. *Global Change Biol.* **4,** 759–772.

van Veen, J. A., Liljeroth, E., Lekkerkek, L. J. A., and van de Geijn, S. C. (1991). Carbon fluxes in plant-soil systems at elevated atmospheric CO_2 levels. *Ecol. Appl.* **1,** 175–181.

van Veen, J. A., Merckx, R., and van de Geijn, S. C. (1989). Plant- and soil-related controls of the flow of carbon from roots through the soil microbial biomass. *Plant Soil* **115,** 43–52.

Vitousek, P. M., and Howarth, R. H. (1991). Nitrogen limitation on land and in the sea: How can it occur? *Biogeochemistry* **13,** 87–115.

Vogt, K. A., Vogt, D. J., and Bloomfield, J. (1998). Analysis of some direct and indirect methods for estimating root biomass and production of forests at an ecosystem level. *Plant Soil* **200,** 71–89.

Walker, B. H., and Steffen, W. L. (eds.) (1996). "Global Change and Terrestrial Ecosystems." Cambridge Univ. Press, Cambridge, UK.

Walker, R. F., Johnson, D. W., Geisinger, D. R., and Ball, J. T. (1998). Growth and ectomycorrhizal colonization of ponderosa pine seedlings supplied different levels of atmospheric CO_2 and soil N and P. *For. Ecol. Manage.* **109,** 9–20.

Wardle, D. A., and Giller, K. E. (1996). The quest for a contemporary ecological dimension to soil biology. *Soil Biol. Biochem.* **28,** 1549–1554.

Wardle, D. A., and Lavelle, P. (1997). Linkages between soil biota, plant litter quality and decomposition. *In* "Driven by Nature. Plant Litter Quality and Decomposition" (G. Cadish and K. E. Giller, eds.), pp. 107–124. CAB International, Wallingford, UK.

Wardle, D. A., Verhoef, H. A., and Clarholm, M. (1998). Trophic relationships in the soil microfood-web: Predicting the responses to a changing global environment. *Global Change Biol.* **4,** 713–727.

Warembourg, F. R., and Billès, G. (1979). Estimating carbon transfer in the rhizosphere. *In* "The Soil–Root Interface" (J. L. Harley and R. S. Russell, eds.), pp. 183–197. Academic Press, London.

Wessén, B., and Berg, B. (1986). Long-term decomposition of barley straw: Chemical changes and ingrowth of fungal mycelium. *Soil Biol. Biochem.* **18,** 53–59.

Whipps, J. M., and Lynch, J. M. (1985). Energy losses by the plant in rhizodeposition. *Ann. Proc. Phytochem. Soc. Eur.* **26,** 59–71.

Whipps, J. M., and Lynch, J. M. (1986). The influence of the rhizosphere on crop productivity. *Adv. Microb. Ecol.* **9,** 187–244.

Zheng, D. W., Bengtsson, J., and Ågren, G. I. (1997). Soil food webs and ecosystem processes: Decomposition in donor-control and Lotka–Volterra systems. *Am. Nat.* **149,** 125–148.

6

Herbivory and Trophic Interactions

Sam J. McNaughton

I. Introduction

Herbivory is a pervasive ecological process, ranging in terrestrial habitats from the extremes of snails (Shachak *et al.*, 1987) and lepidopteran larvae (Wessels and Wessels, 1991) that obtain nourishment from epilithic algae and lichens by grinding the surface off desert rocks, to the conspicuous large mammalian herbivores of African savannas with a manifestly impressive effect on plant biomass (Petrides and Swank, 1965; McNaughton, 1976, 1985; Andere, 1981). Nevertheless, it is a neglected topic in the analysis of terrestrial global productivity (McNaughton *et al.*, 1989, 1991), and the concept of trophic levels has been a subject of continuing controversy, although there is evidence of a developing synthesis (Oksanen, 1991). Instead of being treated as an integral control of primary production, herbivory and trophic relations are more often considered as an adjunct, processes grafted by evolution onto a base of plant growth and yield, customarily considered a suppressant of plant processes and fitness.

Current approaches to estimating global values of primary production, and other ecological processes, are invariably based on the biome concept influentially stated by Clements and Shelford (1939), who were, during their time of publication, the preeminent plant and animal ecologists, respectively, in the United States. They wrote (1939, p. 20) that "the biome or plant–animal formation is the basic community unit; that is, two separate communities, plant and animal do not exist in the same area . . . [but] a great biotic complex of fully developed and developing communities." Still, in application, the fundamental biome concept has been based wholly on vegetation types in terrestrial habitats, so, operationally, biome is vegetation.

Indeed, all global modeling of terrestrial global productivity known to me considers primary production a function of climatic parameters, typically as modified by soils, with higher trophic levels unrepresented (McGuire *et al.*, 1993; Running and Hunt, 1993; Parton *et al.*, 1994). Models of higher spatial resolution, in contrast, reveal that herbivores can have a considerable controlling effect on the expression of climatic and soil constraints in production processes (Coughenour *et al.*, 1984; Holland *et al.*, 1992; Seagle *et al.*, 1992). If herbivory and trophic relations have powerful local effects integral to primary productivity, are those effects, then, so spatially and temporally localized that they are insignificant and can be disregarded on a global basis? Is it sufficient to predict primary production through correlative methods without a comprehension of contributing mechanisms? Or, do localized effects of herbivores cancel one another at large spatial and temporal scales? There is a time-honored conceptual view of ecosystems as donor controlled, operating within a climatic–geological context with life and its processes flowing inexorably from a physical environmental context. Although there can be no doubt that ecosystem processes are abiotically constrained, neither can it be doubted that ecosystem processes feed back to abiotic factors to have profound effects, even on a global scale (Schwartzman and Volk, 1989; Shear, 1991). Recognition is required of the many complex and powerful biotic feedbacks that operate in the biosphere through the trophic web.

II. Herbivores and Trophic Relations in Global Evolutionary Context

A self-sustaining ecosystems must have three attributes: (1) It must be open to energy (most often, the sun); (2) it must contain autotrophic producers (most often, photosynthetic organisms) capable of transforming incoming (usually radiant) energy into C–C bond energy; and (3) it must contain heterotrophic consumers that recycle elements back into the environment in forms that can be utilized by producers. Here is the crux: those consumers can (1) feed entirely on the dead tissues of plants or other organisms, participating in a detritus/litter food web, or (2) feed on the live tissues of plants and other, just killed, organisms, participating in a grazing food web.

In fact, because they depend on live, rather than dead, production, members of a grazing web have temporal and food-quality priority over the detritus web, intervening before energy and materials can flow to the latter. The unused residue from the grazing food web transferred to the detritus web is of substantially lower quality than the material harvested, but the processed material excreted as waste is of much higher quality than the dead tissues of plants. This chapter focuses on food web processes to address the question raised above and considers how grazing and detritus food webs are quantitatively related in terrestrial ecosystems.

The earliest organisms likely were chemoautrophs (lithotrophs) with an inorganically based nutrition in hydrothermal habitats (Wäschtershäuser, 1988; Huber and Wäschtershäuser, 1997). Later, heterotrophic organisms evolved to utilize lithotroph leakage, leading to ecosystems in which nutrients were passed between lithotrophs and heterotrophs, eventually generating feeding processes culminating in recycling by a detritus-like food web.

Columnar stromatolites, the most abundant of Precambrian fossils, are organosedimentary structures built by cyanobacteria (Awramik, 1971). Their decline in abundance and distribution is simultaneous with the appearance of the Metazoa, representing the earliest documented abundance decline and range restriction of primary producers due to consumers. Similarly, the first terrestrial plants likely were cyanobacterial crusts and mats of algae, which, along with detritivores and associated carnivores, constituted the earliest terrestrial ecosystems (Shear, 1991). Herbivores appeared only later, feeding directly on the developing ground cover of cyanobacteria, mosses, and lichens.

Thus, there can be little uncertainty that the earliest food webs were detrital both in sea and on land, but that herbivore–plant interactions have been major ecological processes from near the dawn of the trophic web. Consider the effect that herbivorous dinosaurs of body mass up to 100 tons, with huge nutritional and energetic requirements and resulting enormous appetites, consuming several thousand kilograms of plant tissue per capita daily, must have had on the vegetation that supported them. They undoubtedly profoundly affected energy flow and nutrient cycling in ecological time and acted as major evolutionary forces in geological time. Certainly, the herbivorous dinosaurs must have had a major impact on plants for millions of years, influencing all those plant properties known today to be important in herbivory: palatability, regrowth potential, assimilation capacity, structural strength, phytotoxins, nutritional suitability, and growth form. Should we consider herbivores as only passive, largely negative, participants in energy flow, damaging plants and therefore reducing productive potential? Or should we consider them regulators, influencing primary production intimately and acting as organismal catalysts of ecosystem processes (Chew, 1974; McNaughton, 1976, 1979; Owen and Weigert, 1976; Loreau, 1995), setting limits on the bounds of energy flow and nutrient cycling through direct and indirect effects on resource supply at the food web base, modifying plant physiological status by influencing both environmental factors and producer metabolic processes (McNaughton, 1976, 1979, 1983; Bianchi *et al.*, 1989; Bianchi and Jones, 1991; Jones *et al.*, 1994)? African elephants (*Loxadontia africana*) have ecosystem effects that have been characterized as nearly as pervasive as those of humans (Laws, 1970). It is appropriate, therefore, to regard members of the grazing web as fundamental organizers of photoautotroph-based food webs.

III. Herbivory and the Environment

A. Nutrients and Energy

Plant regrowth after herbivores remove foliage (Wegener and Odasz, 1997) depends on predefoliation reserves and postdefoliation assimilates (de Visser *et al.*, 1997). Considerable emphasis in studies of the physiological response of plants to herbivory has been placed on the role of stored reserves, particularly mobilizible carbohydrates, and in the ability of affected plants to tolerate, and recover from, herbivory (Alberda, 1966; Misra and Singh, 1981; Bahrani *et al.*, 1983; Danckwerts and Gordon, 1987). However, data on carbohydrate and N depletion following herbivory indicate that it generally is not severe in plants in otherwise favorable environments, and is very transitory, with the nutritional and energetic needs of regrowing tissues rapidly met by new assimilate (Caldwell *et al.*, 1981; Valentine *et al.*, 1983). Therefore, sustained depletion of energetic and nutritional supplies is an uncommon consequence of herbivory, more likely to be important if the plant is otherwise stressed by, for example, nutrient (Chapin and McNaughton, 1989) or water (Georgiadis *et al.*, 1989; Simoes and Baruch, 1991) shortage. A dwarf shrub in South Africa severely defoliated (80% of leaf and twig biomass removed) generated regrowth during the first 2 postdefoliation weeks principally from stored reserves (60%), but current assimilate also made a substantial contribution until full assimilate competence was reestablished (van der Heyden and Stock, 1996). Even severely clipped plants attain photosynthetic competence sufficient to meet regrowth requirements in 10 days to 2 weeks, although if it occurs late in the growing season when reserves are acccumulated for new growth in the following year, defoliation can have significantly detrimental, but delayed, effects (Eckardt *et al.*, 1982), because spring growth is initially driven by mobilization of reserves stored the previous growing season (Volonec *et al.*, 1996). Stored reserves utilized to generate foliage rapidly at the outset of the subsequent growing season are a pivotal component of primary productivity in ecosystems dominated by perennial plants (McNaughton, 1976).

Leaf expansion of rain forest trees is positively correlated with leaf N content; more rapidly expanding leaves have higher N concentrations compared to more slowly growing leaves (Kursar and Coley, 1991). This, of course, represents a potentially more nutritious food supply for herbivores. Kursar and Coley (1991) suggest that age-specific N levels and associated leaf expansion can be interpreted as evolutionary consequences of a trade-off between palatability and escape from herbivory through rapid expansion because leaves toughen as they enlarge. In other words, rapid leaf expansion, and therefore rainforest productivity, may have been driven, in part, by evolutionary responses to herbivory.

A fundamental generalization relating herbivory to nutrients is that investment of resources in chemical defenses against herbivory will reduce those available for growth (Feeny, 1976; Rhodes and Cates, 1976; Coley *et al.,* 1985; Gulmon and Mooney, 1986). Within-species genetic variation in chemical defenses has been used to document an inverse relationship between those defenses and the growth rate of a genotype (Coley, 1986; Sagars and Coley, 1995). Similarly, there is convincing evidence that the relative growth rate of radish (*Raphanus sativus*) genotypes declines with higher resistance to a pathogen causing fungal wilt disease (Hoffland *et al.,* 1996). As proportional infection varied from 0 to 80%, relative growth rate declined by half.

An investigation of the genetic basis and cost of defense, in contrast, presented evidence that selection by herbivory favored both resistance and tolerance, via regrowth, to herbivory, with no evident trade-offs (Mauricio *et al.,* 1997). Therefore, an obligate reciprocal resource exchange colimiting defense against, or regrowth after, herbivory clearly is not universal.

A definitive study of the role of stored carbohydrates in regrowth of two bunchgrass species differing markedly in their regrowth capacity following defoliation found that only a very small proportion of regrowth was at the expense of stored assimilates (Richards and Caldwell, 1985). Comparing the carbohydrate budgets of etiolated plants, entirely dependent on stored reserves following defoliation, with plants regrowing in the light revealed that between 89 and 99% of regrowth was derived from current assimilates. Regrowth was unrelated to initial stored nonstructural carbohydrate, total pools, or pool depletion. Thus, this study indicates that defoliated grasses rapidly become photosynthetically competent and are not dependent on major pool drawdown for regrowth (Oesterheld and McNaughton, 1991).

Nutrient supply, in contrast, can be an important element of regrowth potential (McNaughton, 1979; Bazely and Jefferies, 1985). Multifactorial experiments with a heavily grazed dryland C_4 sedge, *Kyllinga nervosa,* from the Serengeti ecosystem, revealed that different components of trophic yield were controlled by different environmental variables (McNaughton *et al.,* 1983). Yield to grazers (the amount clipped off) was controlled principally by nitrogen supply, yield to producers (the residual plant mass) was controlled primarily by water supply, and yield to decomposers (litter and standing dead) was controlled jointly by nitrogen supply and clipping height. However, yields to decomposers and grazers were not simply competitive, with the latter diverting flow from the former, but were partitioned, in part, between additional primary productivity in a compensatory growth response. Experiments with a heavily grazed African grass, *Sporobolus kentrophyllus,* in the Amboseli ecosystem (Georgiadis *et al.,* 1989), varied nitrogen supply, clipping, and the interval between storms (watering interval). Maximum relative aboveground growth stimulation occurred when watering and clipping were both infrequent, but the maximum absolute stimulation occurred under the combined conditions expected under grazing: nitrogen inputs (Bazely and Jefferies, 1985) and moderate defoliation under frequent watering. This pattern is conceptually consistent with relationships between leaf longevity, assimilation capacity, and nitrogen status (Reich *et al.,* 1992; see also Chapter 11, this volume). Reich and colleagues have demonstrated a complex of mutually related traits in species exploiting habitats that are generally resource rich, including high nitrogen concentrations per unit mass, high specific leaf areas, high assimilation rates per unit leaf mass, and short leaf life-span. Moreover, there were evident temporal components to these traits because species from habitats in which resources are temporally variable, but abundant overall, had this complex of traits. That trait complex is entirely consistent with the properties of grasses adapted to high-grazing environments that, on defoliation, convert to higher leaf nitrogen status (McNaughton *et al.,* 1983), greater specific leaf areas, and higher relative growth rates (Oesterheld and McNaughton, 1991).

Africa's most widespread and abundant grass, the midheight *Themeda triandra,* was incapable of completely compensating for herbivory, although yield to grazers was positively related to tillering, the activation of basal meristems (Coughenour *et al.,* 1985a). Yield to grazers was affected by water and nitrogen supply, residual plant yield was affected negatively by clipping height and positively by nitrogen, and decomposer yield was influenced only by clipping height, being reduced by shorter clipping. An African tall grass, *Hyparrhenia filipendula,* responded to defoliation with increased photosynthetic rates and more rapid rates of leaf blade elongation (Coughenour *et al.,* 1985b). Principal factors affecting yield to grazers were clipping height and water supply; plant terminal biomass was regulated by nitrogen and water supplies and yield to decomposers was affected primarily by clipping height and water.

Four African grass species grown under phosphorus deficiency were severely inhibited by defoliation, with reductions in both yield to producers and yield to decomposers (Chapin and McNaughton, 1989), although adequate phosphorus supply allowed regrowth to compensate for tissue removal (McNaughton and Chapin, 1985). These data suggest that rapid nutrient cycling and adequate nutrient supplies from soils are important elements of ecosystem properties that allow plants to cope with herbivory and reduce its effects to a nondetrimental status (Chapin and McNaughton, 1989), although some studies with temperate grasses suggest that soil may be less important in their responses to defoliation (Hicks and Reader, 1995).

Fox and Morrow (1992) found that eucalyptus trees growing on infertile soil were more responsive to herbivore removal than to fertilization. That is, even though the plants were growing in nutritionally poor soils, growth increased in response to herbivore removal, but was unresponsive to nutrient supplementation.

B. Carbon Dioxide

Increasing atmospheric carbon dioxide is a component of global change than can be expected to continue for the foreseeable future (see Chapters 7 and 22, this volume). Elevated CO_2 commonly results in an increase in plant C:N ratio (Lincoln, 1993), principally due to sugar accumulation and a reduction in ribulose bisphosphate carboxylase/oxygenase (rubisco) concentrations (Besford, 1990). This results in decreased food quality for herbivores. Arthropod herbivores may increase consumption in response to lower food quality (Slansky and Feeny, 1977), compensating for reduced quality by increased intake (Fajer, 1989). This could result in increased herbivory accompanying elevated CO_2. Mammals, however, because they eat to fill and then ruminate, spend so much time feeding that it is unlikely that they could increase intake substantially as forage quality declines (Wilsey *et al.*, 1994).

A comparison of grasses from grasslands composed exclusively of C_3 species (Yellowstone, United States), exclusively C_4 plants (Serengeti), and a mixture of photosynthetic types (flooding pampas, Argentina) revealed declines of forage quality at increased CO_2 in Yellowstone and Argentine pampas grasses, but no change in those from the tropical Serengeti ecosystem (Wilsey *et al.*, 1997). Neither was there an interaction between defoliation and CO_2. These data indicate that trophic relations in ecosystems dominated by plants with C_4 photosynthesis may change little, whereas ecosystems in which C_3 species are important may be significantly affected.

A specialist lepidopteran larvae, *Junonia coenia,* feeding on *Plantago lanceolata,* grew more slowly and experienced higher mortality when grown at high CO_2 (Fajer *et al.*, 1989). *Rumex obtusifolius* fed on by *Gastrophysa viridula* larvae at two different consumption and CO_2 levels revealed complicated patterns of plant growth in response to herbivory (Pearson and Brooks, 1996). The low-herbivore treatment reduced leaf area 20–40% and the high treatment, 50–70%, and CO_2 had no effect on consumption. Plants grown in elevated CO_2 without herbivores expressed the common syndrome of trait change: increased growth, higher root/shoot, lower stomatal conductance, and increased photosynthesis. At the end of a 1-month recovery period following herbivore removal, plants grown at high CO_2 concentration had increased plant yields at higher levels of previous herbivory.

These data, though sparse, suggest that herbivory effects will likely differ in ecosystems differing in the photosynthetic pathway of the predominant producers. In C_3-dominated ecosystems, in which herbivores are likely to be principally insects, increasing CO_2 is expected to result in greater per capita consumption by herbivores as they increase intake to compensate for poorer food quality. Whether gross rates of herbivory change, however, will finally depend on the demographic consequences of poor food quality. In C_4-dominated ecosystems, in contrast, in which herbivores are likely to be principally large mammals, there is likely to be little change in either per capita or trophic level consumption because C:N balances are largely insensitive to atmospheric CO_2.

A frequent effect of increased CO_2 is a shift of plant aboveground/belowground allocation balance toward an increase of the latter (Wilsey *et al.*, 1997). This would be expected to result in increased food availability to underground herbivores. The few experimental studies that have been done on belowground herbivores indicate that they can have a pronounced effect on primary productivity (Stanton *et al.*, 1981; Ingham and Detling, 1990). In general, elimination of belowground herbivores tends to have variable and unpredictable effects on plant productivity. Changes in food quality belowground, as aboveground, also can have substantial effects on herbivory (Seastedt *et al.*, 1988) that would be expected to affect soil trophic interactions.

IV. Plant Architecture

A. Geometry

The structure of vegetation is influenced by and influences herbivory and trophic relations in ecosystems (McNaughton and Sabuni, 1988). These reciprocal influences act at both ecological and evolutionary time scales (Whitham and Slobodchikoff, 1981; Coughenour, 1984; Haukioja *et al.*, 1990).

A clear evolutionary trade-off between competitive ability and regrowth competence following herbivory has been documented within a species of Serengeti Plains grass (Hartvigsen and McNaughton, 1995). Within a local population are individuals ranging from tall stature with low relative growth rate to short stature with high relative growth rate. Height is clearly a variable associated with competitive ability in vascular plants, whereas high relative growth rate allows rapid recovery following defoliation (Oesterheld and McNaughton, 1991). When fences were erected in Serengeti grasslands, the tall-statured genotypes became predominant within 5 years (McNaughton, 1989). Similarly, when fences were removed, the short-statured genotypes became predominant in 2–3 years. Briske and Anderson (1992), in contrast, found that *Schizachyrium scoparium* genotypes from grazed habitats were stronger competitors compared to plants lacking a history of grazing, even in the absence of defoliation, because the grazing-adapted genotypes produced abundant tillers that provided a competitive advantage.

The fractal geometry of a legume shrub was modified by livestock grazing in a pattern suggesting that both no grazing and heavy grazing were associated with developmental instability (Escós *et al.*, 1997). Moderate grazing promoted both growth and developmental stability of vegetative structures. This was manifested in increased fractal complexity and greater symmetry of morphological structures. These consequences of moderate grazing were also associated with greater total inclusive fitness, assessed as population rate of change, of grazed shrubs compared with those either heavily grazed or ungrazed.

B. Root–Shoot Balance

Among the enduring generalizations about aboveground herbivore effects on plant productivity is that any increase in plant growth to replace aboveground loss will be at the expense of belowground growth (McNaughton *et al.*, 1997). Most of the evidence for this tenet is derived from short-term pot experiments of many years ago (Jameson, 1963). More recent evidence from field (McNaughton *et al.*, 1997) and laboratory (Oesterheld and McNaughton, 1991) experiments indicate that Serengeti grasslands, at least, among the most heavily defoliated ecosystems known (McNaughton, 1985; McNaughton *et al.*, 1989, 1991), do not suffer reductions in belowground growth as a consequence of aboveground defoliation.

Plant growth and, therefore, ecosystem primary productivity, are consequences of the interaction between plant biomass at the beginning of an assay period (usually the beginning of a growing season) and relative growth rate (RGR) until the time of assay (usually the end of the growing season). Therefore, if foliage area is regenerated after defoliation, foliage RGR must increase (Oesterheld and McNaughton, 1991). Plasticity of allocation must be a component of recovery (Alward and Joern, 1993), and leaf area restoration (McNaughton, 1974; Potter and Jones, 1977; Poorter and Remkes, 1990) becomes of singular functional importance. Oesterheld (1992) studied the regrowth of two perennial grass species from the flooding pampa of Argentina in relation to defoliation intensity. He found that foliage RGR increased linearly in both species in relation to defoliation up to complete removal. Belowground RGR was little affected by defoliation up to 20% removal, but then declined gradually up to the complete defoliation level. In addition, there were striking differences between the species in root RGR. It is evident that a whole panoply of root–shoot interactions due to trophic impact can be observed (McNaughton, 1986). Plant regrowth following herbivory, like plant defense against that herbivory, must be considered with a phytocentric perspective if it is to be properly understood (Coleman and Jones, 1991).

In closing this section, it is important to emphasize that plants with an evolutionary history of strong defoliation, and the communities they compose, are most likely to compensate for its effects, at the system level (Milchunas and Lauenroth, 1993) and at the individual level (Lennartsson *et al.*, 1997). At the former level, the spatial pattern of defoliation is important, with grazing of large patches more likely to increase productivity than grazing of small, isolated patches (Semmartin and Oesterheld, 1996). At the latter level, some plants can sufficiently adjust growth following defoliation so that fitness is enhanced (Paige and Whitham, 1987; Maschinski and Whitham, 1989; Lennartsson *et al.*, 1997).

V. Tri-trophic Interactions

Although an assortment of interrelated models (Hairston *et al.*, 1960; Fretwell, 1977; Oksanen *et al.*, 1981) predict the qualitative nature of trophic cascades (Carpenter *et al.*, 1985) as one or another trophic level is perturbed by natural forces or experimental intervention, very little empirical research has been done in terrestrial habitats (Pimm, 1982, 1991; Howe and Westley, 1988). A fundamental view of alternative organizers of trophic webs is that some may be organized by resource availability to producers, others by predators acting as keystone species (Paine, 1969). In fact, a review of tritrophic studies revealed that none measured aspects relevant to global terrestrial productivity (Price *et al.*, 1980), most concentrating on population dynamics of only one or two of the three levels, and most were single species per trophic level studies. Prevailing evidence from pelagic freshwater ecosystems on their responses to resource addition and predator manipulations does not provide support for the prevailing models linking food web levels to ecosystem limitations (Brett and Goldman, 1997).

A recent experimental test of top-down versus bottom-up control of a terrestrial system was a growth chamber study, but, nevertheless, this is a clear, unequivocal test of trophic cascades (Hartvigsen *et al.*, 1995). Plants were 2-year-old *Populus deltoides* (cottonwood), herbivores were *Tetranychus urticae* (two-spotted mite), and carnivores were *Phytoseiulus persimilis* (predatory mite). Nutrient supply was regulated by fertilization, and plant performance was determined by calculating relative height growth rate based on weekly measurements. Plants responded to fertilization with increased growth only in the absence of herbivores; thus, herbivores reduced performance of all plants, regardless of fertilization, in the absence of predators to keep their numbers low. However, fertilized plants grew more rapidly if both herbivores and predators were present than if both were absent. As in the freshwater ecosystem experiments, the authors concluded that this study characterized a complex that did not fit the dichotomous model of ecosystem control, but an interaction intermediate to the two extreme hypotheses.

VI. Trophic Interactions and Plant Community Composition

The idea that herbivores influence the species composition of plant communities, and even the growth forms within those communities, is an august one in ecology (Clements and Shelford, 1939). To what extent do consumers control the occurrence and abundance of plants so that a trophic web can be considered a self-organizing complex (McNaughton, 1983, 1994)?

A comprehensive review of the subject provided overwhelming evidence that animals have major controlling effects on plant community composition, and that the effects of vertebrates are much more pronounced than are the effects of invertebrates (Crawley, 1989). Some data suggest that intrinsic processes–vegetation interaction (competition, soil modification, etc.) are more important than insect herbivores in successional processes (Gibson and Brown, 1992; Rees and Brown, 1992). However, the effects of tiny herbivores are underestimated because their effects on plants may be comparatively more subtle, though no less far reaching (*e.g.*, Louda, 1982), than the effects of animals that take large, perceptible mouthfuls as they feed; Brown (1984) has presented evidence that arthropods influence the direction of plant succession. Hacker and Bertness (1996) found that aphid attack in the absence of neighboring plants was so intense as to lead to extinction, providing evidence for the importance of neighbors in stabilizing plant community composition in response to herbivory (Atsatt and O'Dowd, 1976; McNaughton, 1978).

Brown and Heske (1990) found that elimination of small mammals from arid ecosystems had a prounounced effect on species compositon. Therefore, even small herbivores must influence the nature of the biomes on which global terrestrial productivity estimates are based. Thus, a question that remains unsatisfactorily resolved in my mind is the extent to which insects and other tiny herbivores regulate the species compositions of the ecosystems that support them. For large mammalian herbivores, on the other hand, I am convinced that they are major controllers of species composition and plant growth form (McNaughton, 1983; Crawley, 1989; Pastor *et al.*, 1993; McNaughton and Banyikwa, 1995), so it is unsound to estimate biome productivity without incorporating that control.

VII. Grazing and Detritus Food Webs in a Global Biome Context

A. Energetics

It is a customary rule of thumb in ecology that herbivores commonly consume less than a quarter of net primary productivity in ecosystems, with the vast majority of production entering detritus food webs directly (Wallace *et al.*, 1997), and many basic ecology textbooks do not treat herbivory as an ex-

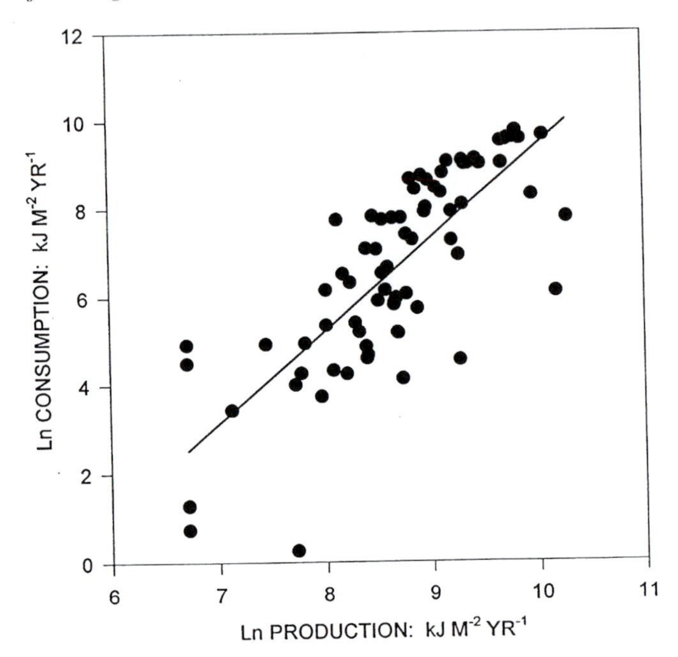

Figure 6-1 Relationship between producer and consumer energy flow in native ecosystems ranging from desert and tundra to tropical forests and grasslands. Production is foliage production and consumption is amount consumed by herbivores. After McNaughton *et al.* (1989, 1991).

plicit process at all but subsume it under two other topics. One of these is energy flow, wherein herbivory is merely recognized as the base of grazing food webs. The other is coevolution, within which herbivory is treated as a driver of plant defenses against herbivores and the evolution of means of circumventing defenses by those herbivores. If less than 25% of production flows through the grazing food web, then over 75% must eventually flow through the detritus food web, so the generally greater attention to decomposition than to herbivory as an ecological process in texts can be justified on the basis of relative magnitude and, therefore, presumed importance.

However, evidence from data syntheses reveals that herbivory is a positive function of primary production in both terrestrial (Fig. 6-1) and aquatic ecosystems (McNaughton *et al.*, 1989, 1991; Oesterheld *et al.*, 1992; Cyr and Pace, 1993; Cebrián and Duarte, 1994). In all these studies, herbivory and net primary productivity are positively related.

The simplest conclusion to draw from these data is that herbivory and, therefore, overall flow to the grazing food web are uncomplicated functions of plant growth. If that conclusion were correct, and plant growth were con-

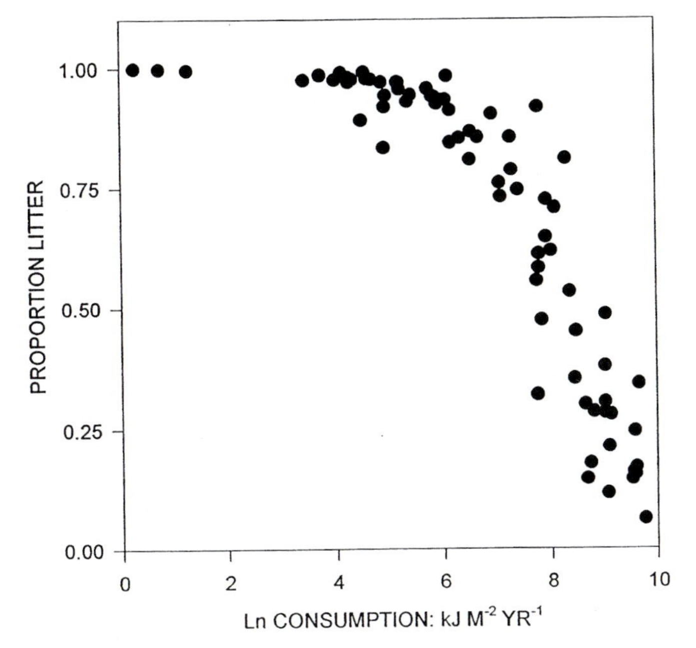

Figure 6-2 Relationship between the consumption of foliage by herbivores and the proportion of foliage production converted to litter. Note that the grazing food web has priority over decomposers in both food quantity and food quality. Data from McNaughton *et al.* (1989, 1991).

trolled principally by physical factors such as climate and soil, estimating global productivity would be a comparatively simple exercise. However, given abundant evidence discussed earlier in this chapter that herbivores have strong controlling effects on plant growth, decreasing it in most documented cases but increasing it in others (McNaughton, 1979, 1983; Maschinski and Whitham, 1989; Milchunas and Lauenroth, 1993), it is more prudent to recognize that trophic processes feed back to influence processes at the base of the web (McNaughton *et al.*, 1989, 1991).

Because herbivory is directly related to primary productivity, what is left over from herbivory, and therefore enters decomposition food webs, must also be related to those two processes in some predictable fashion. Reexamination of the terrestrial data (McNaughton *et al.*, 1989, 1991) reveals a definite relationship between the level of herbivory and the proportion of foliage production flowing into detritus food webs on an annual basis (Fig. 6-2). At the lowest levels of herbivory, typical of desert and tundra biomes, virtually all of the primary production was transmitted to the detritus web. Above a herbivory level of 400 kJ m^{-2} yr^{-1} (approximately 20 g m^{-2} yr^{-1}), however, the importance of direct detrital flow began to decrease sharply, approaching zero at the highest levels recorded. At the highest levels, characteristic of C_4 grasslands and salt marshes, direct flow to detritus food webs is a minor component of ecosystem energy flow. Most of the material entering decomposition webs, then, must have been processed previously by the grazing web. The three-quarters level cited above was achieved at around 1100 kJ m^{-2} yr^{-1}. These data suggest that the significance of detritus food webs is overestimated by traditional rule-of-thumb generalizations in ecology.

B. Coupling: A Synthesis

The patterns above involve many complex mechanisms, ranging from the behavior, body size, and demography of herbivores (Fryxell and Sinclair, 1988; Owen-Smith, 1988), to the abundance of predators supported (Schindler *et al.*, 1997), to the growth form and physiology of plants (McNaughton, 1979; Mooney and Gulmon, 1982). A recent series of synthesis papers provides the means for a conceptual linkage between the phenomenological evidence presented throughout this chapter and the data syntheses of Figs. 6-1 and 6-2, which, taken together, falsify the concept of ecosystems as solely donor controlled as a workable hypothesis.

The chemical properties of plant litter are linked to decomposition (Rastetter *et al.*, 1991) in both grasslands (Schimel *et al.*, 1990) and forests (Mellilo *et al.*, 1982), leading to a general hypothesis coupling decomposition to proximate chemical analyses of decomposing tissues (Bosatta and Ågren, 1991; Ågren and Bosatta, 1996). The fundamental assumption of the first linking hypothesis is that decomposers are principally C limited and that C compounds vary in quality, i.e., fugacity. This is a fundamental concept of long-standing in livestock husbandry, which has developed precisely defined analytical protocols to partition forages and other feedstuffs between readily digestible fractions (largely cell contents), recalcitrant fractions (cell wall), and indigestible components (Van Soest, 1982).

In decomposition, it is proposed (Bosatta and Ågren, 1991; Ågren and Bosatta, 1996) that the higher litter quality is, the greater the growth rate of microbial populations and, therefore, the greater the rate of disappearance of litter, with residual, poorly utilized C fractions accumulating to supplement the soil organic matter pool (Mellilo *et al.*, 1982). A data set of 978 decomposition rates from 19 litter types at 16 locations, coupling litter proximate analysis to rates, substantiates the general hypothesis by documenting a generally close association between residual C at a given time and the C calculated to remain at that time (Ågren and Bosatta, 1996). Thus, litter quality, as evaluated by straightforward chemical partitioning, is directly related to the rate of return of chemicals to the grazing food web.

Grime *et al.* (1997) have proposed that life history traits of adults of a wide variety ($n = 43$) of plant species from the British flora are ordered, in first importance in multivariate analysis, by a matching among high foliar mineral content, high potential RGR, poor yield under nutrient limitation, low tensile strength and high palatability of foliage, and a high rate of decomposition of leaf litter. That is (Coley *et al.*, 1985), to generalize, life history traits associated with high RGR and, therefore, a capacity to recover from damage, together with little protection from damage, are to be found in species of productive environments rich in resources. Alternatively, life history traits associated with low RGR and a poor capacity to recover from damage are found in species from resource-poor habitats with substantial investment in traits that protect them from damage.

Finally, evidence experimentally couples resistance to herbivore damage with resistance to decomposition (Grime *et al.*, 1996). Tissues that are highly palatable to herbivores are also rapidly decomposed. This provides a basis for interpreting Fig. 6-2 in mechanistic terms and unifies two seemingly separate ecological concerns, animal foraging behavior and mineralization. Forage highly palatable to herbivores obviously will be more heavily consumed than less palatable tissues, leading to a reduction in flow to detritus food webs. Taken together with these generalizations, Figs. 6-1 and 6-2 reveal across-biome patterns of relative plant palatability, at the extremes, high in native grasslands and low in desert and tundra. This, of course, is palatability to the *in situ* herbivores, not to some standard herbivore. Therefore, low energy flow through the herbivore trophic level, less than 400–1100 kJ m^{-2} yr^{-1}, is likely due to low resource availability and is credibly associated with increasing levels of plant defenses against herbivory, including both chemicals and structural potency (Grime *et al.*, 1997). Above that threshold, resource limitations on growth are less stringent, plant tissues are less defended from herbivory, and flow to the detritus web is progressively choked off by herbivory. As flow to herbivores increases, we expect also that flow to higher trophic levels will increase and predators will be more abundant.

VIII. Resolving a Dilemma: Remote Sensing, Herbivory, and Trophic Interactions

Given the controlling feedbacks that regutate plant production by higher trophic levels, remote sensing, which determines only standing crop, can only be a very approximate estimate of actual ecosystem primary productivity, particularly above an annual herbivore energy flow of 400–1100 kJ m^{-2} yr^{-1}. However, remote sensing might be corrected for consumption by applying the relationships documented above, allowing a partitioning of the energy capture by plants between grazing and detritus food webs (McNaughton *et al.*, 1989, 1991). Livestock biomass density per level of primary production is about an order of magnitude above unmanaged natural ecosystems (Oesterheld *et al.*, 1992). And it has also been documented that NOAA–AVHRR satellite data are good estimators of rangeland stocking rates (Oesterheld *et al.*, 1997). Therefore, more refined relationships could be established between satellite indices, actual aboveground primary production, belowground production for well-characterized systems that include measures of the effects of herbivores on partitioning, and consumption by herbivores. This would change the perceived disparities between ecosystems differing levels of herbivory, narrowing the range, if not leading to convergence, between ecosystems with abundant herbivores and those with low levels of herbivory.

IX. Conclusions

Herbivores and the trophic web that they support are not merely a parasitic component supported by an abiotic-controlled level of primary productivity independent of their activities. Instead, there are many feedbacks from higher trophic levels to plants that can reduce or increase primary productivity. Projecting terrestrial global productivity accurately into the future will require the incorporation of these feedbacks, but that task does not appear insurmountable. There are clear associations between primary productivity and (1) all functional properties of the herbivore trophic level, consumption, biomass, and production (McNaughton *et al.*, 1989, 1991), (2) partitioning of energy and nutrient flow through the producers between grazing and detritus food webs with an important breakpoint between consumption of 400 and 1100 kJ m^{-2} yr^{-1} (approximately 20–55 g m^{-2} yr^{-1}), and (3) proximate chemical properties of plant foliage. If properly incorporated into remote-sensing theory, these phenomena will increase considerably the accuracy of remotely sensed estimates of terrestrial global productivity.

Acknowledgment

Preparation of this chapter was supported by NSF DEB-9312435. Douglas A. Frank reviewed the manuscript and made helpful suggestions for improvement.

References

Ågren, G. I., and Bosatta, E. (1996). Quality: A bridge between theory and experiment in soil organic matter studies. *Oikos* **76,** 522–528.

Alberda, T. (1966). The influence of reserve substances on dry-matter production after defoliation. *Proc. Xth Int. Grassl. Congr.*, pp. 140–147.

Alward, R. D., and Joern, A. (1993). Plasticity and overcompensation in grass response to herbivory. *Oecologia* **95,** 358–364.

Andere, D. K. (1981). Wildebeest *Connochaetes taurinus* (Burchell) and its food supply in Ambosli Basin. *Afr. J. Ecol.* **19,** 239–250.

Atsatt, P. R., and O'Dowd, D. J. (1976). Plant defense guilds. *Science* **193,** 24–29.

Awramik, S. M. 1971. Precambrian columnar stromatolite diversity: Reflection of metazoan appearance. *Science* **174,** 825–827.

Bahrani, J., Beatty, E. R., and Tan, K. H. (1983). Relationship between carbohydrate, nitrogen contents, and regrowth of tall fescue tillers. *J. Range Manage.* **36,** 234–235.

Bazely, D. R., and Jefferies, R. L. (1985). Goose faeces: a source of nitrogen for plant growth in a grazed salt marsh. *J. Appl. Ecol.* **22,** 693–704.

Besford, R. T. (1990). The greenhouse effect: Acclimation of tomato plants in high CO_2, relative changes in Calvin Cycle enzymes. *J. Plant Physiol.* **136,** 458–463.

Bianchi, T. S., and Jones, C. G. (1991). Density-dependent positive feedbacks between consumers and their resources. *In* "Comparative Analyses of Ecosystems" (J. Cole, G. Lovett, and S. Findlay, eds.), pp. 339–340. Springer-Verlag, New York.

Bianchi, T. S., Jones, C. G., and Shachak, M. (1989). The positive feedback of consumer population density on resource supply. *Trends Ecol. Evol.* **4,** 234–238.

Bosatta, E., and Ågren, G. I. (1991). Dynamics of carbon and nitrogen in the organic matter of the soil: A generic theory. *Am. Nat.* **138,** 227–245.

Brett, M. T., and Goldman, C. R. (1997). Consumer versus resource controls in freshwater pelagic food webs. *Science* **275,** 384–386.

Briske, D. D., and Anderson, V. J. (1992). Competitive ability of the bunchgrass *Schizachyrium scoparium* as affected by grazing history and defoliation. *Vegetatio* **103,** 41–49.

Brown, V. K. (1984). Secondary succession: Insect–plant relationships. BioScience **34,** 710–716.

Brown, J. H., and Heske, E. J. (1990). Control of a desert-grassland transition by a keystone rodent guild. *Science* **250,** 1705–1707.

Caldwell, M. M., Richards, J. H., Johnson, D. A., Nowak, R. S., and Dzurec, R. S. (1981). Coping with herbivory: Photosynthetic capacity and resource allocation in two semiarid *Agropyron* bunchgrasses. *Oecologia* **50,** 14–24.

Carpenter, S. R., Kitchell, J. F., and Hodgson, J. R. (1985). Cascading trophic interactions and lake productivity. *BioScience* **35,** 634–639.

Cebrián, J., and Duarte, C. M. (1994). The dependence of herbivory on growth rate in natural plant communities. *Funct. Ecol.* **8,** 518–525.

Chapin III, F. S., and McNaughton, S. J. (1989). Lack of compensatory growth under phosphorus deficiency in grazing-adapted grasses from the Serengeti Plains. *Oecologia* **79,** 551–557.

Chew, R. M. (1974). Consumers as regulators of ecosystems: An alternative to energetics. *Ohio J. Sci.* **74,** 369–370.

Clements, F. E., and Shelford, V. E. (1939). "Bio-Ecology." Wiley, New York.

Coleman, J. S., and Jones, C. G. (1991). A phytocentric perspective of phytochemical induction by herbivores. *In* "Phytochemical Induction by Herbivores" (D. W. Tallamy and M. J. Raup, eds.), pp. 3–45. Wiley, New York.

Coley, P. D. (1986). Costs and benefits of defense by tannins in a Neotropical tree. *Oecologia* **70,** 238–241.

Coley, P. D., Bryant, J. P., and Chapin III, F. S. (1985). Resource availability and plant anitherbivore defense. *Science* **230,** 895–899.

Coughenour, M. B. (1984). Graminoid responses to grazing by large herbivores: Adaptations, exaptations and interacting processes. *Ann. Mo. Bot. Gard.* **72,** 852–863.

Coughenour, M. B., McNaughton, S. J., and Wallace, L. L. (1984). Modelling primary productivity of perennial graminoids—Uniting physiological processes and morphometric traits. *Ecol. Model.* **23,** 101–134.

Coughenour, M. B., McNaughton, S. J., and Wallace, L. L. (1985a). Responses of an African graminoid (*Themda triandra* Forsk.) to frequent defoliation, nitrogen, and water: A limit of adaptation to herbivory. *Oecologia* **68,** 105–110.

Coughenour, M. B., McNaughton, S. J., and Wallace, L. L. (1985b). Responses of an African tall-grass (*Hyparrhenia filipendula* Stapf.) to defoliation and limitations of water and nitrogen. *Oecologia* **68,** 80–86.

Crawley, M. J. (1989). The relative importance of vertebrate and invertebrate herbivores in plant population dynamics. *In* "Insect-Plant Interactions" (E. A. Bernays, ed.), pp. 45–71.

Cyr, H., and Pace, M. L. (1993). Magnitude and patterns of herbivory in aquatic and terrestrial ecosystems. *Nature (London)* **361,** 148–150.

Danckwerts, J. E., and Gordon, A. J. (1987). Long-term partitioning, storage and remobilization of ^{14}C-assimilated by *Lolium perenne* (cv. Melle). *Ann. Bot.* **59,** 55–66.

de Visser, R., Vianden, H., and Schnyder, H. (1997). Kinetics and relative significance of remobilized and current C and N incorporation in leaf and root growth zones of *Lolium perenne* after defoliation: Assessment by ^{13}C and ^{15}N steady-state labelling. *Plant Cell Environ.* **20,** 37–46.

Eckardt, F. E., Heerfordt, L., Jørgensen, and Vaag, P. (1982). Photosynthetic production in Greenland as related to climate, plant cover and grazing pressure. *Photosynthetica* **16,** 71–100.

Escós, J., Alados, C. L., and Emlen, J. M. (1997). The impact of grazing on plant fractal geometry and fitness in a Mediterranean shrub *Anthyllis cytisoides* L. *Funct. Ecol.* **11,** 66–78.

Fajer, E. D. (1989). The effects of enriched CO_2 atmospheres on plant–insect herbivore interactions: Growth responses of larvae of the specialist butterfly, *Junonia coenia* (Lepidoptera: Nymphalidae). *Oecologia* **81,** 514–520.

Fajer, E. D., Bowers, M. D., and Bazzaz, F. A. (1989). The effects of enriched carbon dioxide atmospheres on plant–insect herbivore interactions. *Science* **243,** 1198–1200.

Feeny, P. (1976). Plant apparancy and chemical defense. *Recent Adv. Phytochem.* **10,** 1–40.

Fox, L. R., and Morrow, P. A. (1992). Eucalypt responses to fertilization and reduced herbivory. *Oecologia* **89,** 214–222.

Fretwell, S. D. (1977). The regulation of plant communities by food chains exploiting them. *Perspect. Biol. Med.* **20,** 169–185.

Fryxell, J. M., and Sinclair, A. R. E. (1988). Seasonal migrations by white-eared kob in relation to resources. *Afr. J. Ecol.* **26,** 17–31.

Georgiadis, N. J., Ruess, R. W., McNaughton, S. J., and Western, D. (1989). Ecological conditions that determine when grazing stimulates grass production. *Oecologia* **81,** 316–322.

Gibson, C. W. D., and Brown, V. K. (1992). Grazing and vegetation change: Deflected or modified succession? *J. Appl. Ecol.* **29,** 120–131.

Grime, J. P., Cornelissen, J. H. C., Thompson, K., and Hodgson, J. G. (1996). Evidence for a causal connection between anti-herbivore defence and the decomposition rate of leaves. *Oikos* **77,** 489–494.

Grime, J. P., Thompson, K., Hunt, R., Hodgson, J. G., Cornelissen, J. H. C., Rorison, I. H., Hendry, G. A. F., Ashenden, T. W., Askew, A. P., Band, S. R., Booth, R. E., Bossard, C. C., Campbell, B. D., Cooper, J. E. L., Davison, A. W., Gupta, P. L., Hall, W., Hand, D. W., Hannah, M. A., Hillier, S. H., Hodkinson, D. J., Jalili, A., Liu, Z., Mackey, J. M. L., Matthews, N., Mowforth, M. A., Neal, A. M., Reader, R. J., Reiling, K., Ross-Fraser, W., Spencer, R. E., Sutton, F., Tasker, D. E., Thorpe, P. C., and Whitehouse, J. (1997). Integrated screening validates primary axes of specialisation in plants. *Oikos* **79,** 259–281.

Gulmon, S. L., and Mooney, H. A. (1986). Costs of defense and their effects on plant productivity. *In* "On the Economy of Plant Form and Function" (T. Givnish, ed.), pp. 681–698. Cambridge Univ. Press, Cambridge.

Hacker, S. D., and Bertness, M. D. (1996). Trophic consequences of a positive plant interaction. *Am. Nat.* **148,** 559–575.

Hairston, N. G., Smith, F. E., and Slobodkin, L. B. (1960). Community structure, population control and competition. *Am. Nat.* **94,** 421–425.

Hartvigsen, G., and McNaughton, S. J. (1995). Tradeoff between height and relative growth rate in a dominant grass from the Serengeti ecosystem. *Oecologia* **102,** 273–276.

Haukioja, E., Ruohomäki, K., Senn, J., Suomela, J., and Walls, M. (1990). Consequences of herbivory in the mountain birch (*Betula pubscens* ssp. *tortuosa*): importance of the functional organization of the tree. *Oecologia* **82,** 238–247.

Hicks, S. L., and Reader, R. J. (1995). Compensatory growth of three grasses following simulated grazing in relation to soil nutrient availability. *Can. J. Bot.* **73,** 141–145.

Hoffland, E., Niemann, G. J., van Pelt, J. A., Pureveen, J. B. M., Eijkel, G. B., Boon, J. J., and Lambers, H. (1996). Relative growth rate correlates negatively with pathogen resistance in radish: The role of plant chemistry. *Plant Cell Environ.* **19,** 1281–1290.

Holland, E. A., Parton, W. J., Detling, J. K., and Coppock, D. L. (1992). Physiological responses of plant populations to herbivory and their consequences for ecosystem nutrient flow. *Am. Nat.* **140,** 685–706.

Howe, H. F., and Westley, L. C. (1988). Ecological relationships of plants and animals. Oxford Univ. Press, Oxford.

Huber, C., and Wächtershäuser, G. (1997). Activated acetic acid by carbon fixation on (Fe, Ni)S under primordial conditions. *Science* **276,** 245–247.

Ingham, R. E., and Detling, J. K. (1990). Effects of root-feeding nematodes on aboveground net primary production in a North American grassland. *Plant Soil* **121,** 279–281.

Jameson, D. A. (1963). Responses of individual plants to harvesting. *Bot. Rev.* **29,** 532–594.

Jones, C. G., Lawton, J. H., and Shachak, M. (1994). Organisms as ecosystem engineers. *Oikos* **69,** 373–386.

Kursar, T. A., and Coley, P. D. (1991). Nitrogen content and expansion rate of young leaves of rain forest species: Implications for herbivory. *Biotropica* **23,** 141–150.

Laws, R. M. (1970). Elephants as agents of habitat and landscape change in East Africa. *Oikos* **21,** 1–15.

Lennartsson, T., Tuomi, J., and Nilsson, P. (1997). Evidence for an evolutionary history of overcompensation in the grassland biennial *Gentianella campestris* (Gentianaceae). *Am. Nat.* **149,** 1147–1155.

Lincoln, D. E. (1993). The influence of plant carbon dioxide and nutrient supply on susceptibility to insect herbivores. *Vegetatio* **104,** 273–180.

Loreau, M. (1995). Consumers as maximizers of matter and energy flow in ecosystems. *Am. Nat.* **145,** 22–42.

Louda, S. M. (1982). Limitation of the recruitment of the shrub *Happlopappus squarrosus* (Asteraceae) by flower- and seed-feeding insects. *J. Ecol.* **70,** 43–54.

Maschinski, J., and Whitham, T. G. (1989). The continuum of plant responses to herbivory: The influence of plant association, nutrient availability and timing. *Am. Nat.* **134,** 1–19.

Mauricio, R., Rausher, M. D., and Burdick, D. S. (1977). Variation in the defense strategies of plants: Are resistance and tolerance mutually exclusive? *Ecology* **78,** 1301–1311.

McGuire, A. D., Joyce, L. A., Kicklighter, D. W., Melillo, J. M., Esser, G., and Vorosmarty, C. J. (1993). Productivity response of climax temperate forests to elevated temperature and carbon dioxide: A North American comparison between two global models. *Climatic Change* **24,** 287–310.

McNaughton, S. J. (1974). Developmental control of net productivity in *Typha latifolia* ecotypes. *Ecology* **55,** 864–869.

McNaughton, S. J. (1976). Serengeti migratory wildebeest: Facilitation of energy flow by grazing. *Science* **191,** 92–94.

McNaughton, S. J. (1978). Serengeti ungulates: Feeding selectivity influences the effectiveness of plant defense guilds. *Science* **199,** 806–807.

McNaughton, S. J. (1979). Grazing as an optimization process: Plant–ungulate relationships in the Serengeti. *Am. Nat.* **113,** 691–703.
McNaughton, S. J. (1983). Serengeti grassland ecology: The role of composite environmental factors and contingency in community organization. *Ecol. Monogr.* **53,** 291–320.
McNaughton, S. J. (1985). Ecology of a grazing ecosystem: The Serengeti. *Ecol. Monogr.* **55,** 259–294.
McNaughton, S. J. (1986). On plants and herbivores. *Am. Nat.* **128,** 765–770.
McNaughton, S. J. (1991). Evolutionary ecology of large tropical herbivores. *In* "Plant–Animal Interactions: Evolutionary Ecology in Tropical and Temperate Regions" (P. W. Price, T. M. Lewinsohn, G. W. Fernandes, and W. W. Benson, eds.), pp. 509–522. Wiley-Interscience, New York.
McNaughton, S. J. (1994). Conservation goals and the configuration of biodiversity. *In* "Systematics and Conservation Evaluation" (P. L. Forey, C. J. Humphries, and R. K. Vane-Wright, eds.), pp. 41–62. Clarendon Press, Oxford.
McNaughton, S. J., and Banyikwa, F. F. (1995). Plant communities and herbivory. *In* "Serengeti II: Dynamics, Management, and Conservation of an Ecosystem" (A. R. E. Sinclair and P. Arcese, eds.), pp. 49–70. Univ. Chicago Press, Chicago.
McNaughton, S. J., Banyikwa, F. F., and McNaughton, M. M. (1998). Root biomass and productivity in a grazing ecosystem: The Serengeti. *Ecology* **79,** 587–592.
McNaughton, S. J., and Chapin III, F. S. (1985). Effects of phosphorus nutrition and defoliation on C_4 graminoids from the Serengeti Plains. *Ecology* **66,** 1617–1629.
McNaughton, S. J., Oesterheld, M., Frank, D. A., and Williams, K. J. (1989). Ecosystem-level patterns of primary productivity and herbivory in terrestrial habitats. *Nature (London)* **341,** 142–144.
McNaughton, S. J., Oesterheld, M., Frank, D. A., and Williams, K. J. (1991). Primary and secondary production in terrestrial ecosystems. *In* "Comparative Analyses of Ecosystems" (J. Cole, G. Lovett, and S. Findlay, eds.), pp. 120–139. Springer-Verlag, New York.
McNaughton, S. J., and Sabuni, G. A. (1988). Large African mammals as regulators of vegetation structure. *In* "Plant Form and Vegetation Structure" (M. J. A. Werger, P. J. M. van der Aart, H. J. During, and J. T. A. Verhoeven, eds.), pp. 339–354. SPB Academic Publ., The Hague.
McNaughton, S. J., Wallace, L. L., and Coughenor, M. B. (1983). Plant adaptation in an ecosystem context: Effects of defoliation, nitrogen, and water on growth of an African C_4 sedge. *Ecology* **64,** 307–318.
Melillo, J. M., Aber, J. D., and Muratore, J. F. (1982). Nitrogen and lignin control of hardwood leaf litter decomposition dynamics. *Ecology* **63,** 621–626.
Milchunas, D. G., and Lauenroth, W. K. (1993). Quantitative effects of grazing on vegetation and soils over a global range of environments. *Ecol. Monogr.* **63,** 327–366.
Misra, G., and Singh, K. P. (1981). Total nonstructural carbohydrates of one temperate and two tropical grasses under varying clipping and soil moisture regimes. *Agro-Ecosystems* **7,** 213–223.
Mooney, H. A., and Gulmon, S. L. (1982). Constraints of leaf structure and function in reference to herbivory. *BioScience* **32,** 198–206.
Nielsen, S. L., Enríquez, Duarte, C. M., and Sand-Jensen, K. (1996). Scaling maximum growth rates across photosynthetic organisms. *Funct. Ecol.* **10,** 167–175.
Oesterheld, M. (1992). Effect of defoliation intensity on aboveground and belowground relative growth rates. *Oecologia* **92,** 313–316.
Oesterheld, M., di Bella, C. M., and Kerdiles, H. (1997). Relation between NOAA–AVHRR satellite data and stocking rate of rangelands. *Ecol. Appl.* **8,** 207–212.
Oesterheld, M., and McNaughton, S. J. (1991). Effect of stress and time for recovery on the amount of compensatory growth after grazing. *Oecologia* **85,** 305–313.
Oesterheld, M., Sala, O. E., and McNaughton, S. J. (1992). Effects of animal husbandry on herbivore carrying-capacity at a regional scale. *Nature (London)* **356,** 234–236.

Oksanen, L. (1991). Trophic levels and trophic dynamics: A consensus emerging? *Trends Ecol. Evol.* **6,** 58–60.

Oksanen, L., Fretwell, S. D., Arruda, J., and Niemela, P. (1981). Exploitation ecosystems in gradients of primary productivity. *Am. Nat.* **118,** 240–261.

Owen, D. F., and Wiegert, R. G. (1976). Do consumers maximize plant fitness? *Oikos* **27,** 488–492.

Owen-Smith, N. (1988). Megaherbivores: The influence of very large size on ecology. Cambridge Univ. Press, Cambridge.

Paige, K. N., and Whitham, T. G. (1987). Overcompensation in response to mammalian herbivory: The advantage of being eaten. *Am. Nat.* **129,** 407–416.

Paine, R. T. (1969). A note on trophic complexity and species diversity. *Am. Nat.* **100,** 65–75.

Parton, W. J., Ojima, D. S., Cole, C. V., and Schimel, D. S. (1994). A general model for soil organic matter dynamics: Sensitivity to litter chemistry, texture and management. *In* "Quantitative Modeling of Soil Forming Processes," pp. 147–167. Soil Science Society of America, Madison, Wisconsin.

Pastor, J., Dewey, B., Naiman, R. J., McInnes, P. F., and Cohen, Y. (1993). Moose browsing and soil fertility in the boreal forests of Isle Royale National Park. *Ecology* **74,** 467–480.

Pearson, M., and Brooks, G. L. (1996). The effect of elevated CO_2 and grazing by *Gastrophysa viridula* on the physiology and regrowth of *Rumex obvusifolius*. *New Phytol.* **133,** 605–616.

Petrides G. A., and Swank, W. G. (1965). Population densities and the range carrying capacity for large mammals in Queen Elizabeth National Park, Uganda. *Zool. Afr.* **1,** 209–225.

Pimm, S. L. (1982). "Food Webs." Chapman and Hall, London.

Pimm, S. L. (1991). "Balance of Nature?" Univ. Chicago Press, Chicago.

Poorter, H., and Remkes, C. (1990). Leaf area ratio and net assimilation rate of 24 wild species differing in relative growth rate. *Oecologia* **83,** 553–559.

Potter, J. R., and Jones, J. W. (1977). Leaf area partitioning as an important factor in growth. *Plant Physiol.* **59,** 10–14.

Price, P. W., Bouton, C. E., Gross, P., McPheron, B. A., Thompson, J. N., and Weis, A. E. (1980). Interaction among three trophic levels: Influence of plants on interactions between insect herbivores and natural enemies. *Annu. Rev. Ecol. Syst.* **11,** 41–65.

Rastetter, E. B., Ryan, M. G., Shaver, G. R., Melillo, J. M., Nadelhoffer, K. J., Hobbie, J. E., and Aber, J. D. (1991). A general biogeochemical model describing the response of the C and N cycles in terrestrial ecosystems to changes in CO_2, climate, and N deposition. *Tree Physiol.* **9,** 101–126.

Rees, M., and Brown, V. K. (1992). Interactions between invertebrate herbivores and plant competition. *J. Ecol.* **80,** 353–360.

Reich, P. B., Walters, M. B., and Ellsworth, D. S. (1992). Leaf life-span in relation to leaf, plant, and stand characteristics among diverse ecosystems. *Ecol. Monogr.* **62,** 365–392.

Rhodes, D. F., and Cates, R. G. (1976). Toward a general theory of plant antiherbivore chemistry. *Recent Adv. Phytochem.* **10,** 168–213.

Richards, J. H., and Caldwell, M. M. (1985). Soluble carbohydrates, concurrent photosynthesis and efficiency in regrowth following defoliation: A field study with *Agropyron* species. *J. Appl. Ecol.* **22,** 907–920.

Running, S. W., and Hunt Jr., E. R. (1993). Generalization of a forest ecosystem model for other biomes. *In* "Scaling Physiological Processes: Leaf to Globe" (J. R. Ehleringer and C. B. Field, eds.), pp. 141–158. Academic Press, San Diego.

Sagers, C. L., and Coley, P. D. (1995). Benefits and costs of defense in a Neotropical shrub. *Ecology* **76,** 1835–1843.

Schimel, D. S., Parton, W. J., Kittel, T. G. F., Ojima, D. S., and Cole, C. V. (1990). Grassland biogeochemistry: Links to atmospheric processes. *Climate Change* **17,** 13–25.

Schindler, D. E., Carpenter, S. R., Cole, J. J., Kitchell, J. F., and Pace, M. L. (1997). Influence of food web structure on carbon exchange between lakes and the atmsophere. *Science* **177,** 248–251.

Schwartzman, D. W., and Volk, T. (1989). Biotic enhancement of weathering and the habitability of Earth. *Nature (London)* **340,** 457–460.

Seagle, S. W., McNaughton, S. J., and Ruess, R. W. (1992). Simulated effects of grazing on soil nitrogen and mineralization in contrasting Serengeti grasslands. *Ecology* **73,** 1105–1123.

Seastedt, T. R. Ramundo, R. A., and Hayes, D. C. (1988). Maximization of densities of soil animals by herbivory: Empirical evidence, graphical and conceptual models. *Oikos* **51,** 243–248.

Semmartin, M., and Oesterheld, M. (1996). Effect of grazing pattern on primary productivity. *Oikos* **75,** 431–436.

Shachak, M., Jones, C. G., and Granot, Y. (1987). Herbivory in rocks and the weathering of a desert. *Science* **236,** 1098–1100.

Shear, W. A. (1991). The early development of terrestrial ecosystems. *Nature (London)* **351,** 283–289.

Simoes, M., and Baruch, Z. (1991). Responses to simulated herbivory and water stress in two tropical C_4 grasses. *Oecologia* **88,** 173–180.

Slansky, F., and Feeny, P. (1977). Stabilization of the rate of nitrogen accumulation by larvae of the cabbage butterfly on wild and cultivated food plants. *Ecol. Monogr.* **47,** 209–228.

Stanton, N. L., Allen, M., and Campion, M. (1981). The effect of the pesticide carbofuran on soil organisms and root and shoot production in shortgrass prairie. *J. Appl. Ecol.* **18,** 417–431.

Valentine, H. T., Wallner, W. E., and Wargo, P. M. (1983). Nutritional changes in host foliage during and after defoliation, and their relation to the weight of gypsy moth pupae. *Oecologia* **57,** 298–302.

van der Heyden, F., and Stock, W. D. (1996). Regrowth of a semiarid shrub following simulated browsing: The role of reserve carbon. *Funct. Ecol.* **10,** 647–653.

Van Soest, P. J. (1982). "Nutritional Ecology of the Ruminant." O & B Books, Corvallis, Oregon.

Volonec, J. J., Ourry, A., and Joern, B. C. (1996). A role for nitrogen reserves in forage regrowth and stress tolerance. *Physiol. Plant.* **97,** 185–193.

Wallace, J. B., Eggert, S. L., Meyer, J. L., and Webster, J. R. (1997). Multiple trophic levels of a forest stream linked to terrestrial litter inputs. *Science* **277,** 102–104.

Wäschtershäuser, G. (1988). Before enzymes and templates: Theory of surface metabolism. *Microbiol. Rev.* **52,** 452–484.

Wegener, C., and Odasz, A. M. (1997). Effects of laboratory simulated grazing on biomass of the perennial Arctic grass *Dupontia fisheri* from Svalbard: Evidence of overcompensation. *Oikos* **79,** 496–502.

Wessels, D. C. J., and Wessels, L.-A. (1991). Erosion of biogenically weathered Clarens sandstone by lichenophagous bag-worm larvae (Lepidoptera: Pyschidae). *Lichenologist* **23,** 283–291.

Whitham, T. G., and Slobodhikoff, C. N. (1981). Evolution by individuals, plant–herbivore interactions, and mosaics of genetic variability: The adaptive significance of somatic mutations in plants. *Oecologia* **49,** 287–292.

Wilsey, B. J., Coleman, J. S., and McNaughton, S. J. (1997). Effects of elevated CO_2 and defoliation on grasses: A comparative ecosystem approach. *Ecol. Appl.* **7,** 844–853.

Wilsey, B. J., McNaughton, S. J., and Coleman, J. S. (1994). Will increases in atmospheric CO_2 affect regrowth following grazing in C_4 grasses from tropical grasslands? A test with *Sporobolus kentrophyllus. Oecologia* **99,** 141–144.

7

Water, Nitrogen, Rising Atmospheric CO_2, and Terrestrial Productivity

Denis Loustau, Bruce Hungate, and Bert G. Drake

I. Introduction

The functioning of plants in terrestrial ecosystems must satisfy different constraints imposed by the physical environment. The prevention of embolism and conservation of internal water constrain stomatal behavior and leaf area indices that plants may sustain. Plant height and canopy structure are controlled by water and nutrients through carbon allocation between roots and leaves. The amount of available nutrients strongly influences net primary production, largely by determining the amount of photosynthetic enzymes, and in turn leaf area, that may be achieved in a given ecosystem. The impacts of stomatal function, leaf area index, and photosynthetic capacity on the net primary production of terrestrial ecosystems vary according to canopy roughness. Increasing atmospheric CO_2 concentration (C_a) usually stimulates carbon uptake and carbon distribution belowground, though the magnitude of these responses varies among ecosystems. Translating increased carbon uptake at the leaf and canopy levels to long-term carbon storage is not straightforward, and, so far there is little experimental verification of a CO_2-driven expansion of carbon pools with long-term storage potential. Nevertheless, evidence to date suggests that carbon uptake by the terrestrial biosphere will increase in concert with rising C_a.

The importance of water and nutrient regimes for net primary production (NPP) and net ecosystem production (NEP) has long been recognized for agricultural crops. A simple plot of annual net primary production vs. rainfall illustrates the dependency of NPP on rainfall for a wide range of agricultural and natural ecosystems (Fig. 7-1), the dependency being stronger for the drier ecosystems than for the more humid ones. A variety of surveys

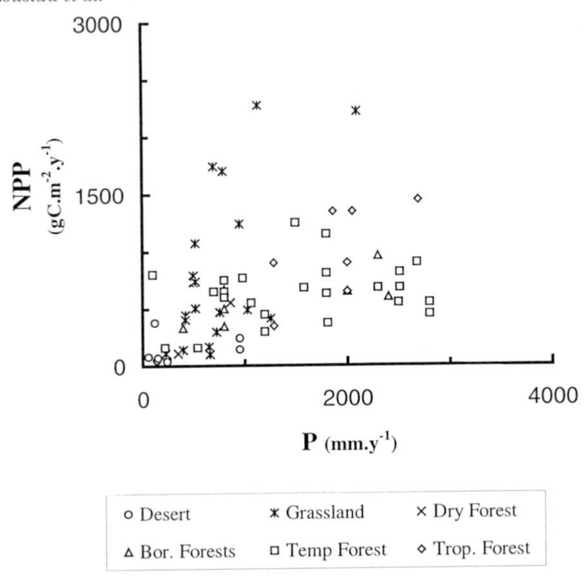

Figure 7-1 Annual net primary production (NPP) and rainfall for a range of terrestrial ecosystems. Data were taken from: Baldocchi and Vogel (1996), Baldocchi *et al.* (1997), Black *et al.* (1996), Breda and Granier (1996), Fan *et al.* (1995), Grace *et al.* (1995), Greco and Baldocchi (1996), Grier *et al.* (1992), Goulden *et al.* (1996), Harrington *et al.* (1995), Kelliher *et al.* (1993), Knapp *et al.* (1993), Long *et al.* (1989), Nizinski and Saugier (1989), Pook (1985), Redmann (1978), Runyon *et al.* (1995), Ryan *et al.* (1994), Schulze (1982), Schulze *et al.* (1996), Valentini *et al.* (1996), Vermetten *et al.* (1994), Waring *et al.* (1995), and Wofsy *et al.* (1993).

and experiments allow a quantitative estimation of the impact of water and nutrient limitations on the NPP and NEP in terrestrial ecosystems. Fertilization and irrigation experiments clearly show the extent to which NPP is limited by water and nutrients in agricultural crops, forests (Linder, 1987), and grasslands (Date, 1973). The role of nitrogen deposition in the enhancement of forest production across northern Europe shows that nutrient limitation affects forest NPP at large scales (Kauppi *et al.*, 1992), though excessive N deposition can reduce NPP in forests by causing soil acidification and losses of base cations (Johnson *et al.*, 1994; Aber *et al.*, 1998). Dendroclimatological studies have demonstrated that the history of drought experienced by various species of temperate trees accounts for some of the annual variation in carbon accumulation by secondary growth in a number of forests (e.g., Becker, 1989; Federer *et al.*, 1989; Becker *et al.*, 1994). A close correlation has been shown between the water balance of forest stands, annual secondary growth, and the carbon isotope ratio in annual ring series of different temperate species (Dupouey *et al.*, 1993; Bert *et al.*, 1997). A number of field experiments show that elevated atmospheric concentration in CO_2 (C_a) stimulates photosynthesis at the leaf and canopy levels (Drake *et al.*, 1997); the degree of stimulation appears to be higher under limiting water conditions and, in some cases, to be lower when nutrient supply is low.

The aim of this chapter is to analyze the effects of water, nitrogen, and C_a on net primary and net ecosystem production of terrestrial ecosystems. In terrestrial ecosystems, carbon photosynthetic assimilation requires absorption of light and CO_2 and thus exposure of leaves to the atmospheric environment. Associated with the entry of CO_2 into the leaf is an output flux of water vapor from the internal (vapor-saturated) leaf tissues to the air, and loss of plant water. As explained further in Section II,A, plants need to maintain their water content within a relatively narrow range, corresponding to water potentials ranging roughly from 0 to -5 MPa. The presence of plants on the land surface demonstrates that plants can successfully conserve water in their internal tissues across a wide range of external water regimes. This implies that plants can replace lost water and adjust their transpiration to water availability. In addition, photochemical energy conversion and biochemical fixation of carbon result from a combination of various enzymatic activities, all demanding a certain amount of nitrogen, phosphorus, and other nutrients. Terrestrial plants must therefore simultaneously satisfy the different constraints imposed by water relations and nutrient requirements. These constraints are exerted on different components of NPP, including leaf area, stomatal function, photosynthetic capacity, and canopy structure. In Section II, we summarize the main constraints involved. In Sections III and IV, we discuss the effects of water regime and C_a on net primary and ecosystem production as mediated through these different components.

II. The Constraints

The conservation of an internal aqueous medium requires that the plant water losses not exceed the amount of water available over a given time period. For a plant, the water available includes water stored in soil, plant reservoirs, and the net input by precipitation. In the vast majority of ecosystems, plants rely solely on rainfall input during the growing season, although this is not always the case—for example, plants can extract water from deep subsoil reserves or riparian ecosystems. The atmospheric variables that govern water transfer through plants can change rapidly during a day; in contrast, soil water availability typically varies on a time scale of days to weeks. Terrestrial plants must therefore control their water loss, which faces both short-term fluctuations of climatic demand and long-term fluctuations in water availability (Cowan, 1982).

The sensitivity of evapotranspiration (E) at the leaf, plant, and canopy levels has been analyzed by Jarvis and MacNaughton (1986), who introduced the useful notion of coupling. They demonstrated that ecosystem E is affected to different degrees by stomatal conductance and leaf area according to the canopy aerodynamic roughness. Water regime should thus differentially constrain the leaf area and stomatal conductance of terrestrial plants, depending on canopy structure. Indeed, it has long been recognized that water regime affects (1) the canopy leaf area, (2) stomatal behavior, and (3) canopy structure. These effects of water regime are described below.

A. Impact of Water and Nitrogen Availability on Leaf Area

1. Water The restriction of leaf size and number by water limitation is documented for a wide range of crop, grass, and tree species. Many experimental studies have reported detailed information on the processes involved in the control of leaf and stem growth by the water regime. Processes vary according to the plant growth type and life form. Storage factors of carbohydrates required for leaf growth—either in seeds or in perennial vegetative plant parts—(Andriani *et al.*, 1991), leaf number (Cavelier *et al.*, 1992), leaf expansion (Zahner, 1968; Van Volkenburgh and Boyer, 1985; Metcalfe *et al.*, 1990; Zhang and Davies, 1990; Belaygue *et al.*, 1996) and leaf life-span (Pook, 1985; Andriani *et al.*, 1991) are all sensitive to water stress. Leaf cell division and elongation rates are affected by water stress (Durand *et al.*, 1995; Lecoeur *et al.*, 1995). This sensitivity explains both the direct effects (e.g., leaf shedding; Tyree *et al.*, 1993) and indirect effects (e.g., initiation of cell number and leaf primordia of future foliage) of drought on the leaf area of individual plants and canopies (Lecoeur *et al.*, 1995). The combination of these control mechanisms with different response times allows plant communities to integrate the effects of the water regime on short and long time scales. Indeed, even if long time series of simultaneous measurements of leaf area index and water balance are scarce, there is some empirical evidence showing the indirect effects of water stress on leaf area, primary production, and plant growth in natural ecosystems (Webb *et al.*, 1983). One example is given in Fig. 7-2.

Canopy evaporation depends on leaf area index, L, through both the net energy absorbed by the canopy, and the canopy conductance, g_C. At the individual plant scale, decreased leaf area reduces transpiration and conserves water, whatever the canopy structure and roughness. At the canopy scale, the response of ecosystem E to leaf area is not linear but reaches a plateau at high leaf area, due to mutual shading between leaves, compensatory effects of soil evaporation, and related effects on radiation absorption and turbulence. Kelliher *et al.* (1995), analyzing the relationship between L and g_C, demonstrated that the response of the maximal canopy conductance to L may show a plateau above a threshold value close to $L = 4$, depending on the net radiation available, vapor pressure deficit, and aerodynamic con-

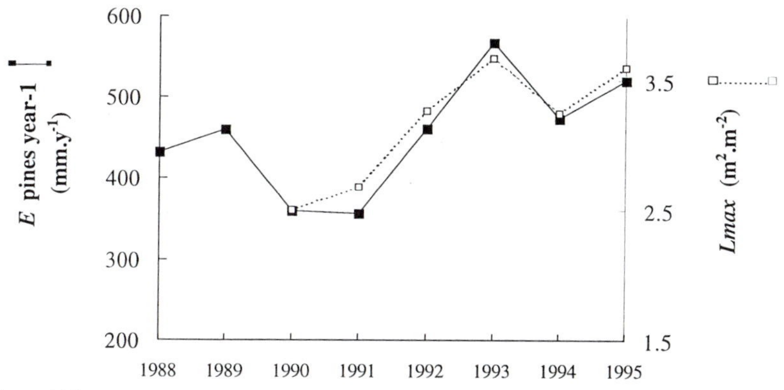

Figure 7-2 Time course of annual maximum of needle area (*L*) and previous year's evapotranspiration (*E*) in a 25-year-old maritime pine forest (EUROFLUX Site FR2, "Bray site," Southwest France). The *L* values were measured by optical methods (Demon system; CSIRO) (P. Berbigier, unpublished results). Estimates of *E* were based on the Penman–Monteith equation and the stand water balance model, published in Loustau *et al.* (1997).

ductance. Any further increase in *L* affects only slightly the maximal canopy conductance. Similarly, in an analysis of the sensitivity of the annual water balance of a three-layer forest canopy, Loustau *et al.* (1997a) showed that a change in tree leaf area affected the tree layer transpiration but has a smaller effect on *E*, because the reduction in pine transpiration is offset by increased soil and understorey evapotranspiration. The dependence of transpiration on leaf area index is thus more sensitive at low values of *L*. Accordingly, the dependence of *L* on site water balance is strongest in dry environments, characterized by unpredictable rainfall, very low values of *L*, and strong seasonality in leaf area and plant life (Nicholson *et al.*, 1990). This is illustrated by Fig. 7-3, which shows variations in *L* along a water availability gradient from deserts, savannas, and grasslands to sclerophyllous forests and rainforests: with increasing water supply, the canopy structure changes from sparse canopies with seasonal vegetation (ephemerals) such as desert, to canopies in patches, e.g., tiger bush, continuous low vegetation (savannas, grasslands), to continuous, tall, multilayered canopies. The relationship between site water balance and leaf area index has also been established for other vegetation types, such as temperate forests (Grier and Running, 1977; Gholz, 1982; Gholz *et al.*, 1990) or mediterranean ecosystems (Poole and Miller, 1981; Rambal and Leterme, 1987).

*2. **Nitrogen Availability*** Photon harvesting, photochemical conversion, and biochemical photosynthetic energy fixation require that leaves contain a certain amount of nitrogen and nutrients incorporated in structural compounds, enzymes, and other metabolic components. The N concentration

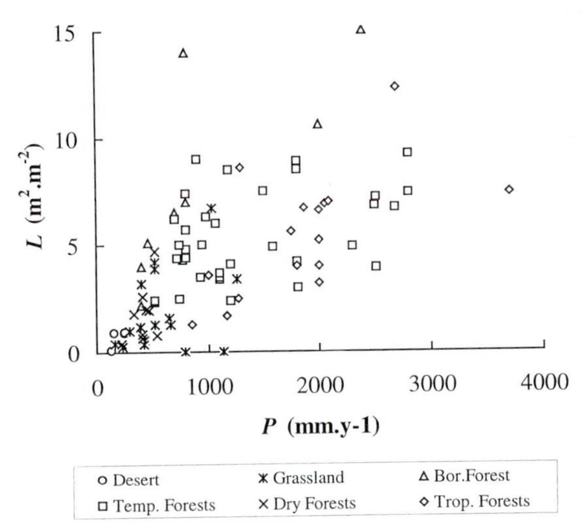

Figure 7-3 Relationship between the annual maximum leaf area index (L) and annual precipitation (P) for terrestrial ecosystems (same data as Fig. 7-1).

in plants is highest in leaves, particularly young leaves, reflecting the nitrogen cost of photosynthesis. This includes nitrogen required for light harvesting and for CO_2 fixation, e.g., for the CO_2 carboxylation enzyme, ribulose bisphosphate carboxylase/oxygenase (rubisco), which constitutes 5–30% of leaf nitrogen (Field, 1991), the proportion generally increasing as nitrogen concentration increases (Evans, 1989). The proportions and activity of nitrogenous compounds in leaves change according to the balance of available resources (Chapin *et al.*, 1987): when carbon acquisition is enhanced (by increasing photosynthesis in elevated Ca, for example), rubisco content and activity decrease, freeing N for other functions (Woodrow, 1994). Similarly, the CO_2 concentrating mechanism in C_4 plants allows a greater photosynthetic capacity per unit N compared to C_3 plants—concentrating intracellular [CO_2] in C_4 plants makes C less limiting, and thus N can be allocated to other functions (Sage and Pearcy, 1987). Similarly, nitrogen in chlorophyll constitutes 15–20% of total leaf nitrogen when plants are grown under high light; this proportion varies inversely with irradiance and can increase to 60% when plants are grown in the shade (Evans, 1989). Plants adapted to shade conditions show greater N investment in light harvesting (Bjorkman, 1981), whereas high-light adapted plants invest relatively more in rubisco (CO_2 fixation) (Seemann *et al.*, 1987).

In response to increasing N supply, production of leaf area increases more than photosynthetic rate per unit leaf (Sage and Pearcy, 1987). For example, the response of coniferous forests to increased nutrient availability was a large increase in leaf area and stem growth with only a small increase in the photosynthetic capacity (Brix, 1981; Linder and Rook, 1984). Similarly, decreasing N supply decreases the rate of leaf expansion more than the rate of photosynthesis per unit leaf area (Evans, 1983; Pettersson and McDonald, 1992). Plant growth rate is therefore not strongly correlated with photosynthetic capacity, but is strongly correlated with the rate of leaf expansion (Potter and Jones, 1977).

Relatively high N partitioning to new leaves should tend to maximize growth, because of the compounding effect on growth of producing new photosynthetic tissue. However, producing new leaves creates additional demand for nitrogen (and other belowground resources). Thus, when nutrients become limiting to growth, plants increase partitioning of resources to roots (Davidson, 1969; Reynolds and D'Antonio, 1996). In response to nitrogen stress, N can be remobilized from shoots and distributed to roots to enhance acquisition of the more limiting resource (Vessey and Layzell, 1987). For carbon allocation, as well, decreasing N availability reduces the rate of leaf expansion, reducing foliar sink strength, and shifting carbon allocation from developing leaves to roots. Conversely, increasing nitrogen availability enhances shoot growth relatively more than root growth. As shown by Vessey and Layzell (1987), coordination of root and shoot growth in response to increasing N supply involves translocation of N between shoots and roots. When leaf N concentration is high, foliar nitrogen is transported as amino acids to roots. The nitrogen the roots do not use is returned to shoots to support growth of new leaves.

The morphology of leaves and roots also changes in response to variation in nitrogen supply. Specific leaf area increases with increasing N supply, and fine root production decreases (e.g., Boot and Mensink, 1990; Linder and Rook, 1984; Fitter and Hay, 1981; Hunt *et al.*, 1985; Fitter, 1985), reflecting a shift in partitioning toward carbon acquisition when nitrogen is abundant. According to several plant growth models, the shift in biomass partitioning in response to increased or decreased nitrogen supply maximizes growth rate (Mooney *et al.*, 1988; Hirose, 1987; Kachi and Rorison, 1989). With increasing nitrogen supply and plant internal nitrogen concentration, leaf weight ratio, specific leaf area, and net assimilation rate all increase, resulting in higher relative growth rate (Hirose, 1988). In fact, plant internal nitrogen concentration alone is a powerful predictor of plant growth rate and primary production (Agren, 1985; Agren and Ingestad, 1987). Nitrogen limitation of net primary production has been demonstrated in many temperate ecosystems by N addition experiments and is inferred in many cases based on carbon:nutrient ratios (Vitousek and Howarth, 1991), because limited nitrogen supply reduces foliar nitrogen concentration relative to other nutrients (Ingestad, 1979; Birk and Vitousek, 1986).

B. The Role of Stomatal Control

The vulnerability of water transport tissues to cavitation implies that vascular plants must control their internal water tensions and transpiration over short time scales (Tyree and Sperry, 1988). They modulate loss of water through stomatal control (Woodward, 1998). The water transfer system of most plants operates under tension, i.e., at negative water potentials (Sperry *et al.,* 1996), and plants must keep their water potential above the point of catastrophic runaway embolism, which varies from above −1.0 MPa for riparian species (Tyree *et al.,* 1994) and some rainforest tree species (Machado and Tyree, 1995) to −10 MPa in less vulnerable species, e.g., *Juniperus* sp. (Sperry and Tyree, 1990). Depending on the value of the soil-to-leaf hydraulic conductance and soil water potential in the rooting zone, vulnerability to cavitation sets an upper limit on water flow through the plant. This constraint must therefore be exerted on stomatal conductance (Tyree and Sperry, 1988; Jones and Sutherland, 1991; Cochard *et al.,* 1996). From this point of view, it is worth noting that the main external variables constraining stomatal function—water availability at the soil–root interface, and leaf-to-air vapor pressure deficit—also determine the water potential difference between the end points of the soil-to-leaf pathway.

In aerodynamically rough canopies, stomatal closure is a very efficient mechanism for adjusting both individual plant transpiration and stand E (Choudhury and Monteith, 1986; Kelliher *et al.,* 1993). This has been shown, e.g., in sclerophyllous Mediterranean vegetation (Tenhunen *et al.,* 1990) or coniferous canopies (Granier and Loustau, 1994; Loustau *et al.,* 1996). In such canopies, atmospheric vapor pressure deficit, D, at the leaf surface is only weakly dependent on plant transpiration, so that plant transpiration and ecosystem E depend strongly on canopy conductance, i.e., on leaf conductance and leaf area. Additionally, the increase in sensible heat flux caused by stomatal closure leads to a vertical expansion of the convective boundary layer (Jacobs and De Bruin, 1992). Incorporation of drier air from above the convective boundary layer dilutes the vapor emitted by the canopy within a larger volume of air, which leads to a positive feedback on D. Indeed, there is increasing evidence showing that stomatal function effectively allows woody plants to operate above their cavitation threshold (Alder *et al.,* 1996), as illustrated in Fig. 7-4 (Cochard *et al.,* 1996).

In a smooth canopy, whole-ecosystem E is less sensitive to leaf area or stomatal conductance because stomatal variation has a negative feedback effect on D at the canopy surface, canceling the impact on E. The transpiration of an individual plant is more sensitive to its leaf area than to its stomatal conductance, because plant transpiration is dominated by the equilibrium term, i.e., the amount of absorbed energy. Additionally, small leaf size confers a high boundary layer conductance, favoring heat dissipation, which is advantageous in dry environments. In aerodynamically smooth canopies,

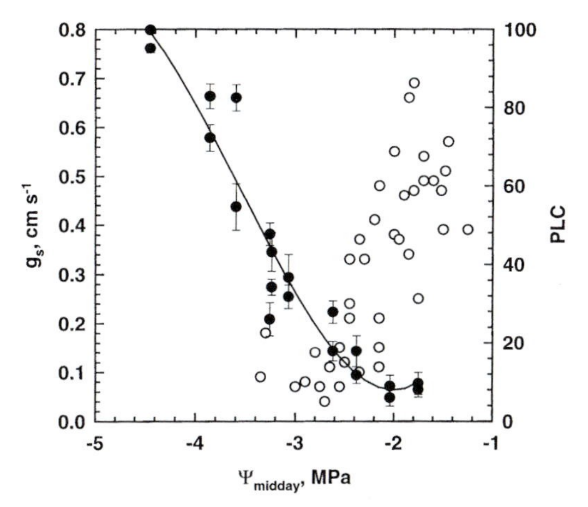

Figure 7-4 Percent loss in conductivity due to xylem embolism (●) and midday stomatal conductance (○) vs. leaf water potential at midday (Ψ_{midday}) in *Quercus petraea*. Vertical error bars represent one standard deviation. Embolism significantly increased in petioles and twigs, $\Psi_{midday} < -2.7$ Mpa, at which point stomatal conductance (g_s) was reduced to about 10% of its maximal value. From Cochard *et al.* (1996), with permission.

the sensible heat flux and, in turn, the height of the convective boundary layer are less sensitive to stomatal conductance. Therefore, plant transpiration is rather regulated by control of the net absorption of radiation by leaves, through such processes as leaf growth, leaf rolling, leaf shedding, and leaf orientation (Dingkuhn *et al.*, 1989). Avoidance of critically low water potentials may also be less important for systems (e.g., grass species) in which the nocturnal repair of embolized vessels may be more common due to positive root pressures (Tyree *et al.*, 1986).

The hydraulic constraint may also restrict species expansion, precluding the survival of a given species in environments that are too dry (or cold), environments according to their hydraulic vulnerability and stomatal function characteristics. Convincing examples may be found in Sperry and Tyree (1990) and Tyree and Cochard (1996), who compared the hydraulic vulnerability of coniferous and *Quercus* species, respectively, and concluded that the differential vulnerabilities to embolism contribute to their geographical distribution.

C. Impact of Water Availability on Canopy Structure

The constraints on L and g_C by the water regime, with related effects on assimilation rate, are, together with temperature, major factors determining the distribution of plant life form over the land surface (Raunkiaer, 1934; Schulze, 1982; Woodward, 1987). This influence has been widely recognized and we shall not add further comments on this point. Although water stress has been shown to affect plant architecture through qualitative processes involved in morphogenetic development, here we will focus on two water-related processes involved in the control of plant height—plant carbon balance and hydraulic architecture.

It is relatively straightforward to understand the potential impacts of the water and nutrient availability on plant height and canopy structure from an analysis of the carbon balance of a single plant, using the pipe model formalism (Makela, 1986; Valentine, 1990). The net production of a leaf over an annual cycle must account for the cost of maintenance and renewal of the tissues that support the water transfer system to the leaf, i.e., sapwood, cambium, and root (Fig. 7-4). As canopy height increases, so does the length of the pathway between roots and leaves and the size of the sapwood connecting them. Water and nutrients affect the amount and distribution of assimilates, and in turn plant and canopy height through different components of the plant carbon balance.

First, as detailed in Section III,B, the time integral of leaf net assimilation rate, and, in turn, the amount of carbohydrates available for stem maintenance, depend on the water regime and nutrient availability. Furthermore, this effect interacts with increasing plant height as a consequence of increased gravitational force and decreased stem conductance (Mencuccini and Grace, 1996). Both of these factors increase the water stress in leaves and impair their photosynthetic production.

Second, the root:shoot ratio is increased under drought conditions because carbon allocation shifts in favor of roots under water- or nutrient-limiting conditions. This is well exemplified by the data obtained by Schulze *et al.* (1996) along an aridity gradient in Patagonia. Consequently, humid and fertile environments, where leaf assimilation is not restricted by drought, allow growth of larger plants and taller canopies, with proportionally less belowground biomass. In these conditions, height growth provides a competitive advantage for light capture. Nutrient-poor or dry environments limit the development of the plant aerial structure and produce larger root systems. The competition for light will not play a major role in canopy structure.

Finally, Mencuccini and Grace (1996) reported convincing evidence that the age-related decrease of soil-to-leaf hydraulic conductance could explain the maximal height reached by a Scots pine stand at Thetford Forest (South England), through a decrease in apical growth rate linked to lack of turgor (Tyree and Ewers, 1991). This may also place a ceiling on the maximal height sustainable in a given environment. Similarly, Margolis *et al.* (1995) derived an expression relating the maximal height as a function of internal hydraulic conductance, leaf surface/sapwood ratio, and variables determining the transpiration of canopy (Whitehead *et al.*, 1984). From this re-

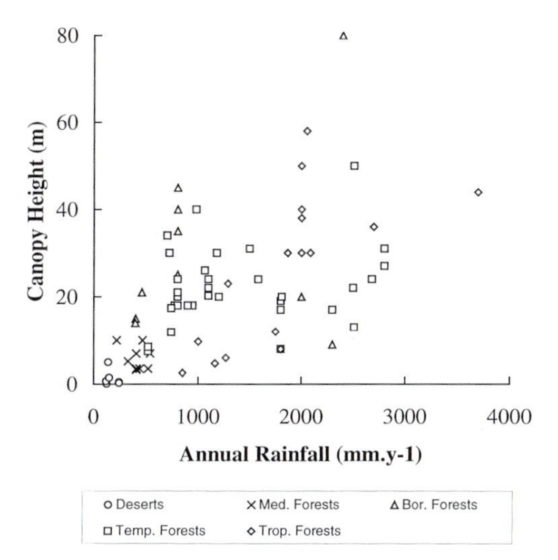

Figure 7-5 Relationship between canopy height and annual rainfall for a range of terrestrial ecosystems. (same data as Fig. 7-1).

lation, Beerling *et al.* (1996) explained some of the effects of drought on the height growth of *Fagus sylvatica* in England.

The relative extent to which carbon balance, hydraulic architecture, and availability of nutrients limit plant height is under debate (Ryan and Waring, 1992; Ryan *et al.*, 1994). The major difficulty when attempting to provide a quantitative estimate of the upper limit in canopy height is the assessment of the carbon balance of the whole plant and canopies over long periods. Some components, e.g., root turnover and respiration costs, are notoriously difficult to estimate and can account for 10–60% of the carbon assimilated. The maintenance and renewal costs of stem transfer tissue expressed per meter length may account for typically 0.5–1.5% of total net assimilation as calculated from Ryan (1990), Ryan *et al.* (1994), or Bosc (1999). This order of magnitude is quite compatible with the hypothesis of control of canopy height by carbon balance: the maximum height sustainable by a canopy would then be around 80 m and a more reasonable estimate will be around 60 m for the most humid part of the world where the tallest forests occur (Fig. 7-5).

III. Effects of Water Regime on Net Primary Production

Leaf area index, stomatal function, and canopy structure differentially affect NPP and NEP of terrestrial ecosystems. Carbon assimilation by terrestrial plants can be regarded as the result of three interrelated processes: light interception and energy conversion by photochemical reactions, diffusion of CO_2 to chloroplastic carboxylation sites, and carbon fixation by carboxylation of ribulose-1,5-bisphosphate (RuBP) (C_3) or phosphoenol pyruvate (PEP) (C_4). Leaf area primarily affects light interception, whereas canopy structure and stomatal function influence the diffusion of CO_2 into leaves. Photochemical energy conversion and carbon metabolism are controlled by the amount and efficiency of enzymes per unit leaf area, which appear to be controlled by nutrient availability. This will be analyzed in the next section.

A. Impact of Leaf Area on NPP

The primary impact of leaf area on net production derives from the interception of photosynthetically active radiation by the canopy. The impact of leaf area on light interception and ecosystem production has been widely documented (Monteith, 1977; Gosse *et al.*, 1986; Cannell *et al.*, 1987), and there is strong empirical evidence that differences in leaf area between ecosystems explain most of the geographical variations in primary production. For instance, Webb *et al.* (1983) demonstrated that a unique linear relationship between maximal annual foliar standing crop and NPP can well describe the variations observed over a range of ecosystems, from desert to coniferous forests, in North America. This relationship has also been established within particular biomes, e.g., coniferous forest (Runyon *et al.*, 1994; McMurtrie *et al.*, 1994), eucalypt forests (Landsberg and Hingston, 1996), and along aridity transects, e.g., *Acacia koa* in Hawai (Harrington *et al.*, 1995). Water regime contributes also to explaining local and temporal variations in NPP through variations in *L*. This is clearly shown by irrigation experiments, e.g., in annual crops such as sunflower and soybean (Cox and Jolliff, 1986; Huck *et al.*, 1986) or lucerne (Durand *et al.*, 1989) or for some forest stands (Linder *et al.*, 1987). Figure 7-6 illustrates the relationship between annual net primary production and leaf area index for a range of ecosystems.

B. Stomatal Control and Related Effects

1. Leaf Level

a. Effects of D Stomatal closure under increasing atmospheric deficit is commonly observed in arid (Schulze *et al.*, 1974; Roessler and Monson, 1985), mediterrannean (Eckardt *et al.*, 1975; Tenhunen *et al.*, 1984), temperate (Beadle *et al.*, 1985), and even tropical (Smith, 1989; Roy and Salager, 1992; Koch *et al.*, 1994; Zotz and Winter, 1996) environments. In the early 1980s, the observed decrease in carbon assimilation accompanying a midday increase in *D* and temperature was attributed to both stomatal and non-stomatal effects (Tenhunen *et al.*, 1984). However, these conclusions have been questioned in the light of observations of drought-induced patchy

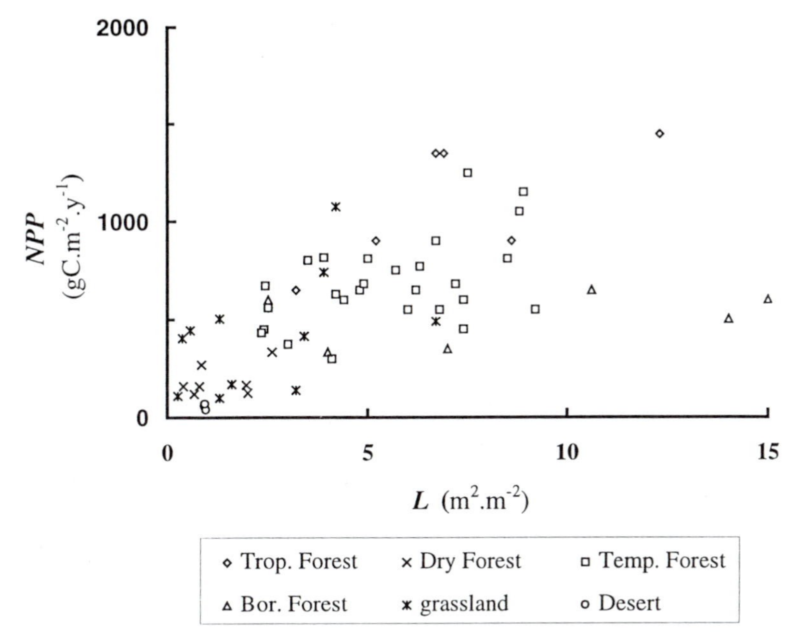

Figure 7-6 Relationship between annual net primary production and leaf area index for the data referenced in Fig. 7-1.

stomatal closure, the consequences of which for the internal CO_2 concentration C_i computations were made explicit by Terashima *et al.* (1988). The effect of stomatal response to D on assimilation rate appears to be primarily a drop in C_i, as can be readily understood from current leaf photosynthesis models (e.g., Farquhar *et al.*, 1980). In particular, it is worth noting that the light limited- and light-saturated photosynthetic rates are not equally affected by internal co_2 concentrations. Stomatal closure affects the light-saturated rate more than the apparent quantum efficiency of carbon assimilation, and therefore decreases the light threshold intensity for saturating photosynthesis.

There is now little evidence that atmospheric drought may produce a substantial change in C assimilation through enzymatic activities. When patchiness effects were taken into account, a decrease in enzymatic activities in the short-term response of leaf assimilation to atmospheric D or mild water stress could not be demonstrated (Sharkey and Seemann, 1989), leading authors to postulate that the effect of D on carbon assimilation could be considered as a purely stomatal limitation (Comstock and Ehleringer, 1993; Dai *et al.*, 1992). The linear decrease of C_i induced by D found in a wide range of plants by Zhang and Nobel (1996) is consistent with this hypothesis.

b. Soil Water Deficit Stomatal closure and the related drop in C_i is the first mechanism that affects carbon assimilation during a period of soil water shortage. However, nonstomatal effects, i.e., a reduction in enzyme activity, are also commonly involved during soil water deficits. Adjustment of photosynthetic enzymatic capacity has a longer time constant, typically hours to days, and requires a longer application time of the primary signal. There is an abundant literature showing a down-regulation of enzymatic activities when photosynthesis is decreased by CO_2 availability, e.g., under drought-induced stomatal closure, in both C_3 (Jones, 1973; Martin and Ruiz-Torres, 1992; Wise *et al.*, 1990; Sharkey and Seeman, 1989; Kaiser, 1987) and C_4 (Du *et al.*, 1996) plants.

Following the work of Kaiser (1987), Cornic *et al.* (1989), and Quick *et al.* (1992), who suppressed artificially the diffusional limitations induced by water stress using CO_2 concentrations as high as 15%, it is now commonly accepted that leaf water stress or dehydration has no effect per se on the photon-harvesting systems and thylakoid enzymes controlling the light reaction processes of photosynthesis, at least in the range encountered under natural conditions (Chaves, 1991; Dreyer *et al.*, 1992; Epron and Dreyer, 1992). However, under high radiation load, stomatal closure can raise leaf temperature to supraoptimal levels. Under such conditions, the capacity of leaves to recycle the excess reducing power through processes such as heat dissipation, the Mehler reaction (e.g., Biehler and Fock, 1996), and photorespiration (Heber *et al.*, 1996) may be overriden. This leads to irreversible damages to the photosynthetic apparatus. In the field, reversible photoinhibition, leading to a decrease in light-saturated rate of photosynthesis, has been shown to occur typically during the afternoon, for grapevine (Correia *et al.*, 1990) and *Quercus cerris* (Valentini *et al.*, 1995), and is attributed to indirect effects induced by high light and temperature on photosystem II (PSII).

A consequence of the stomatal effects on carbon assimilation and transpiration lies in the characteristics of discrimination between stable carbon isotopes by C_3 plants. Farquhar *et al.* (1989) demonstrated that the discrimination rate is proportional to the ratio A/g_s, or intrinsic water use efficiency, and decreases with stomatal closure. This finding opened interesting possibilities for the use of carbon isotope analysis of plants (more particularly, annual rings of trees) to assess fluctuations in water use efficiency and net primary production caused by drought (Dupouey *et al.*, 1993; Livingston and Spittlehouse, 1996; Bert *et al.*, 1997; Walcroft *et al.*, 1997; Nguyen-Queyrens, 1998, Duquesnay *et al.*, 1998).

2. Ecosystem Level

a. Impacts of D A significant decrease in the radiation use efficiency (RUE) concurrent with high vapor pressure deficit values has been observed in a wide variety of terrestrial ecosystems, e.g., tall canopies such as coniferous forests (Fan *et al.*, 1995; Baldocchi and Vogel, 1996; Lamaud *et *al.*, 1997), tropical forests (Grace *et al.*, 1995), and broadleaved temperate forests (Hollinger *et al.*, 1994). Only a few exceptions report the absence of any contribution of *D* to variations in net carbon exchanges over a deciduous forest (Verma *et al.*, 1986). Runyon *et al.* (1994) estimated that *D* reduced the NPP of coniferous ecosytems in the Oregon transect by 10–20%, and a stronger effect of *D* can be presumed in more arid environments. A decrease in the net carbon exchange by increased *D* was also observed for medium-size canopies, such as *Andropogon* tallgrass prairie (Verma *et al.*, 1989, 1992), and even for shorter-statured grasslands (Kim and Verma, 1990; Pettigrew *et al.*, 1990) and crops (Stockle *et al.*, 1990). A positive interaction of the effect of *D* with soil drought is commonly observed (e.g., Verma *et al.*, 1992). The effects of *D* on stomatal conductance and net assimilation can certainly account for a part of the midday (or afternoon) depression in RUE reported in most studies of CO_2 exchanges above rough canopies, either in water-limited or well-watered conditions (e.g., Valentini *et al.*, 1996; Lamaud *et al.*, 1997). However, high *D* occurs simultaneously with high temperature and radiation, which raise leaf and biomass temperatures and increase their respiration rates. Additionally, photosynthesis may also decrease when temperature exceeds an optimal value. These effects are confounded in the above-mentioned decrease of NEP, and cannot be discriminated easily.

Unfortunately, the variety of sites and climate conditions precludes a comprehensive comparison of the impact of *D* on RUE between canopies of different roughness. The impact of *D* on canopy net exchanges may explain some differences in the behavior of different ecosystems, because grassland plant species would be expected to be less sensitive to *D* than tree species (Jarvis, 1985). Because large values of *D* occur mostly under saturating light conditions, the difference RUE of forest and crops might be explained in part by the difference in response of their canopy conductances to atmospheric *D* (Ruimy *et al.*, 1995). The observation that the assimilation rate of uncoupled canopies, e.g., grasslands, does not saturate with increasing photosynthetically active radiation (PAR), as coupled canopies do (Ruimy *et al.*, 1995), suggests that the light saturation point of net assimilation could be reduced more by *D* in rough canopies.

b. Impact of Soil Water Deficit Effects of soil drought on RUE have also been reported for a wide range of ecosystems. A drought-induced decrease in RUE has been demonstrated in most forests, e.g., a canopy of *Pinus pinaster* (Fig. 7-7), *Fagus sylvatica* (Valentini *et al.*, 1996), or mixed deciduous forest (Greco and Baldocchi, 1996). However, this drought impact can be tenuous for temperate forests growing under high rainfall, as in the case of the deciduous forest studied by Goulden *et al.* (1996).

Only the NPP of lowland boreal or flooded forests, where the water table remains close to the soil surface during the growing season, escapes such soil water limitations (Black *et al.*, 1996; Baldocchi *et al.*, 1997). There is some current uncertainty concerning the extent of the effects of soil moisture

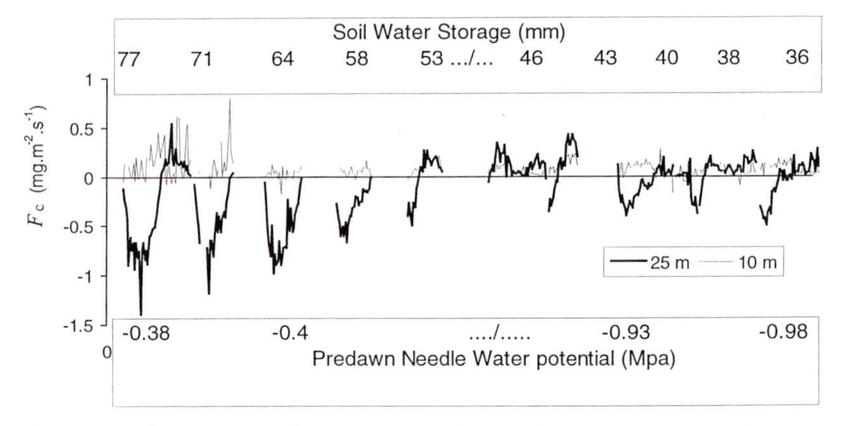

Figure 7-7 Time course of CO_2 fluxes beneath (z = 10 m) and above (z = 25 m) a maritime pine canopy, for selected days covering a range of soil moisture deficits in summer, 1995. Soil water content, predawn needle water potential, and daily sum of downward PAR are given for each day. Flux data are from Lamaud *et al.* (1997).

deficit on net exchange between tropical rainforest and atmosphere (see Chapter 17, this volume), but there is no doubt that soil water can also play a role in these systems, at least where there is a significant dry season (Monteny, 1989). The RUE of grass and savannas is also affected by soil moisture deficit. Estimates of NPP from biomass changes (Knapp *et al.*, 1993; Petersson and Hansson, 1990; Long *et al.*, 1989, 1996) and from continuous measurements of net ecosystem exchanges (Kim and Verma, 1990; Verma *et al.*, 1989, 1992; Redmann, 1978) show a clear decline linked to soil water deficit. Such effects have also been widely documented for agricultural crops, for which some continuous measurements of ecosystem exchanges have been made using crop chambers (Jones *et al.*, 1986), Bowen ratio–CO_2 combined measurements (Baldocchi *et al.*, 1981a, 1983, 1985), or eddy covariance techniques (Baldocchi, 1994). Nevertheless, it should be noted that the growth habits of some adapted species exhibit a high tolerance to water stress and are affected weakly by soil moisture deficit, e.g., alfalfa (Baldocchi *et al.*, 1981b).

C. Impact of Canopy Roughness

An important characteristic of the control of transpiration and assimilation by L or g_C is that the efficiency of control by L (or g_C) varies according to the degree of coupling of the canopy to the atmosphere (Jarvis and MacNaughton, 1986). As pointed out by Jarvis (1985), the effects of a fractional change in canopy conductance on net assimilation are expected to be stronger in tall canopies than in short canopies, because the drop in C_i caused by stomatal closure is partially offset by an increase in C_a in the latter. Figure 7-8 shows the theoretical relationship between g_C and the light-

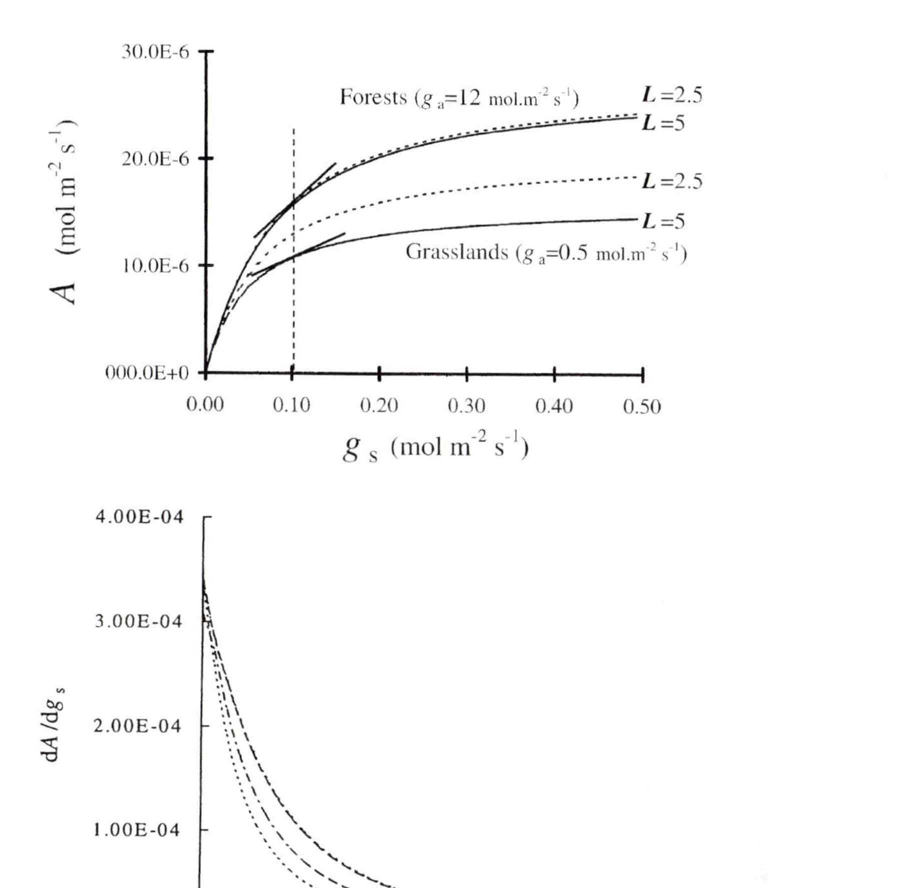

Figure 7-8 Impact of a change in g_s on the light-saturated carbon assimilation of theoretical "big C_3 leaf" canopies differing in leaf area index and aerodynamic conductance, at 25°C and $[CO_2] = 350\ \mu\text{mol mol}^{-1}$ (upper graph). The tangents drawn at $g_s = 100\ \text{mmol m}^{-2}\ \text{s}^{-1}$ give the absolute sensitivities of the forest and grassland to g_s. The lower graph gives the sensitivity of carbon assimilation to g_s. Assimilation was calculated using the model of Farquhar *et al.* (1980). Parameters values are maximal carboxylation rate, $V_{c\ max} = 100\ \mu\text{mol m}^{-2}\ \text{s}^{-1}$; daylight respiration $R_d = 1\ \mu\text{mol m}^{-2}\ \text{s}^{-1}$; CO_2 light-insensitive compensation point, $\Gamma^* = 31\ \mu\text{mol mol}^{-1}$; Michaelis–Menten constant for carboxylation, $K_c = 0.46\ \text{mmol mol}^{-1}$; and oxygenation $K_o = 0.33\ \text{mol mol}^{-1}$.

saturated rate of photosynthesis for short and tall canopies. Both canopies are modeled as a big C_3 leaf. Photosynthesis is calculated from Farquhar *et al.* (1980), internal mesophyll resistance is neglected, and other parameters are given in the legend of the figure. The figure shows clearly that the sensitivity of net assimilation to stomatal conductance is higher for forest canopies than for short, smooth canopies: when C_i drops below 100 μmol mol^{-1} a fractional reduction in stomatal conductance reduces A almost twice as much for a forest than for a short grass. Unfortunately, available data describing the effects of soil water deficit or D on stomatal conductance and carbon assimilation in natural canopies remain scarce, making it difficult to compare the behavior of aerodynamic contrasting canopies. Indeed, there is some experimental evidence showing that the midday CO_2 mixing ratio within a sunlit canopy does not deviate from the reference value by more than 20 μmol mol^{-1} in tall canopies (Buchmann *et al.*, 1996), but can decrease by 100 μmol mol^{-1} in short canopies. Turbulent mixing has been shown to have a positive effect on net assimilation at high irradiance in short C_3 canopies such as rice (Yabuki *et al.*, 1978) or alfalfa (Baldocchi, 1981b), but does not appear to affect the carbon exchange of tall canopies to the same extent. In tall canopies, Buchmann *et al.* (1996) showed that the CO_2 profile between the soil and the top of the canopy was more depleted in a broadleaved, smooth canopy than in a conifer canopy. In addition, Baldocchi and Vogel (1996) observed that D had more severe effects on net assimilation for a boreal coniferous canopy than for a broadleaved, smoother, temperate forest.

IV. Effects of CO_2 on NPP and NEP

The effects of rising C_a on net primary production and net ecosystem production depend on the physiological responses of plants to elevated C_a, how these responses interact with environmental stresses, and feedbacks and constraints that modulate their translation to the ecosystem level. Experimental CO_2 doubling stimulates photosynthesis at the leaf level, and in most cases at the canopy level as well, in both managed and unmanaged ecosystems. Elevated C_a often reduces stomatal conductance and transpiration, and the stimulation of photosynthesis by elevated C_a is generally larger when plants are under water-stressed conditions. In some cases, photosynthetic and growth responses to elevated C_a are smaller in plants under under nutrient stress, but they are rarely eliminated and in many cases do not seem to depend on nutrient stress at all. Furthermore, elevated C_a reduces the requirement for rubisco, contributing to the reduction in foliar nitrogen concentrations in elevated C_a, and this plasticity in plant C:N allows a positive response to elevated C_a in nitrogen-limited systems. In field experiments in herbaceous systems, where enough data are available to construct partial carbon budgets, the enhancements of above- and belowground biomass by elevated C_a are usually smaller than measured increases in CO_2 uptake at the leaf and canopy levels. This discrepancy suggests that much of the extra carbon in elevated C_a is distributed belowground, making it difficult to confirm directly increases in carbon mass in response to elevated C_a. Determining the trajectory of increased NEP beyond the time scales of C_a-enrichment experiments requires considering the distribution of carbon to pools of differing turnover times, and also biogeochemical and atmospheric feedbacks that operate on temporal and spatial scales beyond those of manipulative experiments.

A. CO_2 Stimulation of Photosynthesis: Interactions with Nitrogen

The evidence that elevated C_a stimulates photosynthesis is overwhelming. In experiments conducted in pots with large rooting volumes or with high nitrogen supply, growth in elevated C_a increased photosynthesis 57–58% compared to the rate for plants grown in normal ambient C_a (Table 7-1). While restricted rooting volume and low nitrogen supply reduced the C_a enhancement to 28 and 23%, respectively (Table 7-1), neither eliminated the

Table 7-1 Stimulation of Photosynthesis by Elevated C_a [a]

Attribute (A)	R	Species (n)
A at growth C_a		
Large rv	1.58[b]	45 (60)
Small rv	1.28[c]	28 (103)
High N supply	1.57[b]	8 (10)
Low N supply	1.23[b]	8 (10)
Protein	0.86[b]	11 (15)
[Rubisco]	0.85[b]	11 (8)
Leaf [N]		
High N	0.85[b]	8 (10)
Low N	0.81[b]	22 (39)

[a]Determined as the ratio (R) of the value of the attribute for plants grown in elevated C_a compared to normal ambient C_a in various species and experiments (n), and for plants grown in containers with large (>10 liters) or small (<10 liters) rooting volumes (rv) and under high or low N supply. Also shown are the effects of elevated C_a on protein, rubisco, and leaf N concentrations. After Drake *et al.* (1997).

[b]Means statistically different from 1.0 ($p < 0.01$) by Student's t-test.

[c]Means statistically different from 1.0 ($p < 0.01$) by Mann–Whitney rank sum test for data normality test.

stimulation of photosynthesis by elevated C_a. Acclimation of photosynthesis to elevated C_a clearly reduces photosynthetic capacity (Sage, 1994; Gunderson and Wullschleger, 1994), but rarely enough to compensate completely for the stimulation of the rate by high C_a.

The primary carboxylase of C_3 photosynthesis, rubisco, is the most abundant protein in the biosphere and constitutes as much as 30% of total leaf N mass. In current ambient C_a, light-saturated rates of photosynthesis require large quantities of rubisco (Masle *et al.*, 1993), but elevated C_a reduces this requirement markedly. For example, *Nicotiana tabaccum* transformed with antisense RbcS to produce 13–18% less rubisco photosynthesized and grew more slowly than the wild type when both were grown in ambient C_a, but there was no difference in C gain or growth when both were grown at 80 Pa C_a (Masle *et al.*, 1993), showing the decreased requirement for rubisco at elevated C_a. Similarly, calculations suggest that in doubled C_a, 35% of rubisco could be lost before it would colimit photosynthesis (Long and Drake, 1991). Because of the strong temperature dependence of CO_2 stimulation of photosynthesis, the amount of rubisco required in elevated C_a will decline further with increasing temperature. For example, at 25°C, elevated C_a reduces by 41% the amount of rubisco necessary to maintain a given photosynthetic rate, while at 35°C, elevated C_a reduces this requirement by 58% (Woodrow, 1994). This reduced requirement for rubisco in elevated C_a partly causes the commonly observed reduction in leaf N concentration in elevated C_a (Table 7-1) (Conroy, 1992; Curtis *et al.*, 1992; Hocking and Meyer, 1991; Norby *et al.*, 1986; Wong, 1979).

B. Stimulation of Photosynthesis by Elevated C_a: Interactions with Water

The effect of elevated C_a on photosynthesis and plant growth interacts with water stress (Chaves and Pereira, 1992; Grant *et al.*, 1995). Elevated C_a not only enhances CO_2 availability at the leaf surface, but also reduces stomatal conductance in various species. For 23 species and 29 observations, the average reduction of g_s was 23%, leading to an average reduction in leaf transpiration of 27% (Table 7-2) (see also Field *et al.*, 1995). Responses in trees are highly variable (Curtis, 1996), and in some species there is no response to elevated C_a. Differences among species are at least partly related to growth form, as the stomatal response of coniferous trees to elevated C_a tends to be smaller than the response of herbs, and deciduous trees tend to be intermediate (Saxe *et al.*, 1998). It is also possible that the failure of some species to respond directly to elevated C_a is due to acclimation of stomata to high humidity. For example, stomata of *Xanthium strumarium* grown in a greenhouse in high humidity failed to respond to elevated C_a until given a cycle of chilling stress (Drake and Raschke, 1974).

Elevated C_a increases water use efficiency (WUE), the ratio of net photo-

Table 7-2 Effects of Growth in Elevated C_a [a]

Attribute	R	Species (n)
g_s	0.80[a,b]	28 (41)
E	0.72[b]	35 (81)
C_i/C_a	0.99	26 (33)
L	1.03	8 (12)

[a]Effects are on acclimation of stomatal conductance (g_s), transpiration (E), the ratio of intercellular to ambient CO_2 concentration (C_i/C_a), and leaf area index (L; field-grown species only), using a number of species and studies (n). R is the mean of n observations in various species of the ratio of the attribute in plants grown in elevated C_a compared to that for plants grown in current ambient C_a. After (Drake *et al.*, 1997a).

[b]Means statistically different from 1.0 ($p < 0.01$) by Student's *t*-test.

[c]Means statistically different from 1.0 ($p < 0.01$) by Mann–Whitney rank sum test for data that failed normality test.

synthesis to transpiration, either by reducing g_s, increasing A, or both. In a study of oats, mustard, and two cultivars of wheat, WUE increased 40–100% as the ambient C_a was increased from about 15 to 35 Pa (Polley *et al.*, 1993). In a free-air CO_2 enrichment (FACE) study in wheat, C_a elevated to 55 Pa increased WUE by 76 and 86% in cotton crops, averaged over two full growing seasons (Pinter *et al.*, 1996). Elevated C_a also increased WUE in both C_3 and C_4 wetland species (Arp, 1991) and in annual grasses (Jackson *et al.*, 1994; Hungate *et al.*, 1997a). Water use efficiency determined from gas exchange measurements is increased by elevated C_a in almost every species studied in chambers or greenhouse experiments, and this result has been confirmed in the longer term by growth analysis and carbon isotope discrimination (Jackson *et al.*, 1994; Guehl *et al.*, 1994; Picon *et al.*, 1996).

Elevated C_a can also mitigate plant water stress by improving osmoregulation capacity (Vivin *et al.*, 1996), activating the deoxydative metabolic pathway (Schwanz *et al.*, 1996), and, in some but not all cases, increasing root to shoot ratio (Rogers *et al.*, 1994; but see Norby, 1994). Through these mechanisms, elevated C_a partially alleviates the effects of drought on net assimilation and plant growth (Jackson *et al.*, 1994; Idso and Idso, 1995, Owensby *et al.*,1997). For example, the increase in carbon assimilation by C_a doubling is 2- to 10-fold higher under drought than under well-watered conditions (Idso and Idso, 1995; Guehl *et al.*, 1994; Clifford *et al.*, 1993). So far, there is no indication of an effect of elevated C_a on hydraulic characteristics of plants—e.g., in *Quercus suber, Pinus pinaster,* or *Quercus pubescens*

(H. Cochard, personal communication)—and thus no evidence of associated changes in primary production, although relatively few studies have addressed this issue.

C. Respiration

1. Mechanism of the Direct Response to Elevated C_a Within minutes of a doubling of C_a, respiration often declines by about 20% (reviewed in Drake *et al.*, 1997a). This has been observed in many different tissues—leaves, roots, stems, and even soil bacteria—suggesting that the basic mechanism involves a fundamental aspect of respiration. Elevated C_a reduces the *in vitro* activity of both cytochrome *c* oxidase (Cytox) and succinate dehydrogenase, key enzymes of the mitochondrial electron transport system, by about 20% (Gonzalez-Meler *et al.*, 1996; Palet *et al.*, 1991; Reuveni and Gale, 1985), but has no effect on the activity of the alternative pathway (Gonzalez-Meler *et al.*, 1996). Under experimental conditions in which Cytox controlled the overall rate of respiration in isolated mitochondria, O_2 uptake was inhibited by about 15% (Gonzalez-Meler *et al.*, 1996). Another proposed mechanism for the apparent inhibition of respiration is that elevated C_a stimulates dark CO_2 fixation (Amthor, 1997). However, measurements of the respiratory quotient (consumption of O_2/emission of CO_2) show that this is unlikely, because reduced CO_2 evolution is balanced by an equal reduction of O_2 uptake in elevated C_a (Reuveni *et al.*, 1993).

The possibility that CO_2 inhibition of these enzymes mediates the direct effect of C_a on respiration in plants is supported by measurements on different types of plant organelles and tissues. Doubled C_a reduced O_2 uptake by soybean mitochondria and by extracts from excised shoots of the sedge *Scirpus olneyi* (Gonzalez-Meler *et al.*, 1996). Experiments in which CO_2 efflux was used to measure dark respiration showed that doubling C_a reduced respiration in excised shoots removed from the field to the lab and from intact stands in which respiration was determined in the field on the C_3 sedge, *S. olneyi* (Drake, 1992). The importance of this effect for carbon metabolism of plants and ecosystems is that it apparently occurs at a very fundamental level of organization—the mitochondrial electron transport system. Thus, all respiring tissues are subject to this effect.

2. Acclimation of Respiration to Elevated C_a Elevated C_a could also affect respiration by altering tissue composition. As tissues age, the rate of dark respiration of foliage declines. This occurs as tissue N and protein concentrations decline, indicating a decreased demand for energy to sustain growth and/or maintenance. Thus, the reduction in protein and N concentration of plants grown in elevated C_a (Table 7-1) suggests that rising C_a could reduce growth and maintenance respiration associated with protein turnover (Amthor, 1997; Curtis, 1996; Wullschleger *et al.*, 1992). We reviewed data on measurements of respiration on leaves of 17 species grown in current ambient and elevated C_a. Acclimation of dark respiration was determined by comparison of the rate of CO_2 efflux or O_2 consumption measured on samples of tissue grown in current ambient or elevated C_a at a common background C_a. In our survey of the literature we found no overall difference between the specific rates of respiration of shoots and leaves grown in elevated or ambient C_a. However, some C_3 species—*S. olneyi, Lindera benzoin,* and *Triticum aestivum*—do show acclimation to high C_a, apparently by reducing the activity of enzymatic complexes of the mitochondrial electron transport chain (Cytox and Complex III), resulting in diminished capacity of tissue respiration (Aranda *et al.*, 1995; Azcon-Bieto *et al.*, 1994). Reduction of the activity of these enzymes was not found in the C_4 species, *Spartina patens.*

D. Canopy CO_2 Uptake, Ecosystem Carbon Mass, and Net Ecosystem Production

Based on results from a number of elevated C_a experiments in a variety of ecosystems, there is broad agreement that the stimulation of leaf-level photosynthesis by elevated C_a (Drake *et al.*, 1996, 1997) (Table 7-1) is reflected in increased CO_2 uptake at the canopy level (Drake and Leadley, 1991). For example, even with acclimation of photosynthesis to elevated C_a, in the sedge, *S. olneyi,* elevated C_a stimulated ecosystem carbon uptake in the salt marsh community where *S. olneyi* is dominant (Drake *et al.*, 1996). Increased canopy CO_2 uptake in response to elevated C_a has been observed in field experiments in a number of herbaceous systems, including arctic tundra (Oechel *et al.*, 1994), alpine grassland (Diemer, 1994), annual Mediterranean grassland (Field *et al.*, 1997), tallgrass prairie (Ham *et al.*, 1995), and calcareous grassland (Stocker *et al.*, 1997). In two of these cases, the stimulation of canopy CO_2 uptake was short-lived, disappearing after 3 years in the alpine grassland (Körner *et al.*, 1997) and after 2 years in arctic tundra (Oechel *et al.*, 1994). Photosynthesis in the dominant species in the tundra system rapidly adjusted to elevated C_a in controlled environment studies (Oberbauer *et al.*, 1986; Cook *et al.*, 1998). Low temperatures in tundra and alpine grasslands may reduce the response to elevated C_a (Long and Drake, 1991). Similarly, high temperatures tend to amplify the stimulation of photosynthesis by elevated C_a at both the leaf level (in *Pinus taeda*) (Lewis *et al.*, 1996) and at the canopy level (in cotton and wheat) (Pinter *et al.*, 1996). In summary, in the majority of field experiments conducted to date in both crops and native species, elevated C_a increased canopy CO_2 uptake, and the stimulation was sustained over the duration of the experiment.

Net ecosystem production integrates annual carbon inputs through photosynthesis and annual carbon losses through respiration for a given land area over time (e.g., g C m^{-2} yr^{-1}), so canopy gas exchange measurements, such as those cited above, can be used to estimate NEP and its response to elevated C_a. Usually, however, annual NEP and instantaneous canopy gas exchange differ in time scale. Chamber-based gas exchange measurements are rarely continuous for an entire annual period, because maintaining the measurement system is too costly or could introduce artifacts that alter the fluxes being measured (e.g., altered microenvironment, excluded rainfall), and also because gas exchange rates during certain periods of the year are deemed negligible. Thus, a potential source of error in determinations of NEP from gas exchange measurements is the extrapolation to the nonmeasured periods, including the assumption that net fluxes during certain periods (e.g., winter) are zero. The effects of elevated C_a on NEP can also be determined by comparing carbon stocks in the C_a treatments after a given period of time (Hungate *et al.*, 1997b). This approach, although somewhat simpler in methodology, requires that experiments are of sufficient duration that biologically important differences in NEP yield statistically detectable differences in ecosystem carbon mass (Hungate *et al.*, 1996). Usually, this means ignoring large, relatively inert, ecosystem carbon pools in order to reduce the noise (e.g., excluding soil, or soil below the top 15 cm), which can also yield erroneous estimates of NEP.

There have now been a number of elevated C_a experiments conducted under field conditions in native ecosystems), but there are only a few in which estimates of the effect of elevated C_a on NEP have been reported from canopy gas exchange measurements, inventories of ecosystem carbon stocks, or, most instructive, from both approaches (Table 7-3). These experiments support the general conclusion that elevated C_a increases NEP (with the exception of the arctic tundra study). In cases where enough information is available, the effects of elevated C_a on NEP determined by gas exchange are larger than those determined by carbon inventories, dramatically so in the salt marsh study, but also the case in the alpine grassland.

The stimulation of carbon uptake at the canopy level is only partly accounted for by increases in measured ecosystem carbon pools (Drake *et al.*, 1996; Diemer, 1997; Körner *et al.*, 1997; Canadell *et al.*, 1996), suggesting that much of the extra carbon taken up in elevated C_a is distributed to large carbon pools, where changes are difficult to detect, which, in the case of the herbaceous ecosystems listed in Table 7-3, is most likely the soil carbon pool (Kuikmann *et al.*, 1991; Canadell *et al.*, 1996; Gorissen, 1996; Hungate *et al.*, 1997b). Determining the fate of this carbon is critical for extrapolating to long-term carbon storage in such systems, and presents a challenge to researchers in this area. First, it must be demonstrated unequivocally that carbon is accumulating in soil in these experiments, either by measuring a change in soil carbon mass, or by ruling out other fates of the extra carbon, including leaching or horizontal transfer of labile carbon, or return of carbon to the atmosphere as CO_2 during periods when net CO_2 exchange is

Table 7-3 Estimates of Net Ecosystem Productivity from Elevated C_a Field Experiments[a]

Ecosystem	Plant + litter ($g\,C\,m^{-2}$)		Soil ($g\,C\,m^{-2}$)		ΔNEP_{stocks} ($g\,C\,m^{-2}\,yr^{-1}$)[b]	CO_2 exchange ($g\,C\,m^{-2}\,yr^{-1}$)		$\Delta NEP_{gas\ exch}$ ($g\,C\,m^{-2}\,yr^{-1}$)[c]
	A	E	A	E	$(E - A)/t$	A	E	E − A
Salt marsh[d]	3900	3910	1400	1470	10	1450	2170	720
Alpine grassland[e]	548	595	NR	NR	16	191	233	42
Serpentine grassland[f]	186	246	1994	2086	38	NR	NR	—
Sandstone grassland[f]	418	565	2385	2383	36	NR	NR	—
Tallgrass prairie[g]	850	1223	6290	6365	72	NR	NR	—
Arctic tundra[h]	NR	NR	NR	NR	—	−60	−60	0

[a] *NEP* is estimated, first, from reports of ecosystem carbon stocks (shown are plant + litter, and soil), expressed as the difference in carbon stocks between elevated (E) and ambient (A) C_a treatments divided by the duration of the experiment at the time the measurements of carbon stocks were made (ΔNEP_{stocks}). NR, not reported. *NEP* is also estimated from reports of canopy CO_2 exchange, extrapolated to an annual flux, and the difference between elevated and ambient C_a treatments for one annual period is shown ($\Delta NEP_{gas\ exch}$).

[b] Calculated as the difference in total carbon stocks between elevated and ambient C_a treatments divided by the duration of the experiment.

[c] Calculated as the difference in integrated canopy gas exchange between elevated and ambient C_a treatments.

[d] All data for 1994 season, from Drake *et al.* (1996), after 8 years of experimental treatment.

[e] Körner *et al.* (1997). Canopy gas exchange calculated from the reported 22% stimulation of seasonal C uptake and reported absolute difference of 42 $g\,C\,m^{-2}\,yr^{-1}$ between the ambient and elevated C_a treatments.

[f] Plant and litter C mass from Hungate *et al.* (1997b); 1994 sample, after 3 years of experimental treatment; soil C mass from Hungate *et al.* (1996); 1995 sample, after 4 years of experimental treatment.

[g] Plant C only. Data from website http://spuds.agron.ksu.edu/, assuming plant mass is 50% C; aboveground biomass from 1996; belowground estimated by summing root ingrowth from 1990 to 1995. Soil data from Rice *et al.* (1994) for the third year of treatment.

[h] Oechel *et al.* (1994); calculated for a 60-day growing season using 1985 data.

not being measured. Second, soil carbon comprises several carbon pools of varying turnover times, and plant carbon allocation partly determines carbon distribution among these pools (Parton *et al.*, 1987). Because the potential for carbon storage in a given soil pool depends on its turnover time, shifts in plant carbon allocation in elevated C_a (e.g., increased allocation to fine roots and root exudates) will affect long-term carbon storage. Thus, quantitative predictions of long-term increases in NEP require understanding how elevated C_a affects the distribution of the carbon to pools of varying carbon storage potential.

In field experiments, the quantitative evidence that elevated C_a enhances carbon distribution to soil is indirect, resting on observations that the carbon increments from increased canopy photosynthesis are larger than observed increases in plant biomass. Although some pot studies using soils with low background amounts of carbon (Lutze, 1996) or carbon isotope labels (Ineson *et al.*, 1996; Gorrisen, 1996) provide direct evidence for carbon accumulation in soil, others show that much of the extra carbon distributed belowground in elevated C_a is preferentially allocated to relatively labile pools, limiting carbon accretion in recalcitrant pools with high storage potential (Tate *et al.*, 1995; Hungate *et al.*, 1997a,b). In these cases, elevated C_a may cause a larger increase in carbon turnover than in soil carbon mass (Newton *et al.*, 1995; Tate and Ross, 1997).

V. Interactions between CO_2 and Nutrients

Interactions between rising C_a and nutrients can occur in two ways: (1) the response of plants to elevated C_a can depend on nutrient availability and (2) the responses of plants to elevated C_a can alter nutrient availability, changes that can further modify both NPP and NEP. The effects of elevated C_a are rarely eliminated by nitrogen stress, at least partly because the decreased demand for rubisco in elevated C_a (as discussed above) allows a positive response even without an expansion of plant nitrogen mass. For example, in a 4-year study of a native Australian grass, elevated C_a reduced tissue nitrogen concentration irrespective of the availability of N in the soil, and this was accompanied by accumulation of carbon in the microcosm, although the lower N availability levels reduced the relative effects of C_a on carbon accumulation (Lutze, 1996). Whether growth responses to elevated C_a are relatively smaller under conditions of nutrient limitation remains ambiguous. In some cases, responses to elevated C_a are markedly smaller when nitrogen supply is restricted (Curtis *et al.*, 1994, 1995; McGuire *et al.*, 1995; Saxe *et al.*, 1998; (Table 7-1), whereas other experiments show no evidence for a short-term nitrogen constraint (reviewed in Lloyd and Farquhar, 1996; Idso and Idso, 1995). In native ecosystems, whereas nitrogen addition augments responsiveness to elevated C_a in tallgrass prairie (Owensby *et al.*, 1993), small responses in alpine grasslands are apparently unrelated to nitrogen availability (Körner *et al.*, 1997). The relatively large C_a stimulation of net ecosystem carbon uptake in a C_3 saltmarsh (Drake *et al.*, 1996) is widely attributed to its "nutrient-rich" status, but this explanation seems unlikely given that net primary production in these coastal salt marshes is limited by nitrogen, as demonstrated by marked increases in aboveground production in response to experimental nitrogen addition (Valiela and Teal, 1974; Jefferies, 1977; Jefferies and Perkins, 1977; Kiehl *et al.*. 1997). Decreased growth responses to elevated C_a under restricted phosphorus supply have also been observed in some cases (Cure *et al.*, 1988; Goudriaan and DeRuiter, 1983; Conroy *et al.*, 1988, 1990), and not in others (Conroy, 1992; Israel *et al.*, 1990). More striking than the presence or absence of a putative nutrient constraint on production responses to elevated C_a is the wide variation in responses among different experiments, matched by the variety of conclusions drawn in reviews of the topic (e.g., Idso and Idso, 1994; McGuire *et al.*, 1995; King *et al.*, 1997; Curtis and Wang, 1998). As suggested by Johnson *et al.* (1998), limitation by any growth factor is more likely to be a continuum than a dichotomy of the presence or absence of limitation. When the data are reexamined with a quantitative assessment of the degree of nutrient limitation, a weak dependence of the response to elevated C_a on the degree of nutrient limitation emerges, with smaller responses under more nutrient-limiting conditions (Poorter and Garnier, 1996).

Nutrient constraints on the productivity responses to elevated CO_2 fall into two classes: (1) whether the relative growth response to elevated C_a depends on nutrient availability (which, as described in the preceding paragraph, is controversial), and (2) whether, over longer time scales (10–100 yr), sequestering nutrients in biomass and soils that occurs in concert with greater carbon storage in elevated C_a will diminish the potential for further increases in productivity. Greater carbon uptake in elevated C_a will result in nutrient sequestration that reduces nutrient availability to plants, unless, as C_a continues to rise, N inputs increase, N losses decrease, or distribution of N to ecosystem pools with high C:N ratios increases(Field 1999), particularly, from soils to wood. Widening C:N ratios of plant tissues allow some increased productivity in response to rising C_a, but this is limited by plant stoichiometry. To date, field experiments have not lasted long enough to test whether elevated C_a causes redistribution of nitrogen between ecosystem pools of varying C:N ratios, but there have been a few tests of the effects of elevated C_a on nitrogen inputs and losses.

Plants with direct access to atmospheric nitrogen through nitrogen fixation generally show a relatively larger growth response to elevated C_a compared to nonfixing species, in both the laboratory (Poorter, 1993) and in the field (Sousanna and Hartwig, 1996). Nitrogen fixation is energetically expensive, but it is usually not directly regulated by the availability of photosynthate (Hartwig *et al.*, 1990, 1994; Hunt and Layzell, 1993; Weisbach *et al.*, 1996). Rather, increased growth of N fixers in elevated C_a might be viewed as a realisation of the CO_2 limitation of photosynthesis matched by a ready supply of fixed nitrogen. Nitrogen fixation is a very small part of the annual nitrogen cycle in most ecosystems, but it is nevertheless the major mechanism for the entry of nitrogen. Summed over many years, a sustained increase in nitrogen fixation in high C_a could increase ecosystem N mass enough to partially counteract reduced N availability through sequestration in biomass and soils (Gifford *et al.*, 1996). However, this will be less important in cases where low phosphorus availability restricts the growth of nitrogen-fixing plants (Vitousek and Howarth, 1991). For example, elevated C_a strongly increased growth and aboveground N mass of a legume in a scruboak ecosystem in which soil phosphorus availability was high, but in a calcareous grassland, elevated C_a increased the growth of legumes only with additional phosphorus supply, and had no effect on legume growth under field conditions where phosphorus availability was low (Niklaus *et al.*, 1998).

Changes in the processes that control nitrogen losses from ecosystems will amplify or counteract increased inputs of nitrogen through fixation. Studies to date are inconclusive. Elevated C_a can increase nitrogen losses by increasing carbon supply to rhizosphere denitrifiers (Smart *et al.*, 1997) and by increasing soil moisture and N/N_2O efflux (Arnone and Bohlen, 1998). By contrast, elevated C_a can decrease N losses by increasing N immobilization by microbes and thereby reducing NO efflux (Hungate *et al.*, 1997a). Small changes in N inputs, losses, or distribution will substantially modify the productivity responses to elevated C_a, underscoring the importance of futher experimental and modeling studies of these potential changes.

VI. Interactions between Water Regime and CO_2 Concentrations

C_a affects the two main canopy characteristics determining ecosystem water balance, L and g_s, but with opposite impacts on water balance: the increase in L enhances plant transpiration and ecosystem E, but the reduction in g_s decreases transpiration, depending of canopy roughness. The water balance constraint may allow effects on L to take place in water-limited ecosystems, provided its impact on the water balance is low. No savings in water can be expected in canopies where elevated C_a stimulates increase in L relatively more than it decreases g_s.

In smooth canopies, low atmospheric coupling explains the lack of sensitivity of E to surface conductance (Hileman *et al.*, 1994; Kimball *et al.*, 1995; Bunce *et al.*, 1997). However, an improvement in soil water status due to a decrease in E under doubled C_a has been observed in rice crops (Baker *et al.*, 1997), Mediterranean grasslands (Field *et al.*, 1995), and a C_4–tallgrass prairie, which suggests than reduction in g_s (e.g., Owensby *et al.*, 1997) can overcome the impact of increased L on E. However, our survey shows that L did not increase in any of the long-term field studies of the effects of elevated C_a on crops or native species (Table 7-1). This survey included studies of wheat (*Triticum aestivum*) and cotton in Arizona, where FACE was used to expose the plants to 55 Pa, as well as open-top chamber studies of native species. Elevated C_a (>68 Pa) reduced E compared with normal ambient in the Maryland wetland (Arp, 1991), Kansas prairie (Ham *et al.*, 1995), and California grassland ecosystems (Field *et al.*, 1995). In the wetland ecosystem, E was evaluated for a C_3-dominated and a C_4-dominated plant community. In these two communities, instantaneous values of E averaged 5.5–6.5 and 7.5–8.7 mmol H_20 m^{-2} s^{-1} for the C_3 and C_4 communities, respectively, at present ambient C_a, but at elevated C_a (68 Pa), evapotranspiration (ET) was reduced 17–22% in the C_3 and 28–29% in the C_4 community, indicating the relatively greater effect of elevated C_a on g_s in the C_4 species. In the prairie ecosystem, cumulative ET over a 34-day period in midsummer was 180 kg m^{-2} at present ambient C_a whereas it was 20% less at elevated C_a. In several grassland ecosystems, elevated C_a reduced E to the extent that soil water availability increased (Fredeen *et al.*, 1995; Rice *et al.*, 1994).

In aerodynamically rough canopies, the effect of L on E is potentially large, but the extent to which it could be compensated by effects on g_s will vary among species (Eamus and Jarvis, 1989; Saxe *et al.*, 1998). Also, in contrast to most grass and crop species, the sensitivity of g_s to D or soil drought seems unaffected in species such as sunflower (Bunce, 1993) or *Pinus pinaster* (Picon *et al.*, 1996). The E from forest canopies composed of species unresponsive to C_a would therefore be potentially affected only through changes in leaf area or when root growth enhancement gives access to new sources of soil water. In humid and fertile environments, where a change in L has little effect on ecosystem water balance, the associated impact on radiation interception will be low. In forest exposed to water limitations, the water regime will limit any impact of C_a on L, unless root growth enhancement will allow plants to access new sources of water. However, primary production may be increased through the enhancement of CO_2 availability, as far as plant growth can use the additional assimilated carbon in stem or root growth, depending on other limiting conditions. For species responsive to C_a, such as numerous *Quercus* species, reduction in g_s may allow L to increase with little effect on transpiration. Then C_a can have a larger potential impact on primary production and growth. But still, this improvement will only take place providing that the other possible climatic and trophic limitations of primary production can be overcome.

The reduction in E possibly caused by increased C_a would also alter canopy energy balance and shift some energy loss from transpiration to convective heat loss. This effect has important consequences for climate. Incorporating a model of stomatal response to elevated C_a into a coupled simple biosphere–atmosphere global circulation model (SiB2–GCM) showed that decreased g_s and latent heat transfer will cause a warming of the order of 1–2°C over the continents (Sellers *et al.*, 1996) in addition to warming from the CO_2 greenhouse effect. Implicit in this development is that any loss of photosynthetic capacity, through acclimation, would lead to further decreased g_s.

VII. Interactions between CO_2, Water, and Nitrogen

In arid and semiarid grasslands, increased water availability in elevated C_a can extend the length of the growing season (Fredeen *et al.*, 1995) and increase nitrogen availability (Hungate *et al.*, 1997b), both potentially amplifying the effect of elevated C_a on NPP (Hungate *et al.*, 1997b). However, increased soil moisture in elevated C_a will also stimulate decomposition and thus CO_2 release from soils (Rice *et al.*, 1994). Because appreciable carbon storage in grasslands is largely restricted to soils, the balance of the effects of increased soil water content in elevated C_a is to reduce the effects of elevated C_a on NEP. In forests, greater nutrient mineralization associated with wetter soils could support greater tree growth and thus carbon accumulation in wood, which has a higher C:N ratio than soils (e.g., Shaver *et al.*, 1992). However, as mentioned earlier, elevated C_a tends to cause smaller reductions in g_s in woody species compared to herbs, so this interaction between C_a, water, and nitrogen may be less important in forest ecosystems.

Acknowledgments

DL acknowledges gratefully J. Grace, P. G. Jarvis, and E. Dreyer for their comments and suggestions, and E. Lamaud, P. Berbigier, and H. Cochard, for providing the data used in this analysis.

References

Aber, J., McDowell, W., Nadelhoffer, K., Magill, A., Berntson, G., Kamakea, M., McNulty, S., Currie, W., Rustad, L., and Fernandez, I. (1998). Nitrogen saturation in temperate forest ecosystems. *BioScience* **48,** 921–934.

Agren, G. I. (1985). Theory for growth of plants derived from the nitrogen productivity concept. *Physiol. Plant.* **64,** 17–28.

Agren, G. I., and Ingestad, T. (1987). Root:shoot ratio as a balance between nitrogen productivity and photosynthesis. *Plant Cell Environ.* **10,** 579–586.

Alder, N. N., Sperry, J. S., and Pockman, W. T. (1996). Root and stem xylem embolism, stomatal conductance, and leaf turgor in *Acer grandidendatum* populations along a soil moisture gradient. *Oecologia* **105,** 293–201.

Amthor, J. (1997). Plant respiratory responses to elevated CO_2 partial pressure. "Advances in CO_2 effects research" (J. L. H. Allen, M. H. Kirkham, D. M. Olszyk, and C. E. Whitman, eds.). ASA, CSSA, and SSSA. ASA Special Publication. Madison, Wisconsin.

Andriani, J. M., Andrade, F. H., Suero, E. E., and Dardanelli, J. L. (1991). Water deficits during reproductive growth of soybeans. I. Their effects on dry matter accumulation, seed yield and its components. *Agronomie* **11,** 737–746.

Aranda, X., Gonzàlez-Meler, M. A. *et al.* (1995). Cytochrome oxidase activity and oxygen uptake in photosynthetic organs of *Triticum aestivion* and *Scirpus olneyi* plants grown at ambient and doubled CO_2 (abstract no. 262). *Plant Physiol.* **108,** S-62.

Arnone III, J. A., and Bohlen, P. J. (1998). Stimulated N_2O flux from intact grassland monoliths after two growing seasons under elevated atmospheric CO_2. *Oecologia* **116,** 331–335.

Arp, W. J. (1991). Vegetation of a North American salt marsh and elevated atmospheric carbon dioxide. Ph.D. Thesis. Vrije Universiteit, Amsterdam.

Azcón-Bieto, J., Gonzàlez-Meler, M. A., Dougherty, M. A., and Drake, B. G. (1994). Acclimation of respiratory O_2 uptake in green tissues of field grown native species after long-term exposure to elevated atmospheric CO_2. *Plant Physiol.* **106,** 1163–1168.

Baker, J. T., Allen, Jr., L. H., Boote, K. J., and Pickering, N. B. (1997). Rice responses to drought under CO_2 enrichment. 2. Photosynthesis and evapotranspiration. *Global Change Biol.* **3,** 129–138.

Baldocchi, D. D. (1994). A comparative study of mass and energy exchange rates over a closed C_3 (wheat) and an open C_4 (corn) crop. II: CO_2 exchange and water use efficiency. *Agric. For. Meteorol.* **67,** 291–321.

Baldocchi, D. D., and Vogel, C. A. (1996). Energy and CO_2 flux densities above and below a temperate broad-leaved forest and a boreal pine forest. *Tree Physiol.* **16,** 5–16.

Baldocchi, D. D., Verma, S. B., and Rosenberg, N. J. (1981a). Mass and energy exchanges of a soybean canopy under various environmnetal regimes. *Agron. J.* **73,** 706–710.

Baldocchi, D. D., Verma, S. B., Rosenberg, N. J., Blad, B. L., Garay, A., and Specht, J. E. (1983). Influence of water stress on the diurnal exchange of mass and energy between the atmosphere and a soybean canopy. *Agron. J.* **75,** 543–548.

Baldocchi, D. D., Verma, S. B., and Rosenberg, N. J. (1981b). Environmental effects on the CO_2 flux & CO_2-water flux ratio of alfalfa. *Agric. For. Meteorol.* **24,** 175–184.

Baldocchi, D. D., Verma, S. B., and Rosenberg, N. J. (1985). Water use efficiency in a soybean field: Influence of plant water stress. *Agric. For. Meteorol.* **34,** 53–65.

Baldocchi. D. D., Vogel, C. A., and Hall, B. (1997). Seasonal variation of carbon dioxide exchange rates above and below a boreal jack pine forest. *Agric. For. Meteorol.* **83,** 147–170.

Beadle, C. L., Talbot, H., Neilson, R. E., and Jarvis, P. G. (1985). Stomatal conductance and photosynthesis in a mature Scots pine *Pinus sylvestris* forest: Variation in canopy conductance and canopy photosynthesis. *J. Appl. Ecol.* **22,** 587–596

Becker, M. (1989). The role of climate on present and past vitality of silver fir forests in the Vosges mountains of northeastern France. *Can. J. For. Res.* **19,** 1110–1117.

Becker, M., Nieminen, T. M., and Geremia, F. (1994). Short-term variations and long-term changes in oak productivity in northeastern France. The role of climate and atmospheric CO_2. *Ann. Sci. For.* **51,** 477–492.

Beerling, D. J., Heath, J., Woodward, F. I., and Mansfield, T. A. (1996). Drought–CO_2 interactions in trees: Observations and mechanisms. *New Phytol.* **134,** 235–242.

Belaygue, C., Wery, J., Cowan, A. A., and Tardieu, F. (1996). Contribution of leaf expansion, rate of leaf appearance, and stolon branching in growth of plant leaf area under water deficit in white clover. *Crop Sci.* **36,** 1240–1246.

Bert, D., Leavitt, S. W., and Dupouey J.-L. (1997). Variations of wood ^{13}C and water-use efficiency of *Abies alba* (Mill.) during the last century. *Ecology* **78,** 1588–1596.

Biehler, K., and Fock, H. (1996). Evidence for the contribution of the Mehler–peroxidase reaction in dissipating excess electrons in drought-stressed wheat. *Plant Physiol.* **112,** 265–272.

Birk, E. M., and Vitousek, P. M. (1986). Nitrogen availability and nitrogen use efficiency in loblolly pine stands. *Ecology* **67,** 69–79.

Björkman, O. (1981). Responses to different quantum flux densities. *In* "Physiological Plant Ecology. I. Responses to the Physical Environment" (O. L. Land, P. S. Novel, C. B. Osmond, and H. Ziegler, eds.), pp. 57–107. Springer-Verlag, Berlin.

Black, T. A., Den Hartog, G., Neumann, H. H., Blanken, P. D., Yang, P. C., Russell, C., Nesic, Z., Lee, X., Chen, S. G., Staebler, R., and Novak, M. D. (1996). Annual cycles of water vapour and carbon dioxide fluxes in and above a boreal aspen forest. *Global Change Biol.* **2,** 219–229.

Boot, R. G. A., and Mensink, M. (1990). Size and morphology of root systems of perennial grasses from contrasting habitats as affected by nitrogen supply. *Plant Soil* **129,** 291–299.

Bosc, A. (1999). Etude expérimentale du fonctionnement hydrique et carboné des organes aériens du Pin maritime (*Pinus pinaster* Ait.): Intégration dans un modèle Structure–Fonction appliquè à l'analyse de l'autonomie carbonée des branches de la couronne d'un arbre adulte. Ph.D. Thesis, Université de Bordeaux II, France.

Breda, N., and Granier, A. (1996). Intra- and interannual variations of transpiration, leaf area index and radial growth of a sessile oak stand (*Quercus petraea*). *Ann. Sci. For.* **53,** 521–536.

Brix, H. (1981). Effects of thinning and nitrogen fertilisation on branch and foliage production in Douglas-fir. *Can. J. For. Sci.* **11,** 502–511.

Buchmann, N., Kao, W.-Y., and Ehleringer, J. R. (1996). Carbon dioxide concentrations within forest canopies—Variation with time, stand structure, and vegetation type. *Global Change Biol.* **2,** 421–432.

Bunce, J. A. (1993). Effects of doubled atmospheric carbon dioxide concentration on the responses of assimilation and conductance to humidity. *Plant Cell Environ.* **16,** 189–197.

Bunce, J. A., Wilson, K. B., and Carlson, T. N. (1997). The effect of doubling CO_2 on water use by alfalfa and orchard grass: Simulating evapotranspiration using canopy conductance measurements. *Global Change Biol.* **3,** 81–87.

Canadell, J. G., Pitelka, L. F., and Ingram, J. S. I. (1996). The effects of elevated [CO_2] on plant–soil carbon below-ground: A summary and synthesis. *Plant Soil* **187,** 391–400.

Cannell, M. G., Milne, R., Sheppard, L. J., and Unsworth, M. H. (1987). Radiation interception and productivity of willow. *J. Appl. Ecol.* **24,** 261–278.

Cavelier, J., Machado, J.-L., Valencia, D., Montoya, J., Laignelet, A., Hurtado, A., Varela, A., and Mejia, C. (1992). Leaf demography and growth rates of *Espeletia barclayana* Cuatrec. Compositae a caulescent rosette in a Colombia Paramo. *Biotropica* **24,** 52–63.

Chapin III, F. S., Bloom, A. J., Field, C. B., and Waring, R. H. (1987). Plant responses to multiple environmental factors. *BioScience* **37,** 49–57.

Chaves, M. M. (1991). Effects of water deficit on carbon assimilation. *J. Exp. Bot.* **42,** 1–16.

Chaves, M. M., and Pereira, J. S. (1992). Water stress, CO_2 and climate change. *J. Exp. Bot.* **43,** 1131–1139.

Choudhury, J., and Monteith, J. L. (1986). Implications of stomatal response to saturation deficit for the heat balance of vegetation. *Agric. For. Meteorol.* **36,** 215–225.

Clifford, S. C., Stronach, I. M., Mohamed, A. D., Azam-Ali, S. N., and Crout, N. M. J. (1993). The effects of elevated atmospheric carbon dioxide and water stress on light interception, dry matter production and yield in stands of groundnut (*Arachis hypogaea* L.). *J. Exp. Bot.* **44,** 1763–1770.

Cochard, H., Bréda, N., and Granier, A. (1996). Whole tree hydraulic conductance and water loss regulation in *Quercus* during drought: Evidence for stomatal control of embolism? *Ann. Sci. For.* **53,** 197–206.

Comstock, J., and Ehleringer, J. (1993). Stomatal response to humidity in common bean (*Phaseolus vulgaris*): Implications for maximum transpiration rate, water-use efficiency and productivity. *Aust. J. Plant Physiol.* **20,** 669–691.

Conroy, J. P. (1992). Influence of elevated atmospheric CO_2 concentrations on plant nutrition. *Aust. J. Bot.* **40,** 445–456.

Conroy, J. P., Küppers, M., Küppers, B., Virgona, J., and Barlow, E. W. R. (1988). The influence of CO_2 enrichment, phosphorus deficiency, and water stress on the growth, conductance and water use of *Pinus radiata* D. Don. *Plant Cell Environ.* **11,** 91–98.

Conroy, J. P., Milham, P. J., Bevege, D. I., and Barlow, E. W. R. (1990). Influence of phosphorus deficiency on the growth response of four families of *Pinus radiata* seedlings to CO_2-enriched atmosphere. *For. Ecol. Manage.* **30,** 175–188.

Cook, A. C., Tissue, D. T., Roberts, S. W., and Oechel, W. C. (1998). Effects of long-term elevated [CO_2] from natural CO_2 springs on *Nardus stricta:* Photosynthesis, biochemistry, growth and phenology. *Plant Cell Environ.* **21,** 417–425.

Cornic, G., Le Gouallec, J. L., Briantais, J. M., and Hodges, M. (1989). Effect of dehydration and high light on photosynthesis of two C_3 plants (*Phaseolus vulgaris* L. and *Elastotema repens* (Lour.) Hall f.) *Planta* **177,** 84–90.

Correia, M. J., Chaves, M. M., and Pereira, J. S. (1990). Afternoon depression in photosynthesis in grapevine leaves—Evidence for a high light stress effect. *J. Exp. Bot.* **41,** 417–426.

Cowan, F. (1982). Regulation of water use in relation to carbon gain in higher plants. *In* "Physiological Plant Ecology II—Water Relations and Carbon Assimilation" (O. L. Lange, P. S. Nobel, C. B. Osmond, and H. Ziegler, eds.), pp. 589–612. Springer-Verlag, Berlin–Heidelberg–New York.

Cox, W. J., and Jolliff, G. D. (1986). Growth and yield of sunflower and soybean under soil water deficit. *Agron. J.* **78,** 226–230.

Cure, J. D., Rufty, Jr., T. W., and Israel, D. W. (1988). Phosphorus stress effects on growth and seed yield responses of nonnodulated soybean to elevated carbon dioxide. *Agron. J.* **80,** 897–902.

Curtis, P. S. (1996). A meta analysis of leaf gas exchange and nitrogen in trees grown under elevated CO_2 *in situ. Plant Cell Environ.* **19,** 127–137.

Curtis, P. S., and Teeri, J. A. (1992). Seasonal responses of leaf gas exchange to elevated carbon dioxide in *Populus grandidentata. Can. J. For. Res.* **22,** 1320–1325.

Curtis, P. S., and Wang, X. (1998). A meta-analysis of elevated CO_2 effects on woody plant mass, form, and physiology. *Oecologia* **113,** 299–313.

Curtis, P. S., Zak, D. R., Pregitzer, K. S., and Teeri, J. A. (1994). Above- and belowground responses of *Populus gradidentata* to elevated atmospheric CO_2 and soil N availability. *Plant Soil* **165,** 45–51.

Curtis, P. S., Vogel, C. S., Pregitzer, K. S., Zak, D. R., and Teeri, J. A. (1995). Interacting effects of soil fertility and atmospheric CO_2 on leaf area growth and carbon gain physiology in *Populus* × *euroamericana* (Dode) Guinier. *New Phytol.* **129,** 253–263.

Dai, Z., Edwards, G. E., and Ku, M. S. B. (1992). Control of photosynthesis and stomatal conductance in *Ricinus communis* L. (castor bean) by leaf to air vapour pressure deficit. *Plant Physiol.* **99,** 1426–1434.

Date, R. A. (1973). Nitrogen: A major limitation in the productivity of natural communities, crops and pastures in the Pacific area. *Soil Biol. Biochem.* **5,** 5–18.

Davidson, R. L. (1969). Effect of root:leaf temperature differentials on root:shoot ratios in some pasture grasses and clover. *Ann. Bot.* **33,** 561–569.

Diemer, M. W. (1994). Mid-season gas exchange of an alpine grassland under elevated CO_2. *Oecologia* **98,** 429–435.

Diemer, M. (1997). Effects of elevated CO_2 on gas exchange characteristics of alpine grassland. *Acta Oecologica* **18,** 177–182.

Dingkuhn, M., Cruz, R. T., O'Toole, J. C., and Doerffling, K. (1989). Net photosynthesis, water use efficiency, leaf water potential and leaf rolling as affected by water deficit in tropical upland rice. *Aust. J. Agric. Res.* **40,** 1171–1182.

Drake, B. G. (1992). A field study of the effects of elevated CO_2 on ecosystem processes in a Chesapeake Bay wetland. *Aust. J. Bot.* **40,** 579–595.

Drake, B. G., and Raschke, K. (1974). Prechilling of *Xanthium strumarium* L. reduces net photosynthesis and, independently, stomatal conductance, while sensitizing the stomata to CO_2. *Plant Physiol.* **53,** 808–812.

Drake, B. G., and Leadley, P. W. (1991). Canopy photosynthesis of crops and native plant communities exposed to long term elevated CO_2. *Plant Cell Environ.* **14,** 853–860.

Drake, B. G., Muehe, M. S., Peresta, G., Gonzalez-Meler, M. A., and Matamala, R. (1996). Acclimation of photosynthesis, respiration and ecosystem carbon flux of a wetland on Chesapeake Bay, Maryland to elevated atmospheric CO_2 concentration. *Plant Soil* **187,** 111–118.

Drake, B. G., Long, S. P. *et al.* (1997). More efficient plants: A consequence of rising atmospheric CO_2? *Annu. Rev. Plant Physiol. Plant Mol. Biol.* **48,** 607–637.

Dreyer, E., Epron, D., and Eyog-Matig, O. (1992). Compared sensitivity of photochemical efficiency of PSII to rapid leaf dehydration among 11 temperate and tropical forest tree species differing in their tolerance to drought. *Ann. Sci. For.* **49,** 615–625.

Du, Y. C., Kawamitsu, Y., Nose, A., Hiyane, S., Murayama, S., Wasano, K., and Uchida, Y. (1996). Effects of water stress on carbon exchanges rate and activities of photosynthetic enzymes in leaves of sugarcane *(Saccharum* sp.). *Aust. J. Plant Physiol.* **23,** 719–726.

Dupouey, J.-L., Leavitt, S., Choisnel, E., and Jourdain, S. (1993). Modelling carbon isotope fractionation in tree rings based on effective evapotranspiration and soil water status. *Plant Cell Environ.* **16,** 939–947.

Duquesnay, A., Bréda, N., Stievenard, M., and Dupouey, J.-L. (1998). Changes in tree-ring ^{13}C and water-use efficiency of beech (*Fagus sylvatica* L.) in north-eastern France during the past century. *Plant Cell Environ.* **21,** 565–572.

Durand, J.-L., Lemaire, G., Gosse, C., and Chartier, M. (1989). Analyse de la conversion de l`énergie solaire en matière sèche par un peuplement de luzerne (*Medicago sativa* L.) soumis à un déficit hydrique. *Agronomie* **9,** 599–607.

Durand, J.-L., Onillon, B., Schnyder, H., and Rademacher, I. (1995). Drought effects on cellular and spatial parameters of leaf growth in tall fescue. *J. Exp. Bot.* **46,** 1147–1155.

Eamus, D., and Jarvis, P. G. (1989). The direct effects of increase in the global atmospheric CO_2 concentration on natural and commercial temperate trees and forests. *Adv. Ecol. Res.* **19,** 1–55.

Eckardt, F. E., Heim, G., Methy, M., and Sauvezon, R. (1975). Interception de l'énergie rayonnante, échanges gazeux et croissance dans une forêt méditérranéenne à feuillage persistant (*Quercetum ilicis*) *Photosynthetica* **9,** 145–156.

Epron, D., and Dreyer, E. (1992). Effects of severe leaf dehydration on leaf photosynthesis in *Quercus petraea* (Matt.) Liebl.: Photosystem II eficiency, photoshemical and nonphotochemical fluoresence quenchings and electrolyte leakage. *Tree Physiol.* **10,** 273–284.

Evans, J. R. (1983). Nitrogen and photosynthesis in the flag leaf of wheat as affected by nitrogen supply. *Plant Physiol.* **72,** 297–302.

Evans, J. R. (1989). Photosynthesis and nitrogen relationships in leaves of C_3 plants. *Oecologia (Berlin)* **78,** 9–19.

Fan, S.-M., Goulden, M. L., Munger, J. W., Daube, B. C., Bakwin, P. S., Wofsy, F. C., Amthor, J. S., Fitzjarrald, D. R., Moore, K. E., and Moore, T. R. (1995). Environmental controls on the photosynthesis and respiration of a boreal lichen woodland: A growing season of whole ecosystem exchange measurements by eddy correlation. *Oecologia (Berlin)* **102,** 443–452.

Farquhar, G. D., Von Caemmerer, S., and Berry, J. A. (1980). A biochemical model of photosynthetic CO_2 assimilation in leaves of C_3 species. *Planta* **149,** 78–90.

Farquhar, G. D., Ehleringer, J. R., and Hubick, K. T. (1989). Carbon isotope discrimination and photosynthesis. *Annu. Rev. Plant Physiol. Plant Mol. Biol.* **40,** 503–537.

Federer, C. A., Tritton, L. M., Hornbeck, J. W., and Smith, R. B. (1989). Physiologically based dendroclimate models for effects of weather on red spruce basal-area growth. *Agric. For. Meteorol.* **46,** 159–172.

Field, C. B. (1991). Ecological scaling of carbon gain to stress and resource availability. *In* "Responses of Plants to Multiple Stresses" (H. A. Mooney, W. E. Winner, and E. J. Pell, eds.), pp. 35–65. Academic Press, San Diego.

Field, C. B., Jackson, R. B., and Mooney, H. A. (1995). Stomatal responses to increased CO_2: Implications from the plant to the global scale. *Plant Cell Environ.* **18,** 1214–1225.

Field, C. B., Lund, C. P., Chiariello, N. R., and Mortimer, B. E. (1997). CO_2 effects on the water budget of grassland microcosm communities. *Global Change Biol.* **3,** 197–206.

Fitter, A. H. (1985). Functional significance of root morphology and root system architecture. *In* "Ecological Interactions in Soil" (A. H. Fitter, ed.), pp. 87–106. Blackwell Scientific Publications, Oxford.

Fitter, A. H., and Hay, R. K. M. (1981). "Environmental Physiology of Plants." Academic Press, London.

Fog, K. (1988). The effect of added nitrogen on the rate of decomposition of organic matter. *Biol. Rev.* **63,** 433–462.

Fredeen, A. L., Koch, G. W., and Field, C. B. (1995). Effects of atmospheric CO_2 enrichment on ecosystem CO_2 exchange in a nutrient and water limited grassland. *J. Biogeog.* **22,** 215– 219.

Garnier, E. (1991). Resource capture, biomass allocation and growth in herbaceous plants. *Trends Ecol. Evol.* **6,** 126–131.

Gholz, H. L. (1982). Environmental limits on aboveground net primary production, leaf area and biomass in vegetation zones of the pacific northwest. *Ecology* **53,** 469–481.

Gholz, H. L., Ewel, K. C., and Teskey, R. O. (1990). Water and forest productivity. *For. Ecol. Manage.* **30,** 1–18.

Gifford, R. M., Lutze, J. L., and Barrett, D. (1996). Global atmospheric change effects on terrestrial carbon sequestration: Exploration with a global C- and N-cycle model (CQUESTN). *Plant Soil* **187,** 369–387.

Gonzàlez-Meler, M. A., Ribas-Carbó, M., Siedow, J. N., and Drake, B. G. (1996). The direct inhibition of plant mitochondrial respiration by elevated CO_2. *Plant Physiol.* **112,** 1349–1355.

Gorissen, A. (1996). Elevated CO_2 evokes quantitative and qualitative changes in carbon dynamics in a plant/soil system: Mechanisms and implications. *Plant Soil* **187,** 289–298.

Gosse, G., Varlet-Grancher, C., Bonhomme, R., Chartier, M., Allirand, J. M., and Lemaire, G. (1986). Production maximale de matière sèche et rayonnement solaire intercepté par un couvert végétal. *Agronomie* **6,** 47–56.

Goudriaan, J., and Ruiter, H. E. de (1983). Plant growth in response to CO_2 enrichment, at two levels of nitrogen and phosphorus supply. 1. Dry matter, leaf area and development. *Neth. J. Agric. Sci.* **131,** 157–169.

Goulden, M. L., Munger, J. W., Fan, S.-M., Daube, B. C., and Wofsy, S. C. (1996). Exchange of carbon dioxide by a deciduous forest: response to internannual climate variability. *Science* **271,** 1576–1578.

Grace, J., Lloyd, J., McIntyre, J., Miranda, A., Meir, P., Miranda, H., Moncrieff, J., Massheder, J., Wright, I., and Gash, J. (1995). Fluxes of carbon dioxide and water vapour over an undisturbed tropical forest in south-west Amazonia. *Global Change Biol.* **1,** 1–12.

Granier, A., and Loustau, D. (1994). Measuring and modelling transpiration of maritime pine stands from sap flow data. *Agric. For. Meteorol.* **71,** 61–81.

Grant, R. F., Garcia, R. L., Pinter, P. J., Hunsaker, D., Wall, G. W., Kimball, B. A., and LaMorte,

R. L. (1995). Interaction between atmospheric CO_2 concentration and water deficit on gas exchange and crop growth: Testing of ecosys with data from the Free Air CO_2 Enrichment (FACE) experiment. *Global Change Biol.* **1,** 433–454.

Greco, S., and Baldocchi, D. D. (1996). Seasonal variations of CO_2 and water vapour exchange rates over a temperate deciduous forest. *Global Change Biol.* **2,** 183–197.

Grier, C. C., and Running, S. W. (1977). Leaf area of mature northwestern coniferous forests: Relation to site water balance. *Ecology* **58,** 893–899.

Grier, C. C., Elliott, K. J., and McCullough, D. G. (1992). Biomass distribution and productivity of *Pinus edulis* and *Juniperus monosperma* woodlands of north central Arizona. *Forest Ecol. Manage.* **40,** 331–350.

Guehl, J.-M., Picon, C., Aussenac, G., and Gross, P. (1994). Interactive effects of elevated CO_2 and soil drought on growth and transpiration efficiency and its determinants in two European forest tree species. *Tree Physiol.* **14,** 707–724.

Gunderson, C. A., and Wullschleger, S. D. (1994). Photosynthetic acclimation in trees to rising atmospheric CO_2: A broader perspective. *Photosynth. Res.* **39,** 369–388.

Ham, J. M., Owensby, C. E. *et al.* (1995). Fluxes of CO_2 and water vapor from a prairie ecosystem exposed to ambient and elevated CO_2. *Agric. For. Meteorol.* **77,** 73–93.

Harrington, R. A., Fownes, J. H., Meinzer, F. C., and Scowcroft, P. G. (1995). Forest growth along a rainfall gradient in Hawaii: *Acacia koa* stand structure, productivity, foliar nutrients, and water- and nutrient-use efficiencies. *Oecologia (Berlin)* **102,** 277–284.

Hartwig, U., Boller, B. C., Baur-Hoch, B., and Nosberger, J. (1990). The influence of carbohydrate reserves on the response of nodulated white clover to defoliation. *Ann. Bot.* **65,** 97–105.

Hartwig, U. A., Heim, I., Luscher, A., and Nosberger, J. (1994). The nitrogen-sink is involved in the regulation of nitrogenase activity in white clover after defoliation. *Physiol. Plant.* **92,** 375–382.

Heber, I., Bligny, R., Streb, P., and Douce, R. (1996). Photorespiration is essential for the protection of the photosynthetic apparatus of C_3 plants against photoinactivation under sunlight. *Bot. Acta* **109,** 307–315.

Hileman, D. R., Huluka, G., Kenjige, P. K., Sinha, N., Bhattacharya, N. C., Biswas, P. K., Lewin, K. F., Nagy, J., and Hendrey, G. R. (1994). Canopy photosynthesis and transpiration of field grown cotton exposed to free air CO_2 enrichment (FACE) and differential irrigation. *Agric. For. Meteorol.* **70,** 189–207.

Hirose, T. (1987). A vegetative plant growth model: Adaptive significance of phenotypic plasticity in matter partitioning. *Funct. Ecol.* **1,** 195–202.

Hirose, T. (1988). Modelling the relative growth rate as a function of plant nitrogen concentration. *Physiol. Plant.* **72,** 185–189.

Hocking, P. J., and Meyer, C. P. (1991). Effects of CO_2 enrichment and nitrogen stress on growth, and partitioning of dry matter and nitrogen in wheat and maize. *Aust. J. Plant Physiol.* **18,** 339–356.

Hollinger, D. Y., Kelliher, F. M., Byers, J. N., Hunt, J. E., McSeveny, T. M., and Weir, P. L. (1994). Carbon dioxide exchange between an undisturbed old-growth temperate forest and the atmosphere. *Ecology* **75,** 134–150.

Huck, M. G., Peterson, C. M., Hoogenboom, G., and Busch, C. D. (1986). Distribution of dry matter between shoots and roots of irrigated and nonirrigated determinate soybeans. *Agron. J.* **78,** 807–813.

Hungate, B. A., Jackson, R. B., Field, C. B., and Chapin III, F. S. (1996). Detecting changes in soil carbon in CO_2 enrichment experiments. *Plant Soil* **187,** 135–145.

Hungate, B. A., Chapin III, F. S., Zhong, H., Holland, E. A., and Field, C. B. (1997a). Stimulation of grassland nitrogen cycling under carbon dioxide enrichment. *Oecologia* **109,** 149–153.

Hungate, B. A., Holland, E. A., Jackson, R. B., Chapin III, F. S., Mooney, H. A., and Field, C. B.

(1997b). The fate of carbon in grasslands under carbon dioxide enrichment. *Nature (London)* **388,** 6642, 576–579.

Hunt, S., and Layzell, D. B. (1993). Gas exchange of legume nodules and the regulation of nitrogenase activity. *Annu. Rev. Plant Physiol. Plant Mol. Biol.* **44,** 483–511.

Hunt, E. R., Weber, J. A., and Gates, D. M. (1985). Effects of nitrate application on *Amaranthus powellii* Wats. I. Changes in photosynthesis, growth rates, and leaf area. *Plant Physiol.* **79,** 609–613.

Idso, K. E., and Idso, S. B. (1995). Plant responses to atmospheric CO_2 enrichment in the face of environmental constraints:a review of the past 10 years' research. *Agric. For. Meteorol.* **69,** 153–203.

Ineson, P., Cortrufo, M. F., Bol, R., Harkness, D. D., and Hartwig, U. (1996). Quantification of soil carbon inputs under elevated CO_2 : C_3 plants in a C_4 soil. *Plant Soil* **187,** 345–350.

Ingestad, T. (1979). Nitrogen stress in birch seedlings. II. N, K, P, Ca, and Mg nutrition. *Physiol. Plant.* **45,** 149–157.

Israel, D. W., Rufty, T. W., Jr., and Cure, J. D. (1990). Nitrogen and phosphorus nutritional interactions in a CO_2 enriched environment. *J. Plant Nutr.* **13,** 1419–1433.

Jackson, R. B., Sala, O. E., Field, C. B., and Mooney, H. A. (1994). CO_2 alters water use, carbon gain and yield for the dominant species in a natural grassland. *Oecologia (Berlin)* **98,** 257–262.

Jacobs, C. M. J., and De Bruin, H. A. R. (1992). The sensitivity of regional transpiration to land surface characteristics: Significance of feedback. *J. Climate* **5,** 683–698.

Jarvis, P. G. (1985). Transpiration and assimilation of trees and agricultural crops: The "omega" factor. *In* "Attributes of Trees as Crop Plants" (M. G. R. Cannell and J. E. Jackson, eds.), pp. 441–460. Institute of Terrestrial Ecology.

Jarvis, P. G., and MacNaughton, K. G. (1986). Stomatal control of transpiration: Scaling up from leaf to region. *Adv. Ecol. Res.* **15,** 1–49.

Jefferies, R. L. (1977). Growth responses of coastal halophytes to inorganic nitrogen. *J. Ecol.* **65,** 847–865.

Jefferies, R. L., and Perkins, N. (1977). The effects on the vegetation of the additions of inorganic nutrients to salt marsh soils at Stiffkey, Norfolk. *J. Ecol.* **65,** 867–882.

Johnson, A. H., Schwartman, T. N., Battles, J. J., Miller, R., Miller, E. K., Friedland, A. J., and Vann, D. R. (1994). Acid rain and soils of the Adirondacks. II. Evaluation of calcium and aluminum as causes of red spruce decline at Whiteface Mountain, New York. *Can. J. For. Res.* **24,** 4, 654–662.

Johnson, D. W., Thomas, R. B., Griffin, K. L., Tissue, D. T., Ball, J. T., Strain, B. R., and Walker, R. F. (1998). Effects of carbon dioxide and nitrogen on growth and nitrogen uptake in ponderosa and loblolly pine. *J. Environ. Qual.* **27,** 414–425.

Jones, H. G. (1973). Moderate-term water stress and some associated changes in phototosynthetic parameters in cotton. *New Phytol.* **72,** 1095–1104.

Jones, H. G., and Sutherland, R. A. (1991). Stomatal control of xylem cavitation. *Plant Cell Environ.* **14,** 607–612.

Jones, J. W., Zur, B., and Bennett, J. M. (1986). Interactive effects of water and nitrogen stresses on carbon and water vapor exchange of corn canopies. *Agric. For. Meteorol.* **38,** 113–126.

Kachi, N., and Rorison, I. H. (1989). Optimal partitioning between root and shoot in plants with contrasted growth rates in response to nitrogen availability and temperature. *Funct. Ecol.* **3,** 549–559.

Kaiser, W. M. (1987). Effects of water deficit on photosynthetic capacity. *Physiol. Plant.* **71,** 142–149.

Kauppi, P. E., Mielikaïnen, K., and Kuusela, K. (1992). Biomass and carbon budget of European forests, 1971 to 1990. *Science* **256** 70–74.

Kelliher, F. M., Leuning, R., and Schulze, E. D. (1993). Evaporation and canopy characteristics of coniferous forests and grasslands. *Oecologia (Berlin)* **95,** 153–163.

Kelliher, F. M., Leuning, R., Raupach, M. R., and Schulze, E. D. (1995). Maximum conductances for evaporation from global vegetation types. *Agric. For. Meteorol.* **73,** 1–16.

Kiehl, K., Esselink, P., and Bakker, J. P. (1997). Nutrient limitation and plant species composition in temperate salt marshes. *Oecologia* **111,** 325–330.

Kim, J., and Verma, S. B. (1990). Carbon dioxide exchange in a temperate grassland ecosystem. *Bound. Layer Meteorol.* **52,** 135–142.

Kimball, B. A., Pinter, P. J., Garcia, R. L., Lamorte, R. L., Wall, G. W., Hunsaker, D. J., Wechsung, G., Wechsung, F., and Kartschall, T. (1995). Productivity and water use of wheat under free-air CO_2 enrichment. *Global Change Biol.* **1,** 429–442.

King, A. W., Post, W. M., and Wullschleger, S. D. (1997). The potential response of terrestrial carbon storage to changes in climate and atmospheric CO_2. *Climat. Change* **35,** 199–227.

Knapp, A. K., Fahnestock, J. T., Hamburg, S. P., Statland, L. B., Seastedt, T. R., and Schimel, D. S. (1993). Landscape patterns in soil–plant water relations and primary production in tall grass prairie. *Ecology* **74,** 549–560.

Koch, G. W., Amthor, J. S., and Goulden, M. L. (1994). Diurnal patterns of leaf photosynthesis, conductance and water potential at the top of a lowland rain forest canopy in Cameroon: Measurements from the Radeau des Cimes. *Tree Physiol.* **14,** 347–360.

Körner, C., Diemer, M., Schappi, B., Niklaus, P., and Arnone, J. (1997). III. The responses of alpine grassland to four seasons of CO_2 enrichment: A synthesis. *Acta Oecologica* **18,** 165–175.

Kuikman, P. J., Lekkerkerk, L. J. A., and Veen, J. A. van (1991). Carbon dynamics of a soil planted with wheat under an elevated atmospheric CO_2 concentration. *In* "Advances in Soil Organic Matter Research: Proceedings of a Symposium, Colchester, UK, 3–4 September, 1990," pp. 267–274. Royal Society of Chemistry, Cambridge, UK.

Lamaud, E., Brunet, Y., and Berbigier, P. (1997). Radiation and water use efficiencies of two coniferous forest canopies. Proceedings of the EGS XXI General Assembly, The Hague, 6–10 May, 1996. EGS, in press.

Landsberg, J. J., and Hingston, F. J. (1996). Evaluating a simple radiation/dry matter conversion model using data from *Eucalyptus globulus* plantations in Western Australia. *Tree Physiol.* **16,** 801–808.

Lawler, I. R., Foley, W. J., Woodrow, I. E., and Cork, S. J. (1997). The effects of elevated CO_2 atmospheres on the nutritional quality of Eucalyptus foliage and its interaction with soil nutrient and light availability. *Oeocologia* **109,** 59–68.

Lecoeur, J., Wery, J., Turc, O., and Tardieu, F. (1995). Expansion of pea leaves subjected to short water deficit: Cell number and cell size are sensitive to stress at different periods of leaf development. *J. Exp. Bot.* **46,** 1093–1101.

Lewis, J. D., Tissue, D. T. *et al.* (1996). Seasonal response of photosynthesis to elevated CO_2 in loblolly pine (*Pinus taeda* L) over 2 growing seasons. *Global Change Biol.* **2,** 103–114.

Linder, S. (1987). Responses to water and nutrients in coniferous ecosystems. *In* "Ecological Studies" (E. D. Schülze and H. Zwölfer, eds.), Vol. 61, pp.180–201. Springer-Verlag, Berlin.

Linder, S., and Rook, D. A. (1984). Effects of mineral nutrition on carbon dioxide exchange and partitioning of carbon in trees. *In* "Nutrition of Plantation Forests" (G. D. Bowen and E. K. S. Nambiar, eds.), pp. 211–236. Academic Press, Orlando.

Linder, S., Benson, M. L., Myers, B. J., and Raison, R. J. (1987). Canopy dynamics and growth of *Pinus radiata*. I. Effects of irrigation and fertilisation during a drought. *Can. J. For. Sci.* **17,** 1157–1165.

Livingston, N. J., and Spittlehouse, D. L. (1996). Carbon isotope fractionation in tree ring early and late wood in realtion to intra-growing season water balance. *Plant Cell Environ.* **19,** 768–774.

Lloyd, J., and Farquhar, G. D. (1996). The CO_2 dependence of photosynthesis, plant-growth responses to elevated atmospheric CO_2 concentrations and their interaction with soil nutrient status .1. General-principles and forest ecosystems. *Funct. Ecol.* **10,** 4–32.

Long, S. P., and Drake, B. G. (1991). Effect of the long-term elevation of CO_2 concentration in the field on the quantum yield of photosynthesis of the C_3 sedge, *Scirpus olneyi. Plant Physiol.* **96,** 221–226.

Long, S. P., Garcia Moya, E., Imbamba, S. K., Kamnalrut, A., Piedade, M. T. F., Scurlock, J. M. O., Shen, Y. K., and Hall, D. O. (1989). Primary productivity of natural ecosystems of the tropics: A reappraisal. *Plant Soil* **115,** 155–166.

Long, S. P., Osborne, C. P., and Humphries, S. W. (1996). Photosynthesis, rising atmospheric carbon dioxide concentration and climate change. *In* "Global Change: Effects on Coniferous Forests and Grasslands" (A. I. O. Breymeyer and J. M. Mellilo, eds.), pp. 121–158. John Wiley, New York.

Loustau, D., Berbigier, P., Roumagnac, P., Arrruda-Pacheco, C., David, J. S., Ferreira, M. I., Pereira, J. S., and Tavares, R. (1996). Transpiration of a 64-year-old maritime pine stand in portugal: 1. Seasonal course of water flux through maritime pine. *Oecologia* **107,** 33–42.

Loustau, D., Berbigier, P., and Kramer, K. (1997a). Climatic sensitivity of the water balance of a mature maritime pine forest. *In* "Impacts of Global Change on Tree Physiology and Forest Ecosystems" Proceedings of the international Conference, Wageningen, 26–29 November (G. M. J. Mohren, K. Kramer, and S. Sabaté, eds.), pp. 193–207. Kluwer Academic Publ., Norwell, Massachusetts.

Lutze, J. L. (1996). Carbon and nitrogen relationships in swards of *Danthonia richardsonii* in response to carbon dioxide enrichment and nitrogen supply. Ph.D. Thesis. Australian National University, Canberra.

Machado, J.-L., and Tyree, M. T. (1995). Patterns of hydraulic architecture and water relations of two tropical canopy trees with contrasting leaf phenologies: *Ochroma pyramidale* and *Pseudobombax septenatum. Tree Physiol.* **14,** 219–240.

Makela, A. (1986). Implications of the pipe-model theory on dry matter partitioning and height growth in trees. *J. Theor. Biol.* **123,** 103–120.

Margolis, H., Oren, R., Whitehead, D., and Kaufmann, M. (1995). Leaf area dynamics of conifer forests. *In* "Ecophysiology of Coniferous Forests" (W. K. Smith and T. M. Hinckley, eds.), pp. 123–170. Academic Press, San Diego.

Martin, B., and Ruiz-Torres, N. A. (1992). Effects of water-deficit stress on photosyntheis, its components and component limitations, and on water use efficicency in wheat (Triticum Aestivum L.) *Plant Physiol.* **100,** 733–739.

Masle, J., Hudson, G. S., and Badger, M. R. (1993). Effects of ambient CO_2 concentration on growth and nitrogen use in tobacco (*Nicotiana tabacum*) plants transformed with an antisense gene to the small subunit of rubulose-1,5-bisphosphate carboxylase/oxygenase. *Plant Physiol.* **103,** 1075–1088.

McGuire, A. D., Melillo, J. M., and Joyce, L. A. (1995). The role of nitrogen in the response of forest net primary production to elevated atmospheric carbon dioxide. *Annu. Rev. Ecol. Syst.* **26,** 473–503.

McMurtrie, R. E., Gholz, H. L., Linder, S., and Gower, S. T. (1994). Climatic factors controlling the productivity of pine stands: A model-based analysis. *Ecol. Bull. (Copenhagen)* **43,** 173–188.

Mencuccini, M., and Grace, J. (1996). Hydraulic conductance, light interception and needle nutrient concentration in Scots pine stands and their relations with net primary productivity. *Tree Physiol.* **16,** 459–468.

Metcalfe, J. C., Davies, W. J., and Pereira, J. S. (1990). Leaf growth of *Eucalyptus globulus* seedlings under water deficit. *Tree Physiol.* **6,** 221–228.

Monteith, J. L. (1977). Climate and the efficiency of crop production in Britain. *Philos. Trans. R. Soc. Lond. ser. B* **281,** 277–294.

Monteith, J. L. (1995). A reinterpretation of stomatal response to humidity. *Plant Cell Environ.* **16,** 25–34.

Monteny, B. A. (1989). Primary productivity of a Hevea Forest in the Ivory coast. *Ann. Sci. For.* **46** (suppl.), 502–505.

Mooney, H. A., Kuppers, M., Koch, G., Gorham, J., Chu, C. C., and Winner, W. E. (1988). Compensating effects to growth of carbon partitioning changes in response to SO_2-induced photosynthetic reduction in radish. *Oecologia (Berlin)* **75,** 502–506.

Morison, J. I. L. (1987). Intercellular CO_2 concentration and stomatal response to CO_2. *In* "Stomatal Function" (E. Zeiger, G. D. Farquhar, and I. R. Cowan, eds.), pp. 229–251. Stanford Univ. Press, Stanford, California.

Nelson, C. J. (1988). Genetic associations between photosynthetic characteristics and yield: Review of the evidence. *Plant Physiol. Biochem. (France)* **26,** 543–554.

Newton, P. C. D., Clark, H., Bell, C. C., Glasgow, E. M., Tate, K. R., Ross, D. J., and Yeates, G. W. (1995) Plant growth and soil processes in temperate grassland communities at elevated CO_2. *J. Biogeog.* **22,** 235–240.

Nguygen-Queyrens, A., Fehri, A., Loustau, D., and Guehl, J. M. (1998). Within and between annual ring variations in δ13C values of two contrasting provenances of *Pinus pinaster. Can. J. For. Sci.* **28,** 768–773.

Nicholson, S. E., Davenport, M. L., and Malo, A. R. (1990). A comparison of the vegetation response to rainfall in the Sahel and East Africa using normalized difference vegetation index from NOAA AVHRR. *Climate Change* **17,** 209–242.

Niklaus, P. A., Leadley, P. W., Stöcklin, J., and Körner, C. (1998). Nutrient relations in calcareous grassland under elevated CO_2. *Oecologia* **116,** 67–75.

Nizinski, J. J., and Saugier, B. (1989). A model of transpiration and soil-water balance for a mature oak forest. *Agric. For. Meteorol.* **47,** 1–17.

Norby, R. J. (1994). Issues and perspectives for investigating root responses to elevated atmospheric carbon dioxide. *Plant Soil* **165,** 9–20.

Norby, R. J., Pastor, J. *et al.* (1986). Carbon–nitrogen interactions in CO_2-enriched white oak: Physiological and long-term perspectives. *Tree Physiol.* **2,** 233–241.

Oberbauer, S. F., Sionit, N., Hastings, S. J., and Oechel, W. C. (1986). Effects of CO_2 enrichment and nutrition on growth, photosynthesis, and nutrient concentration of Alaskan tundra plant species. *Can. J. Bot.* **64,** 2993–2998.

Oechel, W. C., Hastings, S. J., Vourlitis, G., Jenkins, M., Riechers, G., and Grulke, N. (1993). Recent change of Artic Tundra ecosystems from a net carbon dioxide sink into a source. *Nature (London)* **361,** 520–523.

Oechel, W. C., Cowles, S., Grulke, N., Hastings, S. J., and Lawrence, B. (1994). Transient nature of CO_2 fertilization in Arctic tundra. *Nature (London)* **371,** 500–503.

Owensby, C. E., Ham, J. M., Knapp, A. K., Bremer, D., and Auen, L. M. (1997). Water vapour fluxes and their impact under elevated CO_2 in a C_4 tall grass prairie. *Global Change Biol.* **3,** 189–195.

Palet, A., Ribas-Carbo, M., Argiles, J. M., and Azcon-Bieto, J. (1991). Short-term effects of carbon dioxide on carnation callus cell respiration. *Plant Physiol.* **96,** 467–472.

Parton, W. J., Schimel, D. S., Cole, C. V., and Ojima, D. S. (1987). Analysis of factors controlling soil organic matter levels in Great Plains grasslands. *Soil Sci. Soc. Am. J.* **51,** 1173–1179.

Pettersson, R., and Hansson, A.-C. (1990). Net primary production of a perennial grass ley (*Festuca pratensis*) assessed with different methods and compared with a lucerne ley (*Medicago sativa*) *J. Appl. Ecol.* **27,** 788–802.

Pettersson, R. McDonald, A. J. S. (1992). Effects of elevated carbon dioxide concentration on photosynthesis and growth of small birch plants at optimal nutrition. *Plant, Cell and Environ.* **15,** 911–919.

Pettigrew, W. T., Hesketh, J. D., Peters, D. B., and Woolley, J. T. (1990). A vapor pressure deficit effect on crop canopy photosynthesis. *Photosynth. Res.* **24,** 27–34.

Picon, C., Guehl, J.-M., and Fehri, A. (1996). Leaf gas exchange and carbon isotope composition responses to drought in a drought-avoiding (*Pinus pinaster*) and a drought tolerant (*Quercus petraea*) species under present and elevated atmospheric CO_2 concentration. *Plant Cell Environ.* **19,** 427–436.

Pinter, P. J. J., Kimball, B. A. *et al.* (1996). Free-air CO_2 enrichment: Responses of cotton and wheat crops. *In* "Carbon Dioxide and Terrestrial Ecosystems" (G. W. Koch and H. A. Mooney. eds.), pp. 215–249. Academic Press, San Diego.

Polley, H. W., Johnson, H. B., Marino, B. D., and Mayeux, H. S. (1993). Increase in C_3 plant water-use efficiency and biomass over glacial to present CO_2 concentrations. *Nature (London)* **361,** 61–64.

Pook, E. W. (1985). Canopy dynamics of *Eucalyptus maculata*. 3. Effects of drought. *Aust. J. Bot.* **33,** 65–80.

Poole, D. K., and Miller, D. C. (1981). The distribution of plant water-stress and vegetation characteristics in Southern California chaparral. *Am. Midl. Nat.* **105,** 32–43.

Poorter, H. (1993). Interspecific variation in the growth response of plants to an elevated ambient CO_2 concentration. *Vegetatio* **104/105,** 77–97.

Poorter, H., and Garnier, E. (1996). Plant growth analysis: An evaluation of experimental design and computational methods. *J. Exp. Bot.* **47,** 1343–1351.

Potter, J. R., and Jones, J. W. (1977). Leaf area partitioning as an important factor in growth. *Plant Physiol.* **59,** 10–14.

Quick, W. P., Chaves, M. M., Wendler, R., David, M., Rodrigues, M. L., Passaharinho, J. A., Pereira, J. S., Adcock, M. D., Leegod, R. C., and Stitt, M. (1992). The effect of water stress on photosynthetic carbon metabolism in four species grown under field conditions. *Plant Cell Environ.* **1,** 25–35.

Rambal, S., and Leterme, C. (1987). Changes in the aboveground structure and resistances to water uptake in *Quercus coccifera* along rainfall gradient. *In* "Plant Responses to Stress. Functional Analysis in Mediterranean Ecosystems. Nato ASI Series," Vol. G15, pp. 191–200 (J. Tenhunen, O. L. Lange, and C. Oechel, eds.). Springer-Verlag, New York.

Raunkiaer, C. (1934). "The Life Form of Plants and Statistical Plant Geography." Oxford Univ. Press, Oxford.

Redmann, R. E. (1978). Seasonal dynamics of carbon dioxide exchange in a mixed grassland ecosystem. *Can. J. Bot.* **56,** 1999–2005.

Reuveni, J., and Gale, J. (1985). The effect of high levels of carbon dioxide on dark respiration and growth of plants. *Plant Cell Environ.* **8,** 623–628.

Reuveni, J., Gale, J., and Mayer, A. M. (1993). Reduction of respiration by high ambient CO_2 and the resulting error in measurements of respiration made with O_2 electrodes. *Ann. Bot.* **72,** 129–131.

Reynolds, H. L., and D'Antonio, C. (1996). The ecological significance of plasticity in root weight ratio in response to nitrogen: Opinion. *Plant Soil* **185,** 75–97.

Rice, C. W., Garcia, F. O., Hampton, C. O., and Owensby, C. E. (1994). Soil microbial response in tallgrass prairie to elevated CO_2. *Plant Soil* **165,** 67–74.

Roessler, P. G., and Monson R. K. (1985). Midday depression in net photosynthesis and stomatal conductance in *Yucca glauca*. *Oecologia (Berlin)* **67,** 380–387.

Rogers, A., Bryant, J. B., Raines, C. A., Long, S. P., Blum, H., and Frehner, M. (1995). Acclimation of photosynthesis to rising CO_2 concentration in the field. Is it determined by source/sink balance? *In* "Photosynthesis: From Light to Biosphere," Vol. V. Proceedings of the Xth International Photosynthesis Congress, Montpellier, France, 20–25 August, 1995, pp. 1001–1004. Kluwer Academic Publishers, Dordrecht, Netherlands.

Roy, J., and Salager, J.-L. (1992). Midday depression of net CO_2 exchange of leaves of an emergent rain forest tree in French Guiana. *J. Trop. Ecol.* **8,** 499–504.

Ruimy, A., Jarvis, P. G., Baldocchi, D. D., and Saugier, B. (1995). CO_2 fluxes over plant canopies and solar radiation: A review. *Adv. Ecol. Res.* **26,** 2–63.

Runyon, J., Waring, R. H., Goward, S. N., and Welles, J. M. (1994). Environmental limits on net primary production and light-use efficiency across the Oregon transect. *Ecol. Appl.* **4,** 226–237.

Ryan, M. G. (1990). Growth and maintenance respiration in stems of *Pinus contorta* and *Picea engelmannii. Can. J. For. Res.* **17,** 472–483.

Ryan, M. G., and Waring, R. H. (1992). Maintenance respiration and stand development in a subalpine lodgepole pine forest. *Ecology* **73,** 2100–2108.

Ryan, M. G., Linder, S., Vose, J. M., and Hubbard, R. (1994). Dark respiration of pines. *Ecol. Bull. (Copenhagen)* **43,** 50–64.

Sage, R. F. (1994). Acclimation of photosynthesis to increasing atmospheric CO_2 the gas exchange perspective. *Photosynth. Res.* **39,** 351–368.

Sage, R. F., and Pearcy, R. (1987). The nitrogen use efficiency of C_3 and C_4 plants. II. Leaf nitrogen effects on the gas exchange characteristics of *Chenopodium album* (L.) and *Amaranthus retroflexus* (L.) *Plant Physiol.* **84,** 959–963.

Saxe, H., Ellsworth, D. S., and Heath, J. (1998). Tansley Review No. 98: Tree and forest functioning in an enriched CO_2 atmosphere. *New Phytol.* **139,** 395–436.

Schulze, E. D. (1982). Plant life forms and their carbon, water and nutrient relations. *In* "Physiological Plant Ecology II—Water Relations and Carbon Assimilation" (O. L. Lange, P. S. Nobel, C. B. Osmond, and H. Ziegler, eds.), pp 616–676. Springer-Verlag, Berlin–Heidelberg–New York.

Schulze, E.-D., and Chapin III, F. S. (1987). Plant specialization to environments of different resource availability. *In* "Potentials and Limitations of Ecosystem Analysis" (E. D. Schulze and H. Zwolfer, eds.), pp. 120–148. Vol. 61. Ecological Studies. Springer-Verlag, Berlin.

Schulze, E.-D., Lange, O. L., Evenari, M., Kappen, L., and Buschbom, U. (1974). The role of air humidity and leaf temperature in controlling stomtatal resistance of *Prunus armeniaca* L. under desert conditions. I. A simulation of the daily course of stomatal resistance. *Oecologia (Berlin)* **17,** 159–170.

Schulze, E.-D., Mooney, H. A., Sala, O. E., Jobbagy, E., Buchmann, N., Bauer, G., Canadell, J., Jackson, R. B., Loreti, J., Oesterheld, M., and Ehleringer, J. R. (1996). Rooting depth, water availability, and vegetation cover along an aridity gradient in Patagonia. *Oecologia (Berlin)* **108,** 503–511.

Schwanz, P., Picon, C., Vivin, C., Dreyer, E., Guehl, J.-M., and Polle, A. (1996). Responses of antioxidative systems to drought stress in pedunculate oak (*Quercus robur* L.) and maritime Pine (*Pinus pinaster* Ait.) as modulated by elevated CO_2. *Plant Physiol.* **110,** 393–402.

Seemann, J. R., Sharkey, T. D., Wang, J., and Osmond, C. B. (1987). Environmental effects on photosynthesis, nitrogen-use efficiency, and metabolite pools in leaves of sun and shade plants. *Plant Physiol.* **84,** 796–802.

Sellers, P. J., Bounoua, L., Collatz, G. J., Randall, D. A., and Dazlich, D. A. (1996). Comparison of radiative and physiological-effects of doubled atmospheric CO_2 on climate. *Science* **271(5254),** 1402–1406.

Sharkey, T. D., and Seemann, J. R. (1989). Mild water stress effects on carbon-reduction-cycle intermediates, ribulose biphosphate carboxylase activity and spatial homogeneity of photosynthesis in intact leaves. *Plant Physiol.* **89,** 1060–1065.

Shaver, G. R., Billings, W. D., Chapin, F. S., Giblin, A. E., Nadelhoffer, K. J., Oechel, W. C., and Rastetter, E. B. (1992). Global change and the carbon balance of arctic ecosystems. *BioScience* **42,** 433–441.

Smart, D. R., Ritchie, K., Stark, J. M., and Bugbee, B. (1997). Evidence that elevated CO_2 levels can indirectly increase rhizosphere denitrifier activity. *Appl. Environ. Microbiol.* **63,** 4621–4624.

Smith, B. G. (1989). The effects of soil water and atmospheric vapour pressure deficit on stomatal behaviour and photosynthesis in the oil palm. *J. Exp. Bot.* **40,** 647–651.

Soussana, J. F., and Hartwig, U. A. (1996). The effects of elevated CO_2 on symbiotic N_2 fixation: A link between the carbon and nitrogen cycles in grassland ecosystems. *Plant Soil* **187,** 321–332.

Sperry, J. S., and Tyree, M. T. (1990). Water stress induced cavitation and embolism in three species of conifers. *Plant Cell Environ.* **13,** 427–436.

Sperry, J. S., Saliendra, N. Z., Pockman, W. T., Cochard, H., Cruiziat, P., Davis, S. D., Ewers, F. W., and Tyree, M. T. (1996). New evidence for large negative xylem pressures and their measurement by the pressure chamber method. *Plant Cell Environ.* **19,** 427–436.

Stocker, R., Leadley, P. W., and Korner, C. (1997). Carbon and water fluxes in a calcareous grassland under elevated CO_2. *Funct. Ecol.* **11,** 222–230.

Stockle, C. O., and Kiniry, J. R. (1990). Variability in crop radiation use efficiency associated with vapour pressure deficit. *Field Crops Res.* **25,** 171–182.

Tate, K. R., and Ross, D. J. (1997). Elevated CO_2 and moisture effects on soil carbon storage and cycling in temperate grasslands. *Global Change Biol.* **3,** 225–235.

Tate, K. R., Parshotam, A., and Ross, D. J. (1995). Soil carbon storage and turnover in temperate forests and grasslands—A New Zealand perspective. *J. Biogeog.* **224**/5, 695–700.

Tenhunen, J. D., Lange, O. L., Gebel, J., Beyschlag, W., and Weber, J. A. (1984). Changes in photosynthetic capacity, carboxylation efficiency and CO_2 compensation point associated with midday stomatal closure and midday depression of net CO_2 exchange of leaves of *Quercus suber. Planta* **162,** 193–203.

Tenhunen, J. D., Serra, A. S., Harley, P. C., Dougherty, R. L., and Reynolds, J. F. (1990). Factors influencing carbon fixation and water use by mediterranean sclerophyllous shrubs during summer drought. *Oecologia (Berlin)* **82,** 381–393

Terashima, I., Wong S.-C., Osmond, C. B., and Farquhar, G. D. (1988). Characterization of nonuniform photosynthesis induced by abscisic acid in leaves having different mesophyll anatomies. *Plant Cell Physiol.* **29,** 385–394.

Tilman, D. (1984). Plant dominance along an experimental nutrient gradient. *Ecology* **65,** 1445–1453.

Tyree, M. T., and Cochard, H. (1996). Summer and winter embolism in oak: Impact on water relations. *Ann. Sci. For.* **53,** 173–180.

Tyree, M. T., and Ewers, F. W. (1991). Tansley Review No. 34. The hydraulic architecture of trees and other woody plants. *New Phytol.* **119,** 345–360.

Tyree, M. T., and Sperry, J. S. (1988). Do woody plants operate near the point of catastrophic xylem dysfunction caused by dynamic water stress?: Answers from a model. *Plant Physiol.* **88,** 574–580.

Tyree, M. T., Fiscus, E. L., Wullschleger, S. D., and Dixon, M. A. (1986). Detection of cavitation in corn under field conditions. *Plant Physiol.* **82,** 597–599.

Tyree, M. T., Cochard, H., Cruiziat, P., Sinclair, B., and Ameglio, T. (1993). Drought-induced leaf shedding in walnut: Evidence for vulnerability segmentation. *Plant Cell Environ.* **16,** 879–882.

Tyree M. T., Kolb, K. J., Rood, S. B., and Patino, S. (1994). Vulnerability to drought-induced cavitation of riparian cottonwoods in Alberta: A possible factor in the decline of the ecosystem? *Tree Physiol.* **14,** 455–466.

Valentine, H. T. (1990). A carbon balance model of tree growth with a pipe model framework. *In* "Process Modelling of Forest Growth Responses to Environmental Stress" (R. K. Dixon, G. A. Ruark, and W. G. Warren, eds.), pp. 229–240. Timber Press, Portland, Oregon.

Valentini, R., Epron, D., de Angelis, P., Matteucci, G., and Dreyer, E. (1995). *In situ* estimation of net CO_2 assimilation, photosynthetic electron flow, and photorespiration in Turkey oak (Q. cerris L.) leaves: diurnal cycles under different levels of water supply. *Plant, Cell and Environment* **18,** 631–640.

Valentini, R., De Angelis, P., Matteucci, P., Monaco, R., Dore, S., and Scarascia-Mugnozza, G. E.

(1996). Seasonal net carbon dioxide exchange of a beech forest with the atmosphere. *Global Change Biol.* **2,** 199–207.

Valiela, I., and Teal, J. M. (1974). Nutrient limitation in salt marsh vegetation. *In* "Ecology of Halophytes" (R. J. Reimold and W. H. Queen, eds.), pp. 547–563. Academic Press, New York.

Van Volkenburgh, E., and Boyer, J. S. (1985). Inhibitory effect of water deficit on maize leaf elongation. *Plant Physiol.* **77,** 190–194.

Verma, S. B., Baldocchi, D. D., Anderson, D. E., Matt, D. R., and Clement, R. J. (1986). Eddy fluxes of CO_2, water vapour, and sensible heat over a deciduous forest. *Bound. Layer Meteorol.* **36,** 71–91.

Verma, S. B., Kim, J., and Clement, R. J. (1989). Carbon dioxide, water vapour and sensible heat fluxes over a tallgrass prairie. *Bound. Layer Meteorol.* **46,** 53–67.

Verma, S. B., Kim, J., and Clement, R. J. (1992). Momentum, water vapour, and carbon dioxide exchange at a centrally located prairie site during FIFE. *J. Geophys. Res.* **97,** 18629–18639.

Vermetten A. W., Ganzefeld, M., Jeuken, L., Hofschreuder, A., and Mohren, G. M. J. (1994). CO_2 uptake by a stand of Douglas fir: flux measurements compared with model calculations. *Agricultural and Forest Meteorology,* **72,** 57–80.

Vessey, J. K., and Layzell, D. B. (1987). Regulation of assimilate partitioning in soybean: initial effects following change in nitrate supply. *Plant Physiol.* **83,** 341–348.

Vitousek, P. M., and Howarth, R. W. (1991). Nitrogen limitation on land and in the sea: How can it occur? *Biogeochemistry* **13,** 87–115.

Vivin, P., Guehl, J. M., Clement, A., and Aussenac, G. (1996). The effects of elevated and water stress on whole plant CO_2 exchange, carbon allocation and osmoregulation in oak seedlings. *Ann. Sci. For.* **53,** 447–459.

Walcroft, A. S., Silvester, W. B., Whitehead, D., and Kelliher, F. M. (1997). Seasonal changes in stable carbon isotope ratios within annual rings of *Pinus radiata* reflect environmnental regulation of growth processes. *Aust. J. Plant Physiol.* **24,** 57–68.

Waring, R., Law, B. E, Goulden, M. L., Bassow, S. L., McCreight, R. W., Wofsy, S. C., and Bazzaz, F. A. (1995). Scaling gross ecosystem production at Harward Forest with remote sensing: A comparaison of estimates from a constrained quantum-use efficiency model and eddy correlation. *Plant Cell Environ.* **18,** 1201–1213.

Webb, W. L., Lauenroth, W. K., Szareck, S. R., and Kinerson R. S. (1983). Primary production and abiotic controls in forests, grasslands, and desert ecosystems in the United States. *Ecology* **64,** 134–151.

Weisbach, C., Hartwig, U. A., Heim, I., and Nosberger, J. (1996). Whole-nodule carbon metabolites are not involved in the regulation of the oxygen permeability and nitrogenase activity in white clover nodules. *Plant Physiol.* **110,** 539–545.

Whitehead, D., Edwards, W. R. N., and Jarvis, P. G. (1984). Relationships between conductiong sapwood area, foliage area and permeability in mature *Picea sitchensis* and *Pinus contorta* trees. *Can. J. For. Res.* **14,** 940–947.

Wise, R. R., Frederick, J. R., Alm, D. M., Kramer, D. M., Hesketh, J. D., Crofts, A. R., and Ort, D. R. (1990). Investigation of the limitations to photosynthesis induced by leaf water deficit in field-grown sunflower (*Helianthus annuus* L.) *Plant Cell Environ.* **13,** 923–931.

Wofsy, S. C., Goulden, M. L., Munger, J. W., Fan, S.-M., Bakwin, P. S., Daube, B. S. C., Bassow, S. L., and Bazzaz, F. A. (1993). Net exchange of CO_2 in a mid-latitude forest. *Science* **260,** 1314–1317.

Wong, S. C. (1979). Elevated atmospheric partial pressure of CO_2 and plant growth. I. Interactions of nitrogen nutrition and photosynthetic capacity in C_3 and C_4 plants. *Oecologia (Berlin)* **44,** 68–74.

Woodrow, I. E. (1994). Optimal acclimation of the C_3 photosynthetic system under enhanced CO_2. *Photosynth. Res.* **39,** 401–412.

Woodward, F. I. (1987). "Climate and Plant Distribution." Cambridge Univ. Press, Cambridge.

Woodward, F. I. (1998). Do plants really need stomata? *J. Exp. Bot.* **49** (special issue), 471–480.

Wullschleger, S. D., Norby, R. J., and Hendrix, D. L. (1992). Carbon exchange rates, chlorophyll content, and carbohydrate status of two forest tree species exposed to carbon dioxide enrichment. *Tree Physiol.* **10,** 21–31.

Yabuki, K., and Aoki, M. (1978). The effect of windspeed on the photosynthesis of rice field. *In* "Ecophysiology of Photosynthetic Productivity" (JIBP Synthesis, Vol. 19) (M. Monsi and T. Saeki, eds.), pp. 152–159. University of Tokyo Press, Tokyo.

Zahner, R. (1968). Water deficits and growth of trees. *In* "Water Deficits and Plant Growth," Vol. 11 (T. T. Kozlowski, ed.), pp. 191–254. Academic Press, New York.

Zhang, J., and Davies, W. J. (1990). Does ABA in the xylem control the rate of leaf growth in soil-dried maize and sunflower plants? *J. Exp. Bot.* **41,** 1125–1132.

Zhang, H., and Nobel P. S. (1996). Dependency of c_i/c_a and leaf transpiration efficiency on the vapour pressure deficit. *Aust. J. Plant Physiol.* **23,** 561–568

Zotz, G., and Winter, K. (1996). Diel patterns of CO_2 exchange in rainforest canopy plants. *In* "Tropical Forest Plant Ecophysiology" (S. S. Mulkey, R. L. Chazdon, and A. P. Smith eds.), pp. 89–113. Chapman and Hall, New York.

8

How Does Biodiversity Control Primary Productivity?

Jacques Roy

I. Introduction

Primary productivity as well as the other ecosystem processes associated with the transfer of energy, water, and nutrient are either directly dependent on or are tightly controlled by the activity of organisms present in the ecosystem. Despite early considerations on the positive impact of diversity on biomass production (Darwin, 1859) or on the stability of populations and ecosystem processes (Elton, 1958), the study of the functional role of biodiversity was not an active field of research. One reason might have been the obvious predominance of abiotic factors such as water and temperature in the control of ecosystem processes (Lieth, 1975). A second reason was the divergent fields of interest of community and ecosystems ecologists, each group having its own perception of diversity (Mooney, 1997). However, with the realization that we are in the midst of a global biodiversity crisis (Pimm *et al.,* 1995) and that the human alteration of ecosystems is substantial and may impair the services ecosystems provide to society (Folke *et al.,* 1996; Vitousek *et al.,* 1997), linking biodiversity and ecosystem functioning has recently received considerable attention (Schulze and Mooney, 1993; Mooney *et al.,* 1995; Chapin *et al.,* 1997, 1998; Tilman, 1999; Schläpfer and Schmid, 1999).

The current alteration of biodiversity has several facets whose impact on ecosystem processes may vary. On one hand there is a decline of biodiversity due to a very high rate of species extinction (Pimm *et al.,* 1995). This decline raises the concern of the functional role of the number of species. What is the shape of the relationship between species number and a given ecosystem process? For how many species does this relationship reach a plateau? Which ecosystem processes are most affected? On the other hand there are changes in floristic composition due to increased disturbance (Thompson, 1994), rising atmospheric CO_2 concentration (Luscher *et al.*, 1998; Leadley *et al.*, 1999; Owensby *et al.*, 1999) and nitrogen deposition (Berendse and Elberse, 1990; Lee and Caporn, 1998), and a growing number of invasive species (Lonsdale, 1999; Dukes and Mooney, 1999). These floristic changes are not at random: specific groups of species are usually selected, often fast-growing species typical of eutrophic disturbed habitats (Thompson, 1994). Species extinction might not be at random either. The question is then: do some species have more impact than others on ecosystem processes? Which ecological traits have these species or groups of species (functional groups) that are responsible for these impacts? Is the number of functional groups rather than the number of species important? Or is the division between species and functional groups arbitrary?

Biodiversity varies spatially, at different scales, from latitudinal changes to changes between local patches. Can we answer some of the above questions using these natural changes in diversity? What are the answers from the experiments conducted recently where biodiversity was specifically manipulated? The research objectives are now to understand the mechanisms relating diversity to ecosystem processes. Some mechanisms have been hypothesized (Mooney *et al.*, 1995) and we now have experimental evidence for some of them.

II. Productivity and Species Diversity at Different Spatial Scales

Biodiversity and productivity both increase as we move from the poles to the equator or, more generally, from stressful low-resource environments to benign, resourceful environments. Taking regional data of net primary productivity (NPP) and species number in different biomes, Lieth (1975) obtained a linear relationship between productivity and species number (Fig. 8-1a). Among the several types of hypotheses suggested to explain the higher number of species in the tropics (Pianka, 1966), the higher productivity, or higher amount of energy available (Wright, 1983; Currie, 1991), is the most persuasive. However, it has been shown to be a complex (Begon *et al.*, 1986; Abrams, 1995) and possibly wrong hypothesis (Tilman and Pacala, 1993).

However, biodiversity does not always increase monotonically with productivity. In many regional surveys, unimodal ("hump-shaped") curves have been found (Fig. 8-1b). Grime (1973, 1979), who first noted this type of relationship, attributed the decline in species diversity in high-productivity environments to increased competitive exclusion. In low-biomass habitats, germination and growth tolerance to stress limit the number of species. Guo

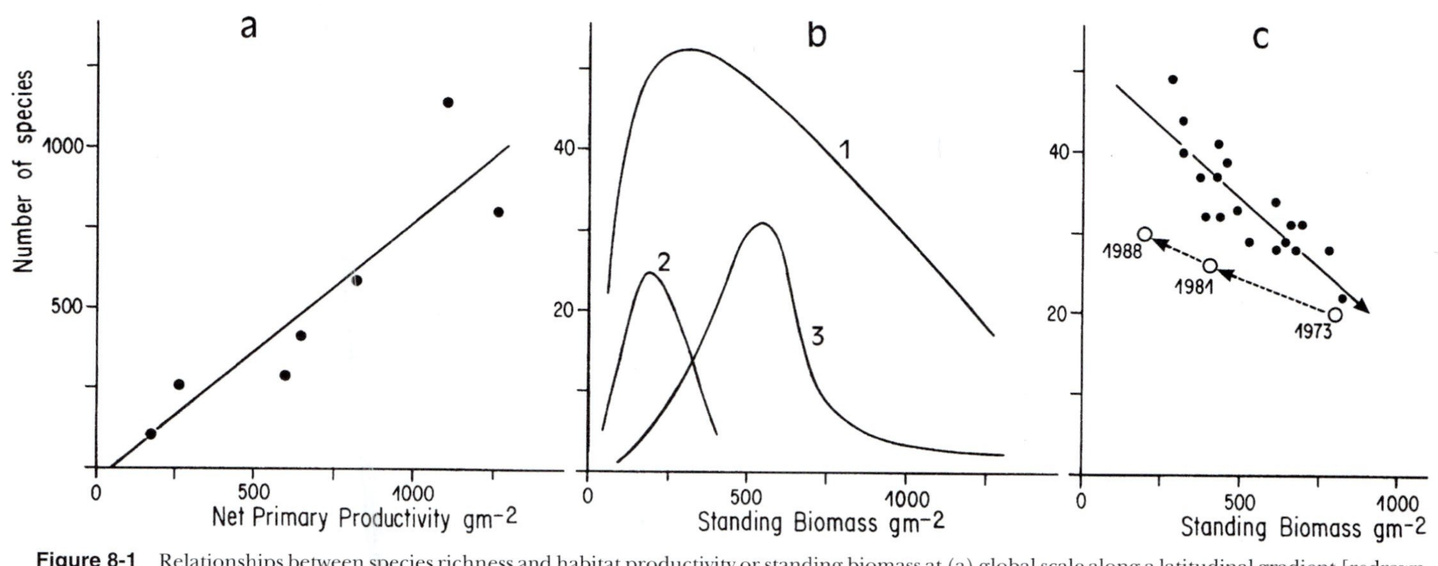

Figure 8-1 Relationships between species richness and habitat productivity or standing biomass at (a) global scale along a latitudinal gradient [redrawn from Lieth (1975)], (b) regional scale in Mediterranean grasslands [1, from Garcia *et al.* (1993), Mediterranean fynbos; 2, from Bond (1983), temperate grassland; 3, from Al-Mufti *et al.* (1977)] and (c) local scale [●, plots with various fertilization treatments for 8 years, from Silvertown (1980), 1862 data; ○, hayfield where fertilization ceased in 1973, from Olff and Baker (1991), field L2].

and Berry (1998) showed how aggregation of microhabitat types extends the environmental gradient and allows the development of a hump-shape relationship. This relationship and its explanation also have some theoretical support (Tilman, 1982; Loreau, 1998a), but alternative explanations have been suggested (Abrams, 1995).

On a given plot, changes in species diversity are observed when productivity is modulated by the manipulation of the resource level. Fertilization, for example, decreases plant species diversity as shown by the Park Grass Experiment at Rothamsted, UK (Silvertown, 1980), whereas high species richness can be recovered by cessation of fertilizer application (Olff and Baker, 1991) (Fig. 8-1c). The mechanism involved is competitive exclusion.

Multiple factors determine species richness in plant communities (Ricklefs and Schluter, 1993) and their relative importance varies with scale, leading to the contrasted relationships between productivity and diversity shown in Fig. 8-1. Locally though, we have good evidence that the resource level modulates plant diversity. Sampling natural ecosystems in order to infer the consequences of the loss of species on productivity and other ecosystem processes is then a hazardous enterprise. Local differences in the intensity and stability of ecosystem fluxes more likely constrain species diversity, rather than the opposite.

III. Impact of Plant Species and Functional Group Diversity on Intensity of Ecosystem Fluxes

Common sense as well as basic agricultural knowledge indicate that some specific species, or groups of species, have a particularly large impact on primary productivity. This is because they represent either the largest proportion of the biota, such as dominant tree species in forest vegetation, or they have biological traits likely to profoundly affect ecosystem processes—for example, nitrogen fixation for the leguminous species. However, identifying keystone species is most of the time not so obvious because they are of many types and are not necessarily important energy transformers (Bond, 1993). Chapin *et al.* (1997) suggested that species most notably affecting ecosystem processes must have traits that (1) modify the availability, capture, and use of soil resources, (2) affect the ecosystem trophic structure, or (3) influence the frequency and severity of disturbance.

Besides the impact of biodiversity through the identity of single species, what is the role of the number of species in the functioning of ecosystems? A few classical ecological studies give some hints to answer this question: during the course of secondary succession there is a change in species number as well as in productivity. In general, plant diversity decreases while productivity increases (Mellinger and McNaughton, 1975; Mooney and Gulmon, 1983). The increase in productivity is often the result of the introduction of a new life form, such as a shrub, suggesting that species identity plays a larger role than species number (Mooney, 1997). There is some evidence from agricultural systems that multicropping, to some extent, increases productivity (Swift and Anderson, 1993; Schläpfer and Schmid, 1999). On average, two species mixtures are 12% more productive than pure stands (Jollife, 1997). For forestry systems, available data suggest that the effects of species diversity are small or absent (Schläpfer and Schmid, 1999).

Specific experiments where diversity was manipulated as an independent factor and in a range relevant to natural communities have been lacking. Designing and analyzing such experiments turned out to be difficult. In some early experiments, number of species was confounded with identity of species and "hidden treatments" were at play (Givnish, 1994; Huston, 1997). Since then, using experimental designs with direct control of diversity, randomly determined species composition, and adequate replication (Tilman, 1997b), several experiments have reported an increase of plant biomass or plant cover with increased number of plant species (Naeem *et al.*, 1996; Tilman *et al.*, 1996, 1997a; Hector *et al.*, 1999). Averaging a trans-European study, Hector *et al.* (1999) describe the relationship as being log-linear, with a reduction of approximately 80 g m^{-2} for each halving of the number of plant species. However, the variability within and between the experiments of this study is very large, giving some latitude to interpret the results. In order to define the number of species at which productivity levels off, I determined the lowest number of species at which 90% of the productivity of the most diverse treatment was obtained. I only used field experiments for which biomass data were available, i.e., Tilman *et al.* (1997a), Hooper (1998), and the 8 experiments described by Hector *et al.* (1999). This determination is tentative, because in some cases it required approximate interpolation between actual treatments and because generally the planned number of species for each treatment is given and not the actual number at the time of biomass measurement. For the majority of experiments, 90% of the productivity of the most diverse treatment was reached with a number of species equal to or less than four species (Fig. 8-2, inset), the average being five species. The curve of Fig. 8-2 was drawn using this information and a doubling of productivity between 1 and 30 species.

The classification of species into functional groups depends on the question being asked and is time and space dependent (Körner, 1993; Gitay and Noble, 1997; Mooney, 1997). With regard to ecosystem productivity, several functional groups have been identified on the basis of species physiology (C_3 vs. C_4 species, N fixers vs. non-N fixers, woody vs. nonwoody species) or of species phenology (early vs. late season species, annual vs. perennial species). Since by definition functional differences between functional groups are larger than functional differences between randomly taken species, it is logically expected that the diversity of functional groups would have a stronger impact than the diversity of species (Walker, 1992, 1995).

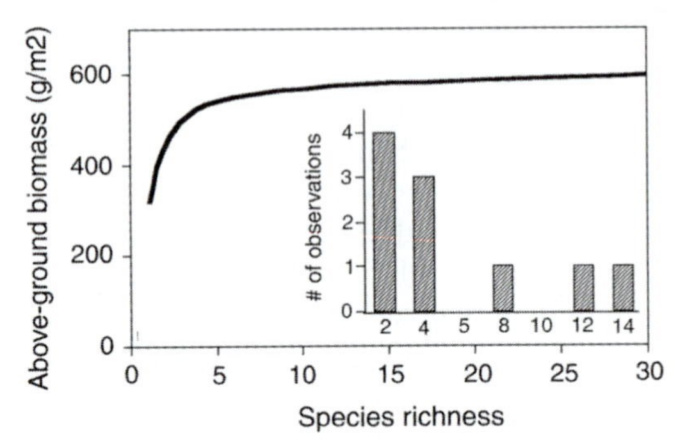

Figure 8-2 Shape of the relationship between aboveground biomass and species richness. The curve is drawn assuming a doubling in biomass between 1 and 30 species and 90% of maximum biomass obtained for five species, the average of the observations shown in the inset. Inset: Plant species richness for which 90% of the productivity of the most diverse treatment is achieved. From data by Tilman *et al.* (1997a), Hooper (1998), and Hector *et al.* (1999).

Tilman *et al.* (1997a) as well as Hector *et al.* (1999) found a positive relationship between the number of functional groups and aboveground biomass. Hector *et al.* (1999), working with three functional groups, found this relationship to be linear, with the removal of one functional group reducing biomass by 100 g m^{-2}. Tilman *et al.* (1997a), with five functional groups, found saturation to be reached with three groups. In that experiment, species diversity effect became nonsignificant once functional diversity was taken into account. However, in both experiments, it was the functional composition (i.e., the identity of the functional group present) that had the larger impact. In Tilman *et al.* (1997a), species and functional diversity together explained 8% of the variance whereas diversity and functional composition together explained 37% of the variance. The legumes and the C_4 grasses were the functional groups with the largest impact on productivity. In Hector *et al.* (1999), species and functional group richness together explained 18% of the variation while species composition explained 39%. The legume *Trifolium pratense* had the largest significant effect. In the single experiment specifically designed to test for the impact of functional group diversity, Hooper (1998) manipulated four functional groups whose suite of relevant characteristics suggested they would use resources in a complementary way. Early season and late season annuals, perennial bunchgrasses, and nitrogen fixers indeed differ in phenology, rooting depth, shoot/root ratio, size per individual, and leaf nitrogen content. However, results show that aboveground biomass did not correlate with increasing functional group richness. Rather the identity of the functional groups present explained much more variance than did richness ($R = 0.72$ vs. $R = 0.13$). The large biomass of perennials in monocultures and the reduction of their contribution in mixture due to the competitiveness of annuals outweighed the effect of complementarity (Hooper, 1998; Hooper and Vitousek, 1997). Strong compensatory effects can be found among functional groups. Removing either C_4 grasses, C_3 annual grasses, all C_3 grasses, or all dicotyledonous weeds, Wardle *et al.* (1999) did not find a strong and consistent impact on the productivity of a perennial grassland ecosystem and attributed it to growth compensatory effects.

The above changes in productivity with plant diversity were generally associated with changes in other ecosystem parameters. Hooper and Vitousek (1998) found that nutrient cycling was more related to the identity of the functional groups than to the number of functional groups. Light penetration as well as soil nitrogen have been reported to decrease with plant diversity, while canopy nitrogen was increasing (Tilman *et al.*, 1997a; Tilman, 1999). This raises the question of the mechanisms through which plant diversity affects ecosystem processes. Are these changes in light and nitrogen the mere consequences of the increase in plant biomass triggered by the occurrence of more productive species? Or is the increase in plant biomass the consequence of an increased resource interception due to complementarity between species at high diversity?

IV. Mechanisms Relating Plant Diversity and Flux Intensity

Species (or genotypes or functional groups) have specific identities, i.e., specific biological traits. As a result, they can profoundly affect ecosystem processes as discussed previously, either because they are dominant species or because they have traits strongly determining resources use, community tropic structure, or ecosystem disturbance (Chapin *et al.*, 1997; Vitousek and Hooper, 1993) (Fig. 8-3).

The main mechanism through which increasing the number of different species in an ecosystem would result in higher productivity was originally hypothesized to be some complementarity between species for resource acquisition (Mooney *et al.*, 1995) or more generally some niche differentiation (Tilman, 1999). It was a logical assumption, but it was soon realized (Aarsen, 1997; Huston, 1997) that increasing the number of species also increased the probability to include the more productive species, which are likely to dominate the community (Fig. 8-3). The modeling of this simple mechanism, called the sampling effect, provided diversity response curves very similar to the one obtained experimentally (Tilman *et al.*, 1997b; Loreau, 1998a). This statistical mechanism is generally thought not to be a mere artifact. However, it is at play in nature only to the extent that community assembling and species extinction occur at random from the regional pool of species, which often is not the case [see Wardle (1999) and van der Heijden *et al.* (1999) for discussion]. A major task in understanding the mechanisms by which biodiversity operates is to separate the sampling effect from other

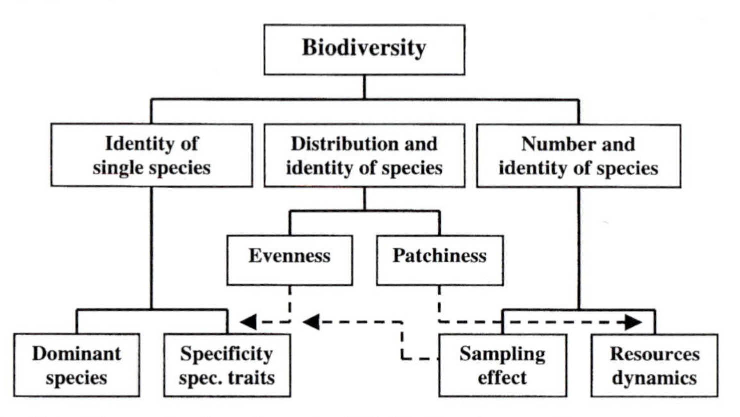

Figure 8-3 The mechanisms through which biodiversity can affects ecosystem processes.

effects. The use of the relative yield total (RYT) of the mixture has been suggested (Hector, 1998) ($RYT = \Sigma_i RY_i$; $RY_i = O_i/M_i$, with O_i the observed yield of species i in the mixture and M_i its yield in monoculture). However, Loreau (1998b) has shown that RYT only allows a test of the null hypothesis that the yield in the mixture can be explained by changes in the relative contributions of the various species in the mixture. Overyielding, a situation whereby the yield of the mixture is larger than the yield of the most productive species in monoculture, is a conservative but unambiguous test, of complementarity effects (Garnier *et al.*, 1997; Loreau, 1998b; Tilman, 1999). Evidence of overyielding was found by Tilman *et al.* (1997a) and Hector *et al.* (1999), suggesting that complementary and/or positive interactions between species occurred and that sampling effect was not the only mechanism at play.

Another component of diversity is the distribution of the species in the community. The evenness of the distribution (the relative proportion of the component species) can affect productivity through two mechanisms (Fig. 8-3). One is the proportion of the most productive species. When the abundance of the most productive species is manipulated to decrease evenness, its proportion as well as community productivity will increase. When the abundance of a less productive species is manipulated to decrease evenness, the proportion of the most productive species as well as community productivity will decrease. Using a simple growth model, the relationship between evenness and productivity has been shown to be linear (Nijs and Roy, 2000). The second mechanism is the complementary and positive interactions between species that are expected to decrease with decreasing evenness. In this case, evenness and productivity are positively related. In the only experiment testing the impact of evenness (Wilsey and Potvin, 2000), a general positive linear relationship has been found between evenness and productivity, mainly driven by root production, suggesting that the second mechanism was dominant. Aboveground biomass, however, showed opposite trends, depending on which species was manipulated, suggesting that the first mechanism also had a role. Patchiness is also expected to affect productivity through a modification of complementary and positive interactions between species (Fig. 8-3). The higher the aggregation of the individuals of the component species, the smaller the frontiers between species where these interactions can take place. Patchiness should then be negatively related to productivity. However, this relationship also depends on the relative strength of inter- and intraspecific competitions.

V. Impact of Plant Diversity on the Stability of Ecosystem Fluxes

The relationship between diversity and stability has been a controversial issue for decades [see review in McNaughton (1993) and Tilman (1999)]. The existence of niche differences between species suggests that more diverse communities would be buffered against environmental variations because the probability that they contain species capable of performing well under each of the different environmental conditions is higher than for less diverse communities (Elton, 1958; McNaughton, 1977). However, some models showed that higher diversity leads to lower stability (May, 1972). As shown by Tilman *et al.* (1996), both were correct: the stability of individual populations decreases with diversity, but the stability of parameters aggregated at the ecosystem level increases with diversity. Aboveground biomass and primary productivity are such aggregated parameters. Field experiments using natural or fertilization-induced variations in species richness have shown that their stability, measured by the temporal variation in biomass, increases substantially with plant diversity (McNaughton, 1977, 1985; Frank and McNaughton, 1991; Tilman *et al.*, 1996). However, in such field experiments, the determinants of biodiversity such as ecological history, resource availability, or species characteristics may best explain the results rather than species diversity per se (Leps *et al.*, 1982; Huston, 1997; Sankaran and McNaughton, 1999). There is some similarity here with the variation across scales of the relationship between biomass and diversity presented in Fig. 8-1. The relationship between diversity and stability derived from studies across environmental gradients are not determined by the same mechanisms as the relationship between diversity and stability obtained by the loss of species in a given ecosystem, and these relationships should not be expected to be similar. After controlling for the potentially confounding variables, however, Tilman (1999) found that the positive relationship between diversity and stability remained significant, with the coefficient of variation of biomass changing typically from 60 to 40% for a 10-fold (2 to 20) increase in plant diversity. Experiments with aquatic microcosms have also shown a positive relationship between diversity and stability (McGrady-Steed *et al.*, 1997; Naeem and Li, 1997). Overall, there are very few experiments relating unambiguously diversity and stability of ecosystem processes, but the relationship is found to be positive. More data exist on the postive impact of diversity on the stability of floristic composition, e.g., Tilman (1997a), Prieur-Richard *et al.* (2000), and Van der Putten *et al.* (2000).

The mechanism through which diversity stabilizes productivity can be seen like species complementarity in time. At one moment species (or a set of species) *X* composes most of the biomass; at another moment, under different prevailing environmental conditions, species (or a set of species) *Y* is the largest producer. Although this change in abundance could be driven solely by the physiological responses to the abiotic environment, competition between species is likely to be involved (McNaughton, 1977; Tilman *et al.*, 1996). However, Doak *et al.* (1998) pointed out that increased stability could well be the result of a simple statistical process called the portfolio effect. Just like diversifying stock holdings brings a less variable total value of the stocks, averaging the randomly fluctuating biomass of many species will provide a more constant total biomass than the same operation done with a few species. Species interactions as well as niche differentiation relevant to the environmental fluctuations do not need to be involved. Tilman *et al.* (1998) showed that the results of Doak *et al.* (1998) are not general but depend on how the variance of the abundance of a species scales with its abundance. The analysis of available data suggests, however, that in nature this scaling is such that the portfolio effect does contribute to stability (Tilman *et al.*, 1998). Additional theoretical studies (Hughes and Roughgarden, 1998; Ives *et al.*, 1999) have shown that the number and strength of competitive interactions have little effect on stability.

Other theoretical studies further analyzed the relationship between stability and diversity. Using the classical approach for the analysis of the local stability of an equilibrium, Loreau and Behera (1999) conclude that diversity (measured as the phenotypic difference between species) can increase or decrease the stability at the level of aggregated ecosystem processes, just as it does at the population level. They also question the relevance of models representing equilibrium deterministic systems and discuss the probability that in the case of a "negative" perturbation such as drought, diversity is likely to increase stability of productivity as seen above, but that in the case of a "positive" perturbation such as nitrogen addition, diversity is likely to decrease stability because of the higher probability for the presence of a species responding strongly to this resource increase. Using a dynamic model where productivity of each species follows a stochastic process in response to environmental fluctuations, Yachi and Loreau (1999) showed that the temporal variance of productivity was reduced at high diversity due to the asynchrony of species responses. When competitive dominance was introduced in the model, they also showed that the temporal mean of productivity was enhanced due to a mechanism similar to the sampling effect.

VII. Conclusions and Perspectives

Much progress has been achieved with regard to our understanding of the role of biodiversity in ecosystem functioning. Most studies confirm the predominant role of species identity on the intensity of fluxes and primary productivity. As a consequence, factors that currently are changing floristic composition (invasions, disturbance, nitrogen deposition, elevated CO_2, global warming) are likely to alter significantly ecosystem processes. The direction of the change is starting to be understood for only some of these factors and much additional work is needed in this area, in particular, for determination of the key species or functional groups whose loss would substantially alter ecosystem processes.

The number of species per se appears to be less important for the intensity of primary productivity than sometimes claimed: saturation is reached with four species in the majority of the studies and part of the observed effect of the number of species is due to the sampling effect, which may not apply fully in nature [see also Grime (1997)]. However, the impacts of diversity on ecosystem processes are far from having been studied thoroughly—in particular, the interactions between plant diversity and soil diversity are numerous and may affect more strongly ecosystem processes other than biomass production. For example, in an experiment manipulating only some dominant species of Mediterranean old-fields (then minimizing the sampling effect), the loss of species had no impact on plant productivity but affected considerably the pool and fluxes of nutrients with a large increase in nitrate leaching in the fall (J. Roy *et al.*, unpublished data). The experiments manipulating litter or soil organism diversity have not been reviewed here [see Brussaard *et al.* (1997), Wall-Freckman *et al.* (1997), Lavelle *et al.* (1997), Wardle and Lavelle (1997), Coûteaux and Darbyshire (1998), and Scheu *et al.* (1999) for reviews]. Many of these studies show strong impacts of specific classes of organisms, either as soil engineers (Jones *et al.*, 1994) or as specific trophic levels (Ingham *et al.*, 1985). Very few experiments have tested the impact of diversity at the species level. van der Heijden *et al.* (1998) have shown a positive impact of the richness of arbuscular mycorrhizal fungi on nutrient capture and productivity. Laakso and Setälä (1999) show no impact of the number of species of animals on plant growth and found an impact of functional groups only for the groups situated at the lowest trophic positions in the decomposer food web.

With experiments in different ecosystem types and with manipulation of diversity at different trophic levels being published, a broader scale and hypotheses-based analysis becomes possible. Schläpfer and Schmid (1999) classified hypotheses according to the ecosystem variable measured and its trophic position with regard to the trophic position of the organism whose diversity was manipulated. Within-trophic effects (mostly impact of plant diversity on plant parameters) were in general positive whereas bottom-up effects (plant diversity impact on herbivores or decomposers) were absent or negative. The manipulation of diversity in trophic levels other than plants affected ecosystem parameters less consistently. This study points to the missing categories of experiment (among their 56 categories, 20 have been so far experimentally tested).

The impact of diversity on the stability of ecosystem functioning has received few experimental tests in comparison to the large potential impact it can have, principally through the "insurance hypothesis." There is no published data on the role of genetic diversity on ecosystem functioning. Although the short term impact of genetic diversity on the intensity of ecosystem processes is probably marginal, its long-term impact on the stability of these processes is expected to be large. Could the opposite stand for functional groups? Their impact on the intensity of ecosystem processes has been shown to be large, but their impact on stability has not been investigated. The presentation of functional group diversity impacting the intensity of ecosystem processes while the number of species within functional groups enhancing ecosystem stability (Walker, 1995) suggests a lower role of functional groups in the maintenance of stability. However, when functional groups are defined according to organisms response to disturbance (Lavorel *et al.*, 1999), at least the identity of functional groups has an impact on stability. The contrasting impact of genotypes, species, and functional groups on the stability and intensity of ecosystem processes is indicated in Fig. 8-4. In future experiments the distinction between genotypes, species, and functional groups could be replaced by a measurement of the pheno-

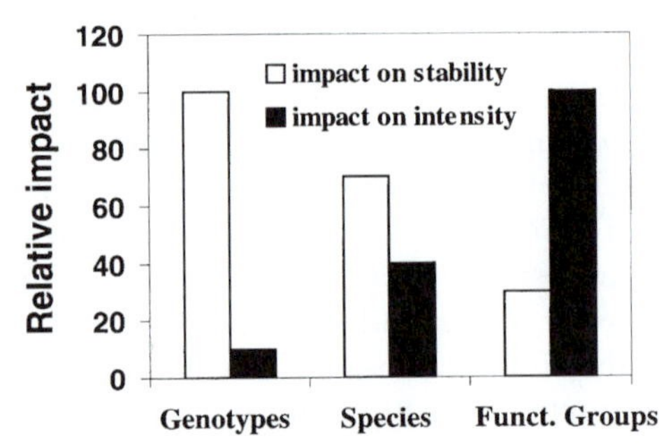

Figure 8-4 Expected relative impact of genotypes, species, and functional groups on the stability and intensity of ecosystem processes.

typic distance between taxa, as is done in some models (Loreau and Behera, 1999; Nijs and Roy, 2000. However, several biological traits often are involved in the relation between diversity and ecosystem processes. In that case, the determination of a synthetic phenotypic distance is not obvious and the genotypes, species, and functional group classifications have some relevance. Stability of ecosystem processes and its control by diversity need to be examined in a larger array of conditions. The response of the different ecosystem processes may not be the same and some perturbations may be more buffered by diversity than others. In particular, patterns of response to abiotic (frost, drought) vs. biotic (herbivory, pathogens) perturbations need to be compared.

In front of the large, rapid, and pervasive environmental changes currently imposed by humans, the biodiversity issue is of great concern for ethical, aesthetic, economic, and ecological reasons. Estimating how large the ecological impact of changes in biodiversity is has been an active and rich area of research during the past decade. It has been demonstrated that different aspects of diversity are not neutral with respect to ecosystem functioning, but major issues are still to be addressed.

Acknowledgments

I thank Sandra Lavorel, Paul Leadley, Michel Loreau, and Hal Mooney for their constructive comments and Pam and Kevin Singh for their warm hospitality during the writing of this paper. Support was provided by CNRS (GDR 1936 DIV-ECO) and by the European Commission, Environment & Climate Framework IV, Terrestrial Ecosystems Research Initiative contract CLUE (ENV4-CT95-0002).

References

Aarsen, L. W. (1997). High productivity in grassland ecosystems: Effected by species diversity or productive species? *Oikos* **80,** 183–184.

Abrams, P. A. (1995). Monotonic or unimodal diversity–productivity gradients: What does competition theory predict? *Ecology* **76,** 2019–2027.

Al-Mufti, M. M., Sydes, C. L., Furness, S. B., Grime, J. P., and Band, S. R. (1977). A quantitative analysis of shoot phenology and dominance in herbaceous vegetation. *J. Ecol.* **65,** 759–791.

Begon, M., Harper, J. L., and Townsend, C. R. (1986). "Ecology: Individuals, Populations and Communities." Blackwell, Oxford.

Berendse, F., and Elberse, W. T. H. (1990). Competition, succession and nutrient availability. *In* "Perspective in Plant Competition" (J. Grace and D. Tilman, eds.), pp. 93–116. Academic Press, San Diego.

Bond, W. (1983). On alpha diversity and the richness of the Cape flora: A Study in Southern Cape fynbos. *In* "Mediterranean-Type Ecosystems: The role of Nutrients" (F. S. Kruger, D. T. Mitchell, and J. V. M. Jarvis, eds.). Springer-Verlag, Berlin.

Bond, W. J. (1993). Keystone species. *In* "Biodiversity and Ecosystem Function" (E. D. Schulze and H. A. Mooney, eds.), pp. 237–253. Springer-Verlag, Berlin.

Brussaard, L., Behan-Pelletier, V. M., Bignell, D. E., Brown, V. K., Didden, W., Folgarait, P., Fragoso, C., Wall-Freckman, D., Gupta, V., Hattori, T., Hawksworth, D. L., Klopateck, C., Lavelle, P., Malloch, D. W., Rusek, J., Söderström, B., Tiedje, J. M., and Virginia, R. A. (1997). Biodiversity an ecosystem function in soil. *Ambio* **26,** 563–570.

Chapin, F. S., Walker, B. H., Hobbs, R. J., Hooper, D. U., Lawton, J. H., Sala, O. E., and Tilman, D. (1997). Biotic control over the functioning of ecosystems. *Science* **277,** 500–504.

Chapin, F. S., Sala, O. E., Burke, I. C., Grime, J. P., Hooper, D. U., Lauenroth, W. K., Lombard, A., Mooney, H. A., Mosier, A. R., Naeem, S., Pacala, S. W., Roy, J., Teffen, W. L., and Tilman, D. (1998). Ecosystem consequences of changing biodiversity. *BioScience* **48,** 45–52.

Coûteaux, M.-M., and Darbyshire, J. F. (1998). Functional diversity amongst soil protozoa. *Appl. Soil Ecol.* **317,** 1–9.

Currie, D. J. (1991). Energy and large-scale patterns of animal- and plant-species richness. *Am. Nat.* **137,** 27–49.

Darwin, C. (1859). "The Origin of Species by Means of Natural Selection." Murray, London.

Doak, D. F., Bigger, D., Harding, E. K., Marvier, M. A., O'Malley, R. E., and Thomson, D. (1998). The statistical inevitability of stability–diversity relationships in community ecology. *Am. Nat.* **151,** 264–276.

Dukes, J. S., and Mooney, H. A. (1999). Does global change increase the success of biological invaders? *Trends Ecol. Evol.* **14,** 135–139.

Elton, C. S. (1958). "The Ecology of Invasions by Animals and Plants." Methuen, London.

Folke, C., Holling, C. S., and Perrings, C. (1996). Biological diversity, ecosystems, and the human scale. *Ecol. Appl.* **6,** 1018–1024.

Frank, D. A., and McNaughton, S. J. (1991). Stability increases with diversity in plant communities: Empirical evidence from the 1988 Yellowstone drought. *Oikos* **62,** 360–362.

Garcia, L. V., Maranon, T., Moreno, A., and Clemente, L. (1993). Above-ground biomass and species richness in a Mediterranean salt marsch. *J. Veg. Sci.* **4,** 417–424.

Garnier, E., Navas, M. L., Austin, M. P., Lilley, J. M., and Gifford, R. M. (1997). A problem for biodiversity–productivity studies: How to compare the productivity of multispecific plant mixtures to that of monocultures? *Acta Oecol.* **18,** 657–670.

Gitay, H., and Noble, I. R. (1997). Functional groups. *In* "Plant Functional Types: Their Relevance to Ecosystem Properties and Global changes" (T. M. Smith, H. H. Shugart, and F. I. Woodward, eds.). Cambridge Univ. Press, Cambridge.

Givnish, T. J. (1994). Does biodiversity beget stability? *Nature (London)* **371,** 113–114.

Grime, J. P. (1973). Competitive exclusion in herbaceous vegetation. *Nature (London)* **242,** 344–347.

Grime, J. P. (1979). "Plant Strategies and Vegetation Processes." John Wiley & Sons, Chichester.

Grime, J. P. (1997). Biodiversity and ecosystem function: The debate deepens. *Science* **277,** 1260–1261.

Guo, Q. F., and Berry, W. L. (1998). Species richness and biomass: Dissection of the hump-shaped relationships. *Ecology* **79,** 2555–2559.

Hector, A. (1998). The effect of diversity on productivity: Detecting the role of species complementarity. *Oikos* **82,** 597–599.

Hector, A., Schmid, B., Beierkuhnlein, C., Caldeira, M. C., Diemer, M., Dimitrakopoulos, P. G., Finn, J. A., Freitas, H., Giller, P. S., Good, J., Harris, R., Hogberg, P., Huss-Danell, K., Joshi, J., Jumpponen, A., Korner, C., Leadley, P. W., Loreau, M., Minns, A., Mulder, P. H., O'Donovan, G., Otway, S. J., Pereira, J. S., Prinz, A., Read, D. J., Sherer-Lorenzen, M., Schulze, E.-D., Siamantziouras, A.-S. D., Sphen, E. M., Terry, A. C. Troumbis, A. Y., Woodward, F. I., Yachi, S., and Lawton, J. H. (1999). Plant diversity and productivity experiments in European grasslands. *Science* **286,** 1123–1127.

Hooper, D. U. (1998). The role of complementarity and competition in ecosystem responses to variation in plant diversity. *Ecology* **79,** 704–719.

Hooper, D. U., and Vitousek, P. M. (1997). The effects of plant composition and diversity on ecosystem processes. *Science* **277,** 1302–1305.

Hooper, D. U., and Vitousek, P. M. (1998). Effects of plant composition and diversity on nutrient cycling. *Ecol. Monogr.* **68,** 121–149.

Hughes, J. B., and Roughgarden, J. (1998). Aggregate communities properties and the strength of species' interactions. *Proc. Natl. Acad. of Sci. USA* **95,** 6837–6842.

Huston, M. A. (1997). Hidden treatments in ecological experiments: Re-evaluating the ecosystem function of biodiversity. *Oecologia* **110,** 449–460.

Ingham, R. E., Trofymow, J. A., Ingham, E. R., and Coleman, D. C. (1985). Interaction of bacteria, fungi and their nematode grazers on nutrient cycling and plant growth. *Ecol. Monogr.* **55,** 119–140.

Ives, A. R., Gross, K., and Klug, J. L. (1999). Stability and variability in competive communities. *Science* **286,** 542–544.

Jollife, P. A. (1997). Are mixed populations of plant species more productive than pure stands? *Oikos* **80,** 595–602.

Jones, C. G., Lawton, J. H., and Schachak, M. (1994). Organisms as ecosystem engineers. *Oikos* **69,** 373–386.

Körner, C. (1993). Scaling from species to vegetation: The usefulness of functional groups. *In* "Biodiversity and Ecosystem Function" (E.-D. Schulze and H. A. Mooney, eds.), pp. 117–140. Springer-Verlag, Berlin.

Laakso, J., and Setälä, H. (1999). Sensitivity of primary production to changes in the architecture of belowground food webs. *Oikos* **87,** 57–64.

Lavelle, P., Bignell, D., Lepage, M., Wolters, V., Roger, P., Ineson, P., Heal, O. W., and Dhillion, S. (1997). Soil function in a changing world: The role of invertebrate ecosystem engineers. *Euro. J. Soil Biol.* **33,** 159–193.

Lavorel, S., Rochette, C., and Lebreton, J.-D. (1999). Functional groups for response to disturbance in Mediterranean old fields. *Oikos* **84,** 480–498.

Leadley, P. W., Niklaus, P. A., Stocker, R., and Korner, C. (1999). A field study of the effect of elevated CO_2 on plant biomass and community structure in a calcareous grassland. *Oecologia* **118,** 39–49.

Lee, J. A., and Caporn, S. J. M. (1998). Ecological effects of atmospheric reactive nitrogen deposition on semi-natural terrestrial ecosystems. *New Phytol.* **139,** 127–134.

Leps, J., Osbornovà-Kosinovà, J., and Rejmànek, M. (1982). Community stability, complexity and species life history strategies, *Vegetatio* **50,** 53–63.

Lieth, H. (1975). Some prospects beyond production measurement. *In* "Primary Productivity fo the Biosphere" (H. Lieth and R. H. Whittaker, eds.), pp. 285–304. Springer-Verlag, New York.

Lonsdale, W. M. (1999). Global patterns of plant invasions and the concept of invasibility. *Ecology* **80,** 1522–1536.

Loreau, M. (1998a). Biodiversity and ecosystem functioning: A mechanistic model. *Proc. Nat. Acad. Sci. U.S.A.* **95,** 5632–5636.

Loreau, M. (1998b). Separating sampling and other effects in biodiversity experiments. *Oikos* **82,** 600–602.

Loreau, M., and Behera, N. (1999). Phenotypic diversity and stability of ecosystem processes. *Theor. Pop. Biol.* **56,** 29–47.

Luscher, A., Hendrey, G. R., and Nosberger, J. (1998). Long-term responsiveness to free air CO_2 enrichment of functional types, species and genotypes of plants from fertile permanent grassland. *Oecologia* **113,** 37–45.

May, R. M. (1972). Will a large complex system be stable? *Nature (London)* **238,** 413–414.

McGrady-Steed, J., Harris, P. M., and Morin, P. J. (1997). Biodiversity regulates ecosystem predictability. *Nature (London)* **390,** 162–165.

McNaughton, S. J. (1977). Diversity and stability of ecological communities: A comment on the role of empiricism in ecology. *Am. Nat.* **111,** 515–525.

McNaughton, S. J. (1985). Ecology of a grazing ecosystem: The Serengeti. *Ecol. Monogr.* **55,** 259–294.

McNaughton, S. J. (1993). Biodiversity and function of grazing ecosystems. *In* "Biodiversity and Ecosystem Function" (E.-D. Schulze and H. A. Mooney, eds.), pp. 361–383. Springer-Verlag, Berlin.

Mellinger, M. V., and McNaughton, S. J. (1975). Structure and function of successional vascular plant communities in Central New York. *Ecol. Monogr.* **45,** 161–182.

Mooney, H. A. (1997). Ecosystem function of biodiversity: The basis of the viewpoint. *In* "Plant Functional Types: Their Relevance to Ecosystem Properties and Global Changes" (T. M. Smith, H. H. Shugart, and F. I. Woodward, eds.), pp. 341–354. Cambridge Univ. Press, Cambridge.

Mooney, H. A., and Gulmon, S. L. (1983). The determinants of plant productivity: Natural vs. man-modified communities. *In* "Disturbance and Ecosystems" (H. A. Mooney and M. Godron, eds.), pp. 146–158. Springer-Verlag, Berlin.

Mooney, H. A., Lubchenco, J., Dirzo, R., and Sala, O. E. (1995). Biodiversity and ecosystem functioning: Basic principles. *In* "Global Biodiversity Assessment" (V. H. Heywood and R. T. Watson, eds.), pp. 279–325. Cambridge Univ. Press, Cambridge.

Naeem, S., and Li, S. (1997). Biodiversity enhances ecosystem reliability. *Nature (London)* **390,** 507–509.

Naeem, S., Hakansson, K., Lawton, J. H., Crawley, M. J., and Thompson, L. J. (1996). Biodiversity and plant productivity in a model assemblage of plant species. *Oikos* **76,** 259–264.

Nijs, I., and Roy, J. (2000). How important are species richness, species evenness and interspecific differences to productivity? A mathematical model. *Oikos* **88,** 57–66.

Olff, H., and Baker, J. P. (1991). Long-term dynamics of standing crop and species composition after the cessation of fertilizer application to mown grassland. *J. Appl. Ecol.* **28,** 1040–1052.

Owensby, C. E., Ham, J. M., Knapp, A. K., and Auen, L. M. (1999). Biomass production and species composition change in a tall grass prairie ecosystem after long term exposure to elevated CO_2. *Global Change Biol.* **5,** 497–506.

Pianka, E. R. (1966). Latitudinal gradients in species diversity: A review of concepts. *Am. Nat.* **100,** 30–46.

Pimm, S. L., Russel, G. J., Gittleman, J. L., and Brooks, T. M. (1995). The future of biodiversity. *Science* **269,** 347–350.

Prieur-Richard, A.-H., Lavorel, S., K. Grigulis and Dos Santos, A. (2000). Plant community diversity and invasibility by exotics: the example of Coniza bonariensis and C. canadensis invasion in Mediterranean annual old fields. *Ecol. Lett.* (in press).

Ricklefs, R. E., and Schluter, D. (1993). "Species Diversity in Ecological Communities: Historical and Geographical Perspectives." Univ. Chicago Press, Chicago.

Sankaran, M., and McNaughton, S. J. (1999). Determinants of biodiversity regulate compositional stability of communities. *Nature (London)* **401,** 691–693.

Scheu, S., Alphei, J. Bonkowski, M., and Jentschke, G. (1999). Soil food web interactions and ecosystem functioning: Experimental approaches with systems of increasing complexity. *In* "Going Underground. Ecological Studies in Forest Soils" (N. Rastin and J. Bauhus, eds.), pp. 1–32. Research Signpost, Trivandrum, India.

Schläpfer, F., and Schmid, B. (1999). Ecosystem effects of biodiversity: A classification of hypotheses and exploration of empirical results. *Ecol. Appl.* **9,** 893–912.

Schulze, E.-D., and Mooney, H. A., eds. (1993). "Biodiversity and Ecosystem Function." Springer-Verlag, Berlin.

Silvertown, J. (1980). The dynamics of a grassland ecosystem: Botanical equilibrium in the Park Grass experiment. *J. Appl. Ecol.* **17,** 491–504.

Swift, M. J., and Anderson, J. M. (1993). Biodiversity and ecosystem function in agricultural systems. *In* "Biodiversity and Ecosystem Function" (D. E. Schulze and H. A. Mooney, eds.), pp. 15–41. Springer-Verlag, Berlin.

Thompson, K. (1994). Predicting the fate of temperate species in response to human disturbance and global changes. *In* "Biodiversity, Temperate Ecosystems, and Global Changes" (T. J. B. Boyle and C. E. B. Boyle, eds.), pp. 61–76. Springer-Verlag, Heidelberg.

Tilman, D. (1982). "Resource Competition and Community Structure," Princeton Univ. Press, Princeton.

Tilman, D. (1997a). Community invasibility, recruitment limitation, and grassland biodiversity. *Ecology* **78,** 81–92.

Tilman, D. (1997b). Distinguishing between the effect of species diversity and species composition. *Oikos* **80,** 185.

Tilman, D. (1999). The ecological consequences of changes in biodiversity: A search for general principles. *Ecology* **80,** 1455–1474.

Tilman, D., and Pacala, S. (1993). The maintenance of species richness in plant communities. *In* "Species Diversity in Ecological Communities: Historical and Geographical Perspectives" (R. E. Ricklefs and D. Schluter, eds.), pp. 13–25. Univ. Chicago Press, Chicago.

Tilman, D., Wedin, D., and Knops, J. (1996). Productivity and sustainability influenced by biodiversity in grassland ecosystems. *Nature (London)* **379,** 718–720.

Tilman, D., Knops, J., Wedin, D., Reich, P., Ritchie, M., and Sieman, E. (1997a). The influence of functional diversity and composition on ecosystem processes. *Science* **277,** 1300–1302.

Tilman, D., Lehman, C. L., and Thomson, K. T. (1997b). Plant diversity and ecosystem productivity: Theoretical considerations. *Proc. Natl. Acad. Sci. U.S.A.* **94,** 1857–1861.

Tilman, D., Lehman, C. L., and Bristow, C. E. (1998). Diversity-Stability relationships: Statistical inevitability or ecological consequence? *Am. Nat.* **151,** 277–282.

van der Hiejden, M. G. A., Klironomos, J. N., Ursic, M., Moutoglis, P., Streitwolf-Engel, R., Boller, T., Wiemken, A., and Sanders, I. R. (1998). Mycorrhizal fungal diversity determines plant diversity, ecosystem variability and productivity. *Nature (London)* **396,** 69–72.

van der Heijden, M. G. A., Klironomos, J. N., Ursic, M., Moutoglis, P., Streitwolf-Engel, R., Boller, T., Wiemken, A., and Sanders, I. R. (1999). 'Sampling effect.' A problem in diversity manipulation? A reply to David A. Wardle. *Oikos* **87,** 408–410.

van der Putten, W. H. Mortimer, S. R., Hedlund, K., Van Dijk, C., Brown, V. K., Leps, J., Rodriguez-Barrueco, C., Roy, J., Diaz Len, T. A., Gormsen, D., Korthals, G. W., Lavorel, S., Santa Regina, I., and Smilauer, P. (2000). Plant species diversity as a driver of early succession in abandoned fields: a multi-site approach. *Oecologia* (in press).

Vitousek, P. M., and Hooper, D. U. (1993). Biological diversity and terrestrial ecosystem biogeochemistry. *In* "Biodiversity and Ecosystem Function" (E.-D. Schulze and H. A. Mooney, eds.), pp. 3–14. Springer-Verlag, Berlin.

Vitousek, P. M., Mooney, H. A., Lubchenco, J., and Melillo, J. M. (1997). Human domination of earth's ecosystems. *Science* **277,** 494–499.

Walker, B. H. (1992). Biodiversity and ecological redundancy. *Conserv. Biol.* **6,** 18–23.

Walker, B. H. (1995). Conserving biological diversity through ecosystem resilience. *Conserv. Biol.* **9,** 747–752.

Wall-Freckman, D., Blackburn, T. H., Brussaard, L., Hutchings, P., Palmer, M. A., and Snelgrove, P. V. R. (1997). Linking biodiversity and ecosystem functioning of soils and sediments. *Ambio* **26,** 556–562.

Wardle, D. A. (1999). Is 'sampling effect' a problem for experiments investigating biodiversity–ecosystem function relationships? *Oikos* **87,** 403–407.

Wardle, D. A., and Lavelle, P. (1997). Linkages between soil biota, plant litter quality and decomposition. *In* "Driven by Nature: Plant Litter Quality and Decomposition" (G. Cadisch and K. E. Giller, eds.), pp. 107–124. CAB International, Wallingford.

Wardle, D. A., Bonner, K. I., Barker, G. M., Yeates, G. W., Nicholson, K. S., Bardgett, R. D., Watson, R. N., and Ghani, A. (1999). Plant removals in perennial grassland: Vegetation dynamics, decomposers, soil biodiversity, and ecosystem properties. *Ecol. Monogr.* **69,** 535–568.

Wilsey, B. J., and Potvin, C. (2000). Biodiversity and ecosystem functioning: importance of species evenness and identity in an old field. *Ecology* **81,** 887–892.

Wright, D. H. (1983). Species-energy theory: An extension of species-area theory. *Oikos* **41,** 496–506.

Yachi, S., and Loreau, M. (1999). Biodiversity and ecosystem productivity in a fluctuating environment: the insurance hypothesis. *Proc. Natl. Acad. Sci. USA* **96,** 1463–1468.

II

Ecosystem Productive Performance

9

Productivity of Arctic Ecosystems

Gaius R. Shaver and Sven Jonasson

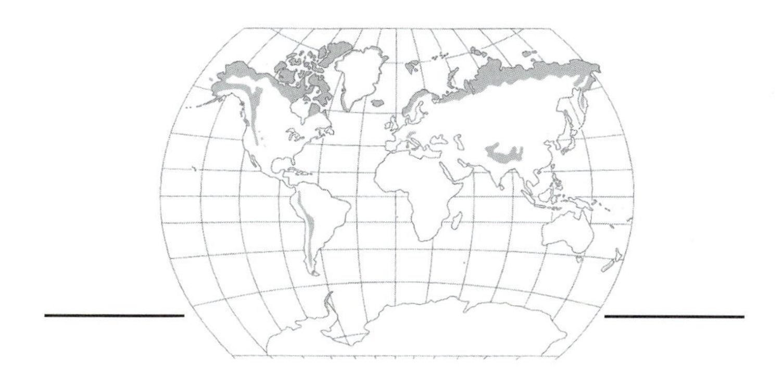

I. Introduction

Over the past 25 years, research on arctic productivity has shifted from a focus on species-level adaptations in an extreme environment to a focus on integrated plant communities and whole ecosystems. Indirect effects of the arctic environment, such as the effects of low temperature on soil resources, have frequently been shown to be of greater long-term importance to productivity than direct effects. In arctic tundras, although there is clear evidence for competitive partitioning of limiting resources among species, it is unclear whether species diversity contributes significantly to total resource uptake or to productivity. Instead, the major long-term controls on both primary production and net ecosystem production appear to be related to long-term inputs and outputs of limiting nutrients such as N and P, and to long-term controls on nutrient mineralization and accumulation of nutrients in major organic matter pools.

Arctic tundras, polar deserts, and semideserts are among the least productive of the world's major ecosystem types. This is not surprising given the extremes of temperature, moisture, light regime, and other features that characterize the arctic environment. Yet, within the Arctic there is also considerable variation in primary production, both within and among the various kinds of arctic vegetation. Furthermore, although primary production is low, large amounts of organic matter have accumulated in the Arctic since deglaciation and changes in net ecosystem production have the potential to contribute significantly to the global C balance. Understanding of controls on primary production and net ecosystem production in the Arctic, and on variation in the responsiveness of different arctic ecosystems to climate change and disturbance, has advanced rapidly over the past two decades. The aim here is to review and summarize these more recent advances, starting in the mid-1970s.

We start in the mid-1970s because the International Biological Programme's Tundra Biome study, which took place from 1969 to 1974, was a watershed event in the development of our understanding of arctic productivity. The Tundra Biome study was significant because it marked a shift from a focus on individual plants and species, and their adaptations to the arctic environment (reviewed by Billings and Mooney, 1968), to a focus on whole plant communities and ecosystems. The Tundra Biome study provided, for the first time, integrated data sets from a wide range of arctic, subarctic, alpine, and antarctic sites in which measurements of primary production and its component processes (e.g., photosynthesis, respiration, nutrient uptake, and allocation) were combined with information on ecosystem carbon and nutrient budgets. Results from many of the main Tundra Biome sites were synthesized in individual volumes, including Wielgolaski (1975), Bliss (1977), Tieszen (1978), Heal and Perkins (1978), Sonesson (1980), and Brown *et al.* (1980); a major international synthesis and comparative analysis was edited by Bliss *et al.* (1981).

Since the mid-1970s, research on primary production in the Arctic has revolved around several main themes that emerged from the Tundra Biome results. One theme is that a narrow focus on the direct effects of environmental factors (especially temperature) on production processes is incomplete, and that indirect effects (such as the effect of temperature on nutrient availability) and interactions between environmental factors must be considered to explain adequately longer term controls and spatial patterns of productivity (Chapin, 1983). A second theme is that primary production, biomass accumulation, and overall element cycling at a given site are affected strongly by the particular species or plant growth forms that occur on the site, in addition to the direct environmental controls (Berendse and Jonasson, 1992; Hobbie, 1995). A third theme, related to both of the others, is that short-term changes in individual carbon and nutrient cycling processes are in the long term constrained by longer term controls on organic matter accumulation and distribution between plants and soil (Shaver *et al.*, 1992).

II. Environment

The total area covered by arctic ecosystems (excluding oceans and ice-covered areas) is about 5.6 million km^2 (Table 9-1) (Bliss and Matveyeva, 1992). Almost all of this area lies north of the Arctic Circle and is snow covered for 9 to 10 months per year. Air temperatures during the growing season are low, usually less than 10°C, and frosts or snowfall can occur at any time. Annual average temperatures are typically from 1 to 15° below freezing (Barry *et al.*, 1981; Maxwell, 1992), leading to the formation of permafrost in most parts of the Arctic. The presence of permafrost means that the soils are both cold and restricted in terms of the volume of unfrozen soil that is available for biological activity. Annual depth of soil thaw varies from 20–30 cm to 1–5 m, depending on topography, vegetation, and parent material. Deep soil drainage is also impeded by permafrost, so arctic soils are often wet and low in oxygen, with a steep temperature gradient from the soil surface to the top of the permafrost (Everett *et al.*, 1981; Kane *et al.*, 1992).

Because they lie north of the Arctic Circle, arctic ecosystems are characterized by continuous daylight for at least some portion of each summer and continuous darkness in midwinter. The period of continuous daylight generally overlaps with the time of transition from continuous snowcover to rapid early-season growth. An important consequence is that the highest solar radiation inputs, which peak in late June, occur before the plant canopy is fully developed and often before annual leaf growth has started (Chapin and Shaver, 1985a). Radiation declines continuously from the start of the growing season. Radiation intensity is never high in comparison with temperate or tropical systems, although daily total inputs can be substantial (e.g., ~23 Mj m^{-2} d^{-1} in June at Point Barrow, Alaska) (Brown *et al.*, 1980). On sunny days at midseason, peak levels of photosynthetically active radiation averaged 1400 μmol photons m^{-2} s^{-1}, with an average daily total of 46 mol photon m^{-2}, at Toolik Lake, Alaska (68°38′N) (Chapin *et al.*, 1995)

The energy balance of arctic ecosystems is tightly linked to the hydrologic cycle, in particular through the thawing and freezing of water at the start and end of the growing season and through the effects of snow cover on net radiation (Brown *et al.*, 1980; Barry *et al.*, 1981; Kane *et al.*, 1992). During the winter the albedo of the snow-covered surface is high, about 80%, shifting to about 20% during the summer. Net radiation increases rapidly with snowmelt, to about 10–15 Mj m^{-2} d^{-1} in June and decreasing to less than half that amount by mid-August at Imnavait Creek, Alaska (Hope and Stow, 1996; Hinzman *et al.*, 1996). During the growing season evapotranspiration is the dominant component of the surface energy budget, because the soils are usually wet (despite low precipitation inputs) due to the presence of permafrost, which prevents deep drainage.

Table 9-1 Primary Production and Organic Matter Stocks in Major Arctic Ecosystem Types[a]

Vegetation type	Net annual production ($g\ m^{-2}$)			Organic matter mass ($kg\ m^{-2}$)			Global total	
	Aboveground	Belowground	Total	Plant	Soil	Total	Area ($\times 10^6$ km)	Organic matter ($\times 10^{15}$ g)
Low Arctic								
Tall shrub	400	600	1000	2.61	0.40	3.01	0.174	0.52
Low shrub	125	250	375	0.77	3.8	4.57	1.282	5.86
Tussock/sedge–dwarf shrub	125	100	225	3.33	29.0	32.33	0.922	29.81
Wet sedge/mire	70	150	220	0.95	38.75	39.70	0.880	34.94
Semidesert	28	17	45	0.29	7.20	7.49	0.358	2.68
High Arctic								
Wet sedge/mire	60	80	140	0.75	21.00	21.75	0.132	2.87
Semidesert	25	10	35	0.25	1.03	1.28	1.005	1.28
Polar desert	0.7	0.3	1	0.002	0.02	0.03	0.847	0.02
Total Arctic							5.600	77.98

[a]Modified from Bliss and Matveyeva (1992) and Oechel and Billings (1992). If organic matter within permafrost is included (i.e., below the seasonally thawed active layer), soil organic matter stocks may be double the amounts shown (Michaelson *et al.*, 1996).

Arctic vegetation is low in stature (usually less than 1 m tall), without trees, and aerodynamically smooth. Often the vegetation canopy is wind sculpted, filling topographic depressions to the level of the top of the winter snow cover; in exposed, wind-blown sites the vegetation may be less than 5 cm tall. Nonetheless, canopy height and foliage area, combined with the energy exchange characteristics of the surface moss layer, are major controls on surface temperature and heat flux into arctic soils, which in turn control soil thaw (Brown *et al.*, 1980; Tenhunen *et al.*, 1992, 1994).

III. Primary Production and Net Ecosystem Production

New measurements of arctic primary production have increased our understanding of its spatial and temporal variation and correlations with environmental factors. Since 1975 the most intensive and detailed studies have been done in Alaskan moist and wet tundras (Batzli, 1980; Miller, 1982; Oechel, 1989; Shaver and Chapin, 1991; Reynolds and Tenhunen, 1995). Other intensively studied sites include Alexandra Fiord on Ellesmere Island, Canada (Svoboda and Freedman, 1994) and Abisko in Sweden (Karlsson and Callaghan, 1996); surveys of primary production across the Canadian High Arctic were completed by Bliss and Svoboda (1984) and Bliss *et al.* (1984). All of these studies have focused on aboveground production, with little or no new information on root production becoming available since the Tundra Biome study.

The broad geographic patterns of primary production in the Arctic (Table 9-1) (see Bliss and Matveyeva, 1992; Oechel and Billings, 1992) range over three orders of magnitude, from high arctic polar deserts ($\sim$1 g m^{-2} yr^{-1}) to low arctic shrub communities ($\sim$1000 g m^{-2} yr^{-1}). Although there is a general trend of decreasing production as latitude increases (and both temperature and length of growing season decrease), at any latitude the variation in annual production may be nearly as great (up to 100-fold) as that between latitudes. Where habitats are similar, as in High Arctic and Low Arctic mires, production is similar (60 vs. 70 g m^{-2} yr^{-1}) (Table 9-1). The reason for this pattern is that local variation in topography, soil moisture, and exposure to wind are dominant controls over the kind of vegetation that can grow in a given site and over the long-term accumulation of organic matter and nutrients that support primary production. Thus primary production in the Arctic varies in a mosaic fashion, with dramatic variation among vegetation types (especially along topographic gradients) (Billings, 1973; Shaver *et al.*, 1996a,b) and increasing frequency of low-productivity vegetation as latitude increases.

Net ecosystem production (NEP), defined as the net change over time in

C or organic matter stocks in the whole ecosystem (vegetation plus soils and consumers), is also highly variable across the Arctic. Although NEP has rarely been measured directly in the Arctic, the 1000-fold range in total organic matter stocks of arctic ecosystems (Table 9-1) indicates that even though both primary production and respiration are low by global standards, small differences in the balance between them can lead to large differences in net accumulation. The arctic ecosystems with the highest net accumulations of organic matter are not those with the highest primary production, but those with the wettest soils where decomposition is limited relatively more than production.

Estimates of arctic NEP calculated from organic matter budgets generally range from 0 to 30 g C m^{-2} yr^{-1} accumulation (Oechel and Billings, 1992), roughly in agreement with early measurements made during the 1970s (Chapin *et al.*, 1980). However, measurements by Oechel *et al.* (1993) indicate large net annual losses of C and organic matter from Alaskan wet and moist tundras, in some years exceeding 200 g C m^{-2}, or nearly 1% of the total organic C stock above the permafrost at these sites. The most likely explanation for these recent observations is that they were made during a series of unusually warm years, perhaps with less precipitation, greater evaporation, and greater soil thaw and drainage than in earlier decades. This explanation is supported by laboratory microcosm studies (Billings *et al.*, 1982, 1983; Johnson *et al.*, 1996), but field experiments have not been completed. It is also uncertain whether such large net losses can continue even if global warming is sustained, due to longer term feedbacks and interactions with the N cycle (McKane *et al.*, 1997a,b).

Most previous estimates of arctic NEP have considered only the net C or organic matter balance of terrestrial components of the landscape, despite the knowledge that many arctic lakes and ponds show consistent net C losses to the atmosphere (Hobbie, 1980). The lakes lose C because they are typically supersaturated with CO_2 with respect to the atmosphere. The most likely source of this excess CO_2 is terrestrial root and soil respiration, which produces CO_2 that is dissolved in soil water and, along with terrestrially fixed dissolved organic C, transported in soil water from land to streams, lakes, and ponds. Kling *et al.* (1991; Kling, 1995) calculated that for the North Slope of Alaska these net losses from aquatic ecosystems may, on a regional basis, be as high as 30% of gross primary production on land. If this is true, previous estimates of NEP based on net terrestrial gas exchange and terrestrial organic matter budgets alone have seriously overestimated landscape-level and regional NEP, and other measurements of net C losses from terrestrial ecosystems have underestimated the net loss (Oechel and Billings, 1992). This is an active area of ongoing research (Johnson *et al.*, 1996).

IV. Environmental Controls, Direct and Indirect

Arctic plants are well adapted to low temperatures, both morphologically and physiologically (Billings and Mooney, 1968). Perhaps in part as a result of this successful adaptation, annual production of arctic plants and vegetation is less sensitive to temperature than to other environmental factors, including light and especially nutrient availability and length of growing season (Chapin, 1983). For vascular plants, water stress is rare in wet or moist tundras although in the High Arctic plants in exposed sites may be directly water limited (e.g., Oberbauer and Miller, 1979; Oberbauer and Dawson, 1992; Gold and Bliss, 1995). Many of these basic patterns of physiological control were well established by the end of the 1970s (reviewed by Chapin and Shaver, 1985a). Subsequent research has focused on CO_2 effects (e.g., Tissue and Oechel, 1987; Oberbauer *et al.*, 1986) and on moss and lichen productivity (e.g., Murray *et al.*, 1989; Tenhunen *et al.*, 1992).

Field experiments have been particularly useful in testing our physiologically based understanding of controls on primary production in the Arctic. Similar experiments, including factorial manipulation of temperature, light, and nutrient availability, have been completed in Alaska, Sweden, and Svalbard (Havström *et al.*, 1993; Wookey *et al.*, 1993–1995; Parsons *et al.*, 1994; Chapin *et al.*, 1995; Chapin and Shaver, 1996; Shaver *et al.*, 2000). A general conclusion from these experiments is that nutrient availability is far more consistently and strongly limiting to annual production than is light or temperature (Callaghan and Jonasson, 1995; Jonasson *et al.*, 1996). For example, although fertilizer addition may double primary production and triple biomass after 3–9 years, when air temperature is increased 2–5°C on average for up to 9 years (with a plastic greenhouse), most species and the vegetation as a whole show only small and usually nonsignificant increases (Chapin *et al.*, 1995). The increases in production that do occur with warming may be explained entirely by increased soil nutrient mineralization in the warmed treatments, and consequent increases in vegetation N mass (Shaver *et al.*, 2000). In some individual species there is a greater warming response in the colder, higher latitudes than at warmer, lower, latitudes (Callaghan and Jonasson, 1995; Jonasson *et al.*, 1996), suggesting greater temperature limitation in the high-latitude site. Similarly, although after 9 years at 50% of ambient light the productivity of Alaskan tussock tundra was reduced by nearly half, this response took several years to develop and contrasted with dramatic increases in production of fertilized plots (Chapin *et al.*, 1995). Consistent, strong nutrient limitation of primary production has also been demonstrated in extensive series of factorial fertilizer experiments (McKendrick *et al.*, 1978, 1980; Shaver and Chapin, 1980, 1986, 1995). Well-drained sites tend to be N limited, whereas poorly drained sites are usually P limited.

There has been only one series of experiments in which CO_2 concentration was manipulated in the field, but it, too, showed little or no long-term effect on primary production, despite an initially strong increase with increased CO_2 (Tissue and Oechel, 1987; Grulke *et al.*, 1990; Oechel *et al.*, 1994). Similar acclimation occurred in laboratory experiments, but there was considerable variation among species in the extent of acclimation (Oberbauer *et al.*, 1986). In general it appears that nutrient limitation restricts the capacity of tundra plants and vegetation to sustain an initial increase in primary production under increased CO_2.

In both field and laboratory experiments, NEP is also strongly nutrient limited and much less affected by temperature increases or decreases in light (Billings *et al.*, 1982–1984; Johnson *et al.*, 1996; McKane *et al.*, 1997a; Shaver *et al.*, 2000). With respect to CO_2 concentration, in both field and laboratory NEP shows a strong initial increase in response to increased CO_2 concentrations, but within weeks (in the laboratory) or 2–3 years (in the field) there is a complete, ecosystem-level acclimation of NEP, again probably due to nutrient limitation (Billings *et al.*, 1982, 1984; Grulke *et al.*, 1990; Oechel *et al.*, 1994). However, increases in temperature interact strongly with both nutrient availability and CO_2 in regulation of NEP, for reasons that are still not entirely clear. In field experiments in Alaska there is a strong *negative* temperature/fertilizer interaction, such that the NEP of warmed and fertilized tundra is significantly lower than that of tundra treated with fertilizer only (McKane *et al.*, 1997a; Shaver *et al.*, 2000). The probable explanation for this is that at high temperatures and high nutrient status, plant respiration is dramatically increased. In contrast, when CO_2 concentrations are increased in combination with an increase in temperature, the initial increase in NEP is sustained for at least 3 years (Oechel and Billings, 1992). It may be that the higher temperatures somehow allow greater plant nutrient uptake when combined with high CO_2, but additional research is needed to explain these temperature/nutrient/CO_2 interactions.

On a landscape and regional basis, NEP is more closely related to soil moisture than any other environmental variable (Tenhunen *et al.*, 1994; Oberbauer *et al.*, 1996), as is organic matter accumulation in general. Among the many factors responsible for this correlation are the strong control by soil moisture over moss and lichen abundance, which is critical in controlling surface temperature, depth of thaw, and evapotranspiration (Tenhunen *et al.*, 1992). A thick moss and/or lichen layer will act to insulate the soil surface, reducing heat flux into the soil. Soil moisture alone is also strongly correlated with depth of thaw and heat flux through the soil, which affect not only soil respiration and decomposition but also plant nutrient uptake and primary production. Laboratory experiments (Billings *et al.*, 1982, 1983; Johnson *et al.*, 1996) indicate that the major impact of soil moisture is on soil respiration, with a strong increase in net C losses if drainage is accompanied by increased temperature but little change under increased temperature only. Modeling studies (McKane *et al.*, 1997b) also support the contention of Oechel *et al.* (1993) that the recent, apparent shift of the North Slope of Alaska from a net C sink to a net source may be due to increased soil drying, increased soil thaw, and increased drainage.

Finally, there is little or no relationship between NEP and primary production in the Arctic, either in comparison of regional patterns (Table 9-1), in field experiments (e.g., Shaver *et al.*, 2000), or in laboratory experiments (Johnson *et al.*, 1996). The arctic ecosystems with the highest primary production are by no means those with the greatest organic matter accumulation rates, nor are those with the lowest primary production. Although the two main components of NEP—primary production and heterotrophic respiration—must ultimately be linked to each other, at least over the time scale of present knowledge that linkage appears quite flexible. Understanding the linkages between primary production and heterotrophic respiration is another key issue for future research.

V. Effects of Species Composition

One of the most striking features of arctic vegetation is the remarkable variation in plant growth form abundance that frequently occurs over very short distances (Billings, 1973; Walker, 1995). This variation in vegetation composition is often correlated with dramatic variation in primary production (Chapin *et al.*, 1980, 1988; Shaver *et al.*, 1996a,b). A number of studies have attempted to determine whether the changes in species composition contribute directly to changes in production, or whether production is determined solely by resource availability and independent of species composition. Research in this area has focused on (1) partitioning of resource uptake among species, particularly uptake of soil nutrients, (2) nutrient use efficiency and C and nutrient turnover within plants, and (3) long-term effects of plants on litter decomposition and nutrient recycling. Comprehensive reviews of these topics are available in several chapters in Chapin and Körner (1995), in Berendse and Jonasson (1992), in Hobbie (1992, 1996), and in Callaghan and Jonasson (1995).

Partitioning of uptake of growth-limiting resources may increase productivity of arctic vegetation if some species can obtain resources in different forms or in different locations or times than other species; the result may be higher total uptake by the community. Indirect evidence of resource partitioning is provided by the common observation of interspecies differences in root distribution, canopy distribution, and phenology (e.g., Shaver and Billings, 1975; Tenhunen *et al.*, 1994; Shaver, 1995). More direct evidence is provided by analyses of $\delta^{15}N$ values in leaves and other tissues, which differ sharply among species and growth forms (Michelsen *et al.*, 1996; Jonasson and Michelsen, 1996; Nadelhoffer *et al.*, 1996). The $\delta^{15}N$ of nonmycorrhizal graminoids such as *Eriophorum* and *Carex* species is typically around +1 to −2, whereas that of mycorrhizal shrubs (*Betula nana, Salix* spp.) is in the −4 to −6 range; ericaceous species (*Ledum, Vaccinium, Empetrum*) may be as low as −8. These data suggest that the major plant forms obtain their N from different sources in the soil, even though most are capable of taking up N in several different chemical forms (Kielland and Chapin, 1992; Kielland, 1994; Leadley *et al.*, 1997). Further evidence for resource partitioning comes from experiments in which removal of species from the community was not compensated by increased growth of remaining species (Fetcher, 1985; Jonasson, 1992).

Species composition may also affect primary production through differences in C and nutrient use. During the 1970s there were several studies that seemed to indicate that rapidly growing plant forms such as deciduous shrubs, which tend to dominate in the most productive arctic sites, were more productive in those sites compared to other growth forms (such as evergreens) because the deciduous shrubs had higher nutrient uptake rates at high levels of nutrient availability and were able to grow more rapidly and thus produce more (e.g., Chapin *et al.*, 1978, 1980). On the other hand, a negative consequence of this more rapid growth was the tendency of deciduous species to produce new tissues with higher N and P content; i.e., they produced less new biomass per unit nutrient, or had lower nutrient use efficiency (Jonasson, 1983; Chapin and Shaver, 1985b). However, later and more detailed analysis of biomass and nutrient budgets in contrasting vegetation types (Shaver and Chapin, 1991) indicated that differences in efficiency of N and P use were negligible at the level of whole plants or whole vegetation, mainly because of the relatively large allocation to woody stems (with low N and P concentrations) by deciduous shrubs, which increased their overall nutrient use efficiency. Furthermore, under most circumstances soil nutrient uptake at the level of the whole vegetation appears to be much more strongly limited by supply to the available pool than by plant uptake capacity, so differences in uptake kinetics among species may have little effect on community productivity (Leadley *et al.*, 1997).

Species composition has also been linked to carbon and nutrient turnover in the Arctic (Chapin *et al.*, 1978, 1980; Schulze and Chapin, 1987), again based on the observation that the most productive sites tend to be dominated by the species with the most rapid growth rates and highest nutrient requirements in high-turnover leaves. The prediction of higher turnover in the most productive sites was based largely on extrapolation from studies of single species and on comparisons of growth and physiology of individual plant parts. Yet, when production:biomass relationships for whole vegetation are compared across a wide range of arctic vegetation types, produc-

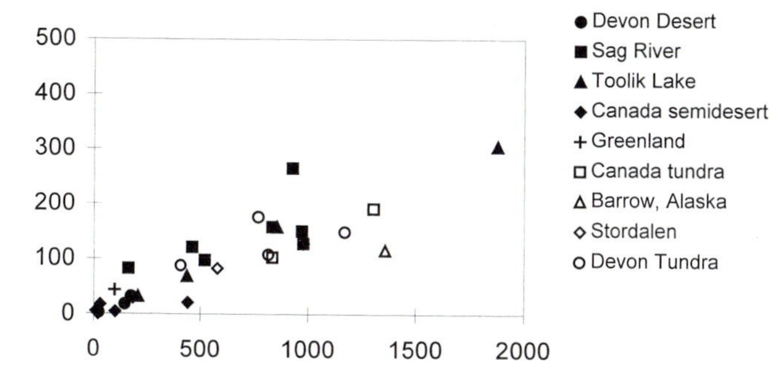

Figure 9-1 Annual net primary production (NPP) above- and belowground (g m^{-2} yr^{-1}) by vascular plants versus live vascular plant biomass (g m^{-2}) in six communities along a single toposequence in northern Alaska (Sag River) (Shaver *et al.*, 1996a,b), plus the four communities studied by Shaver and Chapin (1991) at Toolik Lake, Alaska, and at 21 other polar desert, semidesert, and arctic tundra sites summarized by Wielgolaski *et al.* (1981) (Devon Island, Canada semidesert, Greenland, Canada tundra, Stordalen, and Barrow, Alaska). The correlation between production and biomass is highly significant ($P < 0.001$; $r^2 = 0.77$).

tion:biomass ratios are nearly constant and, if anything, indicate that the sites with the highest biomass turnover are those with the lowest production and biomass (Fig. 9-1) (Shaver *et al.*, 1996a,b). Once again, the reason is that the most productive sites are those with the largest woody biomass, which turns over relatively slowly. Although total C and nutrient flux through the vegetation is greater in the most productive sites, the flux per unit biomass is, if anything, less.

A final mechanism by which species composition might affect both primary production and net ecosystem production is through species-related differences in litter decomposition and nutrient mineralization. The leaf and stem litter produced by arctic plants differs greatly in its relative decomposability, suggesting that soil organic matter accumulation and recycling of nutrients in support of primary production should be strongly affected by species composition of the vegetation (Berendse and Jonasson, 1992; Jonasson and Michelsen, 1996; Hobbie, 1996). The most decomposable litter types are often leaves of graminoids and forbs, with evergreen and deciduous shrub leaves and wood being least decomposable (particularly those with high concentrations of plant secondary compounds). Laboratory studies indicate that soils formed beneath different vegetation types also differ in their respiration and mineralization rates roughly as would be predicted by differences in initial litter quality (Nadelhoffer *et al.*, 1991, 1992; Shaver *et al.*, 1996a,b). In field studies, Jonasson (1983; Jonasson *et al.*, 1996; Jonasson and Michelsen, 1996) observed that overall nutrient turnover in soils and vegetation is related to species composition. Thus, although soil temperature and moisture are probably the dominant factors controlling soil organic matter accumulation and nutrient availability in the Arctic, species-related differences in litter chemistry may contribute significantly.

VI. Short- versus Long-Term Controls

Understanding of controls on productivity of arctic vegetation and ecosystems is complicated by the fact that different components of the production system respond at different rates, and by feedbacks between components. Thus, initial responses to a change in environment are often not sustained, and major long-term controls may not be apparent from short-term experiments (Chapin *et al.*, 1995; Jonasson *et al.*, 1996). Yet, long-term modeling and prediction depends on understanding these changes (Reynolds and Leadley, 1992; Reynolds *et al.*, 1996; McKane *et al.*, 1997b).

Year-to-year variation in primary production in the Arctic is often only poorly correlated with year-to-year variation in weather (Chapin and Shaver, 1985a,b; Shaver *et al.*, 1986), even though average climate is a relatively good predictor of long-term average production within a given vegetation type (Bliss *et al.*, 1981). Although essentially all of the individual physiological processes that support primary production (photosynthesis, nutrient uptake, respiration, growth) are strongly responsive to variation in temperature, light, moisture, and other weather variables, it turns out that at the level of the whole community, production is effectively buffered against annual variation in these variables in at least three different ways:

1. Probably most important, current-year's primary production in most arctic plants is strongly dependent on resources, especially N and P, that were acquired in previous years (Chapin *et al.*, 1986; Chapin and Shaver, 1985a,b, 1988, 1989). Large fractions of these nutrients may have been recycled internally over many years (Jonasson and Chapin, 1985, 1991; Jonasson, 1989; Jónsdóttir and Callaghan, 1989, 1990). Thus, at the individual plant level, annual growth is effectively dependent on resources acquired over several years, so that the effects of weather in the current year are less important than if growth were dependent on concurrent resource uptake.
2. At least in mixed-growth form communities such as moist tussock tundra, not all species respond in the same way to current-year's weather, so that it is unusual for all species to attain maximum or minimum annual growth in the same year. As a result, annual variation in community production is much less than annual variation in growth or production of individual species (Chapin and Shaver, 1985a,b).
3. Because the nutrient supply to plant-available pools depends strongly on mineralization of soil organic matter, rather than on current deposition or fixation (Chapin *et al.*, 1980; Leadley *et al.*, 1997), there is a further

lag in the relationship between current weather and primary production. This lag results from the fact that most net N or P mineralization takes place late in the growing season, after most of the current-year's growth has occurred (Giblin *et al.*, 1991, 1995; Nadelhoffer *et al.*, 1992; Jonasson *et al.*, 1996).

In sum, these three buffering mechanisms act to stabilize annual variation in primary production so that production in a given year is more reflective of long-term climate averages rather than current-year's weather. For much the same reasons, these mechanisms may act to delay the production response to disturbance or climate change that results in sustained, one-directional changes in the plant environment.

Long-term (multiyear) changes in both primary production and NEP are regulated differently from short-term responses to weather within a year or annual variation in weather about climatic means. In addition to controls on fluxes of C and nutrients between vegetation, soil, water, and the atmosphere, long-term changes in production of arctic ecosystems are also constrained by the total amounts of organic matter in vegetation and soil and by stoichiometric constraints on the chemical composition of the organic matter (Shaver *et al.*, 1992; Rastetter *et al.*, 1997; McKane *et al.*, 1997a; Waelbroeck *et al.*, 1997). Current understanding of long-term controls on primary production and NEP in the Arctic can be summarized in a simple conceptual model of carbon–nutrient interactions (Fig. 9-2) (Shaver *et al.*, 1992).

This model structure (Fig. 9-2) helps to clarify several characteristics of the long-term regulation of primary production in arctic ecosystems. Because the C:N and C:P ratios of both primary production and vegetation biomass vary only slightly across a wide range of vegetation types and plant forms (Shaver and Chapin, 1991), large and long-term changes in both production and biomass will require a sustained change in nutrient uptake from the soil. This increase in nutrient uptake could be supplied either from external sources (e.g., N deposition or fixation) or from internal recycling through decomposition of litter and soil organic matter. However, in virtually all arctic ecosystems external inputs are only a small fraction of internal recycling (Chapin *et al.*, 1980; Nadelhoffer *et al.*, 1992), and the external inputs would have to increase several fold to equal only a 5–10% change in internal recycling. Thus, sustained changes in primary production will almost certainly be driven primarily by changes in soil decomposition and nutrient mineralization processes. In most arctic ecosystems the total amounts of potentially mineralizable N and P in the soil are large relative to the amounts annually mineralized, suggesting that large increases in primary production are possible if newly mineralized nutrients can be taken up by the vegetation and not lost by leaching. Exceptions include much of the High Arctic, where there is little soil organic matter available to be miner-

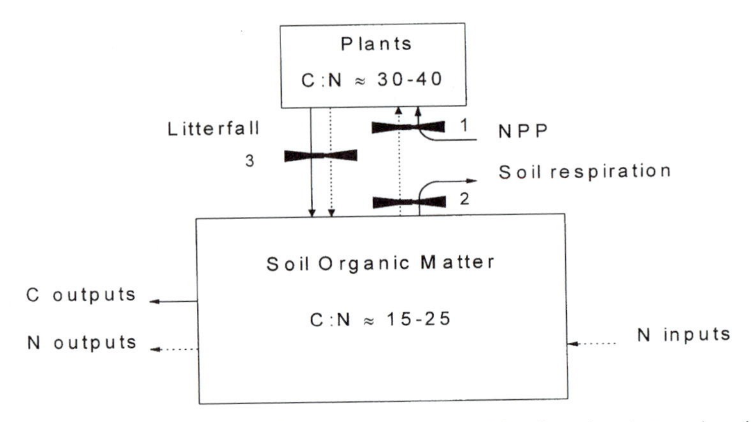

Figure 9-2 The Simple Arctic Model (Shaver *et al.*, 1992) of nutrient interactions in tundras, using C and N as an example. In this model there are two major pools of organic matter in the ecosystem (i.e., plants and soil). Carbon and N fluxes into and out of these pools are indicated by solid and by dashed lines, respectively. The "bow ties" suggest links between C and N fluxes. Bow tie #1 implies that net C uptake by plants (net primary production, NPP) is constrained by the plant's ability to take up N, and vice versa. In most arctic ecosystems, essentially all of the plant's nutrient supply comes from the mineralization of soil organic matter (including recent litter), and that N mineralization is linked to the loss of C in soil respiration. (Dissolved C losses indicated by the arrow at lower left are usually small relative to gaseous exchanges.) Thus, at least in a proximate sense, the overall C balance of arctic ecosystems (i.e., the difference between NPP and soil respiration) is largely determined by C gains associated with N uptake by plants, balanced against C losses associated with N mineralization.

alized and long-term changes in primary production are more tightly linked to N fixation (Gold and Bliss, 1995).

The model (Fig. 9-2) also helps highlight important aspects of the long-term regulation of NEP and total organic matter stocks. Here the key factors are (1) total nutrient content of the ecosystem, (2) the distribution of nutrients between plant and soil pools that differ in their C:nutrient ratios, and (3) the variability of C:nutrient ratios within plant and soil pools (Shaver *et al.*, 1992; Rastetter *et al.*, 1997; McKane *et al.*, 1997a). The total nutrient content of the system is important because inputs and outputs of the principal limiting nutrients (mainly N and/or P) are much more tightly constrained than inputs and outputs of C. Thus changes in NEP and total ecosystem organic matter stocks should be closely linked to changes in total N and/or P stocks. The distribution of nutrients between plants and soil is important because the amount of C associated with each unit of nutrient differs dramatically between these pools. For example, if vegetation has an overall C:N ratio of 30:1 and soil organic matter has a ratio of 15:1, the transfer of one unit of N from soil to plants will be associated with a loss (as soil respiration) of 15 units of C and a gain (as primary production) of 30 units of C, with no net change in ecosystem N stocks. Because in most arctic ecosystems, especially those in the Low Arctic, the amount of N in soil organic matter is many times the amount in vegetation, large, long-term increases in NEP are possible simply by redistributing the N from soils to vegetation. The third factor, changes in C:nutrient ratios within soils or plants, may play a particularly important role in the short-term, reflecting temporary immobilization of excess nutrients released in disturbances. In relatively undisturbed ecosystems, C:nutrient ratios are less variable, particularly in vegetation, than total are nutrient stocks or the distribution of nutrients between plants and soils.

VII. Priorities for New Research

Considerable progress has been made in understanding the controls over primary production and its variation across the arctic landscape. The greatest future need is to understand how long-term changes in primary production are linked to soil processes, and how interactions between vegetation and soil determine NEP. One specific need is for new data on root biomass, root production, and the partitioning of belowground respiration into root and soil components. Basic measurements of NEP are also lacking from many kinds of arctic ecosystems, and information is needed to delineate more clearly the landscape patterns of NEP and its environmental correlations. Controls on soil organic matter accumulation and C:nutrient interactions in arctic soils are another key to improving our understanding.

Perspectives on the role of temperature in arctic ecosystems have changed especially dramatically in the past three decades, from an early focus on adaptations to a low-temperature environment and direct effects of low temperature on individual plants and species. Great progress has been made in understanding of indirect effects of temperature and interactions of temperature with other environmental variables in control of plant processes, but we still do not clearly understand temperature/nutrient, temperature/CO_2, or temperature/soil moisture interactions in soils or in the regulation of NEP.

Acknowledgments

Research on arctic productivity was inspired by W.D. Billings, a great arctic ecologist who died January 4, 1997, during the preparation of this manuscript. The science of ecology owes much to his leadership and his teaching. Research by G.R. Shaver and colleagues was supported by a series of grants from the U.S. National Science Foundation to the Marine Biological Laboratory; the most recent of these were Grants #9019055, #9211775, and #9415411.

References

Barry, R. G., Courtin, G. M., and Labine, C. (1981). Tundra climates. *In* "Tundra Ecosystems: A Comparative Analysis" (L. C. Bliss, O. W. Heal, and J. J. Moore, eds.), pp. 79–114. Cambridge Univ. Press, Cambridge.

Batzli, G. O., ed. (1980). Patterns of vegetation and herbivory in arctic tundra: Results from the Research on Arctic Tundra Environments (RATE) Program. *Arctic Alpine Res.* **12,** 401–518.

Berendse, F., and Jonasson, S. (1992). Nutrient use and nutrient cycling in northern ecosystems. *In* "Arctic Ecosystems in a Changing Climate: An Ecophysiological Perspective" (F. S. Chapin III, R. L. Jefferies, J. F. Reynolds, G. R. Shaver, and J. Svoboda, eds.), pp. 337–358. Academic Press, New York.

Billings, W. D. (1973). Arctic and alpine vegetations: Similarities, differences, and susceptibility to disturbance. *BioScience* **23,** 697–704.

Billings, W. D., and Mooney, H. A. (1968). The ecology of arctic and alpine plants. *Biol. Rev.* **43,** 481–530.

Billings, W. D., Luken, J. O., Mortensen, D. A., and Peterson, K. M. (1982). Arctic tundra: A source or sink for atmospheric carbon dioxide in a changing environment? *Oecologia* **53,** 7–11.

Billings, W. D., Luken, J. O., Mortensen, D. A., and Peterson, K. M. (1983). Increasing atmospheric carbon dioxide: Possible effects on arctic tundra. *Oecologia* **58,** 286–289.

Billings, W. D., Luken, J. O., Mortensen, D. A., and Peterson, K. M. (1984). Interaction of increasing atmospheric carbon dioxide and soil nitrogen on the carbon balance of tundra microcosms. *Oecologia* **65,** 26–29.

Bliss, L. C. (1962). Adaptations of arctic and alpine plants to environmental conditions. *Arctic* **15,** 117–144.

Bliss L. C., ed. (1977). "Truelove Lowland, Devon Island, Canada: A high arctic ecosystem." University of Alberta Press, Edmonton.

Bliss, L. C., and Matveyeva, N. V. (1992). Circumpolar arctic vegetation. *In* "Arctic Ecosystems in a Changing Climate: An Ecophysiological Perspective" (F. S. Chapin III, R. L. Jefferies, J. F. Reynolds, G. R. Shaver, and J. Svoboda, eds.), pp. 59–89. Academic Press, New York.

Bliss, L. C., and Svoboda, J. (1984). Plant communities and plant production in the western Queen Elizabeth Islands. *Holarctic Ecol.* **7,** 325–344.

Bliss, L. C., Heal, O. W., and Moore, J. J., Eds. (1981). "Tundra Ecosystems: A Comparative Analysis." Cambridge Univ. Press, Cambridge.

Bliss, L. C., Svoboda, J., and Bliss, D. I. (1984). Polar deserts, their plant cover and production in the Canadian High Arctic. *Holarctic Ecol.* **7,** 305–324.

Brown, J., Miller, P. C., Tieszen, L. L., Bunnell, F. L., and MacLean, S. F., Jr., eds. (1980). "An Arctic Ecosystem: The Coastal Tundra at Barrow, Alaska." Dowden, Hutchinson, and Ross, Stroudsburg, Pennsylvania.

Callaghan, T. V., and Jonasson, S. (1995). Arctic terrestrial ecosystems and environmental change. *Philos. Trans. R. Soc. Lond. A* **352,** 259–276.

Chapin III, F. S. (1983). Direct and indirect effects of temperature on arctic plants. *Polar Biol.* **2,** 47–52.

Chapin III, F. S., and Körner, C. (1995). "Arctic and Alpine Biodiversity: Patterns, Causes, and Ecosystem Consequences." Springer-Verlag, Ecological Studies Series, Vol. 113. Springer-Verlag, New York.

Chapin III, F. S., and Shaver, G. R. (1985a). Arctic. *In* "Physiological Ecology of North American Plant Communities" (B. Chabot and H. A. Mooney, eds.), pp. 16–40. Chapman and Hall, London.

Chapin III, F. S., and Shaver, G. R. (1985b). Individualistic growth response of tundra plant species to manipulation of light, temperature, and nutrients in a field experiment. *Ecology* **66,** 564–576.

Chapin III, F. S., and Shaver, G. R. (1988). Differences in carbon and nutrient fractions among arctic growth forms. *Oecologia* **77,** 506–514.

Chapin III, F. S., and Shaver, G. R. (1989). Differences in growth and nutrient use among arctic plant growth forms. *Funct. Ecol.* **3,** 73–80.

Chapin III, F. S., and Shaver, G. R. (1996). Physiological and growth responses of arctic plants to a field experiment simulating climatic change. *Ecology* **77,** 822–840.

Chapin III, F. S., Johnson, D. A., and McKendrick, J. D. (1980). Seasonal nutrient movements in various plant growth forms in an Alaskan tundra: Implications for herbivory. *Ecology* **68,** 189–210.

Chapin III, F. S., Shaver, G. R., and Kedrowski, R. A. (1986). Environmental controls over carbon, nitrogen, and phosphorus chemical fractions in *Eriophorum vaginatum* L. in Alaskan tussock tundra. *J. Ecol.* **74,** 167–196.

Chapin III, F. S., Fetcher, N., Kielland, K., Everett, K. R., and Linkins, A. E. (1988). Productivity and nutrient cycling of Alaskan tundra: Enhancement by flowing soil water. *Ecology* **69,** 693–702.

Chapin III, F. S., Jefferies, R. L., Reynolds, J. F., Shaver, G. R., and Svoboda, J., eds. (1992). "Arctic Ecosystems in a Changing Climate: An Ecophysiological Perspective." Academic Press, New York.

Chapin III, F. S., Shaver, G. R., Giblin, A. E., Nadelhoffer, K. J., and Laundre, J. A. (1995). Responses of arctic tundra to experimental and observed changes in climate. *Ecology* **76,** 694–711.

Chapin III, F. S., Hobbie, S. E., and Shaver, G. R. (1997). Impacts of global change on composition of arctic communities: Implications for ecosystem functioning. *In* "Global Change and Arctic Terrestrial Ecosystems" (W. C. Oechel and J. Holten, eds.). Springer-Verlag, New York.

Everett, K. R., Vassilevskaya, V. D., Brown, J., and Walker, B. D. (1981). Tundra and analogous soils. *In* "Tundra Ecosystems: A Comparative Analysis" (L. C. Bliss, O. W. Heal, and J. J. Moore, eds.), pp. 139–180. Cambridge Univ. Press, Cambridge.

Fetcher, N. (1985). Effects of removal of neighboring species on growth, nutrients, and microclimate of *Eriophorum vaginatum*. *Arctic Alpine Res.* **17,** 7–17.

Giblin, A. E., Nadelhoffer, K. J., Shaver, G. R., Laundre, J. A., and McKerrow, A. J. (1991). Biogeochemical diversity along a riverside toposequence in arctic Alaska. *Ecol. Monogr.* **61,** 415–436.

Giblin, A. E., Laundre, J. A., Nadelhoffer, K. J., and Shaver, G. R. (1994). Measuring nutrient availability in arctic soils using ion exchange resins. *Soil Sci. Soc. Am. J.* **58,** 1154–1162.

Gold, W. G., and Bliss, L. C. (1995). Water limitations and plant community development in a polar desert. *Ecology* **76,** 1558–1568.

Grulke, N. E., Reichers, G. H., Oechel, W. C., Hjelm, U., and Jaeger. C. (1990). Carbon balance in tussock tundra under ambient and elevated atmospheric CO_2. *Oecologia* **83,** 485–494.

Havström, M., Callaghan, T. V., and Jonasson, S. (1993). Differential growth responses of *Cassiope tetragona,* an arctic dwarf-shrub, to environmental perturbations among three contrasting high- and subarctic sites. *Oikos* **66,** 389–402.

Heal, O. W., and Perkins, D. F., eds. (1978). "Production Ecology of British Moors and Montane Grasslands." Springer-Verlag, New York.

Hinzman, L., Kane, D. L., Benson, C. S., and Everett, K. R. (1996). Energy balance and hydrological processes in an arctic watershed. *In* "Landscape Function: Implications for Ecosystem Response to Disturbance. A Case Study in Arctic Tundra" (J. Reynolds and J. Tenhunen, eds.), pp. 131–154. Springer-Verlag, New York.

Hobbie, J. E., ed. (1980). "Limnology of Tundra Ponds: Barrow, Alaska." Dowden, Hutchinson and Ross, Stroudsburg, PA.

Hobbie, S. E. (1992). Effects of plant species on nutrient cycling. *Trends Ecol. Evol.* **3,** 336–339.

Hobbie, S. E. (1995). Direct and indirect species effects on biogeochemical processes in arctic ecosystems. *In* "Arctic and Alpine Biodiversity: Patterns, Causes, and Ecosystem Consequences" (F.S. Chapin III and C. Körner, eds.), pp. 213–224. Springer-Verlag, Ecological Studies Series, Vol. 113. Springer-Verlag, New York.

Hobbie, S. E. (1996). Temperature and plant species control over litter decomposition in Alaskan tundra. *Ecol. Monogr.* **66,** 503–522.

Hope, A., and Stow, D. A. (1996). Shortwave reflectance properties of arctic tundra landscapes. *In* "Landscape Function: Implications for Ecosystem Response to Disturbance. A Case Study in Arctic Tundra" (J. Reynolds and J. Tenhunen, Eds.), pp. 155–164. Springer-Verlag, New York.

Johnson, L. C., Shaver, G. R., Giblin, A. E., Nadelhoffer, K. J., Rastetter, E. R., Laundre, J. A., and Murray, G. L. (1996). Effects of drainage and temperature on carbon balance of tussock tundra microcosms. *Oecologia* **108,** 737–748.

Jonasson, S. (1983). Nutrient content and dynamics in north Swedish shrub tundra. *Holarctic Ecol.* **6,** 295–304.

Jonasson, S. (1989). Implications of leaf longevity, leaf nutrient reabsorption, and translocation for the resource economy of five evergreen plant species. *Oikos* **63,** 420–429.

Jonasson, S. (1992). Plant responses to fertilization and species removal in tundra related to community structure and clonality. *Oikos* **63,** 420–429.

Jonasson, S., and Chapin III, F. S. (1985). Significance of sequential leaf development for nutrient balance of the cotton sedge, *Eriophorum vaginatum* L. *Oecologia* **67,** 511–518.

Jonasson, S., and Chapin III, F. S. (1991). Seasonal uptake and allocation of phosphorus in *Eriophorum vaginatum* L. measured by labeling with super (32) P. *New Phytol.* **18,** 349–357.

Jonasson, S., and Michelsen, A. (1996). Nutrient cycling in subarctic and arctic ecosystems, with special reference to the Abisko and Torneträsk region. *In* "Plant Ecology in the Subarctic Swedish Lapland" (P. S. Karlsson, and T. V. Callaghan, eds.), Ecological Bulletins 45, pp. 45–52. Swedish Natural Science Research Council, Stockholm.

Jonasson, S., Lee, J. A., Callaghan, T. V., Havström, M., and Parsons, A. N. (1996). Direct and indirect effects of increasing temperatures on subarctic ecosystems. *In* "Plant Ecology in the Subarctic Swedish Lapland" (P. S. Karlsson, and T. V. Callaghan, eds.), Ecological Bulletins 45, pp. 180–191. Swedish Natural Science Research Council, Stockholm.

Jónsdóttir, I. S., and Callaghan, T. V. (1989). Localized defoliation stress and the movement of ^{14}C-photoassimilates between tillers of *Carex bigelowii*. *Oikos* **54,** 211–219.

Jónsdóttir, I. S., and Callaghan, T. V. (1990). Intraclonal translocation of ammonium and nitrate nitrogen in *Carex bigelowii* Torr. ex Schwein. using ^{15}N and nitrate reductase assays. *New Phytol.* **114,** 419–428.

Kane, D. L., Hinzman, L. D., Woo, M., and Everett, K. R. (1992). Arctic hydrology and climate change. *In* "Arctic Ecosystems in a Changing Climate: An Ecophysiological Perspective" (F. S. Chapin III, R. L. Jefferies, J. F. Reynolds, G. R. Shaver, and J. Svoboda, eds.), pp. 35–58. Academic Press, New York.

Kielland, K. (1994). Amino acid absorption by arctic plants: Implications for plant nutrition and nitrogen cycling. *Ecology* **75,** 2373–2383.

Kielland, K., and Chapin III, F. S. (1992). Nutrient absorption and accumulation in arctic plants. *In* "Arctic Ecosystems in a Changing Climate: An Ecophysiological Perspective" (F. S. Chapin III, R. L. Jefferies, J. F. Reynolds, G. R. Shaver, and J. Svoboda, eds.), pp. 321–336. Academic Press, New York.

Kling, G. W. (1995). Land–water interactions: the influence of terrestrial diversity on aquatic ecosystems. *In* "Arctic and Alpine Biodiversity: Patterns, Causes, and Ecosystem Consequences" (F. S. Chapin III, and C. Körner, eds.), pp. 297–312. Springer-Verlag Ecological Studies Series, Vol. 113. Springer-Verlag, New York.

Kling, G. W., Kipphut, G. W., and Miller, M. C. (1991). Arctic lakes and streams as gas conduits to the atmosphere: Implications for tundra carbon budgets. *Science* **251,** 298–301.

Leadley, P. W., Reynolds, J. F., and Chapin III, F. S. (1997). A model of nitrogen uptake by *Eriophorum vaginatum* roots in the field: Ecological implications. *Ecol. Monogr.* **67,** 1–22.

McKane, R. B., Rastetter, E. B., Shaver, G. R., Nadelhoffer, K. J., Giblin, A. E., and Laundre, J. A. (1997a). Effects of experimental changes in CO_2 and climate on carbon storage in arctic tundra. *Ecology* **78,** 1178–1187.

McKane, R. B., Rastetter, E. B., Shaver, G. R., Nadelhoffer, K. J., Giblin, A. E., and Laundre, J. A. (1997b). Reconstruction and analysis of historic changes in carbon storage in arctic tundra. *Ecology* **78,** 1188–1198.

McKendrick, J. D., Ott, V. J., and Mitchell, G. A. (1978). Effects of nitrogen and phosphorus fertilization on the carbohydrate and nutrient levels in *Dupontia fischeri* and *Arctagrostis latifolia*. *In* "Vegetation and Production Ecology of an Alaskan Arctic Tundra" (L. L. Tieszen, ed.), pp. 509–537. Springer-Verlag, New York.

McKendrick, J. D., Batzli, G. O., Everett, K. R., and Swanson, J. C. (1980). Some effects of mammalian herbivores and fertilization on tundra soils and vegetation. *Arctic Alpine Res.* **12,** 565–578.

Michaelson, G. J., Ping, C. L., and Kimble, J. M. (1996). Carbon storage and distribution in tundra soils of arctic Alaska, U.S.A. *Arctic Alpine Res.* **28,** 414–424.

Michelsen, A., Schmidt, I. K., Jonasson, S., Quarmby, C., and Sleep, D. (1996). Leaf ^{15}N abundance of subarctic plants provides field evidence that ericoid, ectomycorhizal, and non- and arbuscular mycorrhizal species access different sources of soil nitrogen. *Oecologia* **105,** 53–56.

Miller, P. C. (1982). Environmental and vegetational variation across a snow accumulation area in montane tundra in central Alaska. *Holarctic Ecol.* **5,** 85–98.

Murray, K. J., Tenhunen, J. D., and Kummerow, J. (1989). Limitations on *Sphagnum* growth and net primary production in the foothills of the Philip Smith Mountains, Alaska. *Oecologia* **80,** 256–262.

Nadelhoffer, K. J., Giblin, A. E., Shaver, G. R., and Laundre, J. A. (1991). Effects of temperature and organic matter quality on C, N, and P mineralization in soils from six arctic ecosystems. *Ecology* **72,** 242–253.

Nadelhoffer, K. J., Linkins, A. E., Giblin, A. E., and Shaver, G. R. (1992). Microbial processes and plant nutrient availability in arctic soils. *In* "Arctic Ecosystems in a Changing Climate: An Ecophysiological Perspective" (F. S. Chapin III, R. L. Jefferies, J. F. Reynolds, G. R. Shaver, and J. Svoboda, eds.), pp. 281–300. Academic Press, New York.

Nadelhoffer, K. J., Shaver, G. R., Fry, B., Johnson, L., and McKane, R. (1996). ^{15}N natural abundances and N use by tundra plants. *Oecologia* **107,** 386–394.

Oberbauer, S. F., and Dawson, T. E. (1992). Water relations of arctic vascular plants. *In* "Arctic Ecosystems in a Changing Climate: An Ecophysiological Perspective" (F. S. Chapin III, R. L. Jefferies, J. F. Reynolds, G. R. Shaver, and J. Svoboda, eds.), pp. 259–279. Academic Press, New York.

Oberbauer, S. F., and Miller, P. C. (1979). Plant water relations in montane and tussock tundra vegetation types in Alaska. *Arctic Alpine Res.* **11,** 69–81

Oberbauer, S. F., Sionit, N., Hastings, S. J., and Oechel, W. C. (1986). Effects of CO_2 enrichment and nutrition on growth, photosynthesis, and nutrient concentration of Alaskan tundra plant species. *Can. J. Bot.* **64,** 2993–2998.

Oberbauer, S. F., Hastings, S. J., Beyers, J. L., and Oechel, W. C. (1989). Comparative effects of downslope water and nutrient movement on plant nutrition, photosynthesis, and growth in Alaskan tundra. *Holarctic Ecol.* **12,** 324–333.

Oberbauer, S. F., Cheng, W., Gillespie, C. T., Ostendorf, B., Sala, A., Gebauer, R., Virginia, R. A., and Tenhunen, J. D. (1996). Landscape patterns of carbon dioxide exchange in tundra ecosystems. *In* "Landscape Function: Implications for Ecosystem Response to Disturbance. A Case Study in Arctic Tundra" (J. Reynolds and J. Tenhunen, eds.), pp. 223–256. Springer-Verlag, New York.

Oechel, W. C., (1989). Nutrient and water flux in a small arctic watershed: An overview. *Holarctic Ecol.* **12,** 229–237.

Oechel, W. C., and Billings, W. D. (1992). Anticipated effects of global change on the carbon balance of arctic plants and ecosystems. *In* "Arctic Ecosystems in a Changing Climate: An Ecophysiological Perspective" (F. S. Chapin III, R. L. Jefferies, J. F. Reynolds, G. R. Shaver, and J. Svoboda, eds.), pp. 139–168. Academic Press, New York.

Oechel, W. C., Hastings, S. J., Vourlitis, G., Jenkins, M., Reichers, G., and Grulke, N. (1993). Recent change of arctic tundra ecosystems from a carbon sink to a source. *Nature (London)* **361,** 520–523.

Oechel, W. C., Cowles, S., Grulke, N., Hastings, S. J., Lawrence, B., Prudhomme, T., Reichers, G., Strain, B., Tissue, D., and Vourlitis, G. (1994). Transient nature of CO_2 fertilization in arctic tundra. *Nature (London)* **371,** 500–503.

Parsons, A. N., Welker, J. M., Wookey, P. A., Press, M. C., Callaghan, T. V., and Lee, J. A. (1994). Growth responses of four sub-Arctic dwarf shrubs to simulated environmental change. *J. Ecol.* **82,** 307–318.

Rastetter, E. B., McKane, R. B., Shaver, G. R., and Melillo, J. M. (1992). Changes in C storage by terrestrial ecosystems: How C–N interactions restrict responses to CO_2 and temperature. *Water Air Soil Poll.* **64,** 327–344.

Rastetter, E. B., McKane, R. B., Shaver, G. R., and Nadelhoffer, K. J. (1997). Analysis of CO_2, temperature, and moisture effects on carbon storage in Alaskan arctic tundra using a general ecosystem model. *In* "Global Change and Arctic Terrestrial Ecosystems" (W. C. Oechel and J. Holten, eds), pp. 437–451. Springer-Verlag, New York.

Reynolds, J. F., and Leadley, P. W. (1992). Modeling the response of arctic plants to changing climate. *In* "Arctic Ecosystems in a Changing Climate: An Ecophysiological Perspective" (F. S. Chapin III, R. L. Jefferies, J. F. Reynolds, G. R. Shaver, and J. Svoboda, eds.), pp. 413–440. Academic Press, New York.

Reynolds, J., and Tenhunen, J., eds. (1996). "Landscape Function: Implications for Ecosystem Response to Disturbance. A Case Study in Arctic Tundra." Springer-Verlag, New York.

Reynolds, J. F., Tenhunen, J. D., Leadley, P. W., Li, H., Moorehead, D. L., Ostendorf, B., and Chapin III, F. S. (1996). Patch and landscape models of arctic tundra: Potentials and limitations. *In* "Landscape Function: Implications for Ecosystem Response to Disturbance. A Case Study in Arctic Tundra" (J. Reynolds and J. Tenhunen, eds.), pp. 293–325. Springer-Verlag, New York.

Rosswall, T., and Heal, O. W., eds. (1975). Structure and function of tundra ecosystems. Ecological Bulletin No. 20. Swedish Natural Science Research Council, Stockholm.

Schulze, E.-D., and Chapin III, F. S. (1987). Plant specialization to environments of different resource availability. *In* "Potential and Limitations of Ecosystem Analysis" (E.-D. Schulze and H. Swölfer, eds.), pp. 120–148. Springer-Verlag Ecological Studies Series, Vol. 61. Springer-Verlag, Berlin.

Shaver, G. R. (1995). Plant functional diversity and resource control of primary production in Alaskan arctic tundras. *In* "Arctic and Alpine Biodiversity: Patterns, Causes, and Ecosystem Consequences" (C. Körner and F. S. Chapin III, eds.), pp. 199–212. Springer-Verlag Ecological Studies Series, Vol. 113. Springer-Verlag, New York.

Shaver, G. R. (1996). Integrated ecosystem research in northern Alaska, 1947–1994. *In* "Landscape Function: Implications for Ecosystem Response to Disturbance. A Case Study in Arctic Tundra" (J. Reynolds and J. Tenhunen, eds.), pp. 19–34. Springer-Verlag, New York.

Shaver, G. R., and Billings, W. D. (1975). Root production and root turnover in a wet tundra ecosystem, Barrow, Alaska. *Ecology* **56,** 401–410.

Shaver, G. R., and Chapin III, F. S. (1980). Response to fertilization by various plant growth forms in an Alaskan tundra: Nutrient accumulation and growth. *Ecology* **61,** 662–675.

Shaver, G. R., and Chapin III, F. S. (1986). Effect of fertilizer on production and biomass of tussock tundra, Alaska, U.S.A. *Arctic Alpine Res.* **18,** 261–268.

Shaver, G. R., and Chapin III, F. S. (1991). Production/biomass relationships and element cycling in contrasting arctic vegetation types. *Ecol. Monogr.* **61,** 1–31.

Shaver, G. R., and Chapin III, F. S. (1995). Long-term responses to factorial NPK fertilizer treatment by Alaskan wet and moist tundra sedge species. *Ecography* **18,** 259–275.

Shaver, G. R., Chapin III, F. S., and Gartner, B. L. (1986a). Factors limiting growth and biomass accumulation in *Eriophorum vaginatum* L. in Alaskan tussock tundra. *J. Ecol.* **74,** 257–278.

Shaver, G. R., Fetcher, N., and Chapin III, F. S. (1986b). Growth and flowering in *Eriophorum vaginatum:* Annual and latitudinal variation. *Ecology* **67,** 1524–1525.

Shaver, G. R., Nadelhoffer, K. J., and Giblin, A. E. (1991). Biogeochemical diversity and element transport in a heterogeneous landscape, the North Slope of Alaska. *In* "Quantitative Methods in Landscape Ecology" (M. G. Turner and R. H. Gardner, eds.), pp. 105–126. Springer-Verlag, New York.

Shaver, G. R., Billings, W. D., Chapin III, F. S., Giblin, A. E., Nadelhoffer, K. J., Oechel, W. C., and Rastetter, E. B. (1992). Global change and the carbon balance of arctic ecosystems. *BioScience* **42,** 433–441.

Shaver, G. R., Laundre, J. A., Giblin, A. E., and Nadelhoffer, K. J. (1996a). Changes in vegetation biomass, primary production, and species composition along a riverside toposequence in arctic Alaska. *Arctic Alpine Res.* **28,** 363–379.

Shaver, G. R., Giblin, A. E., Nadelhoffer K. J., and Rastetter, E. B. (1996b). Plant functional types and ecosystem change in arctic tundras. *In* "Plant Functional Types" (T. Smith, H. Shugart, and I. Woodward, eds.), pp. 152–172. Cambridge Univ. Press, Cambridge.

Shaver, G. R., Rastetter, E. B., Giblin, A. E., and Nadelhoffer, K. J. (1997). Carbon–nutrient interactions as constraints on recovery of arctic ecosystems from disturbance. *In* "Disturbance and Recovery in Arctic Lands: An Ecological Perspective" (R. M. M. Crawford, ed.), pp. 553–562. Kluwer Academic Publishers, Dordrecht.

Shaver, G. R., Johnson, L. C., Cades, D. H., Murray, G., Laundre, J. A., Rastetter, E. B., Nadelhoffer, K. J., and Giblin, A. E. (2000). Biomass accumulation and CO_2 flux in three Alaskan wet sedge tundras: Responses to nutrients, temperature, and light. *Ecology.*

Sonesson, M., ed. (1980). "Ecology of a subarctic mire." Ecological Bulletin No 30. Swedish Natural Science Research Council, Stockholm. 313 pp.

Svoboda, J., and Freedman, B., eds. (1994). "Ecology of a Polar Oasis: Alexandra Fiord, Ellesmere Island, Canada." Captus Press, Toronto.

Tenhunen, J. D., Lange, O. L., Hahn, S., Siegwolf, R., and Oberbauer, S. F. (1992). The ecosystem role of poikilohydric tundra plants. *In* "Arctic Ecosystems in a Changing Climate: An Ecophysiological Perspective" (F. S. Chapin III, R. L. Jefferies, J. F. Reynolds, G. R. Shaver, and J. Svoboda, eds.), pp. 213–237 Academic Press, New York.

Tenhunen, J. D., Siegwolf, R. A., and Oberbauer, S. F. (1994). Effects of phenology, physiology, and gradients in community composition, structure, and microclimate on tundra ecosystem CO_2 exchange. *In* "Ecophysiology of Photosynthesis." (E.-D. Schulze and M. M. Caldwell, eds.), pp. 431–462. Springer-Verlag Ecological Studies Series, Vol. 100. Springer-Verlag, New York.

Tieszen, L. L., ed. (1978). "Vegetation and Production Ecology of an Alaskan Arctic Tundra." Springer-Verlag Ecological Studies Series, Vol. 29. Springer-Verlag, New York.

Tissue, D. T., and Oechel, W. C. (1987). Response of *Eriophorum vaginatum* to elevated CO_2 and temperature in the Alaskan tussock tundra. *Ecology* **68,** 401–410.

Waelbroeck, C., Monfray, P., Oechel, W. C., Hastings, S., and Vourlitis, G. (1997). The impact of permafrost thawing on the carbon dynamics of tundra. *Geophys. Res. Lett.* **24,** 229–232.

Walker, M. D. (1995). Patterns and causes of arctic plant community diversity. *In* "Arctic and Alpine Biodiversity: Patterns, Causes, and Ecosystem Consequences" (C. Körner and F. S. Chapin III, eds.), pp. 3–20. Springer-Verlag Ecological Studies Series, Vol. 113. Springer-Verlag, New York.

Wielgolaski, F. E. (1972). Vegetation types and plant biomass in tundra. *Arctic Alpine Res.* **4,** 291–305.

Wielgolaski, F. E., ed. (1975). "Fennoscandian Tundra Ecosystems. Part 1: Plants and Microorganisms." Springer-Verlag, New York.

Wielgolaski, F. E., Bliss, L. C., Svoboda, J., and Doyle, G. (1981). Primary production of tundra. *In* "Tundra Ecosystems: A Comparative Analysis" (L. C. Bliss, O. W. Heal, and J. J. Moore, eds.), pp. 187–226. Cambridge Univ. Press, Cambridge.

Wookey, P. A., Parsons, A. N., Welker, J. M., Potter, J. A., Callaghan, T. V., Lee, J. A., and Press, M. C. (1993). Comparative responses of phenology and reproductive development to simulated environmental change in sub-arctic and high arctic plants. *Oikos* **67,** 490–502.

Wookey, P. A., Welker, J. M., Parsons, A. N., Press, M. C., Callaghan, T. V., and Lee, J. A. (1994). Differential growth, allocation, and photosynthetic responses of *Polygonum viviparum* to simulated environmental change at a high arctic polar semi-desert. *Oikos* **70,** 131–139.

Wookey, P. A., Robinson, C. H., Parsons, A. N., Welker, J. M., Press, M. C., Callaghan, T. V., and Lee, J. A. (1995). Environmental constraints on the growth, photosynthesis, and reproductive development of *Dryas octopetala* at a high arctic polar semi-desert, Svalbard. *Oecologia* **102,** 478–489.

10

Productivity of Boreal Forests

Paul G. Jarvis, Bernard Saugier, and E.-Detlef Schulze

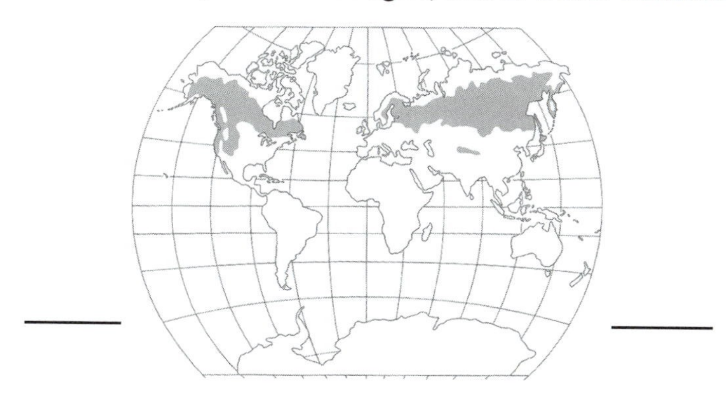

I. Introduction

Boreal forests play an important role in regulating the global carbon cycle and the climate of the Earth. They cover large areas and store large amounts of carbon in biomass and soil organic matter. Having a low albedo, they efficiently trap solar radiation. This is true even in winter, in comparison to snow-covered tundra, which has a much higher albedo. Because boreal forest canopies have high resistances to water vapor transfer, boreal forests have relatively low evaporation rates and act as strong heat sources, much stronger than tundra surfaces.

Increased precipitation and global warming resulting from the enhanced greenhouse effect are expected to be more pronounced at high latitudes, and this will make boreal forests especially sensitive to climate change. It is feared that as a result of this warming boreal forests may release large amounts of CO_2, by tree and "soil" respiration, and of methane, from wetter, low-lying areas. For these reasons, boreal forests have recently received considerable attention through large-scale vegetation–atmosphere interaction experiments such as the Boreal Ecosystem Atmosphere Study (BOREAS) in Canada, the Northern Hemisphere Climate Processes Experiment (NOPEX) in Scandinavia, and the Eurosiberian carbon flux studies in Siberia. Most results quoted in this chapter come from such large-scale programs.

II. Nature and Extent of Boreal Forests

A. Area Occupied

The boreal forests form a nearly continuous band around the northern hemisphere, 500 to 1500 km wide, located largely between latitudes 46°N and 66°N. At the northern limit, the forests grade through taiga into tundra: at the southern boundary they grade into mixed temperate forests or into grasslands. Coniferous boreal forest species require at least 1 month with a mean temperature above 10°C to regenerate so that their geographical extent northward is limited by the July 10°C climatic isotherm and southward by areas with more than 4 months above this temperature (Köppen, 1936). Comparable forests occur only sporadically in the southern hemisphere, where they comprise very different species and are of very different structure, as a result of a more equable climate because of the absence of large land masses. Estimates of the extent of the northern boreal forests vary somewhat among authors, ranging from 12.8 million km^2 (Kurz and Apps, 1993) to 14.7 million km^2 (Bonan and Shugart, 1989), i.e., between 8 and 10% of the Earth's terrestrial surface. Approximately 34% of this area lies in North America, 22% in Europe to the west of the Urals, and 44% in northern Russia, east of the Urals.

B. Importance

Boreal forests have become the focus of special attention recently for several reasons. Because of their large area, they play an important role in global climate, regulating evaporation and energy balance at the surface. They also store large masses of carbon: about 42 Gt C in biomass and 200 Gt C in soil organic matter, representing 7.5 and 13% of the global amounts, respectively. To satisfy the global balance of anthropogenically produced carbon dioxide, it has been suggested that the terrestrial biosphere may store ~2 Gt (C) yr^{-1} (Houghton *et al.*, 1996; Bolin *et al.*, 2000). Inverse modeling of carbon sources and sinks from the latitudinal distribution of atmospheric CO_2 concentrations indicates that a major terrestrial carbon sink is located in ecosystems of the northern hemisphere (Tans *et al.*, 1990; Ciais *et al.*, 1995). It is believed that a significant fraction of this storage occurs in the northern boreal and temperate forests, partly as a result of the fertilizing effects of rising atmospheric CO_2 concentration, partly because recent warming has increased the length of the photosynthetic period (Myneni, 1997), and partly as a result of the global increase in nitrogen deposition (Vitousek, 1994), although it has been argued that most of the increased nitrogen goes into the soil and not into wood, where carbon storage is taking place (Nadelhoffer *et al.*, 1999). However, it is possible that global warming, which is likely to be especially significant in high latitudes (Mitchell *et al.*, 1995), might decrease this storage of carbon, or even reverse it, by stimulating the release of CO_2 in enhanced soil and plant respiration, although this is quite uncertain (see Section X). Furthermore, boreal forests release methane from the many low-lying areas that are saturated with water of low oxygen content. The extension in area of these wet lands, and the increase in the emission of methane with warming, may significantly increase the role of the boreal forests in the greenhouse effect (Roulet *et al.*, 1992).

This review is centered on the net primary productivity (NPP) of North American boreal forests, and includes a discussion of the main biotic and environmental variables that limit this productivity to relatively low values. Because CO_2 assimilation and transpiration of water vapor are regulated in similar ways by leaf area and stomatal conductance, we also deal briefly with energy partitioning and exchange of water. Additional data on biomass and NPP are also given for Eurosiberia, based on two reviews by Shvidenko and Nilsson (1994, 2000), who have made available the large body of information about the Russian and Siberian forests that has been published in Russian. With respect to the European boreal forests, our analysis is based on the summary by the European Environment Agency (Stanners and Bourdeau, 1995). In addition, we include data that were collected during three expeditions to Siberia (Schulze *et al.*, 1995, 1999; Wirth *et al.*, 1999).

C. Climate

The climate of the boreal region is strongly seasonal, characterized by long, very cold winters and short, mild summers with wide annual temperature variation, and also strong contrasts between day and night temperatures. Daily summer temperatures average 12–15°C, whereas winter temperatures may often go down below −30°C. The frost-free season lasts between 50 and 100 days, the growing season about 100 days, and the photosynthetic season is about 180 days; summer daylength varies between 16 and 24 h. The most severe climate occurs in eastern Siberia, where continentality is combined with moderate elevation. There mean monthly temperatures range from −50°C (January) to 15°C (July). Annual precipitation is relatively low, with a range of between 300 and 700 mm. In spite of this, boreal forests are not in general short of water, containing many lakes and streams, where they are on flattish ground, as a result of recent glaciation.

D. Soils

The area presently occupied by boreal forests was covered by glaciers 18,000 yr ago and consequently soils are young. Plant litter decomposes slowly because it is low in nutrients and high in resins, waxes, and lignin; the temperatures are low and the soils are often poorly aerated. This results in low fluxes of mineral nitrogen that limit tree density and forest productivity. There is intense leaching because precipitation exceeds evaporation for a significant part of the year. This results in podzolic soils with a pH around

4, a thick litter layer, and a leaching of organic acids that accumulate in the B horizon. On poorly drained sites, organic matter accumulates at the surface to form a thick humus layer. In the colder areas, permafrost reduces the depth of the rooting zone to less than 1 m.

E. Species

Boreal forests are dominated by coniferous species but also contain poplar, alder, and birch, especially in the southern parts. In North America, the main species are white spruce (*Picea glauca*) on well-drained sites, black spruce (*Picea mariana*) and tamarack (*Larix laricina*) on poorly drained sites, and jack pine (*Pinus banksiana*) on dry sandy soils. Over much of the area of the boreal forest zone in mid-Canada, black spruce covers 40–50% of the land area. In Siberia, there are extensive areas of Siberian larch (*Larix gmelinii*) in the east, and of Scots pine (*Pinus sylvestris*), Siberian fir (*Abies sibirica*), and Siberian spruce (*Picea obovata*) in the west. Large areas in both Canada and Russia have been changed from coniferous species into *Betula* and *Populus* spp. by repeated fires.

F. Fire

The main mode of ecosystem regeneration is through responses to natural fires that recur on a site with an average return time of 100 to 150 yr. Kurz *et al.* (1992) computed the fraction of forested area burning each year. They found an average of 0.6% for the entire Canadian forested area, with variations between 0.24% in the eastern boreal zone and 1.5% in the western boreal zone, which is much drier. Thus the boreal forest appears as a mosaic of relatively even-aged stands often dominated by a single tree species or sometimes up to three species growing together. The landscape also contains numerous lakes of various sizes and bogs and fens devoid of trees because the ground is too wet to supply oxygen to tree roots.

III. Recent Research Programs

The Boreal Ecosystem Atmosphere Study began in summer of 1993 to investigate the exchanges of radiative energy, heat, water, CO_2, and trace gases between the forest and atmosphere in the boreal forest zone in central Canada (Sellers *et al.*, 1995). The study covered an area of 1000 × 1000 km with two main study areas, one near the southern boundary around Prince Albert, Saskatchewan (SSA), and the other near the northern boundary around Thompson, Manitoba (NSA), and a transect in between. In each study area, there were intensive study sites where carbon dioxide, heat, and water vapor flux measurements were made on a total of 10 towers and on four aircraft. Remote-sensing measurements were made from additional aircraft and satellites, and there were supporting measurements within the

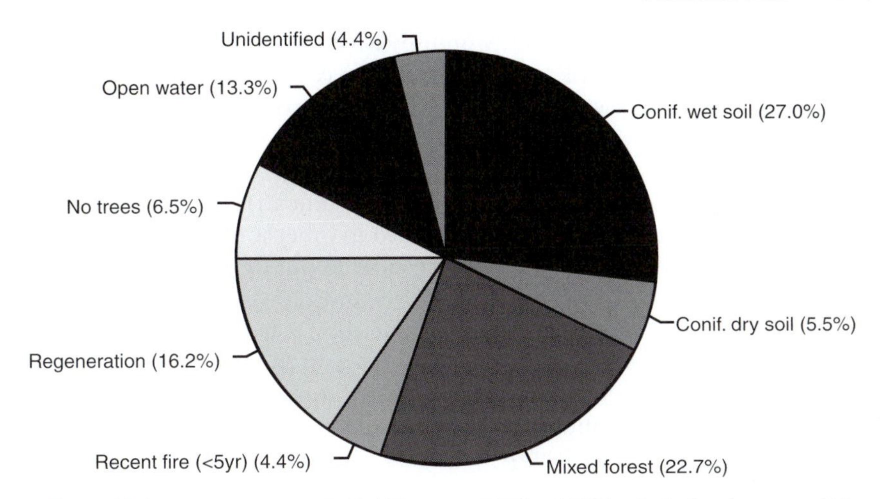

Figure 10-1 Land use on the BOREAS square (1000 × 1000 km in Saskatchewan and Manitoba, Canada (from Steyaert *et al.*, 1997).

stands of production, ecology, physiology, trace gas exchanges, and radiative transfer. Flux towers were installed in extensive stands of the main vegetation types: aspen (*Populus tremuloides*), black spruce (*Picea mariana*), jack pine (*Pinus banksiana*), and on fen/bog sites. Figure 10-1 shows the distribution of the various vegetation types in the BOREAS square, obtained from remote-sensing data taken by Steyaert *et al.* (1997). Open water accounts for 13.3% of the total area and bogs and fens for another 6.5%. Thus forests cover only 80% of the area. Regeneration after fire or harvesting accounts for 20.6% of the area, including 4.4% of recent (<5 yr) postfire regeneration, leading to an estimate of return time of ~100 yr.

The following ecophysiological measurements made on the tower flux sites, together with net ecosystem CO_2 flux above and below the canopy, have enabled evaluation of the various components of the carbon balance: net photosynthesis of the foliage; leaf, trunk and soil respiration; and above- and belowground biomass stock and production. Examples of the resulting carbon budgets will be given here for one of the 10 flux sites in the southern study area (54°00′N, 105°07′W). The remote-sensing and airborne flux measurements have allowed spatial integration of the site data, but this falls outside the present context. Most measurements were made in 1994 and 1996 and some are still being analyzed.

At the time of writing two sites in Eurosiberia have been studied:

1. The *Larix gmelinii* forest in eastern Siberia near Yakutsk (60°51′N, 128°16′E). Deciduous larch covers about 54% of the total forested area in Siberia (Shvidenko and Nilsson, 1994), growing mainly on sandy soils be-

tween the Ural Mountains and the Yenisey River. The CO_2 flux (net ecosystem production; NEP) measurements were made above a 125-yr-old stand. Biomass and growth were measured in a chronosequence of 49- to 380-yr-old stands growing on loamy permafrost soil (Schulze *et al.*, 1995; Hollinger *et al.*, 1997).

2. The *Pinus sylvestris* forest in Central Siberia near Bor (60°44′N, 89°09′E). Pine covers 14% of the forest area in Siberia (Shvidenko and Nilsson, 1994). NEP measurements were carried out in a range of habitats, including 7- and 13-yr-old logging areas, 53- and 215-yr-old stands naturally regenerated after fire, and an area dominated by lichens. In the same region, NEP of a representative bog was also measured (Schulze *et al.*, 1999). Biomass and growth were studied in a chronosequence of stands ranging from 28- to 450-yr-old stands on sandy soil without permafrost (Pergelic Cryochrept) (Wirth *et al.*, 1999).

IV. Biophysical Properties

A. Leaf Area Index

Because broad-leaved and coniferous species are both present in the boreal forest, in BOREAS leaf area index (LAI) was defined in a uniform way as the ratio between half the total leaf surface area and the ground area. However, for comparison with the wider literature, plan area is given here using the factor for conversion to plan area for the coniferous species, ×0.787 for *Picea mariana* (black spruce) and ×0.789 for *Pinus banksiana* (jack pine). Characteristics of three BOREAS stands are given in Table 10-1. Despite very narrow crowns, LAI (on a plan area basis) of the tree canopy is 4.4 in the old black spruce (OBS), because of the extreme density of the foliage in the tree crowns, is less than 2 in the old jack pine (OJP) and reaches 2.3 in the old aspen (OA) stand. However, the deciduous stand of aspen differs from the coniferous stands by the presence of a thick understorey of hazelnut (*Corylus cornuta*), with a LAI reaching 3.3. Under the spruce stands there is a ground flora of sphagnum and feather mosses and a few herbaceous plants with a sparse shrub layer largely of Labrador tea (*Ledum groenlandicum*), and under the pine stands, lichens and a few herbaceous plants with very small LAI. The small LAI of jack pine canopies results from sparse tree crowns with large gaps between crowns of individual trees, caused by limitations in water and nitrogen supply.

The LAI of the Siberian larch and pine forests is generally lower than in the BOREAS stands, ranging between 1 and 2 (Schulze *et al.*, 1999).

B. Radiation and Aerodynamic Characteristics

By comparison with the aspen site, the albedos (α) of the black spruce and jack pine sites are very low (Table 10-2), but are consistent with measure-

Table 10-1 Selected Mean Stand Properties of Three of the BOREAS Sites in the Southern Study Area[a]

Site	Dominant species	Age (yr)	Height (m)	Trees (ha^{-1})	Basal area (m^2 ha^{-1})	DBH (cm)	Sapwood vol. (m^3 ha^{-1})	LAI[b]	LAI[c]	SAI
OBS	*Picea mariana*	115	7.2	6005[d]	29.9	7.1	61	5.6	4.44	0.4
OJP	*Pinus banksiana*	65	12.7	1192	16.9	16.9	57	2.4	1.9	0.2
OA	*Populus tremuloides*	67	20.1	984	33.5	20.5	327	3.3	3.3	0.4

[a] Located at ~54°N, 105°W (from Gower *et al.*, 1997; Ryan *et al.*, 1997); SAI, stem area index; LAI, leaf area index; OBS, old black spruce; OJP, old jack pine; OA, old aspen; DBH, diameter at breast height.

[b] Half surface area basis.

[c] Plan area.

[d] Includes 2% *Pinus banksiana* comprising 8% of the basal area.

Table 10-2 Biophysical Characteristics of Three of the Southern Study Area BOREAS Sites[a]

Parameter	Black spruce	Jack pine	Aspen
α	0.09	0.10	0.15
h (m)	7 to 10	13.5	21
z_0 (m)	1.1	2.1	2.7
d (m)	4.5	5.9	13.2

[a]Black spruce data from Hale (1996), jack pine data from Baldocchi *et al.* (1996), and aspen data from Simpson (1996). α, Albedo; h, stand height; z_0, roughness length; d, displacement of the zero plane of the wind speed profile.

ments over spruce and pine sites in the temperate region (Jarvis *et al.*, 1976). In coniferous stands, roughness length (z_0) is rather large, but zero plane displacement height (d) is remarkably small in comparison to canopy height (h), as a result of the narrow crowns and gaps among the trees.

V. Evaporation, Transpiration, and Climate

Evaporation from the black spruce, jack pine, and aspen stands amounted to 240, 190, and 320 mm, respectively, over the 120-day summer measurement season in 1994. In contrast to the aspen site, where evaporation exceeded 5 mm d^{-1}, evaporation from both the jack pine and black spruce sites did not exceed 2.5 mm d^{-1} (one-third to one-half of the net radiation). In Siberia, Schulze *et al.* (1999) found similar rates, between 1.3 and 2.6 mm d^{-1}.

Energy Partitioning

The average partitioning of available energy, i.e., the total daily net absorbed radiation (R_n), reflects both the dryness of the sites and the area and physiological properties of the leaves (Table 10-3). On the sandy, dry jack pine site, sensible heat transfer predominates and the Bowen ratio (ratio of sensible over latent heat fluxes) frequently exceeds 3 at midday in summer (Baldocchi *et al.*, 1997); on the wet, peaty black spruce site, despite freely available water and a large LAI, the Bowen ratio frequently exceeds 2 at midday (Jarvis *et al.*, 1997), because of very low stomatal conductances. By contrast, the energy balance of the aspen stand is quite different. On the well-drained, loamy aspen site, the latent heat flux accounts for most of the net radiation in the stand and the Bowen ratio is ~0.5 or less, rarely exceeding 1, and very similar to that of fen sites. This is partly attributable to the thick understorey of hazelnut, but mostly results from the high stomatal conductance of the

Table 10-3 Average Partitioning of Available Energy in Summer 1994[a]

Parameter	Black spruce	Jack pine	Aspen
Sensible heat flux	0.49	0.51	0.25
Latent heat flux	0.44	0.36	0.61
Storage	0.04	0.01	0.14
Residual	0.03	0.12	—

[a]Black spruce data for 120 days from Jarvis *et al.* (1997), jack pine data for 120 days from Baldocchi and Vogel (1996), and aspen data from Blanken *et al.* (1997) for daytime only on 4 August.

aspen leaves compared to spruce or pine needles (Black *et al.*, 1996). In the Siberian forests daily-averaged Bowen ratios ranged between 1.4 and 5.2 (Schulze *et al.*, 1999). This accords with many previous studies showing that many coniferous stands usually utilize less than one-half to one-third of R_n in latent heat flux (see Jarvis *et al.*, 1976).

Thus despite the large areas of open water, the boreal forest is a strong heat source during the day. This has been confirmed by measurements of the height of the convective boundary layer (CBL), which reaches 2500 to 3000 m early on sunny days (Barr and Betts, 1994), whereas the height of the CBL does not exceed 1500 m above temperate deciduous forests (Baldocchi and Vogel, 1996). This has a strong influence on the local climate and partly explains the strong contrast between day and night temperature (Sellers *et al.*, 1995).

Transpiration rates of the trees in the stands of jack pine, black spruce, and aspen have been measured as the difference between the water vapor fluxes measured by eddy covariance above and below the tree canopy (Baldocchi *et al.*, 1997) and by a sap flow method (Saugier *et al.*, 1997). For the jack pine site, for example, the tree canopy water vapor flux was found to be 0.76 of the overall above-canopy flux, and seasonal transpiration based on this estimate is plotted in Fig. 10-2, together with the sap flow measurements. There is a good overall agreement between the two sets of measurements except during rainy periods, when the eddy flux estimate includes evaporation of intercepted water whereas the sap flow method does not.

VI. Biomass and Productivity

A. Methods

We are aware of the problems involved in extrapolations from plot data to regions (see Moore, 1996). Often good stands are chosen for ecological stud-

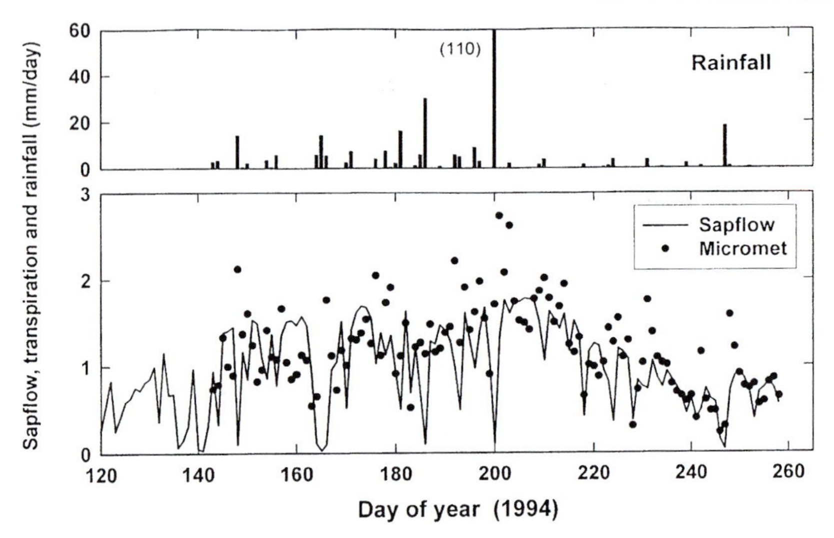

Figure 10-2 Daily transpiration of a stand of jack pine measured by sap flow (continuous line) and as H_2O flux above the canopy (×0.76; see text) at the BOREAS southern study area old jack pine site.

ies, with higher biomass and productivity than the average. Thus we report both detailed studies and large-scale inventories that take into account the full range in variability. For Siberia and Europe, forest inventory data were converted into wood mass and carbon pools using the following average wood densities (dry mass per unit volume): 300 kg m^{-3} for Siberia (Schulze *et al.*, 1995; Wirth *et al.*, 1999), and 400 kg m^{-3} for European conifers and 500 kg m^{-3} for European hardwoods (Kramer, 1988; Körner *et al.*, 1993). In Canada comparable values were measured on the BOREAS sites. The carbon fraction of dry mass was taken as 0.5. The contribution of losses by self-thinning to NPP was estimated from the plot scale data (Schulze *et al.*, 1995; Wirth *et al.*, 1998) and these data were compared with regional estimates by Shvidenko and Nilsson (1994). Root biomass and fine root turnover rates were estimated from Jackson *et al.* (1997) and Steele *et al.* (1997). This analysis is based on as many Russian publications as possible, especially in the field of biomass and productivity measurements (Isaev, 1991).

B. Biomass

The total amount of biomass is quite variable in boreal forests and may amount to only 10 to 30 t ha^{-1} in open lichen woodland and up to 80 to 100 t ha^{-1} in closed southern forests (Larsen, 1980). In a review of biomass and productivity of pine forests all over the world, Gower *et al.* (1994) gave an average aboveground biomass of 73 t ha^{-1} for 58 boreal forest stands. If stands younger than 40 yr are removed, biomass of mature stands averaged

88 t ha^{-1} for the 33 stands. Root biomass was measured in 16 of the mature stands and was found to be on average 36% of aboveground biomass (range 12–50%). Thus total biomass was on average 120 t ha^{-1}. Foliage mass varied between 3.4 and 13.9 t ha^{-1} with an average of 5.9 t ha^{-1} for 33 mature stands. Kurz and Apps (1993) studied the carbon balance of Canadian forests that lie mainly in the boreal zone. They used figures from the national inventory to compute the average forest biomass per eco-climatic zone. This average biomass varied from 12.4 t ha^{-1} of dry matter for the arctic zone to 54 t ha^{-1} for the relatively humid eastern boreal zone (Ontario and Quebec), and was 39 t ha^{-1} for the drier western boreal zone (Alberta, Saskatchewan, and Manitoba). These figures are averages over very large areas, and for that reason are lower than those of Gower *et al.* (1994), which are averages of 58 plots in only 13 sites—four in North America, eight in Europe, and one in Siberian Russia.

Biomass is represented as a function of age in Fig. 10-3 for representative stands from Siberia and Europe. In Siberia, aboveground biomass may reach 110 t ha^{-1} (11 kg m^{-2}) in mature pine stands on moist sites and 80–130 t ha^{-1} (8–13 kg m^{-2}) in good larch stands. For the whole of Siberia, however, the average biomass is much lower and estimated to be 56 t ha^{-1} (5.6 kg m^{-2}) by Shvidenko and Nilsson (1994). This figure is slightly higher than that reported for Canada. For Europe, Stanners and Bourdeau (1995) computed an average figure of 101 t ha^{-1} (10.1 kg m^{-2}).

In Canada, in the southern study area of BOREAS, the total amount of carbon present on the black spruce site is very large but 86% of it is in the soil, whereas around 20% is in the soil on the pine and aspen sites (Table 10-4). The vegetation on the aspen site contains more than twice as much carbon as the vegetation on the coniferous forest sites. Within the vegetation, the bulk of the carbon is in the woody tree stems, branches (including old foliage on the black spruce), and coarse roots, including standing dead trees and detritus on the forest floor (Table 10-4).

Global biomass data for boreal forests are summarized in Table 10-5.

C. Net Primary Productivity

Lieth (1975) reviewed 52 data sets on net primary productivity measured in various biomes as annual biomass increment plus detritus, and found NPP was related to annual temperature and to annual precipitation. When his relationships are applied to the climatic range of boreal forests, resulting annual NPP of dry mass ranges from 2.0 to 11.0 t ha^{-1}. Net annual increment (NAI) in woody biomass was also computed by Kurz and Apps (1993) and found to be quite small, averaging 0.46 t ha^{-1} (dry mass) for the entire Canadian boreal forest and 0.40 t ha^{-1} for the western boreal zone, but NAI is a small fraction of NPP. Annual, aboveground biomass NPP (ANPP) compiled by Gower *et al.* (1994) for 24 boreal pine stands ranged between 2.3 and 7.0 t ha^{-1}, with an average of 4.2 t ha^{-1}.

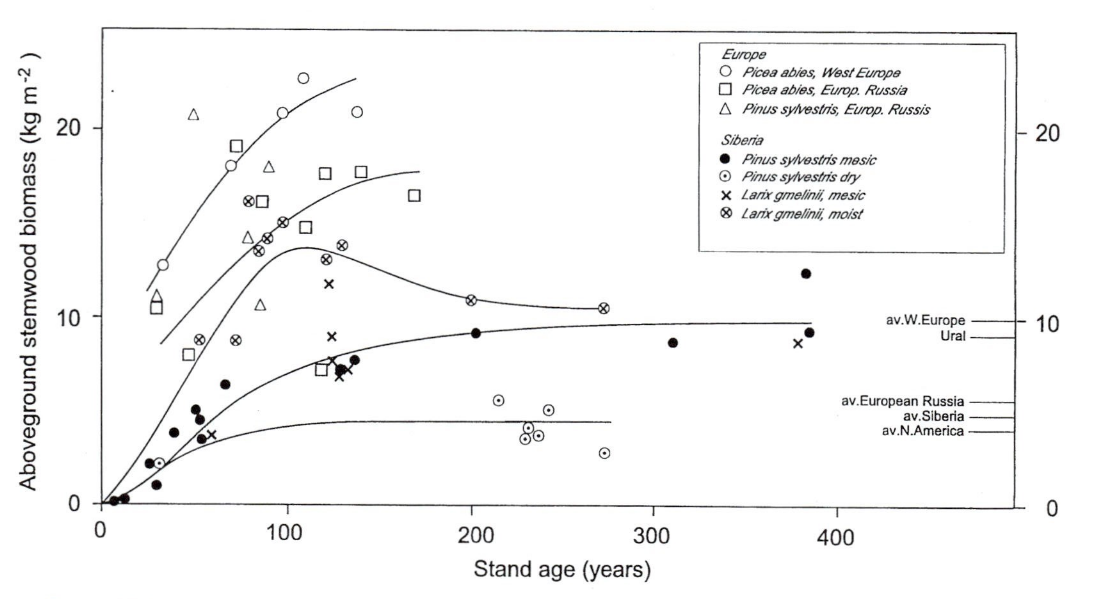

Figure 10-3 Live aboveground biomass in Siberian pine and larch stands as compared to European spruce. Average amounts of biomass in Europe (Stanners and Bourdeau, 1995), Siberia (Shvidenko and Nilsson, 2000), and North America (Botkin and Simpson, 1990).

Table 10-4 Mean Amounts and Distribution of Carbon in Three Southern BOREAS Sites[a]

Amount/distribution	Black spruce	Jack pine	Aspen
C in soil (t ha^{-1})	390.3	14.2	36.0
C in vegetation (t ha^{-1})	67.5	63.7	137.6
Total C in ecosystem (t ha^{-1})	455.9	76.1	170.9
Soil C/total C (%)	85.6	18.7	21.1
C in tree organs (%)			
Tree stem	53.8	40.4	60.1
Tree branch	10.0	9.4	6.5
Tree new foliage	0.7	0.3	0.7
Tree old foliage	6.8	1.3	—
Understorey	0.8	0.2	0.5
Ground moss/lichen	0.9	5.4	—
Detritus standing dead	9.1	8.9	7.0
Detritus forest floor	—	22.8	14.1
Live fine roots (<2 mm)	0.5	0.6	0.2
Live fine roots (2–5 mm)	3.4	1.4	1.9
Dead fine roots (<2 mm)	0.4	0.9	0.3
Dead fine roots (2–5 mm)	1.2	0.8	0.9
Total fine roots	5.5	3.8	3.3
Coarse roots	12.4	10.2	7.7

[a]From Gower *et al.* (1997) and Steele *et al.* (1997).

Table 10-5 Dry Biomass of Boreal Forests[a]

Continent	Area (10^6 km^2)	Biomass density (kg m^{-2})	Regional biomass (Pg)	Source
Siberia	5.97	5.6	33.43	Shvidenko and Nilsson (1994)
European Russia	1.67	9.1	15.20	Pryajnikov and Zamolodchikov (1993)
Western Europe	1.37	10.1	13.84	Stanners and Bourdeau (1995)
North America	4.80	4.6	22.08	Botkin and Simpson (1990)
Total above ground		6.1	84.55	
Total below ground		2.2	30.44	
Total	**13.81**	**8.3**	**114.99**	

[a]Belowground biomass is assumed to be 0.36 of the aboveground biomass. To convert to carbon, divide by 2.0.

Belowground NPP (BNPP) data are harder to come by. There are few reliable values of root productivity because the turnover rate of fine roots is difficult to measure. Of the 24 pine stands quoted by Gower *et al.* (1994), only 13 had simultaneous measurements of root productivity. On average, root productivity was 32% of total (above- plus belowground) NPP, with a few extreme values ranging from 3 to 60%. The very small values reported by Rodin and Bazilevitch (1965) probably failed to account for the turnover of fine roots. The high value of 60% was confirmed by Swedish data on *Pinus sylvestris*. Thus it is likely that roots represent 30–60% of total NPP, the highest figure corresponding to sites of low fertility (low nutrient supply or poor drainage). If roots represent 32% of total productivity on average, this amounts to 47% of ANPP. Thus, for an average annual ANPP of 4.2 t ha^{-1}, there would be an additional 2.0 t ha^{-1} of root production, i.e., 6.2 t ha^{-1} total annual dry biomass production.

Comprehensive new data on ANPP and BNPP of carbon, based on measured biomass increments of the components, are given for the three southern BOREAS sites in Table 10-6. The major components are wood increment, woody and fine root production, and detritus, the largest part of which is made up of overstorey foliage. Total NPP of carbon varies between 1.4 and 3.8 t (C) ha^{-1}, with an intermediate figure of 2.1 for the black spruce (Table 10-6). Gower *et al.* (1997) also present a summary of recent boreal carbon NPP estimates that all include measurements of BNPP, with group means ranging from 2.4 for pines to 4.8 for a mix of deciduous species, and with a mean of 3.4 t (C) ha^{-1} for 10 spruce data sets (Table 10-7).

In Siberia, *Pinus sylvestris* and *Larix* spp. represent 68% of the forested area (Shvidenko and Nilsson, 1994), whereas *Picea* and *Pinus* species cover 51% of the European forest area (Stanners and Bourdeau, 1995). Figure 10-3

Table 10-6 Net Primary Production of Carbon Above- (1994) and Belowground (1994/95) at Three Southern BOREAS Sites[a]

	NPP (kg (C) ha^{-1} yr^{-1})		
Parameter	Black spruce	Jack pine	Aspen
Aboveground increment			
Wood	800	650	1760
Understorey	0	0	530
Detritus (tree and understorey)	660	560	1230
Total ANPP	**1460**	**1210**	**3520**
Total BNPP	**1200**	**1050**	**400**
Overall total NPP	**2660**	**2260**	**3920**
BNPP/NPP	**0.45**	**0.46**	**0.10**

[a]From Gower *et al.* (1997) and Steele *et al.* (1997).

Table 10-7 NPP for Boreal Forests Based on the Sum of Annual Increases in Biomass[a]

Forest	n	ANPP Tree	ANPP Understorey	BNPP	NPP (kg (C) ha^{-1} yr^{-1})	BNPP/NPP	$D/\Delta B$
Evergreen							
Picea sp.	10	1390	960	1080	3420	0.39	1.84
		(280)	(230)	(120)		(0.06)	(0.29)
Pinus sp.	5	1420	100	950	2430	0.40	1.71
		(160)	(50)	(80)		(0.04)	(0.23)
Deciduous							
birch, alder, poplar, larch	7	3640	430	1050	4780	0.21	0.88
		(560)	(160)	(270)	(840)	(0.03)	(0.09)

[a]Biomass (ΔB) and detritus (D) summarized from Gower *et al.* (1997). Numbers in parentheses, 1 SD.

summarizes biomass measurements of these dominant species and gives regional estimates for North America (Botkin and Simpson, 1990), west Europe (Stanners and Bourdeau, 1995), and European Russia as well as Siberia (Shvidenko and Nilsson, 2000).

On mesic sites (Fig. 10-3), *Pinus sylvestris* and *Larix* spp. in Siberia accumulate stem wood biomass at less than half the rate of *Picea abies* spp. in west Europe. *Picea abies* in Europe reaches an aboveground biomass of about 21 kg m^{-2} (210 t ha^{-1}) in 100 yr, whereas pine and larch biomass reached typically 7.5 kg m^{-2} at the same age. In contrast to the situation in Europe, aboveground biomass continues to increase with stand age in Siberian forests. Thus a 380-yr-old pine stand on mesic sites has about the same biomass as a 50-yr-old spruce stand in west Europe. On average, aboveground stem wood biomass density of boreal forests in Siberia and North America is half that of west Europe (Stanners and Bourdeau, 1995; Shvidenko and Nilsson, 1998; Botkin and Simpson, 1990). The biomass of spruce and pine in European Russia, is intermediate between the west European and the Siberian biomass amounts (Vygodskaya *et al.*, 1998; Smirnov and Alekseev, 1967; Smirnov, 1970).

The variation of stand biomass with site quality can be very large. Under favorable conditions the amounts of biomass in larch and pine stands in Siberia may approach those in European Russia (Utkin, 1965; Usolzev, 1985). In this case biomass appears to decrease when stands age beyond a certain threshold, whereas there are dry sites where biomass remains very small irrespective of age. Within European Russia there is also large regional variation, with the largest biomasses in the Ural and Kaukasus regions (Shvidenko and Nilsson, 2000).

To estimate biomass and NPP for large areas such as Canada and Siberia, other methods should be used. From forest inventories it is possible to approximate ANPP by volume growth of stem wood (i.e., NAI), as listed in yield tables, the species-appropriate wood volume, and allometric relations between growth of stem wood and other organs. However, in addition to the standing crop, which is usually measured by forest inventories, rates of tree death and harvesting must be included in estimates of NPP (Schulze, 1982; Lloyd, 1999; Shvidenko and Nilsson, 2000). As stands develop, tree death by self-thinning can even offset growth. Forest management usually manages to keep natural abscission and death as low as possible. The amount of carbon being harvested by thinning operations is included in figures of average total stem wood growth in Europe. In contrast, in Siberia thinning by natural processes reduces stem number with age following a negative exponential curve (Wirth *et al.*, 1999), but dead trees decomposing on site are not included in the inventory (Shvidenko and Nilsson, 2000). Thus, dividing standing biomass by stand age underestimates ANPP.

It is very difficult to quantify the contribution of tree death to total stand growth from inventories unless the change in stand density is known. From plot studies we estimate that in Siberia the amount of wood that is lost by self-thinning is slightly larger than the stand wood increment (WNPP) (Table 10-8). This is in agreement with Shvidenko and Nilsson (2000), who

Table 10-8 Average Figures for NPP of Biomass and Carbon for Siberian Forests[a]

Parameter	NPP, biomass ($g\ m^{-2}\ yr^{-1}$)	Computation	NPP, C ($Pg\ (C)\ yr^{-1}$)
Wood net annual increment (NAI)	36.8	Mean NAI = $0.92\ m^3\ ha^{-1}\ yr^{-1}$; wood density = $400\ kg\ m^{-3}$	0.11
Self-thinning	44.3	1.2 × NAI	0.13
Total wood productivity (WNPP)	**81.1**	—	**0.24**
Needle litter	87.6	1.08[b] × WNPP	0.26
Aboveground productivity (ANPP)	**168.7**	—	**0.50**
Coarse roots	32.4	0.4[c] × WNPP	0.10
Fine roots	112.5 to 244	(1.4 to 3)[d] × WNPP	0.34 to 0.73
Belowground productivity (BNPP)	**144.9 to 276.4**	—	**0.44 to 0.83**
Total net productivity (NPP)	**314 to 445 (281)**	—	**0.94 to 1.33 (0.84)**
BNPP/NPP	**0.46 to 0.62 (0.4)**	—	**0.46 to 0.62 (0.4)**

[a]Total area $5.97 \times 10^6\ km^2$. The estimate for self-thinnings is slightly higher than that of Shvidenko and Nilsson (2000). The carbon fraction of biomass is taken as equal to 0.5 for all plant organs.

[b]The value 1.08 is an average over stands of *Pinus sylvestris* (Wirth *et al.*, 1999) and of *Larix gmelinii* (Schulze *et al.*, 1995).

[c]The value 0.4 comes from Monserud *et al.* (1996).

[d]The value 3 is estimated following Jackson *et al.* (1997). It appears to be an overestimate. Total root productivity represents an average of 40% of total NPP in BOREAS stands (see Table 10-8). The value 1.4 is derived from this estimate in Canadian forests.

estimated that for the whole of Russia tree mortality (913×10^6 m^3 yr^{-1}) is about as high as net stem wood growth (966×10^6 m^3 yr^{-1}). According to Shvidenko and Nilsson (2000), gross rate of stem growth is the sum of net growth and mortality (1879×10^6 m^3 yr^{-1}). By our calculation, ANPP for dry wood biomass (including growth by trees that then died) averages 81 g m^{-2} yr^{-1}, which is only slightly higher than the estimate by Shvidenko and Nilsson (2000), i.e., 72 g m^{-2} yr^{-1} tree growth and mortality for all of Siberia. Our estimate of ANPP, including dead and living wood and litter, of 169 g m^{-2} yr^{-1} is close to the estimate for *Larix* spp. in Central Siberia of 180 g m^{-2} yr^{-1} (Kajimoto *et al.*, 1999). Table 10-8 summarizes our estimates. The main sources of errors are in the fine root productivity, which was based on only a few boreal sites. Total ANPP and NPP are found to be 0.5 Pg (C) yr^{-1} and 1.33 Pg (C) yr^{-1}, respectively, for Siberian forests. The last figure is likely to be an overestimate: using a BNPP/NPP ratio of 0.40, as found in the coniferous Canadian stands (Table 10-7), leads to a total NPP of only 0.84 Pg (C) yr^{-1}.

VII. Production Processes

A. Net Ecosystem Flux

Long-term measurements of CO_2 flux extending from 4 months to 7 yr have now been measured in a number of boreal forest stands and enable annual net fluxes to be estimated, or provide them directly. At the northern black spruce site of BOREAS, measurements have continued since 1993 for over 7 yr (Goulden *et al.*, 1997). A sample of measurements made in 1994 over the jack pine stand by Baldocchi *et al.* (1996) is shown in Fig. 10-4. Half-hourly data have been averaged over about 20-day periods in spring, mid-summer, and end of summer. The fluxes are quite small, with maximum amplitudes between day and night of about 0.3, 0.45, and 0.23 mg m^{-2} s^{-1} (or 7, 10, and 5 μmol m^{-2} s^{-1}) for the three periods, respectively. Maximum net uptake rates were reached in the morning between 1000 and 1200 h, followed by a decrease in CO_2 uptake associated with an increase in atmospheric water vapor pressure deficit (VPD). Similar measurements for black spruce (Jarvis *et al.*, 1997) and aspen (Blanken *et al.*, 1997) in 1996, including the freeze in the fall, part of the winter, and the thaw in the spring, when integrated over the year give the annual net carbon sequestration by the forest (Fig. 10-5). Estimates range from a small annual loss of carbon from old-growth Norway spruce (*Picea abies*) in central Sweden (Lindroth *et al.*, 1998), through zero for the northern black spruce site of BOREAS (Goulden *et al.*, 1997), to a gain of ~0.4 t (C) ha^{-1} for the southern black spruce site of BOREAS (Fig. 10-5), and of 1.3 t (C) ha^{-1} for the aspen site in the south (Black *et al.*, 1996). These figures are all rather smaller than

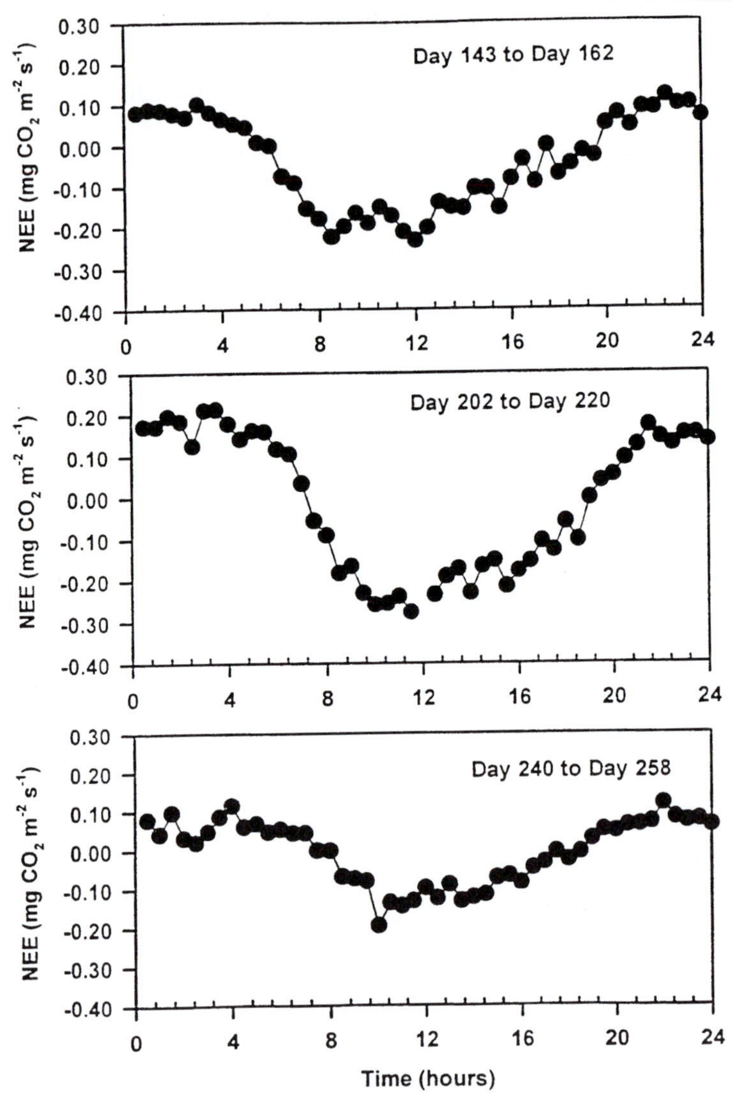

Figure 10-4 Diurnal variations (NEP, net ecosystem flux) in CO_2 flux above a stand of old jack pine in spring, midsummer, and late summer of 1994. Half-hourly measurements by the eddy-covariance method have been averaged over 18-day periods (from Baldocchi *et al.*, 1997).

comparable data from temperate zone forests largely in the range 2.0 to 7.0 t (C) ha^{-1} (Goulden *et al.*, 1996; Greco and Baldocchi, 1996; Valentini *et al.*, 1996; Baldocchi, 1997; Valentini *et al.*, 2000).

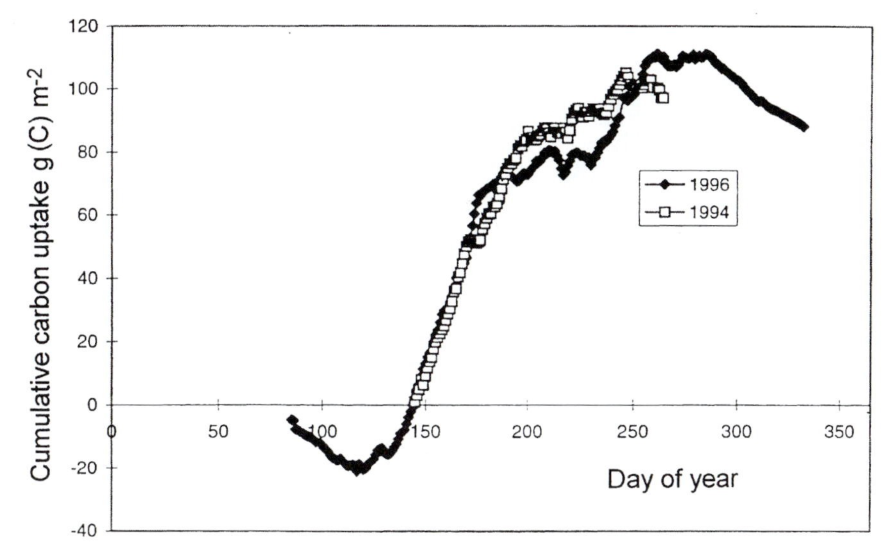

Figure 10-5 The accumulated annual net sequestration of carbon at the old black spruce site in the southern study area of BOREAS.

B. Canopy Photosynthesis

Canopy photosynthesis (A_c) has been measured directly either as the difference between the net ecosystem fluxes measured both above and below the forest canopy, or as the rate of photosynthesis and respiration of branches in long-term branch bags (Dufrêne *et al.*, 1993; Rayment and Jarvis, 1999; Hogg *et al.*, 2000), integrated over the whole canopy.

Rates of canopy net photosynthesis derived from CO_2 fluxes measured above and below the jack pine canopy (Baldocchi and Vogel, 1996) are shown plotted against *absorbed* (above canopy minus below canopy) photosynthetic photon flux density, Q_p, in Fig. 10-6. Again the amplitude is quite small, with maximum rate of canopy photosynthesis of ~10 μmol m^{-2} s^{-1}. Absorbed Q_p never exceeded 1000 μmol m^{-2} s^{-1} because of the relatively sparse canopy, with a maximum LAI below 2 in midsummer.

Rates of photosynthesis and stomatal conductance of jack pine, black spruce, and aspen have also been measured using branch bags. Figure 10-7 shows daytime carbon uptake between spring and fall of four branches of black spruce. Over 90% of the variability in total daytime carbon uptake was accounted for by variations in daily photosynthetic photon flux density and temperature (Rayment and Jarvis, 1999). In spring and fall, assimilation of CO_2 became zero at −6°C and respiration at −8°C. The average accumulated net gain of carbon over the intervening growing season was 192 g (C) m^{-2} (leaf area) (Fig. 10-8). This was corrected for slightly higher tempera-

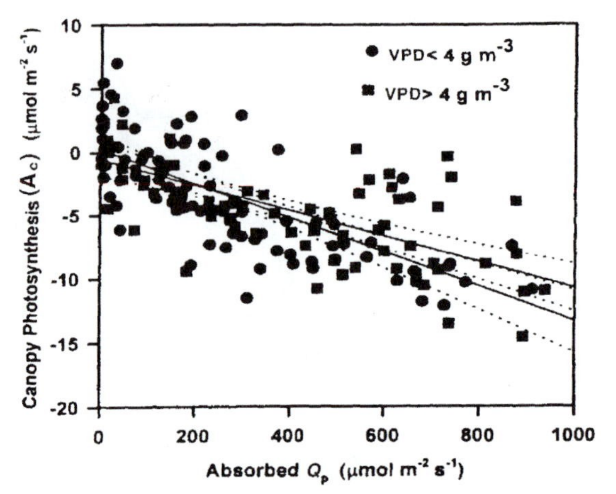

Figure 10-6 The relationship between canopy photosynthesis (A_c) and the absorbed photosynthesis photon flux density (Q_P) for two vapor pressure deficit classes (from Baldocchi and Vogel, 1996).

tures within the branch bags in daytime and for respiration during daytime (addition of 0.6 times the respiration at night corrected to the appropriate daytime temperature) and nighttime (addition of the respiration at night) and scaled up by LAI to give an annual gross carbon gain of 9.3 t ha^{-1}. [The

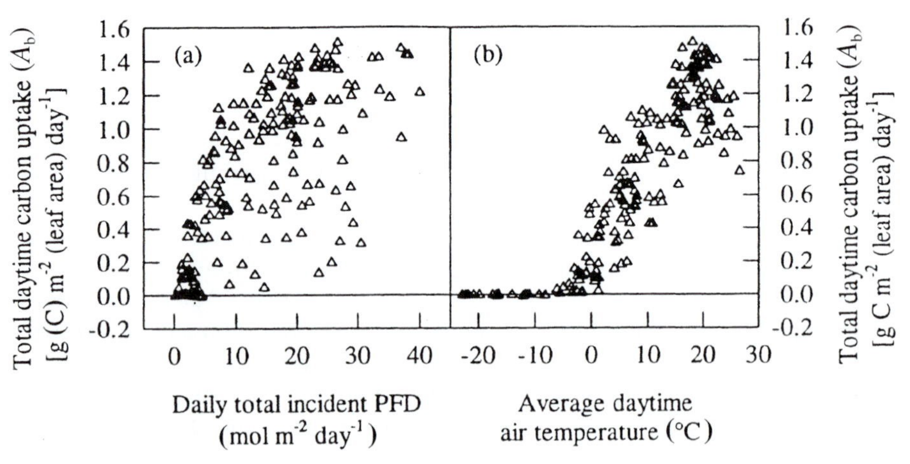

Figure 10-7 The daily daytime carbon uptake of branches of black spruce in the southern study area of BOREAS in relation to (a) daily total incident photosynthetic photon flux density (PFD) and (b) average daytime air temperature within the branch bag (from Rayment and Jarvis, 1999).

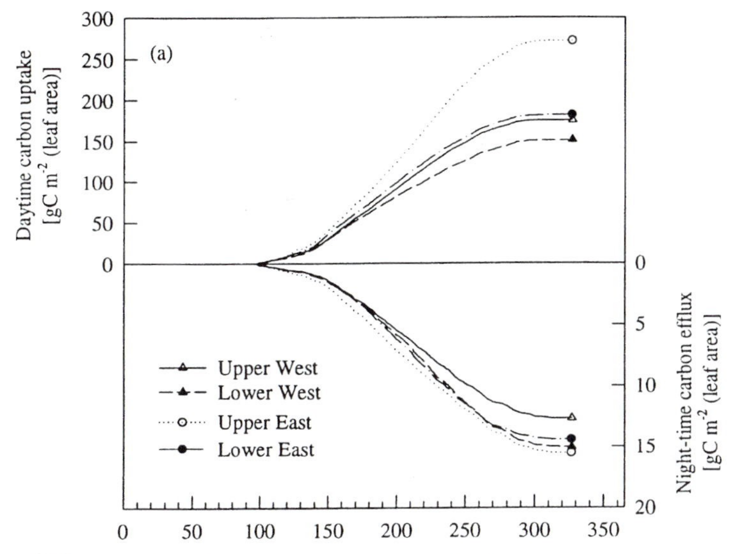

Figure 10-8 Cumulative daytime carbon uptake and nighttime loss measured over the 230 days, during which physiological activity occurred in 1996 on four branches of black spruce at the BOREAS southern study area old black spruce site (from Rayment and Jarvis, 1999).

factor of 0.6 is the presumed proportion of mitochondrial respiration that continues during photosynthesis; see, for example, Lloyd *et al.* (1995).]

Bulk stomatal conductances (g_s) of a branch of jack pine, computed from transpiration measurements, humidity, and leaf/air temperature within the bag, are shown in Fig. 10-9. There is a quite good correlation between g_s and A_b, the net assimilation of the branch, indicating that they respond in a similar manner and that there is relative constancy in the ratio between CO_2 concentrations inside and outside the needles, i.e., C_i/C_a.

Compared with other species of pine reviewed by Teskey *et al.* (1994), jack pine has typical average values of photosynthetic rate but low values of g_s, resulting in C_i/C_a near 0.4. Similar low values of C_i/C_a have been found for *Pinus radiata* (0.39) (Conroy *et al.*, 1988) and *Pinus sylvestris* (0.47) (Troeng and Linder, 1982). By contrast, rates of assimilation and stomatal conductances are much higher in aspen and, to a lesser extent, in black spruce. The low values of g_s and moderate values of A_b enable jack pine to save water without a large sacrifice of carbon uptake, an important feature for a species that grows on sandy soils with small water-holding capacity.

C. Autotrophic Respiration

High autotrophic respiration rates occur in the foliage, aboveground wood, and fine roots (<2 mm), especially during periods of rapid expansion and

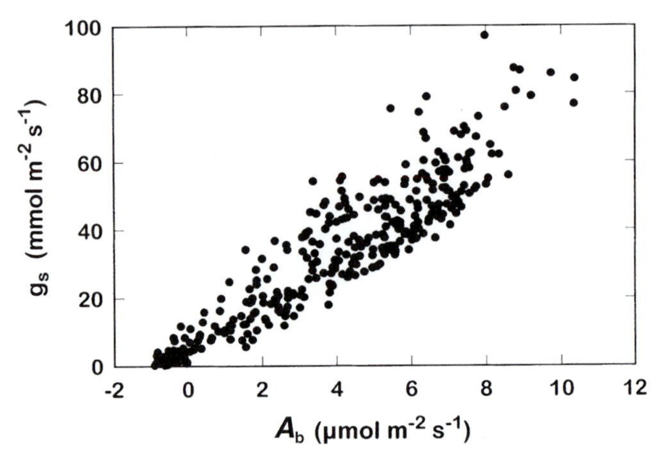

Figure 10-9 The relationship between bulk stomatal conductance (g_s) and net CO_2 assimilation rate (A_b) for a branch of jack pine measured in a branch bag at the BOREAS southern study area old jack pine site (from Saugier *et al.*, 1997).

extension growth and when temperatures are high (Ryan *et al.*, 1997). The highest CO_2 effluxes of >2 μmol m^{-2} s^{-1} occur from black spruce and aspen during midsummer as a result of their large foliar mass and, particularly in aspen, large sapwood volume (Tables 10-1 and 10-4), and, particularly in black spruce, high specific respiration rate. Annual losses of carbon as a result of aboveground respiration are largest from the foliage and, to a lesser extent, the wood and belowground losses from both fine and coarse roots are also appreciable (Table 10-9). The total annual efflux of carbon resulting from autotrophic respiration is largest in aspen, somewhat less in black spruce, and least in the jack pine, in close accordance with their NPP (Table 10-6). Consequently, carbon use efficiency (NPP/GPP) is similar in all three cases.

D. Soil Respiration

The autotrophic root respiration and respiration of the soil heterotrophs feeding on detritus and soil organic matter (SOM) comprise the efflux of CO_2 from the soil, the so-called soil respiration, which has been measured either from low level (~2 m), eddy-covariance measurements of CO_2 efflux or using soil surface chambers in either an open or closed system mode (e.g., Norman *et al.*, 1997; Rayment and Jarvis, 1997). With both approaches, ground vegetation, particularly a thick cover of mosses, and understorey, may deplete the measured efflux. Another problem with the chamber methods is the extreme local heterogeneity of the flux, sometimes over only a meter or two in relation to variation in local microtopography and amount of SOM, so that both intensive and extensive sampling is required. Soil CO_2 ef-

Table 10-9 Mean Annual Autotrophic Respiratory Losses of Carbon at Three Southern BOREAS Sites[a]

Source	Carbon loss (t (C) ha^{-1} yr^{-1})		
	Black spruce	Jack pine	Aspen
Overstorey foliage (night)	0.56	1.29	1.05
Overstorey foliage (day)[b]	1.73	1.30	1.28
Aboveground wood	0.87	0.30	1.23
Understorey wood and foliage[b]	0.02	0.004	1.16
Total aboveground	**3.18**	**2.89**	**4.72**
Fine root (<2 mm)	0.51	0.93	1.00
Fine root (2–5 mm)	0.47	0.11	0.65
Coarse root	0.95	0.47	1.49
Total belowground	**1.93**	**1.51**	**3.14**
Total autotrophic respiration R_a	**5.11**	**4.40**	**7.86**
NPP (from Table 10-6)	**2.66**	**2.26**	**3.92**
GPP (= NPP + R_a)	**7.77**	**6.66**	**11.78**
NPP/GPP	**0.34**	**0.34**	**0.33**

[a]From Ryan *et al.* (1997); showing resultant values for GPP and NPP/GPP.
[b]Modified on the basis that hourly daytime respiration = 0.6 × hourly night respiration at the same temperature.

flux during summer was largest from the black spruce site, followed by the aspen and jack pine sites. Figure 10-10 shows the seasonal course of soil CO_2 efflux at the black spruce site, peaking sharply shortly after midsummer.

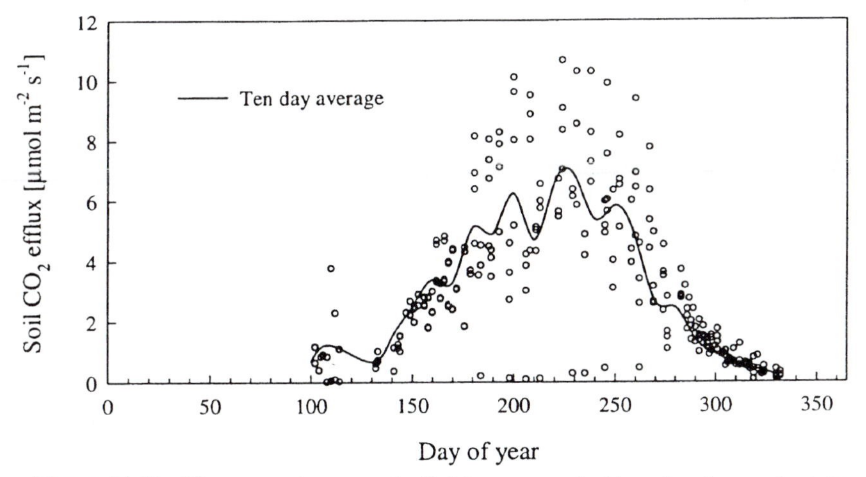

Figure 10-10 The seasonal course of soil CO_2 measured with a chamber method (Rayment and Jarvis, 1997) at the BOREAS southern study area old black spruce site (Rayment and Jarvis, 2000).

Rates of soil respiration in midsummer are very large, reaching 10 μmol m^{-2} s^{-1}, and comparable with the rates of canopy photosynthesis and net ecosystem flux. In general, when the data covered a sufficient range of temperature, the short-term, median Q_{10} was within the range 1.8–2.2. Much larger values result, however, if Q_{10} is calculated from data obtained as the season progresses, such as is shown in Fig. 10-10, because of concurrent changes in the populations of fine roots and microorganisms.

VIII. Stand Annual Carbon Balance

The carbon balance is severely, and very usefully, constrained by the long-term measurement of net ecosystem carbon flux, i.e., $P_e = R_e - P_g$ (Fig. 10-11). In this equation P_e is the long-term sum of CO_2 fluxes measured by eddy covariance, R_e is total ecosystem respiration, computed from nighttime values of CO_2 flux extrapolated to 24 hr using a simple relationship between R_e and air temperature. P_g (gross productivity, the sum of canopy photosynthesis) is calculated as $(R_e - P_e)$, but may also be estimated from direct photosynthesis measurements, using branch bags, for instance (Dufrêne *et al.*, 1993). Stem, branch, leaf, and soil respiration are also directly measurable and their sum should be equal to the ecosystem respiration R_e, the sum

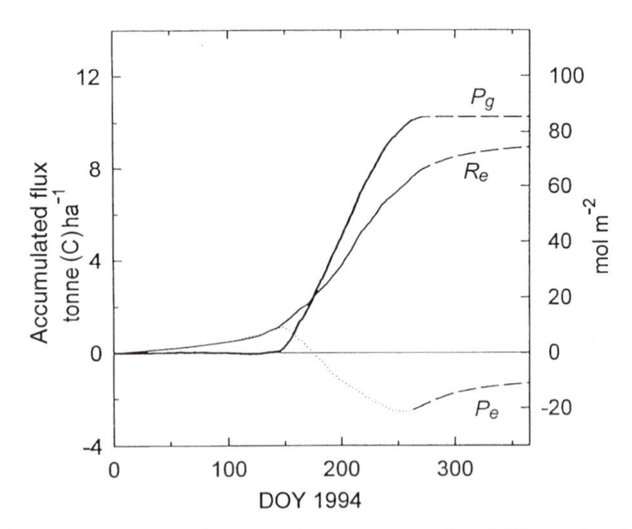

Figure 10-11 The accumulated measured net ecosystem flux (NEP = P_e) and total system respiration (R_e), and the ecosystem gross photosynthesis (GEP = P_g) derived by difference, for the BOREAS old aspen site in 1994. The dashed sections of the lines were estimated from Black *et al.* (1996).

of all autotrophic and heterotrophic respiration. Thus, in principle, the overall carbon gains and losses of the system, and the net difference between them, are independently measurable.

Furthermore, if aboveground growth and detritus production (i.e., ANPP) are known, together with the aboveground autotrophic respiration of wood and foliage, the quantity of carbon going belowground can be obtained as the difference from canopy photosynthesis. This known quantity is used for root growth, root, and microbial respiration, support of mycorrhizae, and fine root turnover (i.e., belowground detritus production. Subtracting root growth, which is probably the best known, from the amount of

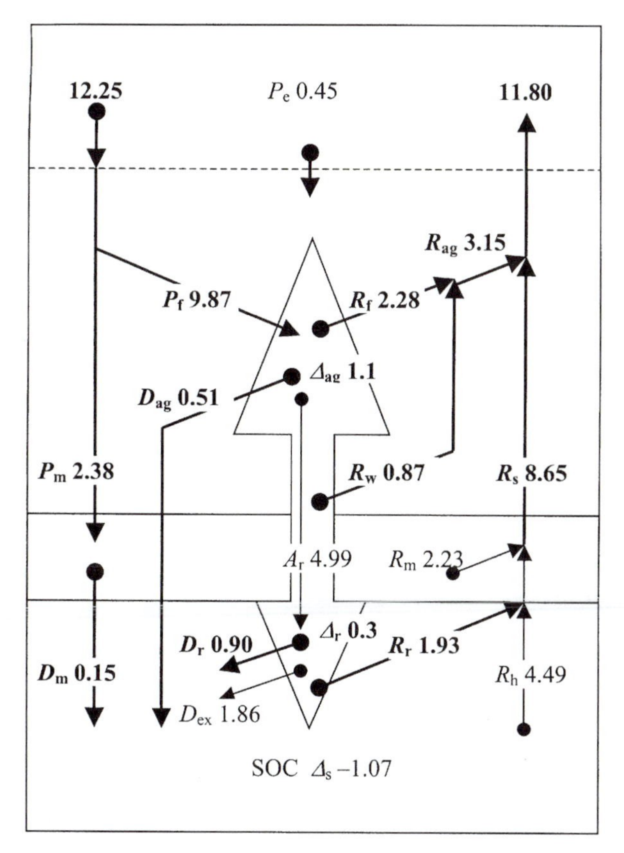

Figure 10-12 The annual carbon balance of the black spruce stand based on data from a number of sources (Ryan *et al.*, 1997, Steele *et al.*, 1997; Gower *et al.*, 1997; Williams and Flanagan, 1998; Rayment and Jarvis, 1999, 2000) constrained by the annual net ecosystem flux (Rayment, 1998; Massheder, 2000) at the BOREAS southern study area old black spruce site.

carbon going belowground, gives a combined estimate of the other processes. This estimate, together with the input of detritus from aboveground, comprises the soil carbon efflux, provided that the soil organic carbon (SOC) is unchanging in amount. The soil carbon efflux is reasonably well known from one or more independent measures, or it can be deduced from canopy photosynthesis, aboveground autotrophic respiration, and net ecosystem flux. It is, therefore, possible to address the key question of whether the SOC is increasing or decreasing in amount and also to pinpoint the measurements that require refinement to determine the reliability of this estimate (Malhi *et al.*, 1999).

The carbon balance for the mature black spruce site in the BOREAS southern study area for 1996, derived as indicated above, is depicted in Fig. 10-12. It can be deduced from this synthesis that, to close the balance, the SOC may be decreasing in amount annually by about 1 t ha^{-1}, possibly as a result of soil warming. Although this is an unusually comprehensive carbon budget, there are inevitably substantial uncertainties attached to the accuracy of the majority of the measurements so that this deduction must be viewed with caution.

IX. Radiation Use Efficiency

Monteith (1977) showed that crop productivity is linearly proportional to intercepted solar radiation and this has been confirmed for a number of tree stands (e.g., Linder, 1985; Wang *et al.*, 1991; Runyon *et al.*, 1994). Utilizing the Monteith relationship, empirical models have been developed to compute plant productivity from intercepted solar radiation to provide regional estimates of NPP directly (Ruimy *et al.*, 1994; Landsberg *et al.*, 1996) or from gross primary productivity (GPP) and a model of respiration (Ruimy *et al.*, 1996). Thus, there is considerable interest in establishing reliable base values of radiation use efficiency (RUE) for different boreal forest types.

Additionally, it has been shown that carbon flux measurements made on a wide range of plant canopies have a more or less linear response to radiation, particularly when measurements are averaged over periods of 1 day or longer (Ruimy *et al.*, 1995); Fig. 10-13A shows the fitted curve through all the forest data and the curve obtained for the jack pine stand data. Clearly, the values of CO_2 flux for this boreal forest stand are among the lowest measured for the same irradiance. The slope of the line in Fig. 10-13B gives an estimate of the RUE of GPP of 0.01 mol (C) per mol (photosynthetically active radiation; PAR), which is equivalent to a RUE of 1.24 g of dry mass (or 0.55 g of C) per MJ of PAR. This value is only one-third of the value of 1.77 g (C) MJ^{-1}(PAR) given by McMurtrie *et al.* (1994) for various pine stands, including a stand of *Pinus sylvestris* in Sweden, and is half the average value

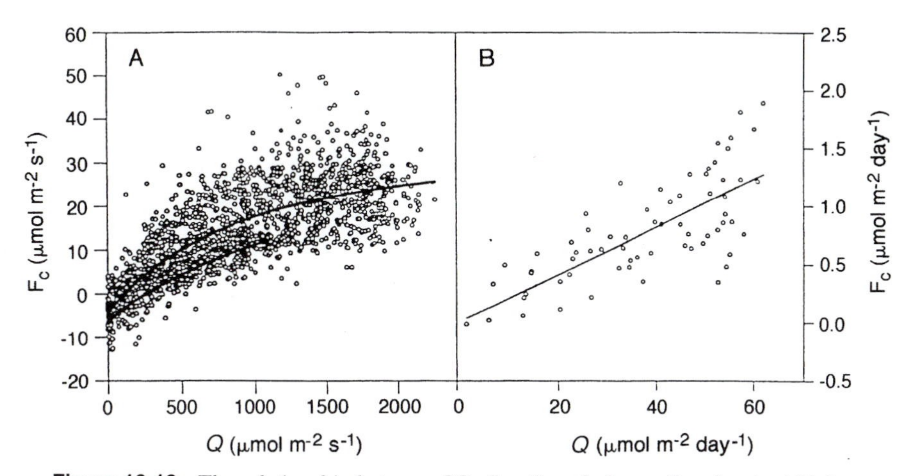

Figure 10-13 The relationship between CO_2 flux F_C and photon flux density (Q) for various types of closed forests (from Ruimy *et al.*, 1995). The upper line is the line of best fit through all the points. The lower line in A is the line of best fit for the *jack pine* stand of Fig. 10-6. B shows the jack pine data integrated on a daily basis.

of RUE for all the various stands of crops, grasslands, and forests reviewed by Ruimy *et al.* (1995).

With some simple assumptions, these figures can be used to provide alternative estimates of NPP. Table 10-10 shows NPP calculated for the jack pine stand, assuming that the stand is photosynthetically active from May 1 to September 15 and that respiration accounts for 60% of GPP. An NPP of dry mass of 380 g m^{-2} is obtained for an intercepted PAR of 766 MJ m^{-2}, i.e., a RUE of 0.5 g MJ^{-1}. For several boreal forests in Quebec, Royer *et al.* (1997) have calculated very similar, low values of RUE for ANPP, ranging

Table 10-10 Estimation of ANPP from APAR for the Jack Pine Stand at the Southern BOREAS Site[a]

Parameter	Value
From CO_2 flux vs. APAR	ε = 0.01 mol (C) mol^{-1} (PPFD)
	= 1.24 g (dm) MJ^{-1} (PAR)
Total solar radiation (1 May–15 Sept)	2280 MJ m^{-2}
Absorbed PAR (APAR = 0.7 · 0.48 · PAR)	766 MJ m^{-2}
(ρ = 0.05; τ = 0.25; f = 0.48)	
GPP (= ε · APAR)	1.24 × 766 = 950 g (dm) m^{-2}
NPP (= 0.4 GPP)	380 g (dm) m^{-2}
RUE (= NPP/APAR)	0.5 g (dm) MJ^{-1} PAR

[a] ρ = canopy reflectance, τ = canopy transmittance, f = fraction of PAR in the global solar radiation, (dm) = dry mass.

from 0.12 to 0.5 g per MJ. In general, whatever the method used, estimates of NPP in boreal forests are relatively low and this can be largely attributed to the small values of LAI and the low photosynthetic capacity resulting from high VPDs and shortage (or in the spruce stands excess) of water and, in particular, shortage of nitrogen. Leaf nitrogen concentrations of black spruce, for example, are in the range 0.6–1.2% of dry mass, i.e., they lie in the commonly accepted, highly deficient range for spruce.

X. Regional Production and Sensitivity to Global Change

It is desirable to have more measurements of the RUE of GPP and NPP of boreal forests to get realistic regional scale maps of the productivity of these forests. Goetz and Prince (1996) made an extensive survey of 29 stands of black spruce and 30 stands of aspen in northeastern Minnesota, using allometric relations to estimate LAI and ANPP. They also estimated *f*, the fraction of PAR intercepted by the stands, from LANDSAT multispectral scanner (MSS) data. The dry mass ANPP estimates were quite variable, ranging from 200 to 1200 g m^{-2} yr^{-1} for aspen and from 40 to 600 g m^{-2} yr^{-1} for black spruce. Most of this variability could be explained by differences in intercepted radiation. Average values for RUE were, respectively, 0.92 g (dry mass) MJ^{-1} (PAR) for aspen and 0.49 g MJ^{-1} for black spruce. Using these values they computed for their area (499 km^2 of aspen and 424 km^2 of black spruce) an average ANPP of 726 g (dry mass) m^{-2} yr^{-1} for aspen and 250 g m^{-2} yr^{-1} for black spruce, figures which are comparable to those that were found in the BOREAS stands in the southern study area (646 g m^{-2} yr^{-1} for aspen and 324 g m^{-2} yr^{-1} for black spruce).

It is possible to predict the radiation interception efficiency of these forests, i.e., the fraction of radiation they intercept during their growing season, and Royer *et al.* (1997) have made an interesting attempt in that direction. They showed a clear seasonal signal of the global vegetation index (GVI) computed from reflectances in the red and near-infrared wavebands, as measured by satellites at low spatial resolution, and demonstrated that the values of GVI for several sites in Alaska, Siberia, Finland, and Quebec were well correlated with the temperature sum above a threshold of 5°C, but with different regression lines for the various sites. However, when the temperature sum was multiplied by an estimate of stand transpiration, they were able to obtain a single regression line with GVI for all sites, thus showing that this approach has good potential for useful generalization.

Global Change

How will boreal forests respond to global change? Relevant variables include increase in wet and dry deposition of nitrogen in several forms, increase in precipitation, increase in temperature, and increase in atmospheric CO_2 concentration. Boreal forests are chronically deficient in nitrogen and consequently are sensitive to nutrient supply, as well as being limited by the low rates of mineralization induced by low soil temperatures. In a long-term experiment at Flakaliden (Sweden, 64°N), full fertilization, applied either as liquid or solid fertilizer during the growing season, has increased NPP of Norway spruce stands by a factor of four to five relative to the unfertilized control plots over the past 12 yr (Berg *et al.*, 1999; Jarvis and Linder, 2000). Previously, it had been widely supposed that growth was entirely constrained by temperature at this latitude.

It is commonly thought that an increase in temperature will stimulate respiration from both vegetation and, particularly, soils more than the rise in CO_2 concentration and temperature will stimulate photosynthesis, and thus will turn boreal forests from carbon sinks into carbon sources in the not too distant future. However, this may be too simplistic a view for three or four reasons. First, warming may increase the duration of the photosynthetic season, as has apparently already been detected from satellite data (Myneni *et al.*, 1997), thus leading to increase in GPP and NPP. Second, "soil" respiration has been found to acclimate downward in response to increase in temperature. For example, in a long-term soil warming experiment in the fertilized and unfertilized Norway spruce stands at Flakaliden, there was no increase in soil respiration as a result of warming the soil by 5°C after 5 years, and, furthermore, the initial period of warming stimulated nitrogen release by soil microorganisms and increased NPP of the trees on the warmed, unfertilized plots by an average of 60% per year (Jarvis and Linder, 2000). Third, the current increase in atmospheric CO_2 is causing an increase in canopy photosynthesis. This effect is predicted by photosynthesis models, especially for tree leaves that present high resistances to CO_2 diffusion in both gaseous and liquid phases, and has now been demonstrated using large tree chambers with climate control and with or without CO_2 enrichment in boreal spruce and pine stands in northern Sweden and Finland (Medlyn *et al.*, 1999). It is becoming increasingly clear that there are more significant feedback processes among temperature, atmospheric CO_2 concentration, nutrition, and precipitation that need to be taken into account in models that purport to predict the likely effects of global change on boreal forests (McMurtrie *et al.*, 2001).

XI. Summary

Boreal forests cover about 13.8 million km^2 between 46°N and 66°N. They have low values of LAI (typically 1 to 3), of albedo (about 0.1 in summer and 0.3 in winter for conifers), and of surface conductance to water vapor. The photosynthetic season is short (4 to 5 months) and canopy photosynthesis is limited by low leaf nitrogen concentration, by low temperatures in spring and fall, and by high VPDs and by soil water stress in summer. This leads to moderate values of GPP, of which respiration removes at least 60%. The resulting NPP is low, typically ranging from 2 to 5 t ha^{-1} yr^{-1}, with about 40% belowground. Growing boreal forests are currently carbon sinks, with NEP in the range −1 to +2.5 t (C) ha^{-1} yr^{-1}, and seem likely to remain so for some time to come on the basis of recent experimental and modelling evidence, although there is uncertainty as to how long.

References

Baldocchi, D. D. (1997). Measuring and modeling carbon dioxide and water vapor exchange over a temperate broad-leaved forest during the 1995 summer drought. *Plant Cell Environ.* **20,** 1108–1122.

Baldocchi, D. D., and Vogel, C. A. (1996). Energy and CO_2 flux densities above and below a temperate broad-leaved forest and a boreal pine forest. *Tree Physiol.* **16,** 5–16.

Baldocchi, D. D., Vogel, C. A., and Hall, B. (1996). Seasonal variation of carbon dioxide exchange rates above and below a boreal jack pine forest. *Agric. For. Meteorol.* **83,** 147–170.

Baldocchi, D. D., Vogel, C. A., and Hall, B. (1997). Seasonal variation of energy and water vapor exchange rates above and below a boreal jack pine forest canopy. *J. Geophys. Res. (BOREAS Special Issue)* **102,** D24, 28939–28951.

Barr, A., and Betts, A. (1994). Preliminary Summary of BOREAS Upper-Air Soundings. Candle Lake and Thompson 1994 Intensive Field Campaigns. BOREAS Report, Saskatoon, Canada.

Berg, J., Linder, S., Lundmark, T., and Elfving, B. (1999). The effect of water and nutrient availability on the productivity of Norway spruce in northern and southern Sweden. *For. Ecol. Manage.* **119,** 51–62.

Black, T. A., den Hartog, G., Neumann, H. H., Blanken, P. D., Yang, P. C., Russell, C., Nesic, Z., Lee, X., Chen, S. G., Staebler, R., and Novak, M. D. (1996). Annual cycles of water vapour and carbon dioxide fluxes in and above a boreal aspen forest. *Global Change Biol.* **2,** 219–229.

Blanken, P. D., Black, T. A., Yang, P. C., Neumann, H. H., Nesic, Z., Staebler, R., den Hartog, G., Novak, M. D., and Lee, X. (1997). Energy balance and canopy conductance of a boreal aspen forest: partitioning overstory and understory components. *J. Geophys. Res. (BOREAS Special Issue)* **102,** D24, 28915–28927.

Bolin, B., Sukumar, R., Ciais, P., Cramer, W., Jarvis, P., Kheshgi, H., Nobre, C., Semenov, S., and Steffen, W. (2000). Global Perspective. *In* "Special Report on Land Use, Land-Use Change, and Forestry" (R. Watson and D. Verrado, eds). Cambridge University Press.

Bonan, G. B., and Shugart, H. H. (1989). Environmental factors and ecological processes in boreal forests. *Annu. Rev. Ecol. Syst.* **20,** 1–28.

Botkin, D. B., and Simpson, L. G. (1990). Biomass of North American boreal forest. *Biogeochemistry* **9,** 161–174.

Ciais, P., Tans, P. P., Trolier, M., White, J. W. C., and Francey, R. J. (1995). A large northern hemisphere terrestrial CO_2 sink indicated by the $^{13}C/^{12}C$ ratio of atmospheric CO_2. *Science* **269,** 1098–1102.

Conroy, J. P., Küppers, M., Küppers, B., Virgona, J., and Barlow, E. W. R. (1988). The influence of CO_2 enrichment, phosphorus deficiency and water stress on the growth, conductance and water use of *Pinus radiata* D. Don. *Plant Cell Environ.* **11,** 91–98.

Dufrêne, E., Pontailler, J. Y., and Saugier, B. (1993). A branch bag technique for simultaneous CO_2 enrichment and assimilation measurements on beech (*Fagus silvatica* L.). *Plant Cell Environ.* **16,** 1131–1138.

Goetz, S. C., and Prince, S. D. (1996). Remote sensing of net primary production in boreal forest stands. *Agric. For. Meteorol.* **78,** 179–179.

Goulden, M. L., Munger, J. W., Fan, S.-M., Daube, B. C., and Wofsy, S. C. (1996). Exchange of carbon dioxide by a deciduous forest: response to interannual climate variability. *Science* **271,** 1576–1578.

Goulden, M. L., Daube, B. C., Fan, S.-M., Sutton, D. J., Bazzaz, A., Munger, J. W., and Wofsy, S. C. (1997). Physiological responses of a black spruce forest to weather. *J. Geophys. Res. (BOREAS Special Issue)* **102,** D24, 28987–28996.

Gower, S. T., Gholz, H. L., Nakane, K., and Badwin, V. C. (1994). Production and allocation patterns of pine forests. *Ecol. Bull. (Copenhagen)* **43,** 115–135.

Gower, S. T., Vogel, J., Norman, J. M., Kucharik, C. J., Steele, S. J., and Stow, T. K. (1997). Carbon distribution and aboveground net primary production in aspen, jack pine and black spruce stands in Saskatchewan and Manitoba, Canada. *J. Geophys. Res. (Boreas Special Issue)* **102,** D24, 29029–29041.

Greco, S., and Baldocchi, D. D. (1996). Seasonal variations of CO_2 and water vapour exchange rates over a temperate deciduous forest. *Global Change Biol.* **2,** 183–197.

Grelle, A. (1997). Long-term water and carbon dioxide fluxes from a boreal forest: Methods and applications. *Acta Univ. Agric. Suec. Sylv.* **28,** 1–80.

Hale, S. E. (1996). Turbulent transport above and within a black spruce forest canopy. Ph.D. Thesis, University of Edinburgh.

Hogg, E. H., Saugier, B., Pontailler, J.-Y., Black, T. A., Chen, W., Hurdle, P. A., and Wu, A. (2000). Responses of trembling aspen and hazelnut to vapor pressure deficit in a boreal deciduous forest. *Tree Physiology* **20,** 725–734.

Hollinger, D. Y., Kelliher, F. M., Schulze, E.-D., Bauer, G., Arneth, A., Byers, J. N., Hunt, J. E., McSeveny, T. M., Kobak, K. I., Milukova, I., Sogachev, A., Tatarinov, F., Varlagin, A., Ziegler, W., and Vygodskaya, N. N. (1997). Forest–atmospere carbon dioxide exchange in eastern Siberia. *Agric. For. Meteorol.* **90,** 291–306.

Houghton, J. T., Meira Filho, L. G., Callander, B. A., Harris, N., Kattenberg, A., and Maskell, K., eds. (1996). "Climate Change 1995: The Science of Climate Change." Cambridge Univ. Press, Cambridge.

Isaev, A. S., ed. (1991). Forecast of the utilization and reproduction of the forest resources by economic regions of the USSR. Vols. 1 and 2, Academy of Sciences of the USSR and the State Forest Committee of the USSR, Moscow [Russian].

Jackson, R. B., Mooney, H. A., and Schulze, E.-D. (1997). A global budget for fine root biomass, surface area, and nutrient contents. *Proc. Natl. Acad. Sci. U.S.A.* **94,** 7362–7366.

Jarvis, P. G., and Linder, S. (2000). Constraints to growth of boreal forests. *Nature* **405,** 904–905.

Jarvis, P. G., James, B. G., and Landsberg, J. J. (1976). Coniferous forest. *In* "Vegetation and Atmosphere (J. L. Monteith, ed.), pp. 171–204. Academic Press, London and New York.

Jarvis, P. G., Massheder, J. M., Hale, S. E., Moncrieff, J. B., Rayment, M., and Scott, S. L. (1997). Seasonal variation of carbon dioxide, water vapour and energy exchanges of a boreal black spruce forest. *J. Geophys. Res. (BOREAS Special Issue)* **102,** D24, 28953–28966.

Kajimoto, T., Matsuura, Y., Sofronov, M. A., Volokitina, A. V., Mori, S., Osawa, A., and Abaimov, A. P. (1999). Above- and belowground biomass and net primary productivity of a *Larix gmelinii* stand near Tura, central Siberia. *Tree Physiology* **19,** 815–822.

Köppen W. (1936). Das geographischen System der Klimate. *In* "Handbuch der Klimatologie" (W. Köppen and R. Geiger, eds.). Gebrüder Bornträger, Berlin.

Körner, C., Schilcher, B., and Pelaez-Riedl, S. (1993). Bestandesaufnahme: Anthropogene Klimaänderungen: Mögliche Auswirkungen auf Österreich—Mögliche Nassnahmen in Österreich. *Österr. Akad. Wissensch. Kap.* **6,** 1–46.

Kramer, H. (1988). "Wald-Wachstumskunde." Paul Parey, Hamburg, Berlin.

Kurz, W. A., and Apps, M. J. (1993). Contribution of Northern forests to the global C cycle: Canada as a case study. *Water Air Soil Pollut.* **70,** 163–176.

Kurz, W. A., Apps, M. J., Webb, T., and MacNamee, P. J. (1992). The carbon budget of the canadian forest sector: Phase I. ENFOR Information Report NOR-X-326, Forestry Canada Northwest Region, Edmonton, Alberta.

Landsberg, J. J., Prince, S. D., Jarvis, P. G., McMurtrie, R. E., Luxmoore, R., and Medlyn, B. E. (1996). Energy conversion and use in forests: The analysis of forest production in terms of radiation utilisation efficiency (ϵ). *In* "The Use of Remote Sensing in the Modelling of Forest Productivity at Scales from the Stand to the Globe" (H. L. Gholz, K. Nakane, and H. Shimoda, eds.), pp. 273–298. Kluwer Academic Publishers, Norwell, Massachusetts.

Larsen, J. A. (1980). "The Boreal Ecosystem." Academic Press, New York.

Lieth, H. (1975). Modeling the primary productivity of the world. *In* "Primary Productivity of the Biosphere" (H. Lieth and R. H. Whittaker, eds.), pp. 237–263. Springer-Verlag, New York.

Linder, S. (1985). Potential and actual production in Australian forest stands. *In* "Research for Forest Management" (J. J. Landsberg and W. Parsons, eds.), pp. 11–35. CSIRO, Australia.

Lindroth, A., Grelle, A., and Morén, A.-S. (1998). Long-term measurements of boreal forest carbon balance reveal large temperature sensitivity. *Global Change Biology* **4,** 443–450.

Lloyd, J. (1999). Current perspectives on the terrestrial carbon cycle. *Tellus* **51B,** 336–342.

Lloyd, J., Grace, J., Miranda, A. C., Meir, P., Wong, S. C., Miranda, H., Wright, I., Gash, J. H. C., and McIntyre, J. (1995). A simple calibrated model of Amazon rainforest productivity based on leaf biochemical properties. *Plant Cell Environ.* **18,** 1129–1145.

Malhi, Y., Baldocchi, D. D., and Jarvis, P. G. (1999). The carbon balance of tropical, temperate and boreal forest. *Plant Cell Environ.* **22,** 715–740.

Massheder, J. M. (2000). "Surface Flux Measurements by Eddy Covariance Over a Black Spruce Stand". Ph.D. Thesis, University of Edinburgh.

McMurtrie, R. E., Gholz, H. L., Linder, S., and Gower, S. T. (1994). Climatic factors controlling the productivity of pine stands: A model-based analysis. *Ecol. Bull. (Copenhagen)* **43,** 173–188.

McMurtrie, R. E., Medlyn, B. E., and Dewar, R. E. (2001). Increased understanding of nutrient immobilisation in soil organic matter is critical for predicting the carbon sink strength of forest ecosystems over the next 100 years. *Tree Physiology* (in press).

Medlyn, B. E., Badeck, F.-W., de Pury, D. G. G., Barton, C. V. M., Broadmeadow, M., Ceulemans, R., de Angelis, P., Forstreuter, M., Jach, E., Kellomäki, S., Laitat, E., Marek, M., Philippot, S., Rey, A., Strassemayer, J., Laitinen, K., Liozon, R., Portier, B., Roberntz, P., Wang, K., and Jarvis, P. G. (1999). Effects of elevated [CO_2] on photosynthesis in European forest species: a meta-analysis of model parameters. *Plant, Cell and Environment* **22,** 1475–1495.

Mitchell, J. F. B., Johns, T. C., Gregory, J. M., and Tett, S. F. B. (1995). Climate response to increasing levels of greenhouse gases and sulphate aerosols. *Nature (London)* **376,** 501–504.

Monserud, R. A., Onuchin, A. A., and Tchebakova, N. M. (1996). Needle, crown, stem, and root phytomass of *Pinus sylvestris* stands in Russia. *For. Ecol. Manage.* **82,** 59–67.

Monteith, J. L. (1977). Climate and the efficiency of crop production in Britain. *Trans. R. Soc. Lond. Ser. B* **281,** 277–294.

Moore, T. R. (1996). The carbon budget of boreal forests: Reducing the uncertainty. *In* "Global Change: Effects on Coniferous Forests and Grasslands" (A. I. Breymeyer, D. O. Hall, J. M. Melillo, G. I. Agren, eds.), pp. 17–40. John Wiley, New York.

Myneni, R. B., Keeling, C. D., Tucker, C. J., Asrar, G., and Nemani, R. R. (1997). Increased plant growth in the northern high latitudes from 1981 to 1991. *Nature (London)* **386,** 698–702.

Nadelhoffer, K. J., Emmett, B. A., Gundersen, P., Kjonaas, O. A., Koopmans, C. J., Schleppi, P., Tietma, A., and Wright, R. F. (1999). Nitrogen deposition makes a minor contribution to carbon sequestration in temperate forests. *Nature (London)* **398,** 145–148.

Norman, J. M., Kucharik, C. J., Gower, S. T., Baldocchi, D. D., Crill, P. M., Rayment, M., Savage, K., and Striegel, R. G. (1997). A comparison of six methods for measuring soil-surface carbon dioxide fluxes. *J. Geophys. Res. (BOREAS Special Issue)* **102,** D24, 28771–28777.

Pryagnikov, A. A., and Zomolodchikov, D. G. (1993). The estimation of carbon stocks in stand phytomass of the European Ural forest zone. *In* "Forests of the Russian Plain," pp. 146–149. Russian Academy of Sciences, Moscow.

Rayment, M. B. (1998). "Carbon and Water Fluxes in a Boreal Forest Ecosystem". Ph.D. Thesis, University of Edinburgh.

Rayment, M., and Jarvis, P. G. (1997). An improved open chamber system for measuring soil CO_2 effluxes in the field. *J. Geophys. Res. (BOREAS Special Issue)* **102,** D24, 28779–28784.

Rayment, M., and Jarvis, P. G. (1999). Seasonal gas exchange of black spruce using an automatic branch bag system. *Can. J. For. Res.* **29,** 1528–1538.

Rayment, M., and Jarvis, P. G. (2000). Temporal and spatial variation of soil CO_2 efflux in a Canadian boreal forest. *Soil Biol. Biochem.* **32,** 35–45.

Rodin, L. E., and Bazilevitch, N. I. (1965). "Production and Mineral Cycling in Terrestrial Vegetation" [Translated by G. E. Fogg]. Oliver & Boyd, London.

Roulet, N., Moore, T., Bubier, J., and Lafleur, P. (1992). Northern fens: Methane flux and climate change. *Tellus* **44B,** 100–105.

Royer, A., Goïta, K., Ansseau, C., Faizoun, A., and Saint, G. (1997). Analysis of boreal forest dynamics using the global vegetation index. *Remote Sensing Rev.* **15,** 265–282.

Ruimy, A., Dedieu, G., and Saugier, B. (1994). Methodology for the estimation of terrestrial net primary production from remotely sensed data. *J. Geophys. Res. (Atmospheres)* **99,** D3, 5263–5283.

Ruimy, A., Jarvis, P. G., Baldocchi, D. D., and Saugier, B. (1995). CO_2 fluxes over plant canopies and solar radiation: A review. *Adv. Ecol. Res.* **26,** 1–68.

Ruimy, A., Dedieu, G., and Saugier, B. (1996). TURC: A diagnostic model of continental gross primary productivity and net primary productivity. *Global Biogeochem. Cycles* **10,** 269–285.

Runyon, J., Waring, R. H., Goward, S. N., and Welles, J. M. (1994). Environmental limits on above-ground production: Observations from the Oregon transect. *Ecol. Appl.* **4,** 226–237.

Ryan, M. G., Lavigne, M. B., and Gower, S. T. (1997). Annual carbon cost of autotrophic respiration in boreal forest ecosystems in relation to species and climate. *J. Geophys. Res. (BOREAS Special Issue)* **102,** D24, 28871–28883.

Saugier, B., Granier, A., Pontailler, J.-Y., Dufrene, E., and Baldocchi, D. D. (1997). Transpiration of a boreal pine forest measured by branch bags, sapflow and micrometeorological methods. *Tree Physiol. (BOREAS Special Issue)* **17,** 511–519.

Schulze, E.-D. (1982). Plant life forms and their carbon, water and nutrient relations. *Encycl. Plant Physiol.* **12B,** 616–676.

Schulze, E.-D., Schulze, W., Kelliher, F. M., Vygodskaya, N. N., Ziegler, W., Kobak, K. I., Koch, H., Arneth, A., Kusnetsova, W. A., Sogachev, A., Issajev, A., Bauer, G., and Hollinger, D. Y. (1995). Above-ground biomass and nitrogen nutrition in a chronosequence of pristine Dahurian *Larix* stands in Eastern Siberia. *Can. J. For. Res.* **25,** 943–960.

Schulze, E.-D., Lloyd, J., Kelliher, F. M., Wirth, C., Rebmann, C., Lühker, B., Mund, M., Milukova, I., Schulze, W., Ziegler, W., Varlagin, A., Valentini, R., Dore, S., Grigoriev, S., Kolle, O., Sogachov, A., and Vygodskaya, N. N. (1999). Productivity of forests in the Eurosiberian boreal region and their potential to act as carbon sink. *Global Change Biol.* **5,** 703–722.

Sellers, P., Hall, F., Margolis, H., Kelly, B., Baldocchi, D., den Hartog, G., Cihlar, J., Ryan, M. G., Goodison, B., Crill, P., Ranson, K. J., Lettenmaier, D., and Wickland, D. E. (1995). The Boreal Ecosystem–Atmosphere Study (BOREAS): An overview and early results from the 1994 field year. *Bull. Am. Meteorol. Soc.* **76,** 1549–1577.

Shvidenko, A., and Nilsson, S. (1994). What do we know about the Siberian forests? *Ambio* **23,** 396–404.

Shvidenko, A., and Nilsson, S. (2000). Phytomass, increment, mortality and carbon budget of Russian forests. *J. Climate Change* (in press).

Simpson, I. J. (1996). Trace gas exchange and the validity of similarity theory in the roughness sublayer above forests. Ph.D.Thesis, University of Guelph.

Smirnov, V. V. (1970). Productivity of stands of subzone broad leaved-spruce forest. *Veg. Resource* **6,** 165–176.

Smirnov, V. V., and Alekseev, V. I. (1967). Productivity of above-ground parts of a 75-year old spruce stand. *Veg. Resource* **3,** 505–512 [russian].

Stanners, D., and Bourdeau, P. (1995). "Europe's Environment; The Dobrís Assessment." European Environment Agency, Copenhagen.

Steele, S. J., Gower, S. T., Vogel, J. G., and Norman, J. M. (1997). Root mass, net primary production and turnover in aspen, jack pine and black spruce forests in Saskatchewan and Manitoba, Canada. *Tree Physiol. (BOREAS Special Issue)* **17,** 577–587.

Steyaert, L. T., Hall, F. G., and Loveland, T. R. (1997). Land cover mapping, fire regeneration, and scaling studies in the Canadian boreal forest with 1 km AVHRR and Landsat TM data. *J. Geophys. Res. (BOREAS Special Issue)* **102,** D24, 29581–29598.

Tans, P. P., Fung, I. Y., and Takahashi, T. (1990). Observational constraints on the global atmospheric CO_2 budget. *Science* **247,** 1431–1438.

Teskey, R. O., Whitehead, D., and Linder, S. (1994). Photosynthesis and carbon gain by pines. *Ecol. Bull. (Copenhagen)* **43,** 35–49.

Troeng, E., and Linder, S. (1982). Gas exchange in a 20-year old stand of Scots pine. II. Variation in net photosynthesis and transpiration within and between trees. *Physiol. Planta.* **54,** 7–14.

Usol'tsev, V. A. (1985). Modelling of structure and dynamic of stand phytomass [russian]. Krasnoyarsk University, Krasnoyarsk, Siberia.

Utkin, A. J. (1965). The forests of central Yakutia. Academia NAUK CCCP, Moscow [russian].

Valentini, R., De Angelis, P., Matteucci, G., Monaco, R., Dore, S., and Scarascia-Mugnozza, G. E. (1996). Seasonal net carbon dioxide exchange of a beech forest with the atmosphere. *Global Change Biol.* **2,** 199–207.

Valentini, R., Matteucci, G., Dolman, A. J., Schulze, E.-D., Rebmann, C., Moors, E. J., Granier, A., Gross, P., Jensen, N. O., Pilegaard, K., Lindroth, A., Grelle, A., Bernhofer, Ch., Grunwald, T., Aubinet, M., Ceulemans, R., Kowalski, A. S., Vesala, T., Rannik, U., Berbigier, P., Loustau, D., Guomundsson, J., Thorgeirsson, H., Ibrom, A., Morgenstern, K., Clement, R., Moncrieff, J., Montagnani, L., Minerbi, S., and Jarvis, P. G. (1999). The carbon sink strength of forests in Europe: novel results from the flux observation network. *Nature* **404,** 861–865.

Vitousek, P. M. (1994). Beyond global warming: Ecology and global change. *Ecology* **75,** 1861–1876.

Vygodskaya, N. N., Milukova, I. M., Tatarinov, F. A., Kusnezova, V. A., Zhukova, V. M., Solnzeva-Elbe, O. N., Gubenko, I. Y., Zavelskaya, N. A., Panfyorov, M. I., Sachaposhnikov, E. S., and Kobak, K. I. (1998). Carbon stock and deposition in phytomass of the forest ecosystems of the Central Forest Reserve (eastern European taiga), **1357,** pp. 1–280. University Göttingen.

Wang, Y.-P., Jarvis, P. G., and Taylor, C. M. A. (1991). PAR absorption and its relation to aboveground dry matter production of Sitka spruce. *J. Appl. Ecol.* **28,** 547–560.

Williams, T. G., and Flanagan, L. B. (1998). Measuring and modelling environmental influences on photosynthetic gas exchange in *Sphagnum* and *Pleurozium. Plant, Cell and Environment* **21,** 555–564.

Wirth, C., Schulze, E.-D., Schulze, W., Ziegler, W., Stünzner-Karbe, D., Milukova, I., Sogachov, A., Varlagin, A., Panfyorov, M., Grigoriev, S., Kusnecova, W., and Vygodskaya, N. N. (1999). Aboveground biomass, structure and self-thinning of prestine boreal Scots pine forests as controlled by fire and competition. *Oecologia* **121,** 66–80.

11

Productivity of Evergreen and Deciduous Temperate Forests

Peter B. Reich and Paul Bolstad

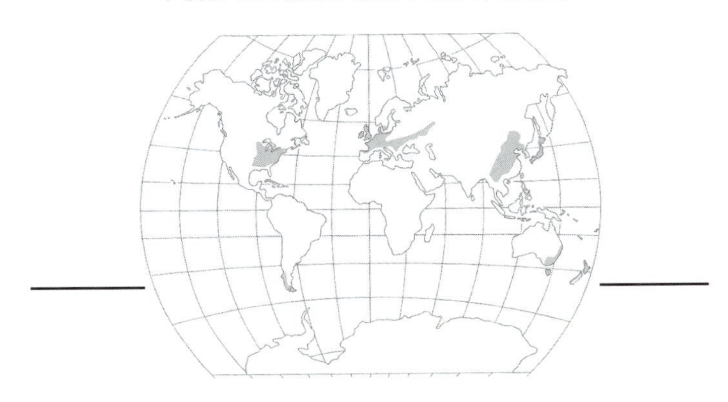

I. Nature and Extent of Temperate Forests

Temperate forests occupy a central position among the world biomes: they occur in regions of intermediate temperature, precipitation, and latitude, and are bracketed geographically by biomes that are more extreme in productivity and diversity: boreal forest and tundra toward the colder high latitudes, grasslands toward the drier climate zones, and tropical forests in warmer regions. Temperate forests make up a substantial proportion of world forests and have generally been subjected to more severe direct and indirect human impact than have other forest types. Temperate forests are important regionally for their ecological services, economic utility, and recreational, tourism, and aesthetic benefits. In a global context, temperate forests are important for their role in the carbon cycle (Tans *et al.*, 1990; Schimel, 1995). However, the degree to which these forests are a net carbon source or sink is controversial.

Although scientists will argue endlessly about logical classifications of vegetation, for simplicity we follow the vegetation classification system used by Melillo *et al.* (1993). Within this scheme, forest vegetation is classified by broad climate zones (tropical, temperate, boreal) and by leaf form and habit (broadleaf vs. needle-leaf forest, evergreen vs. deciduous). Temperate forests occupy 14.2×10^8 ha, or approximately the same area covered by boreal forests or tropical evergreen forests. Thus, temperate forests can be considered one of the three dominant forest types. Generally, a forest is considered temperate if it is found in a climate with a warm summer and a markedly cool or cold winter. For this chapter we consider as temperate those forests found between 3 and 18°C mean annual temperature, with midwinter mean monthly temperatures generally below 8°C and midsummer mean monthly temperatures generally above 18°C.

Temperate forests occupy much of North America, Europe, Asia, and Australia, and a small part of South America. Four different temperate forest types occur: broadleaf deciduous and needleleaf evergreen (the dominant types), and needleleaf deciduous and broadleaf evergreen. All combinations of these also can occur within a given region, landscape, or stand. In some areas one of these forest types may be dominant, although in much of the temperate zone a mixture of at least two types at the landscape scale is more often the rule. Needleleaf deciduous forests are more common in the true boreal zone and broadleaf evergreen forests are more common in Mediterranean and subtropical biomes, and thus do not receive broad coverage in this chapter. Site heterogeneity is usually large in most of the temperate zone, and sites vary as much in fertility, water-holding capacity, and other features locally as they do regionally. Thus, the following generalizations about where we find conifers vs. hardwoods should be considered a rough guide to observed distributional patterns.

Broadleaf deciduous forests typically occupy colder and/or moister parts of the temperate zone compared to broadleaf evergreen forests, which span the ecotones between warm temperate, subtropical, and Mediterranean biomes. For instance, in Japan, moving from north to south within a continuously moist climate, native forests are dominated by mixed conifer–hardwood forests in the cold temperate north, by broadleaf hardwoods in the central region, and in the warm temperate south by mixed deciduous–evergreen or evergreen broadleaf forests. Temperate coniferous forests occupy a wide range of sites, dominating on sandy soils; or in regions with dry summers and mild winters (such as the Pacific Northwestern United States and adjacent parts of Canada), or in montane forests. The causes of these differing distributional patterns appear to be related to the relative advantages among varying sites and climates of differing leaf types, leaf habits, and/or phenologies in terms of annual carbon gain potential (Waring and Franklin, 1979; Reich *et al.*, 1992; Kikuzawa, 1991).

II. Productivity Values

What features distinguish productivity in temperate forests from other biomes, and as well, temperate forests from one another? The answer to this question likely depends on the definition or component of productivity in question. Net ecosystem production (NEP, or P_e) is defined as the net carbon dioxide flux to a forest ecosystem and integrates all ecosystem carbon sources and sinks; gross primary production (GPP, or P_g), i.e., total gross photosynthesis, and total net primary productivity (NPP, or P_n) consider only the autotrophic part of the ecosystem (see Section IV for equations and further details). Aboveground NPP (ANPP) is much easier to measure than either belowground production (and hence NPP) or NEP. Consequently, there are many more estimates of ANPP than NPP or NEP across a much wider range of ecosystems and conditions. Given the paucity of data on belowground NPP, detailed comparisons of productivity across biomes or among temperate forests can only be made for ANPP (these other components are discussed in Section IV).

On average, temperate forests have intermediate levels of ANPP compared to other forested biomes, largely due to differences in growing season length, temperature, and moisture (Landsberg and Gower, 1997). Along gradients of increasing moisture or temperature, ANPP generally increases (Gholz, 1982; Running, 1994). For instance, moving from colder boreal regions through the temperate forest zone toward the equator, ANPP increases, most likely due to a greater number of months in which temperatures can support vegetative activity rather than due to large differences in the peak monthly productivity of subtropical vs. temperate vs. boreal forests. What are the major sources of variation in temperate forest ANPP, and what are the controlling factors? Factors that are often considered important include water, temperature, soil nutrient supply, and forest type.

To address these questions, we assembled a database that includes 197 values of ANPP for closed-canopy temperate forests from 15 countries in Asia, Australia, Europe, and North America (the data set is available on request from the authors). Data are almost entirely for evergreen coniferous and broadleaf deciduous ("hardwood") forests, with only a few data points for deciduous conifers or broadleaf evergreen forests. This data set includes temperate forests that are found from 4 to 17°C mean annual temperature (m.a.t.) and from 350 to 2750 mm mean annual precipitation (m.a.p.), with average values of 8–9°C m.a.t. and 900–1000 mm m.a.p. Stand ages (for the central 95% of the distribution) range from 30 to 200 yr for the hardwoods and from 10 to 300 yr for the conifers, with mean ages of 75 yr for both groups.

In this data set, hardwood and coniferous forests do not differ appreciably in the range or average values for ANPP. Mean (±1 standard deviation) values for evergreen conifers, deciduous conifers, and deciduous hardwoods are 10.1 (4.0), 10.1 (2.9), and 8.8 (3.0) Mg ha^{-1} yr^{-1}, respectively. Variation within each forest type is much larger than their average differences. Using multiple regression analyses, 25% of the variation in ANPP

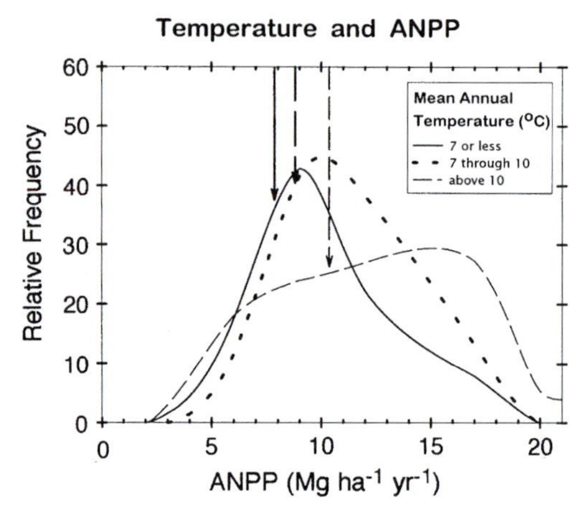

Figure 11-1 Frequency distribution of ANPP for temperate forests from three different temperature ranges (4–7, 7–10, and 10–17°C; $n = 58$ to 63 for each class). Also shown (downward arrows) is the median ANPP for each mean annual temperature class.

among these 197 forests can be explained due to the combination of m.a.t., m.a.p., and stand age (each factor was significant; $P < 0.05$). ANPP increases with m.a.t. and m.a.p., and decreases with stand age. However, given the well-known importance of temperature and precipitation to forest growth, it is somewhat surprising that so much variation (75%) in ANPP is left unexplained. This can be demonstrated by examining the frequency distribution of ANPP by m.a.t. classes (Fig. 11-1). Although forests from warmer climates tend to have higher ANPP on average, a high proportion of such stands have ANPP values that fall well within the distribution for stands in colder areas. In general, other factors beyond temperature or moisture are responsible for large heterogeneity in ANPP in any given temperate climate zone.

Sources of this unexplained variation in ANPP include local variation in soil quality, topographic position (see below), forest type, and experimental error (i.e., inaccuracies in measuring ANPP). Variation in mean ANPP among 22 species is noteworthy as much for the large variation within species as for differences among species (Fig. 11-2). There is no simple trend from coldest to warmest sites, nor between hardwoods versus conifers. The two pine species from the warmest habitats had a mean ANPP significantly greater ($p < 0.05$) than about half of the other species. Four other species with high ANPP (two conifers, two hardwoods) had significantly greater values than one–three of the species (all conifers) with the lowest ANPP. For the 16 species with intermediate ANPP, there were no significant differences in ANPP.

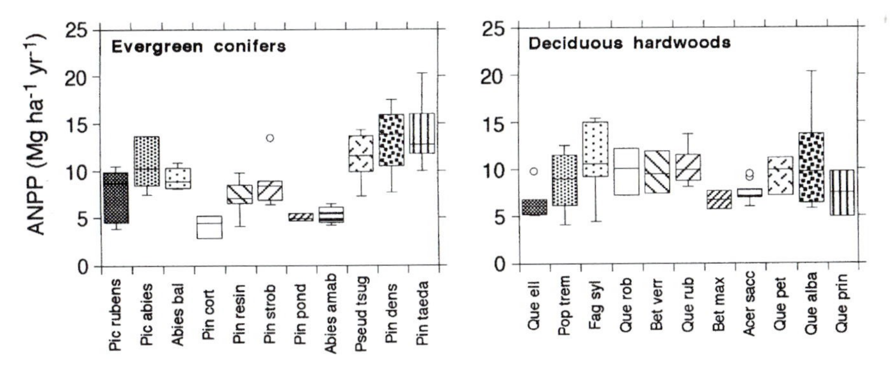

Figure 11-2 Variation in ANPP for 22 temperate deciduous hardwood and evergreen conifer species. Species are arranged left to right from coldest to warmest (by mean annual temperature of studied plots).

III. Controls on ANPP due to Water, Soils, Disturbance, and Vegetation Type

A. Water

Site water status reflects local conditions (soils, topography, evaporative demand) as much as broad m.a.p. patterns, so it is not surprising that variation in m.a.p. among temperate forests corresponds poorly with ANPP (Section II above). However, both manipulative experiments and observational or gradient studies have demonstrated the influence of water supply on ANPP. Irrigation experiments with Scots pine, Douglas fir, and loblolly pine found modest increases in ANPP (Linder, 1987; Linder *et al.*, 1987; Gower *et al.*, 1992) under ambient soil fertility, and heightened increases when irrigation was combined with fertilization. Interannual variation in precipitation had a significant effect on ANPP in five species in a common garden plantation experiment in Wisconsin (Gower *et al.*, 1993), with ANPP being 25% lower during a drought year than in other years. In a comparison of 46 stands in a forested landscape in central North America (Indiana) (Jose and Gillespie, 1996), ANPP increased by 40% across a range of site water availability (Fig. 11-3). From basic physiological principles, it follows that tree productivity will decline when low water availability results in a decline in tree water status sufficient to affect leaf area, leaf gas exchange, or other determinants of productivity. Such declines are highly heterogeneous in time and space, however, making their prediction difficult at local scales.

B. Soils and Nutrients

Both manipulation experiments and observational studies indicate that N is the most common limiting soil nutrient in the temperate forest zone. Fer-

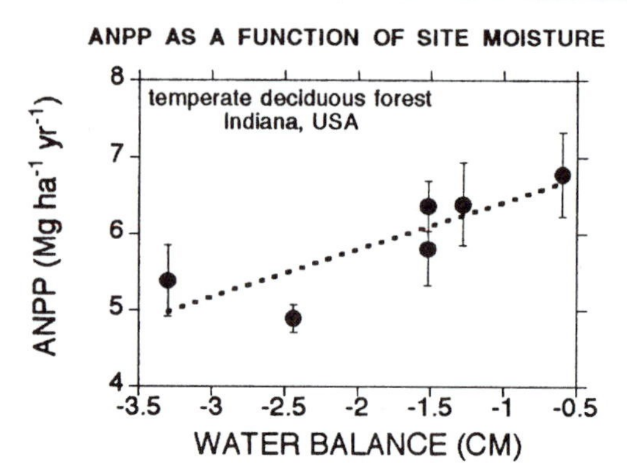

Figure 11-3 Variation in mean ANPP of temperate hardwood forests in Indiana, associated with landscape heterogeneity in site water balance (annual precipitation − potential evapotranspiration). Redrawn from Jose and Gillespie (1996).

tilization studies with N in temperate forests have found increases in annual ANPP of 30% on average (Gower *et al.*, 1992, 1994; Magill *et al.*, 1997), but with enhancement ranging from large to none. The role of soil N supply in influencing ANPP was also highlighted in a comparison of ANPP and soil N availability for 50 forest stands at six sites in central North America (Reich *et al.*, 1997a). The 50 stands included a range of forest types, stand ages, and soil types within a modest regional climate range. ANPP was as much as 80% greater in stands of high soil N availability within broad soil groups. ANPP was highly correlated with annual soil N availability (assessed as *in situ* net N mineralization, Fig. 11-4), supporting the idea of a strong functional relationship between the two. Hardwoods and conifers were apparently similar in this respect. Moreover, ANPP at any given level of soil N availability was further influenced by soil texture. At any given level of annual N mineralization rate, forests on finer textured soils had higher ANPP than did those on sandy soils, suggesting that water-holding capacity of soils interacts with N supply to influence ANPP. ANPP was well predicted by the combination of soil net N mineralization and by soil texture, which together explained 74% of the variation in ANPP for the 48 stands with native species (Reich *et al.*, 1997a).

C. Influence of Forest Type on ANPP via Feedbacks on Soil N Dynamics

It has been hypothesized that feedback controls by species on soil properties (largely soil N availability) may jointly control productivity and species composition. The well-known Blackhawk Island study of Pastor *et al.* (1984) clearly demonstrated large differences in ANPP and net N mineralization among stands varying in litter chemistry, but did so along a combined edaphotopographic and species composition gradient. Thus, hemlock stands on intrinsically poorer soils had a lower soil N mineralization rate and ANPP than did oak stands on better soils—but these data cannot provide an answer as to what are the main controls, soil or species, nor even convincingly demonstrate the existence of strong feedbacks.

A subsequent study by Gower and Son (1992) examined monocultures (five species, two deciduous and three evergreen) planted on similar soils and found feedback effects of tree species (via litter quality and other processes) on soil N mineralization. However, in the regional 50-stand data set mentioned above (Reich *et al.*, 1997a), when conifers and hardwoods were compared on roughly comparable soils and for comparable disturbance histories, there were no significant differences in net N mineralization or ANPP, whereas finer textured Alfisol soils had higher annual N mineralization rates than did sandy Entisol soils, even when similarly vegetated. These differing results (i.e., experimentally observed feedbacks were not detected in a larger observational survey) have yet to be reconciled, but it appears likely that both innate soil properties and vegetation feedbacks on soil processes contribute to variation in soil N supply and ANPP.

D. Productivity at the Ecotone: The Forest, Woodland, Grassland Border

How do ANPP and associated ecosystem measures vary between temperate forests and other neighboring biomes? We can provide one example using a case study at the grassland–forest border. The native vegetation in much of central North America was a mosaic of grassland, savanna, woodland, and forest depending on landscape position, soils, and fire history. In eastern Minnesota, a 32-yr-old fire frequency experiment had created a vegetation continuum from unburned closed-canopy oak forest (>90% tree dominated) to periodically burned mixed woodland (70–30% tree dominated) to frequently burned open savanna (<20% tree dominated) (Peterson and Reich, 2000). Both ANPP and net N mineralization decreased substantially with increasing fire frequency and/or decreasing dominance by oak trees (P. B. Reich *et al.*, 2001) (Fig. 11-5) and these two ecosystem measures were closely correlated.

Why do ecosystem properties and processes vary so much across the forest–grassland continuum? At least three factors work together to create such a sharp gradient in ecosystem processes. First, frequent fire is likely to lead to losses of N from the ecosystem over a 32-yr period (Ojima *et al.*, 1994). Second, the vegetation continuum also "pushes" N cycling in the same direction: litter quality is high in oak trees, lower in C_3 grasses, and lowest in C_4 grasses, and the proportional dominance of the litter pool by trees decreases while increasing for C_4 grasses along the forest to grassland contin-

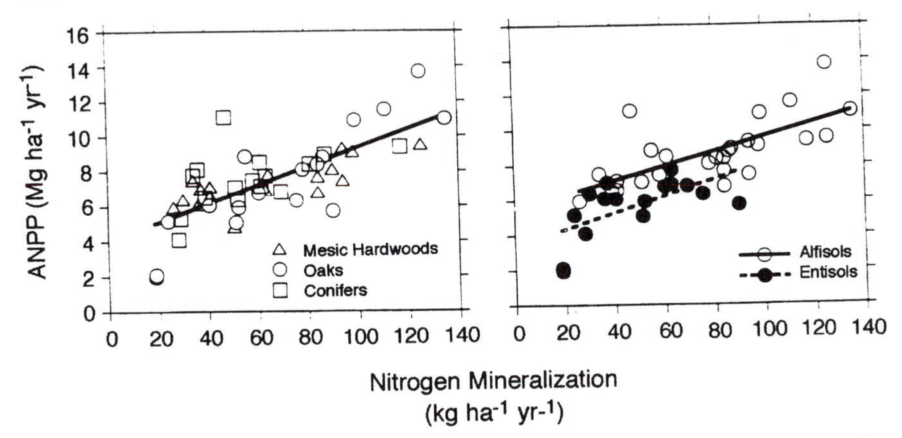

Figure 11-4 Relationships between ANPP and annual net N mineralization rates for broadleaf deciduous oaks, mesic hardwoods, and needleleaf conifers (left) and for stands on fine-textured Alfisol soils and sandy Entisol soils (right). Data based on 50 forest stands in Minnesota and Wisconsin. From Reich *et al.* (1997a).

uum in this experimental landscape. Given the demonstrated feedbacks between litter quality and net N mineralization rate for grasses and trees in this region (Wedin and Tilman, 1990; Gower and Son, 1992), litter quality differences likely contributed to the development of the five-fold gradient in soil N availability across the experimentally developed forest–grassland ecotone. Feedback effects of trees on soil organic matter and moisture relations (Ko and Reich, 1993; K. Wrage and P. Reich, unpublished data) also likely affect decomposition and N mineralization processes, exacerbating grass vs. tree patch differences. Third, if ANPP at least partially determines net N mineralization rates and vice versa, which is likely, then the forest–grassland gradient in the quantity of litter turnover would have gradually increased over 32 yr. In essence, decreasing soil N supply (due to fire and vegetation feedbacks) at the frequently burned, grass-dominated end of the vegetation-type continuum leads to lower ANPP, which provides lesser amounts of a lower quality litter substrate for future mineralization (Reich *et al.*, 2001), and this process repeats itself, year by year. Thus, the combination of varying fire regimes and vegetation properties contributes to dramatic differences in ANPP and related ecosystem processes along the grassland–forest continuum in central North America, and may be representative of such contrasts in other regions.

E. ANPP through Stand Development

It is well recognized that ANPP increases, plateaus, and eventually declines during the life of an even-aged forest stand (Kira and Shidei, 1967; Lands-

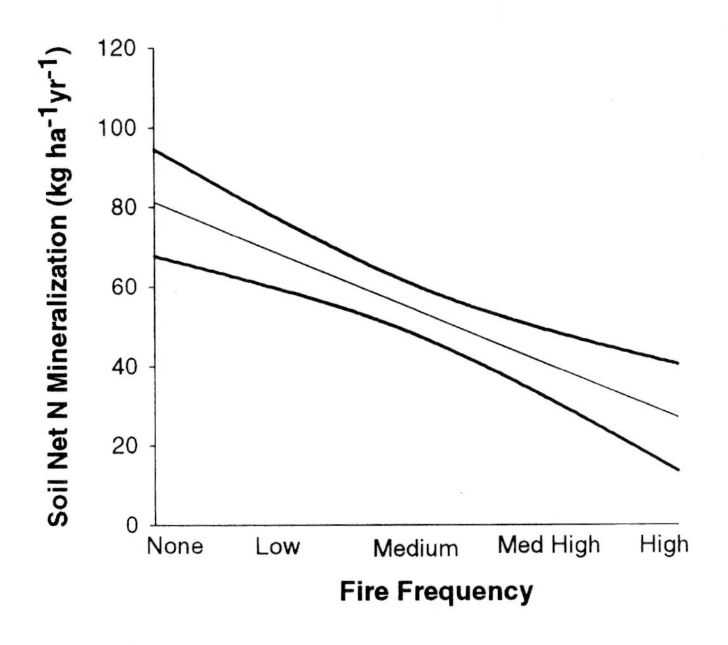

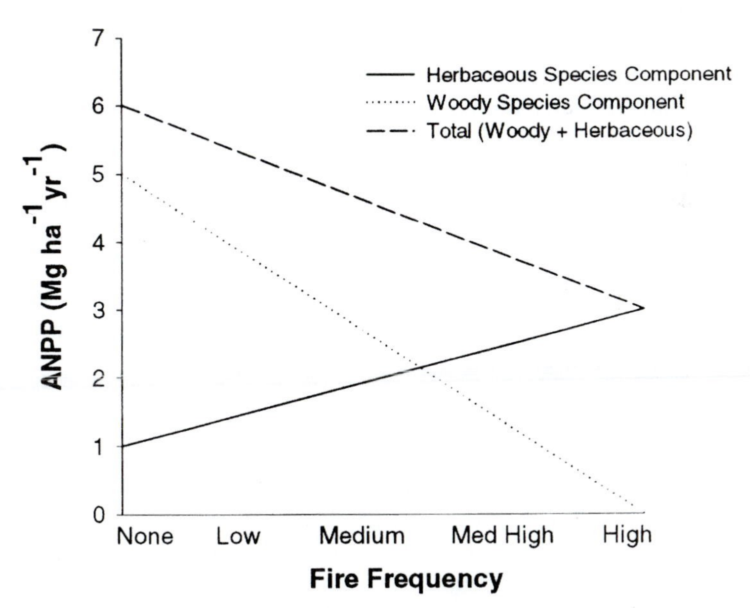

Figure 11-5 Idealized variation in ANPP, the relative contribution to ANPP by woody and herbaceous ecosystem components, and annual net N mineralization rates across a 32-year experimental fire frequency gradient that has caused a vegetation continuum from closed canopy oak forest to semiopen oak savanna to open grassland in eastern Minnesota. Based on data from Pastor *et al.* (1993), Grigal and Homann (1994), Reich *et al.* (1997a), Reich *et al.* (2001).

berg and Gower, 1997). The initial increase is associated with the gradual development of full canopy closure and maximum leaf area index (LAI) (Gholz and Fisher, 1982). Following a peak early in stand development ANPP often declines by as much as 20–50% as stands mature and age (Gower *et al.*, 1996). Given a limited database, these patterns appear to occur equally in temperate forests and in boreal or tropical forests. The mechanisms responsible for the decline are not well understood. Three main hypotheses have been proposed. First, it is possible that as the proportion of photosynthetic vs. respiring tissues changes with age, the balance of total tree respiration vs. net photosynthesis increases (Kira and Shidei, 1967; Ryan and Waring, 1992). Second, N limitation may increase during stand development due to a slowing of decomposition resulting from shifts in litter chemistry and structure as woody detritus becomes an ever greater fraction of annual turnover (Gower *et al.*, 1996). Third, as trees age, increased hydraulic resistance may lead to reduced stomatal conductance and hence reduced canopy photosynthesis (Ryan and Yoder, 1997). Heightened consideration of stand age structure and greater understanding of productivity patterns with stand age should help us better quantify and predict future patterns of temperate forest C cycling at every scale.

F. Genotypic Variation in ANPP: A Pine Is a Pine Is a Pine??

Earlier this century Stein (1922) suggested that "a rose is a rose is a rose." She was referring to human cognition and conceptualization rather than to functional biology, but as ecologists, we often ignore variation within species in modeling patterns of productivity across large climate or biogeographic gradients (Melillo *et al.*,1993; Haxeltine and Prentice, 1996)—in essence, assuming that "a pine is a pine is a pine." Is such an assumption warranted? If not, when might such an assumption cause significant problems in large-scale modeling?

To address this issue we highlight an experiment involving Scots pine populations from a large latitudinal gradient planted in a common garden environment in central Poland. ANPP is higher in populations from the central European part of the range than it is in populations from increasingly cold, northern parts of the range (Fig. 11-6). Why? Northern populations and central populations have comparable net photosynthetic rates, but northern populations have greater proportional biomass distribution to roots both as seedlings (Oleksyn *et al.*, 1992) and as trees (Oleksyn *et al.*, 1998), have a shorter shoot growth period annually due to photoperiodic differences (Oleksyn *et al.*, 1992, 1998), and have greater respiration rates (Reich *et al.*, 1996) than do central populations. Greater proportional biomass distribution to roots is associated with lower tree growth rates (Reich *et al.*, 1992) and lower ANPP (Gower *et al.*, 1992, 1994), whereas greater respiration rates could lead to proportionally greater carbon losses and hence

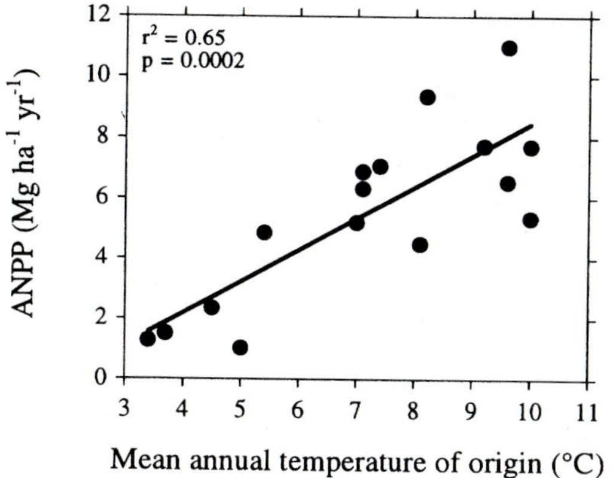

Figure 11-6 Relationship between ANPP of 16 15-yr-old Scots pine populations grown and measured in a common garden environment (52°N; m.a.t. 7°C, in central Poland) and the mean annual temperature of their native environment. Data from Oleksyn *et al.* (2000).

to lower ANPP. It is well recognized that in common environments, cold-adapted tree populations generally grow more slowly than conspecifics from warmer environments: if this is generally also true for NPP, it suggests that biogeographic variation in temperate and boreal NPP is driven by genotypic adaptation as well as by climatic and associated environment differences. Such a possibility is not currently incorporated in most, or probably any, global productivity models.

IV. Component and Total Carbon Flux and Scaling Relations

Carbon flux and pools in temperate forest ecosystems can be categorized by various component sinks, sources, and processes. Most plant tissues are near 50% carbon on a dry mass basis (thus all the biomass production values in Sections II and III can be converted to an approximate C basis by multiplying by 0.5). However, typically 2 to 5 unit masses of carbon must be fixed for every unit mass of living plant tissue found in a temperate forest ecosystem because a proportion of the fixed carbon is lost directly through autotrophic respiration. Over intermediate to longer time scales, system carbon is lost through detritus and ultimately heterotrophic respiration. Growing-season net ecosystem–atmospheric carbon flux is typically negative during daylight hours in temperate forests because canopy assimilation dominates, and flux is typically positive at night because autotrophic and heterotrophic respiration dominate. Although fixed carbon may be transported out through downward leaching beyond the biologically active soil zone, exported through surface flow, immobilized for long periods of time (for example, soil-embedded charcoal from combustion), or removed through herbivory, these losses are typically relatively minor compared to respiration components. In most temperate forests the majority of herbivory is by insects and arthropods, which are effectively sessile; herbivory may greatly increase detrital production and heterotrophic respiration for short periods, on the order of a few weeks to months. (Fire, as a natural or human-caused disturbance, usually significantly increases outward carbon flux and is discussed elsewhere in this chapter.)

Despite more than three decades of concentrated efforts, we are ignorant of many aspects of global temperate forest mass and energy cycles, in large part due to the aforementioned heterogeneity in species composition, structure, and environmental factors. There are at least two important areas where data are particularly sparse and knowledge may be significantly improved. First, belowground components and processes in temperate forest systems are poorly understood relative to aboveground rates and processes. Many aboveground variables are easily measured, e.g., leaves or stems may be enclosed in portable cuvettes, fed ambient air, and carbon flux rates measured without significantly altering the plant component or environment. Allometric relationships may be developed between easy, nondestructive measurements of aboveground plant parts such as stem diameters or leaf size and thickness, and converted to whole-plant or component variables useful in estimating ecosystem processes. Most belowground measurements are significantly more difficult, both to perform and interpret. Roots are intimately mixed with the soil, and any direct measurement technique runs the risk of altering the root environment or damaging root tissue, and hence biasing estimates of related belowground processes. Plants may respond to higher temperatures and drier conditions by allocating more production belowground to compensate for higher fine root turnover. Alternatively, they may shift production toward aboveground components after large nitrogen additions to the system. However, because of a lack of belowground data, we cannot with any certainty quantitatively describe allocation strategies. Although recent work is encouraging, a considerably larger effort is required.

A second major source of uncertainty comes from the small number of ecosystem types for which ecosystem carbon budgets or process-based measurements of productivity are available. Table 11-1 shows that NPP and whole-system flux data are available for forests that dominate less than 10% of the temperate forest zone, and work has been biased toward economically important species (e.g., *Pinus taeda, Pseudotsuga menziesii*), sometimes in plantations outside their native range (e.g., *Pinus radiata, Picea abies*). Many wide-ranging species and types have not been measured. In addition, we have sampled but a small portion of the range of environmental and genet-

Table 11-1 Examples of Total NPP Data for Temperate Forest Ecosystem Types[a]

Forest type	Total NPP (Mg C ha^{-1} yr^{-1})	Belowground as % of total NPP[b]	Ref. and notes
Broadleaf deciduous	19.5	n.r.	Powell and Day (1991); 78-yr-old wet site
Acer–Nyssa	8.4	n.r.	Powell and Day (1991); 78-yr-old wet site
Acer saccharum	8.0	n.r.	Nadelhoffer *et al.* (1985); 35-yr old
Betula spp.	6.3	n.r.	Nadelhoffer *et al.* (1985); 35-yr old
Quercus rubra	9.3	n.r.	Nadelhoffer *et al.* (1985); 35-yr old
Liriodendron tulipifera	8.8	60	DeAngelis *et al.* (1981); 48-yr old
Pinus elliottii	4.7	52	Ewel *et al.* (1987); 7-yr old
	9.6	32	Ewel *et al.* (1987); 27-yr old
Pinus taeda	20.6	38	Kinerson *et al.* (1977); 16-yr old
	7.2	14	Nemeth (1973); 8- to 12-yr old
Pinus resinosa	4.3	n.r	Nadelhoffer *et al.* (1985); >100-yr old
Pinus strobus	6.6	36	Gower *et al.* (1994); 60-yr old
Pseudotsuga menziesii	7.7	53	Vogt (1991); 40-yr old, poor site
	8.3	23	Vogt (1991); 40-yr old, good site
	7.6	41	Vogt (1991); 40-yr old, poor site
	10.1	31	Vogt (1991); 40-yr old, fertilized
	8.5	46	Gower *et al.* (1992); 50-yr old
Abies amabilis	8.9	64	Vogt (1991); 23-yr old
	7.9	80	Vogt (1991); 180-yr old
Chamaecyparis thyoides	11.4	n.r.	Day (1982, 1984); Powell and Day (1991)
Taxodium distichum	10.9	n.r.	Day (1982, 1984); Powell and Day (1991)

[a]Adapted from Vost, 1991.
[b]n.r., Not reported.

ic variation within which each respective ecosystem is found. For example, there have been more whole-system carbon flux and balance studies of *P. mensziesii* than any other species in the biome. Yet work has sampled less than half the range of ages for this species, and reports on young forests (<20 yr) and old forests (>150 yr) are sparse or nonexistent.

A. Net Ecosystem Production

Net ecosystem production (P_e) is defined as the net carbon dioxide flux to a forest ecosystem, and integrates all ecosystem carbon sources and sinks:

$$P_e = P_g - R_a - R_h$$

or

$$P_e = P_n - R_h.$$

Primary components of NEP are net daytime canopy photosynthesis; nighttime leaf, woody tissue, and root respiration; heterotrophic respiration from standing and soil surface detritus; and heterotrophic respiration from microbial breakdown of soil organic matter. Components contributing to C flux from the forest floor (litter, roots, and soil organic matter) are difficult to separate, and this combined forest floor respiration, typically measured directly, is usually much larger than aboveground autotrophic respiration. NEP may be estimated from scaled component measurements (forest floor, woody, and foliar respiration), or from integrated whole-ecosystem flux (eddy covariance). However, the most informative studies have combined both whole-canopy and component flux measurements.

NEP has been measured at relatively few sites to date, so it is difficult to generalize concerning temperate forests. Estimates from one of the best-studied temperate forests (Harvard Forest) range from 1.4 to 2.8 t C ha^{-1} yr^{-1} (Wofsy *et al.*, 1993; Goulden *et al.*, 1996), with significant interannual variability. Changes in annual NEP over 5 yr resulted from climate variation during periods when the forest was particularly sensitive, rather than due to changes in annual mean conditions. Air temperatures in spring and fall (which affected carbon gain by altering the length of the growing season), cloud cover in summer, and snow depth and other factors that affected soil temperatures (and hence soil respiration) in winter were among the most important factors regulated components of net ecosystem exchange (Goulden *et al.*, 1996).

NEP was estimated at 5.3 t C ha^{-1} yr^{-1} for a deciduous forest in a warmer region of the United States, in Tennessee (Greco and Baldocchi, 1996), where hardwood forests typically have higher ANPP as well. Using a lumped parameter functional model (PnET), which was validated against the eddy covariance measurements at Harvard Forest (Aber *et al.*,1996), simulations of NEP for temperate forests in the northeastern United States showed a predicted range of NEP from -1.5 to 3.5 t C ha^{-1} yr^{-1} (Fig. 11-7), with most data points between -0.5 and 2.5 t C ha^{-1} yr^{-1} and a mean of 0.8 t C ha^{-1} yr^{-1} (Aber *et al.*, 1995). It is important to recognize that although these modeled NEP estimates are among the best regional predictions we can make today, they have serious limitations in terms of model structure and assumptions. Given the importance of temperate forest NEP to the global carbon balance, it is likely that measuring and modeling NEP will be a major research activity in the coming decade. In the following sections we outline the components of NEP at increasingly fine levels of scale.

B. Total Net Primary Production

Total NPP (P_n) represents the net productivity of the autotrophic part of the ecosystem:

$$P_n = P_g - R_a,$$

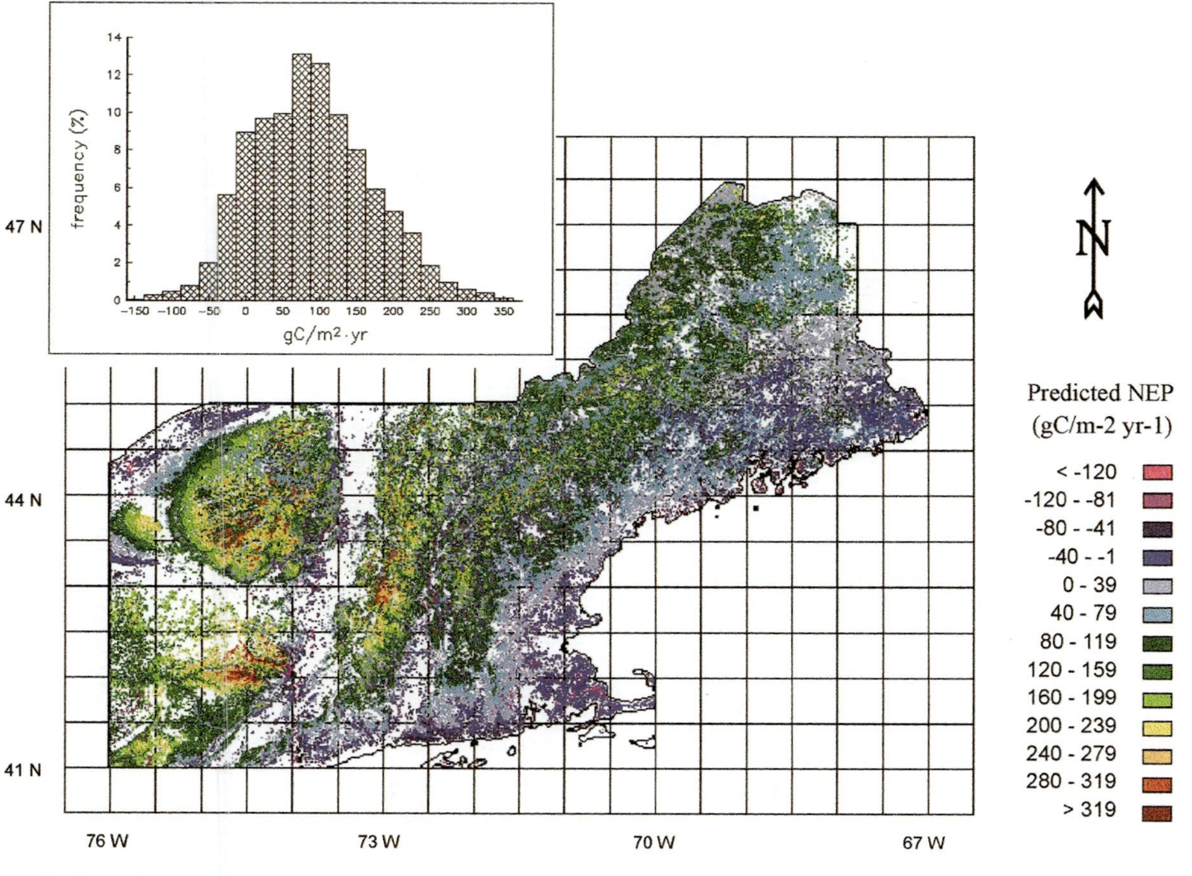

Figure 11-7 A map of the spatial distribution of estimated current net ecosystem production (NEP) for forests in the northeastern United States based on an ecosystem process modeling approach. The insert shows the histogram of all data points. Redrawn from Aber *et al.* (1995).

where P_g (i.e., GPP) is gross primary production and R_a is autotrophic respiration. For simplification, R_a may be split into growth respiration R_g and maintenance respiration R_m. Leaf, branch, stem, and root respiration are the major autotrophic respiration components in temperate forests. Significant data on R_a have been collected during the past decade, particularly for aboveground components, greatly expanding our knowledge of component respiration rates (see Section IV,D). Because of difficulties in estimating GPP and R_a for all components, NPP is typically estimated by

$$P_n = \Delta B + \mathrm{Lf} + \mathrm{wr} + \mathrm{wg},$$

where ΔB is the increase in autotrophic biomass, Lf is litterfall, wr is root production (as biomass), and wg is herbivory. Root production is typically measured via soil cores, ingrowth screens or boxes, or a combination of these and rhizotron technologies (Hendrick and Pregitzer, 1993). Because of the previously described lack of belowground data, total NPP data are available for relatively few temperate forest ecosystem types (Table 11-1), and their accuracy are debated.

C. Gross Primary Production

Gross primary productivity is defined as the total autotrophic carbon fixation during a given time period, usually on an annual basis. There are no good methods for directly estimating whole-system GPP in forest systems because it is impractical to isolate photosynthetic from respiring tissues while making measurements over the whole canopy. GPP may be estimated from leaf photosynthetic measurements combined with canopy-scale models (Reich *et al.*, 1990; Aber *et al.*, 1996), or from daytime whole-canopy flux and storage and component respiration measurements (Ruimy *et al.*, 1996). Intensive field measurements have yielded GPP estimates that range from less than 5 to more than 40 Mg C ha^{-1} yr^{-1} (Kira and Shidei, 1967; Whittaker *et al.*, 1974; Kinerson *et al.*, 1977; Benecke and Nordmeyer, 1982; Gholz *et al.*, 1991; Ryan and Waring, 1992; Ryan *et al.*, 1994). Within broadleaved and deciduous functional types, GPP increases with both growing season and leaf area index, and leaf area duration—consequently, leaf area duration, the LAI times growing season months, is strongly related to GPP (Waring and Schlesinger, 1985). GPP on an instantaneous basis is a measure of the instantaneous net CO_2 exchange rate of the canopy. Typical rates for temperate forests range from 0.9 to 4.3 g C ha^{-1} s^{-1} (or 10 to 45 μmol m^{-2} s^{-1}) at light saturation, based on eddy covariance and modeling analyses (e.g., Verma *et al.*, 1986; Reich *et al.*, 1990; Leuning *et al.*, 1991; Hollinger *et al.*, 1994; Greco and Baldocchi, 1996; Ruimy *et al.*, 1996).

Given the difficulty of direct experimental or observational approaches to quantifying GPP, it is necessary that accurate modeling approaches be developed (see Section VI). Moreover, given the high diversity of species, soils, and climates within and among temperate forest regions of the world, it is equally important that generalized models be developed. For this to occur, we must develop a general understanding of the principles that underlie temperate forest ecosystem production, including the source of production, GPP. In the recent past, there have been substantial advances in the development of interspecific leaf- and canopy-trait scaling relationships (Reich *et al.*, 1991a,b, 1992, 1994b, 1997b; Gower *et al.*, 1993;) and of vertical intracanopy leaf property and microenvironmental scaling relationships (Jarvis and Leverenz, 1983; Caldwell *et al.*, 1986; Hollinger, 1989; Reich *et al.*, 1990, Leuning *et al.*, 1991, 1995; Ellsworth and Reich, 1993) that provide both enhanced understanding of the biological foundation for forest ecosystem GPP, and tools for canopy and ecosystem modeling.

1. Interspecific Canopy-Trait Scaling Forest-forming tree species possess only certain combinations of leaf and canopy properties, ostensibly as a result of trade-offs between leaf- and canopy-scale performance, and juvenile vs. mature tree performance (Reich *et al.*, 1992, 1994a; Gower *et al.*, 1993). In general, species either have leaves with high rates of photosynthetic productivity per unit leaf, but which can be arrayed only in sparse canopies, the opposite (dense canopy of low-productivity leaves), or some intermediate set of traits. This is reflected in (1) the inverse relationship between mean canopy specific leaf area (SLA) and total canopy leaf area or leaf mass; (2) the inverse relationship between photosynthetic rate and total canopy leaf area or mass; (3) the positive relationship between total canopy foliage mass and foliage life-span; and (4) the inverse relationships between canopy production efficiency and either leaf life-span or total canopy foliage mass (Reich *et al.*, 1992, 1994a; Gower *et al.*, 1993).

Because the carbon assimilation that drives GPP and hence NPP is determined by the product of mean leaf photosynthesis multiplied by the size of a canopy, the trade-off between canopy size and leaf performance serves as a major constraint on the range of NPP that can likely be obtained. Species with high photosynthetic rates and sparse canopies (e.g., *Betula* sp., *Populus tremuloides, Larix* sp.) achieve ANPP roughly comparable to that of species with low photosynthetic rates and dense canopies (e.g., *Abies* sp., *Picea* sp.) (Fig. 11-2), and no species has high mass-based photosynthetic rates and a large canopy. The canopy size vs. leaf performance trade-off also makes accurate prediction or modeling of variation in ANPP much more difficult than prediction of variation among forest types in either canopy attributes or leaf attributes, because the latter span well over an order of magnitude, whereas ANPP varies relatively slightly on average among forest types (Fig. 11-2). Nonetheless, successful modeling of NPP will require that we accurately model both leaf and canopy properties, or model surrogate properties that integrate both such values—such as models that use forest type and LAI, with forest type being used to determine leaf-level properties (e.g., Running, 1994; Runyon *et al.*, 1994; Aber *et al.*, 1992).

2. Intracanopy Vertical Scaling One of the most important features of canopies that influences their flux rates involves the well-known attenuation of light vertically (Norman and Jarvis, 1975) and the increasingly well-known parallel vertical scaling of leaf properties (Hollinger, 1989; Ellsworth and Reich, 1993). Leaf structure in particular is plastic and is strongly influenced by the leaf light microenvironment. Thus, leaves located at different positions along a light gradient in a canopy tend to differ predictably in their SLA, providing simple scaling relationships on a leaf area basis (Ellsworth and Reich, 1993) (Fig. 11-8).

This intracanopy vertical scaling has significant implications for our ability to model canopy fluxes. Both top-down models [such as those based on the amount of light intercepted by the canopy multiplied by a light use efficiency factor (Monteith, 1977; Waring *et al.*, 1995)] and bottom-up models [such as PnET (Aber *et al.*, 1996)] incorporate the vertical scaling of light and leaf properties either at a detailed leaf/canopy-layer scale or as part of the features of whole-canopy functioning (i.e., the idea of a canopy light use efficiency).

Top-down light use efficiency models assume that total GPP will be a function of the amount of light intercepted by the entire canopy and the efficiency with which that light is used, with the efficiency either fixed by ecosystem type (Ruimy *et al.*, 1996) or variable, depending on species and environmental conditions (CASA) (Field *et al.*, 1995). Light use efficiency, in turn, is likely to depend on the combined canopy density and photosynthetic attributes (Field *et al.*, 1995; Ruimy *et al.*, 1996), because it has been shown that dense canopies tend to have lower per leaf photosynthetic efficiencies (Reich *et al.*, 1992). Alternatively, models such as PnET (Aber *et al.*, 1996) and its predecessors (Reich *et al.*, 1990) assume that leaf properties vary with average light levels, which are closely related to vertical light attenuation, and that the combination of acclimated leaf properties, the size of canopy layers, and the average light levels at each layer can successfully be used to predict total canopy gas exchange rates.

3. Leaf-Trait Scaling Given that NPP is driven by GPP, which is a product of canopy size multiplied by the average photosynthetic rate of each leaf in the canopy, leaf-level photosynthesis is central to all higher levels of productivity. Leaf photosynthesis is strongly influenced by key leaf traits such as leaf nitrogen concentration, life-span, and structure. These traits vary greatly, often by orders of magnitude, within any given terrestrial ecosystem, including temperate forests (Reich *et al.*, 1995, 1997b, 1998a). Studies from many different ecosystems have identified the existence of interspecific trade-offs at the leaf level between potential photosynthetic productivity on the one hand, and defenses, structural strength, and longevity on the other (e.g., Mooney and Gulmon, 1982; Coley, 1988; Reich *et al.*, 1991b, 1992).

Comparative studies from a variety of species and ecosystems report that

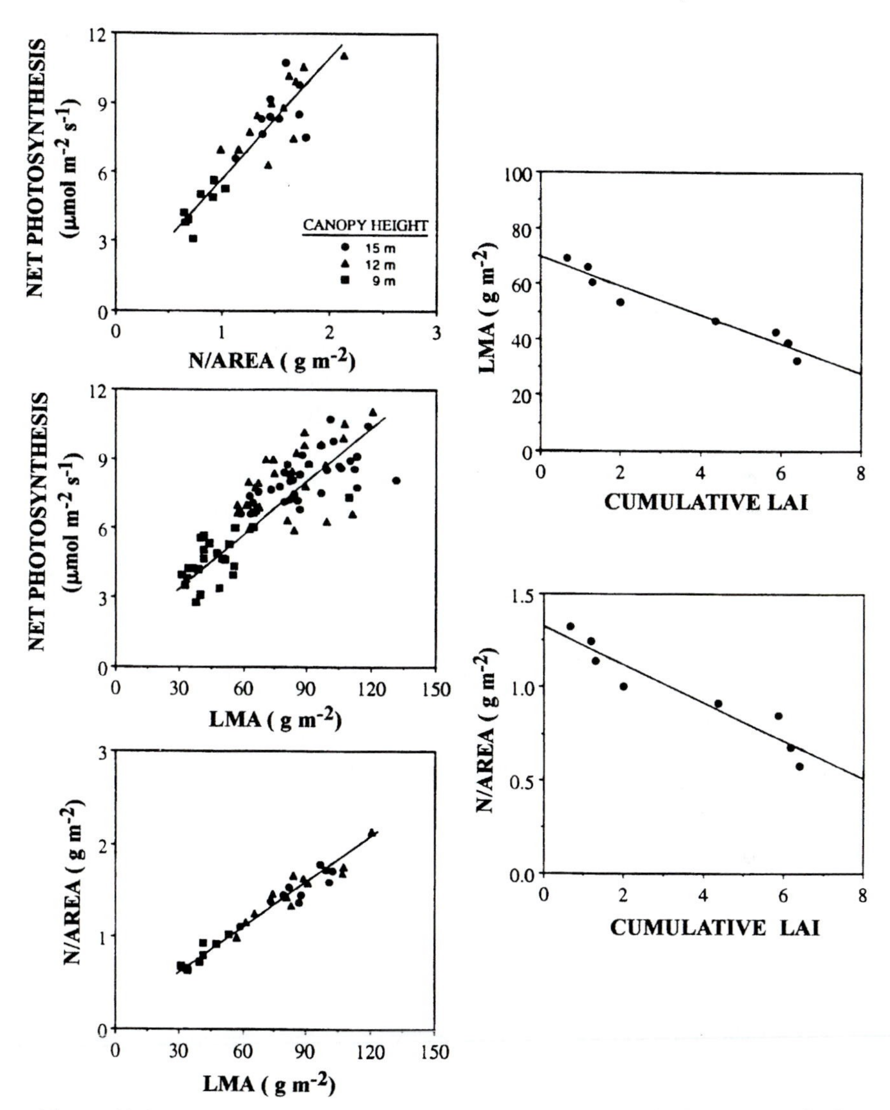

Figure 11-8 Relationships (left) between leaf mass per area and leaf N per area, both regressed against cumulative leaf area index (LAI) from the top to the bottom on a vertical transect through an 18-m tall sugar maple canopy in Wisconsin. Right: Relationships between maximum light-saturated net photosynthesis measured in the field, area-based leaf N, and leaf mass per area (LMA) for sugar maple leaves from three different canopy positions in the same vertical transects. Redrawn from Ellsworth and Reich (1993).

photosynthetic rates and leaf nitrogen levels are usually positively related to each other and negatively related to leaf thickness/density (low SLA) and longevity (Field and Mooney, 1986; Chazdon and Field, 1987; Reich *et al.*, 1991a,b, 1992, 1995; Gower *et al.*, 1993). Other studies have found that these relationships and their proportional scaling are universal at the broadest, continental to global scales (Reich *et al.*, 1997b) because of the limitations placed by biophysical constraints and natural selection leading to ecological trade-offs in leaf traits (Coley *et al.*, 1985; Field and Mooney, 1986; Reich *et al.*, 1992). For instance, maximum photosynthetic capacity can be accurately predicted for temperate forest species from their leaf N and SLA, based on relationships developed among these leaf traits from an independent data set from nontemperate biomes (Fig. 11-9).

This evidence for functional leaf convergence has two major implications. First, any given species tends to have a set of leaf traits that might be located at one end or the other, or someplace along, a multidimensional leaf-trait surface: the ends are represented by leaves of low density, thickness, and longevity, with high rates of net photosynthesis and high mass-based nitrogen concentrations, or the reverse set of traits (Reich *et al.*, 1992, 1997b, 1998a). Second, because relationships among leaf traits are similar among ecosystems and biomes, they have great potential for use in modeling leaf and canopy properties and functioning across a range of scales (e.g., Reich *et al.*, 1990; Aber *et al.*, 1996). For instance, use of a lumped-parameter GPP model based on ecophysiological leaf- and canopy-trait relationships from central North America (Ellsworth and Reich, 1993; Reich *et al.*, 1991a, 1995) successfully predicted the time course of GPP in a forest in eastern North

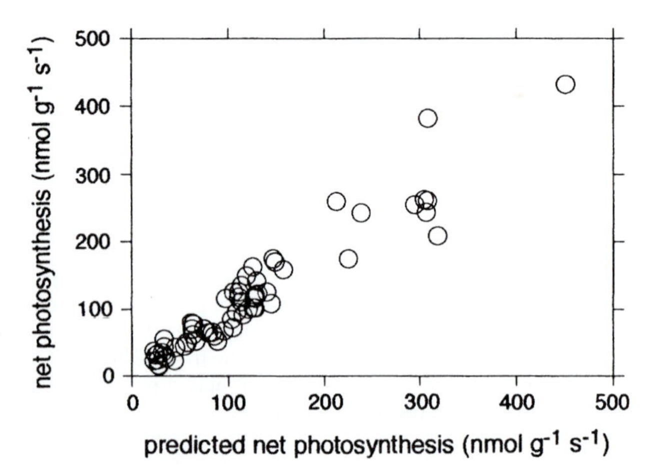

Figure 11-9 Relationship between observed net photosynthetic capacity in the field for temperate forest species and their predicted photosynthetic capacity based on their leaf N, leaf life-span, and SLA, using a model developed independently from nontemperate species. Data from Reich *et al.* (1997b, 1998a), recalculated for this figure.

America over a 4-yr period in a comparison of modeled fluxes vs. eddy covariance estimates (Aber *et al.*, 1996).

D. Respiration

Respiration may be partitioned into that associated with the construction of new tissue (growth respiration) and that required to maintain living tissue (maintenance respiration). These components may be separated by measuring growth rate concurrent with tissue respiration; the slope and intercept of the specific respiration rate vs. specific growth rate plot should be the growth and maintenance respiration coefficients (Amthor, 1989). During periods of no growth, all respiration is considered to be for maintenance. Although there is an increasing body of data for easily accessible plant components, in practice it is often quite difficult to measure growth rates and respiration simultaneously, e.g., for roots or large stems, so direct measurements of growth respiration are quite sparse for some tissue types, particularly in a field setting. It is also difficult to separate autotrophic from heterotrophic respiration belowground, so those will be discussed jointly below.

1. Leaf Respiration There are numerous published data on instantaneous leaf respiration rates for needleleaf and broadleaf deciduous taxa, and this work has established that respiration rates vary by temperature, leaf tissue nitrogen content, species, leaf age, and leaf structural traits (Ryan *et al.*, 1994; Reich *et al.*, 1996, 1998b). Respiration rates at a fixed temperature will scale linearly with nitrogen content both within and across species (Fig.

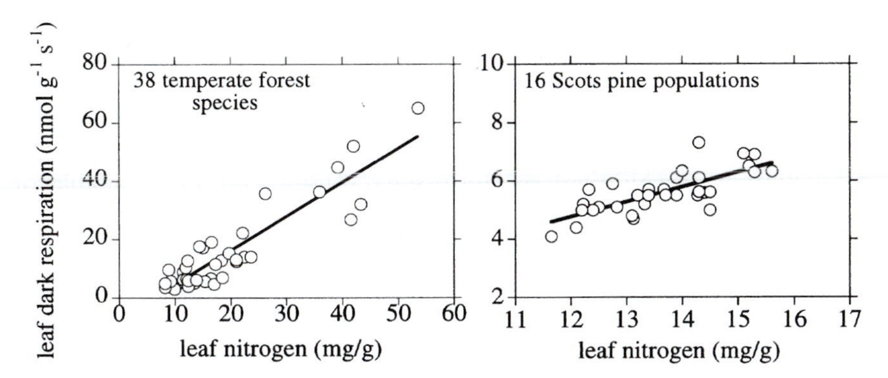

Figure 11-10 Relationships between leaf dark respiration rate (measured at 25°C) and leaf nitrogen for temperate forest species pooled from three areas in the United States (cold temperate forest in Wisconsin, cool temperate montane forest in North Carolina, and warm temperate coastal plain forest in South Carolina) (left) and for Scots pine populations originating across a broad climate range and grown in a common garden in central Poland (right). Data for 38 species are from Reich *et al.* (1998b). Data for 16 Scots pine populations redrawn from Reich *et al.* (1996).

Table 11-2 Cumulative Annual Nighttime Foliar Respiration in Temperate Forests

Forest type	Leaf tissue R (Mg C ha^{-1} yr^{-1})	Total autotrophic R (Mg C ha^{-1} yr^{-1})	Ref. and notes
Mixed deciduous broadleaf	1.81	Not reported	Bolstad *et al.* (1999); high elevation
	3.16	Not reported	Bolstad *et al.* (1999); low elevation
Pinus elliotii	1.35	5.92	Gholz *et al.* (1991); Cropper and Gholz (1993)
Pinus contorta	7.80	35.0	Benecke and Nordmeyer (1982), report in Ryan *et al.* (1994)
	2.30	5.83	Ryan and Waring (1992)
Pinus taeda	6.57	20.68	Kinnerson *et al.* (1977)

11-10). However, there are relatively few integrated whole-canopy estimates of leaf respiration rates over daily to annual time scales (Table 11-2).

*2. **Woody Tissue Respiration*** There are also few reports of woody tissue respiration, an important shortfall given the apparent relative magnitude of this source, which may approach that of total leaf respiration in some systems when integrated over an annual basis (Edwards *et al.*, 1981; Ryan *et al.*, 1994). Respiration rates depend largely on temperature and on woody tissue sapwood biomass and nitrogen content, which are quite variable and in many respects much more difficult to sample than are leaf biomass or nitrogen content (Sprugel and Benecke, 1991). Sapwood biomass and nitrogen content vary by tree size, shape, species, time of year, and structure (e.g., relative branch/bole volumes) in an unknown manner for most species and temperate forest types. Although specific tissue respiration rates are available for many species, few have been collected with sufficient ancillary data for whole-system estimates. Whole-system maintenance woody tissue respiration rates have been reported from 0.5 to 1.6 Mg C ha^{-1} yr^{-1} (Table 11-3).

*3. **Annual Fine Root Production and Respiration*** Comparatively few studies have measured belowground components, and there is considerable variation in observed root biomass and production (Table 11-4). Variation in estimated root production may have many causes. Higher root productivity may be required in nutrient poor soil because decreased nutrient availability drives increased root competition. Intraannual variability, particularly in soil moisture, may cause periodic root mortality and higher root

Table 11-3 Annually Integrated Woody Tissue Maintenance and Growth Respiration for Temperate Forest Stands

Forest type	Woody tissue R_m (Mg C ha^{-1} yr^{-1})	Woody tissue R_g (Mg C ha^{-1} yr^{-1})	Ref. and notes
Mixed *Quercus–Acer*	0.83	0.66	Edwards and Hanson (1996); year 1
Mixed *Quercus–Acer*	1.10	0.94	Edwards and Hanson (1996); year 2
Pinus ponderosa	0.52	Not reported	Gower *et al.* (1993); Ryan *et al.* (1995)
Pinus resinosa	0.70	Not reported	Gower *et al.* (1993); Ryan *et al.* (1995)
Pinus taeda	1.46	Not reported	Cropper and Gholz (1993)
Psuedotsuga–Tsuga	1.62	Not reported	Runyon *et al.* (1994)

turnover rates. Measurement methods and limits also affect estimates; some studies report a 2-mm, others a 3-mm, and yet others a 5-mm size limit for identifying fine roots. Ingrowth cores vs. coring will also add variation, the former by altering the local within-core environment and stimulating production, the latter by subjecting estimates to the inherent among-core variation in root biomass and hence standing crop estimates.

It is quite difficult in practice to separate live root respiration from bulk soil and litter heterotrophic respiration. Soil and litter microorganisms are the source of most forest ecosystem heterotrophic respiration, in part because soil and litter contain a major portion of total ecosystem carbon in temperate regions. However, root respiration may be quite high, particularly during times of high fine-root production. Disturbance inherent in many techniques that try to separate soil from roots may have a priming effect (Hendrick and Pregitzer, 1993), resulting in a stimulation and subsequent increase in heterotrophic respiration, and may damage roots, leading to either an increase or decrease in respiration, depending on the extent of damage and vegetation condition. As with autotrophic respiration, temperature is a major determinant of forest floor respiration rates (Landsberg and Gower, 1997). However, soil moisture, total soil organic carbon, fine root biomass and production, and soil nitrogen status also influence forest floor respiration rates.

4. Forest Floor Respiration Forest floor respiration varies considerably within temperate forest biomes, ranging between at least 2.5 and 14.1 Mg C ha^{-1} yr^{-1} (Raich and Schlesinger, 1992). Soil carbon flux averaged 6.8 Mg C ha^{-1} yr^{-1} under coniferous species ($n = 13$, s.e. $= 0.7$) and 7.3 Mg C ha^{-1} yr^{-1} under broadleaf species ($n = 20$, s.e. $= 0.6$) when considering only na-

Table 11-4 Fine Root Production and Standing Fine Root Biomass[a]

Forest type	Fine root production (Mg C ha^{-1} yr^{-1})	Standing fine root biomass (Mg C ha^{-1})	Ref. and notes
Acer saccharum	2.02	2.14	Nadelhoffer *et al.* (1985); roots <3 mm, mature
	3.25	3.23	Aber *et al.* (1982); <3 mm, mature
	4.04	3.95	Hendrick and Pregitzer (1993); site 1, <2 mm, mature
	3.65	3.44	Hendrick and Pregitzer (1993); site 2, <2 mm
Liriodendron tulipifera	3.37	4.25	Cox *et al.* (1978); roots <5 mm, to 60 cm depth
Mixed broadleaf deciduous	4.94	3.71	Powell and Day (1991); sequential cores
	1.72	3.71	Powell and Day (1991); ingrowth cores
Northern hardwoods	2.54	2.35	Fahey and Hughes (1994); mature forest, <2 mm, mature
	2.40	2.16	Fahey and Hughes (1994); 5-yr-old forest; both ingrown cores
Quercus rubra/Acer saccharum	2.45	2.55	McClaugherty *et al.* (1982); min–max estimates
	4.95	2.55	McClaugherty *et al.* (1982); monthly cores, 30 cm depth, <3 mm
Quercus alba	0.42	2.06	Joslin and Henderson (1987); <2 mm, 22 cm depth
	2.07	1.71	Nadelhoffer *et al.* (1985); N budget
	1.54	N.R.	Aber *et al.* (1985); N budget
Quercus rubra	2.62	1.35	Nadelhoffer *et al.* (1985); N budget
	1.16	N.R.	Aber *et al.* (1985); N budget
Quercus velutina	2.85	1.35	Nadelhoffer *et al.* (1985); N budget
Abies amabilis	3.11	3.05	Harmon *et al.* (1986); Vogt *et al.* (1990); <5 mm, 23 yr
	5.85	7.63	Harmon *et al.* (1986); Vogt *et al.* (1990; 180 yr old
Chamaecyparis thyoides	1.83	3.53	Day (1984), Powell and Day (1991); <3 mm, 57 yr
Pinus elliottii	2.45	5.20	Gholz *et al.* (1986); Ewel *et al.* (1987); <5 mm, 7 yr
	3.09	0.88	Gholz *et al.* (1986); Ewel *et al.* (1987); 27 yr old

(*continues*)

Table 11-4 (*Continued*)

Forest type	Fine root production (Mg C ha^{-1} yr^{-1})	Standing fine root biomass (Mg C ha^{-1})	Ref. and notes
Pinus resinosa	2.10	2.55	McClaugherty *et al.* (1982); <3 mm, 53 yr old
	0.99	2.21	Nadelhoffer *et al.* (1985); <3 mm, 100 yr old
	2.15	2.9	Haynes and Gower (1995); unfertilized, <5 mm, max–min
	1.44	2.4	Haynes and Gower (1995); fertilized
Pinus strobus	1.29	1.86	Aber *et al.* (1982), Nadelhoffer *et al.* (1985); <3 mm
Psuedotsuga menziesii	3.06	2.74	Santantonio and Herman (1985); <5 mm, 70 yr, dry site
	2.96	3.61	Santantonio and Herman (1985); 1 yr, mod. site
	2.26	3.50	Santantonio and Herman (1985); <5 mm, 120 yr, wet site
	3.29	4.94	Vogt (1991); <2 mm, 40 yr old, poor site
	1.18	2.12	Vogt (1991); good site
	2.37	2.64	Vogt (1991); poor site
	4.20	3.99	Gower *et al.* (1992); <5 mm, 50 yr old, control; biomass pre-
	2.17	N.R.	Gower *et al.* (1992); fertilized treatment mean
Taxodium distichum	1.54	1.88	Day (1984), Powell and Day (1991); <3 mm, 86 yr

[a]Adapted from Hendrick and Pregizter (1993) and other literature sources. N.R., not reported.

tive, recently undisturbed forests. Rates for disturbed forests were both above and below similar undisturbed forests (e.g., Weber, 1985; Ewel *et al.*, 1987) and were typified by both rapid increases after canopy opening, which cause increases in soil temperature, and subsequent decline in soil C flux because of lower root biomass and litterfall inputs.

Forest floor carbon dioxide flux is significantly influenced by site soil organic matter (SOM), because larger SOM pools generally allow larger microbial populations. However, annual respiration is often low in the highest C sites, associated with perennially or seasonally saturated soils. Soil/litter carbon amounts depend on a number of factors, including soil type, ecosystem productivity, site moisture and temperature, ecosystem nitrogen status and C:N ratios, plant species/litter quality, topographic position and geology, and time since and type of last disturbance. Most global studies and some regional studies show soil carbon increasing with decreasing temperatures, both across and within biomes (Parton *et al.*, 1987; Houghton, 1995).

V. Human Impacts on Production

A. Disturbance

Humans have had a large direct impact on temperate forests, changing this biome more than perhaps any other of comparable or larger size. Early civilizations developed in Oriental and European temperate regions, and temperate biomes in Europe and North America bore the brunt of the industrial revolution, providing wood for fuel and building materials, and as a raw material in industrial processing. Boreal biomes, although in places heavily exploited, have been spared widespread permanent conversion in vegetation type due to their inappropriateness for intensive agriculture. Conversion from native vegetation to agricultural or urban landcovers has become widespread in tropical biomes only in the second half of this century. Thus temperate forest biomes have had more complete, longer impacts over broader areas than any of the other large forest biomes (Williams, 1989).

Human impacts on temperate forest biomes have been both direct and indirect. Direct impacts include site conversion to other land covers and land uses, including inadvertant effects on the native species composition. Indirect impacts include the introduction or elimination of herbivorous or competing species (Pastor *et al.*, 1993); fire suppression and attendant changes in disturbance regimes (Frelich and Reich, 1995); atmospheric pollution, particularly ozone (Reich *et al.*, 1990; Ollinger *et al.*, 1997), sulfur, and nitrogen compounds (Townsend *et al.*, 1996); changes in the age structure of forest through cutting patterns and practices; and changes in stand age demographics on longer time scales related to historical changes in local, national, and global economies.

The largest direct impacts on temperate forests stem from conversion to other land uses. Precivilization forest extent is unknown; however, direct conversions to other landcovers have been immense. Native forests have been all but eliminated in some countries such as Ireland and Britain, and estimates are that up to one-half the original forest land in the biome has been converted to other uses, primarily agriculture. Thus, forest conversion to agriculture has been of critical importance, with attendant effects on both aboveground stored carbon and cycling, and belowground storage and cycling rates. Large quantities of carbon have been released from temperate forest vegetation and soils into the atmosphere as a result of these conversions (Houghton, 1995; and see Chapter 21, this volume). However, most temperate zone economies have undergone or are in the process of shifts from rural, agricultural economies to urban manufacturing and technologically driven economies. Attendant demographic shifts, increases in agricultural efficiency and agricultural corporatization, and consolidation at both regional and local levels have led to large-scale agricultural abandonment and the reversion of agricultural lands back to forests. Reports based on survey and repeat-measurement growth studies identify an increase in forest area and volumes in both North America and Europe, indicating both are currently net sinks for global atmospheric carbon as aggrading forests accumulate carbon in both plant biomass and soils (Kauppi *et al.*, 1992; Houghton, 1995). By all indications these trends are likely to continue, with significant agricultural land abandonment and conversion for some time to come. As far as it is possible to tell, temperate forest productivity has *on average* not been substantially diminished by human activities to date (see below). Whether temperate forests are more resilient in this way compared to other terrestrial biomes is unclear, but is a possibility that should make us wary of assuming that all biomes could respond to major direct human disturbances in a similar fashion.

Human activities have also indirectly affected temperate forest carbon stores and cycling. Fire exclusion has led to large changes in species composition and carbon storage. Also, local species extirpations (e.g., *Castor canadensis,* the North American beaver) and introductions (e.g., *Lymantra dispar,* the European gypsy moth in North America) have large aggregate impacts on species composition and nutrient and mass cycling rates. However, perhaps the largest indirect human impacts are and will be due to changes in global atmospheric composition and N deposition. Atmospheric N inputs have increased significantly over the past 30 years (Aber *et al.*, 1989), with large increases in per hectare N inputs reported, particularly in Europe (Kauppi *et al.*, 1992). Thus N deposition may be having substantial effects of global patterns of C storage (Townsend *et al.*, 1996). Changes in N availability could have large impacts on species compostion (toward more N-demanding taxa), litter quality and quantity, decomposer populations, and nutrient and mass cycling rates. Increases in atmospheric greenhouse gases may have large impacts through climatic alterations. The magnitude and extent of all of these indirect impacts, but particularly the later two, will best be analyzed with the help of process-based models.

B. Potential Temperate Forest Responses to Global Change

How will changing environmental factors (climate, CO_2, N deposition, land use) affect temperate forest NPP and associated processes? Given the difficulty in addressing these questions empirically at the present time (due to the lack of data), these issues—i.e, ANPP responses to global change factors—are addressed in this section from a conceptual and quantitative modeling framework. It should be noted, however, that ecosystem-scale experiments to assess temperate forest responses to elevated CO_2 are underway at this time in pine and hardwood stands (Duke Forest, North Carolina; Oak Ridge, Tennessee; and Rhinelander, Wisconsin; free-air CO_2 enrichment projects).

Models are one of the only viable methods for predicting climate change response from stand to global scales (Ehleringer and Field, 1993). Although many innovative experimental and measurement approaches have been applied over the past decade [including eddy covariance systems, free-air enriched CO_2 studies, *in situ* field soil temperature and nutrient manipulations, and big bottle experiments; see Jarvis and Dewar (1993)], available resources constrain the scope of these studies to few hectares, species, and climate change factors. Whole-stand measurement systems, although becoming more frequent, are still quite expensive and complicated to assemble and maintain. These systems also have limitations, in that exchanges estimated by eddy covariance are uncertain during some meteorological conditions, e.g., during nighttime stability or rainfall, and the contributing area varies with time and is often ambiguously identified. Eddy covariance systems may be established at a range of heights and concomitant range of ground footprints, but there is a need for estimates at longer temporal and broader spatial scales than current technology can provide. Models allow us to scale our point or stand measurements to larger areas and longer time scales, and aid in the identification of emergent ecosystem properties at these scales.

There is a rich literature describing various approaches for carbon cycle modeling across a range of temporal and spatial scales (Ågren *et al.*, 1991). Most models have been optimized for parts of the carbon cycle and specific biomes, e.g., forest canopy assimilation and transpiration (Wang and Jarvis, 1990), coniferous forest NPP and water balance (Running and Hunt, 1993), or grassland soil carbon cycling (Parton *et al.*, 1987), although few have been applied to a range of ecosystems (Running, 1994; Aber and Federer, 1992; Sellers *et al.*, 1997). This literature includes a number of good reviews of general plant–environment models (e.g., Ågren *et al.*, 1991; Norman, 1982) to which the reader is referred for a detailed treatment. In what follows we will briefly summarize a few key points and provide observations on measurements and approaches that might provide for most rapid progress.

Models may be differentiated by the central driving processes; the selection of this process may in part reflect the disciplinary biases of the developer. Many models developed by ecophysiologists have at their heart a photosynthesis submodel, typically describing carbon assimilation as an asymptotic function of light (Sellers *et al.*, 1997). Temperature and/or plant water status effects may also be included, either directly within the photosynthesis function or as separate functions. Detailed, leaf-level models may integrate transpiration and photosynthesis through submodels of stomatal conductance. Many models may also reflect observations of the strong influence of tissue nitrogen content on maximum photosynthesis. The form of these equations governing the effects of light, nitrogen, and soil and atmospheric moisture depends in part on the spatial scale of the analyses. Nonlinear relationships among these factors are typically observed at the cellular to leaf scale. However, at the stand to larger scales, many relationships tend toward linearity (Aber *et al.*, 1996; Ruimy *et al.*, 1996).

Models developed by atmospheric scientists initially focused on energy and mass balance, treating vegetation as passive, porous layers between soil and atmospheric reservoirs (Washington and Parkinson, 1986). Many current models incorporate parts of both atmospheric and ecophysiological approaches, representing energy and carbon exchange through a set of linked equations. In these, radiation is related to assimilation; however, assimilation is modulated by and related to transpiration rate, which in turn depends on plant water status. Light and moisture status are related to stomatal conductance, effectively linking assimilation and evapotranspiration. Changes in evapotranspiration are thus connected to surface energy balances, and in this manner C assimilation is linked to the overall energy balance (Sellers *et al.*, 1997). Atmospheric–biospheric models have become more tightly linked, with a much more sophisticated and process-based representation of canopy flux, at least for short time scales.

A substantial shift toward embedding more process-level detail into ecosystem models presupposes successively greater knowledge of ecosystem processes, their driving variables, and rates. This is true of carbon cycling and productivity models in general, and carbon flux models in particular. However, current models display our rather limited knowledge of many aspects of terrestrial forest carbon cycling. The greatest gaps occur in our measurements and understanding of belowground carbon stores, flux rates, and cycling, perhaps not surprising given the difficulties, time, and expense required to measure belowground variables. Our understanding of carbon allocation is quite poor, largely due to few (and uncertain) measurements of fine root production, respiration and turnover, and total belowground allocation. Although global trends based on long-term average detrital budgets identify the biome-scale determinants of belowground production (Raich and Nadelhoffer, 1989), these trends do not hold for individual stands (Gower *et al.*, 1996); other factors are more important at this finer scale.

There is also only a small empirical base to support predictions of tissue and soil respiration rates in temperate forests. Most work to date has been performed on economically important conifers, and there are scant data on many tree species that are less important economically but of tremendous ecological dominance and importance. The general form of respiration response has been established, but we currently cannot answer the question "how much do species matter?" with respect to respiration. Leaf respiration rates have been measured for relatively few species, woody tissue rates for fewer, and root respiration rate measurements are rarer still. Compounding this problem is a lack of knowledge concerning biometric relationships, e.g., even if we had good measurements of stem respiration per unit sapwood volume and we knew the magnitude of CO_2 transport upward through the transpiration stream, we have limited abilities to predict whole-tree sapwood volume. Although we expect that developing these data and models is a straightforward process, to date they have been developed for but a few species. Bulk soil respiration rates have been measured for a number of sites; however, the short-term nature of most studies, spatially limited sampling, and the variation in methods used confounds many analyses. The extent and short- and long-term dynamics of acclimation and adaptation in photosynthesis and respiration are not well understood, neither in plants nor for microbial populations. Although total or component respiration may not be important for many questions, inasmuch as significant changes in regional water balance, weather variability, and air and soil temperature may substantially affect root, stem, and leaf allocation and whole-system carbon balance, the collection and analysis of these data should be considered an important research opportunity.

Given the status and shortcomings of current models, what do they suggest about temperate forest responses to global change issues such as elevated CO_2, temperature, N deposition, ozone pollution, or other issues? Running and Nemani (1991) modeled forest ecosystem response to various climate change factors, including a doubling of atmospheric CO_2, a 4°C temperature rise, and 10% increased precipitation, and found predicted changes that depended on the forest type and current climatic conditions. A near doubling of NPP was predicted for forests in western Montana, due largely to increased LAI and photosynthesis. However, increased evapotranspiration was also predicted, significantly affecting regional water flow. NPP was predicted to decrease 5% for pines growing in northern Florida, at the edge of the temperate zone.

Solomon (1986) used a FORET-type mixed species stochastic simulator, and predicted changes in species composition with a 2–7.5°C temperature increase. The model predicted distinct shifts in species composition, with distinct diebacks and complex differential species replacement, with net carbon losses at southern sites and net carbon gains at northern sites in eastern North America.

Model simulations by Aber *et al.* (1995) suggested that neither an unrealistically large (6°C) temperature warming, a 15% decrease in mean annual precipitation, nor a doubling of atmospheric CO_2 had substantial consistent effects on NPP of temperate forests (composition was fixed in the model) throughout a six-state region of the northeastern United States.

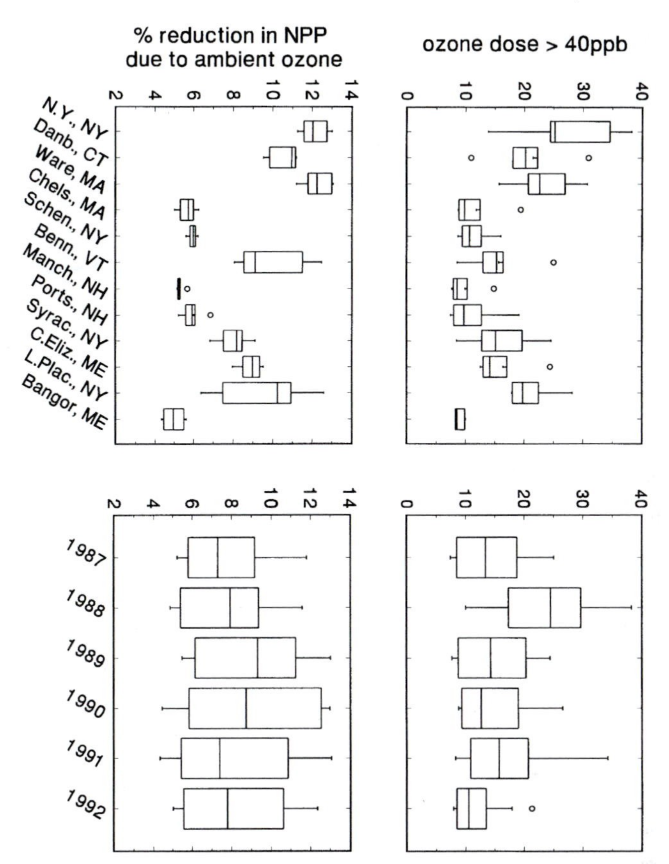

Figure 11-11 Mean annual ambient ozone pollution (summed ozone dose above 40 ppb) and the estimated effect of this pollution on ANPP for 64 locations in the northeastern United States for 1987 through 1992. Left: Ozone levels and simulated responses for 12 of the 64 sites. Right: Averages for all 64 sites for each year from 1987 to 1992. Redrawn from Ollinger *et al.* (1997).

However, all three changes simultaneously resulted in model predictions of increased productivity in all regions and for all forest types because increased water use efficiency associated with elevated atmospheric CO_2 would enable sustained higher canopy photosynthesis during the summer and an extended growing season without development of severe soil water deficits.

For the same region of the northeastern United States, model predictions of NPP responses to ambient ozone pollution levels that occurred from 1987 to 1992 (Ollinger *et al.,* 1997) suggested that forest productivity is currently reduced by 5–10% annually across the region (Fig. 11-11). Despite inter-

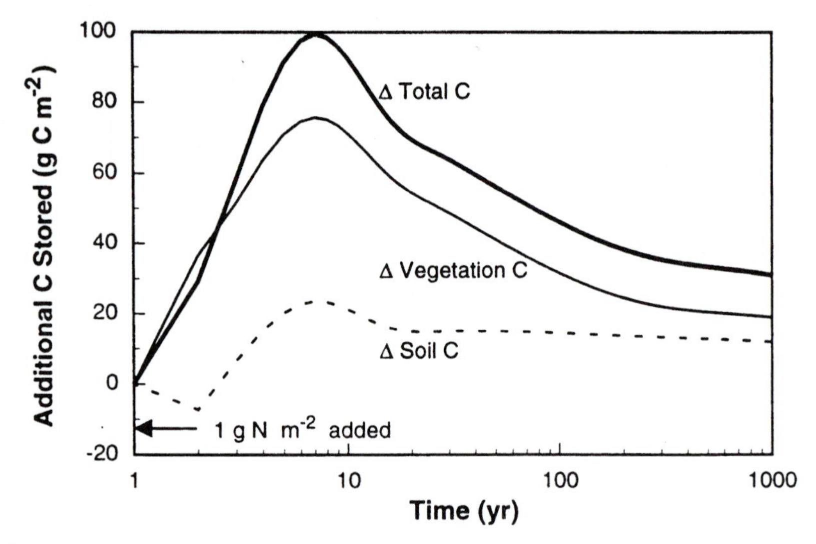

Figure 11-12 Simulated response of undisturbed temperate forest C pools to a unit pulse input of nitrogen (1 g N m^{-2}). From Hudson *et al.* (1994).

nannual variation in the severity of ozone pollution, effects appear to be compensated for by simultaneous interannual variation in precipitation, which is associated with ozone formation. Years with high ozone (e.g., 1988) occur in dry, hot summers where water limitations to ANPP may interact to limit ozone effects. In contrast, sites with greater ozone pollution tended to show greater reductions in NPP from this pollutant compared to sites with lower ozone levels, although interactions within the model clearly suggest that this relationship is not linear.

Hudson *et al.* (1994) developed a linked terrestrial/atmospheric global carbon cycling model to investigate global carbon pools, and to aid in illuminating discrepencies in the measured global carbon balance. They report concurrence with the notion of temperate forests being largely N limited, and identify increased atmospheric N deposition, originating from fossil fuel combustion and other anthropogenic sources, as contributing to large increases in C immobilization in terrestrial vegetation (Fig. 11-12). They conclude that N inputs (and subsequent enhancement of productivity and C storage) may well account for the "missing" C reported in many global C budgets.

VI. Summary

Temperate forests play a major role in the global carbon cycle, their productivity is important to regional economies, and their associated ecosystem, compositional, and structural characteristics are simultaneously a foundation for much of the natural world in temperate climates and of great importance to human society. As illustrated in this chapter we are beginning to better understand the processes that combine to drive productivity and carbon cycling in temperate forests, while still facing enormous challenges both conceptually and quantitatively in understanding those complex interactions and in measuring forest patterns and processes at relevant landscape to regional scales. In the coming century, temperate forests will continue to play an expanding role as demands for fiber, ecological services (such as water supply and stream water quality, carbon storage, and others), other ecological benefits (wildlife, other biodiversity), and other human values (recreation, tourism, aesthetics, spirituality) increase with burgeoning global populations. Given that temperate forests are located in regions that are highly developed (and as a result they have already been almost entirely disturbed by human use and abuse), their future status and extent may well signal whether continued human civilization at high population density is compatible with the sustainability of natural levels of primary production in any ecosystem type or biome.

Acknowledgments

The authors thank D. Ellsworth, A. Gillespie, and J. Oleksyn for insightful reviews of this manuscript and H. Mooney, J. Roy, and B. Saugier for instigating the thought-provoking workshop that has led to this chapter and book.

References

Aber, J. D., and Federer, C. A. (1992). A generalized, lumped-parameter model of photosynthesis, evapotranspiration and net primary production in temperate and boreal forest ecosystems. *Oecologia* **92,** 463–474.

Aber, J. D., Pastor, J., and Melillo, J. M. (1982). Changes in forest canopy structure along a site quality gradient in southern Wisconsin. *Am. Midland Natural.* **108,** 256–265.

Aber, J. D., Melillo, J. M., Nadelhoffer, K. J., McClaugherty, C. A., and Pastor, J. (1985). Fine root turnover in forest ecosystems in relation to quantity and form of nitrogen availablity: A comparison of two methods. *Oecologia* **66,** 317–321.

Aber, J. D., Nadelhoffer, K. J., Steudler, P., and Melillo, J. M. (1989). Nitrogen saturation in northern forest ecosystems. *BioScience* **39,** 378–386.

Aber, J. D., Ollinger, S. V., Federer, C. A., Reich, P. B., Goulden, M., Kicklighter, D. W., Melillo, J. M., Lathrop, R. G., and Ellis, J. M. (1995). Predicting the effects of climate change on water yield and forest production in the northeastern United States. *Climate Res.* **5,** 207–222.

Aber, J. D., Reich, P. B., and Goulden, M. L. (1996). Extrapolating leaf CO_2 exchange to the canopy: A generalized model of forest photosynthesis validated by eddy correlation. *Oecologia* **106,** 267–275.

Ågren, G. I., McMurtrie, R. W., Parton, J. W., Pastor, J., and Shugart, H. H. (1991). State-of-the-art models of production–decomposition linkages in conifer and grassland ecosystems. *Ecol. Appl.* **1,** 118–138.

Amthor, J. S. (1989). "Respiration and Crop Productivity." Springer-Verlag, New York.

Baldocchi, D. D. (1994). An analytical model for computing leaf photosynthesis and stomatal conductance. *Tree Physiol.* **14,** 1069–1079.

Baldocchi, D., Valentini, R., Running, S., Oechel, W., and Dahlman, R. (1996). Strategies for measuring and modeling carbon dioxide and water vapour fluxes over terrestrial ecosystems. *Global Change Biol.* **2,** 159–168.

Benecke, U., and Nordmeyer, A. H. (1982). Carbon uptake and allocation by *Nothofagus solandri* var. *cliffortioides* (Hook. f.) Poole and *Pinus contorta* Douglas ex Loudon ssp. *contorta* at montane and subalpine altitudes. *In* "Carbon Uptake and Allocation in Subalpine Ecosystems as a Key to Management" (R. H. Waring, ed.), pp. 9–21. Forest Res. Lab., Oregon State Univ., Corvallis, Oregon.

Bolstad, P. V., Mitchell, K., and Vose, J. M. (1999). Foliar temperature-respiration functions for eighteen Appalachian broadleaved species. *Tree Physiol.* **19,** 871–878.

Caldwell, M. M., Meister, H.-P., Tenhunen, J. D., and Lange, O. L. (1986). Canopy structure, light microclimate and leaf gas exchange of *Quercus coccifera* L. in a Portuguese macchia: Measurements in different canopy layers and simulations with a canopy model. *Trees* **1,** 25–41.

Chazdon, R. L., and Field, C. B. (1987). Determinants of photosynthetic capacity in six rainforest *Piper* species. *Oecologia* **73,** 222–230.

Coley, P. D. (1988). Effects of plant growth rate and leaf lifetime on the amount and type of anti-herbivore defense. *Oecologia (Berlin)* **74,** 531–536.

Coley, P. D., Bryant, J. P., and Chapin III, F. S. (1985). Resource availability and plant antiherbivore defense. *Science* **230,** 895–899.

Cox, T. L., Harris, W. F., Asmus, B. S., and Edwards, N. T. (1978). The role of roots in biogeochemical cycles in an eastern deciduous forest. *Pedobiologia* **18,** 264–271.

Cropper, W. P., Jr., and Gholz, H. L. (1993). Simulation of the carbon dynamics of a Florida slash pine plantation. *Ecol. Model.* **66,** 231–249.

Day, F. P., Jr. (1982). Litter decomposition rates in the seasonally flooded Great Dismal Swamp. *Ecology* **63,** 670–678.

Day, F. P., Jr. (1984). Biomass and litter accumulation in the Great Dismal Swamp. *In* "Cypress Swamps" (K. C. Ewel and H. T. Odum, eds.), pp. 386–392. Univ. Florida Press, Gainesville, Florida.

DeAngelis, D. L., Gardner, R. H., and Shugart, H. H. (1981). Productivity of forest ecosystems studied during the IBP: The woodlands data set. *In* "Dynamic Properties of Forest Ecosystems" (D. E. Reichle, ed.), pp. 567–672. International Biological Programme 23. Cambridge Univ. Press, London and New York.

Edwards, N. T., and Hanson, P. J. (1996). Stem respiration in a closed-canopy upland oak forest. *Tree Physiol.* **16,** 433–439.

Edwards, N. T., Shugart, H. H., Jr., Mclaughlin, S. B., Harris, W. F., and Reichle, D. E. (1981). Carbon metabolism in terrestrial ecosystems. *In* "Dynamic Properties of Forest Ecosystems" (D. E. Reichle, ed.), pp. 499–536. Cambridge Univ. Press, London.

Ehleringer, J. R., and Field, C. B., eds. (1993). "Scaling Physiological Processes: Leaf to Globe." Academic Press, San Diego.

Ellsworth, D. S., and Reich, P. B. (1993). Canopy structure and vertical patterns of photosynthesis and related leaf traits in a deciduous forest. *Oecologia* **96,** 169–178.

Ewel, K. C., Cropper, Jr., W. P., and Gholz, H. L. (1987). Soil CO_2 evolution in Florida slash pine plantations. I. Changes through time. *Can. J. For. Res.* **17,** 325–329.

Fahey, T. J., and Hughes, J. W. (1994). Fine root dynamics in a northern hardwood forest ecosystem, Hubbard Brook Experimental Forest, NH. *J. Ecol.* **82,** 533–548.

Field, C., and Mooney, H. A. (1986). The photosynthesis–nitrogen relationship in wild plants. *In* "On the Economy of Plant Form and Function" (T. Givnish, ed.), pp. 25–55. Cambridge Univ. Press, Cambridge.

Field, C. B., Randerson, J. T., and Malmstrom, C. M. (1995). Global net primary production: Combining ecology and remote sensing. *Remote Sensing Environ.* **51,** 74–88.

Frelich, L. E., and Reich, P. B. (1995). Spatial patterns and succession in a Minnesota southern-boreal forest. *Ecol. Monogr.* **65,** 325–346.

Gholz, H. L. (1982). Environmental limits on aboveground net primary production, leaf area and biomass in vegetation zones of the Pacific Northwest. *Ecology* **53,** 469–481.

Gholz, H. L., and Fisher, R. F. (1982) Organic matter production and distribution in slash pine (*Pinus elliottii*) plantations. *Ecology* **63,** 1827–1839.

Gholz, H. L., Hendry, L. C., and Cropper, Jr., W. P. (1986). Organic matter dynamics of fine roots in plantations of slash pine (*Pinus elliottii*) in north Florida. *Can. J. For. Res.* **16,** 529–538.

Gholz, H. L., Vogel, S. A., Cropper, Jr., W. P., McKelvey, K., Ewel, K. C., Teskey, R. O., and Curran, P. J. (1991). Dynamics of canopy structure and light interception in *Pinus elliottii* stands, north Florida. *Ecol. Monogr.* **61,** 33–51.

Goulden, M. L., Munger, J. W., Fan, S.-M., Daube, B. C., and Wofsy, S. C. (1996). Exchange of carbon dioxide by a deciduous forest: Response to interannual climate variability. *Science* **271,** 1576–1578.

Gower, S. T., and Son, Y. (1992). Differences in soil and leaf litterfall nitrogen dynamics for five forest plantations. *Soil Sci. Soc. Am. J.* **56,** 1059–1066.

Gower, S. T., Vogt, K. A., and Grier, C. C. (1992). Carbon dynamics of Rocky Mountain Douglas-fir: Influence of water and nutrient availability. *Ecol. Monogr.* **62,** 43–66.

Gower, S. T., Reich, P. B., and Son, Y. (1993). Canopy dynamics and aboveground production of five tree species with different leaf longevities. *Tree Physiol.* **12,** 327–345.

Gower, S. T., Gholz, H. L., Nakane, K., and Baldwin, V. C. (1994). Production and carbon allocation patterns of pine forests. *Ecol. Bull.* **43,** 115–135.

Gower, S. T., McMurtrie, R. E., and Murty, D. (1996). Aboveground net primary production decline with stand age: Potential causes. *Trends Ecol. Evol.* **11,** 378–382.

Greco, S., and Baldocchi, D. B. (1996). Seasonal variation of CO_2 and water vapour exchange rates over a temperate deciduous forest. *Global Change Biol.* **2,** 183–198.

Harmon, M. E., Franklin, J. F., Swanson, F. J., *et al.* (1986). Ecology of coarse woody debris in temperate ecosystems. *Adv. Ecol. Res.* **15,** 133–302.

Harris, W. F., Sollins, P., Edwards, N. T., Dinger, B. E., and Shugart, H.H. (1975) Analysis of carbon flow and productivity in a temperate deciduous forest ecosystem. *In* "Productivity of World Ecosystems," pp. 116–122. National Academy of Science, Washington, D.C.

Haxeltine, A., and Prentice, I. C. (1996). BIOME3: An equilibrium terrestrial biosphere model based on ecophysiological constraints, resource availability, and competition among plant functional types. *Global Biogeochem. Cycles* **10,** 693–709.

Haynes, B. E., and Gower, S. T. (1995). Belowground carbon allocation in unfertilized and fertilized red pine plantations in northern Wisconsin. *Tree Physiol.* **15,** 317–325.

Hendrick, R. L., and Pregitzer, K. S. (1993). The dynamics of fine root length, biomaass, and nitrogen content in two northern hardwood ecosystems. *Can. J. For. Res.* **23,** 2507–2520.

Hollinger, D. Y. (1989). Canopy organization and foliage photosynthetic capacity in a broad-leaved evergreen montane forest. *Funct. Ecol.* **3,** 53–62.

Hollinger, D. Y., Kelliher, F. M., Byers, J. N., Hunt, J. E., McSeveny, T. M., and Weir, P. L. (1994). Carbon dioxide exchange between an undisturbed old-growth temperate forest and the atmsophere. *Ecology* **75,** 134–150.

Houghton, R. A. (1995). Changes in the storage of terrestrial carbon since 1850. *In* "Soils and Global Change" (R. Lal, J. Kimble, E. Levine, and B. A. Stewart, eds.). CRC Press, Lewis Publ., Boca Raton, Florida.

Hudson, R. J. M., Gherini, S. A., and Goldstein, R. A. (1994). Modeling the global carbon cycle: Nitrogen fertilization of the terrestrial biosphere and the "missing" CO_2 sink. *Global Biogeochem. Cycles* **8,** 307–333.

Jarvis, P. G., and Dewar, R. C. (1993). Forests in the global carbon balance: From stand to region. *In* "Scaling Physiological Processes: Leaf to Globe" (J. R. Ehleringer and C. B. Field, eds.), pp. 191–221. Academic Press, San Diego.

Jarvis, P. G., and Leverenz, J. W. (1983). Productivity of temperate deciduous and evergreen forests. *In* "Encyclopedia of Plant Phyhsiology" (O. L. Lange, P. S. Nobel, C. B. Osmond, and H. Ziegler, eds.), New Series 12D, pp. 233–80. Springer-Verlag, Berlin/Heidelberg/New York.

Jose, S., and Gillespie, A. R. (1996). Aboveground production efficiency and canopy nutrient contents of mixed-hardwood forest communities along a moisture gradient in the central United States. *Can. J. For. Res.* **26,** 2214–2223.

Joslin, J. S., and Henderson, G. S. (1987). Organic matter and nutrients associated with fine root turnover in a white oak stand. *For. Sci.* **33,** 330–346.

Kauppi, P., MielikÑinen, K., and Kuusela, K. (1992). Biomass and carbon budget of European forests, 1971 to 1990. *Science* **256,** 70–74.

Kikuzawa, K. (1991). A cost-benefit analysis of leaf habitat and leaf longevity of trees and their geographical pattern. *Am. Nat.* **138,** 1250–1263.

Kinerson, R. S., Ralston, C. W., and Wells, C. G. (1977). Carbon cycling in a loblolly pine plantation. *Oecologia* **29,** 1–10.

Kira, T., and Shidei, T. (1967). Primary production and turnover of organic matter in different forest ecosystems of the western Pacific. *Jpn. J. Ecol.* **17,** 70–87.

Ko, L. J., and Reich, P. B. (1993). Oak tree effects on soil and herbaceous vegetation in savannas and pastures in Wisconsin. *Am. Midland Naturalist* **130,** 31–42

Landsberg, J. J., and Gower, S. T. (1997). "Applications of Physiological Ecology to Forest Management." Physiological Ecology Series. Academic Press, San Diego.

Leuning, R., Wang, Y. P., and Cromer, R. N. (1991). Model simulations of spatial distributions and daily totals of photosynthesis in *Eucalyptus gradis* canopies. *Oecologia* **88,** 494–503.

Leuning, R., Kelliher, F. M., de Prury, D. G. G., and Schulze, E.-D. (1995). Leaf nitrogen, photosynthesis, conductance and transpiration: Scaling from leaves to canopy. *Plant Cell Environ.* **18,** 1183–1200.

Linder, S. (1987). Responses to water and nutrients in coniferous ecosystems. *In* "Potential and Limitations of Ecosystem Analysis" (E.-D. Schulze and H. Zwîlfer, eds.), pp. 180–202. Springer, Berlin.

Linder, S., Benson, M. L., Myers, B. J., and Raison, R. J. (1987). Canopy dynamics and growth of *Pinus radiata.* I. Effects of irrigation and fertilization during a drought. *Can. J. For. Res.* **17,** 1157–1165.

Magill, A. H., Aber, J. D., Hendricks, J. J., Bowden, R. D., Melillo, J. M., and Steudler, P. A. (1997). Biogeochemical response of forest ecosystems to simulated chronic nitrogen deposition. *Ecol. Appl.* **7,** 402–415.

McClaugherty, C. A., Aber, J. D., and Melillo, J. M. (1982). The role of fine roots in the organic matter and nitrogen budgets of two forested ecosystems. *Ecology* **63,** 1481–1490.

Melillo, J. M., Kicklighter, D. W., McGuire, A. D., Moore, B., Vorosmarlty, C. J., and Grace, A. L. (1993). Global climate change and terrestrial net primary production. *Nature (London)* **363,** 234–240.

Monteith, J. L. (1977). Climate and the efficiency of crop production in Britain. *Philos. Trans. R. Soc. Lond. Ser. B.* **281,** 277–294.

Mooney, H. A., and Golmon, S. L. (1982). Constraints on leaf structure and functions in reference to herbivory. *BioScience* **32,** 198–206.

Nadelhoffer, K. J., Aber, J. D., and Melillo, J. M. (1985). Fine roots, net primary production, and nitrogen availability: A new hypothesis. *Ecology* **66,** 1377–1390.

Nemeth, J. C. (1973). Dry matter production in young loblolly (*Pinus taeda* L.) and slash pine (*Pinus elliottii* Engelm.) plantations. *Ecol. Monogr.* **43,** 21–41.

Norman, J. M. (1982). Simulation of microclimates. *In* "Modification of the Aerial Environment of Plants" (B. J. Barfield and J. F. Gerber, eds.), pp. 249–277. American Society of Agricultural Engineers, St. Joseph, MI.

Norman, J. M., and Jarvis, P. G. (1975). Photosynthesis in Sitka spruce (*Picea sitchensis* (Bong.) Carr V. Radiation penetration theory and a test case. *J. Appl. Ecol.* **12,** 839–878.

Ojima, D. S., Schimel, D. S., Parton, W. J., and Owensby, C. E. (1994). Long- and short-term effects of fire on nitrogen cycling in tallgrass prairie. *Biogeochemistry* **24,** 67–84.

Oleksyn, J., Tjoelker, M. G., and Reich, P. B. (1992). Growth and biomass partitioning of populations of European *Pinus sylvestris* L. under simulated 50° and 60°N daylengths: Evidence for photoperiodic ecotypes. *New Phytologist* **120,** 561–574.

Oleksyn, J., Reich, P. B., Chalupka, W., and Tjoelker, M. G. (1998). Differential above- and below-ground biomass accumulation of European *Pinus sylvestris* populations in a 12-year provenance experiment. *Scand. J. For. Res.* **13** (in press).

Oleksyn, J., Reich, P. B., Rachwal, L., Tjoelker, M. G., and Karolewski, P. (2000). Variation in aboveground net primary production of diverse European *Pinus sylvestris* populations. *Trees* (in press).

Ollinger, S. V., Aber, J. D., and Reich, P. B. (1997). Simulating ozone effects on forest productivity: Interactions among leaf-, canopy-, and stand-level processes. *Ecol. Appl.* **7,** 1237–1251.

Parton, W. J., Schimel, D. S., Cole, C. V., and Ojima, D. S. (1987). Analysis of factors controlling soil organic matter levels in Great Plains grasslands. *Soil Sci. Soc. Am. J.* **51,** 1173–1179.

Pastor, J., Aber, J. D., McClaugherty, C. A., and Melillo, J. M. (1984). Aboveground production and N and P cycling along a nitrogen mineralization gradient on Blackhawk Island, Wisconsin. *Ecology* **65,** 256–268.

Pastor, J., Dewey, B., Naiman, R. J., McInnes, P. F., and Cohen, Y. (1993). Moose browsing and soil fertility in the boreal forests of the Isle Royale National Park. *Ecology* **74,** 467–480.

Peterson, D. A., and Reich, P. B. (2000). Fire frequency and stand dynamics in an oak savanna-woodland ecosystem. *Ecol Appl.* (in press).

Powell, S. W., and Day, Jr., F. P. (1991). Root production in four communities in the Great Dismal Swamp. *Am. J. Bot.* **78,** 288–297.

Raich, J. W., and Nadelhoffer, K. N. (1989). Belowground carbon allocation in forest ecosystems: Global trends. *Ecology* **70,** 1346–1354.

Raich, J. W., and Schlesinger, W. H. (1992). The global carbon dioxide flux in soil respiration and its relationship to vegetation and climate. *Tellus* **44B,** 81–89.

Reich, P. B., Ellsworth, D. S., Kloeppel, B. D., Fownes, J. H., and Gower, S. T. (1990). Vertical variation in canopy structure and CO_2 exchange of oak-maple forests: Influence of ozone, nitrogen and other factors on simulated canopy carbon gain. *Tree Physiol.* **7,** 329–345.

Reich, P. B., Ellsworth, D. S., and Walters, M. B. (1991a). Leaf development and season influence the relationships between leaf nitrogen, leaf mass per area, and photosynthesis in maple and oak trees. *Plant Cell Environ.* **14,** 251–259.

Reich, P. B., Uhl, C., Walters, M. B., and Ellsworth, D. S. (1991b). Leaf lifespan as a determinant of leaf structure and function among 23 Amazonian tree species. *Oecologia* **86,** 16–24.

Reich, P. B., Walters, M. B., and Ellsworth, D. S. (1992). Leaf lifespan in relation to leaf, plant and stand characteristics among diverse ecosystems. *Ecol. Monogr.* **62,** 365–392.

Reich, P. B., Koike, T., Gower, S. T., and Schoettle, A. W. (1994a). Causes and consequences of variation in conifer leaf life span. *In* "Ecophysiology of Coniferous Forests" (W. K. Smith and T. M. Hinckley, eds.), pp. 225–254. Academic Press, San Diego.

Reich, P. B., Walters, M. G., Ellsworth, D. S., and Uhl, C. (1994b). Photosynthesis–nitrogen relations in Amazonian tree species. I. Patterns among species and communities. *Oecologia* **97,** 62–72.

Reich, P. B., Kloeppel, B. D., Ellsworth, D. S., and Walters, M. B. (1995). Different photosynthesis–nitrogen relations in evergreen conifers and deciduous hardwood tree species. *Oecologia* **104,** 24–30.

Reich P. B., Oleksyn, J., and Tjoelker, M. G. (1996). Needle respiration and nitrogen concentration in Scots pine populations from a broad latitudinal range: A common garden test with field grown trees. *Funct. Ecol.* **10,** 768–776.

Reich, P. B., Grigal, D. F., Aber, J. D., and Gower, S. T. (1997a). Nitrogen mineralization and productivity in 50 hardwood and conifer stands on diverse soils. *Ecology* **78,** 335–347.

Reich, P. B., Walters, M. B., and Ellsworth, D. S. (1997b). From tropics to tundra: Global convergence in plant functioning. *Proc. Natl. Acad. Sci. U.S.A.* **94,** 13730–13734.

Reich, P. B., Ellsworth, D. S., Walters, M. B., Vose, J., Gresham, C., Volin, J., and Bowman, W. (1999). Generality of leaf trait relationships: a test across six biomes. *Ecology* **80,** 1955–1969.

Reich, P. B., Walters, M. B., Ellsworth, D. S., Vose, J., Volin, J., Gresham, C., and Bowman, W. (1998b). Relationships of leaf dark respiration to leaf nitrogen, specific leaf area, and leaf life-span: A test across biomes and functional groups. *Oecologia* **114,** 471–482.

Reich, P. B., Peterson, D. A., Wrage, K., and Wedin, D. (2001). Fire and vegetation effects on productivity and nitrogen cycling across a forest-grassland continuum. *Ecology* (in press).

Ruimy, A., Jarvis, P. G., Baldocchi, D. D., and Saugier, B. (1996). CO_2 fluxes over plant canopies and solar radiation: A review. *Adv. Ecol. Res.* **26,** 1–51.

Running, S. W. (1994). Testing forest-BGC ecosystem process simulations across a climatic gradient in Oregon. *Ecol. Appl.* **4,** 238–247.

Running, S. W., and Hunt, Jr., E. R. (1993). Generalization of a forest ecosystem process model for other biomes, Biome-BGC, and an application for global-scale models. *In* "Scaling Processes Between Leaf and Landscape Levels" (J. R. Ehleringer and C. Field, eds.), pp. 151–58. Academic Press, San Diego.

Running, S. W., and Nemani, R. R. (1991). Regional hydrologic and carbon balance responses of forests resulting from potential climate change. *Climatic Change* **19,** 349–368.

Runyon, J., Waring, R. H., Goward, S. N., and Welles, J. M.. (1994). Environmental limits on net primary production and light-use efficiency across the Oregon transect. *Ecol. Appl.* **4,** 226–237.

Ryan, M. G. (1991). The effect of climate change on plant respiration. *Ecol. Appl.* **1,** 157–167.

Ryan, M. G. (1995). Foliar maintenance respiration of subalpine and boreal trees and shrubs in relation to nitrogen content. *Plant Cell Environ.* **18,** 765–772

Ryan, M. G., and Waring, R. H. (1992). Maintenance respiration and stand development in a subalpine lodgepole pine forest. *Ecology* **73,** 2100–2108.

Ryan, M. G., and Yoder, B. J. (1997). Hydraulic limits to tree height and tree growth. *BioScience* (in press).

Ryan, M. G., Linder, S., Vose, J. M., and Hubbard, R. M. (1994). Dark respiration of pines. *Ecol. Bull.* **43,** 50–63

Ryan, M. G., Gower, S. T., Hubbard, R. M., Waring, R. H., Gholz, H. I., Cropper, W. I., and Running, S. W. (1995). Stem maintenance respiration of conifer stands in contrasting climates. *Oecologia* **101,** 133–140.

Ryan, M. G., Binkley, D., and Fownes, J. H. (1996). Age-related decline in forest productivity: Pattern and process. *Adv. Ecol. Res.* (in press).

Santantonio, D., and Hermann, R. K. (1985). Standing crop, production, and turnover of fine roots on dry, moderate and wet sites of mature Douglas-fir in western Oregon. *Ann. Sci. For.* **42,** 113–142.

Schimel, D. S. (1995). Terrestrial ecosystems and the carbon cycle. *Global Change Biol.* **1,** 77–91.

Sellers, P. J., Dickinson, R. E., Randall, D. A., Betts, A. K., Hall, F. G., Berry, J. A., Collatz, G. J., Denning, A. S., Mooney, H. A., Nobre, C. A., Sato, N., Field, C. B., and Henderson-Sellers, A. (1997). Modeling the exchanges of energy, water, and carbon between continents and the atmosphere. *Science* **275,** 502–515.

Solomon, A. M. (1986). Transient response of forests to CO_2-induced climate change: Simulation modeling experiments in eastern North America. *Oecologia (Berlin)* **68,** 567–579.

Sprugel, D. G., and Benecke, U. (1991). Measuring woody-tissue respiration and photosynthesis. *In* "Techniques and Approaches in Forest Tree Ecophysiology" (J. P. Lassoie and T. M. Hickley, eds.), pp. 329–355. CRC Press, Boca Raton, Florida.

Stein, G. (1922). "Geography and Plays: Sacred Emily." Four Seas Company, Boston.

Tans, P. P., Fung, I. Y., and Takahashi, T. (1990). Observational constraints on the global atmospheric CO_2 budget. *Science* **247,** 1431–1438.

Townsend, A. R., Braswell, B. H., Holland, E. A., and Penner, J. E. (1996). Spatial and temporal patterns in terrestrial carbon storage due to deposition of fossil fuel nitrogen. *Ecol. Appl.* **6,** 806–814.

Verma, S. B., Baldocchi, D. D., Anderson, D. E., Matt, D. R., and Clement, R. E. (1986). Eddy fluxes of CO_2, water vapour and sensible heat over a deciduous forest. *Bound. Layer Meteorol.* **36,** 71–91.

Vogt, K. (1991). Carbon budgets of temperate forest ecosystems. *Tree Physiol.* **9,** 69–86.

Vogt, K. A., Vogt, D. J., Gower, S. T., and Grier, C. C. (1990). Carbon and nitrogen interactions for forest ecosystems. Proceedings of above- and belowground interactions in forest trees in acidified soils. *Air Pollut. Res. Rep.* **32,** 203–235.

Wang, Y.-P., and Jarvis, P. G. (1990). Description and validation of an array model—MAESTRO. *Agric. For. Meterol.* **51,** 257–280.

Waring, R. H., and Franklin, J. F. (1979). Evergreen coniferous forests of the Pacific Northwest. *Science* **204,** 1380–1386.

Waring, R. H., and Schlesinger, W. H. (1985) "Forest Ecosystems, Concepts and Management." Academic Press, Orlando, Florida.

Waring, R. H., Law, B. E., Goulden, M. L., Bassow, S. L., McCreight, R. W., Wofsey, S. C., and Bazzaz, F. A. (1995). Scaling gross ecosystem production at Harvard Forest with remote sensing: A comparison of estimates from a constrained quantum-use efficiency model and eddy correlation. *Plant Cell Environ.* **18,** 1201–1213.

Weber, M. G. (1985). Forest soil respiration in eastern Ontario jack pine ecosystems. *Can. J. For. Res.* **15,** 1069–1973.

Wedin, D. A., and Tilman, D. (1990). Species effects on nitrogen cycling: A test with perennial grasses. *Oecologia* **84,** 433–441.

Whittaker, R. H., Bormann, F. H., Likens, G. E., and Siccama, T. G. (1974) The Hubbard Brook ecosystem study: Forest biomass and production. *Ecol. Monogr.* **44,** 233–254.

Williams, M. (1989). "Americans and Their Forests." Cambridge Univ. Press, Cambridge.

Wofsy, S. C., Goulden, M. L., Munger, J. W., Fan, S.-M., Bakwin, P. S., Daube, B. C., Bassow, S. L., and Bazzazz, F. A. (1993). Net exchange of CO_2 in a mid-latitude forest. *Science* **260,** 1314–1317.

12

Productivity of Temperate Grasslands

Osvaldo E. Sala

I. Introduction

Grasslands are located in areas with precipitation ranging between 150 and 1200 mm yr^{-1} and temperature between 0 and 25°C (Lieth and Whittaker, 1975). Along a precipitation gradient, in temperate regions, grasslands are located between forests and deserts. Sites with annual precipitation higher than 1200 mm yr^{-1} usually support forests whereas sites receiving less than 150 mm yr^{-1} usually are occupied by deserts. Temperature interacts with precipitation, moving the grassland–forest and grassland–desert boundaries to wetter or drier areas. For example, as temperature and potential evapotranspiration decrease, the grassland–forest boundary occurs at lower precipitation. In the Great Plains of North America where isohyets run in a north–south direction, the boundary between the tallgrass prairie and the temperate forest has a clear SE–NW direction (Barbour and Billings, 1988).

The grassland biome is large, potentially covering an area of 49×10^6 km^2, which is equivalent to 36% of the Earth's surface (Shantz, 1954). This estimate of the grassland area excludes savannas but includes both grass and shrub deserts. The area covered exclusively by grasslands is 15×10^6 km^2, which accounts for 11% of the Earth's surface. There are large expanses of grasslands in North America, South America, and Asia, whereas smaller pieces are found in Europe, Southern Africa, and Australia (Singh *et al.*,

1983). In North America, most of the Great Plains potential natural vegetation is grassland, and encompasses large areas from subtropical Texas in the United States to the boundary with temperate deciduous forest in Canada. In South America, the vast pampas and the Patagonian steppe are considered grasslands. Finally, in Asia, grasslands cover a large region from Ukraine to China.

This chapter is constrained to a discussion of climatically determined grasslands, in contrast with grasslands resulting from human action. Climatically determined grasslands occur in areas where, during at least part of the year, water availability is not enough to support forests, although they receive sufficient precipitation to sustain grasses as the dominant component of vegetation (Lauenroth, 1979). Anthropogenically determined grasslands are usually located in areas where the potential vegetation is forest. These forest areas have been logged, burnt, and sown with grasses and legumes, and consequently transformed into cultivated pastures that do not resemble natural grasslands in their functioning and relationships with the environment. Seldom, cultivated pastures occur in areas where potential vegetation was grassland because, when the transformation is feasible, crops are usually a more economically beneficial option.

Although the vast grassland biome is largely determined by environmental conditions, most of the biome is currently managed in one way or the other. Grasslands have provided food and fiber for our ancestors for millennia and have been at the epicenter of civilization (Stebbins, 1981). Grassland management ranges from pastoralism, which is still common in parts of Africa, to the improvement of cattle and sheep production via the organization of herds in categories, subdivision of ranches into paddocks, development of new water holes, intense veterinary care, and control of predators and parasites (Oesterheld *et al.,* 1992). Management practices rarely include fertilization or irrigation of native grasslands because these practices are reserved for cultivated pastures.

Four major functional types of plants—grasses, shrubs, herbs, and succulents—form the grassland biome. The relative contribution of these four functional types in grasslands depends on the seasonality of precipitation and the soil texture (Sala and Lauenroth, 1993). The two major functional types, grasses and shrubs, have contrasting root patterns. In general, grasses have shallow roots and shrubs have deep root systems. Sites where water availability tends to concentrate in the upper layers of the soil are dominated by grasses, whereas sites where water is predominantly located in deeper layers are dominated by shrubs (Sala *et al.,* 1993). The location of water in the soil depends on soil texture and seasonality of precipitation. The same amount of precipitation penetrates deeper into a soil with a coarse texture and lower water-holding capacity compared to a fine-texture soil with higher water-holding capacity. For example, in Mediterranean ecosystems where the wet and warm seasons do not coincide, soils tend to have water at deeper layers than do regions with a continental climate, where most of the precipitation occurs during the warm season. In Mediterranean ecosystems, during the rainy season, potential evapotranspiration is low and the upper layers stay wet, so when it rains, the upper layers are saturated and water penetrates into deeper soil layers. In contrast, in regions with a continental climate, most of the rain occurs during the warm season when evaporation is high, maintaining the upper layers dry, so when it rains water wets repeatedly the upper layers but rarely reaches deeper layers. Two contrasting examples are the shortgrass steppe in North America and the Patagonian steppe in South America. The shortgrass steppe receives most of the precipitation during the warm season; the wettest layer is located at 5–15 cm of depth, and is dominated by grasses (Sala *et al.*, 1992). In contrast, the Patagonian steppe receives most of the precipitation during fall and winter; the soil layer with the highest probability of being wet during the entire year is located at 80 cm of depth, and primary production is evenly distributed between shrubs and grasses (Paruelo and Sala, 1995).

C_3 and C_4 grasses, differentiated by their photosynthetic pathway, are two major groups within the grass functional type. Physiological differences of the two groups are associated with different ecological characteristics that separate them in space and time. C_4 species, in general, dominate in areas that are warmer and with less available water than C_3 species. Analysis of the distribution of these two types of grasses showed that the proportion of C_3 decreases southward in North America and northward in South America (Paruelo *et al.*, 1998). Many grasslands have both types of grasses but in those cases their phenology and production patterns are separated during the year. For example, the C_3 *Agropyron smithii* dominates the shortgrass steppe during the cool spring but the C_4 grass *Bouteloua gracilis* dominates during late spring and summer (Lauenroth and Milchunas, 1992). Similarly, very little overlap occurs between C_3 and C_4 species in the vast Pampas of South America; productivity of C_3 grasses peaks in early spring to almost disappear in the summer when C_4 grasses reach their maximum (Sala *et al.*, 1981).

Grasslands are utilized for grazing of cattle, sheep, goats, and native animals, all of which produce meat, milk, blood, wool, and hair—important goods for society. In arid and semiarid grasslands, these products represent one of the only ways of harvesting the production of these ecosystems. Grasslands also provide an array of goods and services besides those just mentioned above. These other goods and services, which currently have no market value, include the maintenance of the composition of the atmosphere by sequestering carbon, ameliorating weather, maintaining the genetic library, and conserving the soil. Economic analysis indicates that those services with no market value may in the future exceed the traditional goods and services provided by grasslands. For example, for some grasslands, carbon sequestration may be valued at \$200 ha^{-1}, which can exceed the current price of the land in the region (Sala and Paruelo, 1997).

The objectives of this chapter are twofold: to review the patterns and controls of aboveground productivity in temperate grasslands and to evaluate the impact of expected global change on those patterns. The literature on primary productivity of grasslands is abundant; productivity in grasslands has been measured extensively since the days of the International Biological Programme (Sims and Singh, 1978). The focus here is on those papers that contribute new understanding about the general patterns and controls of productivity in grasslands.

II. Productivity Patterns and Controls

A. Productivity and Precipitation

Grasslands cover a broad range of environmental conditions and consequently show a large range of aboveground net primary productivity (ANPP), from 50 to 800 g m^{-2} yr^{-1}. Precipitation is the major control of aboveground primary production in grasslands at a regional scale (Sala *et al.*, 1988). The United States Department of Agriculture (USDA) Soil Conservation Service developed an exhaustive data set that included estimates of annual average primary production for 9498 sites across the Central Grassland Region of the United States (Joyce *et al.*, 1986). Analysis of the data set indicated that annual precipitation (APPT) was the best predictor of annual primary production and that a straight line was the best model relating precipitation and production (Fig. 12-1) (Sala *et al.*, 1988). Not only was precipitation the best predictor but addition of other variables such as temperature, potential evapotranspiration, or the precipitation/potential evapotranspiration ratio did not improve the estimates of production. The model can be written using the form proposed by Noy-Meir (1973):

$$\text{ANPP (g m}^{-2}\text{)} = 0.6[\text{APPT (mm yr}^{-1}\text{)} - 56],$$

where each term in the equation now has an ecological meaning. The slope 0.6 is the average water use efficiency for grasslands in the region and 56 is the zero-yield intercept or ineffective precipitation (the threshold below which no production can occur).

The model presented in Fig. 12-1 was developed using a large number (9498) of sites, but all from only one continent, North America. How general is this model? McNaughton *et al.* (1993) and Paruelo *et al.* (1998) tested this model against data from other continents. They fitted similar models to data collected for 14 sites located in temperate South America, which is a region with a climate similar to that of the Central Grassland Region, but which belongs to a different biotic realm (Udvardy, 1975) and has a differ-

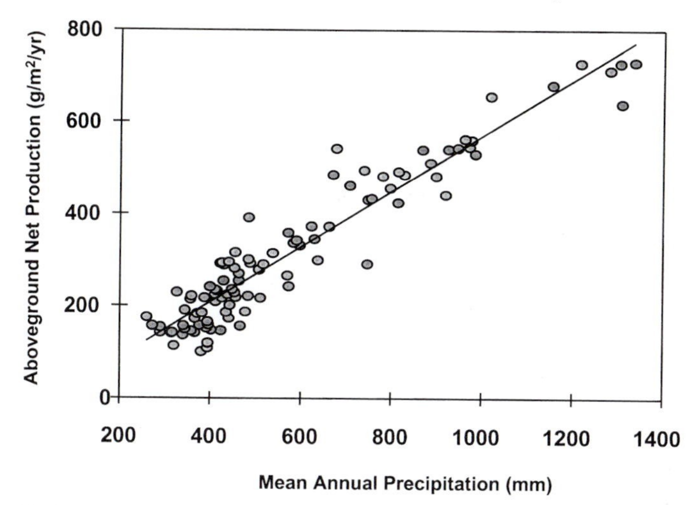

Figure 12-1 Relationship between average aboveground net primary production (ANPP) and mean annual precipitation (APPT) for 100 ecological regions encompassing 9498 sites along the Central Grassland Region of the United States. ANPP (g m^{-2}) = −34 + 0.6 APPT (mm yr^{-1}); $r^2 = 0.90$ and $p < 0.01$. Redrawn from Sala *et al.* (1988).

ent evolutionary grazing history (Sala *et al.*, 1986; Milchunas and Lauenroth, 1993). McNaughton *et al.* (1993) also developed similar models of production for distinct regions using data from 20 sites in the Serengeti ecosystem (McNaughton, 1985), 33 sites in Eastern and Southern Africa (Deshmukh, 1984), and 45 North African locations (Le Houérou and Hoste, 1977). Rodin (1979) reported aboveground primary productivity data for 13 sites in Central Asia ranging in annual precipitation from 99 to 217 mm yr^{-1}. I performed the regression analysis of ANPP versus annual precipitation on the Asian data and obtained a model [ANPP (g m^{-2}) = –30 + 0.59APPT (mm yr^{-1}); $p < 0.01$, $r^2 = 0.63$] that is quite similar to the one reported for the Central Grassland Region of the United States (Fig. 12-1). In all cases there is a remarkable similarity among the models, the efficiencies ranged between 0.48 and 0.85 g m^{-2} mm^{-1} yr^{-1}, encompassing the efficiency of the Central Grassland region and that of other global studies (Lauenroth, 1979; Rutherford, 1980).

B. Productivity and Temperature

Precipitation exerts an overwhelming control over production at the regional scale, although our ecological understanding suggests that other variables should also be important controls of production. Epstein *et al.* (1996), using the database for the Central Grassland Region of the United States, isolated sites along a north–south transect encompassing a broad tempera-

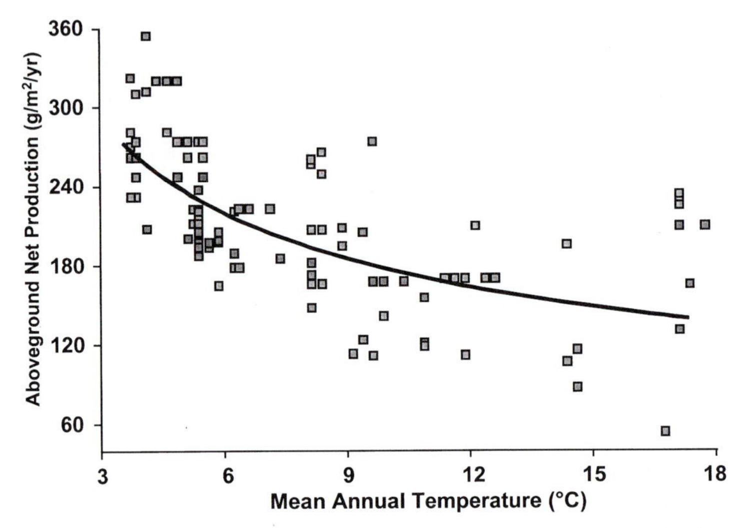

Figure 12-2 Relationship between average aboveground net primary production (ANPP) and mean annual temperature (MAT) for sites along a north–south transect in the Central Grassland Region of North America. ANPP = $456MAT^{(-0.4)}$; $r^2 = 0.4$ and $p < 0.001$. Redrawn from Epstein *et al.* (1996).

ture range (5–18°C) with little variation in precipitation. Once the precipitation variability was eliminated, a temperature effect was visible (Fig. 12-2). Surprisingly, the temperature effect was negative, with production decreasing with increasing temperature. Because biological processes, and certainly those related to plant growth, increase with temperature, the explanation for this pattern is that of an indirect effect on productivity through changes in water availability. As temperature increases, it simultaneously increases the evaporative demand, therefore temperature increases result in a reduction in water availability if precipitation input is maintained constant (Epstein *et al.,* 1996).

C. Productivity and Soil Texture

Soil characteristics also influence primary production in grasslands. Soil texture modifies water-holding capacity, which affects production, but there is an interesting interaction with precipitation (Sala *et al.,* 1988). Analysis of ANPP and soil patterns for 9498 sites in the Central Grassland Region of the United States indicated that production decreases with increasing water-holding capacity for sites with an annual precipitation below 370 mm yr^{-1} and production increases with increasing water-holding capacity at sites with annual precipitation higher than 370 mm (Fig. 12-3). The interaction between water-holding capacity and precipitation, i.e., the inverse texture hypothesis, is based on the ecosystem water balance (Sala *et al.*, 1988). Water losses in grasslands occur via transpiration, deep percolation, and bare soil evaporation and the relative magnitude of the last two pathways varies with precipitation. Bare soil evaporation occurs only from the uppermost layer of the soil. In wetter grasslands, those receiving more than 370 mm yr^{-1}, the major path for water loss is deep percolation and the magnitude of the loss decreases with increasing water-holding capacity. In drier grasslands, receiving less than 370 mm of annual precipitation, precipitation rarely penetrates beyond shallow deep layers and the major path for water loss is bare soil evaporation, which increases with increasing water-holding capacity. A given amount of water penetrates deeper into a coarse-textured soil with low water-holding capacity than into a fine-textured soil and consequently a smaller fraction of water will be located in the uppermost layer where evaporation occurs. In summary, in drier locations the major loss is soil evaporation, which increases with increasing water-holding capacity, whereas in

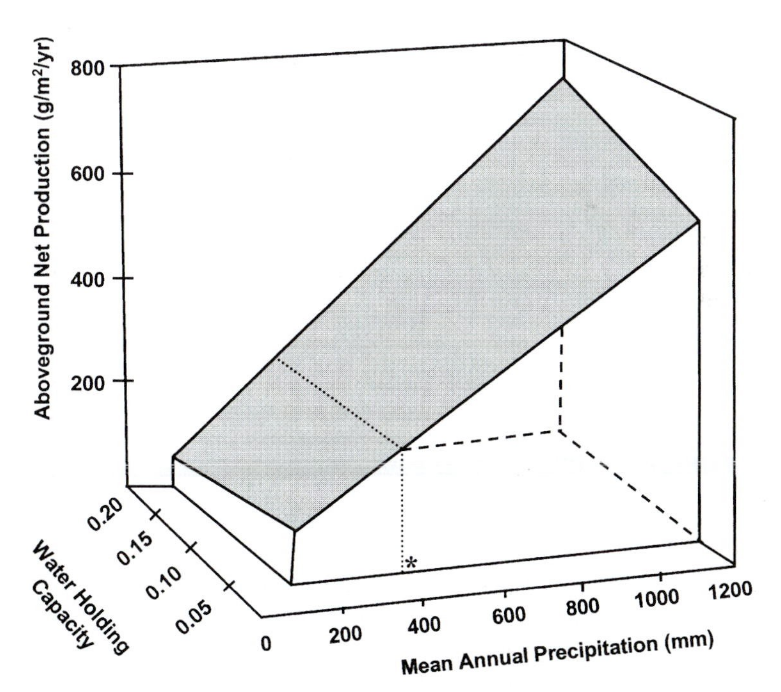

Figure 12-3 Relationship between aboveground net primary production (ANPP), soil water-holding capacity (WHC; proportion of soil dry mass), and mean annual precipitation (APPT) for 9498 sites located in the Central Grassland Region of North America. ANPP = 32 + 0.45APPT − 352WHC + 0.95WHC·APPT; $r^2 = 0.67$ and $p < 0.01$. Redrawn from Sala *et al.* (1988).

wetter sites the major loss is deep percolation, which decreases with increasing water-holding capacity. Consequently, in drier areas, available water for transpiration and plant growth decreases with increasing water-holding capacity and vice versa in wetter sites.

D. Spatial versus Temporal Controls

Up to this point, analyses of patterns and controls of primary production have been based on comparisons among sites. The correlative models were constructed using data on mean production and average precipitation or temperature for different sites. Implicit in the models was the assumption that spatial variability was a good analog of temporal variability and that models developed using spatial data were appropriate tools to predict changes through time. Lauenroth and Sala (1992) analyzed a 52-yr time series of production data for a site, the Central Plains Experimental Range, in the shortgrass steppe of North America. They found that annual precipitation was the variable that accounted for most of the variability in production among years, as precipitation accounted for most of the variability in production among sites. The model for the shortgrass steppe site developed using data for the 52-year period is

$$\text{ANPP (g m}^{-2}\text{ yr}^{-1}) = 56 + 0.13[\text{APPT (mm yr}^{-1})],$$

or, written in the form proposed by Noy-Meir (1973)

$$\text{ANPP (g m}^{-2}\text{ yr}^{-1}) = 0.13[\text{APPT (mm yr}^{-1}) + 430].$$

The striking finding was that the temporal model developed using the time series had a much lower slope compared to the spatial model developed using average data from many sites (Fig. 12-4). In the temporal model, each data point represents the precipitation and production for a different year, but all from the same site, whereas in the spatial model, each data point represents the average production and average annual precipitation for a different site. The large difference between the two models shows that the assumption that it was possible to exchange space for time was incorrect.

The explanation for the difference between the spatial and temporal models is associated with the existence of time lags in the response of ecosystems to changes in water availability. In the 52-yr data set, the lowest production year (1954) corresponds to one of the three driest years during the period. The following year, 1955, had a precipitation slightly above the long-term average; however, production (72 g m^{-2}) was substantially below the long-term production average of 97 g m^{-2}. Production did not reach and pass the long-term average until 1957, which was an extraordinarily wet year. The explanation for the lags in the ability of ecosystems to respond to changes in water availability is related to the inertia of vegetation structure.

Lags occur because what represents the optimal ecosystem structure

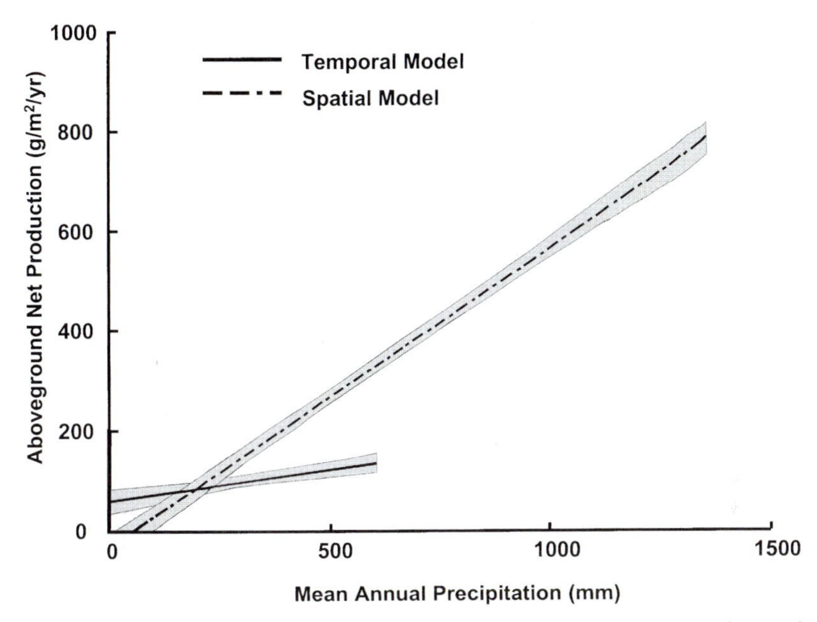

Figure 12-4 Relationship between aboveground net primary production and annual precipitation for a site in the shortgrass steppe during a 52-yr period (solid line), and the model described in Fig. 12-1 (dashed line). Shaded areas represent confidence intervals. Redrawn from Lauenroth and Sala (1992).

changes from year to year. The optimal ecosystem structure is characterized by a leaf area, a density of individuals, and a species composition that result in maximum resource acquisition. The structure of a grassland ecosystem has inertia and does not change immediately with changes in water availability.The different components of the ecosystem structure show different time lags in their ability to respond to increases in water availability. Leaf expansion may respond in a matter of hours after a rainfall event whereas changes in density of individuals or in species composition may take several years. Tilman and Downing (1994) showed how several years were required for species composition in a tallgrass prairie ecosystem to recover after a severe drought. Lags in altering the structure of the ecosystem may be responsible for the differences between the temporal and spatial models. For example, the wettest year during the 52-yr period analyzed for the shortgrass steppe had a precipitation of 588 mm and a production of only 115 g m^{-2}, whereas a site with 588 mm of average annual precipitation has an average production of 318 g m^{-2}. The structure of the ecosystem constrained the production during the wet year.

Burke *et al.* (1997) suggested another explanation for the differences between the spatial and temporal models and the occurrence of lags in the response to changes in water availability. Their modeling exercise using the ecosystem model CENTURY (Parton *et al.*, 1987) suggested that nutrient availability may constrain productivity in wet years. During dry years the model shows higher N mineralization than does the spatial model, because current-year mineralization depends on previous-year production and the resulting active organic matter. Similarly, during wet years N mineralization is lower than expected from the spatial model. Their modeling exercise reproduced the inertial behavior reported in the field data (Lauenroth and Sala, 1992). The nutrient constraint and the structural constraint hypotheses can easily complement each other. Further experimental work is necessary to evaluate the relative importance of these two hypotheses.

The difference between the temporal and spatial models is not restricted to the shortgrass steppe. Knapp *et al.* (1998) found a similar pattern using a 20-yr data set for the tallgrass prairie. Paruelo *et al.* (1999) reviewed data from 11 temperate grasslands (7 in the United States and 4 in Eurasia) with long-term data sets of productivity. They found that the ratio of the slopes of the temporal and spatial models changed along a precipitation gradient from 200 to 1200 mm yr^{-1}; both slopes become similar at intermediate precipitation (475 mm yr^{-1}) but the differences increase toward the dry and wet ends of the gradient. They hypothesized that structural constraints are maximum at the dry end of the gradient and decrease with increasing precipitation. On the contrary, biogeochemical constraints would be maximal at the wettest end of the gradient and decrease with decreasing precipitation. According to this hypothesis, the slope of the temporal model is maximum and equals that of the spatial model at intermediate precipitation values because at that point both the structural and the biogeochemical constraints are minimal.

The temporal model not only had a smaller slope than the spatial model but also accounted for a smaller fraction of the variability in production. A possible explanation for that fact is that different functional groups respond differently to environmental factors that vary from year to year, such as precipitation amount or seasonality. In the tallgrass prairie, grass production increases while forb production decreases with increasing water availability (Knapp *et al.*, 1998). In the Patagonian steppe, which has predominantly winter precipitation, grass production depends on spring and summer precipitation and shrub production depends on the precipitation accumulated during the previous 18 months (Jobbágy and Sala, 2000).

E. Biome Level Productivity

Grasslands occur in all continents and occupy a broad range of temperature, precipitation, and soil conditions. Consequently, estimating production for the entire biome requires taking into account spatial variability of the controls. Accuracy of the estimate depends on the quality of both the information on patterns of the controls as well as the functions relating NPP to the different drivers. Melillo *et al.* (1993) used the terrestrial ecosystem model (TEM) to estimate total NPP (sum of above- and belowground production) for all terrestrial biomes. Although the carbon cycle was the primary interest of their exercise, TEM also simulated the nitrogen cycle and the carbon–nitrogen interactions. The authors divided the grassland biome in two categories, short and tall grasslands, and reported mean NPP values of 428 for short grasslands and 670 g dry matter (DM) $m^{-2} yr^{-1}$ for tall grasslands. The mean NPP for the grassland biome when the two categories were weighted by their areas was 533 g DM $m^{-2} yr^{-1}$. This exercise has since been repeated using 17 different ecosystem models and of course a different set of parameters (Kicklighter *et al.,* 1999). Results of this second exercise were comparable, although higher than those of the first exercise, with values of NPP ranging between 600 and 900 g DM $m^{-2} yr^{-1}$.

Aboveground biomass and primary production are two different concepts; the former is a state variable and the latter is a flow or an ecosystem process. Although conceptually they are different and they even are represented by different units, aboveground biomass and aboveground net primary production in grasslands usually have the same values. Grasslands have an annual turnover of biomass of approximately 1. Consequently, the most common way of estimating aboveground net primary productivity is equating peak biomass with annual productivity (Sala and Austin, 2000). Therefore, mean estimates of biomass for grasslands have an absolute value that is quite similar to those of NPP, although units are different; biomass is represented in mass per unit area whereas productivity also has a unit time.

III. Secondary Productivity Patterns and Controls

A thorough analysis of the literature including 104 sites and encompassing nine different ecosystem types yielded an important conclusion that, at the global scale, there is a highly significant relationship between primary production and herbivore biomass, herbivore consumption, and secondary production (McNaughton *et al.,* 1989). The authors concluded that secondary production, herbivore biomass, and consumption are strong correlates of primary production. Secondary productivity varies linearly with primary productivity, indicating that changes in primary production are directly reflected in changes in secondary productivity. On the contrary, herbivore biomass and consumption vary exponentially with primary production, indicating that as primary production increases, consumption increases more than proportionally and that the fraction of total production consumed increases with increasing primary productivity.

The striking relationship between primary and secondary productivity across all kind of environmental and evolutionary conditions is not unique to terrestrial ecosystems or natural ecosystems. Similar patterns were described for aquatic ecosystems (Cyr and Pace, 1993) and managed ecosystems (Oesterheld *et al.*, 1992). In managed grasslands, analyses of livestock biomass censuses from Argentina and Uruguay showed that there is a tight, exponential relationship between herbivore biomass and primary productivity (Oesterheld *et al.*, 1992). The managed systems showed a relationship with primary production that had the same slope as the relationship found for natural systems but a higher *y* intercept. In managed systems, regardless of socioeconomic differences among regions, herbivore biomass or stocking rate increased exponentially with primary production. The difference in the *y* intercept between the models derived for natural and human-managed systems is an estimate of the effect of animal husbandry on herbivore-carrying capacity at a regional scale.

IV. Grassland Primary Production, Carbon Balance, and Global Change

The carbon balance of an ecosystem depends on inputs and outputs, and for most terrestrial ecosystems, primary production is the major input and soil erosion and decomposition are the major outputs. Four global change drivers will affect the primary production and carbon balance of grasslands: land use change, climate change, changes in the composition of the atmosphere, and biotic exchange or biodiversity change.

Land use change has the largest impact on the carbon balance of grasslands, not only because of the extent of land use change in grassland areas, but because of the effects per unit area. IMAGE 2 (Alcamo, 1994) is a global model of land use dynamics and it predicts large changes in the land use of grasslands for the next century. In the developing world, large areas of native grasslands will be converted into agricultural land, whereas the opposite will occur in North America, where large expanses of cropland will be abandoned. The processes of cultivation and abandonment have opposite effects, but of significantly different magnitude. Consequently, losses due to cultivation would not be compensated equally by an increase in abandoned area. Transformations of grasslands into croplands result in a net carbon flux to the atmosphere, because tillage increases decomposition by breaking soil aggregates and placing organic matter particles that were previously protected in contact with decomposers. Losses during the first 10 to 20 yr of cultivation can be enormous. For example, in the Great Plains of North America, Burke *et al.* (1991) found that agriculture has resulted in carbon losses of 1400 g m^{-2} during this century.

Most of the current climate change scenarios agree on the global warming trend, although there are differences among regions and models regarding the patterns of precipitation change (Kattenberg *et al.*, 1996). A review of the effects of climate change on the carbon balance of grasslands indicated that the expected increase in temperature will result in a decrease of the soil organic matter pool as a result of the increase in decomposition (Sala *et al.*, 1996). The effects of temperature on primary production are likely to be small and indirect, through changes in water availability (see Section II,B). The global circulation models predict decreases in precipitation for most of the temperate grassland region, although for some grasslands they predict small increases. Decreases in precipitation coupled with the increases in temperature will result in a large negative carbon balance. In the small area where precipitation may increase, the increase in production may compensate for the increase in decomposition resulting from the temperature increase.

The direct effects of elevated CO_2 on grassland primary production vary among ecosystems and from year to year. For example, in the annual grasslands of California, elevated CO_2 resulted in a stimulation of productivity in the shallow-soil serpentine ecosystem and a reduction in the deep-soil sandstone (Field *et al.*, 1995). In tallgrass prairie, similar experiments suggested an increase in production of the C_4 grasses during dry years and possibly an accumulation of carbon in surface soils as a result of elevated CO_2 (Owensby *et al.*, 1995).

Finally, our current understanding does not provide a quantitative answer to the question of the effects of changes in biodiversity on primary production. Experimental evidence suggested that production decreases as a result of losses of biodiversity (Mooney *et al.*, 1995a,b; Tilman *et al.*, 1996). However, we do not know yet how general those results are and what is the level of biodiversity change at which production will start to decrease.

The relative importance of the four drivers of global change on the production of grasslands is difficult to assess quantitatively because of the uncertainties in the predictions and in the response of the different grasslands. Burke *et al.* (1991) compared the effects of expected climate change versus the observed changes in land use in the Great Plains of the United States, regarding carbon emissions. They concluded that land use change had a larger effect than did climate change, 1400 versus 200 g m^{-2} of carbon losses. Our difficulty in assessing the relative importance of the four drivers is compounded by the expected change of the relative importance of the drivers. Land use seems the major driver now, but it may be surpassed by biodiversity or climate change in the future.

Acknowledgments

I thank A. T. Austin for her insightful discussions and help in various aspects of this project; J. P. Guerschman significantly contributed to this chapter with very good suggestions. This work was funded by several grants of the InterAmerican Institute for Global Change Research, NSF, CONICET, UBA, and FONCyT.

References

Alcamo, J. (1994). "Image 2: Integrated Modeling of Global Climate Change." Kluwer Academic Publ., Dordrecht.

Barbour, M. G., and Billings, W. D. (1988). North American Terrestrial Vegetation." Cambridge Univ. Press, New York.

Burke, I. C., Kittel, T. G. F., Lauenroth, W. K., Snook, P., Yonker, C. M., and Parton, W. J. (1991). Regional analysis of the Central Great Plains. BioScience **41,** 685–692.

Burke, I. C., Lauenroth, W. K., and Parton, W. J. (1997). Regional and temporal variation in net primary production and nitrogen mineralization in grasslands. Ecology **78,** 1330–1340.

Cyr, H., and Pace, M. (1993). Magnitude and patterns of herbivory in aquatic and terrestrial ecosystems. *Nature (London)* **361,** 148–150.

Deshmukh, I. (1984). A common relationship between precipitation and grassland peak biomass for east and southern Africa. *Afr. J. Ecol.* **22,** 181–186.

Epstein, H., Lauenroth, W., Burke, I., and Coffin, D. (1996). Ecological responses of dominant grasses along two climatic gradients in the Great Plains of the United States. *J. Veg. Sci.* **7,** 777–788.

Field, C., Chapin, F., Chiariello, N., Holland, E., and Mooney, H. (1995). The Jasper Ridge CO_2 experiment: Design and motivation. *In* "Carbon Dioxide and Terrestrial Ecosystems" (G. Koch, and H. Mooney, eds.), pp. 121–145. Academic Press, New York.

Jobbágy, E., and Sala, O. (2000). Controls of grass and shrub aboveground production in the Patagonian steppe. *Ecol. Appl.* **10,** 541–549.

Joyce, L., Chalk, D., and Vigil, A. (1986). "Range Forage Data Base for 20 Great Plains, Southern, and Western States." RM 133. United States Forest Service, Fort Collins, Colorado.

Kattenberg, A., Giorgi, F., Grassl, H., Meehl, G. A., Mitchell, J. F. B., Stouffer, R. J., Tokioka, T., Weaver, A. J., and Wigley, T. M. L. (1996). Climate models—Projections of future climate. *In* "Climate Change: The IPCC Scientific Assessment," pp. 285–358. Cambridge Univ. Press, Cambridge.

Kicklighter, D. W., Bondeau, A., Schloss, A. L., Kaduk, J., and McGuire, A. D. (1999). Comparing global models of terrestrial net primary productivity (NPP): Global pattern and differentiation by major biomes. *Global Change Biol.* **5,** 16–24.

Knapp, A., Briggs, J., Blair, J., and Turner, C. (1998). Patterns and controls of aboveground net primary production in tallgrass prairie. *In* "Grassland Dynamics: Long-Term Ecological Research in Tallgrass Prairie" (A. Knapp, J. Briggs, D. Hartnett, and S. Collins, eds.), pp. 193–221. Oxford Univ. Press, New York.

Lauenroth, W. K. (1979). Grassland primary production: North American grasslands in perspective. *In* "Perspectives in Grassland Ecology. Ecological Studies" (N. R. French, ed.), pp. 3–24. Springer-Verlag, New York, Heidelberg.

Lauenroth, W. K., and Milchunas, D. G. (1992). Short-grass steppe. *In* "Natural Grasslands: Introduction and Western Hemisphere" (R. T. Coupland, ed.), pp. 183–226. Elsevier, Amsterdam.

Lauenroth, W. K., and Sala, O. E. (1992). Long-term forage production of North American shortgrass steppe. *Ecol. Appl.* **2,** 397–403.

Le Houérou, H., and Hoste, C. (1977). Rangeland production and annual rainfall relations in the Mediterranean basin and in the African Sahelo-Sudanian zone. *J. Range Manage.* **30,** 181–189.

Lieth, H., and Whittaker, R. (1975). "Primary Productivity of the Biosphere." Springer-Verlag, New York.

McNaughton, S. J. (1985). Ecology of a grazing ecosystem: The serengeti. *Ecol. Monogr.* **53,** 259–294.

McNaughton, S. J., Oesterheld, M., Frank, D. A., and Williams, K. J. (1989). Ecosystem-level patterns of primary productivity and herbivory in terrestrial habitats. *Nature (London)* **341,** 142–144.

McNaughton, S. J., Sala, O. E., and Oesterheld, M. (1993). Comparative ecology of African and South American arid to subhumid ecosystems. *In* "Biological Relationships between Africa and South America" (P. Goldblatt, ed.), pp. 548–567. Yale Univ. Press, New Haven.

Melillo, J. M., McGuire, A. D. Kicklighter, D. W., Moore III, B., Vorosmarty, C. J., and Schloss, A. L. (1993). Global climate change and terrestrial net primary production. *Nature (London)* **363,** 234–240.

Milchunas, D. G., and Lauenroth, W. K. (1993). Quantitative effects of grazing on vegetation and soils over a global range of environments. *Ecol. Monogr.* **63,** 327–366.

Mooney, H. A., Lubchenco, J., Dirzo, R., and Sala, O. E. (1995a). "Biodiversity and Ecosystem Functioning: Basic Principles. Cambridge Univ. Press, Cambridge.

Mooney, H. A., Lubchenco, J., Dirzo, R., and Sala, O. E. (1995b). "Biodiversity and Ecosystem Functioning: Ecosystem Analyses. Cambridge Univ. Press, Cambridge.

Noy-Meir, I. (1973). Desert ecosystems: Environment and producers. *Annu. Rev. Ecol. Syst.* **4,** 25–52.

Oesterheld, M., Sala, O. E., and McNaughton, S. J. (1992). Effect of animal husbandry on herbivore-carrying capacity at a regional scale. *Nature (London)* **356,** 234–236.

Owensby, C., Ham, J., Knapp, A., Rice, C., Coyne, P., and Auen, L. (1995). Ecosystem-level responses of tallgrass prairie to elevated CO_2. *In* "Carbon Dioxide and Terrestrial Ecosystems" (G. Koch and H. Mooney, eds.), pp. 147–162. Academic Press, New York.

Parton, W., Schimel, D., Cole, C., and Ojima, D. (1987). Analysis of factors controlling soil organic matter levels in great plains grasslands. *Soil Sci. Soc. Am. J.* **51,** 1173–1179.

Paruelo, J. M., and Sala, O. E. (1995). Water losses in the Patagonian steppe: A modelling approach. *Ecology* **76,** 510–520.

Paruelo, J., Jobbagy, E., Sala, O., Lauenroth, W., and Burke, I. (1998). Functional and structural convergence of temperate grassland and shrubland ecosystems. *Ecol. Appl.* **8,** 194–206.

Paruelo, J., Lauenroth, W., Burke, I., and Sala, O. (1999). Grassland precipitation-use efficiency varies across a resource gradient. *Ecosystems* **2,** 64–68.

Rodin, L. (1979). Productivity of desert communities in central Asia. *In* "Arid-Land Ecosystems" (D. Goodall and R. Perry, eds.), pp. 273–298. Cambridge Univ. Press, Cambridge.

Rutherford, M. C. (1980). Annual plant production-precipitation relations in arid and semiarid regions. *S. Afric. J. Sci.* **76,** 53–56.

Sala, O., and Austin, A. (2000). Methods of estimating aboveground net primary production. *In* "Methods in Ecosystem Science" (O. Sala, R. Jackson, H. Mooney, and R. Howarth, eds.), pp. 31–43. Springer-Verlag, New York.

Sala, O. E., and Paruelo, J. M. (1997). Ecosystem services in grasslands. *In* "Nature's Services: Societal Dependence on Natural Ecosystems" (G. C. Daily, ed.), pp. 237–252. Island Press, Washington, D.C.

Sala, O., Deregibus, V., Schlichter, T., and Alippe, H. (1981). Productivity dynamics of a native temperate grassland in Argentina. *J. Range Manage.* **34,** 48–51.

Sala, O. E., Oesterheld, M., Leon, R. J. C., and Soriano, A. (1986). Grazing effects upon plant community structure in subhumid grasslands of Argentina. *Vegetation* **67,** 27–32.

Sala, O. E., Parton, W. J., Lauenroth, W. K., and Joyce, L. A. (1988). Primary production of the central grassland region of the United States. *Ecology* **69,** 40–45.

Sala, O. E., Lauenroth, W. K., and Parton, W. J. (1992). Long term soil water dynamics in the shortgrass steppe. *Ecology* **73,** 1175–1181.

Sala, O. E., Lauenroth, W. K., and Bolluscio, R. A. (1993). Arid and semiarid plant functional types. *In* "Plant Functional Types" (T. M. Smith, H. H. Shugart, and F. I. Woodward, eds.), pp. 217–233. Cambridge Univ. Press, Cambridge.

Sala, O. E., Lauenroth, W. K., and Burke, I. C. (1996). Carbon budgets of temperate grasslands and the effects of global change. *In* "Global Change: Effects on Coniferous Forests and Grasslands" (A. Breymeyer, D. O. Hall, J. M. Mellilo, and G. I. Agren, eds.), pp. 101–119. John Wiley and Sons, Chichester, New York.

Shantz, H. (1954). The place of grasslands in the earth's cover of vegetation. *Ecology* **35,** 142–145.

Sims, P. L., and Singh, J. S. (1978). The structure and function of ten western North American grasslands. III. Net primary production, turnover and efficiencies of energy capture and water use. *J. Ecol.* **66,** 573–597.

Singh, J. S., Lauenroth, W. K., and Milchunas, D. G. (1983). Geography of grassland ecosystems. *Prog. Phys. Geogr.* **7,** 46–80.

Stebbins, G. L. (1981). Coevolution of grasses and herbivores. *Ann. Missouri Bot. Gard.* **68,** 75–86.

Tilman, D., and Downing, J. A. (1994). Biodiversity and stability in grasslands. *Nature (London)* **367,** 363–365.

Tilman, D., Wedin, D., and Knops, J. (1996). Productivity and sustainability influenced by biodiversity in grassland ecosystems. *Nature (London)* **379,** 718–720.

Udvardy, M. (1975). A classification of the biogeographical provinces of the world. IUCN, Rep. 18. Morges, Switzerland.

13

Productivity of Agro-ecosystems

Jan Goudriaan, J. J. Rob Groot, and Peter W. J. Uithol

I. Introduction

The growth of agricultural production worldwide has led to an increasing role of agro-ecosystems within the global carbon cycle. Arable land in the world covers about one-tenth of the 134×10^{12} m^2 of total land area (excluding Antarctica). Much of this land area is bare rock, desert, or covered with ice, and about two-thirds can be considered as covered by vegetation. The total annual carbon flux through all terrestrial ecosystems is estimated at about 50 Pg C (Minami *et al.*, 1993). The objective of this paper is to estimate the contribution of agricultural crops in this total carbon flux. In addition to arable land, grasslands are used for agricultural purposes. They are mainly used for grazing by cattle and sheep, at stocking densities that vary enormously, with the extensively managed rangelands of Australia or Latin America on one end of the scale and the "green carpets" of The Netherlands on the other. Grasslands cover about one-third of the vegetated area and contribute about one-fifth to the global carbon cycle (Minami *et al.*, 1993). For further details on primary productivity of grasslands, reference is made to chapter 12, this volume. Here we present some data on the production of the major crop groups in terms of carbon and nitrogen. The decadal time trend is analyzed, the relation with nitrogen fertilizer production is indicated, and productivity is discussed in relation to other growth-limiting factors.

II. Potential and Actual Net Primary Production

Net primary production (NPP) is generally proportional to global radiation intercepted by green leaves. Although the efficiency of leaves to convert intercepted radiation into dry matter may decline under conditions of shortage of water or nutrients (Bélanger *et al.*, 1992), the main factor causing reduction of growth under poor growing conditions is a reduction in leaf area. The highest production level is simply estimated by using the global average of total incoming global radiation per unit land area (14 MJ m^{-2} d^{-1}), and multiplying that by the maximum radiation use efficiency (1.7 g of dry matter MJ^{-1} of global radiation) reported for optimal growing conditions. This calculation yields an annual carbon fixation of 3.9 kg C m^{-2} yr^{-1}, and at a land area of 134×10^{12} m^2 an absolute upper limit for terrestrial net primary productivity of 520 Pg C yr^{-1}. Although radiation and temperature are positively correlated, low temperatures limit growth in some regions during at least part of the year. If this reduction is estimated at a reduction of 30%, the remainder is 370 Pg C yr^{-1}. Shortage of water causes a much larger reduction. Potential evapotranspiration, averaged worldwide, amounts to 1100 mm yr^{-1}. Precipitation on land is only 730 mm yr^{-1}, of which 260 mm returns as river flow to the ocean [calculated from Schlesinger (1991)], which leaves 470 mm yr^{-1} as actual evapotranspiration on land. If potential evapotranspiration and radiation were not correlated, we could simply use the ratio actual/potential evapotranspiration as a multiplier to find water-limited NPP, yielding 160 Pg C yr^{-1}. In reality, water shortage will especially affect regions with high radiation levels, thus reducing actual radiation use much stronger. This strong interaction explains most of the reduction to the actually observed level of 50 Pg C yr^{-1} (Bolin *et al.*, 1979). Shortage of nutrients will also be important, even though in unexploited ecosystems, nutrient cycling within the growing season allows a much higher NPP at a given nutrient stock than in agricultural ecosystems. Therefore, shortage of nutrients is probably much less a limiting factor to NPP in natural ecosystems than is shortage of water. In agro-ecosystems, nutrients are removed by harvesting, and application of fertilizers to restore soil fertility is common practice. On rangelands this is normally not done. The area of arable land is one-tenth of the total terrestrial land area (see Table 13-1). Agriculture is concentrated in the best climates, pests and diseases are often controlled by biocides, and fertilizers are applied to remedy nutrient deficiencies. Thus, NPP on agricultural land is expected to be much higher than one-tenth of the actual global NPP of 50 Pg yr^{-1}. However, this is not so. Luyten (1995) and Penning de Vries *et al.* (1995) calculated the maximum world food production capacity, if all suitable land for agriculture were cropped, if all available water resources (after subtraction of industrial and urban water requirements) were used, and if multiple cropping was applied where possible. Their calculation was constrained only by availability of water and land suitability, not by present use or vegetation cover. According to their estimates, the land area for agriculture could be as large as 38×10^{12} m^2, of which 25×10^{12} m^2 could be irrigated given the amount of water available

Table 13-1 Global Land Use[a]

Area	Use ($\times 10^{12}$ m^2)
Arable land	13.5[b]
(of which irrigated)	2.44
Permanent crops	1.0
Permanent pastures	33.9
Forest and woodland	41.2
Other (rock, desert, glaciers)	40.8
Inland water bodies	3.5
Total area (excluding Antarctica)	**133.9**

[a]From FAO (1997).
[b]Including fodder crops, temporary meadows for mowing and pastures, and fallow.

for irrigation. Their method applied to our estimated area yields a potential annual NPP of 14.2 Pg C. Actual annual agricultural NPP does not even reach 4 Pg C, as will be shown below. Arable land being uncropped, either bare or fallow during part of the year, must be partly responsible for this low production level.

III. Primary Productivity of Agriculture

Estimates of agricultural NPP have been derived from the Internet-accessible FAOSTAT database (FAO, 1997). A description of agricultural productivity data is given in the FAO Production Yearbooks (FAO, 1996a), whereas for fertilizer production and consumption data reference is made to the FAO Fertilizer Yearbooks (FAO, 1996b). In the FAOSTAT database, 163 different crops for all countries are considered. For each crop–country combination, the harvested area, the agricultural production on this area (harvested product expressed in fresh weight), and yields per unit area are given. For some oil and fiber crops, only production volumes per country are available and for these cases we estimated the cultivated area. For sugar cane the annually harvested area is only two-thirds of the cultivated area, because the length of the growing period is approximately 18 months.

For 1991, we have calculated NPP per country and per crop through a series of crop-specific coefficients, derived from literature (Penning de Vries *et al.*, 1983; Sinclair and De Wit, 1975; van Duivenbooden, 1992; Purseglove, 1968, 1972). Agricultural production was converted to dry matter, using a crop-specific dry matter content. Dry matter production of nonharvested aboveground parts (such as straw, fruit tree trunks) was derived from the harvest index, whereas root production was calculated as a crop-specific fraction of the aboveground production. Crop-specific coefficients for major crops and crop groups are given in Table 13-2. Multiplication of the production figures by their carbon or nitrogen content yields NPP per crop, expressed in either C or N.

Global NPP for the major crop groups is given in Table 13-3. Total NPP of agricultural crops equals 3.5 Pg C yr^{-1}; small grains and coarse grains combined account for 60% of agricultural NPP, whereas oil crop and sugar crops each account for 9%. In Fig. 13-1, the area productivities of the different crop categories, as given in Table 13-3, are plotted versus their cumulative cropping area, ordered in a descending sequence. The area in the graph represents total production. Grains clearly are the most important group. Because their productivity per unit area is not far from the overall average, they are a good representative of all crops. Table 13-3 clearly shows that paddy rice has a higher productivity per unit land area than wheat. This is not surprising in view of the highly intensified growing conditions of paddy rice in South East Asia, where most of the production occurs. Within the group of small grains there is remarkably little variation in the C:N ratio.

From radiation one would expect a much higher production than 3.5 Pg C yr^{-1}, but is this also the case when based on water? As shown above, average evapotranspiration is about 470 mm yr^{-1}. This may be higher in hot climates and lower in cold climates, but this difference is compensated by the higher water requirement per gram of carbon produced in hot climates. A conservative estimate of this requirement may be derived from the ballpark number of 1000 kg of water needed per kilogram of dry matter produced. The total water transpired annually on arable land is 0.5 m times 14×10^{12} m^2, or about 7×10^{15} kg of water. At a transpiration coefficient of 1000, this means a possible NPP of 7×10^{15} g of dry matter, or 3.5 Pg of C. This figure is remarkably close to the actual quantity produced, which means that we are already very close to the maximum production that is possible in terms of local water availability, when water is used at its present efficiency. Although irrigation may increase production locally, worldwide it can only increase agricultural production when it directs water from nonarable land to agricultural land. The alternative is to increase water use efficiency. This is possible by better agricultural technology, which can more than double the water use efficiency. Available water may not be efficiently used when soil fertility limits crop production, and in those situations application of fertilizer may be the simplest way to improve water use efficiency. The data of Luyten (1995) and Penning de Vries *et al.* (1995) can be interpreted as having used a transpiration coefficient of about 200 [where transpiration coefficient is the ratio of water lost through evapotranspiration and dry matter produced (kg kg^{-1})]. At that point the biophysical limit is reached, dictated by diffusion of water vapor and carbon dioxide through the stomata.

Table 13-2 Crop-Specific Coefficients for Major Crops and Crop Groups Used to Estimate NPP and Nitrogen Uptake

Crop	Crop Group	NR[a]	Rel. (%)[b]	DM (%)[c]	HI[d]	Roots[e] (%)	Carbon (%)[f]			Nitrogen (%)[g]		
							Yield	Rest	Roots	Yield	Rest	Roots
Rice, paddy	Small grains	1	(15.7)	88	0.42	15	48	42	43	1.50	0.50	0.75
Other small grains	Small grains	1	(25.1)	85	0.42	15	47	48	47	1.70	0.50	0.75
Maize	Coarse grains	2	(14.7)	85	0.45	15	49	48	47	1.70	0.50	0.75
Sorghum, millet	Coarse grains	2	(4.1)	88	0.27	15	48	48	47	1.70	0.50	0.75
Potatoes	Roots & tubers	3	(1.4)	25	0.70	15	44	48	47	1.36	1.00	0.75
Cassava	Roots & tubers	3	(1.3)	38	0.70	15	44	48	47	0.50	1.00	0.75
Soybeans	Pulses	4	(3.8)	92	0.40	15	52	48	47	4.00	2.00	0.75
Other pulses	Pulses	4	(2.7)	90	0.30	15	47	48	47	2.80	1.70	0.75
Sugar cane	Sugar crops	5	(7.6)	27	0.60	15	48	48	47	1.08	1.00	0.75
Sugar beets	Sugar crops	5	(1.4)	21	0.66	15	44	48	47	0.75	1.00	0.75
Groundnuts	Oil crops	6	(1.5)	95	0.25	15	60	50	47	1.50	0.80	0.75
Coconuts	Oil crops	6	(1.3)	50	0.30	30	60	50	47	1.50	0.80	0.75
Fruit (average)	Fruit	7	(4.0)	19	0.30	25	45	50	47	0.80	0.70	0.75
Cotton	Fibers	8	(2.6)	85	0.33	15	54	50	47	1.00	0.80	0.75
Vegetables (avg.)	Vegetables	9	(2.1)	13	0.45	15	46	47	47	0.80	0.70	0.75

[a]NR, Not reported.
[b]Relative contribution to total carbon production.
[c]Dry matter content.
[d]Harvest index (yield/total aboveground dry matter production).
[e]Root production (as percentage of total aboveground production).
[f]Carbon content harvested products, nonharvested products, and roots.
[g]Nitrogen content of harvested products, nonharvested products, and roots.

Table 13-3 Cultivated Area and Carbon and Nitrogen Productivity for the World's Major Crop Groups for the Year 1991

Crop	Category[a]	Area (10^{12} m^{-2})	NPP (Tg C yr^{-1})	N (Tg N yr^{-1})	NPP/area (g C m^{-2} yr^{-1})	N/area (g N m^{-2} yr^{-1})	C/N (g C/g N)
All crops							
Small grains	1	4.89	1437	29.2	294	6.0	49.2
Coarse grains	2	2.13	660	13.2	309	6.2	50.1
Oil crops + nuts	6	1.00	327	6.1	329	6.1	53.8
Sugar crops	5	0.40	317	6.5	801	16.5	48.5
Pulses	4	1.27	229	10.6	181	8.4	21.6
Fruits	7	0.33	139	2.1	421	6.4	65.7
Roots and tubers	3	0.48	134	2.7	283	5.7	49.7
Fibers	8	0.39	109	1.8	278	4.7	59.8
Vegetables	9	0.34	74	1.2	218	3.4	63.5
Tea, coffee, or tobacco	11	0.19	44	0.7	238	3.8	63.1
Spices	10	0.03	14	0.2	492	7.6	64.5
Others	—	0.07	19	0.3	283	4.4	64.2
Total or (weighted) mean		11.50	3505	74.6	344	6.6	49.6
Subdivision small grains							
Wheat	1	2.23	605	12.3	271	5.5	49.2
Paddy rice	1	1.47	553	11.2	376	7.6	49.4
Remainder	1	1.19	279	5.7	234	4.8	48.9
Total/(weighted) mean		4.89	1437	29.2	294	6.0	49.2

[a]See categories in Fig. 13-1.

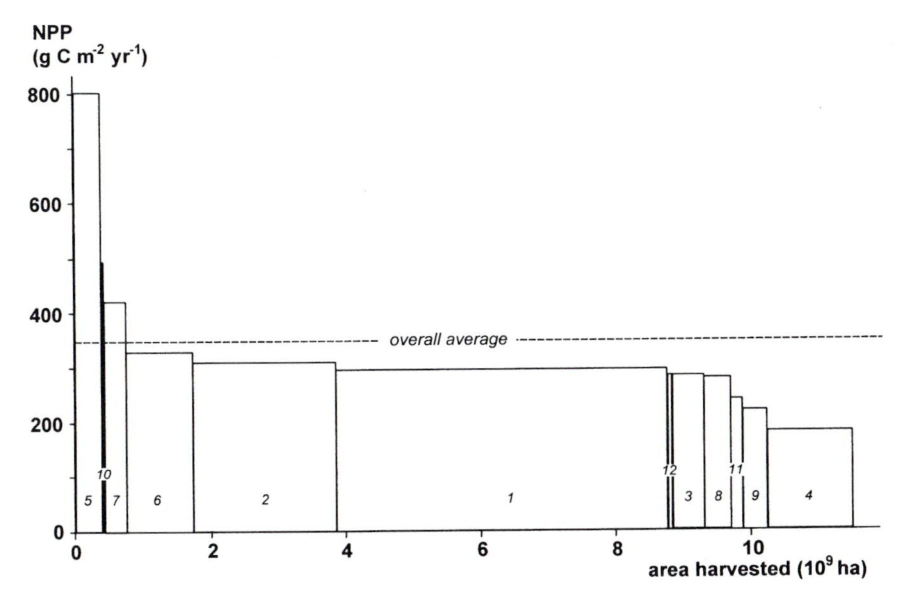

Figure 13-1 Productivities per unit area (height of each rectangle) and cropped areas (width of each rectangle) of the different crop categories, ordered in a descending sequence of productivity (areas in the graph show production per category): (1) small grains, (2) coarse grains, (3) roots and tubers, (4) pulses, (5) sugar crops, (6) oil crops + nuts, (7) fruits, (8) fibers, (9) vegetables, (10) spices, (11) tea, coffee, and tobacco, and (12) rubber.

Such economical use of water can only be achieved under optimal fertilization conditions. Direct evaporation from the soil surface or (worse) from the paddy water in rice fields is ignored. We therefore feel that a transpiration coefficient of 400 is more realistic. Increasing atmospheric CO_2 will definitely improve the water use efficiency (WUE) in practically all conditions, and will continue to act as an improving background.

IV. Time Trends in Productivity, Nitrogen Uptake, and Fertilizer Use

Based on FAO agricultural production statistics (1997) for the period 1960–1991, and applying the conversion coefficients from Table 13-2, time trends in agricultural NPP, nitrogen uptake, and fertilizer use have been derived. Time trends are given as regressions in Table 13-4. The time trend of total NPP on arable land shows an increase at a rate of about 63.1 Tg C yr^{-1} on average (Fig. 13-2). It is a bit worrying that the curve is linear, and not ex-

Table 13-4 Regression Equations over the Period 1960–1991

Equation	Units	Equation number
$C_{total} = 2988 + 63.9\,(t - 1980)$	$Tg\ C\ yr^{-1}$	(a)
$N_{total} = 63 + 1.39\,(t - 1980)$	$Tg\ N\ yr^{-1}$	(b)
$N_{fertilizer} = 60 + 2.61\,(t - 1980)$	$Tg\ N\ yr^{-1}$	(c)
$C_{pulses} = 181 + 4.60\,(t - 1980)$	$Tg\ C\ yr^{-1}$	(d)
$N_{pulses} = 8.3 + 0.228\,(t - 1980)$	$Tg\ N\ yr^{-1}$	(e)
$C_{total} = 1536 + 24.2N_{fertilizer}$	$Tg\ C\ yr^{-1}$	(f)
$N_{total} = 31 + 0.53N_{fertilizer}$	$Tg\ N\ yr^{-1}$	(g)

ponential. Because the total production more than doubled in this period, the relative rate of increase has fallen from 3.5% yr^{-1} in 1960 to about 1.7% yr^{-1} at present (population growth has fallen from 2% yr^{-1} to 1.6% yr^{-1}).

Figure 13-3 shows the time trends in both total nitrogen uptake by agricultural crops and industrially produced nitrogen fertilizer. Both nitrogen uptake and fertilizer production increased linearly, and fertilizer production exceeded total nitrogen uptake from 1981 onward. The decline in fertilizer production from 1987 onward is mainly the result of the changed political situation in the former Soviet Union. In 1991 the total amount of nitrogen incorporated in agriculturally grown dry matter was 75 Tg (Table 13-3), which is only slightly less than the amount of nitrogen fertilizer produced in factories in that year (80 Tg) (FAO, 1996b). The pulses can be considered to fix their own nitrogen, about 11 Tg, so that the remainder, 64 Tg, must have come from external sources. Natural input from precipitation is

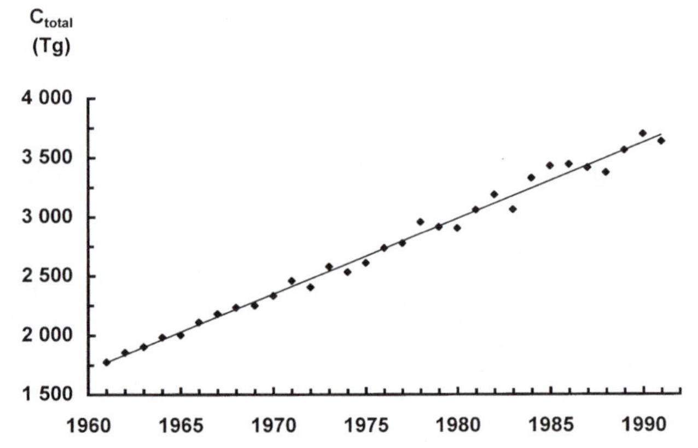

Figure 13-2 Time trend in agricultural net primary production during 1960–1991.

estimated at 1 g N m^{-2} (Schlesinger, 1991) or another 12 Tg for the total arable land area. The sum of leguminous nitrogen fixation and atmospheric input is thus equal to about 23 Tg of nitrogen. This figure is remarkably close to the intercept value of the graph of total nitrogen in the biomass from arable land versus global nitrogen fertilizer production [regression Eq. (g) in Table 13-4; see Fig. 13-4]. The overall conclusion is that the use of industrially produced nitrogen fertilizer has been a major factor in the increase in agricultural net primary production from a "natural" level of 1.5 Pg of C to a level of 3.5 Pg C.

For the stimulatory effect of increasing ambient CO_2 we will use a high value: 0.5% increase per 1% of increased CO_2, which can be translated into an increase of 30% in dry matter production (or NPP) on doubling of the CO_2 concentration. Compared to the rate of increase in net primary productivity of agriculture mentioned above, this effect is rather small on the time scale considered here. Even when using this high estimate for the stimulatory effect of CO_2, it is responsible for a scant 10% of the total increment over this period. It is possible, though, that the effect of increasing CO_2 through improved WUE is underestimated, and that the real CO_2 effect is larger.

V. Who Consumes the Production?

Out of the 3.5 Pg C yr^{-1} produced, about 1.2 Pg is harvested. This portion is taken from the field, transported, stored, and further processed by milling and cooking. The human requirement for food can be derived from the human caloric need, which is about 11 MJ per day per person. This is equivalent to a carbon need of about 320 g C. At a world population of 5.3×10^9 persons (situation in 1991), this means a human carbon consumption of 0.6 Pg C yr^{-1}.

On a world-wide scale, the proportion of grain production fed to livestock has been remarkably stable over the past decades, at a level of about 38% (USDA–ERS, 1993; Myles Mielke, FAO, personal communication). For oil seeds the percentage is higher (about 65%), and for root crops/tuber crops and pulses it is lower (25 and 25–30%, respectively). A rough estimate thus is that livestock consumes 0.5 Pg of harvested carbon; if the same fraction of the harvested product consumed by animals is assumed for the nonharvestable part (straw), the consumption of total C by livestock is 1.5 Pg of C yr^{-1}.

Hence, 1.1 Pg of C out of the 1.2 Pg yr^{-1} harvested is clear: human and animal consumption. Barley, with a production of 0.07 Pg of C yr^{-1} as grain, is almost totally used for beer production, whereas grapes, which form a major portion of the 0.03 Pg of C of the fruits, are almost entirely used for wine.

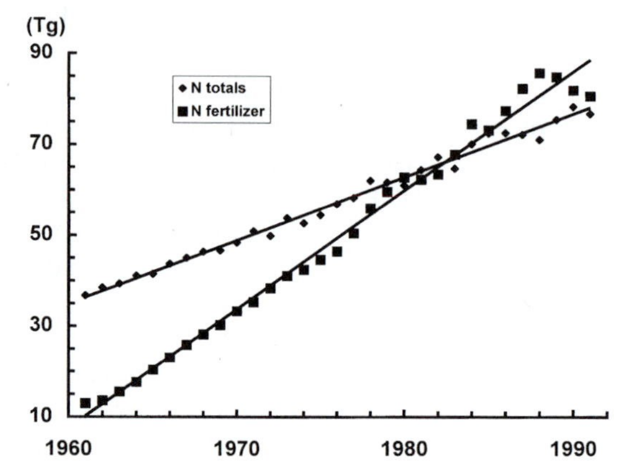

Figure 13-3 Time trend in total crop nitrogen uptake and industrial nitrogen fertilizer production.

Together with other forms of alcohol, these categories may account for the missing 0.1 Pg C per year.

VI. Discussion

In this study, average productivity of agricultural land was found to be 320 g C m^{-2} yr^{-1}, which is low when compared to previous estimates, based on

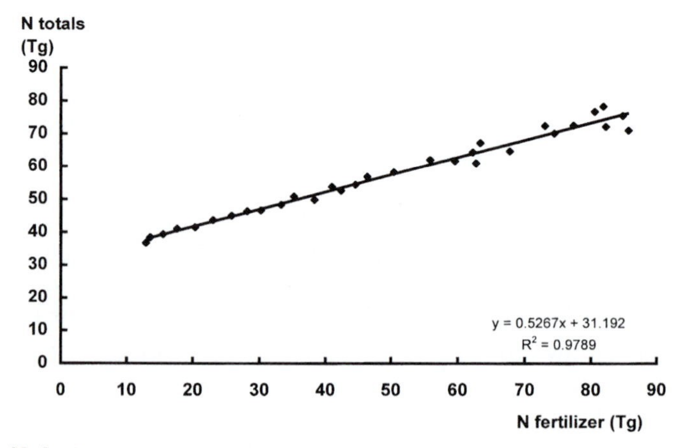

Figure 13-4 The amount of nitrogen in agricultural crops as a function of the amount of industrially produced fertilizer N during 1960–1991.

experimental data from fields, and also low when compared to data for natural ecosystems in similar agroclimatic zones (Schlesinger, 1991; Lieth, 1975). An inherent difficulty in using field data is to find the most representative estimates. Estimates in literature have shown a tendency to become lower over the past few decades, with recognition that the older estimates had been subject to the almost inevitable bias of using too high numbers. The temptation to discard actual observations for alleged problems of disturbance, disease, drought, or whatever has caused a "selective" drift toward higher estimates, whereas in reality these problems are part of normal life.

According to Table 13-1, the area of arable land is 13.5×10^{12} m^2, whereas Table 13-3 shows only 11.5×10^{12} m^2 of cultivated land. This last figure was found as the sum of the harvested area of all crops, so that the fallow land is not included. Because in some areas multiple cropping is practiced, the land area on which global NPP is realized is even smaller. Total arable land as defined by FAO also includes fodder crops, leys (temporary grassland) for mowing and pasture, and fallow. Areas and yields were not available for these categories, because the harvested products do not leave the farm, being consumed by the farmer's own livestock. For the European Community (EC), available statistics allowed estimates of the NPP of fodder crops for the EC (12 states) (Eurostat, 1970–1995). Whereas agricultural NPP (excluding fodder crops) was 320 Tg of C, fodder crops produced 27.5 Tg of C yr^{-1}, less than 9%. Production of fodder crops is related to the existence of intensive animal husbandry systems, thus the share of fodder production in total NPP will be smaller in the rest of the world. The area allocated to fodder crops in China is 2% of the total land area (Simpson *et al.*, 1994), and in the United States it is about 4%. On a global basis we estimate the contribution of fodder crops to agricultural NPP at about 4%.

For some crops in some countries, harvested areas are not reported, and only national production volumes are given. Because the production for these crops is only 1.7% of the total global harvested dry matter production, the total harvested area may be slightly higher than indicated in Table 13-3.

Our estimates are based on FAO statistics, which are not subject to the above biases because they simply account for the totals (though one can have doubts about the absolute accuracy of these statistics). Of course, there are other problems with these statistics, because they may not consider the nonmarketed component of the production. However, in the developed countries of the world, in which the largest share of agricultural production is realized, the nonmarketed component is small or nonexistent. It seems a reasonable guess to assume that actual productivity is bounded by the FAO statistics on the lower end and the field estimates on the upper. The available statistical data seem to be sufficiently accurate to assess ecosystem productivity of agriculture with a much higher accuracy than that of other ecosystems, for which direct production measures are either lacking or highly inaccurate due to the very high spatial variability of these systems. Keeping this in mind, further research or more detailed statistical data will have only a small effect on the accuracy of our calculations.

The study presented here is largely based on production statistics. A complementary agro-ecological study, which relates agricultural productivity to the availability and use of natural resources, would be most welcome. Such a study would require a geographically explicit approach, in which each production situation is characterized by the prevailing ecosystem, its climatic conditions, its soil type and soil fertility, and the availability of water resources for irrigation. Current geographic information system (GIS) techniques and high-resolution climate and soil databases with global coverage (Leemans and Cramer, 1991; FAO, 1996c) allow for geographically explicit estimations of the availability and quality of natural resources. However, the translation of resource availability into (potential) productivity figures requires that the actual distribution of the different ecosystems of the world is known in detail. Present developments in land use and land cover databases with global coverage (IGBP, 1996) will soon make it possible to indicate the location of agricultural land, thus enabling the study of different scenarios for agricultural development (amount of rain forest to be cut, semiarid regions to be irrigated, increased water use efficiency, etc.) and their consequences for the global carbon cycle.

VII. Summary

The area of arable land in the world is 13.5×10^{12} m^2, being one-tenth of the total global land area (excluding Antarctica). Net primary productivity of arable land was derived from FAO agricultural production statistics for the year 1991, through application of crop-specific coefficients (dry matter content, harvest index, root production, carbon and nitrogen content). Total NPP of agricultural crops is estimated at 3.5 Pg C yr^{-1}, i.e., 7% of total NPP of all terrestrial ecosystems. Small and coarse grains together account for 60% of agricultural NPP, and oil crops and sugar crops each account for 9%. From the 3.5 Pg C, 35% (1.2 Pg) is available as harvested product, of which 0.6 Pg is used for human consumption and 0.5 Pg for animal consumption; the remainder is used, among others, for alcohol production. The time trend of total NPP on arable land for the period 1960–1995 shows a linear increase at a rate of about 63 Tg C yr^{-1}. The relative rate of increase in global agricultural productivity has fallen from 3.5% yr^{-1} in 1960 to about 1.7% yr^{-1} at present. It is shown that application of nitrogen fertilizer alone accounts for 60% of the agricultural NPP.

References

Bélanger, G., Gastal, F. and Lemaire, G. (1992). Growth analysis of a tall fescue sward fertilized with different rates of nitrogen. *Crop Sci.* **32,** 1371–1376.

Bolin, B., Degens, E.T., Kempe, S., and Ketner, P. (1979). "The Global Carbon Cycle. SCOPE 13." Wiley and Sons, Chichester.

Eurostat (1970–1995). "Agricultural Statistics of the European Community." EC Information Office, Luxembourg.

FAO (1996a). "Production Yearbook 1995." Vol. 49. FAO, Rome.

FAO (1996b). "Fertilizer Yearbook 1995." Vol. 45. FAO, Rome.

FAO (1996c). "Digital Soil Map of the World and Derived Soil Properties" (data on CD-ROM). FAO-UN, Rome.

FAO (1997). "FAOSTAT Agricultural Database." Internet site: http://apps/fao.org/, March 1997.

IGBP (1996). "Data and Information." Global Change Newsletter No. 27, September. IGBP.

Leemans, R., and Cramer, W. (1991). "The IIASA Database for Mean Monthly Values of Temperature, Precipitation and Cloudiness on a Global Terrestrial Grid." Report RR-91–18. IIASA, Luxembourg.

Lieth, H. (1975). Modeling the primary productivity of the world. *In* "Primary Productivity of the Biosphere" (H. Lieth and R. H. Whittaker, eds.), pp. 237–263. Springer-Verlag, New York.

Luyten, J. C. (1995). "Sustainable world food production and environment." Report 37. AB-DLO, Wageningen.

Minami, K., Goudriaan, J., Lantinga, E. A., and Kimura, T. (1993). Significance of grasslands in emission and adsorption of greenhouse gases. *In* "Proceedings of the XVII International Grassland Congress, 13–16 February 1993, Sessions 27–51," pp. 1231–1237. Palmerston, Hamilton and Lincoln, New Zealand.

Penning de Vries, F. W. T., van Laar, H. H., and Chardon, M. C. M. (1983). Bioenergetics of growth of seeds, fruits and storage organs. *In* "Proc. Symp. on Potential Productivity of Field Crops under Different Environments, 23–26 September 1980," pp. 37–59. International Rice Research Institute, Manila.

Penning de Vries, F. W. T., van Keulen, H., and Rabbinge, R. (1995). Natural resources and limits of food production in 2040. *In* "Eco-regional Approaches for Sustainable Land Use and Food Production" (J. Bouma, A. Kuyvenhoven, B. A. M. Bouman, J. C. Luyten and H. G. Zandstra, eds.), pp. 65–87. Kluwer Academic Press, Dordrecht.

Purseglove, J. W. (1968). "Tropical Crops, Dicotyledons." Longman Ltd., London.

Purseglove, J. W. (1972). "Tropical Crops, Monocotyledons." Longman Ltd., London.

Schlesinger, W. H. (1991). "Biogeochemistry, an Analysis of Global Change." Academic Press, San Diego.

Simpson, J. R., Xu Cheng, X., and Miyazaki, A. (1994). "China's Livestock and Related Agricultural: Projections to 2025." CAB International, Wallingford.

Sinclair, T. R., and de Wit, C. T. (1975). Photosynthesis and nitrogen requirements for seed production by various crops. *Science* **189,** 565–567.

USDA–ERS (1993). "Production, Supply and Demand View." Digital Database, Washington D.C.

van Duivenbooden, N. (1992). "Sustainability in Terms of Nutrient Elements with Special Reference to West-Africa." Report 160. CABO-DLO, Wageningen.

14

Hierarchy and Productivity of Mediterranean-Type Ecosystems

Serge Rambal

I. Nature and Extent of Mediterranean-Type Ecosystems

The five relatively small, isolated regions of the world with Mediterranean-type climates, i.e., parts of the Mediterranean Basin, California, southwestern and southern Australia, central Chile, and southern Africa, are approximately located between latitudes ranging from 30° to 43° north and south. The Mediterranean-type climates can be generally considered as a transition between dry tropical and temperate climates. It is generally accepted that they are characterized by a distinctive annual climatic sequence in which a hot, dry summer alternates with a cool to cold, humid period lasting 5–10 months from fall through winter to spring. There are, however, different opinions about the ranges of annual amounts of rainfall and the extent of summer drought characterizing these climates (Daget, 1977).

Hobbs *et al.* (1995) used areas of 0.28, 0.61, 0.13, and 0.06 Mkm^2 to represent California, Australia, Chile, and South Africa, and Le Houérou (1980) estimated the area of the Mediterranean Basin to be 1.68 Mkm^2, excluding semideserts, deserts, or even extreme deserts. These values yield a total area of about 2.75 Mkm^2.

In these five regions, the vegetation is characterized by the dominance of trees and woody shrubs with small, evergreen, sclerophyllous leaves. These vegetative forms are usually called "garrigue" or "maquis" in France, depending on the nature of the soil substrate, calcareous or siliceous; the terminology is "chaparral" in California, "heath" and "mallee" in Australia, "matorral" in Chile, and "fynbos" in South Africa. This list is not exhaustive and other colloquial names may be used elsewhere to designate short to tall shrublands or woodlands. The plant forms are fairly similar in their physiognomy and have provided the ideal testing ground for the theory of ecological convergence from evolutionary, morphological, and physiological points of view (Cody and Mooney, 1978).

The most apparent differences in Mediterranean-type regions are in soil conditions. Three major categories of soils have been distinguished by Specht and Moll (1983), corresponding roughly to three nutritional plant states: (1) moderately leached soils, with an adequate balance of essential plant nutrients, (2) strongly leached soils or soils derived from nutrient-poor geological substrates, containing low levels of available nutrients, and (3) calcium-rich/high-pH soils, which may induce various nutrient deficiencies. These three categories may be recognized in southern Australia and in South Africa. Many soils of these two regions have been developed from nutrient-poor, very old parent material or from Quaternary infertile siliceous sands. These acidic soils have been heavily weathered and highly leached. The three soil categories also occur around the Mediterranean Sea. In this last case, large areas of calcium-rich/high-pH soils have probably resulted from direct or indirect human impacts on the original forests over the past 3000 to 4000 yr. In California, moderately and strongly leached soils may be found, but only weakly leached soils are found in central Chile.

II. Productivity in Mediterranean-Type Ecosystems

A. Sources of Data

Numerous studies published in the 1970s clearly documented the extreme range of aboveground biomass and productivity that may be present in shrublands and woodlands growing in a Mediterranean-type climate. A shortened summary of these data is presented Table 14-1. Data from before 1975 are largely derived from the pioneering work of Specht and co-workers on Australian mallees and heathlands (Groves and Specht, 1965; Specht, 1966, 1969a,b, 1981; Jones *et al.*, 1969). More than half of the work listed in Table 14-1 was published from 1975 to 1985 under the auspices of the International Biological Programme (IBP). Later work occurred less frequently; on average, about one contribution was made every year.

The nature, content, and accuracy of data from these studies are very varied. Two main categories can be distinguished. First, some data originated

Table 14-1 Bibliographic Review of Biomass and Productivity within Some Mediterranean-Type Ecosystems

Case Study	Formation[a]	Dominant species	Age (yr)	Biomass (g DM m^{-2}) Above-ground	Below-ground	Leaf mass	Leaf area	Plant cover	Litterfall (g DM m^{-2} yr^{-1})	ΔBiomass (g DM m^{-2} yr^{-1})	Leaf N (%)	ANPP (g DM m^{-2} yr^{-1})	NPP (g DM m^{-2} yr^{-1})	Ref.
1	LS, phrygana	*Genista acanthoclada, Thymus capitatus*	40	790	300		0.49		155		0.86[b]			Tsiourlis (1990)
2	LS, phrygana	*Phlomis fruticosa, Euphorbia acanthothamnos*	?	1111	1620[c]	209	1.7		210	202	1.32	412		Margaris (1976)
3	OW, maquis	*Juniperus phoenicea*	60–100	4020	3030	680		0.60	130	240	1.14	370	770	Tsiourlis (1990)
4	LS	*Chamaebatia foliosa*	?	982	491	292					1.77	198	297	Rundel *et al.* (1981)
5	LS, matorral	*Halimium halimifolium*	80	489		69				64				Merino and Martin Vicente (1981)
6	LS, heath	*Phylica cephalantha*	11	1731 ± 336	1397						0.52[b]			Low (1983)
7	LS, heath		5.2	938 ± 96	6630 ± 907[d]		2.5		122	184		306	(6003)	
8	HS, chamise chaparral	*Adenostoma fasciculatum*	2	593		232	1.8				1.90			Rundel and Parsons (1979, 1980)
			6	793		200	1.5				1.35			
			16	1486		286	2.2				1.05			
			37	1400		159	1.2				1.18			
			60+	1370		156	1.2				0.96			
9	OW, mallee	*Eucalyptus* spp., *Acacia aneura*	13	3900		300	0.85					540		Burrow (1976), in Schulze (1982)
			55	7000		700	1.96					330		
			100	6700		240	0.67					240		
10	DW	*Pinus laricio*	15	5280	280	560			264	736	1.41	1000		Ranger (1981)
11	DW	*Pinus pinea*	35	15680	2200	1270			830		0.88	1750	1860	Rapp and Cabanettes (1981)
12	DW	*Quercus suber*	30	4220[e]		230	(2.4)		383	110	1.21	493[e]		Leonardi and Rapp (1993)

(*continues*)

Table 14-1 *(Continued)*

Case Study	Formation[a]	Dominant species	Age (yr)	Biomass (g DM m^{-2}) Above-ground	Biomass (g DM m^{-2}) Below-ground	Leaf mass	Leaf area	Plant cover	Litterfall (g DM m^{-2} yr^{-1})	ΔBiomass (g DM m^{-2} yr^{-1})	Leaf N (%)	ANPP (g DM m^{-2} yr^{-1})	NPP (g DM m^{-2} yr^{-1})	Ref.
13	DW	*Quercus ilex*	1	151		48	0.28				1.50	165		Leonardi and Rapp (1990)
			2	913		278	1.95				1.48	748		
			3	2162		575	4.40				1.50	1248		
			31	15010		800	4.30		313	458	1.01	771		
14	DW	*Pinus laricio*	25	9410		1260	8.1		285	330	0.80	615		Leonardi *et al.* (1988)
15	DW	*Pinus halepensis*	25	4920		210	2.2		120	190	1.12	310		Leonardi and Rapp (1993)
16	DW	*Pinus halepensis*	70	15660		560			393	210	1.23	603		Rapp (1974)
17	LS, coastal sage scrub	*Salvia leucophylla, Artemisia californica*	22	1440		187			199		1.56	355		Gray (1982, 1983)
18	HS, chaparral	*Ceanothus megacarpus*	22	7624		457			801		1.48	1056		
19	HS, chaparral	*Ceanothus megacarpus*	5	2055 ± 357		228 ± 43	0.8				1.65			Schlesinger and Gill (1980)
			12	3708		336 ± 44	1.1		585	265	1.60	850		
			21	6337		465 ± 78	1.6				1.76	850		
20	HS, chamise chaparral	*Adenostoma fasciculatum*	23	2127	1087	438			83		0.53	362	490	Mooney and Rundel (1979)
21	HS	*Banksia ornata, Leptospermum myrsinoides*	13	2650	7600	802	4		395					Maggs and Pearson (1977a,b)
			28	5850	8100	706	4		598					
			85	2078										

22	HS, chaparral	*Adenostoma fasciculatum, Ceanothus greggii*	22	2308	1140		2.6	353			702	1448	Kummerow *et al.* (1981)
23	HS, matorral		15	738			2.0				451	1981	
24	HS, garrigue	*Quercus coccifera*	17	2350	4600	400		228	140	1.35	368		Rapp and Lossaint (1981)
			30	3150		560		261	110		371		
25	DW	*Quercus ilex*	150	26900	5000	654	4.4	384		1.33	650		Lossaint and Rapp (1971)
26	DW	*Quercus ilex*	41	6470 ± 920		550	2.9	428 ± 30	185	1.23	613		Romain (unpublished data)
27	HS, garrigue	*Quercus coccifera*	0.5	253		138	1.0						S. Rambal (unpublished data)
			0.9	713		418	3.0						
			29	1862		425	2.3			1.10			
28	DW	*Quercus ilex, Arbutus unedo*	45	11230		666		228			650		Lledo *et al.* (1992)
29	DW	*Quercus ilex*	0.6	220		110				1.71	220		Rapp *et al.* (1992)
			1.7	580		280				1.50	520		
			40			430		303		1.27	440		

[a] LS, Low (<1 m) shrublands; HS, high (>1 m) shrublands; OW, DW, open and dense woodlands; Δbiomass is the current annual growth increment or an estimate (see text). Data in brackets warn again incertain values. Litter fall, Δbiomass, ANPP and NPP are expressed in g DM m^{-2} yr^{-1}.

[b] Aboveground average.

[c] Estimated in the 0- to 30-cm soil layer.

[d] Estimated in the 0- to 180-cm soil layer.

[e] Neglected understorey species.

from studies on regrowth of the aboveground biomass (B_{ag}) following a major disturbance such as fire or clear-cutting (Kruger, 1977; Rutherford, 1978; Van Wilgen, 1982; Mitchell *et al.*, 1986; Black, 1987). Then there are studies that directly estimated net productivity of vegetation of a given age. In the first case the approach of studying plots with different aged disturbance history usually ignored or minimized spatial variations in productivity resulting from slope orientation or related to altitudinal gradients (see, for example, Schlesinger and Gill, 1978). Problems of data interpretation also occur when the time courses being observed are superimposed on a successional gradient. The first few years of these dynamics are controlled by autosuccessional mechanisms (Hanes, 1971), but as the succession progresses one species becomes dominant and deeply changes the system's productivity. This is the case in *Casuarina pusilla* heathlands, where *Banksia ornata* finally becomes dominant (Groves and Specht, 1965).

In the case of the second type of data, annual net primary productivity (ANPP) is usually based on both estimates of current annual growth increment and litterfall and is very dependent on the weather conditions occurring during the study years. There are many sources of uncertainty, including insufficiently described methodology and failing to take into account spatial variation by utilizing too small samples with no or few replicates. The most frequent error is the overestimation of the current annual growth increment. This is estimated from the ratio of aboveground biomass and stand age.

In both types of study, the belowground biomass is rarely estimated—only in one-third of studies are estimates used. Belowground productivity data are even rarer (Mooney and Gulmon, 1979; Mooney and Rundel, 1979; Rundel *et al.*, 1981; Tsiourlis, 1990) and are subject to great uncertainties (Jones, 1968a,b; Kummerow *et al.*, 1981). Predation is not taken into account, even though it has been suggested that it may be important (Jones, 1968a,b).

B. An Overview of Results

If we analyze the time course of B_{ag} following a disturbance, the separation between low (less than 1 m) compared to high (higher than 1 m) shrublands used in Table 14-1 no longer seems to be apparent. Accordingly, only the distinction between shrublands and woodlands has been retained. This distinction leads to adjustments of Fig. 14-1, which provides an estimate of B_{ag} [in grams of dry matter (DM) per square meter] in relation to time (t) in years:

$$\text{Shrublands:}\quad B_{ag} = 2880(1 - e^{-0.0896t}) \qquad (r^2 = 0.346,\ n = 85,\ P < 0.001),$$

$$\text{Woodlands:}\quad B_{ag} = 26300(1 - e^{-0.00916t}) \qquad (r^2 = 0.565,\ n = 23,\ P < 0.01).$$

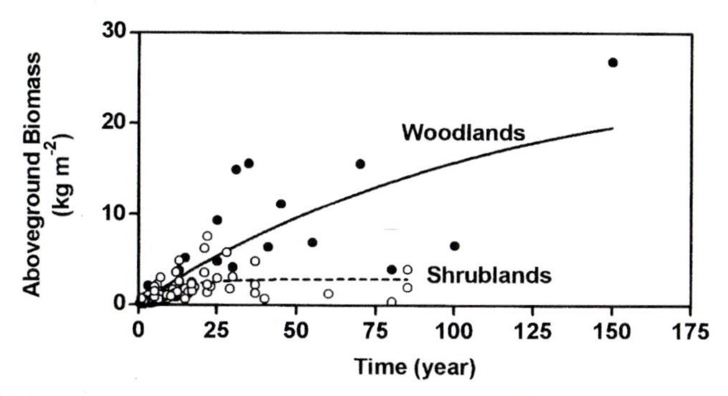

Figure 14-1 Time course of aboveground biomass for Mediterranean-type shrublands and woodlands.

The corresponding half-times are 7.75 and 75.7 yr, respectively. Using the latter equation, the current annual growth increment for a 30-yr-old woodland is 183 g DM m^{-2}.

The factors controlling differences in productivity between both ecosystem are not clearly evident from this review. For example, there is no significant difference in leaf nitrogen content or in amounts of litterfall. On the other hand, there are slight but not significant differences in the leaf ratio (Fig. 14-2), i.e., the ratio between the leaf mass (the mass of carbon-assimilating organs) and the aboveground phytomass (an indicator of the maintenance costs of the ecosystem's carbon balance). The leaf ratio is higher in early stages with shrublands. It then decreases very rapidly and reaches a plateau as B_{ag} increases. A similar pattern is observed with woodlands. The high leaf ratio at low biomass of shrublands undoubtedly explains the more important role of resprouting species compared to the obligate seeders present in some woodlands (e.g., *Pinus* spp.). For modeling purposes we adopt a unique equation:

$$\text{Leaf ratio} = 0.070 + 0.365e^{-0.000589B_{ag}} \qquad (r^2 = 0.590,\ n = 36,\ P < 0.01).$$

The extreme values recorded around the Mediterranean Sea will be used as an example to show the range of variation in B_{ag} that can occur within the isoclimatic region. A 150-yr-old *Quercus ilex* evergreen oak woodland contained nearly 27 kg DM m^{-2} (Lossaint and Rapp, 1971), whereas a *Halimium halimifolium* matorral in Southern Spain (Merino and Martin Vicente, 1981) contained only 489 g DM m^{-2} and a mature phrygana community in Greece contained 790 g DM m^{-2} (Tsiourlis, 1990). The current annual growth increment plus litterfall has been studied in most Mediterranean ecosystems, rather than ANPP. Specht and others reported growth rates in Australia, California, and southern France ranging between 50 and 450 g

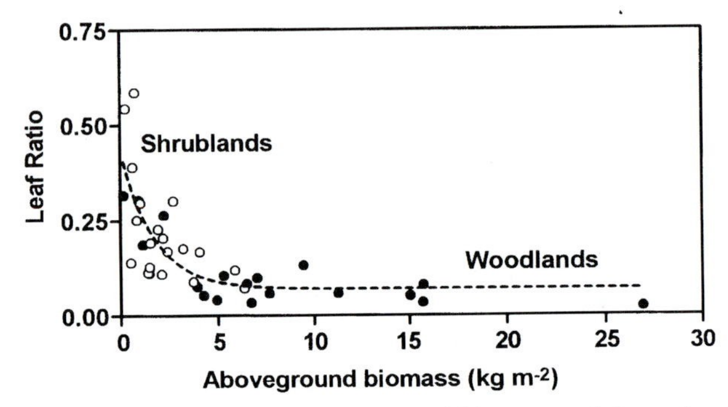

Figure 14-2 Relationships between aboveground biomass and the ratio of leaf mass to aboveground biomass or leaf ratio for both Mediterranean-type shrublands and woodlands.

DM m^{-2} yr^{-1} (Mooney, 1981; Specht, 1969a). Analysis of Table 14-1 hardly extends this range of variation. The lowest value of 40 g DM m^{-2} yr^{-1} was recorded for a *Eucalyptus incrassata* and *Melaleuca uncinata* mallee-broom-bush (Specht, 1966, 1969a, 1981) and the highest was 458 g DM m^{-2} yr^{-1} for a *Q. ilex* woodland (Leonardi and Rapp, 1990). It would thus appear that the contribution of litterfall to the ANPP is the determining factor (Gray, 1982, 1983; Rapp and Cabanettes, 1981). More than 50% of ANPP may be explained by litterfall. As in other plant communities the highest rates of ANPP were recorded in the first four or five postperturbation years, after which they fell rapidly as the community aged (see Leonardi and Rapp, 1990). The production rates in Mediterranean-type ecosystems appear to be only marginally greater than those of adjacent desert-type ecosystems (Hadley and Szarek, 1981).

C. Early Modeling Efforts

Special mention must be made of the pioneering efforts by F. Eckardt and P. C. Miller. They anticipated numerous simulation exercises conducted on productivity from 1985 onward. Eckardt *et al.* (1975) scaled up leaf performances within a 150-yr-old *Q. ilex* woodland canopy and simulated seasonal net assimilation and respiration with both a model of the extinction of photosynthetically active radiation (PAR) and Q_{10}-based models of dark respiration from leaves and stems. The net annual assimilation was 3220 g CO_2 m^{-2}. The maintenance costs for the aboveground parts were estimated to account for more than 51% of this amount. This meant that 1570 g CO_2 m^{-2} or 951 g DM m^{-2} was available for total growth and maintenance of the belowground compartments, a quantity that would seem to be somewhat insufficient when compared to the ANPP of 650 g DM m^{-2} in Table 14-1 (case study 25), the high root:shoot ratio of *Q. ilex* (Canadell and Rodá, 1991), and the fine root production values published by Kummerow and co-workers for Mediterranean-type ecosystems (Kummerow *et al.*, 1981, 1990). They measured values ranging from 700 to 1500 g DM m^{-2} for chaparral, *Quercus coccifera* garrigue, and a Chilean matorral. The belowground compartment was already one of the great unknowns of the ecosystem budget. Miller (1981a), using a more detailed model, Mediterranean Ecosystem Simulator (MEDECS), simulated southern California mixed and chamise chaparral and a Chilean matorral (see case studies 22 and 23, Table 14-1). The annual amounts of assimilated carbon were 1646 and 1107 g DM m^{-2} for mixed and chamise chaparral, respectively. For the Chilean matorral there was little difference in assimilation between polar- and equatorial-facing slopes, 1521 and 1564 g DM m^{-2}, respectively. The total estimated maintenance costs were 54, 52, and 44% and the corresponding values for aboveground growth were 370, 366, 334, and 337 g DM m^{-2}. These values are much greater than measured rates of net aboveground accumulation of 132, 78, and 62, undoubtedly because of the poor simulation of the patterns of allocation of photosynthates to the different plant compartments, and in particular, as in the Eckardt model, to the belowground compartment.

III. Scaling Up Leaf Productivity

Research conducted mainly before the early 1980s also used ecophysiological approaches for identifying the factors controlling productivity at leaf and plant levels (see Miller, 1981a,b). Work carried out since then has continued to look for models to derive canopy performance from that of the leaf. Among the models, the one proposed by Sellers *et al.* (1992) provides a good conceptual framework and has the merit of breaking this procedure down into stages that will be used as a plan in this section. Previous results on canopy optimization have been used by these authors to define the profile of leaf physiological properties within the canopy and yielded a simpler relationship between canopy photosynthesis and remotely sensed spectral vegetation indices that may scale up from top-of-the-canopy (TOC) leaf performance to whole-canopy performance and give good estimates of the area integrals of photosynthesis, even for spatially heterogeneous vegetation cover. Hence,

$$\text{Canopy assimilation} = \text{TOC leaf assimilation} \times \Pi \text{ factor} \times \text{environmental forcing}.$$

Sellers *et al.* (1992) assumed that the leaf performances within the canopy followed the radiation-weighted time-mean PAR and their extinction could be described by a negative exponential of $\bar{k}$ parameter. The Π factor (or canopy PAR use parameter) is the ratio of the fraction of absorbed PAR, FA-

PAR, and $\bar{k}$, where FAPAR = APAR/PAR, APAR being the absorbed PAR. A multiplicative relation is also used by Tenhunen *et al.* (1984) and is suggested in the hierarchy of three organization levels by Field (1991). The continuation of this argument will be introduced by asking four questions that we will attempt to answer in the following sections: (1) Does the leaf performance of Mediterranean plants differ from that of other plants? (2) How is this performance organized through vegetation stands and along resource gradients? (3) Do the patterns of variability within canopies comply with those predicted by optimization theories? (4) What are the time courses of FAPAR in Mediterranean-type ecosystems?

A. Leaf Performances of Species Growing in Mediterranean-Type Ecosystems?

For a wide variety of plants, leaf nitrogen content and photosynthetic capacity are highly correlated. This correlation appears to be a consequence of the limitations on photosynthetic capacity imposed by the levels of the enzyme ribulose-1,5-bisphosphate (RuBP) carboxylase and of the pigment–protein complexes, and of a parallel between levels of leaf nitrogen and of other photosynthetic proteins [see Evans (1989) and Field and Mooney (1986) for comprehensive accounts]. Mediterranean species do not deviate from this rule (Field, 1991), although a distinction between deciduous and evergreen species can be made and interpreted. This distinction, done by Mooney (1981), is in agreement with results of a cost–benefit analysis at leaf level (Mooney and Dunn, 1970). This model was successfully used for two coexisting California oaks (Hollinger, 1984), but does not hold for two other Mediterranean oaks (Damesin *et al.*, 1998a). The latter authors instead propose a model involving nutrient limitation at ecosystem scale (see Aerts, 1995). In any case, the photosynthetic performance of Mediterranean species does not differ particularly from that of species from other biomes. There are many results for the California chaparral and the Chilean matorral (Oechel *et al.*, 1981), the South African fynbos (Mooney *et al.*, 1983; Van der Heyden and Lewis, 1989), and the shrublands and woodlands around the Mediterranean Sea (Tenhunen *et al.*, 1987; Damesin *et al.*, 1998a). Differences have been demonstrated between growth forms (Oechel *et al.*, 1981) or guilds, and between restioid, ericoid, and proteoid species (Van der Heyden and Lewis, 1989), although these do not invalidate the general scheme proposed by Field (1991). These photosynthetic performances may be grouped together within one superclass in the sense of Schulze *et al.* (1994).

Function at the biochemical level of the photosynthetic system, described by Farquhar and von Caemmerer (1982), can be summarized by both maximum carboxylation $V_{c\,max}$ and electron transport rates J_{max}. Mediterranean species also fit within this general scheme if reference is made to the few Mediterranean species included in Wullschleger's (1993) review, although this latter author did not propose a separate grouping for these species. They do not deviate significantly from the empirical linear relation between $V_{c\,max}$ and J_{max} that he observed. Other data support this trend for Mediterranean oaks and *Arbutus unedo* (Hollinger, 1984; Damesin, 1996; Harley and Tenhunen, 1991). The creation of a superclass could also be made on the basis of the discrimination (Δ) against $^{13}CO_2$ during photosynthesis. For the xerophytic woods and scrubs superclass (which includes *Heteromeles arbutifolia, Nerium oleander,* and *Eucalyptus socialis*), Lloyd and Farquhar (1994) gave a surprising low Δ of 12.9‰. By comparison, this discrimination reaches 18.3‰ for cool/cold mixed forests. This distinction in term of Δ values implies a segregation of the long-term estimates of the ratio C_i/C_a between intercellular CO_2 concentration within leaves (C_i) and atmospheric CO_2 (C_a) and therefore of leaf performance and water use efficiency. This approach has been extended by Beerling and Quick (1995), who used it both for superclasses and for plant scale. They estimated $V_{c\,max}$ and J_{max} from Δ and maximum assimilation. What values of Δ should be adopted for Mediterranean species? From a study conducted at 25 stations in southern France with four species of co-occurring trees or shrubs, the following values were obtained: 18.20 $\pm$ 0.65 for *Pinus halepensis,* 18.86 $\pm$ 1.53 for *Quercus pubescens,* 19.50 $\pm$ 0.52 for *Quercus coccifera,* and 19.88 $\pm$ 0.68 for *Quercus ilex* (S. Rambal, unpublished data). Only *Pinus halepensis* deviated significantly from the three other species. A similar distinction was recorded by Williams and Ehleringer (1996) between *Quercus gambelii* and *Pinus edulis.* Values obtained for California oaks (Goulden, 1996) and *Q. ilex* at other sites (Fleck *et al.,* 1996) confirm these orders of magnitude. However, all these values are subject to averaging procedures to overcome great interindividual variation in Δ (Damesin *et al.,* 1997). It would therefore be preferable to turn toward creating functional groups and to taking into account the responses of species to local resources. For example, Williams and Ehleringer (1996) explained the between-site variability in Δ along a summer monsoon gradient in the southwestern United States using a parameter that integrated both the water balance and the climatic demand throughout the growing season. Similarly, Damesin *et al.* (1998b) took into account both within- and between-site variabilities among deciduous and evergreen Mediterranean oaks using the minimum seasonal leaf predawn potential. This response in Δ, and consequently in C_i/C_a, which tends to optimize the use of water resource, can be extended to plant communities growing along a water availability gradient (Stewart *et al.,* 1995).

B. Within-Canopy Variations of Leaf Performances in Mediterranean-Type Ecosystems

Plants acclimate to decreased PAR availability within the canopy by producing a gradient of leaves that are morphologically as well as physiologically distinct. Differences in photosynthetic capacity of leaves exposed to different PAR may arise from variation in both leaf mass per area (LMA) and differential allocation to photosynthetic enzymes as compared to light-harvesting machinery, both of which contribute to variation in area-based leaf nitrogen content (N).

Using an econometric model, Mooney and Gulmon (1979) predicted that decreasing PAR should decrease the level of photosynthetic proteins. Consequently, carbon gain for a whole canopy should be maximized when nitrogen is distributed such that leaves receiving the highest PAR have the highest N. On this basis, Field (1983) developed a biochemically based model of leaf photosynthesis to predict the "optimal" distribution of leaf nitrogen that maximizes daily carbon gain over a Mediterranean drought–deciduous shrub canopy. He assumed that $V_{c\ max}$ and J_{max} are related to mass-based N. From simulation results, he ranked three nitrogen distributions: optimal, uniform, and observed. The expected daily net photosynthesis was greater with the optimal than with the observed nitrogen distribution but greater with the observed than with the uniform. With a close optimization perspective, Gutschick and Wiegel (1988) and Hirose and Werger (1987), among others, suggested that given a fixed amount of nitrogen or carbon available to leaves, plants optimize whole-canopy photosynthesis. They proposed that decreasing leaf N or LMA with depth within the canopy tends to maximize daily photosynthetic carbon gain. The optimal distribution of leaf performance has been shown to depend only on PAR extinction.

However, the observed extinction coefficients of exponential curves describing the changes of leaf N or LMA with depth or leaf area index (LAI) are less than optimum. This difference occurs rather widely, irrespective of the nature of the species (Anten, 1997; Anten *et al.*, 1995). We have compared the extinction coefficients for leaf N, k_N, and for LMA, k_{LMA}, with the extinction of the radiation-weighted time-mean PAR, $\bar{k}$, within a LAI = 4 dense canopy of an evergreen oak largely representative of woodlands growing around the Mediterranean Basin. The calculation of $\bar{k}$ followed Sellers *et al.* (1992). The leaf growth–irradiance history must be considered by integrating over both daylength and the leaf expansion period that occurs approximately between Julian day 100 and 180. The leaf scattering coefficient in the PAR domain is about 12%. We assumed, as Hollinger (1984) did for the evergreen *Q. agrifolia,* that the inclination angles of leaves can be represented by a spherical Poisson model. So, $\bar{k}$ is equal to 0.72 and close to measurements made by Caldwell *et al.* (1986) on nine *Q. coccifera* shrublands of LAI ranging from 2.1 to 8.2 and by Miller (1983) on chamise and mixed chaparrals. Miller wrote "the absorbed irradiance on a daily basis was satisfactorily calculated by using an extinction coefficient of 0.7." However, $\bar{k}$ is far higher than observed for k_N and k_{LMA} (Rambal *et al.*, 1996), 0.222 (see Fig. 14-3). Long-term estimates of C_i/C_a, based on leaf carbon isotope com-

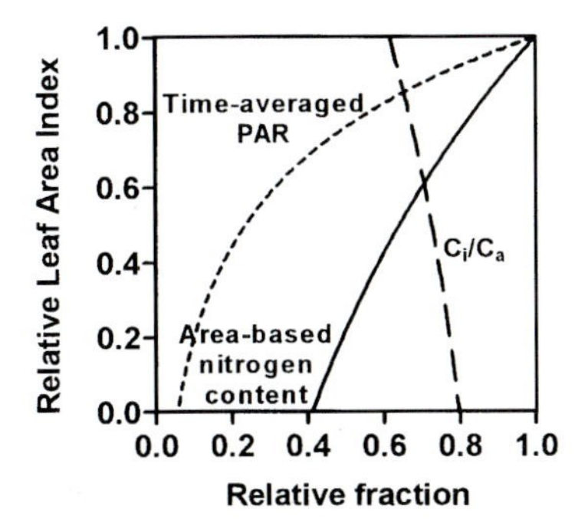

Figure 14-3 Relative variations of time-averaged PAR, area-based nitrogen content, and ^{13}C-based C_i/C_a within a canopy of *Quercus ilex*. The vertical position within the canopy is expressed in relative fraction of total LAI.

positions, were calculated by rearranging the equations originally developed by Farquhar *et al.* (1982), graphed in Fig. 14-3. Inferred from C_i and C_a, the ratio of water use efficiencies (the molar ratio of photosynthesis to transpiration) for upper and lower leaves, which we assumed to experience the same leaf-to-air vapor pressure gradient, was 49%. By comparison, relative cumulative PAR reaching leaves that grow at the bottom of the canopy was only 5% of the full sun radiation.

These last results show that Mediterranean evergreen oak canopies and probably all other Mediterranean-type ecosystems are in some way "nearly" optimal because of nonuniformity in leaf N, and LMA improves canopy carbon gain (Meister *et al.*, 1987). Further studies will be necessary to find out the principle on which this control is based. For Chen *et al.* (1993), leaf nitrogen is continuously partitioned or "coordinated" in such a way to maintain a balance between $V_{c\ max}$ and J_{max}. The nitrogen distribution obtained using this coordination theory is always slightly more uniform than that obtained using optimization theories (i.e., coordinated extinction is less than optimal extinction).

C. What Do We Learn from Gradient Analysis?

We have already shown that k_N and k_{LMA} are site dependent (Rambal *et al.*, 1996). The highest extinction values are reported in the most xeric locations with lowest leaf area indices. In contrast, extinction values in mesic locations are reduced and lower than 0.1. The variation in extinction in relation to site LAI is shown in Fig. 14-4. It can be described by a negative exponential of span 0.425 and parameter 0.267 ($r^2 = 0.766$, $n = 11$, $P < 0.001$).

Changes in extinction are always associated with parallel variations in LMA and even more so in canopy-averaged LMA, $\overline{LMA}$. LMA is an impor-

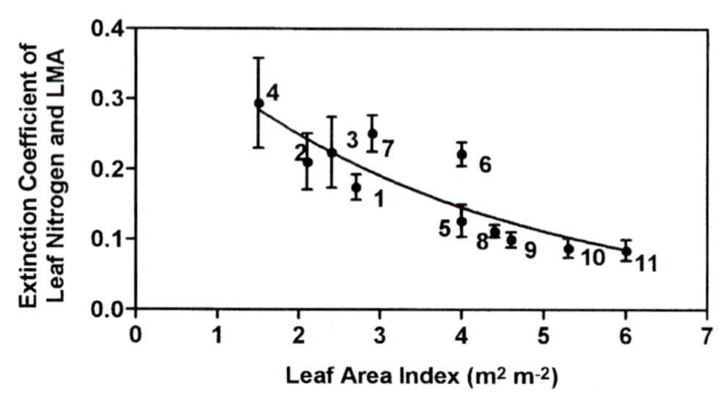

Figure 14-4 Variation of the extinction coefficient of leaf nitrogen and LMA with the site LAI for some Mediterranean evergreen canopies. Numbers refer to locations. Data are for *Quercus coccifera* shrublands (1, Saint-Martin de Londres, France; 2, Montbazin-Gigean, France; 3, La Palme, France; 4, Sierra de la Pila, Spain; 5, Quinta Sao Pedro, Portugal) and *Quercus ilex* woodlands (6, Camp Redon, France; 7, Puechabon, France; 8, Le Rouquet, France; 9, Valley Prades, Spain; 10, Ridge Prades, Spain; 11, La Peyne France) (see description in Rambal *et al.*, 1996).

tant link between carbon, nitrogen, and water budgets because it describes the distribution of leaf biomass relative to leaf area within the canopy. LMA varies with site water availability, and leaf parameters are related to the difference in LMA within and between sites. Changes in LMA are accompanied by changes in the photosynthetic apparatus per unit leaf area and hence by changes in area-based photosynthetic capacity. Heritability of LMA has been demonstrated for evergreen *Eucalyptus* species (Mooney *et al.,* 1978), but significant changes occur between individual plants of the same species. Poole and Miller (1981) observed a decrease in $\overline{\text{LMA}}$ from the driest to the wettest locations in the canopies of the evergreen *Quercus dumosa* growing along a rainfall gradient in southern California. $\overline{\text{LMA}}$ was significantly correlated with the mean rainfall P (millimeters): $\overline{\text{LMA}} = 206 - 0.053P$ ($r^2 = 0.71$, $P < 0.05$). For evergreen *Eucalyptus* spp. forests, Mooney *et al.* (1978) and Specht and Specht (1989) found that $\overline{\text{LMA}}$ decreased with increasing aridity and can be closely related to an evaporative coefficient and to the site LAI. We found the same pattern in relating LMA and LAI for both woodland and shrubland canopies of Mediterranean evergreen oaks (Fig. 14-5). Such an analysis can be extended to complex communities. Pierce *et al.* (1994) applied it to a forest transect in Oregon from dry locations where low LAI *Juniperus occidentalis* or *Pinus ponderosa* communities grow, through to the most mesic locations containing *Tsuga heterophylla* with high leaf area indices. It was also applied by Jose and Gillespie (1996) to mixed-hardwood forest communities of southern Indiana over a range of LAI values varying from 2.8 to 4.5 and for deciduous species ranging from *Quercus prinus* and *Q. alba* to *Fa-*

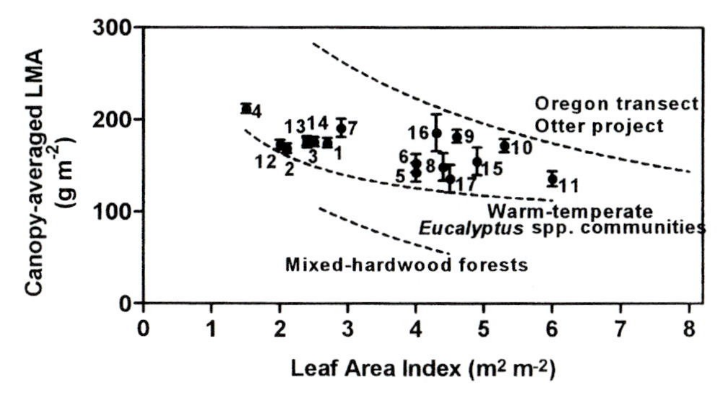

Figure 14-5 Variation of canopy-averaged LMA with site LAI for some Mediterranean evergreen canopies. Also represented are the relationships for the Oregon transect of the OTTER project, a moisture gradient with warm-temperate *Eucalyptus* spp. communities and a transect across mixed-hardwood forest. Numbers 1 to 11 refer to locations in Fig. 14-4. Other data are for *Quercus coccifera* shrublands (12, Grabels, France; 13, Puech du Juge, France; 14, Puech du Juge, France) and *Quercus ilex* woodlands (15, Castelpoziano, Italy; 16, Monte Minardo, Italy; 17, La Castanya, Spain).

gus grandifolia and *Acer saccharum*. These patterns of variation (Fig. 14-5) oppose with those recorded with high LAI canopies by Gower *et al.* (1993) in cold-temperate forests in the transition from *Quercus rubra* stands with an LAI = 4.5 and a $\overline{\text{LMA}} = 82$ g m^{-2} to stands of *Pinus resinosa* or *Picea abies* of LAI 6.2 and 10.2 and of $\overline{\text{LMA}}$ of 263 and 303 g m^{-2}, respectively.

Consequently, LMA as a morphological index could by itself possibly be used as a criterion indicating normal physiological activity and may contribute significantly to a broader application of photosynthesis and productivity models at community and landscape levels.

D. Time Courses of Remotely Sensed Vegetation Indices

The ability to convert TOC spectral vegetation indices to biologically meaningful variables is a key step in making use of the large amounts of satellite information currently available in estimating the productivity of vegetated areas. Quasi-theoretical works provide strong support for using normalized difference vegetation index (NDVI) values to improve our knowledge on the basic processes of ecosystem productivity. Sellers (1987) found that indices based on reflectance in the red and infrared bands are linear indicators of absorbed photosynthetically active radiation (APAR). Sellers *et al.* (1992) demonstrated that spectral indices from low-resolution satellites might provide good area-averaged estimates of photosynthesis.

The relationships between NDVI and APAR, or its fraction FAPAR, have been extensively documented at local scales and for vegetation with simple canopies, such as grasses and crops (Gallo *et al.*, 1985), and more rarely in various natural ecosystems such as coniferous or deciduous forests (Law and Waring, 1994) and especially Mediterranean-type ecosystems (Gamon *et al.*, 1995). These relationships provide a connection between ANPP and NDVI through the empirical model of Monteith (1981) or the process-based model of Sellers *et al.* (1992). There are now sufficient grounds for relating FAPAR to NDVI by a simple linear model (Ruimy *et al.*, 1994). The relationship is independent of pixel heterogeneity, plant leaf area, and variations in leaf orientation and optical properties. On the other hand, the relationship is sensitive to background, atmospheric, and bidirectional effects (Myneni and Williams, 1994) and to the abundance of nongreen biomass (Gamon *et al.*, 1995).

The data set we used was produced as part of the NOAA/NASA Pathfinder AVHRR Land Program and archived at the Distributed Active Archive Center of the Goddard Space Flight Center, Greenbelt, Maryland. It contains global monthly composites of NDVI at 8-km resolution derived from daily data from the NOAA AVHRR Global Area Coverage (GAC). The data covered the period from 1982 to 1993. Only a 10-yr period from 1984 to 1993 was considered here. We digitized polygonal zones of contrasted areas (see Table 14-2) to represent six sites throughout the Mediterranean-type climate regions; southern Australian mallees, South African fynbos, southern Californian chaparral and shrublands, and woodlands of southern France, and one outside the Mediterranean-type region (Mojave Desert) for comparison. For the mallees, we distinguished zones with precipitation between 250 and 750 mm from those where the rainfall supply was greater than 750 mm. These zones were intersected with the NDVI database. For five of the six zones, the patterns of seasonal change in NDVI were very typical (Fig. 14-6). They paralleled the growth of photosynthetic plant material following the rainy season. These changes did, however, reflect complex mechanisms

Table 14-2 Time Course Statistics of the Normalized Difference Vegetation Index Observed in Seven Test Areas

Vegetation	Area	N^a	Min NDVI	Max NDVI	Mean NDVI	CV
Chaparral	Southern California	180	0.142	0.433	0.302	0.199
Desert	Mojave	109	0.024	0.226	0.089	0.510
Garrigues	Southern France	34	0.185	0.504	0.362	0.201
Fynbos	South Africa	289	0.113	0.489	0.303	0.168
Mallee[b]	Southwest Australia	291	0.155	0.583	0.472	0.262
Mallee[c]	Southwest Australia	326	0.092	0.603	0.308	0.479

[a]Number of pixels.
[b]Annual rainfall, >750 mm.
[c]Annual rainfall, <750 mm, >250 mm.

ranging from the growth of new leaves following summer shedding of trees and shrubs to the massive development of a grass understorey in landscapes of open woodlands or shrublands. Only the high-rainfall mallee was distinguished by a different pattern. It showed a relative stability in NDVI at high levels, characteristic of dense stands of evergreen species, interspersed by rapid falls in NVDI values, typical of drought-induced or -mediated partial defoliations. This was the zone that gave the highest mean NDVI values, reaching 0.472. It was followed by the garrigues of southern France (0.362), and by a uniform group consisting of the fynbos, the chaparral, and the dry mallee, whose NDVI values were close to 0.30. By comparison, the value reached in the Mojave Desert was only 0.089. The coefficients of variation were also interesting to analyze. For the chaparral, the garrigue, and the fynbos, they were less than or equal to 20% and clearly demonstrated the buffering effects of plant production in the face of climatic unpredictability. Remember that the variation in the annual rainfall amount reported for the Mediterranean-type climate is of the order of 30% (Rambal and Debussche, 1995). The dry mallee, which stood out from the other zones by its wide CV, is more similar to arid zones.

The strong relationships between NDVI and FAPAR or even canopy structure and chemical content support the use of vegetation indices as indicators of potential photosynthetic production at landscape and larger scales. Evidence indicates that photosynthetic performance and NPP (or ANPP) or water-stressed vegetation cannot be predicted from NDVI alone due to environmental alterations in the radiation use efficiency ε. Additional results illustrated uncoupling between NDVI and photosynthetic productivity due to changing ε (Runyon *et al.*, 1994). The reliability of NDVI as a direct indicator of photosynthetic productivity is particularly suspect in evergreen Mediterranean-type ecosystems, where the seasonal patterns of canopy greenness and photosynthetic activity diverge. Gamon *et al.* (1995) observed in *Q. agrifolia* that NDVI failed to capture the seasonal course in photosynthetic activity of leaf canopy, although environmental limitations and ecophysiological factors affecting both NPP and ε are relatively well understood.

IV. Environmental Limitations on NPP

In the Mediterranean regions the availability of both water and soil nutrients appears to be the major environmental factor affecting the nature and distribution of vegetation, exerting the strongest control on plant productivity.

It is generally difficult to relate soil chemical contents with the availability of nutrients for the plants. However, some general traits occur (see Table

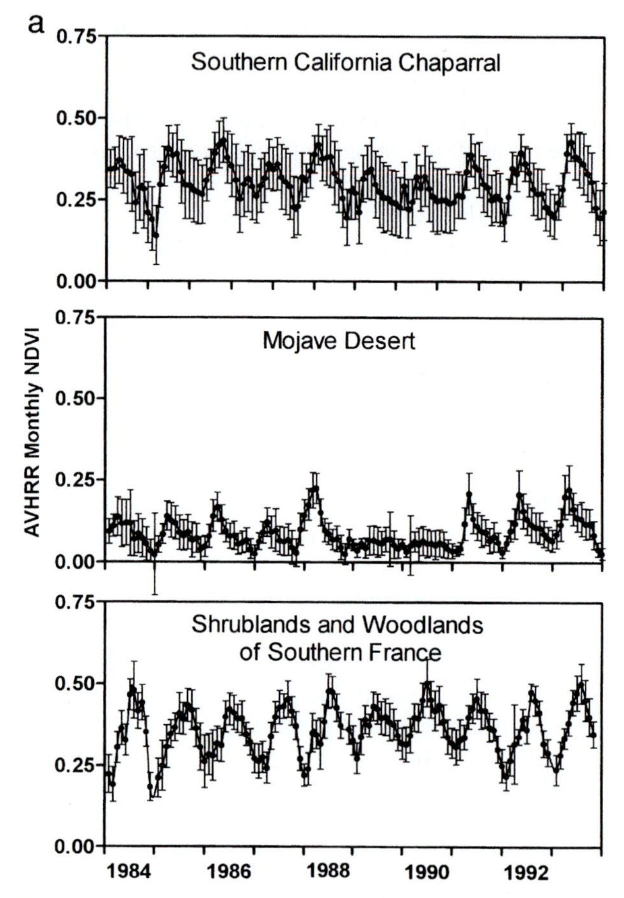

Figure 14-6 (a and b) Time course of area-averaged monthly NDVI with standard deviations from NOAA–AVHRR in the seven test areas.

14-3). Very strongly leached soils characteristic of the Australian heathland and South Africa fynbos show a lack of the major leaf nutrients. Strongly leached soils appear also to have low levels of the major nutrients, but the leaf contents from the mallee samples are not as low in leaf N and P compared to the heathland and fynbos ecosystems. Weakly leached soils of Chile have reasonable levels of N and P whereas calcium-rich / high-pH soils reveal the same pattern. Leaf content of garrigue species show apparently well-balanced nutrient. However, some trace elements may be precipitated and unavailable to plants because of high soil pH.

Studies such as those of Poole and Miller (1975, 1978) showed that some

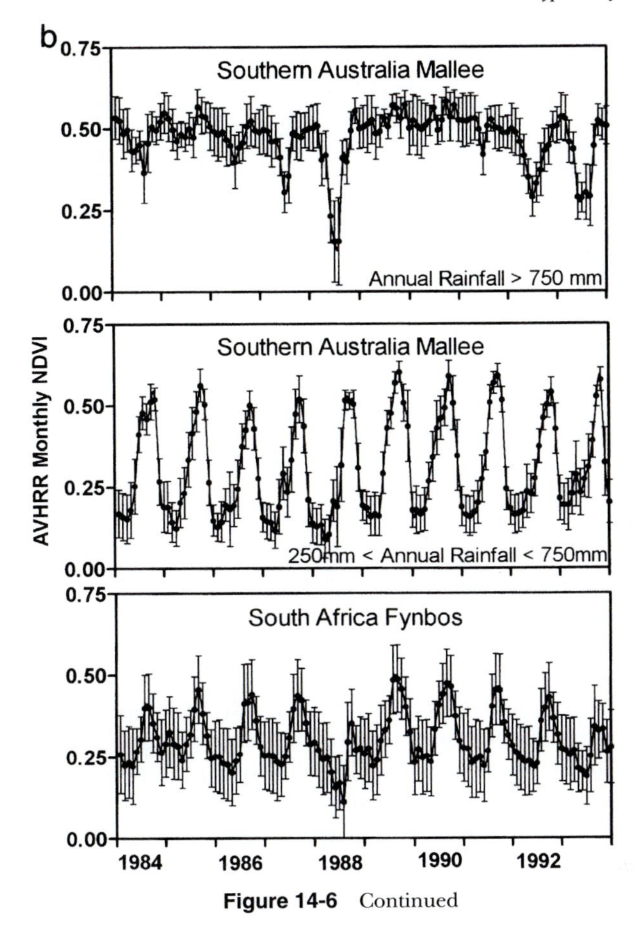

Figure 14-6 Continued

chaparral and matorral species are more sensitive to water stress than others, as indicated by the water potential when stomata close or minimal potential values recorded during drought periods. They assumed that the degrees of drought tolerance, stomatal sensitivity, and carbon-gaining capacity are linked to rooting depth. Thus shallow-rooted species tend to close their stomata at low water potential and to have tissues with the greatest drought tolerance. As a result, Poole and Miller observed clear trends in plant–water relations. Maximum leaf conductances of California evergreen and deciduous species ranged from highest (about 400 mmol m^{-2} s^{-1}) to lowest (about 100 mmol m^{-2} s^{-1}) values with an average of 260 mmol m^{-2} s^{-1} (n = 8). The sequence of species was similar for the lowest observed predawn leaf water potentials, from a low of <-6.5 MPa to a high of -2.0

Table 14-3 Average Values and Standard Deviation of Leaf N and Leaf P[a]

	Overstorey species		Understorey species	
Habitat	Leaf N (%)	Leaf P (%)	Leaf N (%)	Leaf P (%)
Mallee (southwestern and southern Australia)	0.96 ± 0.24	0.054 ± 0.026	1.30 ± 0.75	0.056 ± 0.033
Heathlands (southwestern and southern Australia)	0.79 ± 0.19	0.028 ± 0.012		
Garrigue (southern France)	1.32 ± 0.14	0.101 ± 0.028	1.85 ± 0.46	0.099 ± 0.053
Fynbos (southern Africa)	0.53 ± 0.11	0.029 ± 0.013		
Chaparral (California)	1.04 ± 0.27	0.096 ± 0.018		
Matorral (central Chile)	1.39 ± 0.016	0.17 ± 0.005		

[a]Data for large samples of evergreen species growing in overstorey or understorey canopy position in most Mediterranean-type ecosystems (after Specht and Moll, 1983).

MPa. The rooting depths increase in the same sequence from shallow-rooted to deep-rooted species and may be similarly ordered. The same pattern was observed in the matorral with maximal stomatal conductance ranging from 200 to 60 mmol m^{-2} s^{-1}, associated with variation of predawn potential from −6.0 to −3.5 MPa from shallow- to deep-rooted species.

Other results do not support Poole and Miller's hypothesis. Rhizopoulou and Mitrakos (1990) worked on 11 dominant evergreen species growing around the Mediterranean Basin. They found an average maximal conductance of 320 mmol m^{-2} s^{-1}, close to values reported for some Mediterranean oak species (Acherar and Rambal, 1992). From their results we can derive an opposite relation between conductance and water potential. Larger values of maximal stomatal conductances are associated with lower water stress. This is consistent with work of Davis and Mooney (1986) and Rambal (1993), who concluded that drought tolerance and stomatal sensitivity are not necessarily related to rooting depth, to soil moisture resources available to the plant, or to the degree of seasonal water stress experienced by the plant.

There are a number of other environmental factors that also affect photosynthetic capacity as well as the rate of photosynthesis at any given moment. The temperature of optimum photosynthesis of most Mediterranean-type climate evergreen or broadleaf deciduous species is near 20°C. These plants are characterized by very broad response curves (Oechel *et al.*, 1981) and little acclimation. Except for some ericoid species, the temperature of 20°C at which maximum photosynthesis was recorded in the fynbos remained unchanged during summer and winter (Mooney *et al.*, 1983; Van der Heyden and Lewis, 1990). Because growth requires higher temperatures than does photosynthesis, the winter temperature suppresses growth more than photosynthesis. Apparently growth cannot occur when daily mean temperatures are below 10°C. The patterns of growth are strikingly similar within Mediterranean-type climate areas. Spring growth flushes are the most frequent and for each region a range of flushing periods from late winter to late spring is observed. Kummerow (1983) accurately predicted the onset of stem elongation of four chaparral species with the sum of warming hours greater than 15°C from the day in the preceding fall when soil water was available. With the exception of few summer-growing species, growth ceases when photosynthesis is reduced by summer drought.

V. The Threats of Global Changes

Overpeck *et al.* (1990) wrote "in Northern mid-latitude forests, the chief form of nonanthropogenic catastrophic disturbance is fire." This statement can be easily extended to Mediterranean-type ecosystems too. There is a consensus that the most significant threat to Mediterranean-type ecosystems from global changes would be associated with drying trends, change in rainfall patterns, and seasonality, and that these impacts could be significant, leading to change in fire occurrence (Piñol *et al.*, 1997). Climate, because it affects fuel moisture and ignition probability, is the most useful variable in determining fire behavior and understanding changes in fire frequency. It sets the lower threshold below which fire cannot occur because conditions are too wet for fuel to burn.

Simulations of the future climate with an increased greenhouse effect give variable results, depending on the Global Circulation Model (GCM) used, although some general features appear for the Mediterranean-type climate (Tyson, 1990; Rambal and Hoff, 1998). For instance, the linear trends of the deviation between perturbed minus control runs for UKTR, the fully coupled atmosphere-deep ocean GCM of the Hadley Center for Climate Prediction and Research of the U.K. Meteorological Office, are about +0.035°C yr^{-1} for January and +0.08°C yr^{-1} for July in both southern France and southern Spain. The decreases in annual amount of rainfall range from 0.43 to 1.1 mm yr^{-1} for the same areas and those for the summer period, May to August, from 0.24 to 0.018 mm yr^{-1} (Rambal and Hoff, 1998). GCMs also indicated a systematic shift toward more days of convective rain and less of large-scale rain (Gordon *et al.*, 1992). Gregory and Mitchell (1995) showed in middle latitudes that the average decrease in total annual rainfall is associated with a decrease in both the number of rain days and the frequency of low-rainfall days: "precipitation is distributed over fewer days with relatively larger daily amounts." This means that there will be an increased frequency of dry spells, despite an increase in the number of heavy rain events. In southern France the fire season will lengthen by up to 15 days (Rambal and

Hoff, 1998). As a result, future warmer Mediterranean-type climates will be characterized by more frequent disturbed weather for several reasons. Latitudinal temperature contrasts and atmospheric eddy energy tend to be weakened as climate warms, thus decreasing temperature variability on all time scales (Gregory and Mitchell, 1995). This may lead to longer periods of consecutive hot days without cool breaks during future dry seasons. In addition, the increased water vapor capacity of a warmer atmosphere should be associated with a reduction in stability, leading to increased convection and greater variability in precipitation. As climates become warmer, temperature-driven increases in potential evaporation (Rosenberg *et al.*, 1989) and small decreases in total precipitation combine to make the probability of summer drought, and thus fire, more likely during the warmer season (Johnson and Larsen, 1991; Wotton and Flannigan, 1993).

There is also little doubt that landscape patterning is a synergistic factor, increasing the risk of fire (Whelan, 1995). The climatic threats occurring with short return periods are superimposed on long-term threats resulting from changes in land use practices (Le Houérou, 1991). These changes lead to the homogenization of the mosaic of landscape units and to greater accumulation of B_{ag}, particularly in the northern part of the Mediterranean Sea. Increases in the available fuel biomass can also result from enhanced atmospheric CO_2, but this prediction is yet largely speculative (Oechel *et al.*, 1995). A significant increase in biomass production seems to occur in the very early stage of regrowth (Hättenschwiller *et al.*, 1997). However, it is difficult to quantify the relative importance of fuel or weather on fire behavior (Bessie and Johnson, 1995). The lack of spatial differentiation we observed within natural landscapes in fire frequency seems to result from the behavior of individual fires and the weather patterns under which these fires occur and spread. The fires that shaped present-day Mediterranean landscapes were large and had high rates of spread (see examples in Trabaud and Galtié, 1996; Moreno, 1998). Low-intensity fires burned only small areas and burned under marginal conditions. Large fires are controlled by climate, not fuel loading, and thus they will be difficult, if not impossible, to suppress under global changes.

VI. Summary

Mediterranean-type climate areas cover about 2.75 Mkm^2 in five relatively small, isolated regions of the world. Hot, dry summer alternates with a cool to cold, humid period lasting 5–10 months. The availability of both water and soil nutrients appears to be the major environmental factor affecting the nature and distribution of vegetation and exerting the strongest control on plant productivity. Vegetation is characterized by the dominance of trees and woody shrubs with small, evergreen, sclerophyllous leaves.

The extreme values of aboveground biomass for mature ecosystems range from 489 g DM m^{-2} for a sparse low shrubland to 27 kg DM m^{-2} for dense woodland. The current annual growth increment ranges between 40 and 460 g DM m^{-2} yr^{-1}. More than 50% of ANPP may be explained by litterfall. As in other plant communities, the highest rates of ANPP were recorded in the first four or five postperturbation years, after which they fell rapidly as the community aged. The production rates in Mediterranean-type ecosystems appear to be only marginally greater than those of adjacent desert-type ecosystems.

In any case, leaf photosynthetic performance and discrimination against $^{13}CO_2$ during photosynthesis of Mediterranean species do not differ particularly from that of species from other biomes. Plant–water relations have shown that drought tolerance and stomatal sensitivity are not necessarily related to rooting depth, to soil moisture resources available to the plant, or to the degree of seasonal water stress experienced by the plant. The temperature optimum for photosynthesis in most species is near 20°C, with very broad response curves and little acclimation. Profiles of leaf performance within the canopies suggest that Mediterranean-type ecosystems are in some way "nearly" optimal and improve their canopy carbon gain because of nonuniformity in leaf N and in leaf mass per unit area. However, the extinction coefficients for N and leaf mass per unit area are site dependent. The highest values are reported in the most xeric locations with lowest leaf area indices. In contrast, extinction values in mesic locations are largely reduced. Further studies will be necessary to find out the principle on which this control is based.

The strong relationships between the normalized difference vegetation index and absorbed photosynthetically active radiation support their use as indicators of potential photosynthetic production. Wet Australian mallee showed the highest time-averaged NDVI values (0.472). This was followed by the garrigues of southern France (0.362), and by a uniform group consisting of fynbos, chaparral, and dry mallee with NDVI values close to 0.30. However, evidence indicates the photosynthetic performance and NPP (or ANPP) of water-limited evergreen canopies cannot be predicted from NDVI alone due to environmental alterations in the radiation use efficiency.

There is a consensus that the most significant threat to Mediterranean-type ecosystems from global changes would be associated with drying trends and changes in rainfall patterns and seasonality, and that these impacts could be significant, leading to change in fire occurrence.

Acknowledgments

Helpful comments on the manuscript were provided by J. Roy, H. Mooney, and B. Saugier. This work was partially supported by the MEDEFLU project of the European Community (Contract No. ENV4-CT97-0455).

References

Acherar, M., and Rambal, S. (1992). Comparative water relations of four Mediterranean oak species. *Vegetatio* **99–100,** 177–184.

Aerts, R. (1995). The advantages of being evergreen. *Trends Ecol. Evol.* **10,** 402–407.

Anten, N. P. R. (1997). Modelling canopy photosynthesis using parameters from nondestructive measurements. *Ecol. Res.* **12,** 77–88.

Anten, N. P. R., Schieving, F., and Werger, M. J. A. (1995). Patterns of light and nitrogen distribution in relation to whole canopy carbon gain in C_3 and C_4 mono- and dicotyledonous species. *Oecologia* **101,** 504–513.

Beerling, D. J., and Quick, W. P. (1995). A new technique for estimating rates of carboxylation and electron transport in leaves of C_3 plants for use in dynamic global vegetation models. *Global Change Biol.* **1,** 289–294.

Bessie, W. C., and Johnson, E. A. (1995). The relative importance of fuels and weather on fire behavior in subalpine forest. *Ecology* **76,** 747–762.

Black, C. H. (1987). Biomass, nitrogen, and phosphorus accumulation over a Southern California fire cycle chronosequence. *In* "Plant Response to Stress. Functional Analysis in Mediterranean Ecosystems" (J. D. Tenhunen, F. M. Catarino, O. L. Lange, and W. C. Oechel, eds.), Vol. 15, pp. 445–458. Springer-Verlag, Berlin.

Caldwell, M. M., Meister, H. P., Tenhunen, J. D., and Lange, O. L. (1986). Canopy structure, Light microclimate and leaf gas exchange of *Quercus coccifera* L. in a Portuguese macchia: Measurements in different canopy layers and simulation with a canopy model. *Trees* **1,** 25–41.

Canadell, J., and Rodá, F. (1991). Root biomass of *Quercus ilex* in a montane Mediterranean forest. *Can. J. For. Res.* **21,** 1771–1778.

Chen, J., Reynolds, J. F., Harley, P. C., and Tenhunen, J. D. (1993). Coordination theory of leaf nitrogen distribution in a canopy. *Oecologia* **93,** 63–69.

Cody, M. L., and Mooney, H. A. (1978). Convergence versus nonconvergence in mediterranean-type climate ecosystems. *Annu. Rev. Ecol. System.* **9,** 265–321.

Daget, P. (1977). Le bioclimat méditerranéen: Caractères généraux et modes de caractérisation. *Vegetatio* **34,** 1–20.

Damesin, C. (1996). Relations hydriques, photosynthèse et efficacité d'utilisation de l'eau chez deux chênes méditerranéens caduc et sempervirent cooccurents. Ph.D. thesis, Paris 11, Orsay.

Damesin, C., Rambal, S., and Joffre, R. (1997). Between-tree variations in leaf $\delta^{13}C$ of *Quercus pubescens* and *Quercus ilex* among Mediterranean habitats with different water availability. *Oecologia* **111,** 26–35.

Damesin, C., Rambal, S., and Joffre, R. (1998a). Cooccurrence of trees with different leaf habit: A functional approach on Mediterranean oaks. *Acta Oecol.* **19,** 195–204.

Damesin, C., Rambal, S., and Joffre, R. (1998b). Seasonal and between-year changes in leaf $\delta^{13}C$ in two co-occurring Mediterranean oaks: Relations to leaf growth and drought progression. *Funct. Ecol.* **12,** 778–785.

Davis, S. D., and Mooney, H. A. (1986). Tissue water relations of four co-occurring chaparral shrubs. *Oecologia* **70,** 527–535.

Debano, L. F., and Conrad, C. E. (1978). The effect of fire on nutrients in a chaparral ecosystem. *Ecology* **59,** 489–497.

Eckardt, F. E., Heim, G., Methy, M., and Sauvezon, R. (1975). Interception de l'énergie rayonnante, échanges gazeux et croissance dans une forêt méditerranéenne à feuillage persistant (Quercetum ilicis). *Photosynthetica* **9,** 145–156.

Evans, J. R. (1989). Photosynthesis and nitrogen relationships of C_3 plants. *Oecologia* **78,** 9–19.

Farquhar, G. D., and Caemmerer von, S. (1982). Modeling of photosynthetic response to environmental conditions. *In* "Encyclopedia of Plant Physiology" (O. L. Lange, P. S. Nobel, C. B. Osmond, and H. Ziegler, eds.), Vol. 12B, pp. 549–587. Springer-Verlag, Berlin.

Farquhar, G. D., O'Leary, M. H., and Berry, J. A. (1982). On the relationship between carbon isotope discrimination and the intercellular carbon dioxide concentration in leaves. *Aust. J. Plant Physiol.* **9,** 121–137.

Field, C. (1983). Allocating leaf nitrogen for the maximization of carbon gain: Leaf age as a control on the allocation program. *Oecologia* **56,** 341–347.

Field, C. (1991). Ecological scaling of carbon gain to stress and resource availability. *In* "Response of Plants to Multiple Stresses" (H. A. Mooney, W. E. Winner, and E. J. Pell, eds.), pp. 35–65. Academic Press, San Diego.

Field, C., and Mooney, H. A. (1986). The photosynthesis–nitrogen relationship in wild plants. *In* "On the Economy of Plant Form and Function" (T. J. Givnish, ed.), pp. 25–35. Cambridge Univ. Press, Cambridge.

Fleck, I., Grau, D., SanjosÇ, M., and Vidals, D. (1996). Carbon isotope discrimination in *Quercus ilex* resprouts after fire and tree-fell. *Oecologia* **105,** 286–292.

Gallo, K. P., Daughtry, C. S. T., and Bauer, M. E. (1985). Spectral estimation of absorbed photosynthetically active radiation in corn canopies. *Remote Sensing Environ.* **17,** 221–232.

Gamon, J. A., Field, C. B., Goulden, M. L., Griffin, K. L., Hartley, A. E., Joel, G., Peneulas, J., and Valentini, R. (1995). Relationships between NDVI, canopy structure, and photosynthesis in three Californian vegetation types. *Ecol. Appl.* **5,** 28–41.

Gordon, H. B., Whetton, P. H., Pittock, A. B., Fowler, A. M., and Haylcock, M. R. (1992). Simulated changes in daily rainfall intensity due to the enhanced greenhouse effect: Implications for extreme rainfall events. *Climate Dynam.* **8,** 83–102.

Goulden, M. L. (1996). Carbon assimilation and water-use efficiency by neighboring Mediterranean-climate oaks that differ in water access. *Tree Physiol.* **16,** 417–424.

Gower, S. T., Reich, P. B., and Son, Y. (1993). Canopy dynamics and aboveground production of five tree species with different leaf longevities. *Tree Physiol.* **12,** 327–345.

Gray, J. T. (1982). Community structure and productivity in *Ceanothus* chaparral and Coastal sage scrub of southern California. *Ecol. Monogr.* **52,** 415–435.

Gray, J. T. (1983). Nutrient use by evergreen and deciduous shrubs in Southern California. I. Community nutrient cycling and nutrient-use efficiency. *J. Ecol.* **71,** 21–41.

Gregory, J. M., and Mitchell, J. F. B. (1995). Simulation of daily variability of surface temperature and precipitation over Europe in the current and 2 × CO_2 climates using the UKMO climate model. *Q.J.R. Meteorol. Soc.* **121,** 1451–1476.

Groves, R. H., and Specht, R. L. (1965). Growth of heath vegetation. I. Annual growth curves of two heath ecosystems in Australia. *Aust. J. Bot.* **13,** 261–280.

Gutschick, V. P., and Wiegel, F. W. (1988). Optimizing the canopy photosynthetic rates patterns of investment in specific leaf mass. *Am. Nat.* **132,** 67–86.

Hadley, N. F., and Szarek, S. R. (1981). Productivity of desert ecosystems. *BioScience* **31,** 747–752.

Hanes, T. L. (1971). Succession after fire in the chaparral of southern California. *Ecol. Monogr.* **41,** 27–52.

Harley, P. C., and Tenhunen, J. D. (1991). Modeling the photosynthetic response of C_3 leaves to environmental factors. *In* "Modeling Crop Photosynthesis—From Biochemistry to Canopy" (K. J. Boote and R. S. Loomis, eds.), Vol. 19, pp. 17–39. Crop Science Society of America, Madison.

Hättenschwiler, S., Miglietta, F., Raschi, A., and Körner, C. (1997). Thirty years of *in situ* growth under elevated CO_2: A model for future responses? *Global Change Biol.* **3,** 463–471.

Hirose, T., and Werger, W. J. A. (1987). Maximizing daily canopy photosynthesis with respect to the leaf nitogen allocation pattern in the canopy. *Oecologia* **72,** 520–526.

Hobbs, R. J., Richardson, D. M., and Davis, G. W. (1995). Mediterranean-type ecosystems: Opportunities and constraints for studying the function of biodiversity. *In* "Mediterranean-Type

Ecosystems. The Function of Biodiversity. Ecological Studies 109" (G. W. Davis and D. M. Richardson, eds.), pp. 1–42. Springer-Verlag, New York.

Hollinger, D. Y. (1984). Photosynthesis, water relations, and herbivory in co-occurring deciduous and evergreen California oaks. Ph.D. thesis, Stanford University, Stanford, California.

Johnson, E. A., and Larsen, C. P. S. (1991). Climatically induced change in fire frequency in the Southern Canadian Rockies. *Ecology* **72,** 194–201.

Jones, R. (1968a). Estimating productivity and apparent photosynthesis from differences in consecutives measurements of total living parts of an Australian heathland. *Aust. J. Bot.* **16,** 589–602.

Jones, R. (1968b). The leaf area of an Australian heathland with the contribution of individual species. *Aust. J. Bot.* **16,** 579–588.

Jones, R., Groves, R. H., and Specht, R. L. (1969). Growth of heath vegetation III. Growth curves for heaths in southern Australia: A reassessment. *Aust. J. Bot.* **17,** 309–314.

Jose, S., and Gillespie, A. R. (1996). Aboveground production efficiency and canopy nutrient contents of mixed-hardwood forest communities along a moisture gradient in the central United States. *Can. J. For. Res.* **26,** 2214–2223.

Kruger, F. J. (1977). A preliminary account of aerial plant biomass in fynbos communities of the Mediterranean-type climate zone of the Cape Province. *Bothalia* **12,** 301–307.

Kummerow, J. (1983). Comparative phenology of Mediterranean-type plant communities. *In* "Mediterranean-Type Ecosystems. The Role of Nutrients. Ecological Studies 43" (F. J. Kruger, D. T. Mitchell, and J. U. M. Jarvis, eds.), pp. 300–317. Springer-Verlag, New York.

Kummerow, J., Montenegro, G., and Krause, D. (1981). Biomass, phenology, and growth. *In* "Resource Use by Chaparral and Matorral. A Comparison of Vegetation Function in Two Mediterranean Type Ecosystems" (P. C. Miller, ed.), Vol. 39, pp. 69–96. Springer-Verlag, New York.

Kummerow, J., Kummerow, M., and Trabaud, L. (1990). Root biomass, root distribution and the fine-root growth dynamics of *Quercus coccifera* L. in the garrigue of southern France. *Vegetatio* **87,** 37–44.

Law, B. E., and Waring, R. H. (1994). Remote sensing of leaf area index and radiation intercepted by understory vegetation. *Ecol. Appl.* **4,** 272–279.

Le Houérou, H. N. (1980). L'impact de l'homme et de ses animaux sur la Forât méditeranéenne (2ème partie). *For. Méditerran.* **2,** 155–174.

Le Houérou, H. N. (1991). La Méditerranée en l'an 2050: Impacts respectifs d'une éventuelle évolution climatique et de la démographie sur la Végétation, les écosystèmes et l'utilisation des terres. Etude prospective. *La Météorol.* **36,** 4–37.

Leonardi, S., and Rapp, M. (1990). Production de phytomasse et utilisation des bioéléments lors de la reconstitution d'un taillis de chàne vert. *Acta Oecol.* **11,** 819–834.

Leonardi, S., and Rapp, M. (1993). Forest and agricultural ecosystem productivity in Sicily. *In* "Colloques Phytosociologiques," Vol. 21, pp. 653–667. Camerino.

Leonardi, S., Rapp, M., and La Rosa, V. (1988). Répartition et dynamique de la matière organique dans une forêt de *Pinus laricio* Poir. *Ecol. Mediterran.* **14,** 17–29.

Lledo, M. J., Sanchez, J. R., Bellot, J., Boronat, J., Ibanez, J. J., and Escarre, A. (1992). Structure, biomass and production of a resprouted holm-oak (*Quercus ilex* L.) forest in NE Spain. *Vegetatio* **99–100,** 51–59.

Lloyd, J., and Farquhar, G. D. (1994). ^{13}C discrimination during CO_2 assimilation by the terrestrial biosphere. *Oecologia* **99,** 201–215.

Lossaint, P., and Rapp, M. (1971). Répartition de la matiäre organique, productivité et cycles des éléments minéraux dans des écosystèmes de climat méditerranéen. *In* "Productivité des êcosystèmes Forestiers," pp. 597–617. UNESCO.

Low, A. B. (1983). Phytomass and major nutrient pools in an 11-year post-fire coastal fynbos community. *S. Afr. J. Bot.* **2,** 98–104.

Maggs, J., and Pearson, C. J. (1977a). Litter fall and litter layer decay in Coastal scrub at Sydney, Australia. *Oecologia (Berlin)* **31,** 239–250.
Maggs, J., and Pearson, C. J. (1977b). Minerals and dry matter in coastal scrub and grassland at Sydney, Australia. *Oecologia (Berlin)* **31,** 227–237.
Margaris, N. S. (1976). Structure and dynamics in a phryganic (East Mediterranean) ecosystem. *J. Biogeogr.* **3,** 249–259.
Meister, H. P., Caldwell, M. M., Tenhunen, J. D., and Lange, O. L. (1987). Ecological implications of sun/shade differentiation in sclerophyllous canopies: Assessment by canopy modeling. *In* "Plant Response to Stress. Functional Analysis in Mediterranean Ecosystems" (J. D. Tenhunen, F. M. Catarino, O. L. Lange, and W. C. Oechel, eds.), Vol. 15, pp. 401–411. Springer-Verlag, Berlin.
Merino, J., and Martin Vicente, A. (1981). Biomass, productivity and succession in the scrub of the Donana biological reserve in Southwest Spain. *In* "Components of Productivity of Mediterranean-Climate Regions. Basic and Applied Aspects" (N. S. Margaris and H. A. Mooney, eds.), pp. 197–204. Dr. W. Junk, The Hague.
Miller, P. C. (1981a). "Resource Use by Chaparral and Matorral. A Comparison of Vegetation Function in Two Mediterranean Type Ecosystems," Springer-Verlag, New York.
Miller, P. C. (1981b). Similarities and limitations of resource utilization in Mediterranean type ecosystems. *In* "Resource Use by Chaparral and Matorral. A Comparison of Vegetation Function in Two Mediterranean Type Ecosystems" (P. C. Miller, ed.), Vol. 39, pp. 367–407. Springer-Verlag, New York.
Miller, P. C. (1983). Canopy structure of Mediterranean-type shrubs in relation to heat and moisture. *In* "Mediterranean-Type Ecosystems. The Role of Nutrients." Ecological Studies 43 (F. J. Kruger, D. T. Mitchell, and J. U. M. Jarvis, eds.), pp. 133–166. Springer-Verlag, New York.
Mitchell, D. L., Coley, P. G. F., Webb, S., and Allsopp, N. (1986). Litterfall and decomposition processes in the coastal fynbos vegetation, South-western Cape, South Africa. *J. Ecol.* **74,** 977–993.
Monteith, J. L. (1981). Climatic variation and the growth of crops. *Q. J. R. Soc.* **107,** 749–774.
Mooney, H. A. (1981). Primary production in Mediterranean-climate regions. *In* "Mediterranean-type Shrublands" (F. Di Castri, D. W. Goodall, and R. L. Specht, eds.), Vol. 11, pp. 249–255. Elsevier Scientific Publishing Company, Amsterdam.
Mooney, H. A., and Dunn, E. L. (1970). Photosynthetic systems of Mediterranean-climate shrubs and trees of California and Chile. *Am. Nat.* **104,** 447–453.
Mooney, H. A., and Gulmon, S. L. (1979). Environmental and evolutionary constraints on the photosynthetic characteristics of higher plants. *In* "Topics in Plant Population Biology" (O. T. Solbrig, S. Jain, G. B. Johnson, and P. H. Raven, eds.), pp. 316–337. Columbia University Press, New York.
Mooney, H. A., and Rundel, P. W. (1979). Nutrient relations of the evergreen shrub, *Adenostoma fasciculatum,* in the California chaparral. *Bot. Gaz.* **140,** 109–113.
Mooney, H. A., Ferrar, P. J., and Slatyer, R. O. (1978). Photosynthetic capacity and carbon allocation patterns in diverse growth forms of *Eucalyptus. Oecologia* **36,** 103–111.
Mooney, H. A., Field, C., Gulmon, S. L., Rundel, P., and Kruger, F. J. (1983). Photosynthetic characteristics of South African sclerophylls. *Oecologia* **58,** 398–401.
Moreno, J. M. (1998). "Large Forest Fires." Backhurst Publisher, Leiden, The Netherland.
Myneni, R. B., and Williams, D. L. (1994). On the relationship between FAPAR and NDVI. *Remote Sensing Environ.* **49,** 200–211.
Oechel, W. C., Lawrence, W., Mustafa, J., and Martinez, J. (1981). Energy and carbon acquisition. *In* "Resource Use by Chaparral and Matorral. A Comparison of Vegetation Function in Two Mediterranean Type Ecosystems" (P. C. Miller, ed.), Vol. 39, pp. 151–182. Springer-Verlag, New York.

Oechel, W. C., Hasting, S. J., Vourlitis, G. L., Jenkins, M. A., and Hinkson, C. L. (1995). Direct effects of elevated CO_2 in chaparral and Mediterranean-type ecosystems. *In* "Anticipated Effects of a Changing Global Environment on Mediterranean-type Ecosystems." Ecological Studies 117 (J. Moreno and W. C. Oechel, eds.), pp. 58–75. Springer-Verlag, New York.

Overpeck, J. T., Rind, D., and Goldberg, R. (1990). Climate-induced changes in forest disturbance. *Nature (London)* **343,** 51–53.

Pierce, L. L., Running, S. W., and Walker, J. (1994). Regional-scale relationships of leaf area index to specific leaf area and leaf nitrogen content. *Ecol. Appl.* **4,** 313–321.

Piñol, J., Terradas, J., and Lloret, F. (1997). Climate warming, wildfire hazard, and wildfire occurrence in coastal eastern Spain. *Climatic Change* **38,** 1–13.

Poole, D. K., and Miller, P. C. (1975). Water relations of selected species of chaparral and coastal sage communities. *Ecology* **56,** 1118–1128.

Poole, D. K., and Miller, P. C. (1978). Water related characteristics of some evergreen sclerophyll shrubs in central Chile. *Oecol. Planta.* **13,** 289–299.

Poole, D. K., and Miller, P. C. (1981). The distribution of plant water stress and vegetation characteristics in southern California chaparral. *Am. Midland Nat.* **105,** 32–43.

Rambal, S. (1993). The differential role of mechanisms for drought resistance in a Mediterranean evergreen shrub: A simulation approach. *Plant Cell Environ.* **16,** 35–44.

Rambal, S., and Debussche, G. (1995). Water balance of Mediterranean ecosystems under a changing climate. *In* "Global Change and Mediterranean-Type Ecosystems" (J. M. Moreno and W. C. Oechel, eds.), Vol. 117. Springer-Verlag, New York.

Rambal, S., and Hoff, C. (1998). Mediterranean ecosystems and fire: The threats of global change. *In* "Large Forest Fires" (J. M. Moreno, ed.), pp. 187–213. Backhurst Publisher, Leiden.

Rambal, S., Damesin, C., Joffre, R., Methy, M., and Lo Seen, D. (1996). Optimization of carbon gain in canopies of Mediterranean evergreen oaks. *Ann. Sci. For.* **53,** 547–560.

Ranger, J. 91981). Etude de la minéralomasse et du cycle biologique dans deux peuplements de Pin laricio de Corse, dont l'un a été fertilisé à la plantation. *Ann. Sci. For.* **38,** 127–158.

Rapp, M. (1974). Le cycle biogéochimique dans un bois de pins d'Alep. *In* "Ecologie Forestière. La Forêt: Son Climat, Son Sol, Ses Arbres, Sa Faune" (P. Pesson, ed.), pp. 75–97. Gauthier-Villars, Paris.

Rapp, M., and Cabanettes, A. (1981). Biomass and productivity of a *Pinus pinea* L. stand. *In* "Components of Productivity of Mediterranean-Climate Regions. Basic and Applied Aspects" (N. S. Margaris and H. A. Mooney, eds.), pp. 131–134. Dr. W. Junk, The Hague.

Rapp, M., and Lossaint, P. (1981). Some aspects of mineral cycling in the garrigue of southern France. *In* "Mediterranean-Type Shrublands" (F. di Castri, D. W. Goodall, and R. L. Specht, eds.), Vol. 11, pp. 289–301. Elsevier Scientific Publishing Company, Amsterdam.

Rapp, M., Ed Derfoufi, F., and Blanchard, A. (1992). Productivity and nutrient uptake in a holm oak (*Quercus ilex* L.) stand and during regeneration after clearcut. *Vegetatio* **99–100,** 263–272.

Rhizopoulou, S., and Mitrakos, K. (1990). Water relations of evergren sclerophylls. I. Seasonal changes in the water relations of eleven species from the same environment. *Ann. Bot.* **65,** 171–178.

Rosenberg, N. J., McKenney, M. S., and Martin, P. (1989). Evapotranspiration in a greenhouse-warmed world: A review and a simulation. *Agric. For. Meteorol.* **47,** 303–320.

Ruimy, A., Saugier, B., and Dedieu, G. (1994). Methodology for the estimation of terrestrial net primary production from remotely sensed data. *J. Geophys. Res.* **99,** 5263–5283.

Rundel, P. W., and Parsons, D. J. (1979). Structural changes in chamise (*Adenostoma fasciculatum*) along a fire-induced age gradient. *J. Range Manage.* **32,** 462–466.

Rundel, P. W., and Parsons, D. J. (1980). Nutrient changes in two chaparral shrubs along a fire-induced age gradient. *Am. J. Bot.* **67,** 51–58.

Rundel, P. W., Baker, G. A., and Parsons, D. J. (1981). Productivity and nutritional responses of *Chamaebatia foliolosa* (Rosaceae) to seasonal burning. *In* "Components of Productivity of Mediterranean-Climate Regions. Basic and Applied Aspects" (N. S. Margaris and H. A. Mooney, eds.), pp. 191–196. Dr. W. Junk, The Hague.

Runyon, J., Waring, R. H., Goward, S. N., and Welles, J. M. (1994). Environmental limits on net primary production and light-use efficiency across the Oregon transect. *Ecol. Appl.* **4,** 226–237.

Rutherford, M. C. (1978). Karoo-fynbos biomass along an elevational gradient in the Western Cape. *Bothalia* **12,** 555–560.

Schlesinger, W. H., and Gill, D. S. (1978). Demographic studies of the chaparral shrub, *Ceanothus megacarpus,* in the Santa Ynez mountains, California. *Ecology* **59,** 1256–1263.

Schlesinger, W. H., and Gill, D. S. (1980). Biomass, production, and change in the availability of light, water, and nutrients during the development of pure stands of the chaparral shrub, *Ceanothus megacarpus,* after fire. *Ecology* **61,** 781–789.

Schulze, E.-D. (1982). Plant life forms and their carbon, water and nutrient relations. *In* "Physiological Plant Ecology II. Water Relations and Carbon Assimilation" (O. L. Lange, P. S. Nobel, and O. L. Ziegler, eds.), Vol. 12B. Springer-Verlag, Berlin.

Schulze, E.-D., Kelliher, F. M., Korner, C., Lloyd, J., and Leuning, R. (1994). Relationships among maximum stomatal conductance, ecosystem surface conductance, carbon assimilation rate, and plant nitrogen nutrition: A global scaling exercice. *Ann. Rev. Ecol. System.* **25,** 629–660.

Sellers, P. J. (1987). Canopy reflectance, photosynthesis, and transpiration. II. The role of biophysics in the linearity of their interdependence. *Remote Sensing Environ.* **21,** 143–183.

Sellers, P. J., Berry, J. A., Collatz, G. J., Field, C. B., and Hall, F. G. (1992). Canopy reflectance, photosynthesis and transpiration. III. A reanalysis using improved leaf models and a new integration scheme. *Remote Sensing Environ.* **42,** 187–216.

Specht, R. L. (1966). The growth and distribution of mallee-broombush (*Eucalyptus incrassata–Maleuca uncinata* association) and heath vegetation near Dark Island Soak, Ninety-Mile Plain, South Australia. *Aust. J. Bot.* **14,** 361–372.

Specht, R. L. (1969a). A comparison of the sclerophyllous vegetation characteristic of Mediterranean type climates in France, California, and Southern Australia. I. Structure, morphology, and succession. *Aust. J. Bot.* **17,** 277–292.

Specht, R. L. (1969b). A comparison of the sclerophyllous vegetation characteristic of Mediterranean type climates in France, California, and Southern Australia II. Dry matter, energy, and nutrient accumulation. *Aust. J. Bot.* **17,** 293–308.

Specht, R. L. (1981). Primary production in Mediterranean-climate ecosystems regenerating after fire. *In* "Mediterranean-type Shrublands" (F. Di Castri, D. W. Goodall, and R. L. Specht, eds.), Vol. 11, pp. 257–267. Elsevier Scientific Publishing Company, Amsterdam.

Specht, R. L., and Moll, E. J. (1983). Mediterranean-type heathlands and sclerophyllous shrublands of the world: An overview. *In* "Mediterranean-Type Ecosystems. The Role of Nutrients." Ecological Studies 43 (F. J. Kruger, D. T. Mitchell, and J. U. M. Jarvis, eds.), pp. 41–73. Springer-Verlag, New York.

Specht, R. L., and Specht, A. (1989). Canopy structure in *Eucalyptus*-dominated communities in Australia along climatic gradients. *Acta Oecol. Oecol. Planta.* **10,** 191–213.

Stewart, G. R., Turnbull, M. H., Schmidt, S., and Erskine, P. D. (1995). ^{13}C natural abundance in plant communities along a rainfall gradient: A biological integrator of water availability. *Aust. J. Plant Physiol.* **22,** 51–55.

Tenhunen, J. D., Meister, H. P., Caldwell, M. M., and Lange, O. L. (1984). Environmental constraints on productivity of the Mediterranean sclerophyll shrub *Quercus coccifera. Options Méditerran.* **84,** 33–53.

Tenhunen, J. D., Beyschlag, W., Lange, O. L., and Harley, P. C. (1987). Changes during summer drought in CO_2 uptake rates of macchia shrubs growing in Portugal: Limitations due to photosynthetic capacity, carboxylation efficiency and stomatal conductance. *In* "Plant Response to stress. Functional analysis in Mediterranean Ecosystems" (J. D. Tenhunen, F. M. Catarino, O. L. Lange, and W. C. Oechel, eds.), Vol. 15, pp. 305–327. Springer-Verlag, Berlin.

Trabaud, L., and Galtié, J.-F. (1996). Effects of fire frequency on plant communities and landscape pattern in the Massif des Aspres (southern France). *Landsc. Ecol.* **11,** 215–224.

Tsiourlis, G. M. (1990). Phytomasse, productivité primaire et biogéochimie des écosystèmes méditerranéens phrygana et maquis (île de Naxos, Grèce). Ph.D. thesis, Université Libre, Bruxelles.

Tyson, P. D. (1990). Modelling climatic change in southern Africa: A review of available methods. *S. Afr. J. Sci.* **86,** 318–330.

Van der Heyden, F., and Lewis, O. A. M. (1989). Seasonal variation in photosynthetic capacity with respect to plant water status of five species of the Mediterranean climate region of South Africa. *S. Afr. J. Bot.* **55,** 509–515.

Van der Heyden, F., and Lewis, O. A. L. (1990). Environmental control of photosynthetic gas exchange characteristics of fynbos species representing three growth forms. *S. Afr. J. Bot.* **56,** 654–658.

Van Wilgen, B. W. (1982). Some effects of post-fire age on the above-ground plant biomass of fynbos (Macchia) vegetation in South Africa. *J. Ecol.* **70,** 217–225.

Whelan, R. J. (1995). "The Ecology of Fire." Cambridge Studies in Ecology. Cambridge Univ. Press, Cambridge.

Williams, D. G., and Ehleringer, J. R. (1996). Carbon isotope discrimination in three semiarid woodland species along a monsoon gradient. *Oecologia* **106,** 455–460.

Wotton, B. M., and Flannigan, M. D. (1993). Length of the fire season in a changing climate. *For. Chron.* **69,** 187–192.

Wullschleger, S. D. (1993). Biochemical limitations to carbon assimilation in C_3 plants—A retrospective analysis of the A/C_i curves from 109 species. *J. Exp. Bot.* **44,** 907–920.

15

Productivity of Deserts

James R. Ehleringer

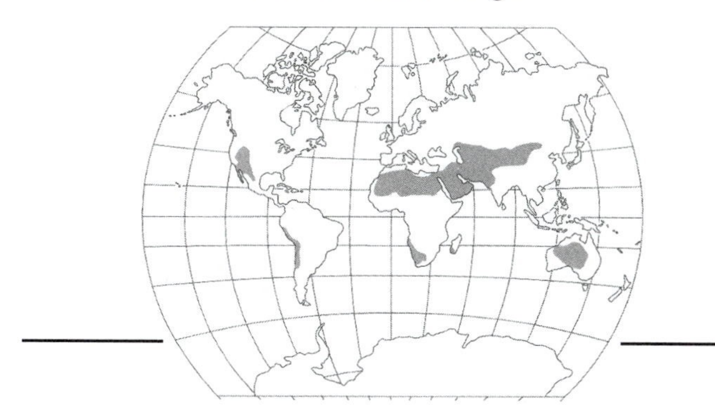

I. Nature and Extent of Deserts

Moving along a geographic gradient of decreasing precipitation from either shrub–woodlands (e.g., Mediterranean climate) or grasslands (see Chapters 14 and 12, this volume), one encounters desert ecosystems, which typically receive less than 250 mm annual precipitation (Fig. 15-1). Arid lands are extensive, occupying nearly 30% of the land surface globally (Noble and Gitay, 1996), and their extent is thought to be increasing annually through desertification into areas previously occupied by arid grasslands (Schlesinger *et al.*, 1990; Kassas, 1995; Bullock and Le Houérou, 1996). Desert ecosystems are typically characterized by extremes, having the lowest standing biomasses, lowest primary productivities, and lowest precipitation inputs. Precipitation is, of course, the principal driver regulating primary productivity rates. Even though rainfall patterns are usually seasonal, these rains come as intermittent pulses, causing desert ecosystems to shift between inactive and active states, depending on soil moisture availability (Noy-Meir, 1985). Only in the higher elevation, cold desert ecosystems of North America, central Asia, and the Middle East do low wintertime temperatures prevent growth and also reduce evaporation, allowing for somewhat longer growth periods in the spring (Caldwell *et al.*, 1977; Goodall and Perry, 1979; Caldwell, 1985).

One feature that distinguishes productivity patterns in desert ecosystems

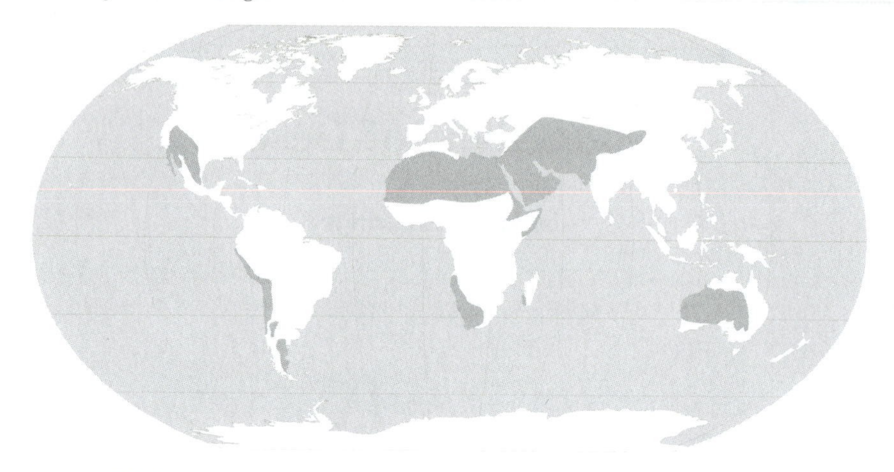

Figure 15-1 Distribution of arid land ecosystems on a global basis. Deserts are typically defined as those nonpolar regions that receive less than 250 mm of precipitation annually. The map is based on information in Logan (1968), McGinnies *et al.* (1977), and Walter and Breckle (1984).

from all others is the increased interannual variability in precipitation. At low precipitation means, both long-term annual and seasonal precipitation (in a biseasonal environment) amounts follow a gamma distribution—not a normal distribution. The consequence is that most years are drier than the arithmetic mean. Additionally, as the mean precipitation amount decreases, the interannual variability increases (Fig. 15-2). The consequence of this is that the drier desert regions are often characterized by several years of below-average precipitation amounts, punctuated by a relatively high precipitation year. Ehleringer (1994) observed that the relative frequency distribution of storm sizes did not differ among North American desert sites receiving different total precipitation amounts. Instead, what differed was simply the number of storm events at a site: annual precipitation was proportional to the total number of storm events in a given year.

II. Standing Biomass and Aboveground Net Primary Productivity Rates

Total standing biomass in desert ecosystems is highly variable, depending on factors such as the seasonal distribution of the precipitation, grazing and human impacts, importance of woody versus herbaceous components, and soil fertility (Noble and Gitay, 1996). A reasonable standing biomass range of 2000–5000 kg ha^{-1} is not uncommon for shrub-dominated desert ecosys-

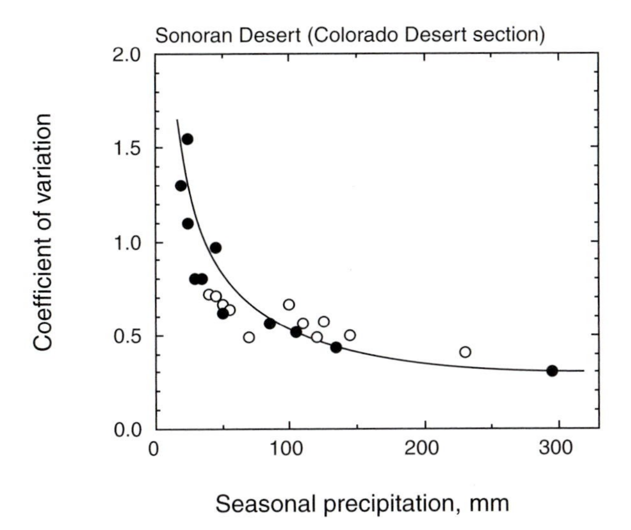

Figure 15-2 A plot of the relationships between long-term mean seasonal precipitation in the Sonoran Desert and the coefficient of variation in that precipitation.

tems (National Academy of Sciences, 1975; Whittaker and Niering, 1975; Goodall and Perry, 1979; Gibbens *et al.*, 1996; Rundel and Gibson, 1996; Schulze *et al.*, 1996). Most of this aboveground mass is associated with woody shrubs, which tend to increase in importance in response to grazing (Schlesinger *et al.*, 1990; Grover and Musick, 1990; Le Houérou, 1990; Pickup, 1996). Overall, the aboveground annual net primary productivity values in desert ecosystems tend to be less than 1500 kg ha^{-1} yr^{-1} (National Academy of Sciences, 1975; Whittaker and Niering, 1975; Goodall and Perry, 1979; Szarek, 1979; Hadley and Szarek, 1981; Rundel and Gibson, 1996), which is low compared to most other ecosystems except possibly grasslands. Much of the interannual variability in primary productivity may be associated with herbaceous components (both annuals and grasses), making it challenging to calculate a general value of net primary productivity for desert ecosystems unless the grazing pressures are well understood. It is possible, however, that remote-sensing approaches may be a valuable approach here; Prince and Goward (1995) and Prince *et al.* (1998) have shown large Normalized Difference Vegetation Index (NDVI) changes in surface spectral characterisitics in response to interannual variations in precipitation.

In desert ecosystems, net primary productivity is typically linearly related to precipitation input (Walter, 1939; Noy-Meir, 1985; Le Houérou, 1984; Sala *et al.*, 1988). This would include productivity of all vegetation components—woody, herbaceous, and annual. Soil nitrogen secondarily limits primary productivity. Linear productivity–precipitation relationships have

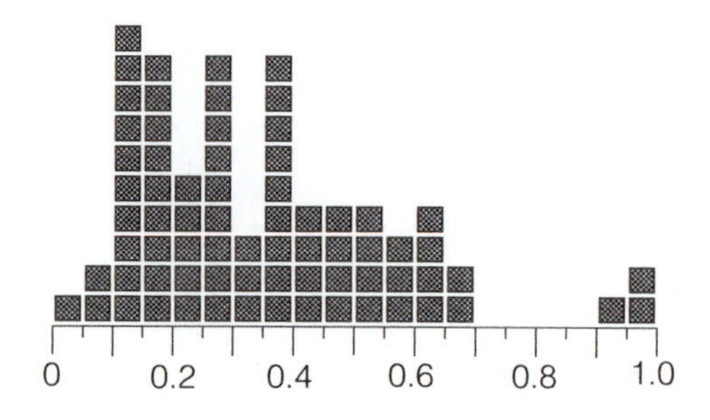

Figure 15-3 Rain use efficiency values for 72 different arid land sites. Data are from Le Houérou *et al.* (1988).

been described in more than 100 separate biomass-based studies throughout arid regions around the globe (Le Houérou, 1984; Le Houérou *et al.*, 1988). Similar tends are also found using satellite-based observations (Tucker *et al.*, 1991; Peters *et al.*, 1993; Prince and Goward, 1995; Prince *et al.*, 1998). Most of the data sets have been produced by measuring the variations in net primary productivity at a particular site from year to year. Le Houérou (1984) proposed the term "rain use efficiency" to describe the slope of the relationship between net primary production and precipitation. He has shown that this linear relationship not only typifies desert regions, but can also be extended into semiarid grassland ecosystems.

Globally, rain use efficiency values range between 0 and 1 g m^{-2} mm^{-1} (Fig. 15-3) (Le Houérou, 1984; Le Houérou *et al.*, 1988). There is a tendency for higher precipitation input sites to have higher rain use efficiencies, but not always. There is an even more evident tendency for disturbed ecosystems to have lower rain use efficiencies (Le Houérou *et al.*, 1988; Varnamkhasti *et al.*, 1995). These include overgrazed sites, regions of soil surface disruption, and areas with wood-harvesting activities. One factor contributing to a decreased rain use efficiency on disturbed sites is likely to be the loss of soil nitrogen, which impacts leaf nitrogen and constrains maximum photosynthetic rates when soil water is available. A second possibility is that in regions with bimodal precipitation patterns, some woody vegetation components may not be equally responsive to winter versus summer moisture inputs, leading to potentially lower rain use efficiencies. These relationships will be discussed further in a later section.

It is perhaps surprising that rain use efficiency values do not show clear differences in C_3- versus C_4-dominated ecosystems (Le Houérou, 1984). Consider, for example, the four major desert regions of western North America (Chihuahuan, Great Basin, Mohave, and Sonoran). These ecosystems differ in C_3/C_4 abundances; rain use efficiency values can be calculated for nongrazed sites based on previously published observations (Whittaker and Niering, 1975; Szarek, 1979; Turner and Randall, 1989). Calculated values range between 0.40 and 0.65 g m^{-2} mm^{-1}, with the Chihuahuan (C_3/C_4, 0.65 g m^{-2} mm^{-1}), Great Basin (C_3/C_4, 0.58 g m^{-2} mm^{-1}), and Mohave (C_3, 0.64 g m^{-2} mm^{-1}) being indistinguishable from each other. One possible explanation for the lack of a relationship between rain use efficiency and photosynthetic pathway is that the more efficient C_4 plants tend to be more summer active, whereas C_3 plants tend to be more spring active (Ehleringer and Monson, 1993). The higher evaporative demand associated with warmer temperatures in the summer would require a greater photosynthetic efficiency in order to attain a rain use efficiency value similar to that of a spring-growing vegetation.

Although both annual and perennial vegetation components contribute to the linear rain use efficiency relationship, it is not evident from regression coefficients that these two vegetation components respond differently to precipitation (Beatley, 1974). Figure 15-3 shows the year-to-year net primary productivity as related to precipitation for Mohave Desert vegetation

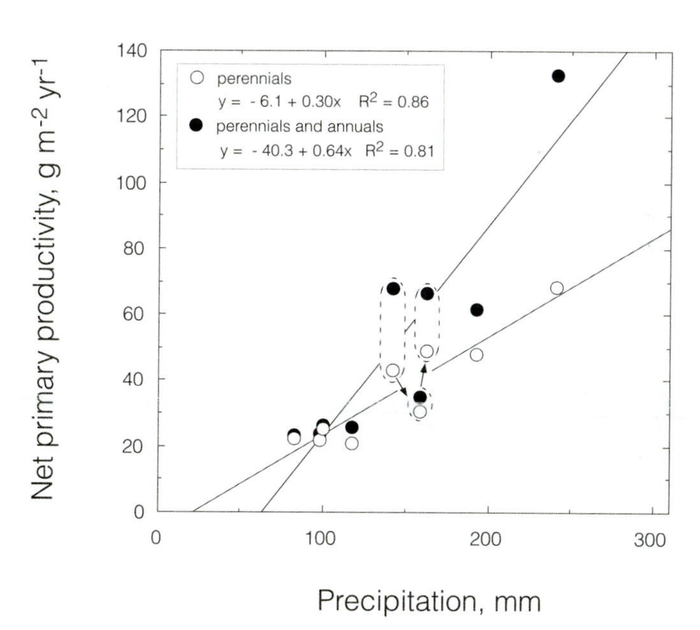

Figure 15-4 The relationship between net primary productivity and precipitation for perennials and all vegetation at the Mohave Desert site, Rock Valley, Nevada. The slope of the regression is known as the rain use efficiency. Arrows and areas enclosed by dashed lines indicate productivity patterns for three successive years. Based on data in Turner and Randall (1989).

in Rock Valley, Nevada. The *x* intercept values indicate that net primary productivity begins at a lower precipitation threshold for perennials than is required to initiate production in annuals. Primary productivity in both annuals and perennials in linearly related to precipitation, yet productivity of annuals is negligible below 100 mm precipitation.

Although the relationship between precipitation and primary productivity is usually linear in desert ecosystems, detailed analyses of year-to-year productivity data often suggest nutrient limitations as a secondary factor. For example, when three successive years of above-average precipitation fell at the Rock Valley site (Fig. 15-4), overall net primary productivity fell in the middle year. This is shown by the circled data in Fig. 15-4, with the arrows showing total net primary productivity from year 1 to year 2 to year 3. Notice that in year 2, the net primary productivity of perennials decreased (well below the regression line) and the productivity of annuals was very low. At the end of any growing season, approximately one-third of the nitrogen may not be recovered and will persist in standing dead plant parts. The likely explanation of the observed pattern in year two is that a significant fraction of the nitrogen within the ecosystem was still in standing dead, leftover from year 1, limiting the capacity of vegetation to photosynthesize even though adequate moisture might have been available. This nutrient limitation was likely removed by year 3, allowing the vegetation productivity to respond as it had in the first year.

III. Seasonality Components of Net Primary Production

Given limited soil moisture availability in deserts and its direct effect on productivity rates, it is essential to understand how different plants will respond to year-to-year changes in the seasonality of precipitation, especially because it is becoming more clear that a macroscale phenomenon, such as El Niño, can influence the duration and amounts of both summer and winter precipitation inputs. How well do desert plants utilize winter-derived and summer, monsoonal moisture inputs?

Cold desert ecosystems of North America and Asia occur at higher elevations. They receive moisture inputs from frontal storms and have a wintertime recharge of deeper soil layers (Gee *et al.*, 1994; Smith *et al.*, 1997), a feature that does not occur in lower elevation, warm desert ecosystems, where plants can be active throughout the winter–spring period. One interesting feature is that many woody species in the cold desert or cold-to-warm desert transition zone in North America tend to use moisture in the upper soil layers only during the spring, but not in the summer (Ehleringer *et al.*, 1991; Donovan and Ehleringer, 1994; Evans and Ehleringer, 1994). This is most evident in evaluating plant responses to summer moisture input, whereby

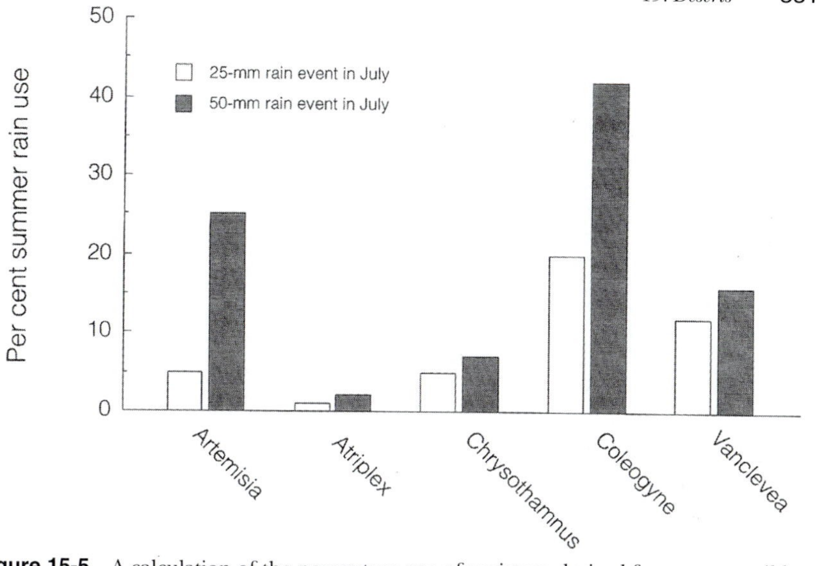

Figure 15-5 A calculation of the percentage use of moisture derived from upper soil layers (summer-derived rain) by five common woody perennials in the Colorado Plateau desert shrub community (Utah). Data are derived from Lin *et al.* (1996) and are based on measured hydrogen isotope ratios of xylem water and a two-member mixing model of deep and shallow soil isotope ratio values of soil water.

monsoon precipitation events often saturate only the upper soil layers. Yet the limited use of monsoonal moisture in the upper soil layers by desert plants is not limited to cold desert species, but can also occur in warm desert woody perennials (Reynolds and Cunningham, 1981; Ehleringer and Cook, 1991).

It is essential to reiterate that soil moisture availability in the upper soil layers is a pulse phenomenon. Because temperatures are warmer in the summer than in the spring, the duration of this pulse will be shorter in the summer than in the spring. Lin *et al.* (1996) have shown that woody perennials in Colorado Plateau deserts have a limited capacity to respond to 25- or 50-mm precipitation events in the summer (Fig. 15-5), although it is known that these plants respond to precipitation events in the spring (Ehleringer *et al.*, 1991). For those woody perennials responding to summer moisture inputs (such as *Coleogyne*), less than 40% of the moisture extracted from the soil and transpired by the shrub is derived from these upper soil layers. Instead it appears that plants are relying to a large degree on moisture stored in deeper soils layers (Thorburn and Ehleringer, 1995; Lin *et al.*, 1996). However, there are life form-dependent differences in the use of moisture in these upper soil layers during the summer. Annuals derive virtually all of their water from

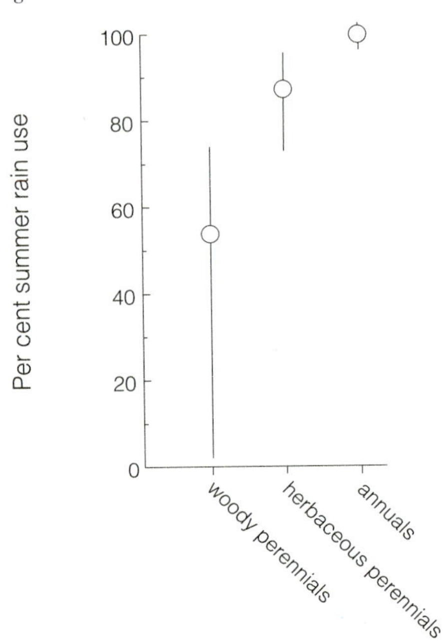

Figure 15-6 The calculated percentage use of moisture derived from summer (August) precipitation events by three life forms on the Colorado Plateau desert shrub community (Utah). Data are means and total ranges for all species within a category. Data are derived from Ehleringer *et al.* (1991).

these surface soil layers, whereas herbaceous perennials derive nearly 85% of their moisture from the shallow depths (Fig. 15-6). On average, only 54% of the water transpired by woody perennials at this site is derived from upper soil layers. Thus, in evaluating the rain use efficiency of a site, an essential factor may be the life form diversity and the capacity to use a specific moisture source. Grazing activities tend to reduce the abundance of annuals and herbaceous vegetation components, which may result in surface-layer moisture being evaporated rather than transpired from a site. It is likely that this relationship explains the observations of reduced rain use efficiencies on impacted arid land sites.

Although the woody shrub *Coleogyne ramosissima* will extract and utilize summer-derived soil moisture, that moisture is not as effective as winter moisture in promoting growth. Ehleringer (unpublished) evaluated the interactions of both competition and water limitations in constraining growth of this dominant shrub. Both the elimination of immediate neighbors and supplemental winter precipitation increased net primary productivity (Fig. 15-7). However, neither supplemental summer precipitation (50 mm) nor

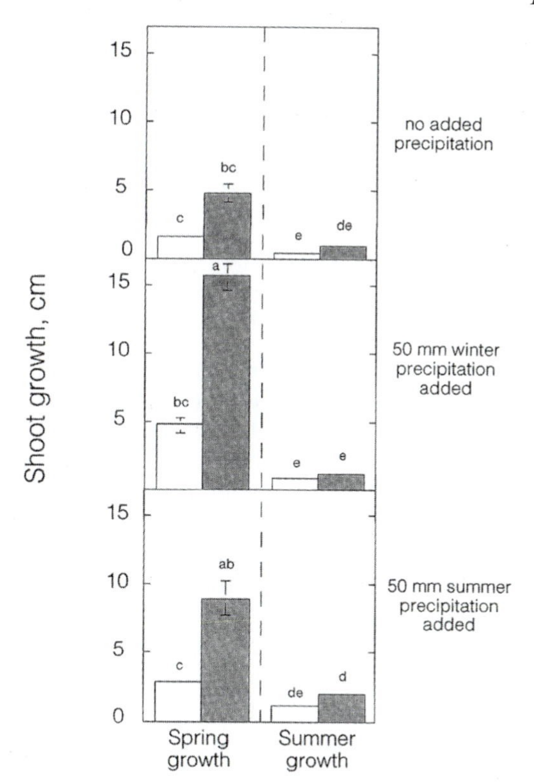

Figure 15-7 The relationships between growth of *Coleogyne ramosissima* and time of the year for treatments receiving supplemental precipitation and/or having their immediate neighbors removed (shaded regions) or neighbors present (unshaded regions). Data are from Ehleringer and Phillips (unpublished).

neighbor removal promoted net primary productivity during the summer period. Thus, it would appear rain use efficiencies in ecosystems dominated by spring-active woody shrub components would decrease under conditions of elevated monsoonal precipitation, particularly if grazing were present to restrict the growth of herbaceous vegetation components.

IV. Human Impacts on Primary Productivity in Desert Regions

The relationships between net primary productivity and human impacts may be generalizable using a simple model illustrating rain use efficiency (Fig. 15-8). Although net primary productivity increases linearly with cu-

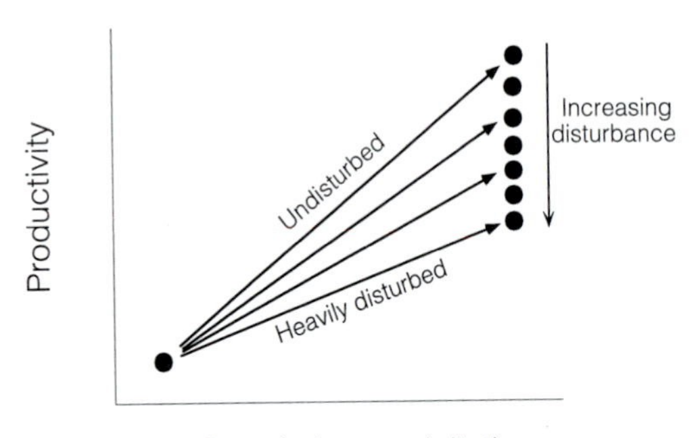

Figure 15-8 A simple model of the relationship between cumulative precipitation and net primary production in desert ecosystems under different levels of soil disturbance.

mulative precipitation inputs through the growing season, human impacts tend to be associated with reductions in the slope of this relationship (i.e., rain use efficiency). The decrease in rain use efficiency is based on two primary factors. First is the reduction in the standing biomass of herbaceous vegetation components through grazing pressures. The second factor is associated with degradation of the land. The causes here may be variable, but it appears that at least in the Colorado Plateau (if not other deserts as well) that breaking up of the biological crust (also known as the cryptobiotic crust) by trampling is perhaps more subtle, but just as important. Arid land ecosystems throughout the world are dominated by a biological crust (Belnap, 1995, 1996), which consists of a diverse mixture of bacteria, fungi, algae, mosses, and lichens. For many desert ecosystems, the biological crust is the primary source of nitrogen input (Evans and Ehleringer, 1993; Evans and Johansen, 1998). Disruption of the crust surface by hooves of cattle, sheep, or goats (or visitors to national recreation areas) results in death of the nitrogen-fixing lichens and algae as they are buried by loosened soil particles (Belnap, 1995, 1996). In the Colorado Plateau desert, where the surface biological crust has been best studied, it is the keystone component influencing both input and loss of nitrogen from the ecosystem. Thus, the biological crust appears to play a central role in affecting primary productivity through its influence on nitrogen availability to higher plants.

Evans and Ehleringer (1994) and Evans and Belnap (1999) provided evidence that soil disturbance was one of the primary factors now decreasing soil nitrogen in desert ecosystems and possibly contributing to desertification (Fig. 15-9). Their model has been evaluated through comparisons of both plant and soil nitrogen on adjacent disturbed and undisturbed sites. Both studies examined the long-term impacts of crust disturbance on

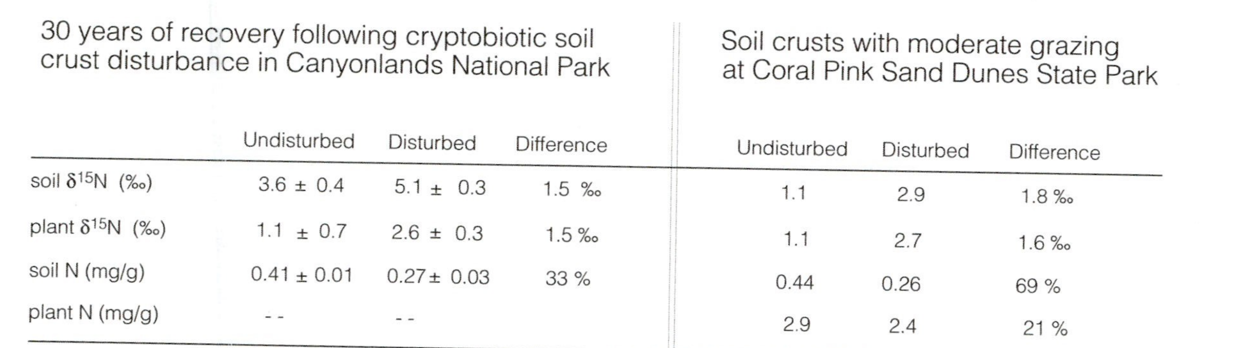

	30 years of recovery following cryptobiotic soil crust disturbance in Canyonlands National Park			Soil crusts with moderate grazing at Coral Pink Sand Dunes State Park		
	Undisturbed	Disturbed	Difference	Undisturbed	Disturbed	Difference
soil $\delta^{15}N$ (‰)	3.6 ± 0.4	5.1 ± 0.3	1.5 ‰	1.1	2.9	1.8 ‰
plant $\delta^{15}N$ (‰)	1.1 ± 0.7	2.6 ± 0.3	1.5 ‰	1.1	2.7	1.6 ‰
soil N (mg/g)	0.41 ± 0.01	0.27 ± 0.03	33 %	0.44	0.26	69 %
plant N (mg/g)	- -	- -		2.9	2.4	21 %

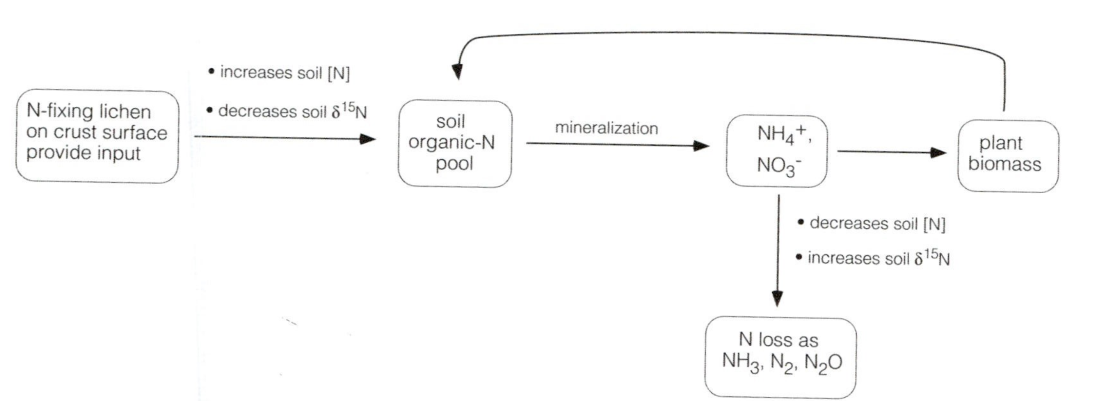

Figure 15-9 Relationships between soil and plant nitrogen isotope ratios and soil and plant nitrogen contents for disturbed and nondisturbed sites within arid land ecosystems on the Colorado Plateau, Utah. Data from Coral Pink Sand Dunes State Park are derived from Evans and Ehleringer (1994); data from Canyonlands National Park are derived from Evans and Belnap (1999).

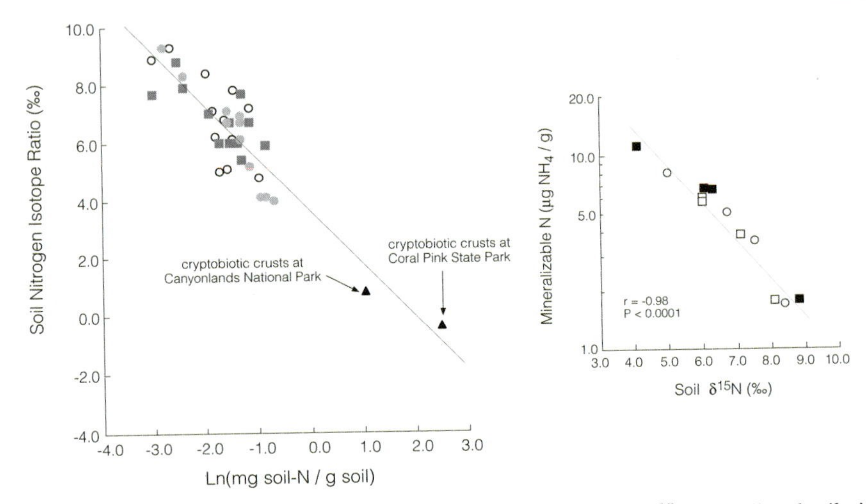

Figure 15-10 The relationships between nitrogen isotope ratio ($\delta^{15}N$) of soil and soil nitrogen content (left) and mineralizable soil nitrogen and nitrogen isotope ratio (right) for soils in arid land ecosystems on the Colorado Plateau. Data are derived from Evans and Ehleringer (1993, 1994).

ecosystem nitrogen relationships by comparing two contrasting ecosystems. At Canyonlands National Park, one site was left undisturbed by cattle (Virginia Park) and the adjacent site (Chessler Park) received light grazing for a decade. Grazing in Chessler Park stopped more than 30 yr ago. Yet recent nitrogen isotope ratio ($\delta^{15}N$) and nitrogen content data indicate that the crusts have not recovered and that the ecosystem continues to lose nitrogen. Chessler Park has only 66% of the soil nitrogen compared to that in the undisturbed Virginia Park. At the Coral Pink Sand Dunes State Park, very light grazing continues on an irregular basis. Note that the disturbed soils are similar to those found in Chessler Park, even though one site is still occasionally disturbed (keeping it from recovering), whereas the other has had 30 yr of recovery time. Both data sets show that soil disturbance and disruption of the biological crusts resulted in substantial decrease in both soil and plant nitrogen. Because plant photosynthesis is directly proportional to leaf nitrogen content (Field and Mooney, 1986; Evans, 1989), disruption of the biological crusts results in very long-term decreases in ecosystem primary productivity. These decreases should be quantifiable in terms of lower rain use efficiencies. At some point, soil nutrition will be impacted sufficiently by disturbance activities such that C_4 species may be expected to out-compete the native C_3 species. This is because photosynthesis is intrinsically more nitrogen-use efficient in C_4 plants, which may translate into a competitive advantage.

The impacts of soil surface disturbance on ecosystem nitrogen cycling and its impact on primary productivity appear to be related to the soil and vegetation $\delta^{15}N$ values. Nitrogen isotope ratios of soils follow a Rayleigh distillation curve (Fig. 15-10), whereby the $\delta^{15}N$ value of the soil N changes in a manner highly correlated with the abundance of nitrogen in the soil. That is, when biological crusts are disturbed, there is a shift in the nitrogen input–loss balance, resulting in a continual loss of nitrogen from the ecosystem. That nitrogen is lost by both dentrification and ammonification processes (Schlesinger, 1997). Because ^{14}N is preferentially lost, the remaining nitrogen is enriched. Nitrogen isotope ratios of nitrogen incorporated into plants reflect the soil enrichment processes (Fig. 15-9).

V. Anticipated Impacts of Global Change on Deserts

Human impacts through grazing activities and land degradation, such as disruption of the biological crusts, are likely to be the dominant factors influencing net primary productivity rates in deserts for some time to come. These stressor impacts may only tend to decrease net primary productivity in desert ecosystems. Nitrogen isotope ratio measurements in plants may be one way of quantifying these long-term ecosystem impacts. Yet, leaf-level ecophysiological responses to elevated carbon dioxide may mitigate these stressors to some degree.

Mooney *et al.* (1991) predicted that water use efficiency would increase under elevated carbon dioxide conditions and that this would increase primary productivity in arid ecosystems. Carbon dioxide acts directly at the stomatal level, reducing stomatal conductance, which results in an increased photosynthesis:transpiration ratio (Knapp *et al.*, 1996). One additional direct, long-term consequence of the reduced stomatal activity will be an extension of the growing season, because under elevated carbon dioxide levels soil moisture is transpired through the vegetation at a slower rate (Ham *et al.*, 1995; Field *et al.*, 1997). In grassland microcosms, Field *et al.* (1997) observed that a doubling of carbon dioxide levels over present-day values reduced evapotranspiration rates by nearly 50%, allowing plant growth substantially longer into the summer drought period.

Although equivalent long-term data on water-use efficiencies are not yet available for desert ecosystems, the same extension of the growing season into the drought period is expected to occur (Fig. 15-11). Modeling studies have calculated the expected impact of an enhanced water use efficiency on seasonal activity and on the increase in leaf area that can be supported by a plant given a reduced transpiration at the single leaf level (Skiles and Hanson, 1994; Neilson, 1995; VEMAP, 1995). Neilson (1995) predicted that as a consequence of elevated carbon dioxide, the leaf area of desert plants in southwestern North America should increase by 25–50% or more (Fig. 15-

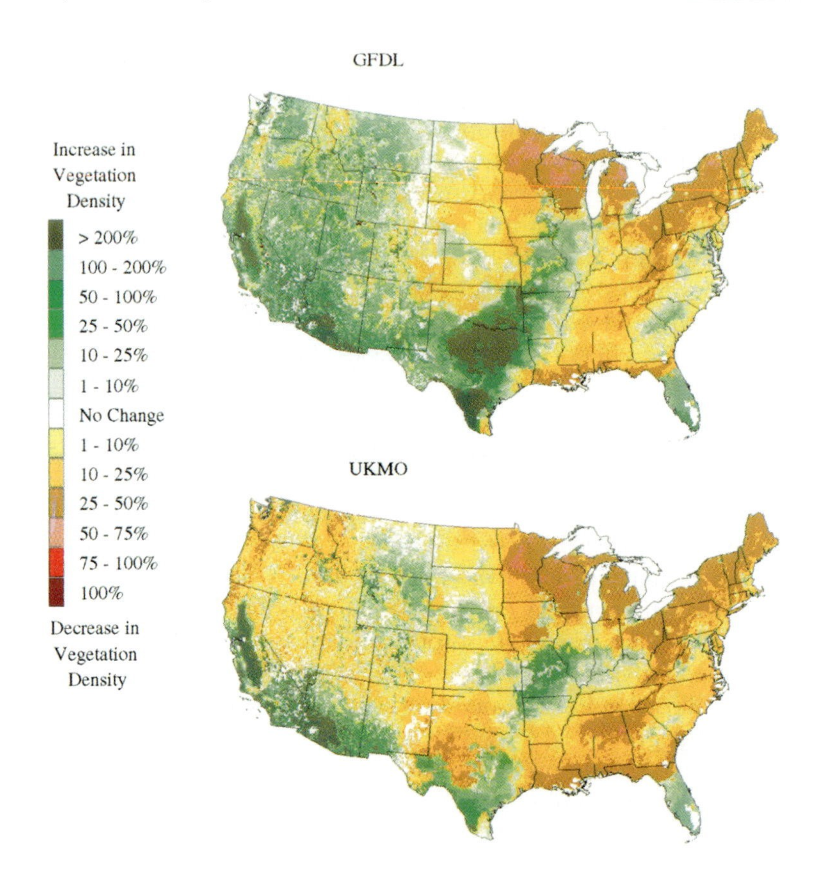

Figure 15-11 Model simulations of the predicted long-term change in vegetation density within the United States in response to a doubling of atmospheric carbon dioxide above present-day values. The Geophysical Fluid Dynamics Laboratory (GFDL) and the United Kingdom Meteorological Office (UKMO) developed two general circulation models. Data are derived from models described in Neilson (1995) and VEMAP (1995). This particular figure first appeared on the back cover of "Our Changing Planet, The FY 1998 U.S. Global Change Research Program," and was a supplement to the President's Fiscal Year 1998 Budget.

11). He refers to this phenomenon as the "greening of the desert," with the expected impact predicted to extend over all arid and semiarid ecosystems in North America. Although global modeling of this phenomenon has not yet be done, it is likely that enhanced water use efficiencies are likely to extend the growing season in most arid ecosystems.

Is there currently strong evidence for broad, climate-induced changes in desert ecosystems? Le Houérou (1996) recently reviewed the available data and concluded that the answer is "no." Although there have been some detectable changes in precipitation of desert regions over the past century, these changes have been region specific and have not shown a consistent, predictable pattern. Desert regions in central Chile and the Sahel have become drier during the past century (Le Houérou, 1996). Yet, in contrast, the deserts of North America may have become wetter during this same period (Nichols *et al.*, 1996). However, there is strong evidence for changes associated with anthropogenic activities, such as those associated with grazing, surface disturbance, and wood harvesting (Le Houérou, 1996). These changes decrease the ability of desert ecosystems to respond to pulse precipitation events. When gas-exchange activities resume following moisture inputs into these impacted regions, it appears that primary productivity usually reflects a reduced rain use efficiency. Yet in satellite-level studies of rain use efficiency in the Sahel, Prince *et al.* (1998) have not been able to detect reduced rain use efficiencies or increased desertification on many of these impacted lands. Future efforts should be directed at resolving this discrepancy, because it lies at the heart of our ability to understand change in primary productivity of desert ecosystems at the global scale.

VI. Conclusions

Primary productivity rates within a desert ecosystem tend to be less than 1500 kg ha^{-1} yr^{-1}, with the productivity rate within a specific location being linear with precipitation inputs. However, the year-to-year primary productivity of a desert ecosystem is challenging to predict, because of high interannual variation in precipitation amounts compounded with human impact (usually grazing related) in many regions. Remote-sensing approaches may be a valuable tool for quantifying productivity in some arid land ecosystems, where vegetation cover is sufficient. Not all perennial plants utilize moisture from the same soil layers, leading to interseasonal differences in the responsiveness of vegetation to biseasonal precipitation inputs. Biological crusts may be a key nitrogen source in many arid regions and disruption of these crusts by grazing is suggested to be a contributing factor to decreased productivity in disturbed arid land ecosystems.

References

Beatley, J. C. (1974). Phenological events and their environmental triggers in Mohave Desert ecosystems. *Ecology* **55,** 856–863.

Belnap, J. (1995). Surface disturbances: their role in accelerating desertification. *Environ. Monitor. Assess.* **37,** 39–57.

Belnap, J. (1996). Soil surface disturbances in cold deserts: Effects on nitrogenase activity in cyanobacterial-lichen soil crusts. *Biol. Fertil. Soils* **23,** 362–367.

Bullock, P., and Le Houérou, H. (1996). Land degradation and desertification. *In* "Climate Change 1995" (R. T. Watson, M. C. Zinyowera, and R. H. Moss, eds.), pp. 171–189. Cambridge Univ. Press, Cambridge.

Caldwell, M. (1985) .Cold desert. *In* "Physiological Ecology of North American Plant Communities" (H. A. Mooney and B. F. Chabot, eds.), pp. 189–212. Chapman and Hall, London.

Caldwell, M. M., White, R. S., Moore, R. T., and Camp, L. B. (1977). Carbon balance, productivity, and water use of cold-winter desert shrub communities dominated by C_3 and C_4 species. *Oecologia* **29,** 275–300.

Donovan, L. A., and Ehleringer, J. R. (1994). Water stress and use of summer precipitation in a Great Basin shrub community. *Funct. Ecol.* **8,** 289–297.

Ehleringer, J. R. (1994). Variation in gas exchange characteristics among desert plants. *In* "Ecophysiology of Photosynthesis" (E.-D. Schulze and M. M. Caldwell, eds.), pp. 361–392. Ecological Studies Series. Springer-Verlag, New York.

Ehleringer, J. R., and Cook, C. S. (1991). Carbon isotope discrimination and xylem D/H ratios in desert plants. "Stable Isotopes in Plant Nutrition, Soil Fertility, and Environmental Studies," pp. 489–497. IAEA, Vienna.

Ehleringer, J. R., and Monson, R. K. (1993). Evolutionary and ecological aspects of photosynthetic pathway variation. *Annu. Rev. Ecol. Syst.* **24,** 411–439.

Ehleringer, J. R., Phillips, S. L., Schuster, W. F. S., and Sandquist, D. R. (1991). Differential utilization of summer rains by desert plants: implications for competition and climate change. *Oecologia* **88,** 430–434.

Evans, J. R. (1989). Photosynthesis and nitrogen relationships in leaves of C_3 plants. *Oecologia* **78,** 9–19.

Evans, R. D., and Belnap, J. (1999). Long-term consequences of disturbance on nitrogen dynamics in an arid grassland ecosystem. *Ecology* **80,** 150–160.

Evans, R. D., and Ehleringer, J. R. (1993). A break in the nitrogen cycles in aridlands: evidence from $\delta^{15}N$ of soils. *Oecologia* **94,** 314–317.

Evans, R. D., and Ehleringer, J. R. (1994). Water and nitrogen dynamics in an arid woodland. *Oecologia* **99,** 233–242.

Evans, R. D., and Johansen, J. R. (1999). Microbiotic crusts and ecosystem processes. *Crit. Rev. Plant Sci.* **18,** 183–225.

Field, C. B., and Mooney, H. A. (1986). The photosynthesis–nitrogen relationship in wild plants. *In* "On the Economy of Plant Form and Function" (T. J. Givnish, ed.). Cambridge Univ. Press, Cambridge.

Field, C. B., Lund, C. P., Chiariello, N. R., and Mortimer, B. E. (1997). CO_2 effects on the water budget of grassland microcosm communities. *Global Change Biol.* **3,** 197–206.

Gee, G. W., Wierenga, P. J., Andraski, B. J., Young, M. H., Fayer, M. J., and Rockhold, M. L. (1994). Variations in water balance and recharge potential at three western desert sites. *Soil Sci. Soc. Am. J.* **58,** 63–72.

Gibbens, R. P., Hicks, R. A., and Dugas, W. A. (1996). Structure and function of C_3 and C_4 Chihuahuan Desert plant communities. Standing crop and leaf area index. *J. Arid Environ.* **34,** 47–62.

Goodall, D. W., and Perry, R. A., eds. (1979). "Arid-land Ecosystems," Vol. 1. Cambridge Univ. Press, Cambridge.

Grover, H. D., and Musick, H. B. (1990). Shrubland encroachment in southern New Mexico, U.S.A.: An analysis of desertification processes in the American Southwest. *Clim. Change* **17,** 305–330.

Hadley, N. F., and Szarek, S. R. (1981). Productivity of desert ecosystems. *BioScience* **31,** 747–753.

Ham, J. M., Owensby, C. E., Coyne, P. I., and Bremer, D. J. (1995). Fluxes of CO_2 and water vapor from a prairie ecosystem exposed to ambient and elevate atmospheric CO_2. *Agric. For. Meteorol.* **77,** 73–93.

Kassas, M. (1995). Desertification: A general review. *J. Arid Environ.* **30,** 115–128.

Knapp, A. K., Hamerlynk, E. P., Ham, J. M., and Owensby, C. E. (1996). Responses in stomatal conductance to elevated CO_2 in 12 grassland species that differ in growth form. *Vegetatio* **125,** 31–41.

Le Houérou, H. N. (1984). Rain use efficiency: A unifying concept in arid-land ecology. *J. Arid Environ.* **7,** 213–247.

Le Houérou, H. N. (1990). Global change: Vegetation, ecosystems, and land use in the southern Mediterranean Basin by the mid twenty-first century. *Israel. J. Bot.* **39,** 481–508.

Le Houérou, H. N. (1996). Climate change, drought and desertification. *J. Arid Environ.* **34,** 133–185.

Le Houérou, H. N., Bingham, R. L., and Skerbek, W. (1988). Relationship between the variability of primary production and the variability of annual precipitation in world arid lands. *J. Arid Environ.* **15,** 1–18.

Lin, G., Phillips, S. L., and Ehleringer, J. R. (1996). Monsoonal precipitation responses of shrubs in a cold desert community on the Colorado Plateau. *Oecologia* **106,** 8–17.

Logan, R. F. (1968). Causes, climates and distributions of deserts. *In* "Desert Biology" (G. W. Brown, ed.), Vol. 1, pp. 21–50.

McGinnies, W. G., Goldman, B. J., and Paylore, P. (1977). "Deserts of the World." University of Arizona Press, Tucson.

Mooney, H. A., Drake, B. G., Luxmoore, R. J., Oechel, W. C., and Pitelka, L. F. (1991). Predicting ecosystem response to elevated CO_2 concentration. *BioScience* **41,** 96–104.

National Academy of Sciences. (1975). "Productivity of World's Ecosystems." National Academy of Sciences, Washington, D.C.

Neilson, R. P. (1995). A model for predicting continental-scale vegetation distribution and water balance. *Ecol. Appl.* **5,** 362–385.

Nichols, M. H., Lane, L. J., and Gibbons, R. (1996). Time series analysis of data for raingauge networks in the Southwest. *In* "Proceedings: Shrubland Ecosystem dynamics in a Changing Environment," pp. 43–47. USDA, Ogden.

Noble, I. R., and Gitay, H. (1996). Deserts in a changing climate: Impacts. *In* "Climate Change 1995. Impacts, Adaptations and Mitigation of Climate Change: Scientific-Technical Analyses" (R. T. Watson, M. C. Zinyowera, R. H. Moss, and D. J. Dokken, eds.), pp. 159–169. Cambridge Univ. Press, New York.

Noy-Meir, I. (1985). Desert ecosystem structure and function. *In* "Ecosystems of the World" (M. Evenari, I. Noy-Meir, and D. W. Goodall, eds.), pp. 92–103. Elsevier, Amsterdam.

Peters, A. J., Reed, B. C., Eve, M. D., and Havstad, K. M. (1993). Satellite assessment of drought impact on native communities of southeastern New Mexico, U.S.A. *J. Arid Environ.* **24,** 305–319.

Pickup, G. (1996). Estimating the effects of land degradation and rainfall variation on productivity in rangelands: An approach using remote sensing and models of grazing and herbage dynamics. *J. Appl. Ecol.* **33,** 819–832.

Prince, S. D., and Goward, S. N. (1995). Global primary production: a remote sensing approach. *J. Biogeogr.* **22,** 815–835.

Prince, S. D., De Colstoun, E. B., and Kravitz, L. L. (1998). Evidence from rain-use efficiencies does not indicate extensive Sahelian desertification. *Global Change Biol.* **4,** 359–374.

Reynolds, J. F., and Cunningham, G. L. (1981). Validation of a primary production model of the desert shrub *Larrea tridentata* using soil-moisture augmentation experiments. *Oecologia* **51,** 357–363.

Rundel, P. W., and Gibson, A. C. (1996). "Ecological Communities and Processes in a Mojave Desert Ecosystem: Rock Valley, Nevada." Cambridge Univ. Press, New York.

Sala, O. E., Parton, W. J., Joyce, L. A., and Lauenroth, W. K. (1988). Primary production of the central grassland region of the United States. *Ecology* **69,** 40–45.

Schlesinger, W. H. (1997). "Biogeochemistry—An Analysis of Global Change." Academic Press, San Diego.

Schlesinger, W. H., Reynolds, J. F., Cunningham, G. F., Huenneke, L. F., Jarrell, W. F., Virginia, R. A., and Whitford, W. G. (1990). Biological concepts in global desertification. *Science* **247,** 1043–1048.

Schulze, E.-D., Mooney, H. A., Sala, O. E., Jobbagy, E., Buchmann, N., Bauer, G., Canadell, J., Jackson, R. B., Loreti, J., Osterheld, M., and Ehleringer, J. R. (1996). Rooting depth, water availability, and vegetation cover along an aridity gradient in Patagonia. *Oecologia* **108,** 503–511.

Skiles, J. W., and Hanson, J. D. (1994). Responses of arid and semiarid watersheds to increasing carbon dioxide and climate change as shown by simulation studies. *Clim. Change* **26,** 377–397.

Smith, S. D., Monson, R. K., and Anderson, J. E. (1997). "Physiological Ecology of North American Desert Plants." Springer-Verlag, Heidelberg.

Szarek, S. R. (1979). Primary production in four North American deserts: indices of efficiency. *J. Arid Environ.* **2,** 187–209.

Thorburn, P. J., and Ehleringer, J. R. (1995). Root water uptake of field-growing plants indicated by measurements of natural-abundance deuterium. *Plant Soil* **177,** 225–233.

Tucker, C. J., Dregne, H. E., and Newcomb, W. W. (1991). Expansion and contraction of the Sahara desert. *Science* **253,** 299–301.

Turner, F. B., and Randall, D. C. (1989). Net production by shrubs and winter annuals in southern Nevada. *J. Arid Environ.* **17,** 23–36.

Varnamkhasti, A. S., Milchunas, D. G., Lauenroth, W. K., and Goetz, H. (1995). Production and rain use efficiency in short-grass steppe: Grazing history, defoliation and water resource. *J. Veg. Sci.* **6,** 787–796.

VEMAP (1995). Vegetation/ecosystem modeling and analysis project: Comparing biogeography and biogeochemistry models in a continental-scale study of terrestrial ecosystem responses to climate change and CO_2 doubling. *Global Biogeochem. Cycles* **9,** 407–437.

Walter, H. (1939). Grasland, Savanne und Busch der ariden Teile Afrikas in ihrer îkologischen Bedingtheit. *Wissen Bot.* **87,** 750–860.

Walter, H., and Breckle, S. W. (1984). "Ecological Systems of the Geobiosphere." Vol. 2. Springer-Verlag, Heidelberg.

Whittaker, R. H., and Niering, W. A. (1975). Vegetation of the Santa Catalina Mountains, Arizona. V. Biomass, production, and diversity along the elevation gradient. *Ecology* **56,** 771–790.

16

Productivity of Tropical Savannas and Grasslands

Jo I. House and David O. Hall

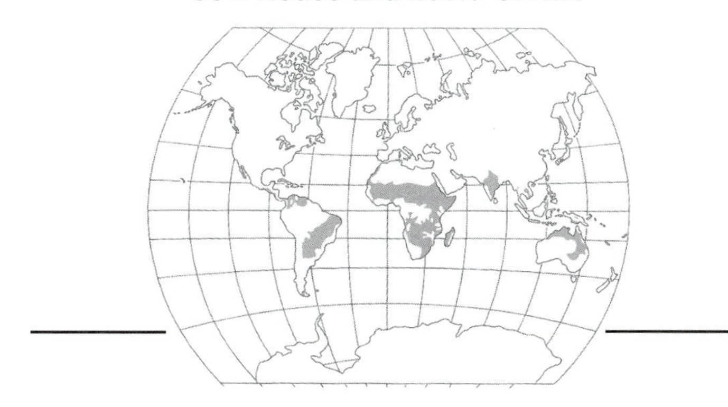

I. Introduction

The savanna biome is diverse, including formations ranging from almost treeless grasslands to more or less closed-canopy woodlands with considerable variation in plant composition, biomass, and net primary productivity (NPP). Savannas cover an extensive area in the tropics, inhabited by a fifth of the human population and supporting the majority of the world's livestock and large mammals. Population pressure and land use changes are high and are likely to increase in the foreseeable future. Savanna areas are increasing due to deforestation and abandoned agriculture, and decreasing due to cultivation and degradation, the balance is probably a decrease. These changes often occur in remote areas and are poorly documented.

Pollen records show that savanna systems are very old, natural systems, characterized by huge heterogeneity, and experiencing rapid changes in plant and animal composition (Menaut *et al.*, 1985). Within this highly dynamic biome, the proportion of trees and grasses can vary enormously, sometimes leading to degradation. Savanna ecosystems and dynamics are currently poorly understood because little attention has been paid to these areas in the past compared to tropical forests or temperate grasslands. Yet it is emerging that savannas have a higher biodiversity (Solbrig *et al.*, 1996a),

Terrestrial Global Productivity

greater productivity (Long *et al.*, 1989, 1992; Scholes and Hall, 1996) and larger impact on global carbon cycles (ORNL, 1998; Hall *et al.*, 1995; Ojima *et al.*, 1993; Scholes and Hall, 1996) than previously realized.

Most research carried out in savanna ecosystems to date has been in the form of short-term site studies of one or two ecological aspects. Some sites have been studied in more detail or over long periods of time, with different emphases, e.g., structure and determinants in Nylsvley, South Africa (Scholes and Walker, 1993), nutrient cycles, fire, and biophysical properties in Lamto, Côte d'Ivoire (Menaut *et al.*, 1999), determinants, fire, and biophysical properties in the Orinocos Llanos, Venezuela (San Jose and Montes, 1997), and hydrology and energy balances in HAPEX-Sahel, Niger (Goutorbe *et al.*, 1997). Due to the great heterogeneity of the savanna biome from site to site, and even from year to year, individual studies cannot provide a representative picture of the whole biome.

More recently there has been a shift from individual studies to syntheses under specifically designed collaborative research programs such as the "Responses of Savannas to Stress and Disturbance" (RSSD) 10-year program set up in 1983 (Frost *et al.*, 1986; Walker, 1987; Walker and Menaut, 1988; Solbrig, 1990); "The Primary Productivity and Photosynthesis of Semi-Natural Ecosystems of the Tropics and Sub-Tropics" program, UNEP 1984 (Long *et al.*, 1989, 1992); the International Geosphere–Biosphere Programme (IGBP) terrestrial transects, which include the Kalahari Transect, the North Australian Terrestrial Transect (NATT), and SALT (Savanna in the Long Term) Ivory Coast to Mali (Koch *et al.*, 1995); The Miombo Network (http://miombo.gecp.virginia.edu); SAFARI 2000 (Southern African Fire–Atmosphere Research Initiative, http://safari.gecp.virginia.edu); and the SCOPE tree–grass modeling group (http://www.nceas.ucsb.edu, search under projects, tree-grass). Over the years, several good synthesis reviews have been published on savannas (Bourlière and Hadley, 1970; Huntley and Walker, 1982; Bourlière, 1983; Sarmiento, 1984; Tothill and Mott, 1985; Cole, 1986; Werner, 1991; Young and Solbrig, 1993; Scholes and Hall, 1996; Solbrig *et al.*, 1996a; Scholes and Archer, 1997).

II. Definition

The classification of "savanna" is inexact and has been based on different, often subjective, criteria. Savannas are most commonly defined as tropical seasonal ecosystems with a more or less continuous herbaceous cover and a discontinuous cover of trees or shrubs in varying proportions (Frost *et al.*, 1986). It is this coexistence and close interaction of herbaceous and woody species that makes savannas unique and complex. The proportions of woody:grass biomass are highly variable and subject to environmental and management changes within broad climatic and edaphic constraints (Scholes and Archer, 1997).

The one constant and defining characteristic of savannas is the seasonality of rainfall: "Tropical savannas, woodlands and grasslands can be defined as those formations constrained by water rather than temperature with an annual dry season of sufficient duration and intensity to cause woody plants to shed their leaves and grasses to dry out, thus providing dry fuel for periodic fires" (Huntley and Walker, 1982). Authors often distinguish between "humid" and "arid" savannas based on a rainfall amount—typically of 600 mm (Walker, 1985), or length of dry season whereby 2.5–5 months is classified as "humid" and 5–7.5 months is "arid"; in some areas the dry season may even last 9 months (Solbrig *et al.*, 1996b). Seasonality may be bimodal with long and short rainy seasons. Fire and grazing are also integral elements in natural and managed savannas.

Savannas form a continuum between tropical forests and grasslands and have often been classified as either one or the other in the past. The distinction between what is forest, grassland, and different structural savanna types can only be set with arbitrary limits and descriptions such as those defined by Scholes and Hall (1996):

Forests: Complete tree canopy cover and three or more overlapping vegetation strata
Woodlands: 50–100% tree canopy cover, and a graminaceous layer
Savannas: 10–50% cover by woody plants and well-developed grass
Grasslands: Less than 10% tree cover

Different authors and classification systems vary as to whether "savanna" includes dense woodlands or treeless tropical grasslands. Some authors use the term "tropical grasslands" to include "the mixed grass and tree communities of the savanna and savanna forest" (Long *et al.*, 1992), and others use "rangelands" in the same context. In this chapter we use the term "savanna" to include the whole range from treeless grasslands to closed-canopy woodlands (with a graminaceous layer) unless otherwise stated.

Savanna Formations

A number of distinctive savanna formations exist, and some common definitions (scientific and vernacular) are listed below (Menaut, 1983; Sarmiento, 1984).

Grass savannas or grasslands: Without woody species taller than the herbaceous stratum
Bushlands/shrublands: Low-statured trees (<3 m), <2% = shrub, 2–15% = bush savanna
Dwarf shrubland: Woody plants <1 or 3 m tall (depending on author)
Wooded grasslands, savanna grasslands, savanna woodland: 2–15% tree cover

Savanna parkland: Mosaic patches of trees
Thicket/scrub forest: Low-statured, multistemmed woodlands or forests, often impenetrable and clumped, with absent or discontinuous grass layer
Tiger savanna: Tree patches in linear bands parallel to slope contours
Bushveld, lowveld, thornveld: Different savanna formations in Southern Africa
Steppes: Savanna grasslands, sometimes with woody component

III. Extent

Savannas occupy a wide range of environments. Rainfall can vary from 200 to 2000 mm, and tends to be more erratic at the arid end of the scale, often occurring as short storm events (Walker 1985; Solbrig *et al.,* 1996b). Temperatures vary from hot tropical temperatures with little seasonality to subtropical temperatures that can approach frost in the winter. Soils vary from fertile to infertile. Typically, savannas in high rainfall areas have poor soils because wet areas with good soils tend to tropical rainforests. Similarly, arid savanna areas tend to have nutrient-rich soils, because low rainfall plus poor soils leads to desert vegetation. However, there are exceptions to these rules.

The extent of savannas is uncertain due to the variation in classification of this biome and the sparsity of data in these areas. Table 16-1 shows a range of estimates. Whittaker and Likens (1973, 1975) estimated savanna cover in 1950 to be 15 Mkm^2 using a modified UNESCO scheme. Atjay *et al.* (1979) used the same classification as Whittaker and Likens for their estimate of 22.5 Mkm^2, which incorporates updated reports and vegetation maps as well as a consideration of human interference such as the cutting of forest areas and subsequent formation of secondary or derived savannas. The Olson *et al.* (1983) map reflects better knowledge of classification, mapping, and ecosystem change, and the tropical and temperate tree-grass classes included in Table 16-1 represent a quarter of the global land surface. Scholes and Hall (1996) used the digitized version of this map published by the United States Environmental Protection Agency (U.S. EPA) for their estimate of 16.1 Mkm^2 using the four savanna classes they considered tropical. However, 74.5% of the "warm or hot grasslands" class is within the tropics giving a total of 27.6 Mkm^3 (almost a fifth of the global land surface). Temperate tree-grass mixes occur in the warmer areas of North America (5 Mkm^2) (McPherson, 1997) Mediterranean Europe, and parts of Russia and Asia.

A. Africa

Savannas form a semicircle around the western central rainforest areas, bordered by the desert zones to the north and south, across a variety of soil conditions with rainfall ranging from 200 to 1800 mm. *Broadleaved savannas* are found in the subhumid interior plateau on old, highly weathered, infertile soils. *Fineleaved savannas* are typical of the low-lying, semiarid regions. The

Table 16-1 Previous Estimates of Area, Biomass, and NPP of Tropical Savannas and Grasslands[a]

Source	Vegetation type	Area (Mkm^2)	Biomass (Pg DM)	Biomass (kg DM m^{-2})	NPP (Pg DM yr^{-1})	NPP (g DM m^{-2} yr^{-1})
Whittaker and Likens (1973, 1975)	**Savannas** (includes tropical grasslands)	15	60	4 (0.2–15)	13.5	900 (200–2000)
Atjay *et al.* (1979)	Dry savanna thorn forest	3.50	52.5	15.0	4.55	1300
	Low tree/shrub savanna	6.00	45	7.5	12.60	2100
	Dry thorny shrubs	7.00	35	5.0	8.40	1200
	Grass-dominated savanna	6.00	13.2	2.2	13.80	2300
	Total	**22.50**	**145.7**	**6.5**	**39.35**	**1749**
Olson *et al.* (1983)	Tropical dry forest and woodland (32)	4.7	73.3	15.6	6.00	1271
	Tropical savanna and woodland (43)	6.7	44.9	6.7	7.33	1091
	Succulent and thorn woods (59)	4.0	35.6	8.9	3.56	889
	Semi-arid woodland or low forest (48)	0.9	11.1	11.1	0.89	977
	Warm or hot shrub and grasslands (41)	17.3	50.2	2.9	15.56	899
	Total (temperate and tropical)	**37.3**	**247.1**	**6.6**	**36.22**	**972**
Scholes and Hall (1996)	Drought-deciduous woodlands	4.6 (4.2)	34.4	8.3	5.2	1263 (462–1789)
	Savanna	6.7 (6.0)	15.1	2.5	8.6	1426 (681–1941)
	Succulant and thorn woods	3.9 (3.1)	1.4	2.5	2.7	856 (289–1370)
	Eucalyptus and *Acacia* woodlands	0.9 (0.5)	7.8	2.5	0.4	733 (186–1242)
	Total (tropical)	**16.1 (13.9)**	**58.7**	**3.6 (0.9–21)**	**16.9**	**1216 (440–4135)**
This study	Tropical dry forest and woodland	4.7	64.7	13.7	6.0	1263
	Tropical savanna and woodland	6.7	31.8	4.7	9.6	1426
	Succulent and thorn woods	4.0	23.4	5.8	3.4	856
	Semi-arid woodland or low forest	0.9	6.3	7.0	0.7	733
	Warm or hot shrub and grasslands	11.2	32.4	2.9	10.1	899
	Total (tropical)	**27.6**	**158.5**	**5.75**	**29.7**	**1078**

[a]DM, Dry matter; Mg = 10^{16} g; Pg = 10^{15} g. Used carbon content of dry matter of 0.45. Olsen *et al.* (1983) categories include tropical at temperate tree-grass mixes (not montane) Scholes and Hall areas, based on Olsen *et al.* (1983) classes 32, 43, 48, and 59, estimated undisturbed area (i.e., still under natural vegetation, not urban/agricutural/degraded) shown in brackets and used for biomass and NPP calculations, although corrected totals shown here due to inconsistency in use of land areas in original paper. NPP calculated using the relationship between WAI and total NPP shown in Fig. 16-3 and discussed in text. This study land areas from Olson *et al.* (1983), tropical portion of "warm or hot shrub and grassland" (between 30°N and 30°S) calculated by Dale Kaiser, ORNL, USA. Biomass average of Table 16-3 based on Scholes and Hall (1996) and Olson *et al.* (1983).

northern *Sudan-type* savannas are open xerophytic grasslands with scattered deciduous trees forming a transition with the Saharan desert vegetation. The *Guinea-type* savanna woodlands form the transition with the evergreen moist forests. The eastern African savannas, with bimodal rainfall typically totalling less than 700 mm, are dominated by herbaceous vegetation with some shrubs or scattered trees. The southern savanna area is known as "*Miombo*" woodland due to its distinctive tree species. (Menaut *et al.*, 1985; Backéus, 1992; Scholes and Walker, 1993; Solbrig, 1996).

B. Australia

Australian savannas occur in two distinctive climatological areas—the cooler, wetter east coast and the warmer, drier north coast. Rainfall is mostly below 1000 mm, soils tend to be poor, fires frequent, and diversity high. Australian savannas include monsoon, tropical, and subtropical tallgrass communities along coastal areas with *Eucalyptus* woodlands; the midgrass savannas on clay soils with *Acacia harpophylla* (Brigalow) and associated *Eucalyptus* woodlands; tussock and hummock grasslands (Mitchell grass); and *Acacia* shrublands (mulga and gidgee pastures) (Mott *et al.*, 1985; Braithwaite, 1990; Burrows *et al.*, 1990, McKeon *et al.*, 1990; Solbrig, 1996).

C. South America

The South American savannas tend to be wetter, with less contrasting seasonal variation, and very nutrient-poor soils, often high in aluminum. The largest savanna type, the cerrado, covers 1.8 Mkm^2 within Brazil and includes the pure grassland campo limpo, through the low open woodland of the true *cerrado,* to the closed dry-forest formations of the cerradao. In Colombia/Venezuela the llanos del Orinoco area is grassland with scattered trees (San Jose and Montes, 1989). Flooded savannas occur in Brazil and Bolivia. The *caatinga* region of Brazil and the *chaco* region of Paraguay/Bolivia/Argentina (which can suffer frost) are often not considered savannas but do fall into the descriptions used by African ecologists. Other smaller areas of distinct savanna types are scattered through South America, Central America, and Cuba (Sarmiento, 1983; Medina and Silva 1990; Solbrig, 1996).

D. Asia

Savannas are mostly "secondary" or "derived" and are formed by deforestation, abandoned cultivation, and burning, being maintained by repeated grazing, harvesting, and burning. Savannas are fairly extensive in India and Sri Lanka, with continued forest clearing increasing their extent, although many areas are under threat from agriculture (Misra, 1983; Yadava, 1990; Backéus, 1992; Pandey and Singh, 1992). Savannas are not so common in South East Asia. Thailand, Laos, Cambodia, and Vietnam have an open deciduous dipterocarp forest with the ground covered by grasses, similar to some vegetation types in Africa, which form a transition between dense forest and shrub savanna. Treeless savannas occur in many places but occupy only a small area (Blasco, 1983; Stott, 1990).

IV. Plant Composition: Structural and Functional Variability

A. Savanna Structure: Tree–Grass Mix

The relative abundance of woody and herbaceous species is highly dependent on environmental conditions and seasonal and interannual variations. Plant available moisture (a function of climate, soil type, and topography) seems to be the key determinant of the tree–grass balance, but structure, function, and species composition are also altered by available nutrients, fire, and herbivory, and these are all discussed later (Walker and Noy-Meir, 1982; Menaut *et al.*, 1985; Solbrig, 1990; Scholes and Hall, 1996; Solbrig *et al.*, 1996b; Scholes and Archer, 1997; Scholes *et al.*, 1997).

Within this naturally dynamic ecosystem, small environmental/management changes may be buffered in the short term, but in the long term can lead to shifts in species and even functional types. Larger changes can lead to rapid transformations in vegetation structure, which may go beyond certain thresholds into a new structural domain, from which it is hard to return (Menaut *et al.*, 1985). For example, fire exclusion and heavy grazing promote increases in tree cover, grasses are suppressed, and there is no fire, thereby leading to dense thickets resistant to fire. Conversely, beyond a certain high fire and low grazing threshold large amounts of grass lead to regular hot fires, destroying all seedlings eventually excluding trees.

B. Plant Function

Functionally, arid and humid savannas are very different (Walker, 1985). Arid savannas have more affinity with semidesert vegetation in their adaptation to water limitation, and growth is less predictable and responds closely to rainfall events. Humid savanna vegetation has more adaptations for fire and low nutrients and trees are functionally more similar to forest vegetation. Savanna vegetation also shows adaptations for herbivory (particularly thorns) and light/shade. Many plants have underground stems and complex root systems to cope with high environmental stresses. Table 16-2 captures the main functional attributes that reflect current environmental conditions for southern Africa (Scholes *et al.*, 1997).

Grasses with a C_4 photosynthetic pathway tend to be dominant, particularly in hotter environments (Huntley, 1982; Menaut *et al.*, 1985). Grasses in wetter African sites tend to be perennial and tufted, with fewer larger seeds, whereas in more arid sites they are rhizomatous or stoloniferous, producing

Table 16-2 Broad Plant Functional Types Found in African Savannas[a]

Plant type	Phenology	Water	Herbivory	Nutrients	Plant functional type common name
Woody plants Long lived (>5 years); high fraction of secondary growth	Drought deciduous	(Mostly fine-leaved)	Thorny (hydrolyzable tannins)	Fast growth and nutrient uptake; may be N-fixing	Thorn shrub/tree
		(Mostly broad-leaved)	Indigestible (condensed tannins)	Inherently slow growth and uptake capacity	Broadleaved shrub/tree
	Evergreen	Sclerophyllous	Indigestible (fiber and tannins)	Inherently slow growth and uptake capacity	Evergreen shrub/tree
		Succulent	Often palatable, sometimes tannins, resins, or alkaloids		Succulent shrub/tree
		Phreatophytic	Tall, or unpalatable		Phreatophyte
	Wet season deciduous	Water storage organ	Toxic	Inherently slow growth	Geoxylic suffrutex
Graminoids Short lived (<2 yr per tiller); monocots, buds basal; mainly vegetative reproduction		Xerophytes	High fiber	Roots sheaths common on sandy soil	Wire grasses
		Mesophytes	Low digestibility	Seasonally low N	Tuft grasses
			Lawn-forming	Continuously adequate N	Creeping grasses
Forbs As above, but dicots with terminal buds; mainly sexually reproducing	Ephemeral/annual	Avoid drought as seeds	Toxic (alkaloids, etc.)		Ephemeral forbs
	Perennial	Drought deciduous	Often palatable but hairy	N-fixing	N-fixing forbs
				Non-N-fixing	Perennial non-N-fixers
Geophytes	Antiseasonal	Water storage bulb	Toxic		Geophytes

[a]From Scholes *et al.* (1997).

many small seeds (Scholes *et al.*, 1997). Some may be fairly deep rooted with some roots even approaching 2 m. Annual grasses tend to have shorter roots compared to perennials and are more competitive in upper soil layers, responding faster to low and erratic rainfall patterns where there is not so much deep percolation. On the other hand, perennials can respond more quickly to the first rains, mobilizing root stores before annuals start their growth from seed. On the whole, grasses are able to respond quickly to rainfall events with rapid leaf flush and high productivity, but leaf loss during dry periods reduces losses due to respiration.

Trees have access to deep water, enabling prerain leaf growth by up to 3 months before the first rains (Tybirk *et al.*, 1992), and late leaf shedding, which enables a longer growing season compared to grasses (Scholes and Walker, 1993). It was initially thought that trees essentially used deeper soil water than grasses, and this resource partitioning allowed the coexistence of these two life forms (Walter, 1971). However, it is now known that some trees, particularly in more arid areas, have shallow roots that extend laterally (particularly on clay soils where percolation is low) and may not have many deep roots (e.g., Menaut, 1983; Knoop and Walker, 1985; Belsky *et al.*, 1989; Coughenour *et al.*, 1990; Le Roux and Bariac, 1998). Besides water, this takes advantage of nutrient concentration in upper soil layers, including that washed out from fire ash in the early rains before grasses become established. In arid areas, deciduous trees are more common because leaf loss reduces water stress. Deciduous trees in humid savannas are thought to have most of their roots in the upper soil layers, thus they shed their leaves in the dry season. Evergreen trees have extensive root systems, enabling them to use water from deep soil layers, thus maintaining relatively high transpiration and photosynthetic rates during the dry season (Vareschi, 1960).

There are few empirical data on allocation of primary production in savanna plants. Herbaceous plants are short lived and almost all aerial parts die back after reaching maturity, to be replaced at the start of the next growing season. Perennials allocate more primary production to belowground structures, dropping their leaves and increasing root production at the onset of the dry season. Annuals, on the other hand, have much higher seed production and may produce seed several times throughout the growing season to avoid drought, fire, and herbivory. Woody plants invest more production in long-lived structures (the bole, branches, and coarse roots). They also use a higher proportion of primary production for maintenance respiration (Illius *et al.*, 1996). The fraction of NPP allocated belowground for all plants is affected by nutrient and water availability, fire, herbivory, and competition, increasing with stress and disturbance. Belowground allocation is typically 40–80% in grasses compared with 20–60% in trees (Scholes and Hall, 1996).

C. Tree–Grass Interactions

Interactions, both competitive and facilatory, between tree and grass components are very complex, variable, and poorly understood (Scholes and Archer, 1997). Trees and grasses both compete for limited resources, and grass biomass/NPP typically decreases as tree biomass/density increases. Several studies show a concave curvilinear change (Donaldson and Kelk, 1970; Walker *et al.,* 1972; Beale, 1973; Dye and Spear, 1982; Scanlan and Burrows, 1990; Teague and Smit, 1992). The inverse relationship between an index of "treeness" and grass production shows the steepest decline when woody biomass is low (Fig. 16-1). It has been shown at one site that the degree of curvature decreases as the productive potential of the site (i.e., water and nutrient availability) increases, suggesting that it is related to the degree of resource competition (Scanlan and Burrows, 1990). Convex relationships also occur (Aucamp *et al.,* 1983). Trees can create a more suitable microhabitat for grass species, and low tree densities have been seen to increase grass production in arid areas by providing shade (reducing evaporation) and nutrient concentration due to leaf litter and root decay and feces of sheltering animals and birds (Weltzin and Coughenour, 1990; Belsky *et al.,* 1989, 1993; Pugnaire *et al.,* 1996).

In most situations, mature trees out-compete grasses for light, water and nutrients, yet grasses out-compete small shrubs and tree seedlings (reducing establishment) and they increase the likelihood of fires, which kill small trees (Knoop and Walker, 1985; Scholes and Archer, 1997). This competitive asymmetry can lead to structural instability (Scholes and Hall, 1996). Often some degree of tree clumping takes place, adding further complexity, with conditions often very different between the undercanopy and intercanopy areas (Weltzin and Coughenour, 1990; Belsky, 1989, 1993; Veetas, 1992; Mordelet and Menaut, 1995).

D. Plant Composition and Ecosystem Productivity

This structural diversity raises the question of how the mixture of trees and grasses, and changes in this mixture, will affect overall ecosystem productivity. The theory that the total NPP of a site remains fixed as the vegetation mix changes (an assumption in some top-down NPP models) seems unlikely, but because most studies measure only the NPP of one component, it is difficult to draw conclusions. It seems probable that some combination of woody and herbaceous biomass leads to higher ecosystem productivity, compared to having one component alone, as is the assumption of some multispecies agroforestry studies. This is partly because of facilitation, and also that trees and grasses can take advantage of different resources spatially and temporally (seasonally).

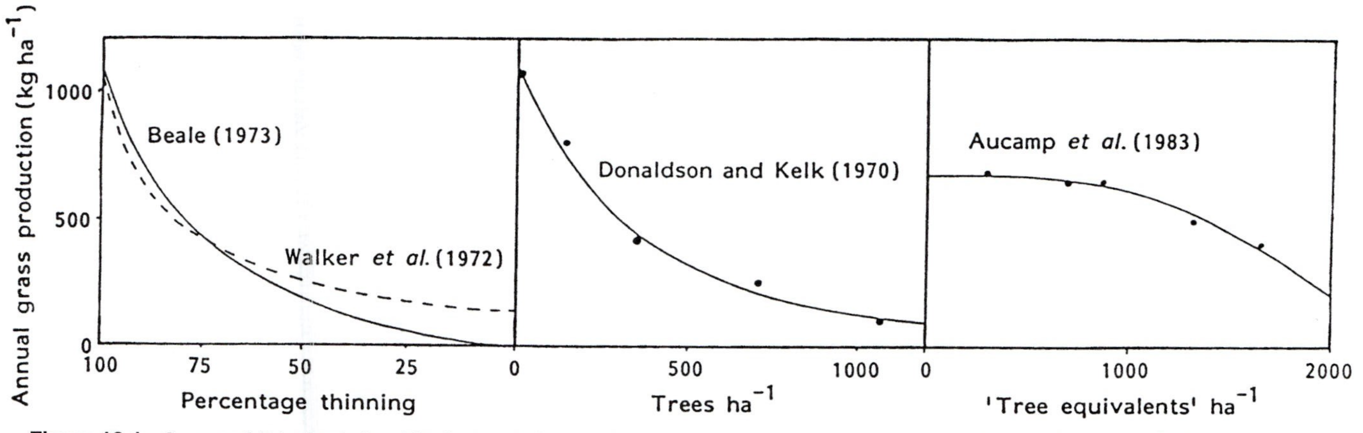

Figure 16-1 Some published relationships between the quantity of trees in a savanna and the annual grass aboveground NPP. Almost all the studies show concave inverse relations; the inverse convex relationship is unusual (reproduced from Scholes and Hall, 1996).

V. Estimates of Biomass and Productivity

There is a paucity of data on the biomass and productivity of different types of savannas and grasslands under the considerable range of climatic and soil conditions they experience. Of the studies that have measured savanna biomass and/or productivity, very few have been sufficiently long term to capture the interannual variations (which can be considerable), few have measured both tree and grass components, even fewer have measured roots, and almost none have looked at insectivorous and microbial production, although herbivorous production is sometimes considered. Thus there are no estimates for total ecosystem biomass/productivity. This chapter considers plants only.

Table 16-1 presents several estimates of biomass each based on different published data. Biomass values reported for individual savanna sites and compiled in Scholes and Hall (1996) are shown in Table 16-3, with the original class names and groupings, although the calculations have been redone and totals differ slightly from the original paper. Due to the paucity of root data, total biomass was calculated from average aboveground biomass using estimates belowground biomass percentage, ranging between 20% and 30% in different savanna types. Jackson *et al.* (1996, 1997) reported a higher average belowground biomass of 40% for tropical savannas and grasslands in their compilation and analysis of root data for all global biomes.

It is difficult to match up the biomass classes given by Scholes and Hall (1996) in Table 16-3 with the land use classes of Olsen *et al.* (1983) in Table 16-1. Scholes and Hall (1996) assigned their *woodlands* biomass class (originally 83 t/ha) to the land class *drought–deciduous woodlands.* In fact, the original Olson *et al.* (1983) description of *tropical dry forest and woodland* class includes drought deciduous forest types such as Australian Brigalow as well as the Miombo woodlands, therefore it seems more reasonable to combine these two classes in Table 16-3. Scholes and Hall (1986) assigned their *savannas* biomass class value (originally 25 t/ha) to all other land classes as it included examples from all of them. The Scholes and Hall (1996) biomass values are all lower than the Olson *et al.* (1983) values which were also based on reported data. For the purposes of this chapter it seems reasonable to take an average of the two sets of estimates, using the re-calculated Scholes and Hall (1996) values in Table 16-3. This gives an average biomass value for the savanna biome of 5.75 kg DM m^{-2}, or a total biomass of 158.5 Pg (DM). Assuming 30% of biomass in roots, the average belowground biomass is therefore 1.73 kg DM m^{-2}, and aboveground is 4.02 kg DM m^{-2}. The Jackson *et al.* (1996, 1997) analysis of root data calculated a total (tree and grass) root biomass of 1.4 kg m^{-2} (of which fine roots account for 0.99 kg m^{-2}). Using his value of 40% root biomass this would equal a total biomass of 3.5 kg m^{-2}, much lower than the other estimates.

Table 16-3 Biomass Reported for Tropical Grasslands and Savannas[a]

Site	Source	Aboveground biomass (t/ha)			Roots	Aboveground biomass	Belowground biomass	Total biomass
		Wood	Tree leaf	Grass	(t/ha)	Total (t/ha)	(%)	(t/ha)
Drought-deciduous forests								
Kurukshutra, Inda	Rajvanshi & Gupta (1985)	53.12	3.29	1.33		57.74		
Brigalow, Australia *Dalbergia sissoo*	Moore *et al.* (1967)	158.1	11	0	40.7	169.1		
Oro Forest, Nigeria *Acacia harpephylla*	Fatubarin (1984)	132				132		
Mean total aboveground biomass						**119.6**	**20**	**149.5**
Woodlands (all *Brachystegia* spp.) = Miombo woodlands								
Kasapa, Zaire	Malaisse *et al.* (1975)	101.7	4.82			106.52		
Makaholi, Zimbabwe	Ward & Cleghorn (1964)	44.6	2.29			46.89		
Zambia		40				40		
Marondera, Zimbabwe	Frost (in Scholes & Hall, 1996)	39.73	2.74	0.1		42.57		
Dukwe, Botswana	Tietema (1993)	81.9				81.9		
Mean total aboveground biomass						**63.9**	**25**	**84.8**
Average of above two classes = Olson *et al.* (1983) "tropical dry forest and woodland" class				**117.1**				
Savannas								
Sengwa, Zimbabwe	Guy (1981)	22.5	0.53			23.3		
Cobar, Australia	Harrington & Johns (1982)	36	3.2	1		40.2		
Nylsvley, S. Africa *Burkea africana*	Scholes & Walker (1983)	15.54	0.748	1.5	7.1	17.788		
Waterberg, Namibia *B. africana*	Rutherford (1982)	22.3				22.3		
Klaserie, S. Africa *Colophospermum Mopane*	Scholes (1988)	20.84	0.8	0.5		22.14		

(*continues*)

Table 16-3 *(Continued)*

Site	Source	Aboveground biomass (t/ha)			Roots (t/ha)	Aboveground biomass Total (t/ha)	Belowground biomass (%)	Total biomass (t/ha)
		Wood	Tree leaf	Grass				
Klaserie, S. Africa *Acacia nigrescens*	Scholes (1988)	5.05	0.664	1		6.714		
Klaserie, S. Africa *Combretum apic.*	Scholes (1988)	10.5	0.759	0.5		11.759		
Kruger, S. Africa *Combretum apiculatum*	Dayton (1978)	15.4	1.5	0.5		17.4		
SE Zimbabwe *Colophospermum Mopane*	Kelly & Walker (1976)	19.7				19.7		
Hwange, Zimbabwe *Terminalia sericea*	Rushworth (1978)	3.65	3.21	1.23	34.4	8.09		
Khakhea, Botswana *Acacia/Combretum*	Tietema (1993)	8.05				8.05		
Kang, Botswana	Tietema (1993)	18.02				18.02		
Morwa Forest, Botswana *A. tortilis*	Tietema (1993)	17.5				17.5		
Morwa Hill, Botswana *Croton/Combretum*	Tietema (1993)	30.87				30.87		
Dikeletsane, Botswana *Acacia/Combretum*	Tietema (1993)	32.97				32.97		
Mean total aboveground biomass (incorporates several Olsen *et al.* savanna classes						**19.8**	**30**	**28.2**

[a]Table re-compiled from Scholes & Hall (1996) Table 4.4, totals, re-calculated, hence values differ slightly. Categories and groupings assigned by Scholes & Hall (1996) according to their judgement.

Woody biomass production is nonlinear, declining with maturity (which can be in the range of 30 to 100 yr) (Scholes and van der Merwe, 1996). It is not possible to estimate average annual production from tree rings because growth may be negligible in drought years, and certain regions experience two growing seasons in some years. Therefore production is typically estimated from change in biomass alone, which is an underestimation because it does not take account of death, removals, and turnover. Litter traps fail to account for dead branches. Removals for fuelwood, building materials, browsing, etc. are common but are poorly recorded in these largely "unmanaged" areas.

Estimates of tropical grassland production vary by almost fivefold, depending on the techniques employed (Long *et al.*, 1989). Herbaceous NPP has generally been estimated using annual maximum standing crop (aboveground), usually at the end of the growing season [see Bourlière and Hadley (1970)—figures for 22 tropical grasslands, and Rutherford (1978)]. Importantly, this method does not account for belowground productivity, the effects of grazing and trampling, mortality before or growth after peak standing crop has been attained, the differences in time at which species attain their peak standing crop (especially mixtures of perennials and annuals), or litter turnover (Rutherford, 1978; Solbrig, 1996). Milner and Hughes (1968) proposed a method for the International Biological Programme (IBP) that measures positive increments in aboveground live biomass [see Singh and Joshi (1979) for a review of 21 studies in tropical grasslands in India and Africa], but this still leads to underestimates.

A more robust methodology was developed under a United Nations Environment Program (UNEP) project for measuring NPP in tropical grasslands (Long *et al.*, 1989, 1992). This measured monthly increments in aboveground live and dead biomass, roots, and monthly decomposition rates for standing dead material, litter, and roots. NPP was defined as the sum of net monthly increases in live biomass plus losses due to death and decomposition. NPP at three of these grassland sites was two to five times higher than that obtained using the standard IBP procedure, which ignores mortality, and two to ten times higher than previous methods, which ignore belowground NPP (Long *et al.*, 1989, 1992).

Measurements of worldwide "grassland" NPP have been incorporated into an Internet database site managed by the U.S. Oak Ridge National Laboratory (ORNL) Distributed Active Archive Center (http://www-eosdis.ornl.gov/npp/npp—home.html) (Scurlock *et al.*, 1999; see Chapter 18, this volume). The sites are often mixed tree–grass systems, although for many of the studies only grass NPP was measured.

Belowground production can be as high as or higher than aboveground production, but it is difficult to measure because it is hard to distinguish between live and dead roots and different species, thus assumed root:shoot ratios are often used. However, the ratio is highly variable, depending on the vegetation formation, maturity, prevailing environmental conditions, and disturbances. For example, in the UNEP study, belowground NPP varied from 40 to 70% of total NPP in different years at some sites (Long *et al.*, 1992). Although root:shoot ratios tend to increase with stress, severe drought or frequent fire can lead to negative belowground NPP (Long *et al.*, 1992). Root turnover and exudates are rarely considered. Menaut and Cesar (1979) observed root turnover up to 100% in an open humid savanna and 70% in wooded savannas.

Some estimates of NPP in the savanna biome as a whole can be found in Table 16-1. Whittaker and Likens (1973, 1975) used measured production and phytomass values and extrapolated them by their estimated area. Atjay *et al.* (1979) used this as a basis for their own study, but they evaluated more recent data on NPP and took account of the role of organic matter. They cite improved root production data as the main factor accounting for their much higher NPP value. In the assessment of Atjay *et al.*, savanna has the second highest share of total production after forest systems. Lieth (1973) suggested mean production of 800 g m^{-2} yr^{-1}, which Long *et al.* (1992) multiplied by 3.5 to take account of methodological underestimates to give a value of 2800 g m^{-2} yr^{-1}. Both Olson *et al.* (1983) and Scholes and Hall (1996) give much lower estimated rates of productivity. It is not clear where Olson *et al.* obtained their figures, but the Scholes and Hall values were derived as follows: Rather than apply one estimate of NPP (based on studies at a small number of sites) to an entire vegetation class that incorporates a range of environmental conditions, Scholes and Hall (1996) used data from individual studies to develop a relationship between water availability and NPP, and then applied this to a number of points in each class. Table 16-4, compiled by Scholes and Hall (1996), lists a series of NPPs reported for tropical savannas and grasslands. To be included, studies must have measured NPP over a period of at least a year, and used the sum-of-positive-increments method plus some assessment of losses.

Several studies reported belowground NPP for tree and/or grass components, enabling the calculation of a relationship between total NPP and aboveground NPP for the trees and grasses:

$$\text{total NPP} = 1.01(\text{aboveground NPP}) + 853.$$

This equation predicts that belowground production accounts for between 5 and 70% of the total NPP, the proportion decreasing with increasing productive potential of the site (Scholes and Hall, 1996). However, this equation would seem to have an unrealistic intercept and is driven by an outlier—the floodplain site Manaus in Brazil, which can be considered an exception due to high water and nutrient availability, although it does show the potential of grassland vegetation. This relationship is redrawn in Fig. 16-2 with-

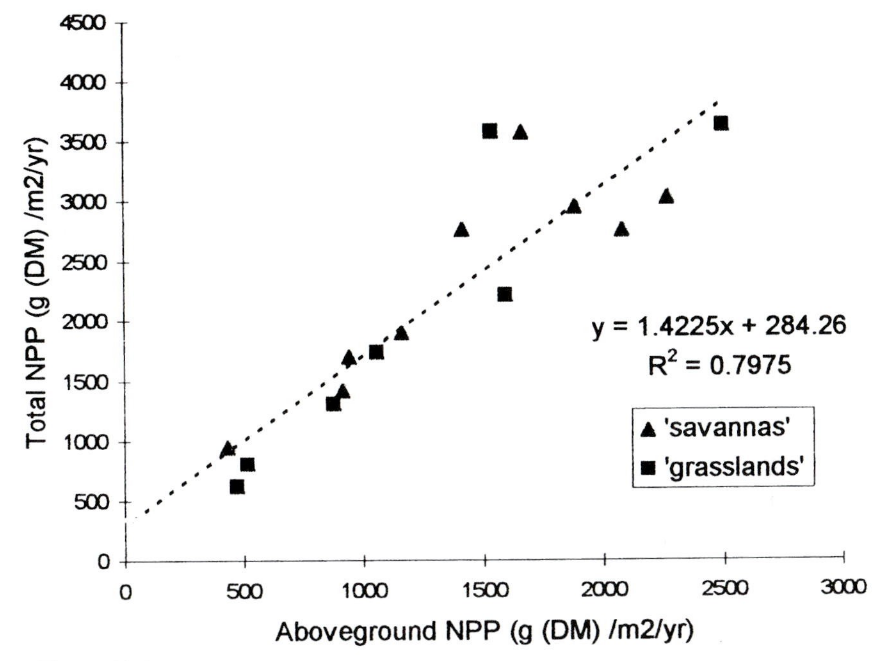

Figure 16-2 The relationship between total NPP and aboveground NPP. [Based on Scholes and Hall (1996, Table 4.2), recompiled in this chapter as Table 16-3]. Excluding "plantations" and outlier site Manaus, Brazil; DM, dry matter

out the outlier, and, sticking to the simple linear relationship, the equation becomes

$$\text{total NPP} = 1.42(\text{aboveground NPP}) + 284.$$

Scholes and Hall used this equation to calculate total NPP for all the sites, then plotted total NPP (tree and grass, above- and belowground) against a calculated water availability index (WAI) based on the monthly ratio of rainfall to evaporation at each site (equivalent to the number of days per year without water stress), presented in Fig. 16-3. The derived relationship was then applied to a large number of randomly selected locations in each of the vegetation classes to calculate a more representative average NPP, which is presented in Table 16-1. Although water and nutrient limitations do constrain savanna productivity, the inherent capacity for production by savanna plants is comparable to forest systems, i.e., they can produce just as much or more biomass per unit of water available. Referring to Fig. 16-3, if the upper range of data is projected to a water availability of 365 days (i.e., no water limitation), the predicted annual NPP is 4000 g m^{-2}, which is well within the average-to-high range for other natural ecosystems (Scholes and Hall, 1996).

Table 16-4 Primary Production Reported for Tropical Grasslands and Savannas[a]

Site	Latitude	Longitude	Aboveground NPP ($g\ DM\ m^{-2}\ yr^{-1}$)			Belowground NPP ($g\ DM\ m^{-2}\ yr^{-1}$)			Total NPP ($g\ DM\ m^{-2}\ yr^{-1}$)	Source
			Trees	Grass	Total	Trees	Grass	Total		
Savannas										
Lamto, Côte d'Ivoire	6°13′N	5°02′W	633	1450	2083	37	633	670	2753	Menaut and Cesar (1979)
			277	1610	1887	23	1040	1063	2950	
			137	1280	1417	13	1330	1343	2760	
			55	1610	1665	5	1900	1905	3570	
Mokwa, Nigeria	9°18′N	5°04′E	378	—	—	—	—	—	—	Collins (1977)
Nylsvley, South Africa	24°42′S	28°42′E	282	157	439	187	325	512	951	Scholes and Walker (1993)
Niona, Senegal	14°18′N	6°0′W	—	225	—	—	—	—	—	De Ridder *et al.* (1982)
Uttar Pradesh	24°18′N	83°0′E	520	645	1165	60	675	735	1900	Pandey and Singh (1992)
			230	715	945	30	720	750	1695	(estimates for area
			430	490	920	40	460	500	1420	protected from grazing)
Grasslands										
Towoomba, South Africa	24°50′S	28°15′E	—	141	—	—	—	—	—	Donaldson *et al.* (1984)
Makaholi, Zimbabwe	19°48′S	30°48′E	—	146	—	—	—	—	—	Ward and Cleghorn (1964)
Nuanetsi, Zimbabwe	21°24′S	30°48′E	—	261	—	—	—	—	—	Kelly *et al.* (undated)
Klong Hoi Kong, Thailand	6°0′N	100°56′E	—	1595	—	—	625	—	2220	Long *et al.* (1992)
Charleville, Australia	26°24′S	146°12′E	—	520	—	—	290	—	810	Christie (1978)

Nairobi, Kenya	1°0′N	36°49′E	—	881	—	—	431	—	1312	Long *et al.* (1989)
Udaipur, India	25°30′N	72°24′E	—	180	—	—	—	—	—	Vyas *et al.* (1972)
Jodphur, India	26°18′N	73°06′E	—	108	—	—	—	—	—	Gupta *et al.* (1972)
Delhi, India	28°54′N	77°12′E	—	798	—	—	—	—	—	Vashney (1972)
Ejura, Ghana	7°30′N	3°30′E	—	870	—	—	—	—	—	Greenland and Nye (1959)
Kurukshetra, India	9°58′N	76°51′E	—	2500	—	—	1131	—	3631	Rajvanshi and Gupt (1985)
Calaboza, Venezuela	8°48′N	67°27′W	—	478	—	—	146	—	624	Sarmiento (1984)
Manaus, Brazil	3°20′S	60°0′W	—	9418	—	—	507	—	9925	Long *et al.* (1989)
Montecillos, Mexico	19°28′N	92°28′W	—	1063	—	—	678	—	1741	Long *et al.* (1989)
Lamto, Coto d'Ivoire	6°13′N	5°02′W	—	1540	—	—	2040	—	3580	Menaut and Cesar (1979)
Plantations (mostly exotic trees)										
Kurukshetra, India	29°58′N	76°51′E	1547	248	—	—	—	—	—	Rajvanshi and Gupt (1985)
Gandi Nagar, India	23°12′N	77°06′E	2270	—	—	757	—	—	3027	Gurumurti *et al.* (1986)
Mudigere, India	13°12′N	75°36′E	1872	—	—	—	—	—	—	Swaminath (1988)
Nigeria	10°30′N	7°18′E	1337	—	—	—	—	—	—	Kadeba and Aduavi (1985, 1986)

[a]Table recompiled from Scholes and Hall (1996; Table 4.2). Terms "savannas" and "grasslands" refer to categories originally separated by Scholes and Hall (1996); DM, dry matter. To be included, studies must have measured NPP over a period of at least a year, and used the sum-of-positive-increments method plus some assessment of losses (it was not always possible to determine from the sources whether losses included turnover or herbivory, or litterfall only). Long *et al.* (1989, 1992) figures used sum-of-positive-increment results rather than UNEP method results (see text), because this fits better with the rest of the data quoted.

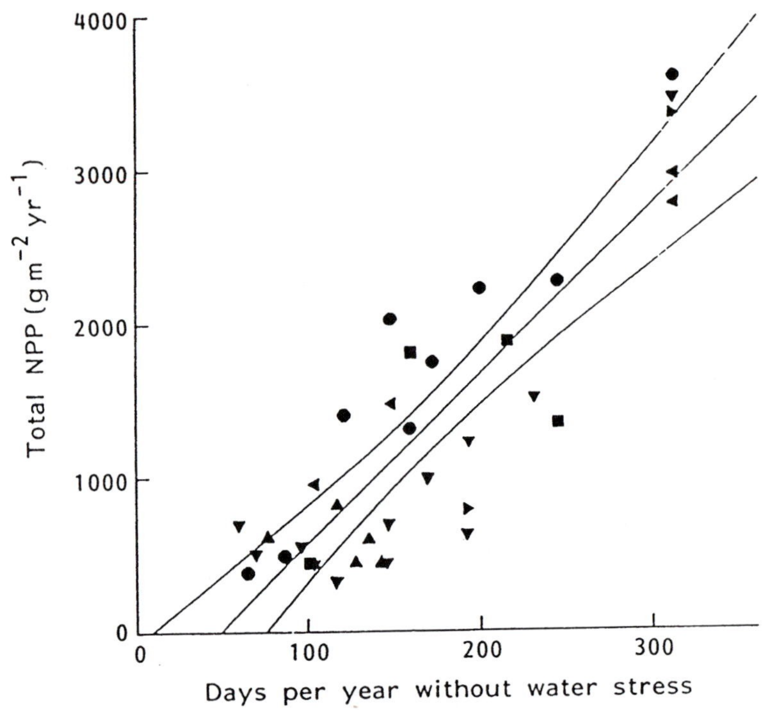

Figure 16-3 The relationship between total NPP and an index of water availability for tropical grasslands and tree–grass mixtures, with 95% confidence limits. The water availability index (WAI) can be thought of as the number of days per year on which growth is not water limited. The derivation of WAI is discussed in the text. The least-squares best-fit line has the equation NPP = 11 × WAI − 539 ($n = 37, F = 108.6, p < 0.001, r^2 = 0.75$). The data are from Scholes and Hall (1996), and a number of other studies that did not satisfy the criteria for inclusion, for instance, because they refer to regions rather than sites. The symbols denote different structural types: (●) grasslands; (▲) derived grasslands; (▼) grassy savannas; (◀) savannas; (▶) woodlands; (■) plantations of exotic trees in savanna areas (reproduced from Scholes and Hall, 1996).

VI. Biophysical Properties, Fluxes, and Efficiencies

Biophysical properties, fluxes, and efficiencies vary enormously and are radically affected by the high seasonality of savannas and fire. The few sites where such measurements have been made in any detail are highlighted in Table 16-5.

A. Leaf Area Indices

The leaf area index (LAI) is highly seasonal, e.g., in Lamto the LAI increases from 0 after fire in January to a peak of 4.0 in November (Le Roux *et al.*,

1997). Values for Venezuela and the UNEP sites also represent the complete seasonal ranges, whereas the Niger study captured part of the wet and dry seasons. The Nylsvley study incorporated trees with measured LAI under canopies (tree + grass, 2.25) and between canopies (grass only, 0.6), giving an overall LAI of 1.2 (Scholes and Walker, 1993). The Brazil study also measured tree LAI but figures seem relatively low (Miranda *et al.,* 1997). Thus, according to this limited data set, the range in arid savannas is 0 to 1.7 (or 2.3 if trees are included), and the range in humid savannas is 0 to 4.8 (and could be higher with the inclusion of trees).

B. Albedo and Radiation

Albedo is a function of underlying soil as well as vegetation, and fires can darken the soil surface and affect its spectral qualities. Albedo in Lamto was 0.07 over bare black soil, and increased over open areas with low shrub coverage from 0.08 to a peak of 0.23, with an overall mean annual albedo of 0.194 (albedo in more shrubby areas was similar) (Le Roux *et al.,* 1994). In contrast, bare soil albedo at Niger was 0.4, greater than the average vegetation albedo of 0.2 (Braud *et al.,* 1997). Similarly in Nylsvley, with a light soil, albedo was at its lowest value of 0.11 over peak biomass, and rose to 0.15 at the end of the dry season, with an annual average of 0.12. San Jose (1992) found that dry season albedo was lower in recently burned areas 0.08 compared to unburned areas 0.12, because the smoother burned surfaces trapped radiation more efficiently (vegetation scattered it). In spite of this, net radiation was also lower due to higher emission of long-wave radiation as a consequence of higher daytime mean temperature in the surface of the bare soil.

C. Roughness, Conductance, and Resistance

Spatial heterogeneity between different patches of savanna vegetation causes a discontinuity in aerodynamic, radiative, thermal, and moisture features. This is exacerbated by fire because air flows from the smooth and hot recently burned savanna (bare soil) to a rougher and cooler unburned savanna (vegetative surface). San Jose (1992) showed that the roughness length and aerodynamic conductance were far lower in burned patches (Table 16-5), and that mixing due to thermal turbulence in the burned savanna was three times greater than that due to roughness-generated turbulence in the unburned savanna. In Brazil and Venezuela (site 2) surface conductance was higher with more leaves in the canopy, and the degree of coupling (Ω) was higher, indicating less coupling between canopy and atmosphere (San Jose *et al.,* 1998; Miranda *et al.,* 1997) (Ω approaches 1 in well-watered and aerodynamically smooth canopies, where the transpiration rate is driven by radiation, and approaches 0 over aerodynamically rough

Table 16-5 Biophysical Properties,

	Lamto, Ivory Coast (humid, open shrubby)				Venezuela, site 1 (humid, grasslands)			Venezuela, site 2[6] (humid, grasslands)	
	Start wet season[1]			Other[2,3]	Dry season[4]		Wet[5]	Wet season	
	Feb	Mar	Apr		Burned	Unburned		Cultivated	Natural
Rainfall (annual mm)			1210			1257		1050	
Total production ($gm/m^2/yr$)			2700						
Herbaceous biomass only (g/m^2)						600–1000		675	433
LAI site average									
Herbaceous layer	0.8	1.5	1.9	4.0		0.08–0.68		4.4	1.7
Trees									
Radiation balance									
Incoming solar radiation ($MJ/m^2/y$)									
($MJ/m^2/d$)					20.6	20.6	15.8	15.7	15.7
max (W/m^2)					720	720	762		
Net radiation, daily ($MJ/m^2/d$)					11.2	11.9	12.0	8.4	7.4
max (W/m^2)					630	550	550	410	360
Mean daytime albedo α	0.08–0.23, av. 0.19, burnt soil 0.0				0.08	0.12		0.3	0.3
Roughness and conductance									
Vegetation height (m)			Shrubs 2–6, grass 1			Grass max 1.8		0.6	1.2
Roughness length Z_0 (m)					0.0013	0.039			
Zero displacement d (m)					0.0070	0.196			
Aerodynamic cond. ($mol/m^2/s$) max					0.068[b]	1.8[b]		4.0[b]	3.2[b]
Canopy conductance ($mol/m^2/s$) max						0.039[b]			
Surface conductance ($mol/m^2/s$) max								1.0[b]	0.6[b]
Coupling coefficient Ω						0.17		0.29	0.17
CO_2 fluxes[c]									
Leaf assimilation ($\mu mol\ CO_2/m^2/s$)				19–33					
Canopy assimilation A_c ($MJ/m^2/day$)							0.138		
max ($\mu mol\ CO_2/m^2/s$)	19.0	23.3	24.7						
Soil respiration ($MJ/m^2/day$)							−0.036		
max ($\mu mol\ CO_2/m^2/s$)	−6.6	−8.9	−9.6						
Net ecosystem flux F, ($MJ/m^2/day$)							0.102		
Daytime peak ($\mu mol\ CO_2/m^2/s$)	19.0	23.3	24.7						
Nighttime peak ($\mu mol\ CO_2/m^2/s$)									
Efficiencies (g/MJ)[d]	Early	Mature	Annual[2]						
Net prod (g)/incident rad (g/MJ)	1.58	1.06	0.85						
Net prod (J)/incident rad (%)							0.7		
NPP/IPAR total	3.38	2.28	1.82						
NPP/APAR total = RUE	3.84	2.56	2.04						

References: [1]Le Roux and Mordelet (1995); [2]Le Roux *et al.* (1997); [3]Le Roux *et al.* (1994); [4]San Jose (1992); [5]San Jose *et al.* (1991); [6]San Jose *et al.* (1998); [7]Miranda *et al.* (1997); [8]assorted papers in Goutorbe *et al.* (1997), Hansan *et al.* (1998); [9]Scholes and Wlaker (1993); [10]chapters in Long *et al.* (1992). Where possible original values have been converted to standared format/units.

[a]Values here for grass and dicotyledons.

[b]Calculated from the original data as inverse of resistances in s/m; and at sea level and 25°C, g $mol/m^2/s$ = 0.04 g mm/s (Jones, 1992).

[c]CO_2 flux notes: Venezuela values calculated from original table for 30th July.

[d]Efficiency notes: Lamto original values converted AG biomass, and OPAR = 0.467 R_s from values in paper. For other sites used IPAR = 0.45 Rs. RUE = radiation use efficiency, IPAR and APAR incident and absorbed hotosynthetically active radiation. Kenya and MExico values recalculated from original data; Thailand and Amazon figures from graphs of cumulative NPP of cumulative NPP against cumulative.

canopies, where transpiration is driven by canopy to air saturation deficit). Although roughness values in Niger were similar to those for the unburned Venezuelan grassland site (Braud *et al.*, 1997), the sparse tree crowns of the Brazilian cerrado vegetation form a more aerodynamically rough surface.

Fluxes, and Efficiencies

Brazil[7] (humid, dense scrub)		Niger[8] (arid, regen. shrub)	Nylsvley[9] (semiarid, mixed S. Africa annual)	UNEP sites[10]			
				Kenya (arid, grassland)	Mexico (arid, grassland)	Thailand (humid, grassland, unburned)	Amazon (wet, grassland, floodplain)
Dry	Wet	Wet/dry 8 weeks		All annual measurement			
1500			623	800	580	2000	2100
			951	1292	1803	2036	9930
474[a]	503[a]						
			1.2				
0.3[a]	0.5[a]	1–1.25	0.6	0–3.1	0–1.7	0.7–2.0	2.3–4.75
0.4	1.0		2.25				
			7316	7190	6741		4400
			20	19.7	18.5	9.9–20.6	12.0
700	950						
			11.3				
650	950						
		0.2 (0.4 soil)	0.11–0.15, av. 0.12				
Trees 9		Shrub 1–3	Tree 6; grass 0.75				
1.2		0.07					
6.3		0.38					
0.15	0.4						
0.17	0.32						
			25	27	23		31
4	15	15					
−2	−5	−5					
		0.99	0.13	0.18	0.27	0.66	2.30
				0.32	0.80		
		2.21	0.29	0.40	0.60	1.47	5.11

D. Carbon Dioxide Fluxes

Several studies have been carried out on leaf photosynthetic rate (or CO_2 flux) of savanna grass species, showing a range of 15 to 33 μmol CO_2 m^{-2} s^{-1} across a range of dry and humid sites, with results showing a high variability (Le Roux and Mordelet, 1995; and see Table 16-5). In Nylsvley alone, measurements for one grass species varied by 13-fold, dependent partly on techniques used and local conditions, although for further calculations Scholes and Walker (1993) used an overall average of 25 μmol CO_2 m^{-2} s^{-s}. Some measurements were also made of tree leaf photosynthesis at Nylsvley and, on average, the C_3 woody plants had 18% lower rates than the C_4 grasses, which are better adapted to high irradiation and heat and water stress.

Primary productivity is much more closely linked to canopy CO_2 flux than to leaf flux, yet there are even fewer savanna studies on this. At Lamto, Le Roux and Mordelet (1995) measured CO_2 canopy fluxes at the beginning of the wet season after a January fire. Despite low N at this site, the peak net canopy CO_2 flux/assimilation F (measured above the canopy) was high at 24 μmol CO_2 m^{-2} s^{-1} for a LAI of only 1.9 not long after the fire. They compare this to net canopy fluxes of 27 μmol CO_2 m^{-2} s^{-1} for LAIs of 7 and 8 in Amazonian and Malaysian rainforests (Fan *et al.*, 1990; Aoki *et al.*, 1975), sustaining the emerging opinion that the primary productivity of tropical savannas could be close to that of tropical forests (Atjay *et al.*, 1979; Gifford, 1980; Long *et al.*, 1989). Net canopy fluxes measured in Niger (LAI 1–1.25) (Hannan *et al.*, 1998) were similar to those found in the wet season in Brazil (LAI 0.5) (Miranda *et al.*, 1997) at 15 μmol m^{-2} s^{-1} into the canopy during the day and -5 μmol m^{-2} s^{-1} out at night. During the wet season in Brazil, the fluxes were higher and although vegetation was found to be a CO_2 sink in the wet season of up to 0.15 mol m^{-2} d^{-1}, it was a source of CO_2 for a brief period at the height of the dry season (Miranda *et al.*, 1997). San Jose *et al.* (1991) measured CO_2 fluxes in the wet season in Venezuela and found that the net community assimilation ranged from 6.6 to 7.9 g (DM) m^{-2} d^{-1}, which was similar to the mean community growth rate they calculated from biomass measurements (2.8–6.9 g (DM) m^{-2} d^{-1}).

E. Production Efficiency

Several different methods are used to calculate efficiency, and some of these are represented in Table 16-5. Radiation use efficiency (RUE) is calculated as total NPP/IPAR (incident photosynthetically active radiation) and we have tried to convert all values to this. IPAR can be assumed to be a fixed proportion of incident radiation (R_s), usually between 45 and 50%. The range of RUE is 0.4 to 1.8 taking the annual results only (i.e., not Niger) and ignoring the outlier Brazil. RUE is partly related to rainfall, with the average across the three semiarid sites being 0.43, and the two humid sites 1.65. Many arid species have a lower conversion efficiency because lower stomatal conductance is an adaptation to water stress. The Lamto results show how efficiency declines as the stand matures. In Thailand, one year of fire reduced RUE from 1.47 to 1.35 and a second year of fire caused it to drop further to only 0.17 (assuming IPAR = $0.45R_s$).

VII. Environmental Determinants

As noted earlier, the amounts of moisture and nutrients available to savanna plants determine the structure of savannas and their NPP. The primary determinants of both of these factors are climate and soil type, rainfall and soil texture being particularly important. In low-rainfall areas, deep, sandy soils allow more water to infiltrate to greater depths, reducing runoff and evaporation and increasing moisture availability. However, where rainfall is high, sandy soils can lose water through drainage and runoff, often leaching nutrients. Clay soils have a higher water-holding capacity and nutrient retention, leading to a higher NPP in moist and mesic systems, but are less advantageous in arid areas due to low infiltration (Dodd and Lauenroth, 1997).

Moisture and nutrient availability in turn influence the occurrence and intensity of fire and herbivory, which are secondary determinants modifying savanna vegetation (Walker and Noy-Meir, 1982; Frost *et al.*, 1986; Solbrig, 1990; Scholes and Hall, 1996). At a more local scale, differences in topography, geomorphology, and management lead to further differences in structure and floristic composition (Solbrig, 1996). Many of the factors controlling savannas are interdependent, interacting positively or negatively on each other either simultaneously or sequentially. The relative importance of individual factors shifts with time and location, leading to the dynamic mosaic of savanna types and NPP. Light is generally sufficient in the tropical savannas except when canopies become dense.

A. Water Availability

Water availability is widely considered the most important controlling factor. Rainfall determines the supply of water, but the amount that is subsequently available to plants (its *effectiveness*) depends on drainage and storage (soil texture, depth, and topography) and losses due to evaporation and evapotranspiration (climatic conditions, vegetation cover, etc.). Annual potential evaporation substantially exceeds annual rainfall, therefore rainfall is, in a sense, less effective than in temperate climates. Rainfall and evapotranspiration are highly seasonal and therefore annual means can be misleading.

Rainfall mostly occurs as short-duration, high-intensity convectional storms. Variability is high, increasing with aridity, and is a primary cause of compositional change. Rainfall may be bimodally distributed and the timing affects the species mix, survivability, extent, and overall growth (Pandey and Singh, 1992; Veenendaal *et al.*, 1996). Many areas regularly experience drought and have developed a range of adapted and tolerant species. Plants close their stomata to reduce water loss during dry periods, which also restricts CO_2 uptake and hence productivity. In some savanna areas flooding is common and can lead to a long-term reduction in production due to soil erosion (Rutherford, 1978).

Many studies in arid areas have shown a high correlation between productivity and rainfall. However, NPP can often be more closely related to the length of the wet and dry seasons than to annual rainfall, because this is a better indicator of water availability. Scholes and Hall (1996) calculated that semiarid savannas, where the duration of the growing season is about 100 days, have a total (above- and belowground tree and grass) NPP of around 500 g (DM) m^{-2} yr^{-1}, whereas very moist savannas (300 growing days per year) have an NPP of around 3000 g (DM) m^{-2} yr^{-1} (Fig. 16-3).

B. Nutrient Availability

Savanna soils tend to have low cation-exchange capacity (CEC), very low phosphorous and nitrogen contents, and high aluminium and iron contents (Solbrig, 1996). More than half of the tropical savanna soils are derived from old, highly weathered acid crystalline igneous rock, leading to leached sandy soils with low fertility and CEC. There are also extensive areas of basic igneous rock forming base-rich clays with more favorable nutrient content and retention (Solbrig, 1996). Local conditions, age, and history lead to a complex regional distribution. Soil catenas are common with dense woodlands on sandy upper areas, through mixed scrub on shallow midslope soils, to open tall-tree savanna on deep fine-textured soils at slope bottoms (Walker, 1985). Iron pans often occur in arid areas (Menaut *et al.*, 1985).

Nutrient mineralization, transport, and root uptake are all dependent on soil water content. Thus it has been suggested that water availability may actually be controlling plant growth by its influence on nutrient availability, and that, certainly in some areas, nutrients rather than water may be the major production constraint. This is borne out by the high carbon:nutrient ratio of many savanna plants, various fertilization experiments, and comparisons of NPP on adjacent soils with different fertility status (Scholes and Hall, 1996).

High rainfall can leach soils, causing a gradient from arid/eutrophic to moist/dystrophic savannas, seen clearly in Africa (Huntley,1982). Arid, sandy savannas have a slow-release, low-nutrient stock favoring perennials with low growth rate over a long period. Clay soils release nutrients in pulses after rain, favoring annuals with rapid regrowth for short periods. In arid savannas, litter production plays a major role in nutrient cycling whereas moist savannas tend to have a short, closed nutrient cycle bound closely to root decomposition, so the amount of nutrients in the soil at any instant may be misleading (Abbadie *et al.* 1992). Nutrients are released steadily throughout the year in humid areas, favoring perennials, although there is still a nutrient pulse at the beginning of the rainy season as rain washes the nutrients from litter decomposition and fire ash into the soil (Menaut *et al.*, 1985).

Low nitrogen and phosphorous availability constrain many savanna ecosystems, yet little is known in detail of the nutrient dynamics of tropical savannas (Breman and De Wit, 1983; Medina,1987; Menaut *et al.*, 1985; Solbrig *et al.*, 1996a). Most nitrogen is lost through pyrodenitrification with frequent fires and an accumulation of decomposition-resistant charcoal. Leguminous plants are common, but may fix little nitrogen on low-fertility soils, possibly due to a deficiency of cofactors (Zietsman *et al.*, 1988). Nitrogen reallocation from leaves can be high, thereby retaining nitrogen stocks before leaf fall and fire.

Earthworms can process up to 70% of soil organic matter in upper horizons (Lavelle *et al.*, 1983; Menaut *et al.*, 1985). Termites are more dominant in drier climates and are very efficient secondary consumers, probably more important than the herbivores, generally consuming around 30% of litter biomass, but this can reach 70%. Termite biomass often averages 10 g m^{-2} of fresh weight with maximums up to 50 g m^{-2}, which is comparable to large herbivores (Wood and Sands, 1978). Nutrients (and seeds) become concentrated in large termite mounds. Herbivory also transforms and concentrates nutrients (Scholes and Walker, 1993).

C. Fires

Fires are inevitable with the buildup of dead grass during the dry season. "Natural" (lightning-induced) fires occur every 1 to 3 years in humid savannas and perhaps 1 in 20 years in arid savannas (Frost, 1985; Walker, 1985). Fires are deliberately set or restricted by humans as part of vegetation management, which has a history of at least 50,000 yr^{-1} in Africa (Menaut *et al.*, 1985). In Africa, 25–50% of the total land surface is burned annually in the "Sudan Zone" (arid) and 60–80% in the "Guinea Zone" (humid) (Menaut *et al.*, 1991). Fire is considered necessary to the functioning of savannas, which have evolved with it, preventing bush encroachment, removing dead material, and recycling nutrients (Troloppe, 1982; Walker and Noy-Meir, 1982). Fire prevention leads to organic matter and litter accumulation and increased tree density, often resulting in the long term in a forested area without grasses. Although fire leads to local and short-term heterogeneity, creating a mosaic structure of open grassland that carries fire, and clumps of trees that lack the herbaceous fuel load to do so, it also maintains regional and long-term homogeneity and stability (e.g., maintaining a sharp boundary between savanna and forest) (Menaut *et al.*, 1985, 1990).

The incidence, intensity, and impact of fire depends on the amount of fuel (grass) present, the prevailing environmental conditions, and thus the time (season) of burning (Trollope, 1982, 1984; Menaut *et al.*, 1993, 1995). Fire removes aboveground herbaceous vegetation and small/immature woody plants, and changes the microclimate and nutrient status. This creates opportunities for seedling development, enhanced reproduction, and fresh growth. Mature trees are not usually affected but total woody biomass is controlled in the long term by reducing recruitment. Underground plant parts are only affected by very hot fires, although the root:shoot ratio increases in regularly burnt areas (Troloppe, 1984). Humid plants show many fire adaptations such as withdrawal of plant parts underground, increased (fire-resistant) seed production, and fire-induced phenology.

Fire-induced mortality of plant populations is often extremely low (0–10%) (Frost, 1985). Early-season fires burn 20–25% of the aboveground herbaceous biomass, and late fires burn 60–95% (Menaut *et al.*, 1991). Because much of the total system NPP is belowground or in tree leaves, less than 20% of the total NPP is lost in annually burnt savannas (Scholes and Walker, 1993), and only 5% in triennially burned savannas (Scholes and Hall, 1996). In the long term, frequent burning tends to increase herbaceous production in humid regions by perhaps 20–50% while decreasing it in arid regions (San Jose and Medina, 1975; Menaut *et al.*, 1993). In fact, the NPP of annually burnt humid savannas is comparable to that of rain forests under similar climatic and edaphic conditions (Scholes and Hall, 1996).

Fire releases carbon, nitrogen, and other elements, which are partly returned to the soil as ash. The long-term effect on nutrient budgets is unclear. It is suggested that savannas that have experienced frequent fires for thousands of years may have been driven to a low-equilibrium nutrient status, but on an annual basis losses of N and C are balanced by annual inputs. Incorporation of charcoal residues in the soil may lead to carbon accumulation in the long term (Menaut *et al.*, 1993).

D. Herbivory

Savannas support large numbers of herbivores, both grazers and browsers. Browsers consume various parts of woody plants and grazers prefer herbaceous biomass, although diets are often mixed and depend on food availability. Insects account for an equal, or even greater, proportion of herbivory compared to mammals, yet they are infrequently studied (Lamotte, 1982; Anderson and Lonsdale, 1990); mammals have greater impact on savanna structure because they trample, urinate, defecate, and trash plants (Skarpe, 1991). Large herbivores, such as elephants, kill trees, thus opening up woodlands and turning savanna into grasslands; giraffes keep trees at a lower, fire-sensitive height (Pellew, 1983). Large concentrations of grazers can cause degradation and erosion of grasslands by severe, long-term removal of the herbaceous cover. Grazing favors tree seedlings, removing competition from grasses and reducing fuel available for fire (Skarpe, 1990).

At the species level, herbivory changes composition and physiognomy, favors annual grasses and increases annual and perennial forbs (Illius *et al.*, 1996). At the individual plant level, changes in quantity, chemical composition, and physiognomy either promote or deter different herbivores. Infertile soils support vegetation with a nitrogen content less than 1%, which is below the threshold for ruminant consumption for most of the year, and less than 10% of NPP is typically consumed compared to 80% on fertile soils (Drent and Prins, 1987; Scholes and Walker, 1993). Some trees can be entirely defoliated by browsing and produce a new crop of leaves in the same season, almost doubling leaf production (Rutherford, 1978).

Grazing has contradictory effects, depending on intensity and local conditions. Moderate grazing removes dead plant material, reduces shade, recycles nutrients, and improves seed dispersal and germination, promoting palatable species, NPP, and high vegetation cover. For example, Pandey and

Singh (1992) found that prevention of grazing in a dry savanna resulted in a decline of NPP (and species diversity), whereas moderate grazing stimulated aboveground NPP by 4–45% and reduced belowground NPP by 25–65%. Intense grazing and trampling result in low plant cover, high mortality, low soil nutrients, low infiltration, decline in water availability, and higher erosion, reducing NPP, particularly in arid and infertile areas (McNaughton, 1983). As the number of humans living in savanna areas has increased, so has the number of domestic livestock, putting pressure on savanna systems (Skarpe, 1991).

VIII. Human Influence

Human-induced land use change has a greater effect on savanna structure and NPP than any other environmental changes. Savannas generally have low human population density because their capacity to support intensive agriculture and livestock grazing is poor; however, population pressure and land use changes are likely to be greater here than in other biomes over the coming decades. The agropastoral communities in savannas are highly dependent on the natural vegetation, which is vulnerable to degradation. Strong social and economic forces will continue to force the conversion of forests to savannas and grasslands to pasture or agriculture fields, although tourism is supporting conservation of large areas of savannas in Africa.

Cattle rearing has been occurring for more than 6000 yr in India and Africa and was sustainable before more recent increases in population, changing social practices, colonialization, war, extensive commercial ranching, etc. Shifting cultivation has also been practiced for thousands of years, generally in areas with rainfall greater than 700 mm, but this is being affected by reduced fallow periods, larger commercial operations, and increased use of fertilizer and irrigation, which increases short-term productivity but can lead to long-term depletion of resources and environmental degradation. South American savannas are rapidly being transformed for soybean cultivation and cattle raising. Australian savannas are likely to remain as grazing lands as alternative economic land uses are limited (Young and Solbrig, 1993; Solbrig, 1996, J. S. Scanlan, personal communication).

Humans influence savannas by managing fire, wildlife, domestic livestock, and wood and grass removals. Intensive livestock grazing, exclusion of indigenous browsers, and fire prevention lead to rapid bush encroachment. In areas of high population and charcoal production, intensive wood harvesting leads to decreased woody biomass, which, combined with high grazing, leads to erosion. The same happens when fire frequency is increased and the animal load is too high (Young and Solbrig, 1993).

IX. Climate Change

Changes in CO_2 concentration, rainfall, and temperature will all affect NPP, fire frequency, etc. Few studies have been carried out on the effects of climate change on savanna ecosystems, there are no true free-air/carbon dioxide enrichment experiments, and few CO_2 studies are as yet complete (R. J. Scholes, personal communication). In tropical savannas the woody biomass is C_3 and most of the herbaceous biomass is C_4. C_3 plants have a greater response to CO_2 enrichment in terms of increased productivity, thus it has been postulated that trees will have an advantage, and, in fact, woody species have been seen to increase in savanna areas although more local changes may be the cause (Polley *et al.*, 1996). However, this is only part of the story. Partial stomatal closure in CO_2-rich environments allows better water and nitrogen use efficiency, which is very important in certain savanna areas constrained in either or both resources (Gifford *et al.*, 1990); Owensby *et al.* (1999) have shown an increase in production in C_4 grasslands due to this effect, and some preliminary results from trials in grasslands in the United States show C_4 grasses actually doing better than C_3. Increased temperatures will lead to increased respiration; however, CENTURY model simulations suggest that increased photosynthesis will exceed increased respiration, resulting in a net carbon sink (Hall *et al.*, 1995). All of the above changes will modify soil carbon storage, which accounts for 80% of the total ecosystem organic carbon in tropical savannas (Scholes and Hall, 1996).

Savannas already have a great impact on carbon cycles due to fires, fuelwood, and land use changes. Amthor *et al.* (1998) found the savanna biome to have the highest potential for carbon gain and among the highest potential for carbon loss. Scurlock and Hall (1998) cautiously propose that this biome may already constitute an annual sink of about 0.5 Pg C, as supported by recent eddy covariance measurements over Brazilian *cerrado* (Miranda *et al.*, 1997) (Table 16-5). Savanna burning releases between 0.87 Pg C yr^{-1} (Scholes and Hall, 1996; extrapolation of Menaut *et al.*, 1991) and 1.66 Pg C yr^{-1} (Scholes and Hall, 1996; extrapolation of Hao *et al.*, 1990). Fire suppression and the resultant increase in tree density could store an extra 30 g C m^{-2} annually in soils (Scholes and Van der Merwe, 1996).

Introducing deep-rooted African grasses and legumes to South American grasslands has increased soil carbon storage, mostly in the form of dead roots, which, if applied throughout Latin America's savanna areas could store 0.1–0.3 Pg C per annum (Fisher *et al.*, 1994). In Venezuela (San Jose *et al.*, 1998) maximum aboveground yield was four times higher for African grasses than native savanna. However, the continuing pressure on these lands and their subsequent degradation are likely to lead to an increase of carbon loss in the future. These human-induced changes will far outweigh any impacts due to climate change.

X. Summary

Tropical savannas (including grasslands) form one of the world's most extensive biomes, covering 27.6 Mkm^2. These areas are experiencing significant population pressure and land use changes, and are vulnerable to rapid land degradation. This will have large impacts on global NPP predictions and climate change, thus it is important to improve our understanding of them. The proportion of trees and grasses is highly variable in space and time, yet very little is understood about the complex dynamic processes and interactions controlling them. There is a great paucity of even basic data for these highly heterogeneous ecosystems because relatively few studies have been carried out in this biome in the past. Additionally there is a need for large-scale syntheses of existing data to improve knowledge and understanding. Some are now underway, and it is becoming increasingly clear that these areas have a greater productivity, higher biodiversity, and larger impact on global carbon cycles than previously realized (Long *et al.*, 1989; Solbrig *et al.*, 1996; Scholes and Hall, 1996). The current best estimate for productivity ranges from 733 to 1426 g (DM) m^{-2} yr^{-1} for different savanna formations, average 1078 g (DM) m^{-2} yr^{-1} (Scholes and Hall, 1996), with total annual production in savannas and grasslands at 29.7 Pg (DM) yr^{-1} (about a quarter of global NPP), although the data these calculations are based on are underestimates (e.g., Long *et al.*, 1989, 1992).

Acknowledgments

Many thanks to Bob Scholes (CSIR, South Africa), Xavier Le Roux (INRA, Clermont-Ferrand, France) Jonathan Scurlock (ORNL, USA), and Joe Scanlan (Department Natural Resources, Qu., Australia) for providing information and making corrections to the manuscript, and Dale Kaiser and Sonja Jones (ORNL, USA) for calculating tropical percent of Olson *et al.* (1983) categories. Sadly David Hall passed away in August 1999 before this chapter was published. His knowledge and love of savannas was only surpassed by his eagerness to learn and teach.

References

Abbadie, L., Mariotti, A., and Menaut, J.C. (1992). Independence of savanna grasses from soil organic-matter for their nitrogen supply. *Ecology* **73,** 608–613.

Anderson, A. N., and Lonsdale, W. M. (1990). Herbivory by insects in Australian tropical savannas: A review. *J. Biogeogr* **17,** 433–444.

Aoki, M., Yabuki, K., and Koyama, H. (1975). Micrometeorology and assessment of primary production of a tropical rainforest in West Malaysia. *Global Carbon Cycle* **31,** 115, 124

Atjay, G. L., Ketner, P., and Duvigneaud, P. (1979). Terrestrial primary production and phytomass. *In* "The Global Carbon Cycle, SCOPE 13" (B. Bolin, E. T. Degens, S. Kempw, and P. Ketner, eds.), pp. 129–181. John Wiley & Sons, Chichester.

Aucamp, A. J., Dankwerts, J. E., Teague, W. R., and Venter, J. J. (1983). The role of *Acacia karoo* in the false thornveld of the Eastern Cape. *J. Grassl. Soc. S. Afr.* **8,** 151–154.

Backéus, I. (1992). Distribution and vegetation dynamics of humid savannas in Africa and Asia. *J. Veg. Sci.* **3,** 345–356.

Beale, I. F. (1973). Tree density effects on yields of herbage and tree components in south-west Queensland mulga (*Acacia aneura* F. Meull.) scrub. *Trop. Grassl.* **7,** 135–142.

Belsky, A. J., Amundson, R. G., Duxbury, J. M., Riha, S. J., Ali, A. R., and Mwonga, S. M. (1989). The effects of trees on their physical, chemical and biological environments in a semi-arid savanna in Kenya. *J. Appl. Ecol.* **26,** 1005–1024.

Belsky, A. J., Mwonga, S. M., Amundson, R. G., Duxbury, J. M., and Ali, A. R. (1993). Comparative effects of isolated trees on their undercanopy environments in high- and low-rainfall savannas. *J. Appl. Ecol.* **30,** 143–155.

Blasco, F. (1983). The transition from open forest to savanna in continental Southeast Asia. *In* "Tropical Savannas" (F. Bourlière, ed.), pp. 167–181. Elsevier, Amsterdam.

Bourlière, F. (1983). "Tropical Savannas. Ecosystems of the World," Vol. 13. Elsevier, Amsterdam.

Bourlière, F., and Hadley, M. (1970). The ecology of tropical savannas. *Annu. Rev. Ecol. Syst.* **1,** 125–152.

Braithwaite, R. W. (1990). Australia's unique biota: Implications for ecological processes. *J. Biogeogr.* **17,** 347–354.

Braud, I., Bessemoulin, P., Monteny, B., Sicot, M., Vandervere, J. P., and Vauclin, M. (1997). Unidimensional modelling of a fallow savannah during the HAPEX-Sahel experiment using SiSPAT model. *J. Hydrol.* **188–189,** 912–945.

Breman, H., and Dewit, C. T. (1983). Rangeland productivity and exploitation in the Sahel. *Science* **221,** 1341–1347.

Burrows, W. H., Carter, J. O., Scanlan, J. C. and Anderson, E. R. (1990). Management of savannas for livestock production in north-east Australia: Contrasts across the tree–grass continuum. *J. Biogeogr.* **17,** 503–512.

Christie E. K. (1978). Ecosystem processes in semi-arid grasslands. I. Primary production and water use of two communities possessing different photosynthetic pathways. *Austr. J. Agric. Res.* **29,** 773–787.

Cole, M. M. (1986). "The Savannas: Biogeography and Geobotany." Academic Press, Orlando.

Collins, N. M. (1977). Vegetation and litter production in a southern Guinea savanna, Nigeria. *Oecologia,* **28,** 163–175.

Coughenour, M. B., Ellis, J. E., and Popp, R. G. (1990). Morphometric relationships and development patterns of *Acacia tortilis* and *Acacia reficiens* in Southern Turkana, Kenya. *Bull. Torrey Bot. Club* **117,** 1, 8–17.

De Ridder, N., Stroosnijder, L., Cisse, A. M., and van Keulen, H. (2000). "Productivity of Sahelian Rangelands. Vol. 1" Dept. of Soil Science and Plant Nutrition, University of Wageningen, The Netherlands.

Dodd, M. B., and Lauenroth, W. K. (1997). The influence of soil texture on the soil water dynamics and vegetation structure of a shortgrass steppe ecosystem. *Plant Ecol.* **133,** 13–28.

Donaldson, C. H., and Kelk, D. M. (1970). An investigation of the veld problems of the Molopo area. 1. Early findings. *Proc. Grassl. Soc.* **5,** 50–57.

Donaldson, C. H., Rethmann, G., and Grossman, D. (1984). Long-term nitrogen and phosphorous application to veld. *J. Grasslands Soc. South Africa* **1,** 27–32.

Drent, R. H., and Prins, H. H. T. (1987). The herbivore as prisoner of it's food supply. *In* "Disturbance in Grassland," (J. van Andel, ed.) pp. 131–147. Dr. W. Junk, Dordrecht. pp. 131–147.

Dye, P. J., and Spear, P. T. (1982). The effects of bush clearing and rainfall variability on grass yield and composition in south-west Zimbabwe. *Zimbabwe J. Agric. Res.* **20,** 103–118.

Fan, S. M., Wofsy, S. C., Bakwin, P. S., Jacob, D. J., and Fitzjarrald, D. R. (1990). Atmosphere–biosphere exchange of CO_2 and O_3 in the central-amazon-forest. *J. Geophys. Res.-Atmos.* **95,** 16851–16864.

Fatubarin, A. (1984). Biomass estimates for the woody plants in a savanna ecosystem in Nigeria. *Trop. Ecol.* **25,** 208–213.

Fisher, M. J., Rao, I. M., Ayarza, M. A., Lascano, C. E., Sanz, J. I., Thomas, R. J., and Vera, R. R. (1994). Carbon storage by introduced deep-rooted grasses in the South American savannas. *Nature (London)* **371,** 236–238.

Frost, P. G. H. (1985). The response of savanna organisms to fire. *In* "Ecology and Management of the World's Savannas," (J. C. Tothill and J. J. Mott, eds.), pp. 232–237. Australian Academy of Science, Canberra.

Frost, P., Menaut, J. C., Solbrig, O. T., Swift, M., and Walker, B. H. (1986). "Responses of Savannas to Stress and Disturbance: A Proposal for a Collaborative Programme of Research." Biology International Special Issue, No.10. International Union of Biological Sciences, Paris.

Gifford, R. (1980). Carbon storage in the biosphere. *In* "Carbon Dioxide and Climate," (G. Peearman, ed.), pp. 167–181. Australian Academy of Sciences, Canberra.

Gifford, R. M., Cheney, N. P., Noble, J. C., Russel, J. S., Wellington, A. B., and Zammit, C. (1990). *In* "Australia's Renewable Resources: Sustainability and Global Change," (R. M. Gifford and M. M. Barson, eds.), pp. 151–187. Bureau of Rural Resources Proceedings No.14. Resource Assessment Commission, Queen Victoria Terrace, Parkes ACT 2600.

Goutorbe, J. P., Dolman, A. J., Gash, J. H. C., Kerr, Y. H., Lebel, T., Prince, S. D., and Stricker, J. N. M. (1997). HAPEX-SAHEL, Special Issue. *J. Hydrol.* **188–189.**

Greenland, D. J., and Nye, P. H. (1959). Increases in carbon and nitrogen contents of tropical soils under natural fallows. *J. Soil Sci.* **10,** 284–299.

Gupta, R. K., Sharm, S., and Sharm, S. K. (1972). Aboveground productivity of grasslands at Jodphur, India. *In* "Tropical Ecology with Emphasis on Organic Production" (P. M. Golley and F. B. Golley, eds.), pp. 75–93. University of Georgia, Athens.

Gurumurti, K., Bhandari, H. C. S., and Dhawan, M. (1986). Studies of yield, nutrients and energy conversion efficiency in energy plantations of *Acacia nilotica J. Tree Sci.* **5,** 36–42.

Guy, P. R. (1981). Changes in biomass and productivity of woodlands in the Sengwa wildlife research area, Zimbabwe. *J. Appl. Ecol.* **18,** 507–519.

Hall, D. O., Ojima, D. S., Parton, W. J., and Scurlock, J. M. O. (1995). Response of temperate and tropical grasslands to CO_2 and climate change. *J. Biogeogr.* **22,** 537–547.

Hanan, N. P., Kabat, P., Dolman, A. J., and Elbers, J. A. (1998). Photosynthesis and carbon balance of a Sahelian fallow savanna. *Global Change Biol.* **4,** 523–538.

Hao, W. M., Lui, M. H., and Crutsen, P. J. (1990). Estimates of annual and regional release of CO_2 and other trace gasses to the atmosphere from fires in the tropics based on FAO statistics for the period 1975–1980. *In* "Fire in Tropical Biota: Ecosystem Processes and Global Challenges" (J. G. Goldammer, ed.), pp. 440–462). Ecological Studies 84. Springer-Verlag, Berlin.

Harrington, G. N., and Johns, G. G. (1990). Herbaceous biomass in a Eucalyptus savanna woodland after removing trees and/or shrubs. *J. Appl. Ecol.* **27,** 775–787.

House, J. *et al.* (2000). Addressing issues in savanna tree-grass dynamics. *J. Biogeography* (submitted.)

Huntley, B. J. (1982). Southern African savannas. *In* "Ecology of Tropical Savannas" (B. J. Huntley and B. H. Walker, eds.), pp. 101–119. Springer Verlag, Berlin.

Huntley, B. J., and Walker, B. H. (1982). "Ecology of Tropical Savannas." Ecological Studies 42. Springer-Verlag, Berlin.

Illius, A. W., Derry, J., and Gordon, I. J. (1996). "Components, Processes and Dynamics of Semi-arid Grazing Systems: A Review of Currrent Knowledge. ODA, UK.

Jackson, R. B., Canadell, J., Ehleringer, J. R., Mooney, H. A., Sala, O. E., and Schulze, E. D. (1996). A global analysis of root distributions for terrestrial biomes. *Oecologia* **108,** 389–411.

Jackson, R. B., Mooney, H. A., and Schulze, E. D. (1997). A global budget for fine root biomass, surface area, and nutrient contents. *Proc. Natl. Acad. Sci. U.S.A.* **94,** 7362–7366.

Jones, H. G. (1992). "Plants and Microclimate." Cambridge Univ. Press, Cambridge.

Kadeba, O., and Aduayi, E. A. (1985). Litter production, nutrient recycling and litter accumulation in *Pinus caribaea* Morelet var *hondurensis* stands in the northern Guinea savanna of Nigeria. *Plant and Soil,* **26,** 197–206.

Kadeba, O., and Aduayi, E. A. (1986). Dry matter production and nutrient distribution in a *Pinus caribaea* stand in a sub-humid tropical savanna site. *Oikos,* **46,** 237–242.

Kelly, R. D., Schwimm, W. F., and Barnes, D. L. (no date). Bush clearing trial at Nuanetsi. Anonymous *Annual report of the Rhodesian Dept of Specialist Services,* Harare, Zimbabwe, about 1975, in Scholes & Hall (1996).

Kelly, R. D., and Walker, B. H. (1976). The effects of different forms of land use on the ecology of a semi-arid region in south-eastern Rhodesia. *J. Ecol.* **64,** 553–576.

Knoop, W. T., and Walker, B. H. (1985). Interactions of woody and herbaceous vegetation in a southern African savanna. *J. Ecol.* **73,** 235–253.

Koch, G. W., Vitousek, P. M., Steffen, W. L., and Walker, B. H. (1995). Terrestrial transects for global change research. *Vegetatio* **121,** 53–65.

Lamotte, M. (1982). Consumption and decomposition in tropical grassland ecosystems at Lamto, Ivory Coast. *In* "Ecology of Tropical Savannas" (B. J. Huntley and B. H. Walker, eds.), pp. 415–430. Ecological Studies 42. Springer-Verlag, Berlin.

Lavelle, P. (1983). The soil fauna of tropical savannas. II. The earthworms. *In* "Tropical Savannas," (F. Bourlière, ed.), pp. 485–504. Elsevier, Amsterdam.

Le Roux, X., and Bariac, T. (1998). Seasonal variations in soil, grass and shrub water status in a West African humid savanna. *Oecologia* **113,** 456–466.

Le Roux, X., and Mordelet, P. (1995). Leaf and canopy CO_2 assimilation in a West African humid savanna during the early growing-season. *J. Trop. Ecol.* **11,** 529–545.

Le Roux, X., Polcher, J., Dedieu, G., Menaut, J.-C., and Monteny, B. A. (1994). Radiation exchanges above West African moist savannas: Seasonal Patterns and comparison with a GCM simulation. *J. Geophys. Res.* **99,** 25,857–25,868.

Le Roux, X., Gauthier, H., Bégué, A., and Sinoquet, H. (1997). Radiation absorption and use by humid savanna grassland: Assessment using remote sensing and modelling. *Agric. For. Meteorol.* **85,** 117–132.

Lieth, H. (1973). Primary production: Terrestrial ecosystems. *J. Human Ecol.* **1,** 303–332.

Long, S. P., Garcia Moya, E., Imbamba, S. K., Kamnalrut, A., Piedade, M. T. F., Scurlock, J. M. O., Shen, Y. K., and Hall, D. O. (1989). Primary productivity of natural grass ecosystems of the tropics: A reappraisal. *In* "Ecology of Arable Land" (M. Clarholm and L. Bergstrom, eds.), pp. 9–20. Kluwer Academic Publishers.

Long, S. P., Jones, M. B., and Roberts, M. J. (1992). "Primary Productivity of Grass Ecosystems of the Tropics and Sub-Tropics." Chapman & Hall, London.

Malaisse, F., Freson, R., Goffinet, G., and Malaisse-Mousset, M. (1975). Litterfall and litter breakdown in Miombo. *In* "Tropical Ecology Systems: Trends in Terrestrial and Aquatic Research" (F. B. Golley and E. Medina, eds.), pp. 137–152. Springer Verlag, New York.

McKeon, G. M., Day, K. A., Howden, S. M., Mott, J. J., Orr, D. M., Scattini, W. J., and Weston, E. J. (1990). North Australian savannas: Management for pastoral production. *J. Biogeogr.* **17,** 355–372.

McNaughton, S. J. (1983). Compensatory plant growth as a response to herbivory. *Oikos* **40,** 329–336.

McPherson, G. (1997). "Ecology and Management of North American Savannas." Univ. Ariz. Press, Arizona.

Medina, E. (1987). Nutrients. Requirements, conservation and cycles of nutrients in the herbaceous layer. Anonymous. "Determinants of Tropical Savannas," pp. 39–66. IUBS, Paris.

Medina, E., and Silva, J. F. (1990). Savannas of northern South America: A steady state regulated by water–fire interactions on a background of low nutrient availability. *J. Biogeogr.* **17,** 403–413.

Menaut, J.-C. (1983). The vegetation of African savannas. *In* "Tropical Savannas" (F. Bourlière, ed.), pp. 109–149. Elsevier, Amsterdam.

Menaut, J.-C., and Cesar, A. (1979). Structure and primary productivity of Lamto savannas, Ivory Cost. *Ecology* **60**(6), 1197–1210.

Menaut, J. C., Barbault, R., Lavelle, P., and Lepage, M. (1985). African savannas: Biological systems of humification and mineralization. *In* "Ecology and Management of the World's Savannas," (J. C. Tothill and J. J. Mott, eds.). Australian Academy of Science, Canberra.

Menaut, J.-C., Gignoux, J., Prado, C., and Clobert, J. (1990). Tree community dynamics in a humid savanna of the Côte-d'Ivoire: Modelling the effects of fire and competition with grass and neighbours. *J. Biogeogr.* **17,** 471–481.

Menaut, J.-C., Abbadie, L., Lavenu, F., Loudjani, P., and Podaire, A. (1991). Biomass burning in West African savannas. *In* "Global Biomass Burning: Atmospheric, Climatic and Biospheric Implications" (J. Levine, ed.), pp. 133–149. MIT Press, Cambridge.

Menaut, J.-C., Abbadie, L., and Vitousek, P. M. (1993). Nutrient and organic matter dynamics in tropical ecosystems. *In* "The Ecological, Atmospheric and Climatic Importance of Vegetation Fires," (P. J. Crutzen and J. C. Goldammer, eds.), pp. 215–231. John Wiley & Sons Ltd.

Menaut, J.-C., Abbadie, L., and Lepage, M. (2000). "Lamto: A Savanna Ecosystem." Ecological Studies. Springer-Verlag, New York (in press)

Milner, C., and Hughes, R. E. (1968). "Methods for the Measurement of Primary Production of Grassland." IBP Handbook No. 6. Oxford Blackwell, London.

Miranda, A. C., Miranda, H. S., Lloyd, J., Grace, J., Francey, R. J., McIntyre, J. A., Meir, P., Riggan, P., Lockwood, R., and Brass, J. (1997). Fluxes of carbon, water and energy over Brazilian cerrado: An analysis using eddy covariance and stable isotopes. *Plant Cell Environ.* **20,** 315–328.

Misra, R. (1983). Indian savannas. *In* "Tropical Savannas, Ecosystems of the World," (F. Bourlière, ed.), Vol. 13., pp. 151–166. Elsevier, Amsterdam.

Moore, A. W., Russel, J. S., and Coaldrake, J. E. (1967). Dry matter and nutrient content of a subtropical semi-arid forest of *Acacia harpophylla* F. Muell. (Brigalow). *Austr. J. Bot.* **15,** 11–24.

Mordelet, P., and Menaut, J.-C. (1995). Influence of trees on above-ground production dynamics of grasses in a humid savanna. *J. Veg. Sci.* **6,** 223–228.

Mott, J. J., Williams, J., Andrew, M. H., and Gillison, A. N. (1985). Australian savanna ecosystems. *In* "Ecology and Management of the World's Savannas," (J. C. Tothill and J. J. Mott, eds.), pp. 56–82. Australian Academy of Science, Canberra.

Ojima, D. S., Dirks, B. O. M., Glenn, E. P., Owensby, C. E., and Scurlock, J. M. O. (1993). Assessment of carbon budget for grasslands and drylands of the world. *Water Air Soil Pollut.* **70,** 95–109.

Olson, J. S., Watts, J. A., and Allison, L. J. (1983). "Carbon in Live Vegetation of Major World Ecosystems." Report ORNL-5862. Oak Ridge National Laboratory, Oak Ridge, Tennessee.

ORNL Ecosystems Analysis Group (1998). *Terrestrial ecosystem responses to global change: A research strategy.* ORNL/TM-1998/27, Environmental Sciences Division Publication No. 821. Oak Ridge National Laboratory, Tennessee.

Owensby, C. E., Ham, J. M., Knapp, A. K., and Auen, L. M. (1999). Biomass production and species composition change in a tallgrass prairie ecosystem after long-term exposure to elevated atmospheric CO_2. *Global Change Biol.* **5,** 497–506.

Pandey, C. B., and Singh, J. S. (1992). Rainfall and grazing effects on net primary productivity in a tropical savanna, India. *Ecology* **73**(6), 2007–2021.

Pellew, R. A. P. (1983). The impacts of elephant, giraffe and fire upon the *Acacia tortilis* woodlands of the Serengeti. *Afr. J. Ecol.* **21,** 41–74.

Polley, W. H., Johnson, H. B., Mayeux, H. S., and Tischier, C. R. (1996). Are some of the recent changes in grassland communities a response to rising CO_2 concentrations? *In* "Carbon Dioxide, Populations and Communities," (C. Komer and F. A. Bazzaz, eds.), pp. 177–195. Academic Press, San Diego.

Pugnaire, F. I., Haase, P., and Puigdefábregas, J. (1996). Facilitation between higher plant species in a semiarid environment. *Ecology* **77**(5), 1420–1426.

Rajvanshi, R., and Gupta, S. R. (1985). Biomass, productivity and litterfall in a tropical *Dalbergia sissoo* Roxb. forest. *J. Tree Sci.* **4,** 73–78.

Rushworth, J. E. (1978). Kalahari sand scrub: something of value. Rhod. Sci. News, **12**(8), 193–195.

Rutherford, M. C. (1978). Primary production ecology in southern Africa. *In* "Biogeography and Ecology of Southern Africa," (M. J. A. Werger, ed.), Chap. 15, pp. 621–659. Dr. W. Junk, The Hague.

Rutherford, M. C. (1982). Woody plant biomass distribution in Burkea african savannas. *In* "The Ecology of Tropical Savannas, Ecological Studies 42" (B. J. Huntley and B. H. Walker, eds.), pp. 120–141. Springer-Verlag, Berlin.

San Jose, J. J. (1992). Mass and energy transfer within and between burned and unburned savanna environments. *Int. J. Wildland Fire* **2**(4), 153–160.

San Jose, J. J., and Medina, E. (1975). Effects of fire on organic matter production in a tropical savanna. *In* "Tropical Ecological Systems," (F. B. Golley and E. Medina, eds.), pp. 251–264. Springer-Verlag, Berlin.

San Jose, J. J., and Montes, R. (1989). An assessment of regional productivity: The *Trachypogon* savannas at the Orinoco Llanos. *Nature Resourc.* **25,** 1, 5–18.

San Jose, J. J., and Montes, R. A. (1997). Fire effect on the coexistence of trees and grasses in savannas and the resulting outcome on organic matter budget. *Interciencia* **22**(6), 289–298.

San Jose, J. J., Montes, R., and Nikonova-Crespo, N. (1991). Carbon dioxide and ammonia exchange in the *Trachypogon* savannas of the Orinoco Llanos. *Ann. Bot.* **68,** 321–328.

San Jose, J. J., Bracho, R., and Nikonova, N. (1998). Comparison of water transfer as a component of the energy balance in a cultivated grass (*Brachiaria decumbens* stapf.) field and a savanna during the wet season of the Orinoco Llanos. *Agric. For. Meteorol.* **90,** 65–79.

Sarmiento, G. (1983). The savannas of tropical America. *In* "Tropical Savannas, Ecosystems of the World" (F. Bourlière, ed.), Vol. 13, pp. 246{endash}288. Elsevier, Amsterdam.

Sarmiento, G. (1984). "The Ecology of Neotropical Savannas." Harvard Univ. Press, Cambridge.

Scanlan, J. C., and Burrows, W. H. (1990). Woody overstory impact on herbaceous understory in *Eucalyptus* spp. communities in central Queensland. *Austr. J. Ecol.* **15,** 191–197.

Scholes, R. J. (1988). "Response of three semi-arid savannas on contrasting soils to removal of the woody element," Ph.D. thesis. University of the Witwatersand, Johannesburg, South Africa.

Scholes, R. J., and Archer, S. R. (1997). Tree–grass interactions in savannas. *Annu. Rev. Ecol. Syst.* **28,** 517–544.

Scholes, R. J., and Hall, D. O. (1996). The carbon budget of tropical savannas, woodlands and grasslands. *In* "Global Change: Effects on Coniferous Forests and Grasslands, SCOPE," (A. I. Breymeyer, D. O. Hall, J. M. Melillo, and G. I. Agren, eds.), Vol. 56, pp. 69–100. Wiley, Chichester.

Scholes, R.J., and van der Merwe, M.R. (1996). Sequestration of carbon in savannas and woodlands. *Environ. Profession.* **18,** 96–103.

Scholes, R. J., and Walker, B. H. (1993). "An African Savanna—Synthesis of the Nylsvely Study." Cambridge Univ. Press, Cambridge.

Scholes, R. J., Pickett, G., Ellery, W. N., and Blackmore, A. C. (1997). Plant functional types in African savannas and grasslands. *In* "Plant Functional Types," (T. M. Smith, H. H. Shugart, and F. I. Woodward, eds.), Chap. 13, pp. 255–268. Cambridge Univ. Press, Cambridge.

Scurlock, J. M. O., and Hall, D. O. (1998). The global carbon sink: A grassland perspective. *Global Change Biol.* **4,** 229–233.

Scurlock, J. M. O., Cramer, W., Olson, R. J., Parton, W. J., and Prince, S. D. (1999). Terrestrial NPP: Towards a consistent data set for global model evaluation. *Ecol. Appl.* **9,** 913–919.

Singh, J. S., and Joshi, M. C. (1979). Primary production. *In* "Grassland Ecosystems of the World: Analysis of Grassland and their Uses" (R. T. Coupland, ed.), pp. 197–218. IBP 18. Cambridge Univ. Press, London.

Skarpe, C. (1990). Shrub layer dynamics under different herbivore intensities in an arid savanna, Botswana. *J. Appl. Ecol.* **27,** 873–885.

Skarpe, C. (1991). Impact of grazing in savanna ecosystems. *Ambio* **20,** 351–356.

Solbrig, O. T. (1990), "Savanna Modelling for Global Change." Biology International Special Issue, No. 24. International Union of Biological Sciences, Paris.

Solbrig, O. T. (1996). The diversity of the savanna ecosystem. *In* "Biodiversity and Savanna Ecosystem Processes—A Global Perspective," Ecological Studies (O. T. Solbrig, E. Medina, and J. F. Silva, eds.), Vol. 121. Springer-Verlag, Berlin.

Solbrig, O. T., Medina, E., and Silva, J. F. (1996a). "Biodiversity and Savanna Ecosystem Processes—A Global Perspective." Ecological Studies, Vol. 121. Springer-Verlag, Berlin.

Solbrig, O. T., Medina, E., and Silva, J. F. (1996b). Determinants of tropical savannas. *In* "Biodiversity and Savanna Ecosystem Processes—A Global Perspective," Ecological Studies (O. T. Solbrig, E. Medina, and J. F. Silva, eds.), Vol. 121. Springer-Verlag, Berlin.

Stott, P. (1990). Stability and stress in the savanna forests of mainland South-East Asia. *J. Biogeogr.* **17,** 373–383.

Swaminath, M. H. (1988). Studies on the response of fast-growing forestry species for biomass production under irrigation. *My Forest,* **24,** 117–123.

Teague, W. R., and Smit, G. N. (1992). Relations between woody and herbaceous components and the effects of bush-clearing in southern African savannas. *J. Grassl. Soc. S. Afr.* **9**(2), 60–71.

Tietema, T. (1992). Possibilities for the management of indigenous woodlands in southern Africa: a case study from Botswana. *In* "The Ecology and Management of Indigenous Forests in Southern Africa, Proceedings of an International Symposium, Zimbabwe, 27–29 July 1992" (G. D. Pearce and D. J. Gumbo, eds.), pp. 134–142. Zimbabwe Forestry Commission & SAREC.

Tothill, J. C., and Mott, J. J. (1985). "Ecology and Management of the World's Savannas." Australian Academy of Science, Canberra.

Trollope, W. S. W. (1982). Ecological effects of fire in South African savannas. *In* "Ecology of Tropical Savannas," (B. J. Huntley and B. H. Walker, eds.), pp. 292–306. Ecological Studies, No. 42. Springer-Verlag, Berlin.

Trollope, W. S. W. (1984). Ecological effects of fire in South African ecosystems. *In* "Fire in Savanna," (P. V. Booysen and N. M. Tainton, eds.), pp. 149–176. Springer-Verlag, Berlin.

Tybirk, K., Schmidt, L. H., and Hauser, T. (1992). Notes and records on the dynamics of soil seed banks and tropical ecosystems. *Afr. J. Ecol.* **32,** 327–330.

Vareschi, V. (1960). Observaciones sobre la transpiration de arboles llaneros, durante le epoca de sequia. *Bol. Soc. Venez. Cienc Nat.* **21,** 128–134. [Quoted from Le Roux & Bariac (1998).]

Varshney, C. K. (1972). Productivity of Delhi grasslands. *In* "Tropical Ecology with Emphasis on Organic Production" (P. M. Golley and F. B. Golley, eds.), pp. 27–42. University of Georgia, Athens.

Veenendaal, E. M., Ernst, W. H. O., and Modise, G. S. (1996). Effect of seasonal rainfall pattern on seedling emergence and establishment of grasses in a savanna in south-eastern Botswana. *J. Arid Environ.* **32,** 305–317.

Veetas, O. R. (1992). Micro-site effects of trees and shrubs in dry savannas. *J. Veg. Sci.* **3,** 337–344.

Vyas, L. N., Garg, R. K., and Agarwal, S. K. (1972). Net aboveground production in the monsoon vegetation at Udiapur. *In* "Tropical Ecology with Emphasis on Organic Production" (P. M. Golley and F. B. Golley, eds.), pp. 95–99. University of Georgia, Athens.

Walker, B. H. (1985). Structure and function of savannas: An overview. *In* "Ecology and Management of the World's Savannas" (J. C. Tothill and J. J. Mott, eds.), pp. 83–92. Australian Academy of Science, Canberra.

Walker, B. H. (1987). "Determinants of Tropical Savannas." IRL Press Ltd., Oxford, UK.

Walker, B. H., and Menaut, J.-C. (1988). "Research Procedure and Experimental Design for Savanna Ecology and Management." International Union of Biological Sciences and UNESCO Man and the Biosphere Programme. CSIRO Printing Centre, Melbourne.

Walker, B. H., and Noy-Meir, I. (1982). Aspects of stability and resilience of savanna ecosystems. *In* "Ecology of Tropical Savannas," (B. J. Huntley and B. H. Walker, eds.), pp. 556–590. Ecological Studies No. 42. Springer-Verlag, Berlin.

Walker, J., Moore, R. M., and Robertson, J. A. (1972). Herbage response to tree and shrub thinning in *Eucalyptus populnea* shrub woodlands. *Aust. J. Agric. Res.* **23,** 405–410.

Walter, H. (1971). "Ecology of Tropical and Subtropical Vegetation." Oliver & Boyd, Edinburgh, UK.

Ward, H. K., and Cleghorn, W. B. (1964). The effects of ringbarking trees in *Brachystegia* woodland on the yield of veld grasses. *Rhod. Agric. J.* **61,** 405–410.

Weltzin, J. F., and Coughenour, M. B. (1990). Savanna tree influence on understory vegetation and soil nutrients in northwestern Kenya. *J. Veg. Sci.* **1,** 325–334.

Werner, P. A. (1991). "Savanna Ecology and Management: Australian Perspectives and Intercontinental Comparisons." Blackwell Science, London.

Whittaker, R. H., and Likens, G. E. (1973). Carbon in the biota. *In* "Carbon and the Biosphere" (G. M. Woodwell and E. V. Pecan, eds.), pp. 281–302. AEC Symposium Series 30. NTIS US Dept. of Commerce, Springfield, Virginia.

Whittaker, R. H., and Likens, G. E. (1975). The biosphere and man. *In* "Primary Productivity of the Biosphere" (H. Leith and R. H. Whittaker, eds.), pp. 305–328. Ecological Studies No. 14. Springer-Verlag, Berlin.

Wood, T. G., and Sands, W. A. (1978). The role of termites in ecosystems. *In* "Production Ecology of Ants and Termites" (M. V. Brian, ed.), pp. 245–292. Cambridge Univ. Press, Cambridge.

Yadava, P. S. (1990). Savannas of north-east India. *J. Biogeogr.* **17,** 385–394.

Young, M. D., and Solbrig, O. T. (1993). *The World's Savannas: Economic Driving Forces, Ecological Constraints and Policy Options for Sustainable Land Use.*" Man and the Biosphere, Vol. 12. Parthenon Publishing Group, Carnforth, UK.

Zietsman, P. C., Grobbelaar, N., and van Rooyen, N. (1988). Soil nitrogenase activity of the Nylsvley Nature Reserve. *S. Afr. J. Bot.* **54,** 21–27.

17

Productivity of Tropical Rain Forests

John Grace, Yadvinder Malhi, Niro Higuchi, and Patrick Meir

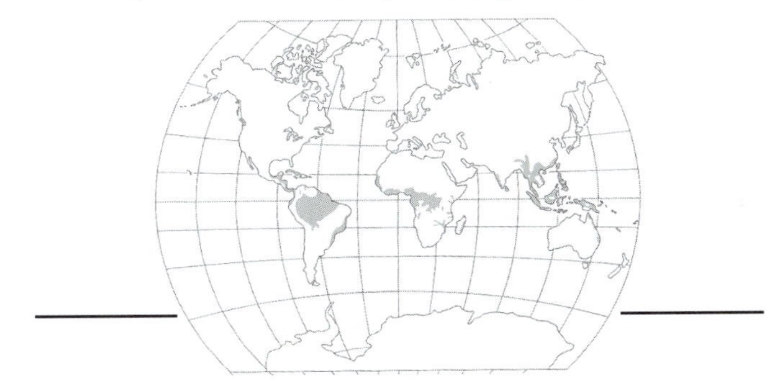

I. The Nature and Extent of Tropical Rain Forests

Tropical rain forests exist in a broad band across the Earth's warm, moist equatorial regions. They are characterized by their great stature (exceeding 30 m), a wide range of life forms (including many trees with buttresses, thick-stemmed climbers, and herbaceous epiphytes), and a large number of tree species (30–300 woody species per hectare). Worldwide, tropical rain forest is being converted to agriculture or being degraded at slightly less than 1% per year (FAO, 1997; Dixon *et al.*, 1994). The CO_2 emitted during this land conversion in the tropics accounts for almost all of the estimated 1.6–2.4 Gt of carbon that is transferred globally from vegetation to atmosphere each year as a result of changes in land use (Schimel, 1995; Houghton, 1999; Fearnside, 2000). Despite the importance of tropical rain forests as a store of carbon, their role in the carbon cycle is not well understood because they are extensive, variable, and generally more difficult to study than other vegetation types.

A. Climate

Tropical rain forests form the natural biome in climates where the mean temperature of the coldest month is 18°C or more, and the monthly rainfall is usually greater than 100 mm (Whitmore, 1990). However, this climatological definition excludes some tropical montane areas, and thus an annual variation of less than 5°C in monthly mean temperature is sometimes used as a more encompassing definition. Where there are several dry months (less than 60 mm rainfall) every year, tropical semideciduous (or monsoon) forests exist. These two types are sometimes bracketed under the term tropical moist forests, but here we shall use the term rain forest to cover lowland evergreen, semideciduous, and montane forests.

B. Area Occupied

Tropical rain forests occupy about 17.5×10^{12} m^2 of land, corresponding to 12% of the terrestrial surface, and may contain as much as 55% of the carbon stored in the world as terrestrial biomass (Whittaker and Likens, 1975; Taylor and Lloyd, 1992). Of the three major blocks of tropical rain forest in the world, the South and Central American block is the largest, about 9.2×10^{12} m^2 (FAO, 1993). Most of this lies within the Orinoco and Amazon basins, with smaller regions extending from the Western Andes up to Mexico, and on the coastal mountains of Brazil. Africa has about 5.3×10^{12} m^2, centered on the Congo basin and extending along the West African coast. Asia has around 3.1×10^{12} m^2, centered on southeast Asia, with outliers in northern Australia, India, and southern China. Within this biome there can be significant variation in vegetation type. For example, in Amazonia the vegetation varies from lowland semideciduous forest in the east and south, through lowland evergreen forest in the center to montane forest in the west, but there are extensive areas of dwarf heath forest on white sands in the upper Rio Negro and bamboo forest in southwest Brazil

C. Canopy Architecture

Classical studies by forest botanists emphasize the structural diversity of the tropical rain forest, which arises from the large influence of the tallest trees and wide spectrum of life forms. Such forests are often said to be vertically "stratified," with a distinct understorey and an emergent layer. Measured profiles of biomass or leaf area density are rather rare (Meir *et al.*, 2000), providing only weak evidence for a layered structure (Koike and Syahbuddin, 1993). However, considering the stratification of life forms, some authorities consider there to be as many as five layers (Whitmore, 1990; Richards, 1996). As well as vertical structure, there is considerable horizontal heterogeneity. Forests are really spatial mosaics of structural phases (Whitmore, 1989). When old trees fall over the gaps are themselves heterogeneous, comprising a "key hole" shape in which the region occupied by the upturned root is quite different from the region of the fallen canopy. The gap is filled by young trees, many of which are derived from seed that has been dormant in the soil or from seedlings that have survived the dense shade, often for several years. The trees colonizing the gaps are usually fast-growing "pioneers" that are succeeded later by trees that grow tall enough to reach the top of the canopy, as well as understorey species with special adaptations to shade. Gaps may be small (caused by the shedding of parts of a tree) or large (caused by several trees falling). Approximately 1–2% of trees die each year (Phillips and Gentry, 1994).

D. Microclimate

Irrespective of whether the canopy is vertically stratified, the canopy height and foliage density are sufficient to cause vertical differentiation in microclimate (Shuttleworth, 1989). This is most evident in the vertical profile of radiation (Chazdon, 1988), but there are important variations in the concentrations of water vapor and other gases that are less obvious but may also be important (Culf *et al.*, 1999). They occur because the lower part of the canopy is largely decoupled from the top of the canopy and from the atmosphere as a whole, particularly at night (Kruijt *et al.*, 2000). Even in the daytime when turbulent transport is effective, there is some degree of decoupling: the air temperatures are much lower at the base of the canopy, and so the air remains nearly water saturated all day. In contrast, at the top of the canopy the air temperature rises and the vapor pressure deficit may increase to as much as 20 mmol mol^{-1} in the late afternoon (Grace *et al.*, 1995a), sufficient to cause xylem water potentials lower than −2 MPa (Roberts *et al.*, 1996). The soil surface is a substantial source of CO_2. The CO_2 concentration in the canopy usually builds up at night in stable conditions, sometimes reaching 550 ppm throughout the vertical profile (Kruijt *et al.*, 1996; Buchmann *et al.*, 1997). The concentration then falls rapidly in the morning, as turbulent transport and photosynthetic activity increase (Kruijt *et al.*, 1996). In this process, significant reassimilation of respired carbon occurs (Medina and Minchin, 1980; Lloyd *et al.*, 1996; Kruijt *et al.*, 1996). Lloyd *et al.* (1996) used the isotopic signature of C to infer that less than 10% of the CO_2 molecules of respiratory origin were re-assimilated.

E. Leaf Characteristics

The area of leaves per area of ground is termed the leaf area index (LAI). For tropical rain forests the LAI has been determined in several studies, either by direct determination or by indirect methods. Direct determination is labor intensive and involves intensive sampling of leaves. It is susceptible to the usual random sampling errors, which are likely to be large when small plots (<1 ha) are used. Indirect methods based on optical principles (Welles and Cohen, 1996) may effectively sample a larger area, but are prone to systematic errors because they contain assumptions that are only approximately met. Examples of LAI values in the literature for undisturbed tropical rain forests are 7.5 (Saldarriaga, 1985), 5.2 (Jordan and Uhl 1978), and 5.7 (McWilliam, 1993). Roberts *et al.* (1996) give data for three forests in the Brazilian Amazon, ranging from 4.4 in Rondonia to 5.7–6.6 in Manaus.

The leaves of tropical rain forests are typically medium-sized (size range a few centimeters, corresponding to mesophylls), with long stalks and leaf margins that are untoothed (Richards, 1996). The leaves tend to be smaller in the drier forest types. Specific leaf areas vary from 5 m^2 kg^{-1} for sunlit leaves at the top of the canopy to over 20 m^2 kg^{-1} at the base (Roberts *et al.*, 1996). At the top of the canopy, leaves have a nitrogen content of 100–200 mmol m^{-2} and a phosphorus content of 1–2 mmol m^{-2}, falling to half this value in the lower canopy (Lloyd *et al.*, 1995). Maximal stomatal conductances also vary with position in the canopy, in the range 100–500 mmol m^{-2} s^{-1}, with some exceptional values of 1000 or more in large-leaved species (Roberts *et al.*, 1990, 1996). Körner (1994) reviewed what is now a quite extensive literature, and found that the maximum leaf conductance was similar in all forest types. The means and standard deviations from his work are as follows: coniferous forests, 234 ± 99; temperate deciduous forests, 190 ± 71; and humid tropical forests, 249 ± 133 mmol m^{-2} s^{-1}. Diurnal trends are also similar for most forest types (Shuttleworth, 1989), with a tendency to open rapidly in the early morning and to reach a maximum before noon.

Photosynthetic rates of rain forest trees on the basis of area of leaf are typical of C_3 species, rarely exceeding 15 μmol m^{-2} s^{-1}, except in pioneer species (Medina and Klinge, 1983; Riddoch *et al.*, 1991a; Roberts *et al.*, 1996). There is strong evidence that the rate of photosynthesis of leaves from mature rain forests, expressed on a leaf area or leaf mass basis, is less responsive to the supply of nitrogen than is the case for crop plants or for early successional tropical vegetation (Reich *et al.*, 1994; Tanner *et al.*, 1998). Laboratory-based studies reveal that many species have the capacity to acclimate to changes in light and shade (e.g., Riddoch *et al.*, 1991b).

The carbon isotopic signal $\delta^{13}C$, is −28 to −34%, typical for C_3 leaves (see Medina and Minchin, 1980; Lloyd and Farquhar, 1996; Buchmann *et al.*, 1997).

F. Root Characteristics

Most studies have found that the mass of roots per area of land declines exponentially with depth (Jackson *et al.*, 1996). In a tropical rain forest, irrespective of whether the forest is evergreen or deciduous, about 70% of the roots are usually within the upper 0.3 m of soil, but there may be additional deep roots that tap the water at great depth (Nepstad *et al.*, 1994).

Most of the water and nutrient uptake by vegetation is achieved by the fine roots (defined as roots ≤2 mm in diameter) or by the extensive mycorrhizal hyphae that are attached to those fine roots. Tropical forests are usually found on highly weathered soils, and the soils often contain a smaller fraction of the nutrient stock than does the aboveground biomass (Jordan, 1982; Zinke *et al.*, 1986; Proctor, 1987; Bruijnzeel, 1991). Thus, the internal nutrient cycling of a tropical forest may be "tighter" than for grasslands and temperate forests. Jackson *et al.* (1997) present a biome-by-biome distribution of fine roots, obtained from published data. Tropical evergreen forest has about 5.7 t biomass ha^{-1} of fine roots, of which 3.3 t are alive. These authors introduce the idea of live root area index (LRAI; by analogy with leaf area index), and suggest that tropical evergreen forest may have a LRAI of 7.4. Corresponding values for grassland types are an order of magnitude higher.

The mass or area of roots in any forest ecosystem is not static but is likely to vary over time scales of weeks to years, in relation to the supply of nutrients and water. This is shown most graphically by the use of ingrowth cylinders (Cuevas and Medina, 1988). In this technique, cylindrical holes are made in the forest soil and are filled with vermiculite, which is imbibed with water or nutrient solution. For nonimbibed cylinders, the 3- to 4-month ingrowth of fine roots was equivalent to 3.1 t biomass ha^{-1} yr^{-1}, but the rate increased threefold when the imbibing solution contained N, P, K, or Ca. The fine roots are clearly dynamic and turn over rapidly, with important implications for the carbon cycle. At a site in Eastern Amazonia Trumbore *et al.* (1995) showed a very high allocation of carbon to roots beyond 1 m deep (19 t C ha^{-1} yr^{-1}), much higher than litterfall (4.6 t C ha^{-1} yr^{-1}), and estimated that 15% of the carbon in deep soil has a turnover time of decades or less.

G. Canopy Biophysical Properties

Models of water, carbon dioxide, and energy exchange, such as the Simple Biosphere (SiB) model, require parameters to represent the biophysical properties of each biome (Sellers *et al.*, 1989, 1997). In this section we briefly review the parameter values that have been determined from tropical rain forest.

The radiation balance is discussed by Shuttleworth (1989) and Gash and Shuttleworth (1991). In the Brazilian Amazon the incoming solar radiation varies somewhat over the course of the year, but the monthly mean is 15–18 MJ m^{-2} d^{-1}. Tropical rain forests have a very low short-wave reflectance, in the range 0.12–0.14, much lower than the pasture that often replaces forests in Brazil (0.17–0.19), and comparable to coniferous forests (Gash and Shuttleworth, 1991; Culf *et al.*, 1996). There is some evidence of seasonality, with the highest values of reflectance coinciding with the period when the soil moisture is driest (Culf *et al.*, 1996).

The relationship between solar radiation and net radiation is usually quite linear, and net radiation is in the range 9–11 MJ m^{-2} d^{-1}. The evapotranspiration rate, expressed as an energy flux, has been found to be approximately 70% of the net absorbed radiation, with rather little seasonal variation (Shuttleworth *et al.*, 1984). The roughness length z_0 and the zero-plane displacement d have been derived from wind profiles, and for a 35-m-tall forest they were practically the same in two studies of the forest near Manaus, Brazil: $z_0 = 2.0$–2.2 m and $d = 29$–31 m (Shuttleworth, 1989). For many kinds of vegetation, it has been possible to relate z_0 and d to the height of the vegetation h, and often z_0/h is 0.1 and d/h is 0.6 (Monteith and Unsworth, 1990). Thus, for tropical rain forests the roughness length may be smaller than expected from other studies and the zero-plane displacement may be larger. This may be related to the unusual canopy structure of tropical rain forests, in relation to that of many other forest types, with its dense undercanopy and sparse emergent layer.

The aerodynamic conductance is a linear function of wind speed, and under neutral and unstable meteorological conditions it is in the range 0–4 mol m^{-2} s^{-1} (Grace *et al.*, 1995a). The canopy stomatal conductance is generally smaller, in the range 0–1 mol m^{-2} s^{-1}. Thus, gas exchange of the canopy as a whole is highly responsive to stomatal behavior. At nighttime, however, the atmosphere often becomes stably stratified, and aerodynamic conductance is then close to zero, and carbon dioxide accumulates in, and above, the forest canopy (Culf *et al.*, 1999).

II. Production Values

A. Biomass of Undisturbed Forest

The aboveground stocks of carbon in tropical rain forest are notoriously difficult to estimate (Brown *et al.*, 1989; Gillespie *et al.*, 1992; Higuchi *et al.*, 1994; Foster Brown *et al.*, 1995; Deans *et al.*, 1996; Alves *et al.*, 1997; Fearnside, 1997). The accuracy of the estimate relies on the unbiased selection of sample plots. In the past, such plots have often been placed in especially luxuriant vegetation, have been poorly replicated, and smaller than the 1-ha minimum usually recommended in forest mensuration (Philip, 1994). The estimation of biomass from forest mensurational data rests on the use of regression equations that relate the diameter (and sometimes height) of the tree to its biomass, the relationships having been determined on a relatively small number of sample trees. The belowground fraction of biomass has rarely been measured and seems to be in the range of one-fifth to one-third of the aboveground fraction in tropical rain forests (Jackson *et al.*, 1996; Deans *et al.*, 1996).

The appropriate figure for the global average carbon stocks in undisturbed tropical rain forests, from contemporary data (Table 17-1), is about 150 t C ha^{-1} aboveground and 35–50 t C ha^{-1} belowground, making a total of 185–200 t C ha^{-1}. Assuming the midrange figure of 192 t C ha^{-1} to apply over the entire 17.5×10^{12} m^2, this implies a global total carbon stock of 336 Gt C, which is more than half of the estimated global biomass stock of 610 Gt C (Schimel, 1995). This figure is more likely to be an overestimate than an underestimate, because researchers may have selected especially luxuriant sites for evaluation.

Table 17-1 Carbon Stocks in Tropical Rain Forests of the World[a]

Rain forest	Leaf area index	Specific leaf area ($m^2\ kg^{-1}$)	Leaf	Wood	Total aboveground	Soil carbon	Ref.
Manaus, Brazil *terra firme* (38 m)			4.6	198	203		Klinge (1976)
Magdalena Valley, Columbia (30 m)			4.5	158	163		Fölster *et al.* (1976)
San Carlos de Rio Negro, Venezuela, *terra firme* (30 m)	(1) 5.2 (2) 7.5	6.5	4.0 4.9	163 112	167 117		Jordan and Uhl (1978), Saldarriaga (1985)
Venezuela, tall Caatinga	5.1	7.8	5.4	128	134		Klinge and Herrera (1983)
El Verde, Puerto Rico	6.6	6.1	5.4	108	114		Jordan (1971)
Manaus, Brazil, dense *terra firme* forest (30 m)	5.7	9.0	3.1	134	137		McWilliam *et al.* (1993)
Manaus, dense *terra firme* forest				158	203		Carvalho *et al.* (1995)
South of Pará, Brazil *terra firme*					93		Higuchi *et al.* (1994)
Rondonia, Brazil					114–171		Foster Brown *et al.* (1995)
15 sites in Malaysia					127–223		Brown *et al.* (1989)
20 sites in Cameroon					119–157		Brown *et al.* (1989)
14 sites in Sri Lanka					76–110		Brown *et al.* (1989)
Pasoh Forest, Malaysia			212	3.9	216	165	Oikawa (1985)
Primary forest, Western Brazilian Amazon					145–247		Alves *et al.* (1997)
Tropical wet forest, 12 sites worldwide						210 ± 9	Post *et al.* (1982)
Tropical moist forest, 162 sites worldwide						115 ± 12	Post *et al.* (1982)
Global average					225		Whittaker and Likens (1975)
Global average					218	124	Taylor and Lloyd (1992)
Global average (data in this review)					152 ± 40		

[a]Expressed as aboveground biomass C per area of land. Units are t C ha^{-1}; obtained from published data on dry biomass by assuming that biomass is 50% carbon (Mathews, 1993).

B. Gross and Net Productivity

The terms gross primary productivity (GPP; P_g and net primary productivity (NPP; P_n) refer to the gross and net rates at which carbon is assimilated by vegetation, and are usually expressed on time scales of months or years. P_g is the rate at which carbon is fixed by photosynthesis of the vegetation as a whole, and P_n is what remains of the photosynthesis after plant respiration R_a:

$$P_n = P_g - R_a.$$

These terms have units of dry mass per area per time, or of carbon per area per time. Traditionally, P_n is measured indirectly within sample plots by sampling and weighing (1) the biomass to determine the increment over the course of the year and (2) the litter that the plants have shed during that period; P_n is then estimated as the sum of the increment plus the accumulated litter. This procedure is subject to sampling errors and cannot be easily applied to the belowground productivity, because it is impractical to measure the shedding of roots and the exudation of organic compounds by the roots, even though both of these may be very large (Sanantonio and Grace, 1987). Thus, to obtain a figure for belowground P_n, various assumptions become necessary and consequently major uncertainties in P_n may arise. For example, the death of roots and rhizodeposition have generally been overlooked (see Clark *et al.*, 2000a).

Reviews of the net primary productivity of tropical rain forests have been presented by UNESCO (1978) and Medina and Klinge (1983), and some of those data are reproduced in Table 17-2. The estimates are based on "traditional" harvesting methods and usually do not contain allowance for belowground processes. Thus they are likely to be underestimates of the real value of P_n. The highest values are for evergreen forests in climatic zones where the dry season is very short or nonexistent, and they exceed the 11 t C ha^{-1} yr^{-1} of Whittaker and Likens (1975), which is often used as a mean value in considerations of global productivity. Recently, Clark *et al.* (2000b) reviewed data from 39 sites, estimating lower and upper limit for each one. They found a mean lower limit of 7.1 and a mean upper limit of 13.0 t C ha^{-1} yr^{-1}. Another relevant data set is that of litterfall, tabulated in UNESCO (1978). Litterfall is of course only one component of P_n. Out of 42 studies cited, 20 exceed 5.0 t C ha^{-1} yr^{-1} and three exceed 10 t C ha^{-1} yr^{-1}. In view of these major uncertainties, much of the older literature is likely to underestimate P_n.

C. Eddy Covariance Data

The productivity of vegetation may also be assessed by micrometeorological measurement of CO_2 fluxes. Early attempts by Allen *et al.* (1972) and Yabuki and Aoki (1978) used the flux-gradient approach, which had been developed in the 1960s and used successfully over field crops. Eddy covariance

Table 17-2 Net Primary Productivity of Tropical Rain Forests[a]

Forest type	Method	Location	NPP ($t\ C\ ha^{-1}\ yr^{-1}$)	Ref.
Equatorial	Harvesting	Yangambi, Zaire	16	Bartholemew *et al.* (1953)
Equatorial	Harvesting	Thailand	15.5	Kira *et al.* (1964)
Subequatorial	Harvesting	Ivory Coast	7.5–8.5	Lemée *et al.* (1975)
Seasonal rain forest	Harvesting	Ivory Coast	6.5	Müller and Nielsen (1965)
Lowland Dipterocarp	Harvesting	Pasoh, Malaysia	15	Whitmore (1984)
Seasonal rain forest	Harvesting	Kade, Ghana	12.5	Greenland and Kowal (1960)
Evergreen forest	Harvesting	San Carlos, Venezuela	6	Jordan and Escalante (1980)
Lowland Dipterocarp	Harvesting	Pasoh, Malaysia	—	Kira (1978)
Tropical rain forest	Modeling	—	10.5	Taylor and Lloyd (1992)
Tropical rain forest	Modeling	—	6.8	Lloyd and Farquhar (1996)
Evergreen rain forest	Modeling	Manaus, Brazil	15.6	Current study (Malhi, Higuchi, and Grace)
Average of those above ± standard deviation			11.2 ± 4.2	
Possible best estimate of global average, and including fine roots (see text)			18.1[b]	Current study

[a]As estimated by traditional harvesting methods, by modeling, and in the present study at Manaus. Units are $t\ C\ ha^{-1}\ yr^{-1}$ obtained from published data on dry biomass by assuming that biomass is 50% carbon (Mathews, 1993).

[b]Clark (2000b) gives a range of 7.1–13.0 $t\ C\ ha^{-1}\ yr^{-1}$, using a different set of assumptions.

has now become the standard micrometeorological technique for investigating fluxes of carbon and water vapor over vegetation (Baldocchi, 1996; Miranda *et al.*, 1997). It is a very direct method and may be run continuously over months or even years. An important advantage of this approach is that the data provide insights into the climatological controls of production processes, because the fluctuations over hours and days may be related to weather patterns. Micrometeorological techniques measure the net ecosystem flux of carbon, F_c, which is the net rate at which the entire ecosystem is exchanging carbon with the atmosphere. It includes not only autotrophic (plant) respiration but respiration from the microbial breakdown of dead organic matter and respiration of animals (*i.e.*, heterotrophic respiration):

$$F_c = P_g - R_a - R_h,$$

where the term R_h is the rate at which carbon is being released by heterotrophic respiration. The errors associated with the technique have been discussed in the literature (Moncrieff *et al.*, 1996). The most important errors appear to be associated with topographic variation, and for this reason it is usual to restrict measurement to flat sites with good fetch. It is important to conduct supporting measurements of in-canopy profiles of CO_2, to enable the biotic flux to be calculated from the eddy flux measured above the canopy. It is also highly recommended that when F_c is being measured by eddy covariance, chambers on plant and soil surfaces should be operated simultaneously to obtain estimates of soil fluxes R_a and R_m. This enables P_n to be found, at least in principle. However, in practice is it hard to partition the total flux of carbon dioxide from the soil into separate plant-derived and microbial-derived components (autotrophic and heterotrophic respiration) and therefore there is always uncertainty in the relative magnitudes of R_a and R_m. Nevertheless, the combination of chamber and micrometeorological techniques provides information about the sensitivity of the constituent fluxes to climatological variables, and such information may form the basis of a useful model.

Eddy covariance measurements of carbon dioxide and water vapor flux demonstrate the sensitivity of ecosystem carbon fluxes to variations in solar irradiance, vapor pressure deficit, and temperature. Typical eddy covariance data from tropical rain forest are shown as a diurnal cycle in Fig. 17-1. The water vapor fluxes are used to calculate the canopy stomatal conductance, using the Penman–Monteith equation. Stomatal conductance usually increases rapidly after sunrise to attain 0.5–1.0 mol m^{-2} s^{-1}, and then gradually declines (Shuttleworth, 1989; Dolman *et al.*, 1991; Grace *et al.*, 1995a). This trend is seen not only from the whole-canopy stomatal response, but also from leaf-scale observations made with porometers inside the canopy (Roberts *et al.*, 1990; McWilliam *et al.*, 1996).

The CO_2 flux at night indicates an ecosystem "dark respiration" of 5–8 μmol m^{-2} s^{-1}, most of which is derived from the soil (Meir *et al.*, 1996). Soon after sunrise, the ecosystem achieves its "light compensation point" when photosynthetic gains are equal to respiratory losses. Thereafter, the rate of uptake increases until some time before solar noon, when net uptake is between 10 and 25 μmol m^{-2} s^{-1}. As the stomatal conductance declines there is a consequent decline in the rate of photosynthesis, so that the afternoon rate is generally less than that of the morning. The relationship between net ecosystem exchange of carbon dioxide and irradiance is nonlinear on an hourly basis, partly as a result of light saturation of sunlit leaves and partly the tendency of the stomata to shut during the second half of the day (Grace *et al.*, 1995a).

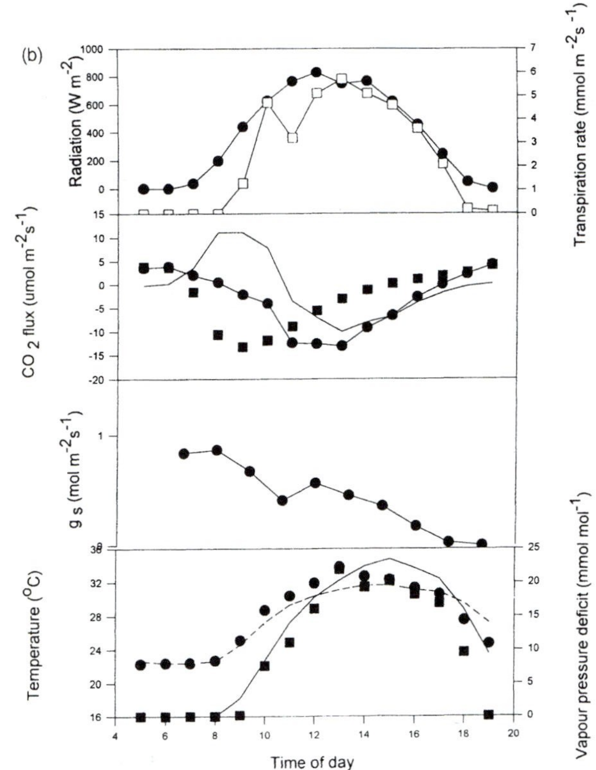

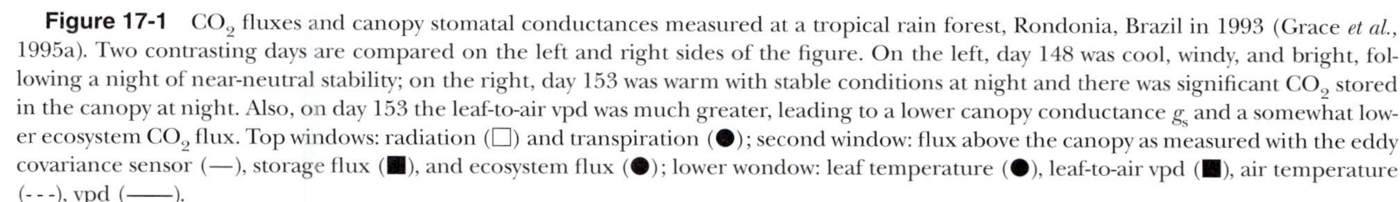

Figure 17-1 CO_2 fluxes and canopy stomatal conductances measured at a tropical rain forest, Rondonia, Brazil in 1993 (Grace *et al.*, 1995a). Two contrasting days are compared on the left and right sides of the figure. On the left, day 148 was cool, windy, and bright, following a night of near-neutral stability; on the right, day 153 was warm with stable conditions at night and there was significant CO_2 stored in the canopy at night. Also, on day 153 the leaf-to-air vpd was much greater, leading to a lower canopy conductance g_s and a somewhat lower ecosystem CO_2 flux. Top windows: radiation (□) and transpiration (●); second window: flux above the canopy as measured with the eddy covariance sensor (—), storage flux (■), and ecosystem flux (●); lower wondow: leaf temperature (●), leaf-to-air vpd (■), air temperature (- - -), vpd (——).

Models of ecosystem carbon and water vapor flux may be fitted to eddy covariance data (da Rocha *et al.*, 1996; Lloyd *et al.*, 1995; Sellers *et al.*, 1997). Such models incorporate the Farquhar model of C_3 photosynthesis (Farquhar *et al.*, 1980), and various empirical models of stomatal response to the environment and heterotrophic respiration. Lloyd *et al.* (1995) fitted such a model to a subset of the data collected from Rondonia during Anglo-Brazilian campaigns in the wet and dry seasons 1992/1993. It was possible to estimate P_g to be 24 t C ha^{-1} yr^{-1} for a semideciduous forest in Rondonia, and Malhi *et al.* (1998) estimate a value of 30.4 t C ha^{-1} yr^{-1} for lowland evergreen forest near Manaus in central Amazonia. Previous estimates of P_g have been made from a knowledge of net primary productivity and ecosystem respiration (Müller and Nielsen, 1965; Yoda, 1978; Oikawa, 1985). In Côte d'Ivoire, Müller and Nielsen (1965) obtained a P_g of 25 t C ha^{-1} yr^{-1} but at Pasoh forest in Malaysia the corresponding figure was approximately double this value (Oikawa, 1985).

The ecosystem respiration measured by eddy covariance includes heterotrophic respiration, and is therefore not especially useful as a means of calculating P_n from GPP. One method of estimating autotrophic respiration is to use theoretical or measured plant respiration rates (Lehto and Grace, 1994). Respiration rate in the dark is divided into maintenance respiration (required for protein turnover and transport) and growth respiration (the respiration associated with biosynthesis of biochemical constituents from CO_2). Theoretical values for these terms are similar to those measured in laboratory experiments, although there is some difficulty in applying the concepts to woody tissue because an unknown fraction of it is "dead" and therefore needs to be discounted from the biomass. Calculations for the Rondonia forest suggest that the respiratory flux from a forest with a leaf turnover of 0.24 yr^{-1} and a stem turnover of 0.03 yr^{-1} is about 70% of the gross primary productivity, implying a P_n of 7.2 t C ha^{-1} yr^{-1}. This figure is similar to the P_n estimated from various recent modeling studies of tropical rain forests (Raich *et al.*, 1991; Wang and Polglase, 1995; Lloyd and Farquhar, 1996). It is interesting to compare such rates with those observed for the increase in biomass of the regrowth forest that follows abandonment of farmland, although this figure is of course less than P_n because it does not allow for production of litter. Nonetheless, it is similar, varying between 3 and 7 t C ha^{-1} yr^{-1} but declining over a few decades (Brown and Lugo, 1990; Saldarriaga and Luxmore, 1991; Honzak *et al.*, 1996; Alves *et al.*, 1997).

III. Components of Production

We can obtain new insights into the allocation of production within a system by combining gas-exchange measurements of photosynthesis and respiration with harvesting data on biomass increments and litterfall. A significant degree of variability between forest sites should be expected. Kira (1978) and Whitmore (1984) present attempts at such calculations for lowland evergreen dipterocarp rain forest at Pasoh, Malaysia, and here we present a current case study for an Amazonian *terra firme* at Cuieiras (near Manaus), Brazil (see Fig. 17-2).

In Fig. 17-2, the exchanges between the ecosystem and the atmosphere are derived from eddy covariance data (Malhi *et al.*, 1998) and the stocks of carbon come from a 10-yr study by the BIONTE project (N. Higuchi *et al.*, unpublished) at three 1-ha sample plots. Leaf, trunk, and soil respiration were measured at another rain forest site in Rondonia, Brazil (Meir *et al.*, 1996) and have been scaled up to provide estimates for the Cuieiras site. Flows between compartments are either directly measured (as, for example the rate of litter fall), or inferred by assuming that the compartment is at a steady state (an assumption that may not be correct).

In the Cuieiras forest, carbon constitutes 48% of dry-weight living biomass (Carvalho *et al.*, 1995). The aboveground biomass was estimated from diameter measurements using allometric relations determined by destructive harvesting of 319 trees at a nearby site, and falls within the range reported in other studies (UNESCO, 1978; McWilliam *et al.*, 1993). Leaf and wood litter pools were directly measured, and the woody litter pool (including large trunks) is significantly larger than that reported in other studies. The quantities of fine roots and large roots were estimated from a fine root:large root:shoot ratio measured by Klinge and Rodrigues (1973) at a nearby site. Total soil carbon was not measured at this site and the average value found in a number of tropical forest studies is used (Post *et al.*, 1982).

Carbon Flows

The details in the assumptions used to build the carbon flow scheme are necessarily crude, but do provide a useful insight into the nature of productivity into the system. The flux data imply a gross primary productivity (incorporating photorespiration) of 30.4 t C ha^{-1} yr^{-1} (Malhi *et al.*, 1998). Of this, 3.7 t are allocated into new leaf and twig growth to replace the (measured) litterfall, and 4.1 t are lost in growth and maintenance respiration. Herbivory (a flow of carbon from leaves to soil via animal guts) was not measured and is not included in this study, although Kira (1978) guessed it may be as large as 2.3 t for the Malaysian rain forest.

A residue of 22.6 t C passes into trunk wood, where an estimated 3.9 t are respired. Mean mortality at the site was 6.7 trees yr^{-1}, or 1.1%, on the lower end of the 1–2% range summarized by Phillips and Gentry (1994). At the BIONTE site there was a measured net increase of aboveground biomass at a rate of 1.7 t C ha^{-1} yr^{-1}. Whether this is part of a natural disturbance–recovery cycle or a response to environmental change is as yet unclear and will be discussed in future work. Overall, new trunk growth accounts for 5.0 t C, leaving a residue of 13.7 t that is allocated belowground. Both large and fine

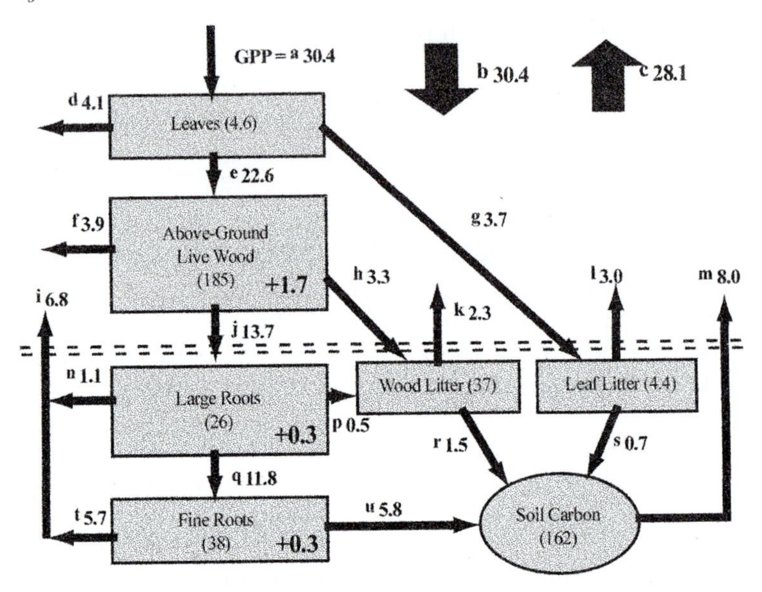

Figure 17-2 Estimates of carbon fluxes at an undisturbed Amazonian *terra firme* forest at Cuieiras (near Manaus), Brazil, based on harvesting of dry mass (Higuchi *et al.*, 1994) and eddy covariance data (Malhi *et al.*, 1998; 1999). Stocks of carbon are shown in parentheses in the boxes as t C ha^{-1}. Fluxes are shown as arrows in t C ha^{-1} yr^{-1}. The explanation of each flux is as follows, coded by the superscript written against each value: a and b, gross primary productivity; c, total ecosystem respiration; d, leaf respiration; e, export from leaves; f, wood respiration; g, litterfall; h, mortality; i, total root respiration; j, transport to the belowground organs; k, microbial respiration from dead wood; l, microbial respiration from dead leaves; m, microbial respiration from soil organic matter; n, coarse root respiration; p, root mortality; q, fine root production; r, decay of wood to form soil organic matter; s, decay of litter to form soil organic matter; t, fine root respiration; shedding of fine roots.

roots are assumed to have a net increment in proportion to aboveground biomass, and some large roots are lost through tree mortality. Large root respiration per unit of carbon is guessed as twice the rate of trunk respiration. Whatever the details of the various assumptions and calculations, to balance the carbon flow we must conclude that a large proportion (here 11.8 t, or 39%) of P_g must be allocated to fine root (and exudate) turnover and respiration. The partitioning of this carbon between production and turnover is unclear and here for simplicity we have guessed at equipartitioning. We have assumed no net increment in litter and soil carbon pools, whereas elsewhere we have made calculations based on the assumption of an increasing pool of soil carbon (Malhi and Grace, 2000).

Having completed a carbon flow diagram for the system, we can now arrive at new estimates for P_n, total autotrophic respiration R_a, and heterotrophic respiration R_h. The numerical value of P_n is evidently strongly dependent on the assumption of partitioning between fine root respiration and production, so we also quote the figures for the alternative assumptions of 75% fine root respiration and 75% fine root production. Thus we estimate an P_n of 15.6 t C ha^{-1} yr^{-1} (range 12.7–18.4), R_a of 14.8 t C ha^{-1} yr^{-1} (range 17.7–12.0), and R_h of 13.3 t C ha^{-1} yr^{-1} (10.4–16.2). In terms of percentage of G_p, the values are P_n, 51% (42–61%), R_a, 49% (58–39%); and R_h, 44% (34–53%).

This value of P_n (42–61% of GPP) is larger than the range of 6–14 t C ha^{-1} yr^{-1} obtained through allometry (see above). The difference is due to the fact that fine root turnover is difficult to sample and is usually neglected. If the fine root turnover is neglected in our calculation, we estimate an P_n of 9.8 t C ha^{-1} yr^{-1} (32% of P_g), within the usual range, though higher than the values found in western Amazonia. This may also be compared to the 11 t C ha^{-1} yr^{-1} quoted by Whittaker and Likens (1975) and the 10.5 t C ha^{-1} yr^{-1} from the model of Taylor and Lloyd (1992). Whether fine root turnover and exudate loss should be included in definitions of P_n will no doubt be debated elsewhere in this book; what these calculations demonstrate is how important an agreed definition is to calculation of the magnitude of P_n.

We have estimated "most likely values" of above- and belowground P_n by taking the best estimates from the Manaus site (Fig. 17-2) and scaling them according to the ratio of the above- and belowground estimated productivity, hence adjusting for the fact that the Manaus site has a somewhat lower traditionally estimated productivity than does an average tropical forest. For the upper and lower ranges we have estimated that two standard deviations in aboveground biomass are equal to 80 t C ha^{-1}. This method provides an estimate of the mean P_n as 18.1 t C ha^{-1} yr^{-1}, with upper and lower limits of 8.5 and 27.5 t C ha^{-1} yr^{-1}.

This carbon flow calculation also allows us to estimate how much of the C efflux from the soil surface is derived from plant respiration R_a as opposed to heterotrophic respiration R_h. We estimate a ratio R_a:R_h of 0.34:0.66, with a range between 0.48:0.52 and 0.15:0.85. Obviously the exact assumption about fine root behavior has an important impact on the ratio, but it seems likely that heterotrophic respiration is the larger component. We estimate a total soil respiration of 20.1 t C ha^{-1} yr^{-1}, equivalent to a mean soil CO_2 efflux of 5.31 μmol CO_2 m^{-2} s^{-1}.

IV. Estimate of Global Productivity

Tropical forests are highly productive and cover a vast area; hence we can expect them to account for a large proportion of total global productivity. We can attempt to estimate their contribution by multiplying the value for

P_n obtained above by the total area of tropical forest. This does not allow for significant heterogeneity but, until a wider range of gas-exchange and allometric studies are undertaken, it is difficult to assess this variability. The lowland rain forest described above is the predominant type in South America, and the review presented by Medina and Klinge (1983) suggests that African forests have similar productivity, whereas Asian rain forests may be more productive.

Assuming tropical rain forests occupy 17.5×10^{12} m^2, and taking the value of P_n derived above as 18.1 t C ha^{-1} yr^{-1}, we estimate the total P_n of tropical forests to be 32 Gt C yr^{-1} [about one-half of the total terrestrial production suggested by Schimel (1995)]. Again assuming a scaling factor to adjust the Manaus P_g to that which might apply to an average tropical rain forest, we estimate a global P_g of 67 Gt C yr^{-1}. The atmosphere contains approximately 765 Gt C as carbon dioxide; we therefore conclude that tropical forests might process 8.2% of the atmospheric CO_2 stock every year as gross production, and 4.2% as net production. The key difference between this calculation of P_n and previous estimates is that we are now attempting to account for fine root turnover and exudate loss. If we ignore this term, the calculated P_n would be somewhat lower.

V. Human and Environmental Impacts

Grace *et al.* (1996) used the model of Lloyd *et al.* (1996) to explore the probable effect of changing the temperature and CO_2 concentration on the fluxes of CO_2 over the rain forest in Rondonia. In the model, small changes in temperature have a large influence on ecosystem respiration, but only a small effect on photosynthesis (Fig. 17-3). In the measurement year (1992–1993) photosynthesis exceeded respiration by about 1 t C, but warmer years would lead to a smaller accumulation or even a loss of carbon. Over the next century, it is expected that the CO_2 concentration will increase. Despite experimental studies on the impact of elevated CO_2 on young trees in artificial (Körner and Arnone, 1992) or natural environments (Würth *et al.*, 1998), the response of tropical forests to elevated CO_2 concentration is not well understood. For forest ecosystems, we must resort to models to make estimates of impacts, because large-scale fumigation studies are not practical. Several models have examined impacts of future climates on the productivity and sink strength of tropical and temperate rain forests (Taylor and Lloyd, 1992; Wang and Polglase, 1995; Lloyd and Farquhar, 1996). Assuming that the photosynthetic system does not down-regulate, and that other factors such as the availability of N and water do not become limiting, we may expect a near-linear increase in GPP of 0.5% per year as the CO_2 concentration increases, which should more than offset the increase in ecosys-

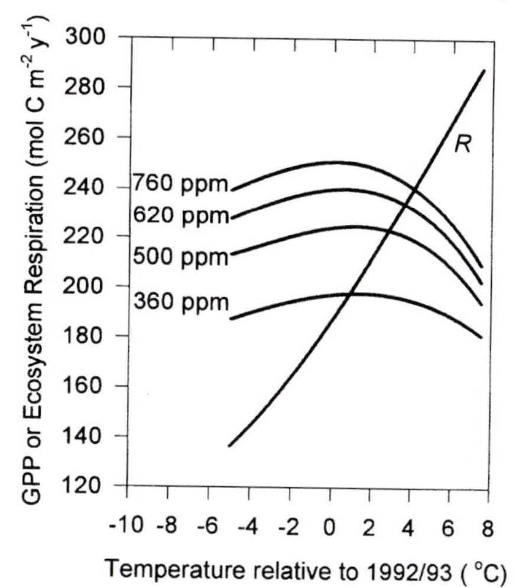

Figure 17-3 The sensitivity of the Rondonia rain forest to changes in temperature and CO_2 concentration, estimated from the model of Lloyd *et al.* (1995) fitted to the data set of Grace (1995a). The model has been rerun several times after subtracting or adding a temperature change to the climatological data. The graph provides a rough estimate of the difference between gross primary productivity (solid lines) and ecosystem respiration (broken lines) that may be expected to be observed as the temperature and CO_2 concentrations rise, assuming no acclimation or floristic changes. Redrawn from Grace *et al.* (1996).

tem respiration caused by the increase in temperature. Such reasoning and such models support the view that forest ecosystems may currently be global sinks for carbon, and may become even stronger sinks in the next few decades as a result of elevated CO_2 and enhanced deposition of nitrogen from anthropogenic sources. It is well known that N deposition rates have increased in northern Europe and the United States as a result of anthropogenic production of NO_x and NH_4 (Pitcairn *et al.*, 1995), and that plant growth has generally increased in northern latitudes (Myeni *et al.*, 1997). It has also been estimated that deposition for tropical areas has increased, too, though less dramatically. Galloway *et al.* (1995) suggest that a preindustrial deposition rate to the Brazilian Amazon and southeastern Asia of 0.28–1.40 kg ha^{-1} yr^{-1} may have (approximately) doubled. Townsend *et al.* (1996) estimate that fossil fuel-derived nitrogen may have stimulated carbon sinks by 0.44 to 0.74 Gt C yr^{-1}. In the temperate zone there is evidence from growth rings of an increase in forest productivity (Badeau *et al.*, 1996). In a study of forestry inventories across the tropics, Phillips *et al.* (1998) concluded that there has been a net increase in tree biomass in the neotropics since the 1970s, corresponding to a net carbon sink of 0.45 Gt C yr^{-1}. Their data hint that there may be a different pattern of change between neotropics and paleotropics, with most of the biomass change being localized in South America. If verified, this may reflect the impact of differing climatic regimes (for example, Asian forests are more influenced by a maritime climate), or the greater degree of low-level human disturbance in the more densely populated neotropics.

The turnover rate of tropical forests may have been increased. Phillips and Gentry (1994) and Phillips *et al.* (1994) report mortality and recruitment of mature tropical forests to have risen from 1% in the 1950s to over 2% in the late 1980s. It is widely held that P_n is currently increasing as a result of elevated CO_2 and N deposition: for example, the model of Lloyd and Farquhar (1996) suggests an increase from 5.6 in preindustrial times to 7.1 at present, reaching 9.3 t C ha^{-1} yr^{-1} around 2100, when the CO_2 concentration becomes 450 ppm.

Some uncertainty still exists in the mean turnover time of tropical forests, a factor that is relevant when assessing the response of the forests to disturbance and global change. The carbon flow diagram in Fig. 17-2 suggests a mean turnover time of 50–100 yr, but recent carbon dating in central Amazonia suggests that some large individual trees may be up to 1400 yr old (Chambers *et al.*, 1998). As such, they would represent significant long-term, low-productivity reservoirs of carbon.

VI. Research Needs

A. Comprehensive Studies of Inputs, Outputs, and Flows in Intact and Disturbed Ecosystems

It has rarely been possible to study the tropical forest as an ecosystem, and particularly to quantify the inputs and outputs in a comprehensive way. The wet/dry deposition of elements and particles to forests in the temperate zone has been studied in considerable detail, and the research has been motivated by the need to understand the impact of acid deposition. For tropical forests the inputs have not been studied systematically, though they can be estimated from atmospheric concentrations and roughness. Of special interest is the rate at which the forest recaptures the nutrients released from within the region, and the extent to which growth is currently being stimulated by such inputs of N and P. The belowground processes are much harder to study than those aboveground, and as previously mentioned, such matters as fine root turnover and root exudation have not been investigated sufficiently. The leakage of carbon from forests via drainage waters to rivers has not been measured: data from the mouth of the Amazon suggest that it may on average be about 0.1 t C ha^{-1} $year^{-1}$, but that discounts any labile carbon lost in transit in the river (Richey *et al.*, 1991).

B. Forests as Global Carbon Sinks

There is a need to corroborate the recent observations that forests are turning over more rapidly than hitherto, and that undisturbed forests in the tropics are carbon sinks (Phillips and Gentry, 1994; Grace *et al.*, 1995b). These issues can be addressed in several ways, one of which is by the analysis of the very large data sets that foresters have collected from permanent sample plots (Phillips *et al.*, 1998); another approach to the question of the ecosystem carbon balance is simply to establish a network of eddy covariance sites throughout the tropics. On a larger scale, an understanding of the forest as a carbon sink requires further application of satellite remote sensing to estimate losses from burning, and the extent of regrowth and agrading areas. The current satellite sensors (NOAH–AVHRR and Synthetic Aperture Radar such as ERS and JERS) have been disappointing, because the signal usually saturates at a relative low biomass density, making it extremely difficult to rely on satellite data for surveillance purposes.

C. Temporal Changes

It appears that even evergreen tropical rain forests are strongly seasonal in their metabolism, exhibiting a seasonal pattern of albedo (Culf *et al.*, 1966), CO_2 assimilation (Malhi *et al.*, 1998), and litterfall. It is not known how this seasonal pattern is determined, although the seasonality in phenology has often been pointed out (Longman and Jenik, 1987; Richards, 1996). On longer time scales, there is likely to be considerable interannual variability as a result in patterns of rainfall and temperature (Lui *et al.*, 1994; Richards, 1996; Kindermann *et al.*, 1996). These patterns cannot be studied in growth rings, as they can in temperate trees, but should be detectable using growth bands and eddy covariance sensors.

D. Impact of Environmental Change

Currently, the long-term impact of climate warming and elevated CO_2 on productivity and biodiversity can only be estimated from mathematical models (Wang and Polglase, 1995; McKane *et al.*, 1995) or from microcosms (Körner and Arnone, 1992). The task of predicting the impact of environmental change on tropical rain forests is made more challenging by the large number of species, the variable soil conditions, and the contrasting response of species to the environment. Early successional species, for example, have been shown by some studies to have stronger photosynthetic and growth responses to nitrogen and phosphorus compared to late successional species (Reich *et al.*, 1995; Huante *et al.*, 1995a,b), whereas others, working in a different geographical region, have found contrary results in some cases (Veenendaal *et al.*, 1996). There is a need to develop new approaches to this difficult problem, perhaps using large-scale experimentation and observation. One aspect of environmental change that has received attention is the influence of forest edges that are created during logging and burning. An experimentally fragmented landscape spanning 20 × 50 km near Manaus, Brazil was used to create a replicated series of patches of 1, 10, and 100 ha. Recent data from these patches show a loss of up to 36% of the biomass within 100 m of the edge (Laurance *et al.*, 1998).

E. The Way Ahead through Interdisciplinary Studies

It has become clear that further progress in understanding the controls on net primary productivity, and the related quantity, the net ecosystem productivity, will require close collaboration between disciplines. Studies at the leaf and stand scale, using ecophysiological and eddy covariance techniques, are advancing our understanding of the temporal changes (hours to months). However, these approaches, used in isolation, are not sufficient to define the total production for the whole biome. The data from such studies are nevertheless vital for parameterization of models, which can be verified against independent data at key sites. Models such as those of Lloyd and Farquhar (1996) and Williams *et al.* (1998) are sufficiently process-based as to be applicable to any forest, and modelers generally await data from the field to enable their models to be parameterized. Thereafter, scaling up to whole regions and biomes still requires remotely sensed data on the distribution of land surface cover, as well as the use of interpolated climatological data from the ground or from global circulation models to drive the models. Inventories of forest plots are likely to be increasingly important (Phillips *et al.*, 1998). It is currently not clear how variable tropical forests will turn out to be in terms of their photosynthetic and respiratory capacities, or their sensitivity to climatic change. Probably the most important questions relate to how the species composition of the rain forest will change as the global environment warms, and how such "acclimation" will influence the capacity of the system to continue to store carbon into the millenium.

Acknowledgments

We thank Deborah Clark for making available her work before it had been published. We acknowledge the financial support provided by the Natural Environmental Research Council, through its TIGER (Terrestrial Initiative in Global Environmental Research) program (award number GST/02/605), through awards GR9/3172 & GR3/11706; and continuing support from the European Commission's DG-XII program.

References

Allen, L. H., Lemon, E., and Muller, J. (1972). Environment of a Costa Rican forest. *Ecology* **53,** 102–111.

Alves, D. S., Soares, J. V., Amaral, S., Mello, E. M. K., Almeida, A. S., da Silva, O. F., and Silveira, A. M. (1997). Biomass of primary and secondary vegetation in Rondonia, Western Brazilian Amazon. *Global Change Biol.* **3,** 451–462.

Badeau, V., Becker, M., Bert, G. D., Dupouey, J. L., Lebourgeois, F., and Picard, J. F. (1996). *In* "Long-Term Growth Trends of Trees: Ten Years of Dendrochronological Studies in France" (H. Spiecker, K. Milikaïnen, M. Köhl, and J. P. Skovsgaard, eds.), pp. 167–181. Springer-Verlag, Berlin.

Baldocchi, D., Valentini, R., Running, S., Oechel, W., and Dahlman, R. (1996). Strategies for measuring and modelling carbon dioxide and water vapour fluxes over terrestrial ecosystems. *Global Change Biol.* **3,** 159–168.

Bartholemew, W. V., Meyer, J., and Laudelot H. (1953). "Mineral Nutrient Immobilisation Under Forest and Grass Fallow in Yangambi (Belgian Congo) Region, with Some Preliminary Results on the Decomposition of Plant Material on the Forest Floor." Ser. Sci. 57, pp. 1–27. INEAC, Bruxelles.

Brown, S., Gillespie, A. J. R., and Lugo, A. E. (1989). Biomass estimation methods for tropical forests with applications to forest inventory data. *For. Sci.* **35,** 881–902.

Brown, S., and Lugo, A. E. (1990). Tropical secondary forests. *J. Tropic. Ecol.* **6,** 1–32.

Bruijnzeel, L. A. (1991). Nutrient input–output budgets of tropical ecosystems: A review. *J. Tropic. Ecol.* **7,** 1–24.

Carvalho, J. A., Santos, J. M., Santos, J. C., Leitão, M. M., and Higuchi, N. (1995). A tropical rainforest clearing experiment by biomass burning in the Manaus region. *Atmosph. Environ.* **29,** 2301–2309.

Chambers, J. Q., Higuchi, N., and Schimel, J. P. (1998). Ancient trees in Amazonia. *Nature (London)* **391,** 135–136.

Chazdon, R. L. (1988). Sunflecks and their importance to forest understory plants. *Adv. Ecol. Res.* **18,** 1–63.

Clark, D. A., Brown, S. Kicklighter, D. W., Chambers, J. Q., Thomlinson, J. R., and Ni, J. (2000a). Measuring net primary production in forests: concepts and field methods. *Ecol. Appl.* (in press).

Clark, D. A., Brown, S. Kicklighter, D. W., Chambers, J. Q., Thomlinson, J. R., Ni, J., and Holland, E. A. (2000b). NPP in tropical forests: an evaluation and synthesis of existing field data. *Ecol. Appl.* (in press).

Cuevas, E., and Medina, E. (1988). Nutrient dynamics in Amazonian forest ecosystems. II. Fine root growth, nutrient availability and leaf litter decomposition. *Oecologia* **68,** 466–472.

Culf, A. D., Esteves, J. L., Marques Filho, A., de O., and da Rocha, H. R. (1996). Radiation temperature and humidity over forest and pasture in Amazonia. *In* "Amazonian Deforestation and Climate" (J. H. C Gash, C. A. Nobre, J. M. Roberts, and R. L. Victoria, eds.), pp. 175–191. Wiley, Chichester.

Culf, A. D., Fisch, G., Malhi, Y., Costa, R. C., Nobre, A. D., Marques, A. D., Gash, J. H. C. and Grace, J. (1999). Carbon dioxide measurements in the nocturnal boundary layer over Amazonian forest. *Hydrology and Earth System Sciences* **3,** 39–53.

da Rocha, H. R., Sellers, P. J., Collatz, G. J., Wright, I. R., and Grace, J. (1996). Estimate of water vapour and carbon exchanges in an Amazonian rain forest by the SiB2 model. *In* "Amazonian Deforestation and Climate" (J. H. C Gash, C. A. Nobre, J. M. Roberts, and R. L. Victoria, eds.), pp. 459–472. Wiley, Chichester.

Deans, J. D., Moran, J., and Grace, J. (1996). Biomass relationships for tree species in regenerating semi-deciduous tropical moist forest in Cameroon. *For. Ecol. Manage.* **88,** 215–225.

Dixon, R. K., Brown, S., Houghton, R. A., Solomon, A. M., Trexler, M. C., and Wisniewski, J. (1994). Carbon pools and flux of global ecosystems. *Science* **263,** 185–190.

Dolman, A. J., Gash, J. H. C., Roberts, J., and Shuttleworth, W. J. (1991). Stomatal and surface conductance of tropical rainforest. *Agric. For. Meteorol.* **54,** 303–318.

FAO (1997). State of the World's Forests 1997. FAO, Rome, Italy.

FAO (1993). FAO Forestry Paper 112. FAO, Rome, Italy.

Farquhar, G. D., Caemmerer, S. von, and Berry, J. A. (1980). A biochemical model of photosynthetic CO_2 assimilation in leaves of C_3 species. *Planta* **149,** 79–90.

Fearnside, P. M. (1997). Greenhouse gases from deforestation in Brazilian Amazonia: net committed emissions. *Climate Change* **35,** 321–360.

Fearnside, P. M. (2000). Global warming and tropical land-use change: greenhouse gas emissions from biomass burning, decomposition and soils in forest conversion, shifting cultivation and secondary vegetation. *Climate Change* **46,** 115–158.

Fölster, H., de las Salas, G., and Khanna, P. (1976) A tropical evergreen forest site with perched water table, Magdelena valley, Colombia. Biomass and bioelement inventory of primary and secondary vegetation. *Oecologia Planta.* **11,** 297–320.

Foster Brown, I., Martinelli, L. A., Thomas, W. W., Moreira, M. Z., Ferreira, C. A. C., and Victoria, R. A. (1995). Uncertainty in the biomass of Amazonian forests: An example from Rondonia, Brazil. *For. Ecol. Manage.* **75,** 175–189.

Galloway, J. N., Schleshinger, W. H., Levy II, H., Michaels, A., and Schnoor, J. L. (1995). Nitrogen fixation: Anthropogenic enhancement–environmental response. *Global Biogeochem. Cycles* **9,** 235–252.

Gash, J. H. C., and Shuttleworth, W. J. (1991). Tropical deforestation: Albedo and the surface-energy balance. *Climatic Change* **19,** 123–133.

Gillespie, A. J. R., Brown, S., and Lugo, A. E. (1992). Tropical forest biomass estimation from truncated stand tables. *For. Ecol. Manage.* **48,** 69–87.

Grace, J., Lloyd, J., McIntyre, J., Miranda, A. C., Meir, P., Miranda, H., Moncrieff, J. M., Massheder, J., Wright, I. R., and Gash, J. (1995a). Fluxes of carbon dioxide and water vapour over an undisturbed tropical rainforest in south-west Amazonia. *Global Change Biol.* **1,** 1–12.

Grace, J., Lloyd, J., McIntyre, J., Miranda, A. C., Meir, P., Miranda, H., Nobre, C., Moncrieff, J. M., Massheder, J., Malhi, Y., Wright, I. R., and Gash, J. (1995b). Carbon dioxide uptake by an undisturbed tropical rain forest in South-West Amazonia 1992–1993. *Science* **270,** 778–780.

Grace, J., Malhi, Y., Lloyd, J., McIntyre, J., Miranda, A. C., Meir, P., and Miranda, H. S. (1996). The use of eddy covariance to infer the net carbon dioxide uptake of Brazilian rain forest. *Global Change Biol.* **2,** 209–218.

Greenland, D. L., and Kowal, J. L. (1960). Nutrient content of the moist tropical forest of Ghana. *Plant Soil* **12,** 154–174.

Higuchi, N., dos Santos, J. M., Imanaga, M., and Yoshida, S. (1994). Aboveground biomass estimate for Amazonian dense tropical moist forest. *Mem. Faculty Agric. Kagoshima Univ.* **30,** 43–54.

Honzak, M., Lucas, R. M., do Amaral, I., Curran, P. J., Foody, G. M., and Amaral, S. (1996). Estimation of leaf area index and total biomass of tropical regenerating forests: A comparison of methodologies. *In* "Amazonian Deforestation and Climate" (J. H. C Gash, C. A. Nobre, J. M. Roberts, and R. L. Victoria, eds.), pp. 365–382. Wiley, Chichester.

Houghton, R. A. (1999). The annual net flux of carbon to the atmosphere from changes in land use 1850–1990. *Tellus* **51B,** 298–313.

Huante, P, Rincon, E., and Chapin III, F. S. (1995a). Responses to phosphorus of contrasting successional tree-seedling species from the tropical deciduous forest of Mexico. *Funct. Ecol.* **9,** 760–766.

Huante, P., Rincon, E., and Acosta, I. (1995b). Nutrient availablity and growth rate of 34 woody species from a tropical deciduous forest in Mexico. *Funct. Ecol.* **9,** 849–858.

Jackson, R. B., Canadell, J., Ehleringer, J. R., Mooney, H. A., Sala, O. E., and Schulze, E.-D. (1996). A global analysis of root distributions for terrestrial biomes. *Oecologia* **108,** 389–411.

Jackson, R. B., Mooney, H. A., and Schulze, E.-D. (1997). A global budget for fine root biomass, surface area, and nutrient contents. *Proc. Natl. Acad Sci.* (submitted).

Jordan, C. F. (1971). Productivity of a tropical forest and its relation to a world pattern of energy storage. *J. Ecol.* **59,** 127–142.

Jordan, C. F. (1982). The nutrient balance of an Amazonian tropical rain forest. *Ecology* **63,** 647–654.

Jordan, C. F., and Escalante, G. S. (1980). Root productivity in an Amazonian rain forest. *Ecology* **61,** 14–18.

Jordan, C. F., and Uhl, C. (1978). Biomass of a *terra firme* forest of the Amazon Basin. *Oecolog. Planta.* **13,** 387–400.

Kindermann, J., Würth, G., and Kohlmaier, G. H. (1996). Interannual variation of carbon exchange fluxes in terrestrial ecosystems. *Global Biogeochem. Cycles* **10,** 737–755.

Kira, T. (1978). Community architecture and organic matter dynamics in tropical lowland rain forests of Southeast Asia with special reference to Pasoh forest, West Malaysia. *In* "Tropical Trees as Living Systems" (P. B. Tomlinson and M. H. Zimmermann, eds.). Cambridge Univ. Press, Cambridge.

Kira, T., Ogawa, H., Yoda, K., and Ogino, K. (1964). Primary production by a tropical rain forest in Southern Thailand. *Bot. Mag. Tokyo* **77,** 428–429.

Klinge, H. (1976). Bilanzierung von Hauptnährstoffen im Ökosystem tropischer Regenwälder (Manaus)-Vorläufige Daten. *Biographica* **7,** 59–76.

Klinge, H., and Herrera, R. (1983). Phytomass structure of natural plant communities on spodosols in southern Venezuela: The tall Amazon Caatinga forest. *Vegetatio* **53,** 65–84.

Klinge, H., and Rodrigues, W. A. (1973). Biomass estimation in a central Amazonian rain forest, *Acta Cientif. Venez.* **24,** 225–237.

Koike, F., and Syahbuddin. (1993). Canopy structure of a tropical rain forest and the nature of an unstratified upper layer. *Funct. Ecol.* **7,** 230–235.

Körner, C. (1994). Leaf diffusive conductances in the major vegetation types of the globe. *In* "Ecophysiology of Photosynthesis" (E.-D. Schulze and M. M. Caldwell, eds.), pp. 463–490. Springer-Verlag, Berlin.

Körner, C., and Arnone, J. A. (1992). Responses to elevated carbon dioxide in artificial tropical ecosystems. *Science* **257,** 1672–1673.

Kruijt, B., Lloyd, J., Grace, J., McIntyre, J. A., Farquhar, G. D., Miranda, A. C., and McCracken, P. (1996). Sources and sinks of CO_2 in Rondonia tropical rain forest. *In* "Amazonian Deforestation and Climate" (J. H. C Gash, C. A. Nobre, J. M. Roberts, and R. L. Victoria, eds.), pp. 331–352. Wiley, Chichester.

Kruijt, B., Malhi, Y., Lloyd, J., Nobre, A. D., Miranda, A. C., Pereira, M. G. P., Culf, A., and Grace, J. (2000). Turbulence statistics above and within two Amazonian rain forest canopies. *Boundary-Layer Meteorology* **94,** 297–331.

Laurance, W. F., Laurance, S. G., Ferreira, L. V., Rankin-de Merona, J-M., Gascon, C., and Lovejoy, T. E. (1998). Biomass collapse in Amazonia Forest Fragments. *Science* **278,** 1117–1118.

Lehto, T., and Grace, J. (1994). Carbon balance of tropical tree seedlings: a comparison of two species. *New Phytol.* **127,** 455–463.

Lemée, G., Huttel, C., and Bernhard-Reversat, F. (1975). Recherches sur l'écosystème de la forêt sub-équatoriale de basse Côte d'Ivoire. *La Terre Vie (Paris)* **29,** 169–264.

Lloyd, J., and Farquhar, G. D. (1996). The CO_2 dependence of photosynthesis and plant growth in response to elevated atmospheric CO_2 concentration and their interrelationship with soil nutrient status. 1. General principles. *Funct. Ecol.* **10,** 4–32.

Lloyd, J., Grace, J., Miranda, A. C., Meir, P., Wong, S. C., Miranda, H., Wright, I., Gash, J. H. C., and McIntyre, J. (1995). A simple calibrated model of Amazon rainforest productivity based on leaf biochemical properties *Plant Cell Environ.* **18,** 1129–1145.

Lloyd, J., Kruijt, B., Hollinger, D. Y., Grace, J., Francey, R. J., Wong, S. C., Kelliher, F. M., Miranda, A. C., Farquhar, G. D., Gash, J. H. C., Vygodskaya, N. N., Wright, I. R., Miranda H. S., and Schulze, E. D. (1996). Vegetational effects on the isotopic composition of atmospheric CO2 at local and regional scales- theoretical aspects and a comparison between rain-forest in Amazonia and a boreal forest in Siberia. *Aust. J. Plant Physiol.* **23,** 371–399.

Longman, K. A., and Jenik, J. (1987). "Tropical Forest and Its Environment." Longman, London.

Lui, W. T. H., Massambani, O., and Nobre, C. (1994). Satellite recorded vegetation response to drought in Brazil. *Int. J. Climatol.* **14,** 343–354.

Malhi, Y., and Grace, J. (2000). Tropical forests and atmospheric carbon dioxide. *Trends Ecol. Evol.* **15,** 332–337.

Malhi, Y., Nobre, A., Grace, J., Kruijt, B., Pereira, M., Culf, A., and Scott, S. (1998). Carbon dioxide transfer over a Central Amazonian rain forest. *J. Geophys. Res.* **D24,** 31593–31612.

Malhi, Y., Baldocchi, D., and Jarvis, P. G. (1999). The carbon balance of tropical, temperate and boreal forests. *Plant Cell & Environment* **22,** 715–740.

Mathews, G. (1993). "The Carbon Content of Trees." Forestry Commission, Technical Paper 4. HMSO, London.

McKane, R. B., Rastetter, E. B., Melillo, J. M., Shaver, G. R., Hopkinson, C. S., and Fernandes, D. N. (1995). Effects of global change on carbon storage in tropical forests of South America. *Global Biogeochem. Cycles* **9,** 329–350.

McWilliam, A.-L. C., Roberts, J. M., Cabral, O. M. R., Leitao, M. V. B. R., de Costa, A. C. L., Maitelli, G. T., and Zamparoni, C. A. G. P. (1993). Leaf area index and above-ground biomass of *terra firme* rain forest and adjacent clearings in Amazonia. *Funct. Ecol.* **7,** 310–317.

McWilliam, A.-L. C., Cabral, O. M. R., Gomes, B. M., Esteves, J. L., and Roberts, J. M. (1996). Forest and pasture leaf-gas exchange in south-west Amazonia. *In* "Amazonian Deforestation and Climate" (J. H. C Gash, C. A. Nobre, J. M. Roberts, and R. L. Victoria, eds.), pp. 265–285. Wiley, Chichester.

Medina, E., and Klinge, H. (1983). Productivity of tropical forests and tropical woodlands. *In* "Physiological Plant Ecology IV, Encyclopedia of Plant Physiology, Volume 12B" (O. L. Lange, P. S. Nobel, C. B. Osmond, and H. Zeigler, eds.), pp. 281–303. Springer-Verlag, Berlin.

Medina, E., and Minchin, P. (1980). Stratification of δ^{13} C values of leaves in Amazonian rain forests. *Oecologia* **45,** 377–378.

Meir, P., Grace, J., Miranda, A. C., and Lloyd, J. (1996). Soil respiration measurements in the Brazil forest and cerrado vegetation during the wet season. *In* "Amazonian Deforestation and Climate" (J. H. C Gash, C. A. Nobre, J. M. Roberts, and R. L. Victoria, eds.), pp. 319–330. Wiley, Chichester.

Meir, P., Grace, J., and Miranda, A. C. (2000). Photographic method to measure the vertical distribution of leaf area density in forests. *Agr. Forest Meteorol.* **102,** 105–111.

Miranda, A. C., Miranda, H. S., Lloyd, J., Grace, J., Francey, R. J., McIntyre, J. A., Meir, P., Riggan, P., Lockwood, R., and Brass, J. (1997). Fluxes of carbon, water and energy over Brazilian cerrado: An analysis using eddy covariance and stable isotopes. *Plant Cell Environ.* **20,** 315–328.

Moncrieff, J. B., Malhi, Y., and Leuning, R. (1996). The propagation of errors in long-term measurements of land-atmosphere fluxes of carbon and water. *Global Change Biol.* **2,** 231–240.

Monteith, J. L., and Unsworth, M. H. (1990) "Principles of Environmental Physics." Arnold, London.

Müller, M. H., and Nielsen, J. (1965). Production brute pertes par respiration et production nette dans la forêt ombrophile tropicale. *Foresel Forsogs. Danmark* **29,** 69–160.

Myneni, R. B., Keeling, C. D., Tucker, C. J., Asrar, G., and Nemani, R. R. (1997). Increased plant growth in the northern high latitudes from 1981 to 1991. *Nature (London)* **386,** 698–702.

Nepstad, D. C., de Carvalho, C. R., Davidson, E. A., Jipp, P. H., Lefebvre, P. A., Negreiros, G. H., da Silva, E. D., Stone, T. A., Trumbore, S. E., and Vieira, S. (1994). The role of deep roots in the hydrological and carbon cycles of Amazonian forests and pastures. *Nature (London)* **372,** 666–669.

Oikawa, T. (1985). Simulation of forest carbon dynamics based on a dry matter production model. 1. Fundamental model structure of a tropical rainforest ecosystem. *Bot. Mag. Tokyo* **98,** 225–238.

Philip, M. S. (1994) "Measuring Trees and Forests." CAB International, Wallingford, UK.

Phillips, O. L., and Gentry, A. H. (1994). Increasing turnover through time in tropical forests. *Science* **263,** 954–958.

Phillips, O. L., Hall, P., Gentry, A. H., Sawyer, S. A., and Vásquez, R. (1994). Dynamics and species richness of tropical rain forests. *Proc. Natl. Acad. Sci. U.S.A.* **91,** 2805–2809.

Phillips, O. L, Malhi, Y., Higuchi, N., Laurance, W. F., Nuñez V. P., Vásquez M. R., Laurance, S. G., Ferreira, L. V., Stern, M., Brown, S., and Grace, J. (1998). Changes in the carbon balance of tropical forests: evidence from long-term plot data. *Science* **282,** 439–442.

Pitcairn, C. E. R., Fowler, D., and Grace, J. (1995). Deposition of fixed atmospheric nitrogen and foliar nitrogen content of bryophytes and *Calluna vulgaris* (L) Hull. *Environ. Pollut.* **88,** 193–205.

Post, W. M., Emanuel, W. R., Zinke, P. J., and Stangenberger, A. G. (1982). Soil carbon pools and world life zones. *Nature (London)* **298,** 156–159.

Proctor, J. (1987). Nutrient cycling in primary and old secondary rain forests. *Appl. Geogr.* **7,** 135–152.

Raich, J. W., and Nadelhoffer, K. J. (1989). Below-ground carbon allocation in forest ecosystems: Global trends. *Ecology* **70,** 1346–1354.

Raich, J. W., Rastetter, E. B., Melillo, J. M., Kicklighter, D. W., Steudler, P. A., Peterson, B. J., Grace, A. L., Moore, B., and Vorosmarty, C. J. (1991). Potential net primary productivity in South America: Application of a global model. *Ecol. Appl.* **1,** 399–429.

Reich, P. B., Walters, M. B., Ellsworth, D. S., and Uhl, C. (1994). Photosynthesis-nitrogen relations in Amazonian tree species. *Oecologia* **97,** 62–72.

Reich, P. B., Ellsworth, D. S., and Uhl, C. (1995). Leaf carbon and nutrient assimilation and conservation in species of different succcessional status in an oligotrophic Amazonian forest. *Funct. Ecol.* **9,** 65–76.

Richards, P. W. (1996). "The Tropical Rain Forest," 2nd Ed. Cambridge Univ. Press, Cambridge.

Richey, J. E., Victoria, R. L., Salati, E., and Forsberg, B. R. (1991). The biogeochemistry of a major river system: The Amazon case study. *In* "Biogeochemistry of Major World Rivers" (E. T. Degens, S. Kempe, and J. E. Richey, eds.), pp. 57–74. Wiley, New York.

Riddoch, I., Grace, J., Fasehun, F. E., Riddoch, B., and Ladipo, D. O. (1991a). Photosynthesis and successional status of seedlings in a tropical semi-deciduous rainforest in Nigeria. *J. Ecol.* **79,** 491–504.

Riddoch, I., Lehto, T., and Grace, J. (1991b). Photosynthesis of tropical tree seedlings in relation to light and nutrient supply. *New Phytologist* **119,** 137–147.

Roberts, J., Cabral, O. M. R., and de Aguiar, L. F. (1990). Stomatal and boundary layer conductances in an Amazonian terra firme rain forest. *J. Appl. Ecol.* **27,** 60–69.

Roberts, J. M., Cabral, O. M. R., da Costa, J. P., McWilliam, A. L.-C., and Sá, T. D. A. (1996). An overview of the leaf area index and physiological measurements during ABRACOS. *In* "Amazonian Deforestation and Climate" (J. H. C Gash, C. A. Nobre, J. M. Roberts, and R. L. Victoria, eds.), pp. 287–306. Wiley, Chichester.

Saldarriaga, J. G. (1985). Forest succession in the upper Rio Negro of Columbia and Venezuela. Ph.D. Thesis. Univ. of Tennessee, Knoxville.

Saldarriaga, J. G., and Luxmore, R. J. (1991). Solar energy conversion efficiency during succession of a tropical rain forest in Amazonia. *J. Trop. Ecol.* **91,** 233–242.

Saldarriaga, J. G., and West, D. C. (1986). Holocene fires in the northern Amazon basin. *Quatern. Res.* **26,** 358–366.

Santantonio, D., and Grace, J. C. (1987). Estimating fine-root production and turnover form biomass and decomposition data: A compartment-flow model. *Can. J. For. Res.* **17,** 900–908.

Schimel, D. S. (1995). Terrestrial ecosystems and the global carbon cycle. *Global Change Biol.* **1,** 77–91.

Sellers, P. J., Shuttleworth, W. J., Dorman, J. L., Dalcher, A., and Roberts, J. M. (1989). Calibrating the Simple biosphere model for Amazonian tropical forest using field and remote sensing data. Part I: Average calibration with field data. *J. Appl. Meteorol.* **28,** 728–759.

Sellers, P. J., Dickinson, R. E., Randall, D. A., Betts, A. K., Hall, F. G., Berry, J. A., Collatz, G. J., Denning, A. S., Mooney, H. A., Nobre, C. A., Sato, N., Field, C. B., and Henderson-Sellers,

A. (1997). Modeling the exchanges of energy, water and carbon between continents and the atmosphere. *Science* **275,** 502–509.

Shuttleworth, W. J. (1989). Micrometeorolgy of temperate and tropical forest. *Philos. Trans. R. Soc. London Ser. B* **324,** 299–334.

Shuttleworth, W. J., Gash, J. H. C., Lloyd, C. R., Moore, C. J., Roberts, J., de O. Marques, A., Fisch, G., de P Silva, V., Ribeiro, M. N. G., Molion, L. C. B., de Abreu Sa, L. D., Nobre, J. C., Cabral, O. M. R., Patel, S. R., and de Moraes, J. C. (1984). Eddy correlation measurements of energy partitioning for Amazon forest. *Q. J. R. Meteorol. Soc.* **110,** 1143–1162.

Tanner, E., Vitousek, P. M., and Cuevas, E. (1998). Experimental investigation of nutrient limitation of forest growth on wet tropical mountains. *Ecology* **79,** 10–22.

Taylor, J. A., and Lloyd, J. (1992). Sources and sinks of atmospheric CO_2. *Aust. J.Bot.* **40,** 407–418.

Townsend, A. R., Braswell, B. H., Holland, E. A., and Penner, J. E. (1996). Spatial and temporal patterns in terrestrial carbon storage due to deposition of fossil fuel nitrogen. *Ecol. Appl.* **6,** 806–814.

Trumbore, S. E., Davidson, E. A., Barbosa de Canargo, P., Nepstad, D. C., and Martinelli, L. A. (1995). Belowground cycling of carbon in forests and pastures. *Global Biogeochem. Cycles* **9,** 515–528.

UNESCO (1978). Gross and net primary production and growth parameters. *In* "Tropical Forest Ecosystems," pp. 233–248. UNESCO, Paris.

Veenendaal, E. M., Swaine, M. D., Lecha, R. T., Walsh, M. F., Abebrese, I. K., and Owusu-Afriyie, K. (1996). Response of West African forest tree seedlings to irradiance and soil fertility. *Funct. Ecol.* **10,** 501–511.

Wang, Y. P., and Polglase, P. J. (1995). The carbon balance in the tundra, boreal and humid tropical forests during climate change—Scaling up from leaf physiology and soil carbon dynamics. *Plant Cell Environ.* **18,** 1226–1244.

Welles, J. M., and Cohen, S. (1996). Canopy structure measurement by gap fraction analysis using commercial instrumentation. *J. Exp. Bot.* **47,** 1335–1342.

Whitmore, T. C. (1984). "Tropical Rain Forests of the Far East." Clarendon Press, Oxford.

Whitmore, T. C. (1989). Canopy gaps and the two major groups of forest trees. *Ecology* **70,** 536–538.

Whitmore, T. C. (1990). "An Introduction to Tropical Rain Forests." Clarendon Press, Oxford.

Whittaker, R. H., and Likens, G. E. (1975). The biosphere and man *In* "Primary Productivity of the Biosphere" (R. H. Whittaker and G. E. Likens, eds.), pp. 305–328. Springer-Verlag, Berlin.

Williams, M., Malhi, Y., Nobre, A. D., Rastetter, E. B., Grace, J., and Pereira, M. G. P. (1998). Seasonal variation in net carbon dioxide exchange and evapotranspiration in a Brazilian rain forest: A modelling analysis. *Plant Cell Environ.*

Würth, M. K. R., Winter, K., and Körner, C. (1998). *In situ* responses to elevated CO_2 in tropical forest understory plants. *Funct. Ecol.* (in press).

Yabuki, K., and Aoki, M. (1978). Micrometeorological assessment of primary production rate of Pasoh Forest. *Malay Nat. J.* **30,** 281–289.

Yoda, K. (1978). Respiration studies in Pasoh forest plants. *Malay Nat. J.* **30,** 259–279.

Zinke, P. J., Strangenberger, A. G., Post, W. M., Emanuel, W. R., and Olson, J. S. (1986). "Worldwide Organic Soil Carbon and Nitrogen Data." ORNL/CDIC-18 NDP-018. Carbon Dioxide Information Center, Oak Ridge, Tennessee.

III

Global Productivity

18

Determining Present Patterns of Global Productivity

Wolfgang Cramer, Richard J. Olson, Stephen D. Prince, Jonathan M. O. Scurlock, and Members of the Global Primary Production Data Initiative[1]

I. Introduction

Net primary productivity (NPP), the flux of carbon between atmosphere and vegetation, is a key quantity of the global carbon cycle and hence highly relevant to the assessment of impacts from and feedbacks to climatic change. The balance between atmospheric concentrations of carbon dioxide and storage of carbon on land and in the oceans is directly affected by changes that may occur in net ecosystem productivity (NEP), which, in turn, is the difference between total annual NPP and heterotrophic respiration (see Chapters 1, 2, and 3, this volume). Estimating present and future NPP and NEP is crucial for the scientific underpinnings of current policy-making (IGBP Terrestrial Carbon Working Group, 1998), but the models on which these estimates are based are in need of improvement, particularly in the important aspect of validation with field observations of NPP.

NPP is a longer term cumulative net flux of carbon in vegetation, and is difficult to measure directly at any scale. NPP estimates therefore must rely

[1]Members include Alberte Bondeau, Michael Coughenour, Gérard Dedieu, Tagir Gilmanov, Stith T. Gower, Kathy Hibbard, David W. Kicklighter, William J. Parton, Wilfred M. Post, David Price, and Larry Tieszen.

on indirect measurements combined with models that relate these measurements to the overall flux. Early quantitative assessments of global NPP were based on results such as those of the International Biological Programme (IBP), during which estimates of biomass and productivity were obtained for a relatively large number of sites around the world. To extrapolate from these sites to the land surface of the planet, a regression model was constructed, using annual means of temperature and rainfall as predictors (the MIAMI model) (Lieth, 1975), resulting in a good fit to the underlying data. From this model, the first estimate of global total NPP was derived. Perhaps more importantly, the MIAMI model also provided a powerful illustration of the role of climate in determining NPP.

Nevertheless, several key problems in assessing carbon fluxes remain after this analysis. First, the quality of the underlying data is crucial for such a model—several studies have shown that earlier assessments, for some biomes, appear to have undervalued total NPP, due to within-season turnover of biomass and underestimation of belowground productivity (Long *et al.*, 1989). Second, even the broad sampling that has occurred so far is inadequate for estimation of the total NPP range in different environments, mostly due to the few measurements of belowground NPP, but also due to the large regions that are insufficiently studied, notably managed or otherwise altered systems. Third, scaling from the level of measurement (leaf, canopy, or ecosystem) to the level of assessment (global) involves assumptions about other influences on productivity that have not been measured, e.g., the heterogeneity of the land surfaces. Fourth, the total carbon balance (NEP) is dependent not only on NPP, but also on (heterotrophic) soil respiration and the effects of harvest or natural disturbance, which are difficult to assess (see Chapter 22, this volume). Finally, future NPP is likely to respond to increased CO_2 concentrations in the atmosphere (Mooney *et al.*, 1999); from this follows that environmental conditions outside the present range of possible sampling strategies may occur, rendering the regression approach unsuitable, at the very least because of changing CO_2 concentrations.

To overcome some of these problems, process models that have been developed use a broad range of measurements to produce, among other variables, better estimates of NPP (Cramer *et al.*, 1999a). Among these models, three major types can be distinguished, based on their data requirements: (1) models that interpret changes in (remotely sensed) radiative properties of the land surface in terms of regional plant productivity (remote-sensing-based models) (Fischer *et al.*, 1996); (2) models that derive flux estimates directly from the simulation of ecosystem processes in terms of physiological processes driven by environmental variables, in some cases with a map of vegetation types (ecosystem process models) (e.g., Raich *et al.*, 1991; Haxeltine and Prentice, 1996); and (3) models that interpret atmospheric CO_2 measurements in terms of the total fluxes between the atmosphere and land biosphere, human industrial system, and oceans (deconvolution models) (see Chapter 19, this volume). Hybrids between these model categories exist as well. Most models require data on actual NPP, for parameterization, calibration, or validation. Direct validation of global models is impossible, nevertheless they gain plausibility through comparison with observational data sets (Rastetter, 1996).

II. The Nature of Available NPP Observations

A. Types of Measurements

NPP in land ecosystems is estimated using several different methods. Much recent attention has been focused on the capacity to measure directly the flux of carbon dioxide at the vegetation–atmosphere interface, using micrometeorological techniques or chambers that include complete trees with some understorey vegetation. These techniques are important for the development of process understanding in well-studied ecosystems. They provide data at a scale of a few to hundreds of square meters. They are currently not, however, capable of yielding data directly suitable to broad-scale aggregations, because heterogeneous environments cannot be sampled, nighttime fluxes are uncertain (Denmead *et al.*, 1996), and they cannot separate NPP and soil CO_2 fluxes—in other words their direct measurement is of NEP.

Another way to measure NPP at the ecosystem scale is by destructive measurement of the biomass of vegetation (usually done in grasslands only). These measurements are very labor intensive, because frequent samplings are required to avoid the omission of any biomass that might be lost in shed plant parts or by herbivory, as well as the sampling of above- and belowground parts. This technique covers few square meters, although an appropriate sampling layout can extend this to representative samples of larger areas. It is the only technique that permits actual belowground sampling, but this is nevertheless difficult and time-consuming, because all roots need to be recovered and separated from other soil components and live and dead roots are not easily distinguished from each other. As a result, biomass and productivity belowground are often estimated rather than measured, using simple multipliers, which themselves may not be based on statistically sound relationships (Singh *et al.*, 1975; Nadelhoffer and Raich, 1992).

The third type of measurement involves estimating yield in ecosystems managed for production, such as forests, rangelands, or agricultural crops. In this case, the yields of economically valuable products are related statistically to total biological productivity. The advantage of this approach is the availability of yield data for many crops over long time spans. It is also particularly suitable for regional applications (integrating from a few to hundreds of square kilometers), because most yield statistics intrinsically relate to relatively large, potentially homogeneous regions, e.g., counties or forests. The main drawback of these data is the difficulty of estimating the noneconomically significant biomass increments that are not included in the reported yields. In addition there are sometimes problems in access to data owing to commercial privacy considerations. Another type of criticism that is sometimes leveled at these data is the fact that they represent the actual NPP and not the potential NPP in the absence of human management; in actuality this is, of course, more a commentary on certain models that cannot allow for the impact of human management. Probably the safest broad-scale, estimates of changing carbon storage based on observations are from forest yield statistics (e.g., Kurz and Apps, 1995).

B. Availability of Data

Despite a growing number of such field studies, worldwide NPP data have not been extensively reviewed since the synthesis of Lieth and Whittaker (1975). Ecosystem-level data compilations existing today can be seen as a hierarchy of data sets of differing quality with, at the one extreme, a limited number of (maybe 10–20) high-quality sites, such as the First ISLSCP Field Experiment (FIFE) site in Kansas, where many factors controlling NPP are relatively well-understood and measured extensively. At the other extreme are some large data collections containing up to thousands of entries, for which many only a limited set of variables is known (e.g., litterfall, tree or crop growth), and for which the protocols used in the observations are often vague. An example of such a collection is the Osnabrück NPP Data Set, a modified and expanded version of the original data set developed for the MIAMI model, presently containing around 720 sites in many different biomes. A goal for ongoing efforts of the Global Primary Production Data Initiative (GPPDI; see below) is to develop, between these extremes, a set of around 100–300 relatively well-characterized study sites for which a time series of NPP data is available, together with driving variables such as climate and the principal site characteristics. Site characteristics include factors such as climate, vegetation type, and management history.

C. Existing Data Compilations—The GPPDI Database

An extensive database of NPP and biomass measurements has been established at Oak Ridge National Laboratory (ORNL; Oak Ridge, Tennessee) as part of the GPPDI, in a database designed according to NASA guidelines (Scurlock *et al.*, 1999).[2] The aim is to assemble the most useful existing data, not to initiate new field data collections, although this is in no way to be discouraged. Data from some intensive study sites and from extensive data col-

[2]These data are described in greater detail and can be accessed at http://www.eosdis.ornl.gov/npp/npp—home.html.

lections are being included, with an emphasis on grasslands, boreal forests, and tropical forests. An example is the grassland data set assembled for use with the CENTURY and SAVANNA plant–soil ecosystem models (Breymeyer and Melillo, 1991; Coughenour, 1992; Parton *et al.*, 1993, 1995), which is in the process of being expanded using data from other grassland study sites. Other data come from the International Biological Programme—particularly for those biomes often underrepresented, such as tundra—as well as other existing data compilations (e.g., Lieth and Box, 1972; Esser, 1992, 1997; Gholz *et al.*, 1994). Presently, additional data are being sought from circumpolar boreal forest studies, worldwide litterfall studies, and forest, crop, and rangeland inventories from the United States and the former USSR (Olson *et al.*, 1997).

Gradually, the GPPDI archive is being expanded using the sources listed in the sections below. Data for over 1600 sites have been obtained from a variety of sources (Table 18-1). The main selection criterion for including published NPP data was adequate documentation of the measurements and completeness of NPP budget. Sites in grasslands, boreal forests, and tropical forests were classified into one of three classes, depending upon completeness of NPP data and ancillary site and stand characteristics. Class 1 sites contain above- and belowground biomass and NPP data. Studies that calculated belowground NPP as a fraction of aboveground biomass or NPP were classified as Class 2, although in some studies it was unclear what methodology was used to estimate belowground NPP. Class 2 stands lack belowground NPP data. Class 3 stands lacked complete NPP budgets or had very sketchy information on NPP methodology and/or site characteristics. The evaluation of sites was performed by the working groups meeting at the U.S. National Center for Ecological Analysis and Synthesis (NCEAS, Santa Barbara, California) as part of the GPPDI project.

Table 18-1 Source Data Archived by the Global Primary Productivity Data Initiative[a]

Source	No. of sites
Osnabrück Data Set (Esser *et al.*, 1997)	679
IBP Woodlands (De Angelis *et al.*, 1981)	117
TEM (McGuire *et al.*, 1992)	16
OTTER (Runyon *et al.*, 1994)	6
Superior National Forest (Hall *et al.*, 1989)	63
Chinese Forests (Ni and Zhang, personal communication)	690
Miscellaneous Intensive Study Sites—ORNL NPP Database	47
Total	1618

[a]Figure 18-1 contains only the 1540 data points where actual NPP values and site coordinates are available.

Table 18-2 Vegetation Data Archived by the Global Primary Productivity Data Initiative[a]

Vegetation type	No. of data points	Mean NPP	Standard deviation	Minimum	Maximum
Tundra	42	99	87	1	427
Desert	32	216	277	9	902
Grassland	101	353	387	3	1680
Boreal forest	297	449	225	18	1550
Tropical forest	48	765	490	116	2013
Subtropical forest	213	773	302	274	1756
Temperate forest	369	813	331	120	1755
Wetland	76	886	754	130	3920
Savanna	14	916	567	154	1700
Combined	1192	639	419	1	3920

[a]Summary statistics for net primary productivity (NPP) (g C m^{-2} yr^{-1}) for sites within biomes as defined by the original source of NPP data.

The Class 3 sites are often classified by biome type; however, there may be inconsistencies between the sources of data as to the criteria that were used to define biome types. Summary statistics for the Class 3 sites (Table 18-2) show a pattern of mean NPP level consistent with published tables and also show the great range of variability within each biome. The presently available 1618 sites (Table 18-1) provide, after removal of uncertain locations or total NPP numbers, 1540 data points that are unevenly distributed throughout the world (Fig. 18-1).

D. Other Data Sets

A compilation of biomass/productivity data from the former USSR was prepared by Bazilevich (1993) with an extensive list of literature sources. The database contains estimates of plant biomass and productivity components for a wide range of natural and managed ecosystems in different zones of the former USSR. At least one-time estimates of aboveground NPP (ANPP) are presented for most records. Expert estimates of root biomass and belowground NPP (BNPP) are provided, using information from "intensive" sites.

The IBP Woodlands Data Set consists of contributions from 117 international forest research sites, all but a few associated with projects that participated in the IBP. The data were collected in the 1960s and early 1970s and compiled into a single data set at the Oak Ridge National Laboratory (De Angelis *et al.*, 1981). Representatives of almost every kind of forest ecosystem are present (Burgess, 1981). A hierarchical scheme was used to assign a forest type to each site based on the climate, life form, behavior, and status of the site. Included are sites of the following types: tropical (26 sites), Mediterranean (3 sites), temperate (55 sites), and boreal (33 sites).

Figure 18-1 Map of the points where NPP data are presently available in the GPPDI data base (for sources, see Table 18-1).

A data set of rangeland production values is being derived from data at the U.S. Department of Agriculture Natural Resources Conservation Service (NRCS). The data set is based on a large number of end-of-season biomass estimates for areas of the United States that have the potential to be used as rangeland, regardless of their present land use. Additional attributes include slope, texture, available water capacity, soil organic matter, hydric soils, and permeability. Production estimates are provided for a range of conditions (wet, normal, and dry years) by map unit, with more detailed information available for different soil phases. Another source of rangeland productivity data, also collected by the NRCS, is the Range Site Descriptions database. This database consists of about 12,000 measurements of aboveground peak-season biomass values. The data have also been used to calculate statistics for 128 Major Land Resource Areas (MLRA) covering the central United States (Sala *et al.*, 1988).

E. Data from Large-Scale Measurement Campaigns

1. FIFE The First ISLSCP (International Satellite Land Surface Climatology Project) Field Experiment Project was conducted on a prairie in Kansas in 1987–1988 to measure the biophysical processes controlling the fluxes of radiation, moisture, and carbon dioxide between the land surface and the atmosphere and to develop remote-sensing methodologies for observing these processes (Hall *et al.*, 1989; Sellers and Hall, 1992).

2. HAPEX-Sahel The HAPEX-Sahel (Hydrologic and Atmospheric Pilot Experiment in the Sahel) international measurement campaign acquired field, aircraft, and satellite data for a $1° \times 1°$ study site in the West African Sahel in Niger during the 1991–1992 field campaign to study the energy, water, and carbon balances of the semiarid Sahel (Prince *et al.*, 1995).

3. OTTER The Oregon Transect Ecosystem Research Project estimated major fluxes of carbon, nitrogen, and water in forest ecosystems along a 300-km transect in 1990–1992, in Oregon, using an ecosystem-process model driven by remotely sensed data (Waring and Peterson, 1994).

4. BOREAS The Boreal Ecosystem–Atmosphere Study Project investigated the interactions between the boreal forest and the atmosphere in 1994 and 1996 in Canada using surface, airborne, and satellite observations to characterize biological and physical processes that govern the exchange of energy, water, heat, carbon, and trace gases (Sellers *et al.*, 1995).

5. LBA The Large-Scale Biosphere–Atmosphere Experiment in Amazonia, which is currently in its starting phase, will be an intensive scientific investigation of the tropical rain forest of Brazil and portions of adjacent countries, using intensive remote-sensing techniques and ground-based experiments to investigate the atmosphere–biosphere–hydrosphere dynamics of this large tropical region (LBA Science Planning Group, 1996).

E. Data from Long-Term Intensive Study Sites

An increasing number of long-term studies are being undertaken in which NPP and other biological variables, as well as environmental variables, are being measured. These studies often involve carefully designed, multiscale measurement techniques, including the leaf, branch, canopy, community, and landscape scales. One of the oldest of these is at Harvard Forest, Massachusetts (Goulden *et al.*, 1996) and others are being added. The Fluxnet[3] program and its regional components (e.g., Ameriflux,[4] Euroflux,[5] and Medeflu) are one source of integration of these sites and the NASA EOS validation program is another.

F. Forest Inventory, Crop Yield, and Rangeland Production Data

Significant economic activities such as agriculture, forestry, and animal production depend on components of NPP and long-standing monitoring activities that make regular field measurements of the resource production and report the results, often for large areas of the land surface. Some regions of the world, such as the midwest in the United States, are almost entirely sown to relatively few crops, for which statistics are available for many years. These represent highly suitable data sets for large-area estimation of NPP. As noted above, the chief difficulties in adapting these data sets for the purpose of global-scale NPP measurement lie in the fact that not all of the total NPP is monitored. In the case of cereal crop yield it is the grain volume that is reported, in forestry the wood volume of merchantable timber, and for rangeland the aboveground standing crop of palatable species at the end of the growing season. In spite of these restrictions, the data are usually collected according to explicit sampling schemes that are intended to estimate values for large areas of the land surface—often up to administrative counties. No measurements of "natural" land cover have been attempted at this spatial scale, nor are any other data available continuously for up to 100 years, and thus there is considerable motivation to make the investment needed to estimate the missing components of NPP and thereby benefit from these archives. In the case of rangelands, and more so in some forests, the degree of human modification of the ecosystem is significantly less, allowing inventory data to be used to validate all types of NPP models, including those that estimate the NPP of a hypothetical, unmanaged land cover.

[3]Internet site: http://www-eosdis.ornl.gov/FLUXNET.
[4]Internet site: http://cdiac.esd.ornl.gov/programs/ameriflux/.
[5]Internet site: http://www.unitus.it/eflux/euro.html.

G. Satellite Data

Current satellite sensors measure neither biomass nor NPP directly, although some early correlation models were based on the partial but strong relationship (in some biomes) of NPP with absorbed photosynthetically active radiation (FPAR) (Prince, 1991). The launch of satellites that carry radiometers capable of remotely sensing both the visible and near-infrared radiation reflected from the Earth's surface has, however, enabled the presence and activity of vegetation to be measured repetitively and with global coverage (Goward and Dye, 1996). The Advanced Very High Resolution Radiometer (AVHRR) sensor, carried on the NOAA-series of operational meteorological satellites, provides red and near-infrared radiances with continuous and global coverage at 1- and 8-km resolution at a frequency of at least 4 days out of every 9. In spite of the coarse spatial resolution of its measurements, the AVHRR is favored because of the relatively high temporal repetition of measurement, which increases the chance of cloudfree data, and also enables the dynamics of vegetation canopies to be tracked reasonably closely. Vegetation is detected using a spectral vegetation index, often the normalized difference vegetation index (NDVI), which is a combination of the two spectral bands. NDVI is a near-linear estimator of the proportion of photosynthetically active radiation absorbed by plants and can also be used to estimate leaf area index (LAI); however, the NDVI–LAI relationship saturates at intermediate LAI values, and thus LAI is not the most effective means of exploiting satellite data in NPP models. The thermal channels on the AVHRR allow other environmental variables such as air temperature and vapor pressure deficit to be estimated simultaneously with FPAR (Goward and Dye, 1996). Modeling frameworks that can fully exploit satellite observations have been reviewed by Prince (1991) and these have been incorporated in an operational NPP model (Prince and Goward, 1995). Other models that exploit satellite observations of FPAR alone have also been developed (e.g., Potter *et al.*, 1993; Ruimy *et al.*, 1996).

In addition to FPAR, the seasonal pattern of NDVI largely defines the phenology of the vegetation (Reed *et al.*, 1994) and indicates the length and other characteristics of the growing season. Start and end of the growing season can be defined by assuming a threshold value (Moulin *et al.*, 1997). A study using 10 years of NDVI data has shown an increase of the length of the growing season in the northern hemisphere, assumed to be related to the late winter/early spring warming of these areas (Myneni *et al.*, 1997). A monthly global data set of FPAR from the AVHRR at 1° longitude/latitude resolution exists for the years 1987–1988 (ISLSCP) (Sellers *et al.*, 1994). Data from this series of satellites have been collected and archived since 1981, and several NDVI time series with different spatial and temporal resolutions, and different levels of processing, are available [e.g., GVI, Tarpley *et al.* (1984);

AVHRR–NDVI Pathfinder, Agbu and James (1994); CESBIO–NDVI, Berthelot *et al.* (1994)].

Current and foreseeable satellite sensors with adequate temporal resolution for NPP measurement all have relatively coarse spatial resolution (EOS MODIS will have a maximum resolution of 250 m) and thus place special constraints on the types of field data that are needed for model validation. Unlike many ecophysiological models that are essentially one dimensional, this class of model is fundamentally spatial, and validation data are needed that have the same spatial dimensions.

III. Assessing Productivity at the Continental to Global Scale

A. Field Data for Ecosystem-Level NPP Estimation

Various methods are used to estimate ANPP and BNPP, with some being more suitable for vegetation communities of small stature (grasslands, tundra, agriculture crops) as opposed to forests. One of the simplest approaches to estimate ANPP for grasslands is to measure peak standing biomass and multiply it with some factor to account for the difference between maximum and minimum biomass. A more complex method implies the UNEP Project formula (Long *et al.*, 1989):

$$\mathrm{ANPP} = \sum(\Delta B + \Delta D + D \bullet r_{\mathrm{D}}),$$

where B is aboveground live biomass, D is aboveground dead matter, and r_{D} is the decomposition rate for dead matter. Methods such as these can only indirectly account for grazing or harvest, which can be important in prairie and agriculture ecosystems. The spatial and temporal variation in above- and belowground biomass components may be large, therefore these methods often fail to provide the statistically desirable degree of replication for even simple systems.

ANPP of woody vegetation communities is commonly determined using the equation

$$\mathrm{ANPP} = \Delta B + L + H,$$

where L is the annual litter/detritus production (commonly measured using litter traps) and H is herbivory or other harvest. Aboveground woody biomass (e.g., stem and branch) and coarse root biomass are commonly estimated from allometric equations that correlate biomass to an independent variable such as diameter or basal area at breast height (1.3 m). Empirical data suggest that these allometric relationships are reasonably constant within a species and, in the case of coarse root biomass, among species. Allometric equations can also be used to estimate total foliage mass and new foliage mass, which can then be used to calculate annual new foliage production. Herbivory is commonly ignored in forest ecosystems because it generally comprises $<10\%$ of NPP.

Methods used to estimate belowground NPP include assuming BNPP to be a constant fraction of ANPP, calculating the difference between maximum and minimum of fine root mass, measuring root ingrowth into cores, or measurement of increments of fine root mass from sequential root cores (Jackson *et al.*, 1997). Generally, the first two methods tend to underestimate BNPP. There are only very few sites for which reliable NPP estimates for both above- and belowground components exist. Therefore, there is a need to screen all data being used and to identify reproducible and reliable methods for estimation of the less well-known variables (Clark *et al.*, 2000; Kelly *et al.*, 2000).

B. Scaling from Site Observations to Large Regions

Measurements of NPP and ecosystem state variables that are used to estimate NPP are, for practical reasons, connected to small patches of vegetation, usually 0.1 ha or less. Productivity varies, however, in space and over time across a wide range of scales. For the assessment of fluxes over large areas and at an annual time-step, explicit scaling methods are required.

Regression models such as the MIAMI model or the later developments of mechanistic ecosystem-process models all connect local observations to local environmental data. Their application to a broad geographical grid depends on the existence of environmental data that adequately represent the conditions that occur in each grid cell. It is important to recognize that different kinds of issues may arise in this process. The forcing variables, typically climatic and other environmental data sets, must be adequate to represent conditions in a grid cell as opposed to a point location. The environmental data provided for the location of observation must be consistent with the conditions at the exact site where the modeled vegetation was growing, so they represent the conditions actually experienced by the modeled vegetation. The total environmental space in which vegetation to be modeled occurs must be sampled so that the successful performance of the model in the total range of conditions that may occur is assured. Potential problems of nonlinearity of model behavior across a range of driving variable values are thereby avoided. The environmental data and associated NPP values must be representative not only of the average conditions in each grid cell but also their frequency distributions.

An example in which the first requirement comes into play would be a mountainous landscape with a broad range of elevations. The environmental data used to run the model and the field observation of NPP must be for the same elevation. To meet the second requirement, it would be important that, somewhere in the data set, observations of NPP and associated environmental variables were available for all elevations in this landscape. Although there are no absolute criteria for judging the fulfillment of these requirements, there are methods to make a qualitative assessment, for example, a graphical display of NPP observation sites in environmental space where the existence of observations for all environmental data combinations can be examined. Finally, no biogeochemical model is useful for global application unless it is coupled with an appropriate data set of driving variables that meet the third requirement. Forcing environmental variables are needed that can adequately represent the range and frequency of NPP results throughout each grid cell. The failure to recognize the issues involved in driving a one-dimensional model at a large grid-cell resolution leads to inadequate "validation."

A frequently used spatial resolution for data used in global models is 0.5° longitude/latitude [e.g., the Global Ecosystems Database; see Kineman and Ohrenschall (1992)], which provides a global grid that is a compromise between the shortage of data for many parts of the Earth's surface and the finer-scale heterogeneity in many regions. Being a compromise, this grid resolution nevertheless cannot escape from errors due to either insufficient resolution (in areas with high heterogeneity) and insufficient accuracy (in areas with poor data). The scaling from point to region is accomplished through the application of the model and is not directly dependent on the existence of a spatially comprehensive observational database of NPP. Its validity, however, depends on the comprehensive evaluation of the model in all relevant environmental conditions plus the necessary fields of environmental data.

Finally, the need for evaluation of the models and the possession of adequate environmental data sets applies to all environmental factors, including those that are not explicit in the model but that implicitly constrain the model to more or less limited sets of conditions. For example, the effect of water-logging on root growth, the presence of toxic elements, or shallow soils may not be explicitly allowed for in the model, rather their absence is assumed. Unless such conditions are known to be absent, or are explicitly incorporated in the model and the necessary data exist on their occurrence in the sites to which the model is to be applied, then the results will be unreliable. It is practically impossible to eliminate this problem because it would require that the whole Earth's surface be surveyed and so they will always be a source of unknown error in applications of this category of model to large areas , especially the entire planet.

Models with implicit considerations of spatial averaging, such as some remote-sensing-based models, which have a resolution defined by the sensor's footprint, require a more direct consideration of spatial representativeness of NPP observations. In the case of models such as GLO-PEM (Prince and Goward, 1995), in which the driving variables are all derived from remotely sensed observations and are therefore implicitly spatial estimates, the emphasis is shifted from the spatial representativeness of the environmental variables to that of the NPP field measurements, which need to be estimated at the scale of the remotely sensed observations. In other models that use both remotely sensed and ground measurements for some environmental variables, both NPP and environmental data representativeness issues need attention.

In practical terms, the degree to which an NPP observation is representative for the grid cell in which it is made must be assessed and, if not, corrections need to be made to the observation to make it better represent the average. Such corrections could in many cases be based on the application of prognostic biogeochemical models such as TEM or BIOME3, using methods as discussed above. The associated assumptions need to be clearly documented, however. Two kinds of errors may occur in the use of biogeochemical models in this mode:

1. The biogeochemical model may, for some reason, fail in the spatial extrapolation from the point(s) to the grid cell.
2. The biogeochemical model may contain crucial components that are based on the same assumptions as the remote-sensing model that the resulting data are used to evaluate, i.e., circularity is introduced and the test itself may be invalid.

In summary, it has been shown that, in the case of process models that are essentially one-dimensional, validation at points using NPP observations is relatively straightforward, but validation at the scale of the grid cell to which they are subsequently applied is also necessary, because the properties of the sets of forcing variables used to drive the models are very demanding, as discussed above, and are rarely fully met. In the case of models that depend on intrinsically spatial forcing variables, such as those that use remotely sensed observations, entirely or partly, no validation is possible without NPP observations for entire grid cells. Thus, for all types of NPP models, field observations at the grid-cell scale are essential. These are being carried out by some large campaigns, such as BOREAS in Canada or LBA in Brazil, but the amount of effort required is obviously very large and therefore only possible in a few regions.

C. Scaling from Short-Term Observations to Longer Term Assessments

Not only are NPP observations connected to specific, small patches of vegetation, but also to specific, often short, time periods. Furthermore, modeling of biogeochemical processes almost always relates to significantly broader temporal as well as spatial units, hence it is necessary to scale from the observation to the modeling units in time as well as in space. The temporal scale of a model is important because the pools to which carbon is allocated have different turnover rates. NPP estimates may be based on, for example, fine root turnover, tree leaf production, or just biomass increments. The estimate at the fine temporal scale (e.g., with fine roots) may not scale up to estimates at the coarse temporal scale because of rapid turnover, which must be modeled. This is analogous to using fine or coarse-scale measuring devices in space. If the scaling was simple, then a scaling factor could be developed and applied to the annual wood increment in all ecosystems. However, the scaling is not simple, and so a constant scaling factor is unlikely to exist. A basic goal in assembling an archive of NPP observations should be to maintain the relation between the NPP and environmental data.

D. Applying NPP Models at the Broad Scale

Estimating NPP at the broad scale and/or over longer periods of time is now routinely being done using global models (Cramer *et al.*, 1999a). Figure 18-2 shows an example for such an estimate, taken as an average of 16 different models. These models basically drive process-based estimates of NPP at a large number of points with climatic information, and data on actual vegetation, soils, and atmospheric CO_2 and in some cases nitrogen. Apart from the site- or region-specific tests discussed above, the quality of these estimates is assessed on the basis of an overall carbon balance estimation and the comparison of results with data from CO_2 concentration measurements. The carbon balance considered includes anthropogenic and natural emissions, soil respiration, and the fluxes between the ocean and the atmosphere, and uses either forward modeling or inverse deconvolution techniques (Prentice *et al.*, 2000; see Chapter 19, this volume). Discrepancies between the flux estimated on the basis of atmospheric measurements and that from process models do not necessarily come from inappropriate models—a significant problem remains also in the (environmental) driving data. Because ecosystem fluxes are spatially highly heterogeneous, also the driving data need to be presented at fairly high spatial resolution. Climate and soils data sets, for example, have only limited quality at very high spatial resolution, and important nutrient fluxes such as nitrogen deposition are hardly quantifiable as global fields.

Efforts are underway to improve the quality of such data sets (Cramer *et al.*, 1999b). New and greatly enhanced data sets are becoming available that will greatly improve the quality of assessments, in particular based on remote-sensing data for actual land cover (such as DISCover) (Loveland and Belward, 1997) or on better aggregations of existing field observations (Scholes *et al.*, 1995). In addition, there is a need to produce databases with enough *temporal* resolution to investigate the seasonal and interannual vari-

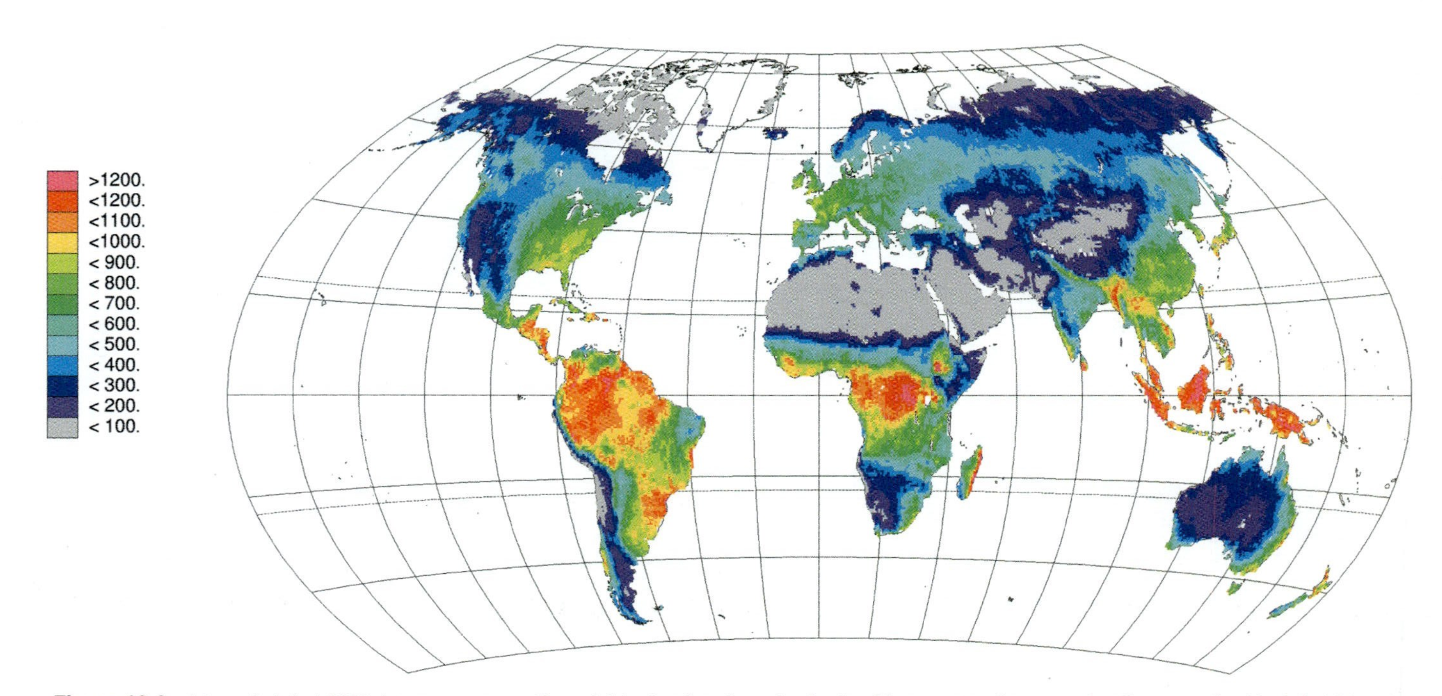

Figure 18-2 Map of global NPP (average across all models), simulated on the basis of long-term climate and soil texture by 16 global biosphere models (model average) at 0.5° longitude/latitude resolution (from Cramer *et al.*, 1999a).

ability of the total carbon flux, because this quantity appears to be well suited for comparison with seasonal and interannual changes in the CO_2 atmospheric growth rate (Heimann *et al.,* 1998).

IV. Outlook: Development of an Improved Database

Progress in understanding the global carbon cycle has been seriously inhibited by the lack of a high-quality data set based on field observations of ecosystem variables with which to calibrate, parameterize, and evaluate global models (Cramer *et al.,* 1996). Net primary productivity is a key ecosystem variable, yet no agreed methodology presently exists for extrapolating sparse and sometimes uncertain field observations of NPP to produce a consistent database representative of major worldwide vegetation and environmental regimes. Under the auspices of the Global Primary Production Data Initiative, an activity endorsed by the International Geosphere–Biosphere Programme Data and Information System (IGBP-DIS), a number of collaborating research groups are now making progress in (1) compiling field site measurements of NPP and associated environmental data, (2) agreeing on consistent standards for cross-site comparisons and scaling up from point measurements to grid cells, and (3) investigating new ways to relate NPP observations to other data sources, such as satellite data. These developments are promising. A consistent data set now exists that will soon become the de facto standard for twentieth century NPP observations. Together with data from experiments and global observing systems, these are the foundation for a higher level of confidence in assessments of future NPP and hence the overall carbon balance. Without such a capacity, Earth system science will be inhibited to provide input into a policy-making progress, which is, for the first time in history, seriously considering the global nature of the human environment.

References

Agbu, P. A., and James, M. E. (1994). "The NOAA/NASA Pathfinder AVHRR Land Data Set User's Manual." Goddard Distributed Active Archive Center, NASA, Goddard Space Flight Center.

Bazilevich, N. I. (1993). "Biological Productivity of Ecosystems of Northern Eurasia." Nauka, Moscow, Russia.

Berthelot, B., Dedieu, G., Cabot, F., and Adam, S. (1994). Estimation of surface reflectances and vegetation index using NOAA/AVHRR: Methods and results at global scale. *In "Physical Measurements and Signatures in Remote Sensing,"* pp. 33–40. *6th International Symposium,* 17–21 January, 1994: ISPRS/CNES, Val d'Isère, France.

Breymeyer, A., and Melillo, J. M. (1991). Global Climate Change: The effects of climate change on production and decomposition in coniferous forests and grasslands. *Ecol. Appl.* **1,** 111.

Burgess, R. L. (1981). Physiognomy and phytosociology of the international woodlands research sites. *In* "Dynamic Properties of Forest Ecosystems" (D. E. Reichle, ed.), pp. 1–35. Cambridge Univ. Press, Cambridge.

Clark, D. A., Brown, S., Kicklighter, D., Chambers, J. Q., Thomlinson, J. R., and Ni, J. (2000). Evaluation and synthesis of existing field NPP data: Tropical forests. *Ecol. Appl.* (in press).

Coughenour, M. B. (1992). Spatial modeling and landscape characterization of an African pastoral ecosystem: A prototype model and its potential use for monitoring drought. *In* "Ecological Indicators." D. H. McKenzie, D. E. Hyatt, and V. J. McDonald, eds.), Vol. 1. pp. 787–810. Elsevier, London, New York.

Cramer, W., Kicklighter, D. W., Bondeau, A., Moore, B., III, Churkina, G., Nemry, B., Ruimy, A., Schloss, A. L., and participants of the Potsdam NPP Model Intercomparison (1999a). Comparing global models of terrestrial net primary productivity (NPP): Overview and key results. *Global Change Biol.* **5** (Suppl. 1), 1–15.

Cramer, W., Leemans, R., Schulze, E.-D., Bondeau, A., and Scholes, R. (1999b). Data needs and limitations for broad-scale ecosystem modelling. *In* "The Terrestrial Biosphere and Global Change: Implications for Natural and Managed Ecosystems" (B. H. Walker, W. L. Steffen, J. Canadell, and J. S. I. Ingram, eds.), IGBP Book Series, Vol. 4., pp. 88–105. Cambridge Univ. Press, Cambridge.

Cramer, W., Moore, B., III, and Sahagian, D. (1996). Data needs for modelling global biospheric carbon fluxes - lessons from a comparison of models. *IGBP Newslett.* **1996,** 13–15.

De Angelis, D. L., Gardner, R. H., and Shugart, H. H. (1981). Productivity of forest ecosystems studied during IBP: The woodlands data set. *In* "Dynamic Properties of Forest Ecosystems" (D. E. Reichle, ed.), pp. 567–672. Cambridge Univ. Press, Cambridge.

Denmead, O. T., Raupach, M. R., Dunin, F. X., Cleugh, H. A., and Leuning, R. (1996). Boundary layer budgets for regional estimates of scalar fluxes. *Global Change Biol.* **2,** 255–264.

Esser, G. (1992). Osnabrück biosphere model—Structure, construction, results. *In* "Modern Ecology" (G. Esser and D. Overdieck, eds.), pp. 679–709. Elsevier Science Publ., Amsterdam.

Esser, G., Lieth, H. F. H., Scurlock, J. M. O., and Olson, R. J. (1997). "Worldwide Estimates of Net Primary Productivity Derived from pre-1982 Publications," TM-13485. Oak Ridge National Laboratory, Oakridge, Tenn.

Fischer, A., Louahala, S., Maisongrande, P., Kergoat, L., and Dedieu, G. (1996). Satellite data for monitoring, understanding and modelling of ecosystem functioning. *In* "Global Change and Terrestrial Ecosystems" (B. Walker and W. Steffen, eds.), Vol. 2. pp. 566–591. Cambridge University Press, Cambridge.

Gholz, H. L., Linder, S., and McMurtrie, R. (1994). Environmental constraints on the structure and productivity of pine forest ecosystems: A comparative analysis. *Ecol. Bull.* **43.**

Goulden, M. L., Munger, J. W., Fan, S. M., Daube, B. C., and Wofsy, S. C. (1996). Measurements of carbon sequestration by long-term eddy covariance—Methods and a critical evaluation of accuracy. *Global Change Biol.* **2,** 169–182.

Goward, S. N., and Dye, D. (1996). Global biospheric monitoring with remote sensing. *In* "The Use of Remote Sensing in Modeling Forest Productivity" (H. L. Gholtz, K. Nakane, and H. Shimoda, eds.), pp. 241–272. Kluwer Academic Publishers, Dordrecht, The Netherlands.

Hall, F. G., Sellers, P. J., MacPherson, I., Kelly, R. D., Verma, S., Markham, B., Blad, B., Wang, J., and Strebel, D. E. (1989). FIFE: Analysis and results—A review. *Adv. Space Res.* **9,** 275–293.

Haxeltine, A., and Prentice, I. C. (1996). BIOME3: An equilibrium biosphere model based on ecophysiological constraints, resource availability and competition among plant functional types. *Global Biogeochem. Cycl.* **10,** 693–709.

Heimann, M., Esser, G., Haxeltine, A., Kaduk, J., Kicklighter, D. W., Knorr, W., Kohlmaier, G. H., McGuire, A. D., Melillo, J., Moore, B., III, Otto, R. D., Prentice, I. C., Sauf, W., Schloss, A., Sitch, S., Wittenberg, U., and Würth G. (1998). Evaluation of terrestrial carbon cycle models through simulations of the seasonal cycle of atmospheric CO_2: "First results of a model intercomparison study. *Global Biogeochem. Cycl.* **12,** 1–24.

IGBP Terrestrial Carbon Working Group (1998). The terrestrial carbon cycle: Implications for the Kyoto Protocol. *Science* **280,** 1393–1394.

Jackson, R. B., Mooney, H. A., and Schulze, E.-D. (1997). A global budget for fine root biomass, surface area and nutrient contents. *Proc. Natl. Acad. Sci. U.S.A.* **94,** 7362–7366.

Kelly, R. H., Parton, W. J., Day, K. A., Jackson, R. B., Morgan, J. A., Scurlock, J. M. O., Tieszen, L. L., Gill, R. A., Castle, J. V., Ojima, D. S., and Zhang X. S. (2000). Using simple environmental variables to estimate belowground productivity in grasslands. *Ecol. Appl.* (in press).

Kineman, J. J., and Ohrenschall, M. A. (1992). "Global Ecosystems Database Version 1.0: Disc A, Documentation Manual." USDOC/NOAA National Geophysical Data Center, Boulder, CO.

Kurz, W. A., and Apps, M. J. (1995). An analysis of future carbon budgets of Canadian boreal forests. *Water Air Soil Pollut.* **82,** 321–331.

LBA Science Planning Group (1996). "The Large Scale Biosphere–Atmosphere Experiment in Amazonia (LBA). Concise Science Plan." Centro de Previsão de Tempo e Estudos Climàticos (CPTEC).

Lieth, H. (1975). Modelling the primary production of the world. *In* "Primary Productivity of the Biosphere" (H. Lieth and R. H. Whittaker, eds.), Vol. 14, pp. 237–263. Springer-Verlag, Berlin and New York.

Lieth, H., and Box, E. (1972). Evapotranspiration and primary productivity: C. W. Thornthwaite Memorial Model. *Publ. Climatol.* **25,** 37–46.

Lieth, H., and Whittaker, R. H., eds. (1975). "Primary Productivity of the Biosphere." Springer-Verlag, Berlin.

Long, S. P., Garcia Moya, E., Imbamba, S. K., Kamnalrut, A., Piedade, M. T. F., Scurlock, J. M. O., Shen, Y. K., and Hall, D. O. (1989). Primary productivity of natural grass ecosystems of the tropics: A reappraisal. *Plant Soil* **115,** 155–166.

Loveland, T. R., and Belward, A. S. (1997). The IGBP-DIS global 1km land cover data set, DISCover: First results. *Int. J. Remote Sensing* **18,** 3289–3295.

McGuire, A. D., Melillo, J. M., Joyce, L. A., Kicklighter, D. W., Grace, A. L., Moore III, B., and Vörösmarty, C. J. (1992). Interactions between carbon and nitrogen dynamics in estimating net primary productivity for potential vegetation in North America. *Global Biogeochem. Cycl.* **6,** 101–124.

Mooney, H. A., Canadell, J., Chapin, F. S., III, Ehleringer, J., Körner, C., McMurtrie, R., Parton, W. J., Pitelka, L., and Schulze, E.-D. (1999). Ecosystem physiology responses to global change. *In* "The Terrestrial Biosphere and Global Change: Implications for Natural and Managed Ecosystems" (B. H. Walker, W. L. Steffen, J. Canadell, and J. S. I. Ingram, eds.), IGBP Book Series, Vol. 4, pp. 141–189. Cambridge Univ. Press, Cambridge.

Moulin, S., Kergoat, L., Viovy, N., and Dedieu, G. (1997). Global scale assessment of vegetation phenology using NOAA/AVHRR satellite measurements. *J. Clim.* **10,** 1154–1170.

Myneni, R. B., Keeling C. D., Tucker C. J., Asrar, G., and Nemani, R. R. (1997). Increased plant growth in the northern high latitudes from 1981 to 1991. *Nature (London)* **386,** 698–702.

Nadelhoffer, K. J., and Raich, J. W. (1992). Fine root production estimates and belowground carbon allocation in forest ecosystems. *Ecology* **73,** 1139–1147.

Olson, R. J., Scurlock, J. M. O., Cramer, W., Prince, S. D., and Parton, W. J. (1997). "Global Primary Production Data Initiative Workshop: From Sparse Field Observations to a Consistent Global Dataset on Net Primary Production." IGBP-DIS 16.

Parton, W. J., Scurlock, J. M. O., Ojima, D. S., Gilmanov, T. G., Scholes, R. J., Schimel, D. S., Kirchner, T., Menaut, J.-C., Seastedt, T. R., Garcia Moya, E., Kamnalrut, A. and Kinyamario, J. I. (1993). Observations and modeling of biomass and soil organic matter dynamics for the grassland biome worldwide. *Global Biogeochem. Cycl.* **7,** 785–809.

Parton, W. J., Scurlock, J. M. O., Ojima, D. S., Schimel, D. S., and Hall, D. O. (1995). Impact of climate change on grassland production and soil carbon worldwide. *Global Change Biol.* **1,** 13–22.

Potter, C. S., Randerson, J. T., Field, C. B., Matson, P. A., Vitousek, P. M., Mooney, H. A., and Klooster, S. A. (1993). Terrestrial ecosystem production—A process model based on global satellite and surface data. *Global Biogeochem. Cycl.* **7,** 811–841.

Prentice, I. C., Heimann, M., and Sitch, S. (2000). The carbon balance of the terrestrial biosphere: Ecosystem models and atmospheric observations. *Ecol. Appl.* (in press).

Prince, S. D. (1991). A model of regional primary production for use with coarse-resolution satellite data. *Int. J. Remote Sensing* **12,** 1313–1330.

Prince, S. D., and Goward, S. N. (1995). Global net primary production: A remote sensing approach. *J. Biogeogr.* **22,** 815–835.

Prince, S. D., Kerr, Y. K., Goutorbe, J.-P., Lebel, T., Tinga, A., Bessemoulin, P., Brouwer, J., Dolman, A. J., Engman, E. T., Gash, J. H. C., Hoepffner, M., Kabat, P., Monteny, B., Said, F., Sellers, P., and Wallace, J. (1995). Geographical, biological and remote sensing aspects of the Hydrologic Atmospheric Pilot Experiment in the Sahel (HAPEX-Sahel). *Remote Sensing Environ.* **51,** 215–234.

Raich, J. W., Rastetter, E. B., Melillo, J. M., Kicklighter, D. W., Steudler, P. A., Peterson, B. J., Grace, A. L., Moore, B., III, and Vörösmarty, C. J. (1991). Potential net primary productivity in South America: Application of a global model. *Ecol. Appl.* **1,** 399–429.

Rastetter, E. B. (1996). Validating models of ecosystem response to global change. *BioScience* **46,** 190–198.

Reed, B. C., Brown, J. F., Vanderzee, D., Loveland, T. R., Merchant, J. W., and Ohlen, D. O. (1994). Measuring phenological variability from satellite imagery. *J. Veg. Sci.* **5,** 703–714.

Ruimy, A., Dedieu, G., and Saugier, B. (1996). TURC: A diagnostic model of continental gross primary productivity and net primary productivity. *Global Biogeochem. Cycl.* **10,** 269–286.

Runyon, J., Waring, R. H., Goward, S. N., and Welles, J. M. (1994). Environmental limits on net primary production and light-use efficiency across the Oregon transect. *Ecol. Appl.* **4,** 226–237.

Sala, O., Parton, W. J., Joyce, L. A., and Lauenroth, W. K. (1988). Primary production of the central grassland region of the United States: Spatial pattern and major controls. *Ecology* **69,** 40–45.

Scholes, R. J., Skole, D., and Ingram, J. S. I. (1995). "A Global Database of Soil Properties: Proposal for Implementation." IGBP DIS 10.

Scurlock, J. M. O., Cramer, W., Olson, R. J., Parton, W. J., and Prince, S. D. (1999). Terrestrial NPP: Toward a consistent data set for global model evaluation. *Ecol. Appl.* **9,** 913–919.

Sellers, P., Hall, F., Margolis, H., Kelly, B., Baldocchi, D. D., Den Hartog, J., Cihlar, J., and Ryan, M. (1995). The Boreal Ecosystem–Atmosphere Study (BOREAS): An overview and early results from the 1994 field year. *Bull. Am. Meteorol. Soc.* **76,** 1549–1577.

Sellers, P. J., and Hall, F. H. (1992). FIFE in 1992: Results, Scientific Gains, and Future Research directions. *J. Geophys. Res.* **97,** 19091–19109.

Sellers, P. J., Tucker, C. J., Collatz, G. J., Los, S. O., Justice, C. O., Dazlich, D. A., and Randall, D. A. (1994). A global 1 degrees-by-1 degrees NDVI data set for climate studies. Part 2. The generation of global fields of terrestrial biophysical parameters from the NDVI. *Int. J. Remote Sensing* **15,** 3519–3545.

Singh, J. S., Lauenroth, W. K., and Sernhorst, R. K. (1975). Review and assessment of various techniques for estimating net aerial primary production in grasslands from harvest data. *Bot. Rev.* **41,** 181–232.

Tarpley, J. D., Schneider, S. R., and Money, R. L. (1984). Global vegetation indices from the NOAA-7 meteorological satellite. *J. Clim. Appl. Meteorol.* **23,** 491–494.

Waring, R. H., and Peterson, D. L. (1994). Oregon Transect Ecosystem Research (OTTER) Project. *Ecol. Appl.* **4,** 210.

19

Integrating Global Models of Terrestrial Primary Productivity

Philippe Ciais, Pierre Friedlingstein, Andrew Friend, and David S. Schimel

I. Introduction

In order to understand the global relevance of basic ecological and physiological processes, their role in biogeochemistry and climate, and the impacts of human-induced changes in the Earth's biosphere, terrestrial net primary productivity (NPP) must be studied globally. Global terrestrial biosphere models (TBMs) aim to understand and eventually predict the cycling of the major elements (C, N) within vegetation and soils, as well as their fluxes with other reservoirs. TBMs can calculate the global fluxes of CO_2 and H_2O between ecosystems and the atmosphere, two key components of the climate system. Because the spatial aggregation of existing and planned *in situ* flux measurements does not allow full global coverage, and given the (large) heterogeneity of most land ecosystems, TBMs are an important technique for extrapolating our understanding of terrestrial processes to the regional and global level. Alternatively, one can use atmospheric measurements together with global atmospheric transport models (TMs) in order to attempt to infer indirectly the distribution of the terrestrial fluxes. This chapter is subdivided into three parts. First, we make an overall presentation of the different types of TBMs and of their range of application. Second, we review how the major processes that control the functioning of the biosphere are incorporated in TBMs. Third, we discuss how global data sets, namely, remote-sensed data and atmospheric CO_2 and other tracer measurements, can be used to constrain indirectly the NPP distribution.

II. Different Nature and Purposes of Global Models

Numerous TBMs have been developed for various purposes. More than 20 are currently available and greatly differ in complexity and flexibility. Model complexity ranges from simple correlation aimed to estimate mean annual NPP, to full physiology that can estimate instantaneous leaf-level photosynthesis. Flexibility of these models is also variable; input requirements and output delivery clearly define their fields of application. Several intercomparison projects have been carried out in order to better understand the strengths and weaknesses of models used to estimate NPP at the global scale: the VEMAP project (VEMAP, 1995) and the Potsdam comparison [see Cramer *et al.* (1999) and companion papers]. Our present goal is to review the advances in NPP modeling and the differences in conceptual approaches.

A. The Statistical Approach

The first major impetus for NPP modeling followed the International Biosphere Programme (IBP) in the early 1960s, when NPP measurements from more than 50 locations around the world had been systematically compiled (Whittaker and Likens, 1973; Duvigneaud, 1974; Rodin *et al.*, 1975; Ajtay *et al.*, 1979). The precursors of the NPP models correlated these NPP data with relevant climatic variables. Rosenzweig (1968), followed by Lieth and Box (1972), correlated NPP with actual evapotranspiration. The best known example of this statistical approach was developed by Lieth (1975), who proposed the Miami model, first introduced at the Second Biological Congress of the American Institute of Biological Sciences, held in Miami, Florida, in 1971. The Miami model estimates NPP from mean annual precipitation (P) and surface temperature (T) regressions, assuming a Liebig's law: the most limiting environmental resource (precipitation or temperature) controls the productivity level:

$$\mathrm{NPP} = \min[f(T), f(P)].$$

The conceptual simplicity of this NPP model and its reasonable input data requirement still make it extremely attractive. Several groups included the Miami NPP equations into more complex global TBMs to address major issues relevant to the global carbon cycle, such as the biospheric activity at the Last Glacial Maximum in SLAVE (Friedlingstein *et al.*, 1992) and HRBM (Esser and Lautenschlager, 1994), the biospheric carbon uptake during the industrial period in HRBM (Esser, 1987) and SLAVE (Friedlingstein *et al.*, 1995), or the response of the biosphere to climate variability (Dai and Fung, 1993; Post *et al.*, 1997). However, statistical TBMs do not provide any "process-based" representation of the physiological mechanisms that control NPP, such as photosynthesis, autotrophic respiration, and allocation of carbon and nutrients. Rather, they use the empirical correlation of climatic variables with NPP and hence provide an extrapolation to the distribution of NPP over the globe.

B. The Physiological Approach

Farquhar *et al.* (1980) proposed a biochemical model of CO_2 assimilation in C_3 plants, now widely accepted for modeling at a range of scales. This model, later refined by Collatz *et al.* (1991) and extended to C_4 plants (Collatz *et al.*, 1992) has the advantage of representing a realistic physiological description of photosynthesis in terms of gas exchange, which yet avoids treating explicitly the complex biochemical reactions that occur during the carboxylation of CO_2. Figure 19-1 illustrates the physiological control on the fluxes of carbon, water, and energy exchanged between the leaf and the atmopshere.

Although relatively simple at first glance (assimilation is the minimum of two or three functions), the calculation of assimilation requires the knowledge of several variables that need to be specified or estimated via a set of equations. The assimilation is, among others, a function of leaf internal CO_2 concentration, which calls for the calculation of stomatal conductance, which may depend on relative humidity and assimilation. Also, scaling up Farquhar's equations established at the leaf level, from leaf to ecosystem, is

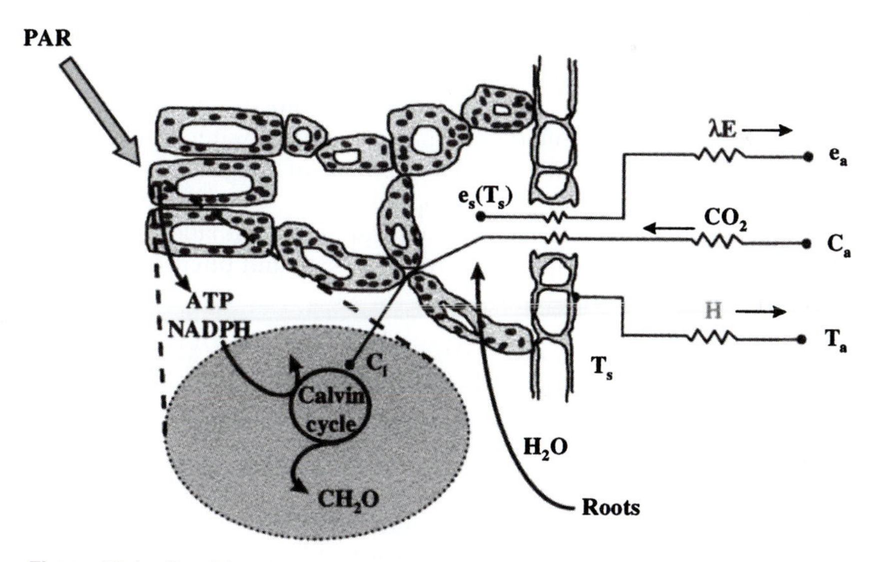

Figure 19-1 Simplified diagram of physiological controls on heat, water vapor, and CO_2 transfer between the leaf and the atmosphere. CO_2 fixation and assimilation, by way of the Calvin cycle, depend on the leaf interior CO_2 concentration (mainly controlled by stomatal conductance) and on the enzyme metabolism (controlled by light supply).

a nontrivial exercise. A growing number of TBMs successfully developed, at the global scale, a photosynthesis model derived from the Farquhar–Collatz formulations: BIOME-BGC (Running and Hunt, 1993), CARAIB (Warnant *et al.*, 1994), LSM (Bonan, 1995), DOLY (Woodward *et al.*, 1995), SiB2 (Sellers *et al.*, 1996a,b), IBIS (Foley *et al.*, 1996); BIOME3 (Haxeltine and Prentice, 1996a,b), and HYBRID (Friend *et al.*, 1997). Generally, these models couple photosynthesis and stomatal conductance to estimate leaf-level photosynthesis. Then, leaf photosynthesis is integrated through the canopy, usually assuming a light attenuation through the canopy, to calculate gross primary productivity (GPP). In order to compute NPP, these models also need to estimate autotrophic respiration (R_a) and deduce the NPP as the difference between GPP and R_a. A more detailed description of some of the processes that are relevant to include in physiological TBMs is given in Section III.

C. Resource-Limitation Models

As an alternative to describing ecosystem-level photosynthesis as an extrapolation of the leaf gas-exchange equations, an important number of ecosystem models have been developed during the past two decades that use the resource or multiple-resource limitation paradigm: TEM (Raich *et al.*, 1991), CENTURY (Parton *et al.*, 1987), CASA (Potter *et al.*, 1993), FBM (Ludeke *et al.*, 1994), DEMETER (Foley, 1994), SDBM (Knorr and Heimann, 1995), SYLVAN (Kaduk and Heimann, 1996), and TURC (Ruimy *et al.*, 1996). Although these models do not have a common generic NPP equation (such the Lieth or the Farquhar equation) they do, however, share some common assumptions about the resources required for photosynthesis. In these models, NPP (or GPP) is controlled by light, temperature, water, nutrients, CO_2, and, in a few cases, disturbances. A typical model of this group is TEM, which calculates GPP as the product of an optimal GPP (GPP_{max}) by several factors, accounting for light [photosynthetically active radiation (PAR)], carbon dioxide (C_i), temperature (T), nutrient (N), and phenological limitations (Φ)

$$GPP = GPP_{max}\, f(PAR)\, f(C_i)\, f(T)\, f(N)\, f(\Phi).$$

Other models use a similar approach, where NPP or GPP is scaled down from an optimal value whenever the resources become limiting (FBM, DEMETER, CENTURY). In a similar fashion, several models (CASA, SDBM, TURC) emphasize radiation as a limiting resource and use the light use efficiency concept introduced by Monteith. These models estimate NPP as the product of absorbed photosynthetically active radiation by light use efficiency (LUE). This latter expresses the efficiency of plants that convert absorbed radiation into carbon. In the TURC model, LUE is constant for photosynthesis, but also depends on temperature and biomass through the plant respiration terms. In other models (CASA, SDBM), the LUE is scaled down by water and temperature constraints. These models, usually called diagnostic models, often make use of remote-sensing information, such as the normalized difference vegetation index (NDVI) further mentioned in Section IV, to derive PAR (Sellers, 1985, 1987) and finally NPP, in some cases obtained as the difference between GPP and R_a.

D. The Issue of Spatial and Temporal Scales

As mentioned before, all models are not intended to estimate NPP at the same temporal and spatial scale. Statistical models are mainly aimed to represent climatic NPP, usually at an annual time scale. Although some models (e.g., SLAVE, HRBM) start with an annual NPP based on correlation, and then estimate seasonal NPP patterns, assuming NPP has the same seasonality as some other climatic variable (e.g., evapotranspiration). Correlative models can be used to investigate biospheric activity, assuming a climatic steady state. Apart from today's mean climate, their typical fields of application are equilibrium simulations under different climate conditions: future 2 × CO_2 climate scenarios, LGM, mid-Holocene, etc. Whenever we move from this scope, e.g., to estimate the diurnal, seasonal, or interannual changes in NPP, there is no reason to expect these models to give any sensible results. For example, under a global warming scenario, with higher surface temperature, the Miami model only suggests increased NPP in regions where the calculated NPP is controlled by temperature (high latitudes) and no change in NPP in precipitation-controlled NPP regions. Obviously the reality may be more complex—high leaf temperatures may inhibit photosynthesis, changes in water and nitrogen cycling feedback to NPP, altered phenology, and allocation of resources may also affect NPP [see McGuire *et al.* (1993), for an interesting comparison of HRBM, a Miami-derived model, with with TEM, a mechanistic model]. Also, correlative models generally do not account for cultivation, or C_3 vs. C_4 plants. Spatially, correlative models should be expected to give realistic results at relatively large scale, where local climate inhomogeneities due to elevation, orientation, etc. have been smoothed (typically to 1 × 1 degree).

The inherent limitation of diagnostic models based on remote-sensed data is that the NDVI time series extends only from 1980 to the present. A remote-sensing-based model is, by essence, unable to predict next year's NPP, or its preindustrial value. Studies of the more distant past, which are often a way to validate model sensitivity to climate change, are also impossible for remote-sensing-driven models. However, the spatial resolution of these models is theoretically as high as the remote-sensed data resolution (of the order of 1 km). Because these models are using a vegetation index data as a major input variable, they can better account for cultivation, land cover change, or deforestation. If the NDVI product was clean of atmospheric and satellite signal degradation (which with today's instruments it is not), these models would also likely provide an excellent diagnostic of interannual NPP variation.

III. Modeling Biospheric Processes

A. Processes at the Plant/Stand Level

In this section we examine the range of approaches being used for modeling the physiological components of global NPP. These components consist of net foliage carbon assimilation that we call here gross primary productivity, maintenance and growth respiration (R_a), the spatial distribution of physiological capacity within the canopy, stomatal conductance, nitrogen uptake, growth, and phenology.

1. Gross Primary Productivity Gross primary productivity is defined here as the net daytime flux of carbon produced by foliage over a given area of ground surface. It is thus the net balance of photosynthesis, photorespiration, and "dark" respiration. NPP is the balance of GPP and maintenance and growth respiration. Some models, such as the Frankfurt Biosphere Model (FBM) (Lüdeke *et al.*, 1994) and the Terrestrial Ecosystem Model (TEM) (Melillo *et al.*, 1993), use aggregated, empirical approaches to predicting GPP. In the FBM, a vegetation-type calibrated rate of GPP is modified by independent relationships with radiation, temperature, and soil water. The approach in TEM is similar, where a vegetation-type prescribed rate of maximum GPP is modified by independent relationships with radiation, temperature, a combined CO_2 and water availability function, and nitrogen availability.

In contrast, most other models predict the uptake of CO_2 using formulations ultimately based on the biochemistry of photosynthesis. The photosynthesis of plants is calculated using the biochemical formulations of Farquhar and co-workers (e.g., Farquhar *et al.*, 1980). This biochemical formulation of GPP implicitly allows for the influence of temperature, radiation, and internal leaf CO_2 concentration on photosynthesis. For C_4 plants, the biochemical approach developed by Collatz *et al.* (1992) is used. In most models, the rate of daytime foliage respiration is estimated directly from the maximum velocity of carboxylation by rubisco, which is itself proportional to temperature. However, in the model of Woodward *et al.* (1995), foliage respiration is estimated from the rate of nitrogen uptake.

2. Maintenance and Growth Respiration In all biochemical-based models, NPP is obtained by subtracting the rate of maintenance respiration in the nonfoliage at all times, the rate of maintenance respiration in the foliage at night, and growth respiration from GPP. Maintenance respiration is usually assumed to be a direct function of mass and temperature. In some models this relationship is refined to depend on sapwood mass in trees (e.g., Foley *et al.*, 1996) and nitrogen content of different tissue types (e.g., Friend *et al.*, 1997). Growth respiration is often calculated as a fixed fraction of NPP (generally taken to be 25 or 30 %).

3. Canopy Physiology The vertical distribution of leaf area and foliage physiological properties is an important component of many global NPP models. Sellers *et al.* (1992) have shown that if leaf photosynthesis decreases as photosynthetically active radiation through canopy, then canopy photosynthesis is proportionnal to leaf photosynthesis. Therefore, because leaf photosynthesis is a nonlinear function of PAR, so is canopy photosynthesis. This has the consequence that it is only necessary to make one calculation of photosynthesis on each time step (e.g., for the top of the canopy), and the result can be scaled to the whole canopy using a canopy PAR use parameter. This parameter is approximated as the time-mean fraction of incident PAR absorbed by the canopy divided by the PAR-flux extinction coefficient. This method of scaling to the canopy is used in the models of Friend *et al.* (1997), Sellers *et al.* (1992), and Woodward *et al.* (1995). Foley *et al.* (1996) and Haxeltine and Prentice (1996a) also use an optimization approach to simplify the calculation of canopy photosynthesis (Haxeltine and Prentice, 1996b). Warnant *et al.* (1994) use prescribed values of biochemical capacity for different biomes. However, the assumption of Sellers *et al.* (1992) has been proved to be at least incorrect: leaf photosynthetic rates typically decrease through canopy depth from 1 to 0.4 whereas PAR decreases from 1 to 0.05 or less in dense canopies. Also, canopy measurements with chambers or micrometeorological methods have shown that canopy photosynthesis–PAR relationships are more linear than leaf counterparts (Ruimy *et al.*, 1996). It follows that the way to take canopy photosynthesis as a linear function of PAR (as in TURC or CASA) is not correct either, but it may be closer to reality in certain cases.

4. Stomatal Conductance The effect of water supply on NPP is usually interpreted as an effect on stomatal conductance, and hence CO_2 supply to rubisco. Stomatal conductance must be known in order to estimate internal leaf CO_2 concentration and evapotranspiration; the latter is required to estimate soil water content. Most models use approaches based on the Ball–Berry model of Ball (1988) and Ball *et al.* (1987). The Ball–Berry model is an empirical equation, relating stomatal conductance to the rate of net photosynthesis, relative humidity, and ambient CO_2 concentration; it includes two empirical parameters. The model of Woodward *et al.* (1995) uses this relation, but also includes an additional multiplier to allow for the effect of soil moisture deficiency. Leuning (1995) developed the Ball–Berry approach by adding a CO_2 compensation point and replacing relative humidity by water vapor saturation deficit, resulting in the requirement for three empirical parameters. Foley *et al.* (1996) use the Leuning relation to calculate stomatal conductance, but they also allow for the effect of moisture stress by assuming a direct effect of soil moisture on photosynthesis. Haxeltine and Prentice (1996a) enable closure of the photosynthesis/stomatal conductance equation by assuming a fixed ratio between ambient and internal leaf CO_2 concentrations. This is equivalent to the Ball–Berry approach because a direct effect of humidity is not included. However, an effect of soil moisture stress is allowed for by imposing a limiting stomatal conductance function if potential evaporation exceeds moisture availability. In contrast, Friend *et al.* (1997) use the approach of Jarvis (1976), as implemented by Stewart (1988), but including an additional effect of photosynthetic capacity on maximum stomatal conductance and an effect of ambient CO_2. Again, this approach is empirical, but uses independent response functions to temperature, radiation, vapor pressure deficit, soil moisture, and CO_2.

5. Nutrient Limitation Nutrient limitation is a key control on productivity of many ecosystems (Vitousek and Howarth, 1991). Most of the physiological processes discussed in the preceding sections capture how water, temperature, and light influence photosynthesis and respiration. In the short-term, nutrient limitation can be thought of in similar terms, applying empirical (Woodward *et al.*, 1995) or Michaelis–Menten type kinetics. These approaches either focus on concentrations of ammonium and nitrate in soil solutions or assume a static nutrient-supplying capacity of the soil. Such approaches can work on short time scales (<years) or under steady-state conditions. On longer time scales, nutrient availability results of interactions between organic matter dynamics—controlling the sequestration of nutrients into inert organic matter and input and losses of nutrients (Parton *et al.*, 1993; Pastor and Post, 1986; Vitousek and Howarth, 1991; Holland *et al.*, 1997). Changes to NPP on decadal–millennial time scales are greatly influenced by nutrient cycling and may dominate long-term trends in NPP (Vitousek *et al.*, 1993). Because of the demonstrated importance of nutrients on successional to longer time scales (Pastor and Post, 1986; Vitousek *et al.*, 1993), many global models of NPP also include the biogeochemical cycles of at least one limiting nutrient (Schimel *et al.*, 1996, 1997a,b; VEMAP, 1995). Representing nutrient limitation may be done via empirical relationships of nutrient supply to soil properties (Woodward *et al.*, 1995) or by explicit simulation of biogeochemistry (Schimel *et al.*, 1997a; Friend *et al.*, 1997). In the CENTURY model, soil reaches an equilibrium with its final C and N content, as a function of climate and soil texture. In other models, the C:N ratios are prescribed. The inclusion of nutrients can fundamentally alter the response of models to global change. In carbon–water models, potential responses to

CO_2 may be large (20–40% NPP enhancement at doubled CO_2), but warmer temperatures may reduce this effect by increasing respiration. In carbon–water–nutrient models this sensitivity is altered (Schimel *et al.*, 1997a). Nutrients reduce the effect of CO_2 alone, but, because warming increases nutrient cycling rates, the effects of CO_2 plus climate change on NPP are larger than the effects of CO_2 alone. Since the inclusion of nutrients in a model can change the sign of an interaction (from negative to positive), evaluating the effects of nutrient vs. biophysical constraints on NPP is crucial.

6. Growth and Phenology The partitioning of net carbon gain to different plant parts (e.g., foliage, stem, and roots) is calculated using empirical coefficients in many global NPP models (e.g., Foley *et al.*, 1996; Lüdeke *et al.*, 1994; Post *et al.*, 1997). In some models, allometric relationships are used to constrain the relative biomasses of the different parts. Other models apply a more sophisticated approach to calculating allocation to foliage (Woodward *et al.*, 1995; Haxeltine and Prentice, 1996a; Lüdeke *et al.*, 1994; Friend *et al.*, 1997), using optimization criteria to predict leaf area index. In the model of Woodward *et al.* (1995), leaf area index is limited by long-term carbon and water constraints. Models can also adopt a more functional approach, where allocation pattern results from evolved responses that partition carbon investments in order to maximize acquisition of the most limiting resources. Root allocation would be invested when water or nutrient resource is limiting growth, whereas aboveground allocation (stem and leaf) would result from light limitations (Friedlingstein *et al.*, 1999).

The carbon constraint takes the form of restricting leaf biomass to that which can be maintained by NPP, can be synthesized from net assimilation, and does not result in a net carbon loss from the foliage at the bottom of the canopy. Once the maximum LAI is found that satisfies these constraints, then leaf area can be further restricted to ensure that moisture loss does not exceed precipitation. In the model of Lüdeke *et al.* (1994), phenological phases are used to define different allocation strategies to leaf mass with time. Leaf mass is maximized within the constraint that sufficient carbon must be allocated to nonleaf mass for support. The model of Friend *et al.* (1997) allocates carbon to foliage using allometric relationships between foliage and structural, fine root, and storage tissues. An additional limitation is imposed by reducing leaf area if the annual carbon balance at the base of the canopy is negative.

The phenological status of cold deciduous trees is usually modeled empirically using critical temperature thresholds (Foley *et al.*, 1996; Haxeltine and Prentice, 1996a) or accumulated growing degree days (Friend *et al.*, 1997). Dry deciduous trees in the model of Foley *et al.* (1996) lose their leaves during the two least productive months of the year. In the model of

Haxeltine and Prentice (1996a), it is assumed that the leaves of dry deciduous trees are not displayed when the soil moisture content in the rooting zone falls below 20% of the available water-holding capacity.

B. Multiple Resource Limitation of NPP: A Synthesis

It is evident that models embodying fundamentally different assumptions have predicted, with significant though varying degrees of success, global patterns of NPP. An example of the global patterns of model estimated NPP is given in Fig. 19-2. Some models emphasize control by radiation (Uchijima and Seino, 1985), by temperature and moisture (Lieth, 1975; Hunt *et al.*, 1996; Bonan, 1993; Sellers *et al.*, 1996a,b), by nutrients (Melillo *et al.*, 1993), and by multiple factors (Schimel *et al.*, 1997a). Empirical evidence reveals that although global patterns of NPP are indeed well correlated with biophysical limitations (Uchijima and Seino, 1985; Lieth, 1975), many if not most ecosystems respond to added nutrients by increasing NPP (see Section III,A,5).

Ecosystem models suggest a hypothesis to explain why biophysical (water and energy) and nutrient limitations might emerge as dominating in different types of studies. In the VEMAP intercomparison (VEMAP, 1995), all of the biogeochemical models produced spatial correlation between NPP and evapotranspiration, as expected from the hypothesis of biophysical dominance (Schimel *et al.*, 1997a). They also showed strong correlation between NPP and nitrogen availability, consistent with nutrient availability being a key control (Fig. 19-3). However, they also showed strong correlation between evapotranspiration (ET) and nitrogen availability, a correlation not expected on mechanistic grounds (ET is at best indirectly related to nitrogen turnover). In further analyses using CENTURY, which simulates the inputs and losses of nitrogen, it became apparent that, in CENTURY, the NPP–ET–nitrogen correlation emerges because water and energy are key controls over nitrogen inputs and losses (through wet deposition, nitrogen fixation, leaching, and trace gas losses). As water and energy availabilities increase, both carbon and nutrient fluxes increase. As organic matter accumulates the amount of N that is stored in organic forms and recycled increases, supporting higher NPP. Thus, on the time scales of ecosystem development, N availability and biophysical limitation should be correlated, a hypothesis consistent with the apparently conflicting observations supporting both biophysical and nutrient dominance (Schimel *et al.*, 1997a,b). This correlation applies on time and space scales over which ecosystem N stocks change substantially. Although research on patterns of NPP (e.g., this volume), nutrient inputs (Galloway *et al.*, 1995; Hedin *et al.*, 1995), and nutrient losses (Andreae and Schimel, 1989) goes on, this hypothesis implies that these processes must be studied together in order to understand large-scale, long-term ecosystem dynamics.

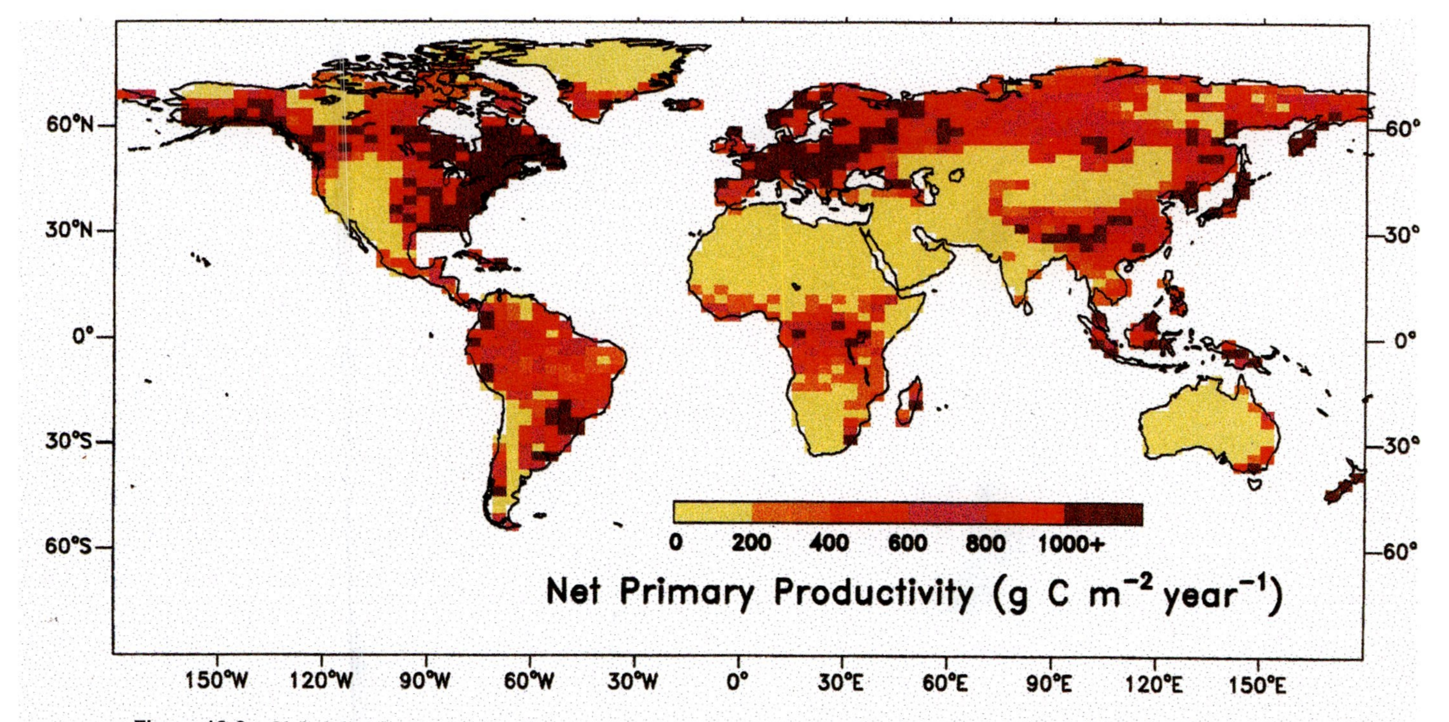

Figure 19-2 Global distribution of the net primary productivity as calculated by the global model HYBRID 3.0. (Friend *et al.*, 1997).

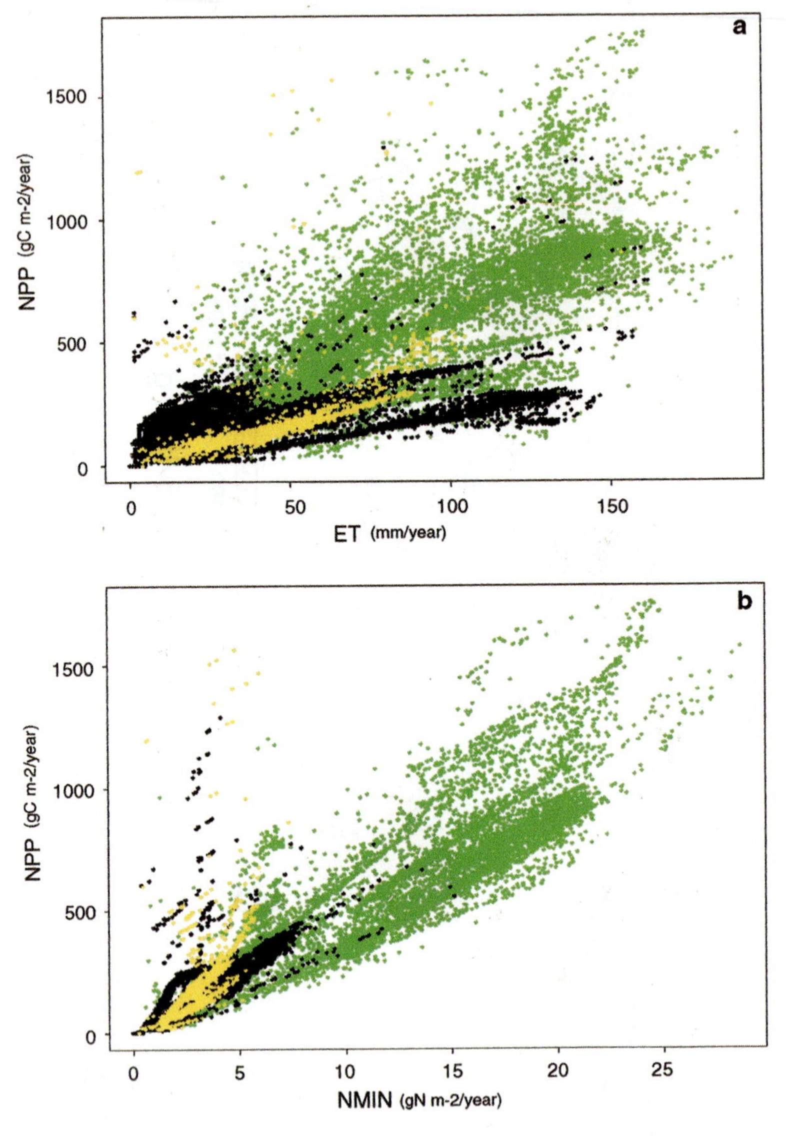

Figure 19-3 **(a)** Simulated NPP vs. ET from a global integration of the Century model. The results are from a global equilibrium simulation under preindustrial conditions and for natural vegetation only. Each point represents a 0.5° × 0.5° grid cell. Green points are forest ecosystems, yellow indicates grasslands, and black points are mixed ecosystem types, which contain both woody and herbaceous vegetation (savannas and woodlands). **(b)** Simulated NPP vs. nitrogen availability (N mineralization) from the same integration as in a. Taken together, the results suggest a tendancy for water/energy and nutrients to become colimiting under steady-state conditions (Schimel *et al.*, 1996).

Additionally, disturbances, especially novel disturbances not found in a particular ecosystem's history, can also change or eliminate these correlations. Although this hypothesis requires careful testing using observational and experimental data, it provides a framework for assessing the importance of different types of limitations of NPP. Models intended for different applications on different time and space scales will necessarily emphasize different aspects of the regulation of NPP, and modeling decisions can be made based on consistent theory instead of arguments arising from different historical approaches to the NPP problem.

IV. Indirect Constraints on NPP from Remote-Sensed and Atmospheric Observations

A. Remote Sensing

The basis for remote sensing of vegetation is the sharp contrast in reflectivity between visible and the near-infrared caused by the optical properties of the chlorophyll molecule. By contrast, nonvegetated surfaces have relatively constant reflectivity across this wavelength region. The NDVI is a normalized difference of the red and near-infrared reflectances, and it is a proxy for vegetation amount. The advantages of the NDVI are that it is simple, provides some resistance to atmospheric and other artifacts, and has been shown in many empirical studies to be quantitatively related to leaf area and the interception of photosynthetically active radiation. The NDVI has many limitations. The only global time series is from the series of NOAA/AVHRR instruments, designed as meteorological instruments. As a consequence, the AVHRR uses a clumsy spatial resampling scheme to reduce data volume, and is not calibrated for long-term stability. In addition, many of the AVHRR instruments were flown for much longer than their design lifetimes, leading to drift in their overpass times. As a consequence, the AVHRR NDVI record is very difficult to use for time series analyses; although considerable effort has been put into corrections, substantial uncertainty remains (Myneni *et al.*, 1997; Braswell *et al.*, 1997). The NDVI does provide considerable information on spatial patterns of vegetation, and on the seasonal cycle of vegetation growth and senescence.

The NDVI has been used in a number of analyses of NPP and the terrestrial carbon balance, following on the pioneering work of Fung *et al.* (1987), and is now used as a key input in many TBMs, as well as in direct analyses of processes influencing ecosystems (Hunt *et al.*, 1996; Malmstrom *et al.*, 1997; Braswell *et al.*, 1997; Myneni *et al.*, 1997). Although there are problems with the NDVI's precision and stability there is no doubt that it contains significant information about the seasonal and spatial evolution of ecosystems, and likely that it contains significant, though less precise, information on interannual time scales. Models incorporating information from comprehensive global observations have substantial advantages for "diagnostic" studies—studies trying to understand the mechanisms underlying an observed pheomenon (Malmstrom *et al.*, 1997). As satellite measurements of ecosystems improve, with both improved instruments (VEGETATION, POLDER, MODIS, MISR) and improved algorithms (Braswell *et al.*, 1996; Privette *et al.*, 1996), models incorporating satellite information should become more accurate and precise (Running, 1994). The distributions of natural vegetation are in principle predictable from climate, disturbance regime, and biotic interactions, but the distributions and characteristics of managed vegetation (crops and forest plantations) reflect human decision making. The ability of satellite measurements to map natural and managed vegetation is a critical contribution of remote sensing to monitoring ecosystem processes and NPP.

B. Atmospheric Measurements

Atmospheric CO_2 brings some information on the biospheric carbon fluxes. Given the limited number of CO_2 measurement sites in the worldwide effort undertaken to monitor "baseline" atmospheric changes and given the fact that the atmospheric transport mixes rapidly CO_2 emitted over different ecosystems, the atmospheric diagnostic is a top-down coarse scale method to estimate the terrestrial fluxes at global and regional scales, (regional being not finer than a continent). Although NPP cannot be inferred as a separate quantity, net ecosystem productivity (NEP = NPP + R_h) is constrained by atmospheric measurements of CO_2, $^{13}C/^{12}C$ in CO_2, and O_2/N_2 in air. Additionally, GPP could be inferred from the global distribution of the $^{18}O/^{16}O$ ratio in CO_2. Table 19-1 summarizes which carbon fluxes are constrained by each tracer. Estimating accurately the net ecosystem productivity is crucial because only half of the anthropogenic CO_2 emissions (1.5 ppm yr^{-1} or 3 Gt of carbon on average) ends up accumulating into the atmosphere, increasing the Earth's radiative budget. The other half is being stored in the world's oceans and land ecosystems, but large uncertainties remain regarding which reservoir is the most important carbon storage. A predominant ocean uptake would represent a much safer case for humanity because once an excess of anthropogenic CO_2 is dissolved into sea water, it is not likely to return to the atmosphere anytime soon and a large fraction of it will remain sequestered in the deep sea. Conversely, a carbon storage into the land biosphere is more transient in nature because the turnover of carbon in most ecosystems hardly exceeds a few decades and because the land biosphere is increasingly disturbed by human activity.

*1. **Atmospheric Transport Models*** Atmospheric transport models (TMs) solve the continuity equation over a three-dimensional grid given a set of

Table 19-1 Atmospheric Tracers Related to Photosynthesis and Respiration[a]

Tracer	Biospheric flux	Constraint
CO_2	Seasonal cycle	Seasonal NPP + R_h
	Latitudinal gradient	NPP + R_h + ocean
	Interannual variations	NPP + R_h + ocean
$^{13}C/^{12}C$	Seasonal cycle	Seasonal NPP + R_h
	Latitudinal gradient	NPP + R_h
	Interannual variations	Year-to-year anomalous changes in NPP + R_h
O_2/N_2	Seasonal cycle	Seasonal NPP + R_h + seasonal ocean biotic fluxes
	Latitudinal gradient	NPP + R_h
	Interannual variations	Year-to-year changes in NPP + R_h
$^{18}O/^{16}O$	Seasonal cycle	GPP
	Latitudinal gradient	GPP + R_a + R_h
	Interannual variations	Not studied

[a]The second column indicates which biospheric flux can be inferred from the measurements of each tracer. The term "ocean" is net air–sea exchange CO_2 flux. In all cases, the sum NPP + R_h includes the effect of possible disturbances induced by land use, fires, insects, etc., and can be termed as NBP (net biome productivity). Fossil fuels emissions are supposed to be the best known fluxes and their effect on the atmospheric concentration fields can thus be separated *a priori* from the least known biospheric fluxes.

prescribed surface CO_2 fluxes and deliver as an output a simulation of the global CO_2 concentration field. The global wind fields used to parameterize the atmospheric transport originate from climate models or from weather forecast models, the latter offering the advantage to simulate synoptic-scale atmospheric CO_2 transport with the same timing as those observed in the real world. Transport models are far from perfect in the way they describe the atmosphere dynamics, especially in the lowest troposphere, which is critical for matching their predictions with surface CO_2 concentration measurements. An additional source of uncertainty is the parameterization of convective mixing, which can differ strongly among models. Measurements of "inert" tracers emitted by sources that are relatively well known improve our present understanding of the large-scale atmospheric mixing processes and help to design more realistic transport models. For instance, ^{222}Rn, with a radioactive decay half-life of 3.8 days, is adapted to constrain vertical mixing, whereas ^{85}Kr and SF_6 provide insights on the meridional transport intensity. A first intercomparison exercise of 10 different TMs for CO_2 has yet shown substantial differences among models (Rayner and Law, 1995).

2. Global CO_2 Measurements Network Long-term measurements of atmospheric CO_2 at many locations around the globe constrain the patterns of the surface CO_2 fluxes. Even far away from the source areas, and despite vigorous atmospheric mixing, small but persistent gradients in atmospheric CO_2 among different measurement sites pertain to surface fluxes. At present, the Scripps, NOAA/CMDL, CSIRO/GASLAB, NIEE, and a few European laboratories operate global air sampling programs comprising a total of roughly 60 "clean air" sites around the world, which document the worldwide geographical distribution and rate of increase of atmospheric CO_2 (Keeling *et al.*, 1989a; Conway *et al.*, 1994; Francey, 1996). Most of these data are compiled into free-access databases (World Meteorological Organization, 1996).

The global atmospheric CO_2 network essentially consists of remote ocean sites that sample "baseline" air masses given the fact that (1) the air-to-sea flux of CO_2 in the vicinity of such stations is generally of smaller magnitude than the fluxes exchanged over productive land ecosystems, (2) over the oceans, the boundary layer thickness varies less than over the continents, the continental planetary boundary layer (PBL) dynamics being furthermore coupled with the diurnal and seasonal surface fluxes, and (3) *in situ* sampling strategies aim at screening out local sources by taking air from the "ocean sector" under windy conditions. An achievable target for common measurement precision between all different groups is 0.1 ppm. Regular intercomparisons coordinated by the WMO/GAW program have shown that background CO_2 measurements on the same samples from two separate laboratories are unlikely to differ by more than 0.5 ppm. Yet, 0.5 ppm is enough to correspond to significant differences in the inferred CO_2 fluxes (a global atmospheric increase of that amount is equivalent to a source of 1 Gt.

The time–latitude variations of atmospheric CO_2 inferred from flask data of the NOAA/CMDL global air sampling network are shown in Fig. 19-4. It is observed that the peak-to-peak amplitude of the seasonal cycle of atmospheric CO_2 increases from zero around the equator, to approximately 15 ppm at high northern latitudes. The southern hemisphere seasonal cycle is opposite in phase with the northern hemisphere and does not exceed 1–2 ppm in amplitude. Figure 19-4 also indicates an average positive north minus south meridional CO_2 gradient, primarily reflecting the accumulation of fossil CO_2 over the northern midlatitude continents. Finally, the mean growth rate of CO_2 exhibits large year-to-year variations.

3. The Seasonal Cycle of Atmospheric CO_2 The seasonal cycle of atmospheric CO_2 provides a constraint on photosynthesis and respiratory processes. The distribution of NPP + R_h as calculated *a priori* by TBMs is prescribed into a TM to generate a model CO_2 concentration field, which is testable against the observed CO_2 seasonal cycle at various stations (Heimann *et al.*, 1998). The air–sea exchange and the fossil fuel source of CO_2 contribute very little to the CO_2 seasonality in the northern hemisphere (Randerson *et al.*, 1997), but can be significant at high southern latitudes,

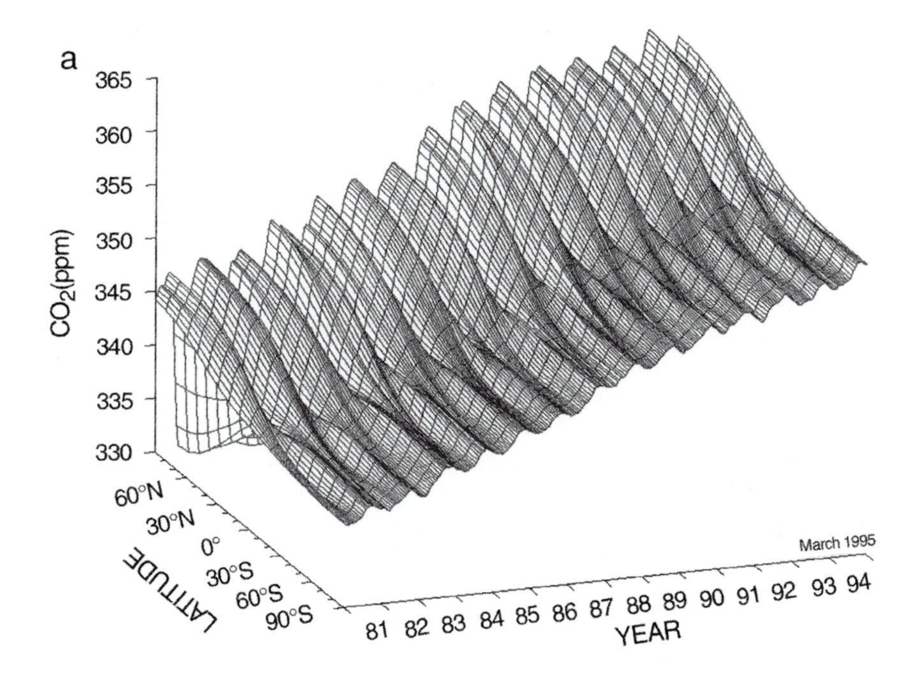

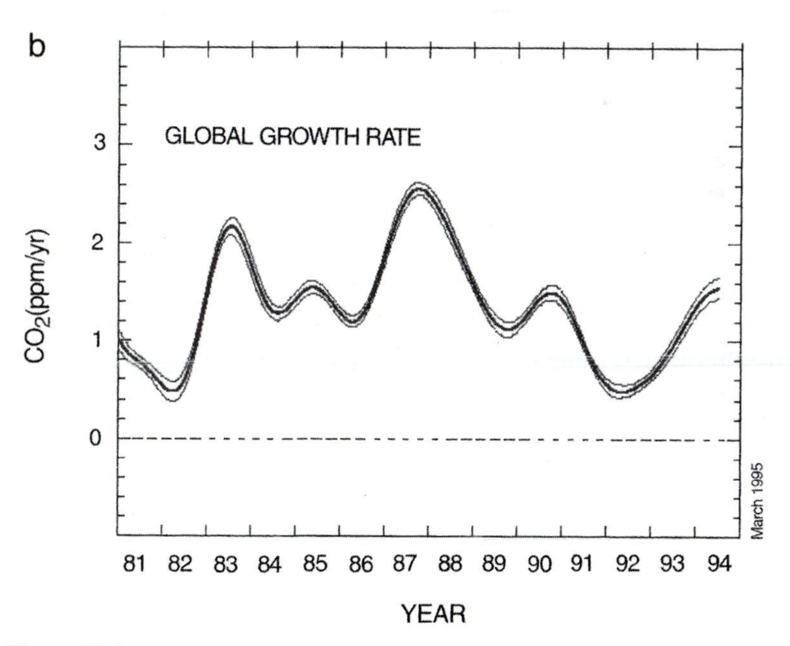

Figure 19-4 (**a**) Atmospheric CO_2 measurements from the NOAA/CMDL global air sampling network gridded as a function of latitude and time. (**b**) Carbon dioxide growth rate variations since 1982 as a function of latitude and time. Data courtesy of NOAA/CMDL Carbon Cycle Group.

where the seasonal peak-to-peak amplitude of CO_2 is nevertheless very small (Erickson *et al.*, 1996; Monfray *et al.*, 1996).

4. Zonally Averaged CO_2 Gradient The relatively well known (±10%) fossil CO_2 emissions are distributed mainly over the northern industrialized regions, and produce a positive interhemispheric CO_2 gradient in the atmosphere. The exact value of this latitudinal "fossil" gradient depends naturally on which TM is used to predict it, but most studies agree on the fact that it is on the order of 6 ppm, i.e., two times higher than the observed one. Closing the CO_2 budget thus requires a "missing" sink in the oceans or the biosphere that would reabsorb roughly half of the fossil CO_2 north of the equator. Based on that rationale, Tans *et al.* (1990), relying on a global estimate of the air–sea flux based on oceanographic measurements, infer a complementary boreal biospheric sink of 2.2 Gt yr^{-1} north of 30°N in order to match the observed latitude CO_2 gradient.

In addition to the meridional CO_2 gradient, the global CO_2 data set indicates significant differences in longitude, with generally sharp constrasts between oceans and continents, despite poor coverage of the interior of the continents. Three-dimensionnal inverse techniques can thus be applied to the atmospheric data to infer the spatial and seasonal patterns of the CO_2 fluxes and the pertaining uncertainties (Enting *et al.*, 1995).

5. $^{13}C/^{12}C$ Ratios in CO_2 The ^{13}C isotope composition of CO_2 (usually expressed as $\delta^{13}C$) traces the partitioning of the carbon sources and sinks between the oceans and the terrestrial biosphere. Photosynthesis strongly discriminates the heavier isotope ^{13}C against the lighter ^{12}C species, whereas air–sea exchange negligibly alters the $^{13}C/^{12}C$ isotope ratio of CO_2 in the air. Consequently, photosynthesis increases the $\delta^{13}C$ value of atmospheric CO_2 whereas respiration decreases $\delta^{13}C$ by simply diluting CO_2 issued from plant material depleted in ^{13}C into the ambient atmosphere. Ultraprecise $\delta^{13}C$ measurements provide a fingerprint of the biospheric net fluxes, the sum of NPP and R_h. A biotic uptake of 1 Gt C on land will shift up the global atmospheric $\delta^{13}C$ value by 0.05‰, compared to the experimental precision on isotopic measurements, which is at best of 0.01‰.

Keeling *et al.* (1989b) analyzed the CO_2 latitudinal gradient, together with the $\delta^{13}C$ latitudinal gradient inferred from six stations in the Pacific ocean, and concluded that the terrestrial ecosystems north of 30°N could not absorb more than 0.6 Gt C yr^{-1}, a finding opposite to the finding of Tans *et al.* (1990). Since the early 1990s, $\delta^{13}C$ measurements progressively have been expanded up to about 40 sites of the NOAA/CMDL global network, allowing to use isotopes as a more powerful constraint on the latitude and time changes of the biotic fluxes. Data collected in 1992–1993, during a pronounced northern hemisphere cooling induced by the Pinatubo eruption, suggest that a biotic uptake of CO_2 of 3.5 Gt C yr^{-1} established at that time at northern midlatitudes (Ciais *et al.*, 1995) which resumed in 1994 (P. P. Tans, personal communication). Whether this strong and apparently short-lived biotic uptake of CO_2 reflects in an increase in NPP or a decrease in R_h cannot be told from the $\delta^{13}C$ measurements.

4. O_2/N_2 Abundance in the Air Further development of experimental techniques has enabled measurements with an accuracy better than 0.01‰ (10 per Meg) of the O_2/N_2 abundance in air (Keeling, 1995). Atmospheric O_2 is exchanged stoechiometrically with CO_2 by photosynthesis and respiration and thus has biospheric fluxes opposite in sign to the carbon fluxes. This also holds for marine biology, but unlike CO_2, which is converted to carbonate and bicarbonate ions once dissolved in the sea, O_2 produced by photosynthesis and consumed by respiration of the marine biota is directly exchanged with the atmosphere, thus giving rise to a large seasonal cycle in the atmosphere over the most productive ocean gyres (e.g., Southern Ocean) (Bender *et al.*, 1996). Yet, given the fact that long-term incremental changes in the size of the marine biota reservoir are unlikely to occur, the O_2/N_2 ratios in air separate the net ocean vs. terrestrial carbon uptake. Keeling *et al.* (1996) infer from O_2/N_2 measurements during 1991–1994 a (large) terrestrial uptake of 2 ± 0.9 Gt yr^{-1} north of the equator, which confirms results obtained by the $\delta^{13}C$ method.

5. $^{18}O/^{16}O$ Ratios in CO_2 The $^{18}O/^{16}O$ ratio in atmospheric CO_2 offers the attractive potential to infer separately the gross photosynthesis (GPP) and respiration ($R_a + R_h$) fluxes. The process of isotopic exchange between CO_2 and H_2O, well known in laboratory studies, has two applications in ecosystems: CO_2 exchanges ^{18}O with two isotopically distinct water pools in the biosphere, the evaporating leaf water and the soil moisture. Water in leaves is isotopically enriched with respect to soil moisture during daytime photosynthesis due to fractionation of water isotopes during evapotranspiration (Craig and Gordon, 1965; Förstel *et al.*, 1975; Yakir *et al.*, 1994). The exchange between CO_2 and leaf water concerns CO_2, which enters the chloroplasts and is retrodiffused back to the atmosphere without being reduced by photosynthesis. This retrodiffused CO_2 flux, a function of the ratio C_i/C_a, is roughly twice the GPP for C_3 plants (Farquhar *et al.*, 1993). On the other hand, CO_2 emitted belowground by roots and decomposers reacts isotopically with surface soil moisture, which has an average $\delta^{18}O$ close to the one of meteoric water, i.e., lower than the one of leaf water. Thus, photosynthetic CO_2 has an isotopic label distinct from that of soil-respired CO_2, allowing separation of both processes.

By combining the output of TBMs for the GPP and the global distribution of $\delta^{18}O$ in meteoric water (Rozanski *et al.*, 1993), it is possible to map the global biospheric source of $\delta^{18}O$, which can then be prescribed into TMs and subsequently compared to the atmospheric $\delta^{18}O$ observations. Regular measurements of $\delta^{18}O$ at several monitoring sites have been carried on concomitantly to $\delta^{13}C$, but $\delta^{18}O$ data from "wet sites" (i.e., tropical sites) are not reliable because of isotopic reaction between CO_2 and water within the sampling container (Gemery *et al.*, 1996). Yet "dry site" $\delta^{18}O$ observations bring up two particularly interesting indirect constraints to the biotic fluxes: (1) a large permanent minimum of atmospheric $\delta^{18}O$ at high northern latitudes and (2) a pronounced seasonal cycle over the northern hemisphere, which lags behind the atmospheric CO_2 seasonal cycle by about 2 months and has a minimum in October. Global atmospheric simulations of $\delta^{18}O$ in CO_2 demonstrate that respiration exerts a dominant control on the atmospheric $\delta^{18}O$ signal (Ciais *et al.*, 1997a,b). Since the annual mean integral of total respiration (R_a + R_h) equals the GPP mean within a few percent, there is hopes for the further use of $\delta^{18}O$ to constrain the gross CO_2 fluxes on land.

C. Linking the NDVI and Atmospheric Records to Year-to-Year and Secular Changes in NPP

1. CO_2 Changes The time series of atmospheric CO_2 and of NDVI spanning the past 15 years indicate short-term (interannual) changes (1) in CO_2 abundance, with maxima that are roughly in phase with the global temperature variations and with El Niño occurrences (Gaudry *et al.*, 1990; Keeling *et al.*, 1989a) and (2) in the CO_2 seasonal amplitudes, whose maxima lag the temperature maxima by about 2 yr (Keeling *et al.*, 1996). Additionally, there is a long-term mean increase in the seasonal cycle amplitude over the past 30 yr (Keeling *et al.*, 1996; Randerson *et al.*, 1997), consistent with an increase in the NDVI amplitude measured by the NOAA/AVHRR sensors since 1981 (recently reported) (Miyeni *et al.*, 1997).

2. $\delta^{13}C$ Changes It is tempting to use year-to-year fluctuations in atmospheric $\delta^{13}C$ to infer changes in terrestrial carbon fluxes. But the only two long $\delta^{13}C$ records covering the past decade show contradictory results (Francey *et al.*, 1995; Keeling *et al.*, 1995). First, it is important to point out that one cannot infer the absolute value of the net ecosystem production from $\delta^{13}C$ measurements, because of "disequilibrium" effects, which induce a systematic (although poorly known) correction to the $\delta^{13}C$ trend (Tans *et al.*, 1993). Yet, relative changes in NEP should be inferrable from very accurate $\delta^{13}C$ time series (Fig. 19-5).

3. Interannual CO_2 Fluctuations Figure 19-4 indicates large positive and negative anomalies of the CO_2 growth rate around a mean value of about 1.5 ppm yr^{-1}. Such changes can have an oceanic or a biospheric origin, and the mechanisms that cause them still remain to be better understood. Yet, it seems that the ocean alone cannot account for the magnitude of the observed CO_2 fluctuations (Winguth *et al.*, 1994), suggesting that the land biosphere plays a major role in the year-to-year CO_2 variability. When the

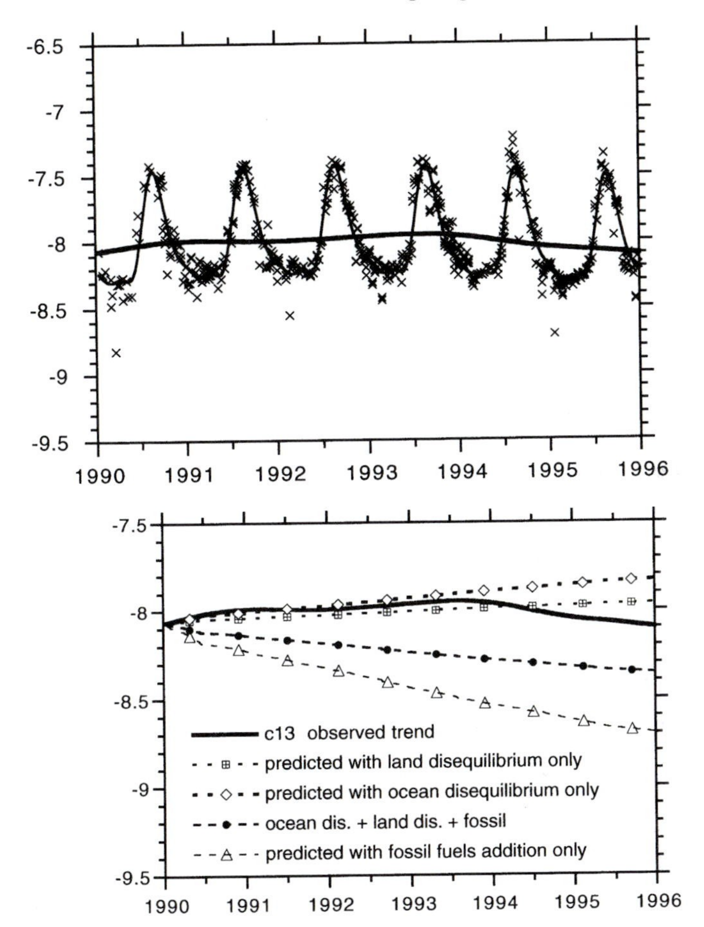

Figure 19-5 Illustration of a "double deconvolution" based on ^{13}C measurements. **(a)** $\delta^{13}C$ measurements on flask air samples at Point Barrow, Alaska (74°N) made by NOAA/CMDL and the University of Colorado/INSTAAR. Crosses, Flask samples; thin line, four harmonics fit to the seasonal cycle; thick line, average long-term trend in $\delta^{13}C$. **(b)** The observed $\delta^{13}C$ trend is then compared to simple model predictions that treat the atmosphere as one single compartment. Annual fossil CO_2 emissions would produce a more negative trend than observed, whereas the "disequilibrium terms" produce positive annual $\delta^{13}C$ shifts. Any positive (or negative) difference between the slope of the observed trend and the slope of the predicted trend with fossil and disequilibria only translates into a carbon uptake (or release) by the land biota. More complex models use that principle but also contain a description of the atmospheric transport.

records of temperature and NDVI are detrended, significant interannual correlations appear, as apparent in Fig. 19-6, with immediate positive global correlation of the CO_2 growth rate with temperature, followed by lagged (1–3 yr) anticorrelation (Braswell *et al.*, 1997). NDVI in northern ecosystems

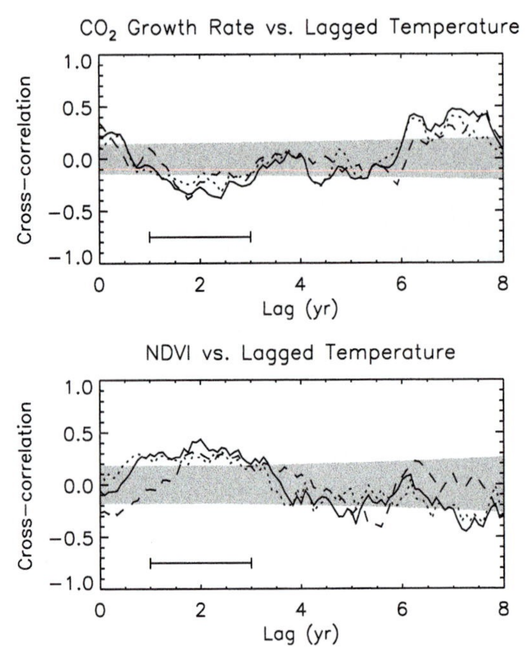

Figure 19-6 Cross-correlation function between lagged temperature and CO_2 growth rate (upper panel) and NDVI (lower panel). The solid line represents the average area-weighted values, the dashed line the southern hemisphere average, and the dotted line the northern hemisphere average. The correlations are statistically significant outside the shaded area. After Braswell *et al.*, (1997).

increases in warm years, whereas arid tropical NDVI declines in warm years. Significant lagged effects in NDVI also appear often as a correlation of the opposite sign to the immediate correlation (Braswell *et al.*, 1997).

During the 1980s, the observed CO_2 and $\delta^{13}C$ fluctuations were used to separate the global oceanic and terrestrial fluxes, with conflicting results (Francey *et al.*, 1995; Keeling *et al.*, 1995). Keeling's data (composite of measurements at four or five sites) show that El Niño events in 1983 and 1988 are associated with a large CO_2 release from the biosphere. In contrast, Francey's data (mostly from a very carefully measured Tasmanian site) conclude no significant year-to-year anomaly in the terrestrial CO_2 fluxes. This discrepancy directly stems from differences in the two $\delta^{13}C$ time series, whereas the isotopic "deconvolution" methods used in both studies are essentially similar. In the early 1990s, both data sets are in agreement with an enhanced terrestrial global carbon uptake. Furthermore, this anomalous land uptake has essentially occurred at middle to high northern latitudes (Ciais *et al.*, 1995; Keeling *et al.*, 1996). This finding, based on $\delta^{13}C$ and independently confirmed by the O_2/N_2 approach, likely relates to the cooling induced by the Pinatubo eruption in June, 1991. However, there is no clear understanding of the terrestrial processes that yielded a brutal augmentation of the NEP

If the biosphere is really driving the quasi-decadal CO_2 changes, the co-variance of temperature with the CO_2 abundance anomalies implies that the CO_2 growth rate, and hence the NEP, are lagging the temperature by roughly 2 yr (Braswell *et al.*, 1997). Additionally, the years with stronger NEP generally coincide with an enhancement of the seasonal cycle amplitude, arising from both an increase in respiration (evident from the positive CO_2 anomaly) and an increase in northern hemisphere NPP (evident from the increase in NDVI there) (Braswell *et al.*, 1997). Keeling *et al.* (1996) further suggest that at this time, NPP does probably exceed R_h. Providing realistic simulations of the CO_2 interannual variability is a very challenging topic for TBMs, although the lags between climate and CO_2 outlined above suggest that TBMs suitable for that exercise should include processes that operate on multiple time scales (allocation, soil carbon, nutrients dynamics effects on NPP, etc.). A study using the FBM model suggests that changes in NPP anomalies rather than in R_h have been causing the CO_2 variations observed during the past 15 years, the most important source of variability being tropical forest ecosystems (Kindermann *et al.*, 1996). Results from CENTURY and Braswell's study contradict this (Schimel *et al.*, 1996).

4. Decadal CO_2 Changes and Long-Term Average Terrestrial Uptake The classic approach that is followed by the IPCC group (IPCC, 1995) is to infer the role of the global biosphere in the global CO_2 budget by the difference between the known anthropic sources and the ocean uptake as estimated by ocean models. This yields a mean terrestrial net sink of 1.5 ± 1.5 Gt yr^{-1}. In parallel, Battle *et al.* (1996) analyzed O_2/N_2 in air preserved in the Antarctic firn dating from the 1980s and estimate a globally "neutral" biosphere, implying that the overall terrestrial carbon fluxes are on average close to zero during the last decade. Taking a commonly admitted land use source of 1.6 ± 1 Gt yr^{-1} (Houghton *et al.*, 1987) would then yield a residual carbon uptake on land of about 1.6 Gt yr^{-1} during the 1980s, a value very close to that of the IPCC.

Keeling *et al.* (1996), using their longest CO_2 record of Mauna-Loa (Hawaii) and Point Barrow (Alaska), have shown that the amplitude of the CO_2 seasonal cycle has been augmenting by roughly 20% over the past 30 yr, while the timing of the CO_2 draw-down of CO_2 in spring and early summer has advanced by 6–7 days. This finding echoes the satellite observations of an increase by approximately 10% in the growing season value of the NDVI index north of 35°N since NOAA/AVHRR measurements became available in 1981, which tentatively suggests enhanced biospheric activity in the boreal hemisphere. Randerson *et al.* (1997) further support this idea. They show that one could only satisfy *both* the amplitude constraint *and* a global terrestrial sink compatible with the IPCC value if most of the land uptake occurs north of the tropics.

Although the uncertainties on the long-term mean terrestrial carbon fluxes and on their location remain quite large, there is increasing evidence of a carbon storage in land ecosystems over the past decade, with NPP exceeding on average R_h based on the NDVI trends. A possible explanation is that climate change has favored plant growth by increasing the length of the growing season at middle and high northern latitudes (Myneni *et al.*, 1997). These findings put some important constraints on TBM simulations of the decadal behavior of the land CO_2 fluxes. First, one must develop a better understanding of the links between the NDVI-inferred 10% increase in photosynthesis over the past decade, and (1) the recent warming, which is especially pronounced at high latitudes during the spring time, (2) the ongoing CO_2 rise at 0.4% per year, and (3) the role of nutrients in triggering a plant's response to CO_2 and climate. Second, one must better quantify the factors that control the strength of R_h and its phasing with respect to NPP.

V. Concluding Remarks

The global integration of NPP is one of the major achievements of carbon cycle research over the past years. Models of various types, levels of complexity, and comprehensiveness have been developed with the aim of predicting not only NPP, but also other carbon and sometimes water fluxes exchanged between the biosphere and the atmosphere. In the context of understanding the future behavior of terrestrial productivity during the ongoing anthropic perturbation of the carbon and nutrient cycles, it is fundamental to rely on models that are validated against independent global data sets. With that respect, satellite measurements of NDVI offer an unsurpassed tool to monitor the seasonal, interannual, and decadal variability of the vegetation. It is reasonable to believe that longer time series of NDVI, as well as better sensors, will refine the constraints brought on by NDVI. Additionally, measurements of atmospheric CO_2 and related tracers provide indirect estimates of the terrestrial carbon fluxes. At present, inversions of the atmospheric measurements are performant to infer the CO_2 budget as a function of latitude, and to a certain extent at the continental level. Improving the CO_2 monitoring over the interior of the continents (using aircraft vertical profiles) will help to better estimate the spatial patterns of the biospheric fluxes.

References

Ajtay, G., Ketner, P., and Duvigneaud, P. (1979). Terrestrial primary production and phytomass. *In* "The Global Carbon Cycle," SCOPE 13 (B. Bolin, E. Degens, S. Kempe, and P. Ketner, eds.), pp. 129–181. John Wiley and Sons, Chichester.

Andreae, M. O., and Schimel D. S., eds. (1989). "Exchange of Trace Gases between Terrestrial Ecosystems and the Atmosphere." John Wiley and Sons, New York.

Ball, J. T. (1988). An analysis of stomatal conductance. Ph.D. Thesis, Stanford University, Stanford, California.

Ball, J. T., Woodrow, I. E., and Berry, J. A. (1987). A model predicting stomatal conductance and its contribution to the control of photosynthesis under different environmental conditions. *In* "Progress in Photosynthesis Research," Vol. 4, pp. 221–224 (J. Biggins, ed.). Nijhoff, Dordrecht.

Battle, M., Bender, M, Sowers, T., Tans, P. P., Butler, J. H., Elkins, J. W., Ellis, J. T., Conway, T., Zhang, N., Lang, P. M., and Clarke, A. D. (1996). Atmospheric gas concentrations over the past century measured in air from firn at the South Pole. *Nature, (London)* **383,** 231–235.

Bender, M., Ellis, T., Tans, P. P., Francey, R. J., and Lowe, D. (1996). Variability in the O_2/N_2 ratio of southern hemisphere air 1991–1994: Implications for the carbon cycle. *Global Biogeochem. Cycles* **10**(1), 9–21.

Bonan, G. B. (1993). Physiological derivation of the observed relationship between net primary production and mean annual air temperature. *Tellus* **45B,** 397–408.

Bonan, G. (1995). Land–atmosphere CO_2 exchange simulated by a land surface process model coupled to an atmospheric general circulation model. *J. Geophys. Res.* **100,** 2817–1831.

Braswell, B. H., Schimel, D. S., Privette, J. L., Moore, III, B., Emery, W. J., Sulzman, E. W., and Hudak, A. T. (1996). Extracting ecological and biophysical information from AVHRR optical measurements: An integrated algorithm based on inverse modeling. *J. Geophys. Res.* **101,** 23,335–23,348.

Braswell, B. H., Schimel, D. S., Linder, E., and Moore III, B. (1997). The response of global terrestrial ecosystems to interannual temperature variability. *Science* **278,** 870–872.

Ciais, P., Tans, P. P., Trolier, M., White, J. W. C., and Francey, R. J. (1995). A large northern hemisphere terrestrial CO_2 sink indicated by the $^{13}C/^{12}C$ ratio of atmospheric CO_2, *Science,* **269,** 1098–1102.

Ciais, P., Denning, S., Tans, P. P., Berry, J. A., Randall, D., Collatz, J. G., Sellers, P. J., White, J. W., Trolier, M., Meijer, H. J., Francey, R., Monfray, P., and Heimann, M. (1977a). A three dimensionnal synthesis study of δ18O in atmospheric CO_2. Part 1. Surface fluxes, *J. Geophys Res.* **102** (D15), 5857–5871.

Ciais, P., Tans, P. P., Denning, S., Francey, R., Trolier, M., Meijer, H. J., White, J. W., Berry, J. A., Randall, D., Collatz, J. G., Sellers, P. J., Monfray, P., and Heimann, M. (1997b). A three dimensionnal synthesis study of δ18O in atmospheric CO_2. Part 2. Simulations with the TM2 transport model. *J. Geophys. Res.* **102** (D5), 5873–5883.

Collatz, G, Ball, J., Grivet, C., and Berry, J. (1991). Physiological and environmental regulation of stomatal conductance, photosynthesis and transpiration: A model that includes a laminar boundary layer. *Agric. For. Meteorol.* **54,** 107–136.

Collatz, G. J., Miquel, R.-C., and Berry, J. A. (1992). Coupled photosynthesis-stomatal conductance model for leaves of C_4 plants. *Aust. J. Plant Physiol.* **19,** 519–538.

Conway, T. J., Tans, P. P., Waterman, L. S., Thoning, K. W., Kitzis, D. R., Masarie, K. A, and Zhang, N. (1994). Evidence for interannual variability of the carbon cycle from the National Oceanic and Atmospheric Administration/Climate Monitoring and Diagnostic Laboratory Global Air Sampling Network. *J. Geophys. Res.* **99** (D11), 22831–22855.

Craig, H., and Gordon, A. (1965). Deuterium and oxygen-18 variations in the ocean and the marine atmosphere. *In* "Stable Isotopes in Oceanic Studies and Paleotemperatures," (E. Tomgiorgi, ed.) pp. 9–130. L.o.g.a.N. Science, Pisa.

Cramer, W., Kicklighter, D., Bondeau, A., Moore III, B., Churcina, G., Nemry, B., Ruimy, A., and Schloss, A. (1999). Comparing global models of terrestrial net primary productivity (NPP): Overview and key results. *Global Change Biol.* **5,** 1–15.

Dai, A. and Fung, I. (1993). Can climate variability contribute to the missing CO_2 sink?, *Global Biogeochem. Cycles* **7,** 599–609.

Duvigneaud, P., (1974). "La Synthese Ecologique." Doin, Paris.

Enting, I. G., Trudinger, C. M., and Francey, R. J. (1995). A synthesis inversion of the concentration and $\delta^{13}C$ of atmospheric CO_2. *Tellus* **47B,** 35–52.

Erickson, D., Rasch, P. J., Tans, P. P., Friedlingstein, P., Ciais, P., Maier-Reimer, E., Six, K., Fischer, C. A., and Walters, S. (1996). The seasonal cycle of atmospheric CO_2: A study based on the NCAR community climate model (CCM2). *J. Geophys. Res.* **101** (D10), 15079–15097.

Esser, G. (1987). Sensitivity of global carbon pools and fluxes to human and potential climatic impacts. *Tellus,* **39,** 245–260.

Esser, G. and Lautenschlager, M. (1994). Estimating the change of carbon in the terrestrial biosphere from 18000 BP to present using a carbon cycle model. *Environ. Pollut.* **83,** 45–53.

Farquhar, G. D., von Caemmerer, S., and Berry, J. A. (1980). A biochemical model of photosynthetic CO_2 assimilation in leaves of C_3 species. *Planta* **149,** 78–90.

Farquhar, G. D., Lloyd, J., Taylor, J. A., Flanagan, L. B., Syvertsen, J. P., Hubick, K. T., Wong, S. C., and Ehleringer, R. (1993). Vegetation effects on the isotope composition of oxygen in atmospheric CO_2. *Nature, (London)* **363,** 439–443.

Foley, J. (1994). Net primary productivity of the terrestrial biosphere: The application of a global model. *J. Geophys. Res.* **99,** 20773–20783.

Foley, J. A., Prentice, I. C., Ramankutty, N., Levis, S., Pollard, D., Sitch, S., and Haxeltine, A. (1996). An integrated biosphere model of land surface processes, terrestrial carbon balance, and vegetation dynamics. *Global Biogeochem. Cycles* **10,** 603–628.

Förstel, H., Putral, A., Schleser, G., and Leith, H. (1975). The world pattern of oxygen-18 in rainwater and its importance in understanding the biogeochemical oxygen cycle. *In* "Isotope Ratios as Pollutant Source and Behavior Indicators." IAEA, Vienna.

Francey, R. J., Tans, P. P., Allison, C. E., Enting, I. G., White, J. W. C., and Trolier, M. (1995). Changes in oceanic and carbon uptake since 1982. *Nature (London)* **373,** 326–330.

Francey, R. J., Steele, L. P., Langenfields, R. L., Lucarelli, M. P., Allison, C. E., Beardsmore, D. J., Coram, S. A., Derek, N., De Silva, F. R., Etheridge, E. D., Fraser, P. J., Henry, R. J., Turner, B., Welch, E. D., Spencer, D. A., and Cooper, L. N. (1996). Global atmospheric sampling laboratory (Gaslab): Supporting and extending the Cape Grim trace gas programs. *Baseline* **1993,** 8–29.

Friedlingstein, P., Delire, C., Muller, J.-F., and Gerard, J.-C. (1992). The climate induced variation of the continental biosphere: A model simulation of the Last Glacial Maximum. *Geophys. Res. Lett.* **19,** 897–900.

Friedlingstein, P., Fung, I., Holland, E., John, J., Brasseur, G., Erickson, D., and Schimel, D. (1995). On the contribution of the biospheric CO_2 fertilization to the missing sink. *Global Biogeochem. Cycles* **9,** 541–556.

Friedlingstein, P., Joel, G., Field, C. B., and Fung, I. Y. (1999). Towards an allocation scheme for global terrestrial carbon models. *Global Change Biol.* **5,** 755–770.

Friend, A. D., Stevens, A. K., Knox R. G., and Cannel, M. G. (1997). A process-based, terrestrial biosphere model of ecosystem dynamics (Hybrid v3.0). *Ecol. Model.* **95,** 249–287.

Fung, I. Y., Tucker, C. J., and Prentice, K. C. (1987). Application of advanced very high resolution radiometer to study atmosphere-biosphere exchange of CO_2. *J. Geophys. Res.* **92,** 2999–3015.

Galloway, J. N., Schlesinger, W. H., Levy II, H., Michaels, A., and Schnoor, J. L. (1995). Nitrogen fixation: Anthropogenic enhancement-environmental response. *Global Biogeochem. Cycles* **9,** 235–252.

Gaudry, A., Monfray, P., Polian, G., and Lambert, G. (1990). The 1982–83 El Niño: A 6 billion ton CO_2 release. *Tellus,* **42B,** 62–75.

Gemery, P. A., Trolier, M., and White, J. W. C. (1996). Oxygen isotope exchange between carbon dioxide and water following atmospheric sampling using glass flasks. *J. Geophys. Res.* **101** (14), 14415–14420.

Harley, P. C., Thomas, R. B., Reynolds, J. F., and Strain, B. R. (1992). Modelling photosynthesis of cotton grown in elevated CO_2. *Plant Cell Environ.* **15,** 271–282.

Haxeltine, A. and Prentice, I. C. (1996a). BIOME3: An equilibrium terrestrial biosphere model based on ecophysiological constraints, resource availability, and competition among plant functional types. *Global Biogeochem. Cycles* **10,** 693–709.

Haxeltine, A. and Prentice, I. C. (1996b). A general model for the light-use efficiency of primary production, *Funct. Ecol.* **10,** 551–561.

Hedin, L. O., Armesto, J. J., and Johnson, A. H. (1995). Patterns of nutrient loss from unpolluted, old-growth temperate forests: Evaluation of biogeochemical theory. *Ecology* **76,** 493–509.

Heimann, M., Esser, G., Haxeltine, A., Kaduk, J., Kicklighter, D. W., Knorr, W., Kohlmaier, G. H., McGuire, A. D., Melillo, J., Moore III, B., Otto, R. D., Prentice, I. C., Sauf, W., Schloss, A., Sitch, S., Wittenberg, U., and Würth, G. (1998). Evaluation of terrestrial carbon cycle models through simulations of the seasonal cycle of atmospheric CO_2: First results of a model intercomparison study. *Global Biogeochem. Cycles* **12,** 1–24.

Holland, E. A., Braswell, B. H., Lamarque, J.-F., Townsend, A., Sulzman, J. M., Müller J.-F., Dentener, F., Brasseur, G., Levy II, H., Penner, J. E., and Roelofs, G. (1997). Variations in the predicted spatial distribution of atmospheric nitrogen deposition and their impact on carbon uptake by terrestrial ecosystems. *J. Geophys. Res.* **102,** 15,849–15,866.

Houghton, R. A., Boone, R. D., Fruci, J. R., Hobbie, J. E., Mellilo, J. M., Palm, C. A., Peterson, B. J., Shaver, G. R., Woodwell, G. M., Moore, B., Skole, D. L., and Myers, N. (1987). The flux of carbon from terrestrial ecosystems to the atmosphere in 1980 due to changes in land use: Geographic distribution of the global flux. *Tellus* **39B,** 122–139.

Hunt, E. R. J., Piper, S. C., Nemani, R., Keeling, C. D., Otto, R., and Running, S. W. (1996). Global net carbon exchange and intra-annual CO_2 concentrations predicted by an ecosystem process model and a three-dimensional atmospheric transport model. *Global Biogeochem. Cycles* **10,** 431–456.

IPCC (1995). "Climate Change 1994. Radiative Forcing of Climate Change. An Evaluation of the IPCC IS92 Emissions Scenarios." Cambridge Univ. Press, Cambridge.

Jarvis, P. G. (1976). The interpretation of the variations in leaf water potential and stomatal conductance found in canopies in the field. *Philos. Trans. R. Soc. Lond. Ser. B* **273,** 593–610.

Kaduk, J., and Heimann, M. (1996). A prognostic phenology scheme for global terrestrial carbon cycle models. *Climate Res.* **6,** 1–19.

Keeling, C. D., Whorf, T. P., Wahlen, N., and Vander Plicht, J. (1995). Interannual extremes in the rate of rise of atmospheric carbon dioxide since 1980. *Nature (London)* **375,** 666–670.

Keeling, R. (1995). The atmospheric oxygen cycle: The oxygen isotopes of atmospheric CO_2 and O_2 and the O_2/N_2 ratio. *Rev. Geophys.* 1253–1262.

Keeling, C. D., Bacastow, R. B., Carter, A. F., Piper, S. C., Whorf, T. P., Heimann, M., Mook, W. G., and Roeloffzen, H. A. (1989a). A Three-dimensional model of atmospheric CO_2 transport based on observed winds: 1. Analysis of observational data. *In* "Aspects of Climate Variability in the Pacific and Western Americas, Geophysical Monograph 55." Washington, D.C.

Keeling, C. D., Piper, S. C., and Heimann, M. (1989b). A three-dimensional model of atmospheric CO_2 transport based on observed winds: 4. Mean annual gradients and interannual variations. *In* "Aspects of Climate Variability in the Pacific and Western Americas, Geophysical Monograph 55." Washington.

Keeling, R., Piper, S. C., and Heimann, M. (1996). Global and hemispheric CO_2 sinks deduced from changes in atmospheric O_2 concentration. *Nature (London)* **381,** 218–221.

Kindermann, J., Wurth, G., Kohlmaier, G. H., and Badeck, F. W. (1996). Interannual variation of carbon exchange fluxes in terrestrial ecosystems. *Global Biogeochem. Cycles* **10**(4), 737–755.

Knorr, W. and Heimann, M. (1995). Impact of drought stress and other factors on seasonal land biosphere CO_2 exchange studied through an atmospheric tracer transport model. *Tellus,* **47,** 471–489.

Leuning, R. (1995). A critical appraisal of a combined stomatal-photosynthesis model for C_3 plants. *Plant Cell Environ.* **18,** 339–355.

Lieth, H. (1975). Modelling the primary productivity of the world. *In* "Primary Productivity of the Biosphere" (H. Lieth and R. B. Whittaker, eds.), pp. 237–263. Springer-Verlag, New York.

Lieth, H., and Box, E. (1972). Evapotranspiration and primary productivity. *Publ. Climatol.* **25,** 37–46.

Lüdeke, M. K. B., Badeck, F., Otto, R. D., Hager, C., Donges, S., Kinderman, J., Wurth, G., Lang, T., Jagel, V., Klaudius, A., Ramge, P., Habermehl, S., and Kohlmaier, G. H. (1994). The Frankfurt Biosphere Model. A global process oriented model for the seasonal and longterm CO_2 exchange between terrestrial ecosystems and the atmosphere. *Climate Res.* **4,** 143–166.

Malmstrom, C. M., Thompson, M. V., and Field, C. B. (1997). Interannual variation in global-scale net primary production: Testing model estimates. *Global Biogeochem. Cycles* **11,** 367–392.

Melillo, J. M., McGuire, A. D., Kicklighter, D. W., Moore III, B., Vorosmarty, C. J., and Schloss, A. L. (1993). Global climate change and terrestrial net primary production. *Nature (London)* **363,** 234–240.

McGuire, A., Joyce, L., Kicklighter, D., Melillo, J., Esser, G., and Vorosmarty, C. (1993). Productivity response of climax temperate forests to elevated temperature and carbon dioxide: A North American comparison between two global models. *Climat. Change* **24,** 287–310.

Monfray, P., Ramonet, M., and Beardsmore, D. (1996). Longitudinal and vertical gradient over the subtropical/subantarctic oceanic CO_2 sink. *Tellus* **48B,** 445–446.

Myneni, R. B., Keeling, C. D., Tucker, C. J., Asrar, G., and Nemani, R. R. (1997). Increased plant growth in the northern high latitudes from 1981 to 1991. *Nature (London)* **386,** 698–702.

Parton, W. J., Scurlock, J., Ojima, D., Gilmanov, T., Scholes, D., Schimel, D., Kirchner, T., Nemaut, J. C., Seastedt, T., Garcia Moya, E., Kamnalrut, A., and Kinymario. J. (1993). Observation and modeling of biomass and soil organic matter dynamics for the grassland biome worldwide. *Global Biogeochem. Cycles* **7,** 785–809.

Parton, W., Schimel, D., Cole, C., and Ojima, D. (1987). Analysis of factors controlling soil organic matter levels in Great Plains grasslands. *Soil Sci. Soc. Am. J.* **51,** 1173–1179.

Pastor, J., and Post, W. M. (1986). Influence of climate, soil moisture and succession on forest carbon and nitrogen cycles. *Biogeochemistry* **2,** 3–27.

Post, W. M., King, A. W., and Wullschleger, S. D. (1997). Historical variations in terrestrial biospheric carbon storage. *Global Biogeochem. Cycles* **11,** 99–109.

Potter, C., Randerson, J., Field, C., Matson, P. A., Vitousek, P., Mooney, H., and Klooster, S. (1993). Terrestrial ecosystem production: A process model based on global satellite and surface data. *Global Biogeochem. Cycles* **7,** 811–841.

Privette, J. L., Emery, W. J., and Schimel, D. S. (1996). Inversion of a vegetation reflectance model with NOAA AVHRR data. *Remote Sensing Environ.* **58,** 187–200.

Raich, J., Rastetter, E., Melillo, J., Kicklighter, D., Steudler, P., Peterson, B., Grace, A., Moore, B., and Vorosmarty, C. (1991). Potential net primary productivity in South America: Application of a global model. *Ecol. Appl.* **4,** 399–429.

Randerson, J. T., Thomson, M. V., Conway, T. J., Fung, I. Y., and Field, C. B. (1997). The contribution of terrestrial sources and sinks to trends in the seasonal cycle of atmospheric carbon dioxide. *J. Geophys. Res.* **11**(4), 535–560.

Rayner, P. J., and Law, R. M. (1995). "Transcom'95: A comparison of modelled responses to prescribed CO_2 sources." *CSIRO Aust. Div., Atmos. Res. Tech. Pap. No. 36,* pp. 1–82.

Rodin, L., Bazilevich, N., and Rozov, N. (1975). Productivity of the world's main ecosystems. *In* "Productivity of World Ecosystems" (D. E. Reichle, J. F. Franklin, and D. W. Goodall, eds.), pp. 13–26. National Academy of Science, Washington, D.C.

Rosenzweig, M. L. (1968). Net primary productivity of terrestrial communities: Prediction from climatological data. *Am. Naturalist* **102,** 67–74.

Rozanski, K., Araguas-Araguas, L., and Gonfiantini, R. (1993). Isotopic patterns in modern global precipitation. *In* "Climate Change in Continental Isotopic Records" (P. K. Swart, K. C. Lohmann, J. MasKenzie, and S. Savin, eds.), pp. 1–37. AGU, Washington, D.C.

Ruimy, A., Dedieu, G., and Saugier, B. (1996). TURC—Terrestrial uptake and release of carbon by vegetation, a diagnostic model of continental gross primary productivity and net primary productivity. *Global Biogeochem. Cycles* **10,** 269–285.

Running, S. W. (1994). Testing Forest-BGC ecosystem process simulations across a climatic gradient in Oregon. *Ecol. Appl.* **4,** 238–247.

Running, S., and Hunt, E. (1993). Generalization of a forest ecosystem to other biomes, BIOME-BGC, and an application for global-scale models. *In* "Scaling Physiological Processes: Leaf to Globe" (J. Ehleringer and C. Field, eds.), pp. 141–158. Academic Press, San Diego.

Schimel, D. S., Braswell, B. H., McKeown, R., Ojima, D. S., Parton, W. J., and Pulliam, W. (1996). Climate and nitrogen controls on the geography and time scales of terrestrial biogeochemical cycling. *Global Biogeochem. Cycles* **10,** 677–692.

Schimel, D. S., Braswell, B. H., and Parton, W. J. (1997a). Equilibration of the terrestrial water, nitrogen, and carbon cycles. *Proc. Nat. Acad. Sci. U.S.A.* **94,** 8280–8283.

Schimel, D. S., VEMAP Participants, and Braswell, B. H. (1997b). Continental scale variability in ecosystem processes: Models, data, and the role of disturbance. *Ecol. Monogr.* **67,** 251–271.

Sellers, P. (1985). Canopy reflectance, photosynthesis and transpiration. *Int. J. Remote Sensing* **6,** 1335–1372.

Sellers, P. (1987). Canopy reflectance, photosynthesis and transpiration. II. The role of biophysics in the linearity of their interdependence. *Remote Sensing Environ.* **21,** 143–183.

Sellers, P. J., Berry, J. A., Collatz, G. J., Field, C. B., and Hall, F. G. (1992). Canopy reflectance, photosynthesis, and transpiration. III. A reanalysis using improved leaf models and a new canopy integration scheme. *Remote Sensing Environ.* **42,** 187–216.

Sellers, P., Randall, D., Collatz, G., Berry, J., Field, C., Dazlich, D., and Zhang, C. (1996a). A revised land-surface parameterization (SiB2) for atmospheric GCMs. Part I: Model formulation. *J. Climate* **9,** 676–705.

Sellers, P. J., Los, S. O., Tucker, C. J., Justice, C. O., Dazlich, D. A., Collatz, G. J., and Randall, D. A. (1996b). A revised land surface parameterization (SiB2) for atmospheric GCMs. Part II: The generation of global fields of terrestrial biophysical parameters from satellite data. *J. Climate* **9,** 706–737.

Stewart, J. B. (1988). Modelling surface conductance of pine forest. *Agric. For. Meteorol.* **43,** 19–35.

Tans, P. P., Fung, I. Y., and Takahashi, T. (1990). Observational constraints on the global atmospheric CO_2 budget. *Science,* **247,** 1431–1438.

Tans, P., Berry, J. A., and Keeling, R. F. (1993). Oceanic ^{13}C data: A New window on CO_2 uptake by the oceans. *Global Biogeochem. Cycles* **7,** 353–368.

Trolier, N., White, J. W., Tans, P. P., Nasarie, K. A., and Gemery, P. A. (1996). Monitoring the isotopic composition of atmospheric CO_2: Measurements from the NOAA global air sampling network. *J. Geophys. Res.* **101,** 25897–259116.

Uchijima, Z., and Seino, H. (1985). Agroclimatic evaluation of net primary productivity of natural vegetations. I. Chikugo model evaluating net primary productivity. *J. Agric. Meteorol.* **40,** 343–352.

VEMAP (1995). Vegetation/ecosystem modeling and analysis project (VEMAP): Comparing biogeography and biogeochemistry models in a continental-scale study of terrestrial ecosystem responses to climate change and CO_2 doubling. *Global Biogeochem. Cycles* **9,** 407–438.

Vitousek, P. M., and Howarth, R. W. (1991). Nitrogen limitation on land and in the sea: How can it occur? *Biogeochemistry* **13,** 87–115.

Vitousek, P. M., Walker, L. R., Whiteaker, L. D., and Matson, P. A. (1993). Nutrient limitations to plant growth during primary succession in Hawaii Volcanoes National Park. *Biogeochemistry* **23,** 197–215.

Warnant, P., Francois, L., Strivay, D., and Gerard, J.-C. (1994). CARAIB: A global model of terrestrial biological productivity. *Global Biogeochem. Cycles* **8,** 255–270.

Whittaker, R. and Likens, G. (1973). Carbon in the biota. *In* "Carbon and the Biosphere" (G. Woodwell and E. Pecan, eds.). U.S. Dept. of Commerce, Springfield, Virginia.

Winguth, A. M. E., Heimann, M., Kurz, K. D., Maier-Reimer, E., Mikolajewicz, U., and Segschneider, J. (1994). El Niño-southern oscillation related fluctuations of the marine carbon cycle. *Global Biogeochem. Cycles* 39–62.

Woodward, F. I., Smith, T. M., and Emanuel, W. R. (1995). A global land primary productivity and phytogeography model. *Global Biogeochem. Cycles* **9,** 471–490.

World Meteorological Organization (1996). WMO WDCGG data report No. 11, GAW data.

Yakir, D., Berry, J. A., Giles, L., and Osmond, C. B. (1994). Isotopic heterogeneity of water in transpiring leaves: Identification of the component that controls $\delta^{18}O$ of atmospheric CO_2 and O_2. *Plant Cell Eniron.* **7,** 73–80.

20

Reconstructing and Modeling Past Changes in Terrestrial Primary Productivity

Joël Guiot, I. Colin Prentice, Changhui Peng, Dominique Jolly, Fouzia Laarif, and Ben Smith

I. Introduction

We have no real proxy records of primary productivity and carbon storage for individual locations, but we can reconstruct net primary productivity (NPP) from pollen data by statistical methods, assuming a relationship between NPP and taxonomic composition. This approach, practicable in regions with a lot of data, is illustrated for Europe, but it has some limitations. Several parallel approaches have attempted to estimate the global terrestrial carbon storage of the Last Glacial Maximum: the large differences between these approaches prove the difficulty of the task. To mitigate this lack of data, we propose an approach based on model simulations, controlled by data. Waiting more sophisticated earth system models, we use BIOME3 asynchronously coupled with a climate model to simulate final vegetation distributions and NPP. The role of paleodata is then to check the vegetation distributions. For 6000 yr BP, data have already been synthesized using objective methods as part of the global BIOME 6000 project for several continents. Not surprisingly, the agreement between data and models is imperfect, although many aspects are quantitatively correct and some recent studies have shown significant improvements due to the inclusion of biogeophysical feedbacks by asynchronous coupling. This approach appears to be a powerful way to check the ability of the models to predict the impact of global warming on vegetation.

Pollen grains and spores are disseminated by plants and are preserved during hundreds of thousands of years when buried in sediments (primarily lakes, peat bogs, and marine sediments), recording in this manner successive states of the surrounding vegetation. Palynologists (pollen analysts) measure the abundance of the different taxa identified in each level of the sediment and establish an assemblage or pollen spectrum for each stage of the past. The fluctuations of these assemblages through the sedimentary profile, represented in the form of a pollen diagram, are used to reconstitute the vegetation history near the sampling site.

Pollen data do not provide direct records of the past net primary productivity (NPP) or biomass of plants or ecosystems. However, past NPP patterns can be inferred by making use of pollen and other paleoecological data that record vegetation composition, then using a modeling approach to link vegetation composition with NPP. In this chapter we briefly review the approaches that have been used to reconstruct vegetation composition and NPP for various times in the past.

Pollen data have been used to map spatial patterns of natural climatic change (e.g., Webb and Bryson, 1972; Huntley and Prentice, 1988; Prentice *et al.*, 1991; Cheddadi *et al.*, 1996) using various empirical methods, all of which ultimately have relied on the use of modern analogs. Pollen data have also been used in a similar way to map changing patterns of NPP and carbon storage in vegetation and soil (Peng *et al.*, 1995a,b). For climate as well as for carbon, these empirical methods have limitations. They do not easily handle "no-analog" climates and vegetation, they take no account of changing CO_2 concentrations, they are ill-determined for many regions of the world where the paleodata are relatively sparse (including substantial areas in the intertropical zone), and they cannot say anything about NPP on the large areas of exposed continental shelves during glacial times.

An alternative approach better suited to global analysis is based on climate model simulations, reserving the pollen data as a test of the simulated vegetation composition, while relying on other aspects of the modeling procedure to generate independent estimates of physical climate and biogeochemical quantities. Ideally, past climate, vegetation, and ecosystem processes (including NPP) would be simulated by interactively coupled models, forced only by the primary changes in boundary conditions between different epochs (e.g., orbital changes between 6000 yr BP and present, ice-sheet and atmospheric composition changes since the last glacial maximum, 18,000 yr BP; see Fig. 20-1). Because fully interactive (atmosphere–ocean–biosphere) models are not yet available, however, we are currently obliged to compromise, e.g., by using prescribed sea-surface temperatures (SSTs) and either one-way forcing or (better) equilibrium asynchronous coupling (Claussen, 1997; Texier *et al.*, 1997) between an atmospheric general circulation model (AGCM) and a global ecosystem model.

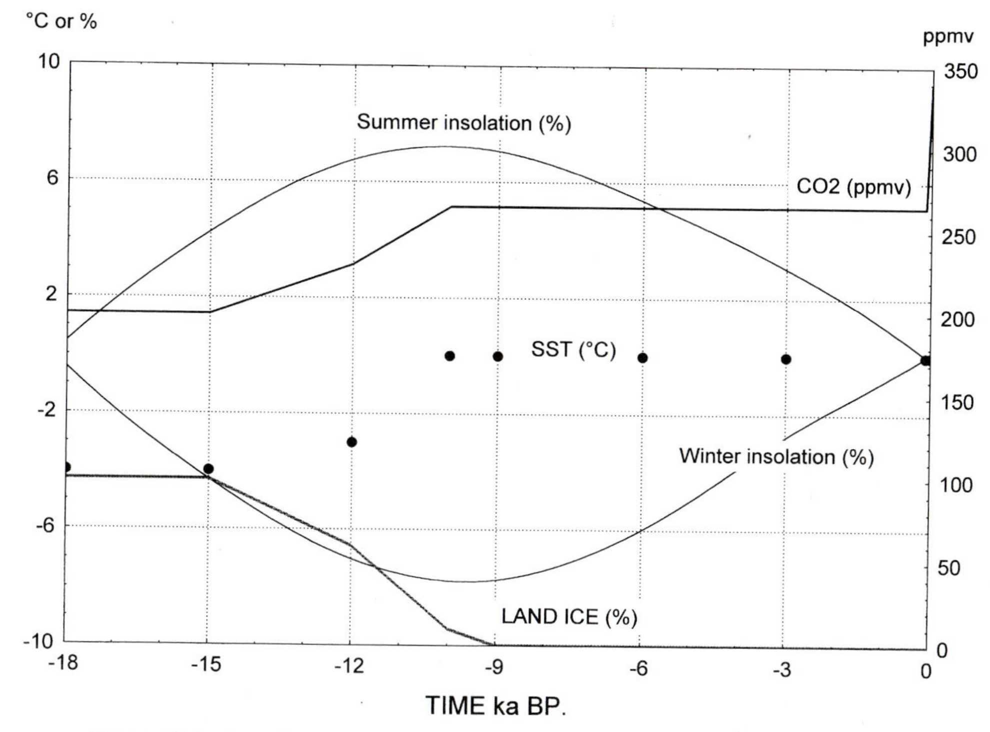

Figure 20-1 Boundary conditions used by the COHMAP simulations (COHMAP members, 1988) for the last 18,000 years. The insolation curves are given in percent difference from the present radiation. The land ice is given in percent of 18,000 yr BP ice volume. The global sea surface temperature (SST) is given as departures from present; atmospheric CO_2 concentration is in parts per million per volume (ppmv). Prior to 15,000 yr BP, the main forcing is high land ice, low SST, and low CO_2; they progressively change up to 9000 yr BP, when the main forcings are high summer insolation and low winter insolation; at 6000 yr BP, the CO_2 concentration, SST, and land ice values are similar to the modern values; CO_2 concentration drastically increased during the past century.

The one-way forcing approach is illustrated here using BIOME3 (Haxeltine and Prentice, 1996) to simulate vegetation distributions and NPP as a function of CO_2 and a climate field derived from an AGCM simulation with the CCM1 model (Kutzbach *et al.*, 1999). The role of the paleodata in this approach is to check the simulated vegetation distributions. For 6000 yr BP, global data are being synthesized for this purpose using objective methods as part of the global BIOME 6000 project (Prentice and Webb, 2000).

II. Reconstruction of Vegetation Composition and NPP from Paleodata

A. Vegetation Reconstruction

The recent development of online pollen databases in a common format at regional, subcontinental, and global scales has spurred advances in the methodology for reconstruction of vegetation composition, carbon storage, and NPP. We begin by summarizing current approaches used for the reconstruction of vegetation composition.

1. Qualitative analysis based on the changing abundances of major taxa. This is still the standard level of analysis in the literature presenting primary pollen data. For example, the Massif Central (France) has been intensively studied (Beaulieu *et al.*, 1988) to establish the vegetation changes since the last glacial period. The earliest Holocene warming (ca. 10000 ^{14}C-yr. BP) is described as being characterized by a *Pinus–Betula* forest that was replaced by a *Corylus–Quercus–Ulmus* forest at about 9500 yr BP. Later, after 5000 yr BP, *Fagus* and *Abies* became dominant. This kind of description has the disadvantage of subjectivity, and interpretations can vary significantly among authors.
2. Taxon calibration approach: pollen percentages for major taxa in modern pollen samples from surface sediments are calibrated against measures of dominance in the contemporary vegetation (Webb *et al.*, 1981; Delcourt *et al.*, 1983, 1984; Prentice, 1982, 1986). The same calibration is then applied to percentages of each taxon in sequences of fossil pollen assemblages to quantify the changes in dominance through time (Delcourt and Delcourt, 1984) . The method is limited by the fact that it does not take into account the less dominant taxa, and by the potential for large statistical errors when multiplying up the number for taxa that have low pollen representation coefficients (Parsons and Prentice, 1984).
3. Modern analog methods: here the fossil pollen assemblages are compared to modern pollen assemblage using dissimilarity measures (Prentice, 1982; Overpeck *et al.*, 1985; Huntley, 1990) and the closest modern assemblages are considered as analogs. The method relies on the strong assumption that the appropriate analogs exist today, which is not always so.
4. Multivariate statistical biome reconstruction: another way to use the modern pollen data is to classify the modern samples according to biomes, then use multivariate analysis to develop discriminant functions for biomes (Peng *et al.*, 1994). The efficiency of this method still depends on the past biomes all existing today and having similar composition.
5. Non-statistical biome reconstruction: in the method developed by Prentice *et al.* (1996), each pollen taxon is assigned to one or several plant

functional types (pft), which are broad classes of plants defined by stature, leaf form, phenology, and climatic thresholds, and the biomes are defined by the pfts they may contain. An affinity index is computed as the sum of the square roots of percentages of all the taxa potentially present in the biome. These affinity indices are compared, and the biome for which the index value is maximum is attributed. This approach circumvents the no-analog problem through the use of *a priori* biome definitions and the classification of taxa into pfts.

Figure 20-2 presents non-statistical biome reconstruction from pollen data at 6000 yr BP (Prentice *et al.,* 1996) for Europe. The reconstruction method was tested using more than 2000 modern pollen assemblages. Only the sites common to both periods (present and 6000 yr BP) are shown. Comparison with the modern (potential) vegetation shows that, 6000 years ago, the various midlatitude forest belts extended toward the north due to generally warmer conditions. In particular, the deciduous forest biome was extended to the north and east (implying warmer winters) and to the south (implying wetter growing-season conditions). In northern Europe the taiga biome was displaced far to the east, allowing a north–south zonation among more oceanic vegetation types.

B. NPP and Carbon Storage Reconstruction

When the biomes are reconstructed for a given period, several methods can be applied to deduce biogeochemical characteristics associated with these biomes.

1. Simple Empirical Estimates The most straightforward method is to use mean values derived from modern databases. Olson *et al.* (1985), for example, compiled a global database of ecosystems and provided typical values for carbon storage in vegetation, and Post *et al.* (1982) provided comparable information for soil carbon. Several empirical reconstructions of the Last Glacial Maximum (LGM) carbon storage globally or regionally have been done on this basis (e.g., Adams *et al.,* 1990; Branchu *et al.,* 1993; Van Campo *et al.,* 1993; Maslin *et al.,* 1995; Crowley, 1995). However, many approximations and questionable assumptions are involved in applying these values globally under a changed climate, especially because atmospheric CO_2 concentration also changed between the LGM and present.

2. Simple Empirical Models Peng *et al.* (1995a,b) (Fig. 20-3) used simple empirical models to calculate changes in NPP and carbon storage based on pollen data from Europe. Three values of NPP were obtained independently, from mean annual temperature, annual precipitation, and actual evapotranspiration, on the basis of the original MIAMI model (Lieth, 1975) and the Montreal model (Lieth and Box, 1972). NPP was taken to be the low-

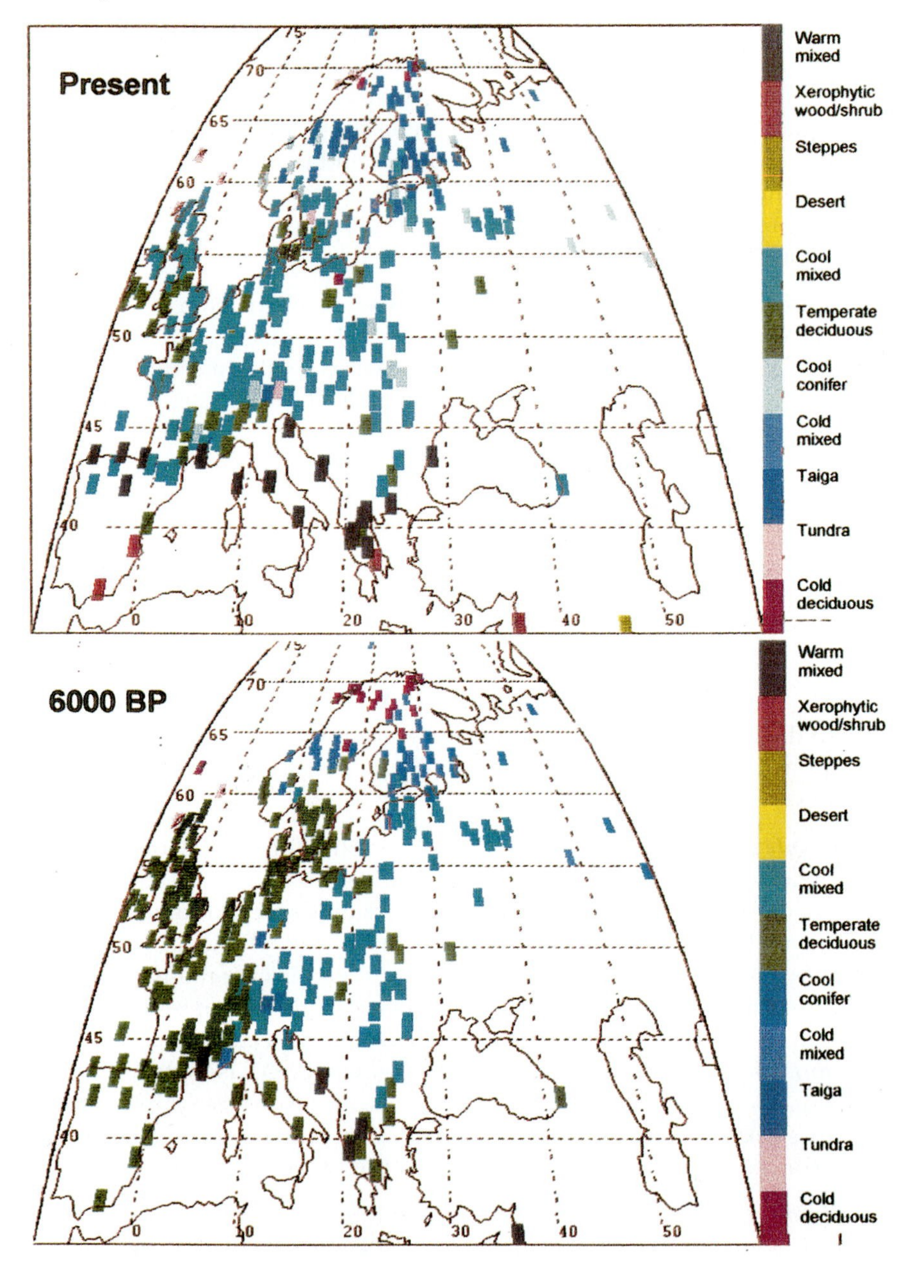

Figure 20-2 Reconstruction of biomes in Europe for 6000 yr BP and present (after Prentice *et al.*, 1996). Each point represents a pollen site.

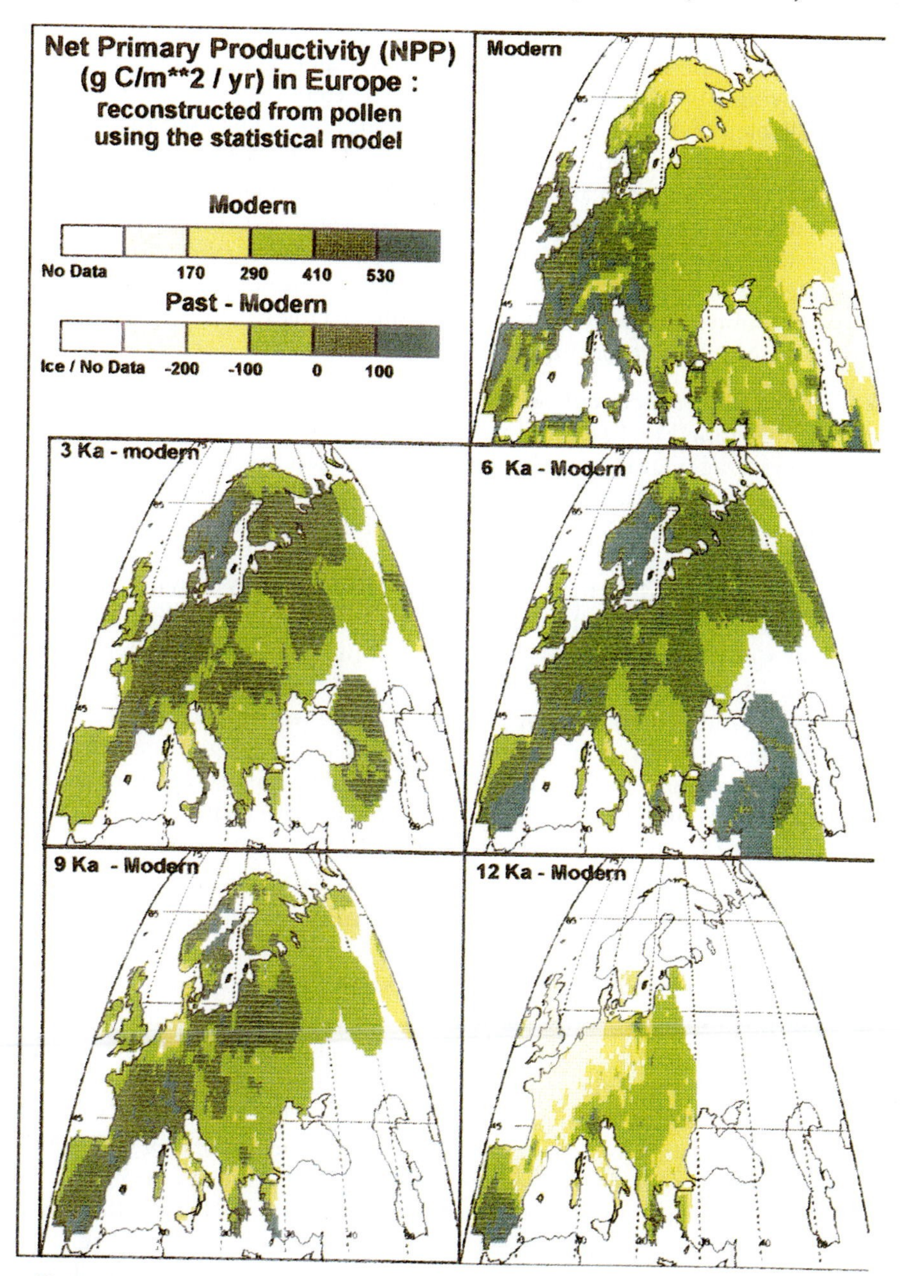

Figure 20-3 Spatial variation of NPP (in g C m^{-2} yr^{-1}) for the present reconstructed from surface pollen data by Peng *et al.* (1995b) and spatial variations of NPP anomalies (past–modern) from 12,000 yr BP to 3000 yr BP by 3000-yr steps (all ages refer to ^{14}C ages). See Fig. 20-1. to understand the climatic forcing for these periods.

est of the three values. Soil respiration was also modeled as an empirical function of temperature and moisture (Schlesinger, 1977; Ewel *et al.*, 1987; Gordon *et al.*, 1987). For Europe, Peng *et al.* (1998) selected 117 observations collected by Raich and Schlesinger (1992) and expressed the soil respiration as a nonlinear function of the same climate variables used to estimate NPP. Biomass was calculated by an empirical relationship from NPP and an assumed mean stand age of the ecosystem (Esser, 1991); soil carbon storage was expressed as a function of potential and actual evapotranspiration (Meentemeyer *et al.*, 1985). Pollen data were used to reconstruct both the paleobiome distribution and the climatic variables required to drive the various empirical relationships.

3. Simple Biogeochemical Models The Osnabrück Biosphere Model (OBM) developed by Esser and modified by Peng is a simple "box model" of carbon storages and fluxes that has also been used to estimate carbon storage in vegetation and soils for Europe (Peng *et al.*, 1995c) and China (Peng and Apps, 1997). Here NPP is still based on the empirical Miami model (Lieth, 1975) but takes into account soil type and CO_2 fertilization through experimentally derived factors (the soil factor corrects the "Miami" productivity according to the soil type and the CO_2 factor is empirically derived from physiological data to take into account the physiological effect of fertilization). Litter production and decomposition and soil organic carbon production and decomposition are functions of mean annual temperature, total annual precipitation, atmospheric CO_2 concentration, soil fertility class, and vegetation type. As before, pollen data were used to provide the vegetation and climatic inputs and the Vostok ice core (Barnola *et al.*, 1987) provided the time sequence of atmospheric CO_2 concentration.

III. Global Simulation of Vegetation and NPP

A. Simulations Using AGCMs and Biome Models

In all of the applications discussed so far, the paleodata have been used as the primary information to drive the models. This approach carries with it the necessity of interpolation among sites for which paleodata are available. Such interpolation is justified in regions of space and time with a high paleodata density, such as Europe during the Holocene. It is much more questionable for less well-studied regions, such as South America, or for any region during the LGM. The alternative approach for global analysis starts from paleoclimate simulation. Several authors (Kutzbach and Guetter, 1980; Joussaume and Taylor, 1995; Kutzbach *et al.*, 1999) have achieved some success in simulating spatial patterns of climate anomalies for key times during the recent geological past (see also Wright *et al.*, 1994). Much effort has focused on two key times, namely, 6000 yr BP and the LGM.

Some attempts have been made to reconstruct NPP and carbon storage based on paleoclimate simulations and the use of biogeographical (biome) models to deduce vegetation distributions from the simulated paleoclimate anomalies. Some of these attempts have included simple empirical schemes in which mean annual temperature and precipitation are used to predict biome distributions (e.g., Prentice and Fung, 1990; Friedlingstein *et al.*, 1992, 1995). Others have used the global BIOME model of Prentice *et al.* (1992) (e.g., Prentice *et al.*, 1993; Esser and Lautenschlager, 1993). In the BIOME model, the distribution of pfts is predicted based on known or hypothesized physiological constraints, and biomes arise as combinations of potentially codominant pfts. The most important practical difference between BIOME and simpler empirical schemes is that BIOME responds to aspects of seasonality that affect plant distribution, e.g., winter temperatures and precipitation distribution. Once biome distributions have been obtained, the results have been used with a range of schemes to estimate NPP or carbon storage, similar to those used in empirically based studies. For example, Prentice *et al.* (1993) used a simple look-up table based on the data sets of Olson *et al.* (1985) and Post *et al.* (1982), whereas Friedlingstein *et al.* (1995) and Esser and Lautenschlager (1993) used carbon flux and storage models similar to the above-mentioned Osnabrück Biospheric Model. Esser and Lautenschlager (1993) in particular drew attention to the potentially large effect of low atmospheric CO_2 concentration on carbon storage.

Empirical and model-based estimates of global changes in carbon storage from the LGM to preindustrial Holocene times have varied widely, both within and between the two categories of estimate (see, e.g., Crowley, 1995). Figure 20-4 illustrates perfectly these variations going from 1350 to 0 Pg C (as departues from the modern storage): in general, estimates done from data are lower than those done from models and marine data. Discussion of the reasons for these variations would be beyond the scope of this chapter. However, we consider all published results on this topic to date as exploratory at best. Published empirically based estimates have major problems due to incomplete LGM data and the necessity of interpolation between sparse data points, whereas model-based estimates (1) have not been performed with state-of-the-art physiological models (e.g., that take into account the interactions of CO_2 and temperature effects on photosynthesis, and (2) have so far been insufficiently tested against what limited paleodata are available for the LGM.

B. Global Simulations with BIOME3

As a further exploratory contribution to the yet-to-emerge science of "paleobiogeochemistry," we present simulation results obtained by a one-way coupling of an AGCM (CCM1) to the equilibrium biosphere model BIOME3 (Haxeltine and Prentice, 1996) for the period around 6000 yr BP. This

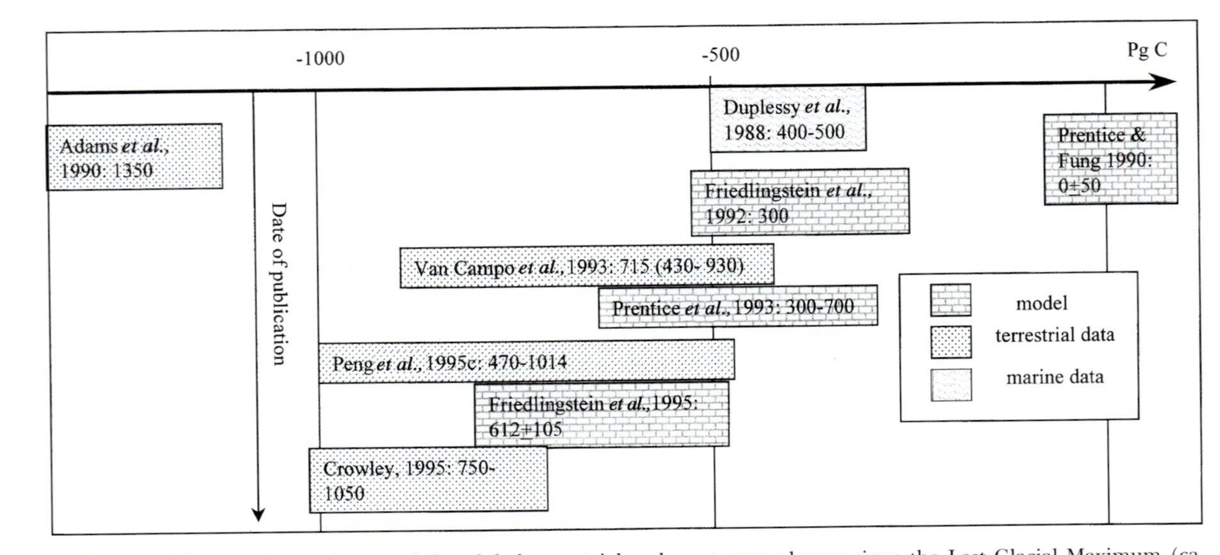

Figure 20-4 Variation estimates of the global terrestrial carbon storage change since the Last Glacial Maximum (ca. 18,000 BP ^{14}C). The widths of the rectangles represent the uncertainty of these estimates. The reference value (0) is the modern storage. The *y* axis represents the publication date.

period is in some ways more tractable than the LGM, because the CO_2 concentration was similar to that of preindustrial time, and because more paleodata is available. The AGCM simulation (Kutzbach *et al.*, 1999) was carried out with the NCAR CCM1, coupled to a mixed-layer interactive model for surface ocean conditions. Anomalies (differences from the control simulation at each grid point) for mean monthly temperature, precipitation, and sunshine fraction were interpolated to a 0.5° grid and added to the Cramer–Leemans climatology that underlies BIOME3, thus providing a new climate field with which to drive BIOME3. Differences in mean coldest-month temperature were also added to the field of absolute minimum temperatures used by BIOME3.

BIOME3 interactively couples vegetation distribution and biogeochemistry. The initial basis of BIOME3 [like the earlier BIOME model of Prentice *et al.* (1992)] is a global set of pfts whose geographical ranges are ultimately controlled by ecophysiological constraints such as cold tolerance and seasonality requirements. Using a coupled carbon and water flux model and an optimization algorithm (Haxeltine and Prentice, 1997) to determine photosynthetic capacity, the model then calculates the maximum sustainable leaf area index (LAI) and NPP for each climatically permitted pft, as a function of moisture and carbon limitations. Differences in physiology, phenology and rooting depth between the PFTs result in differences in their performance in the carbon and water flux model. Herbaceous plants are assumed to utilize either the C_3 or C_4 photosynthetic pathway, depending on temperature and ambient CO_2, whereas woody plant types always use the C_3 pathway, thus allowing seasonal partitioning between C_3 and C_4 species. Competition among pfts is simulated by using NPP as a competitiveness index: the dominant plant type is that with the highest NPP, except where grasses are outcompeted because moisture availability favors trees. The outputs include NPP and LAI for all the pfts, total LAI, and total NPP. For each grid cell, BIOME3 also predicts the biome type according to a classification scheme based on the dominant pft and its LAI and NPP. The model has been tested against global data sets of potential natural vegetation, NPP, and remotely sensed "greeness" as a correlative of LAI.

Figure 20-5 shows the potential distribution of present biomes as simulated by BIOME3, using a CO_2 concentration of 340 ppmv and the simulated distribution for 6000 yr BP using the preindustrial CO_2 value of 280 ppmv and the CCM1 climate anomalies as input (Kutzbach *et al.*, 1999). Comparison of these maps for Europe shows extension at 6000 yr BP of the xeric vegetation in the Mediterranean area, an extension to the north of the temperate deciduous and boreal forest and a reduction of tundra. These changes are the consequence of a climate warming in the north and a drying in southern Europe. Another noticeable change is the decreased desert area in northeast Africa and Arabia, due to an extension of the monsoonal influence.

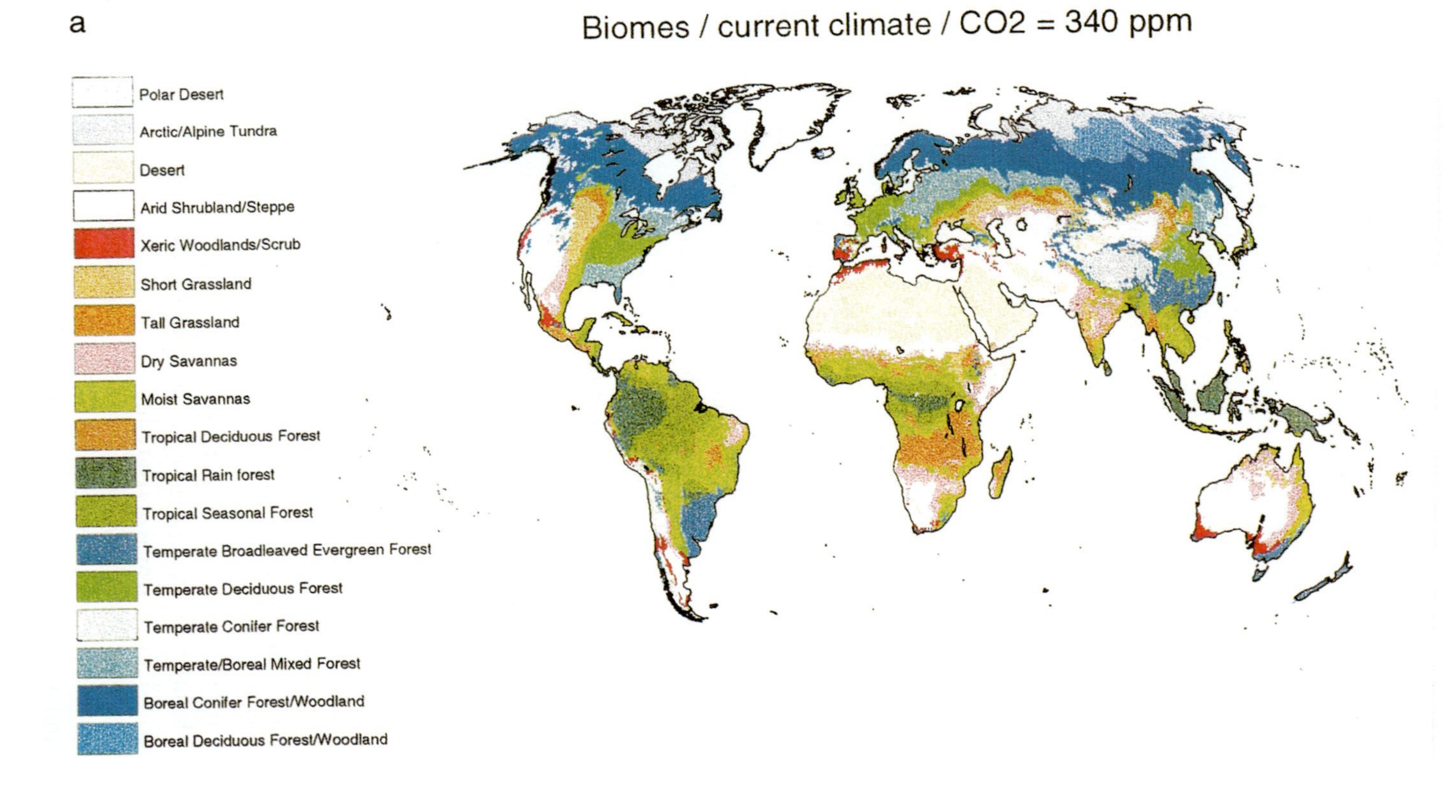
a
Biomes / current climate / CO2 = 340 ppm
Polar Desert
Arctic/Alpine Tundra
Desert
Arid Shrubland/Steppe
Xeric Woodlands/Scrub
Short Grassland
Tall Grassland
Dry Savannas
Moist Savannas
Tropical Deciduous Forest
Tropical Rain forest
Tropical Seasonal Forest
Temperate Broadleaved Evergreen Forest
Temperate Deciduous Forest
Temperate Conifer Forest
Temperate/Boreal Mixed Forest
Boreal Conifer Forest/Woodland
Boreal Deciduous Forest/Woodland

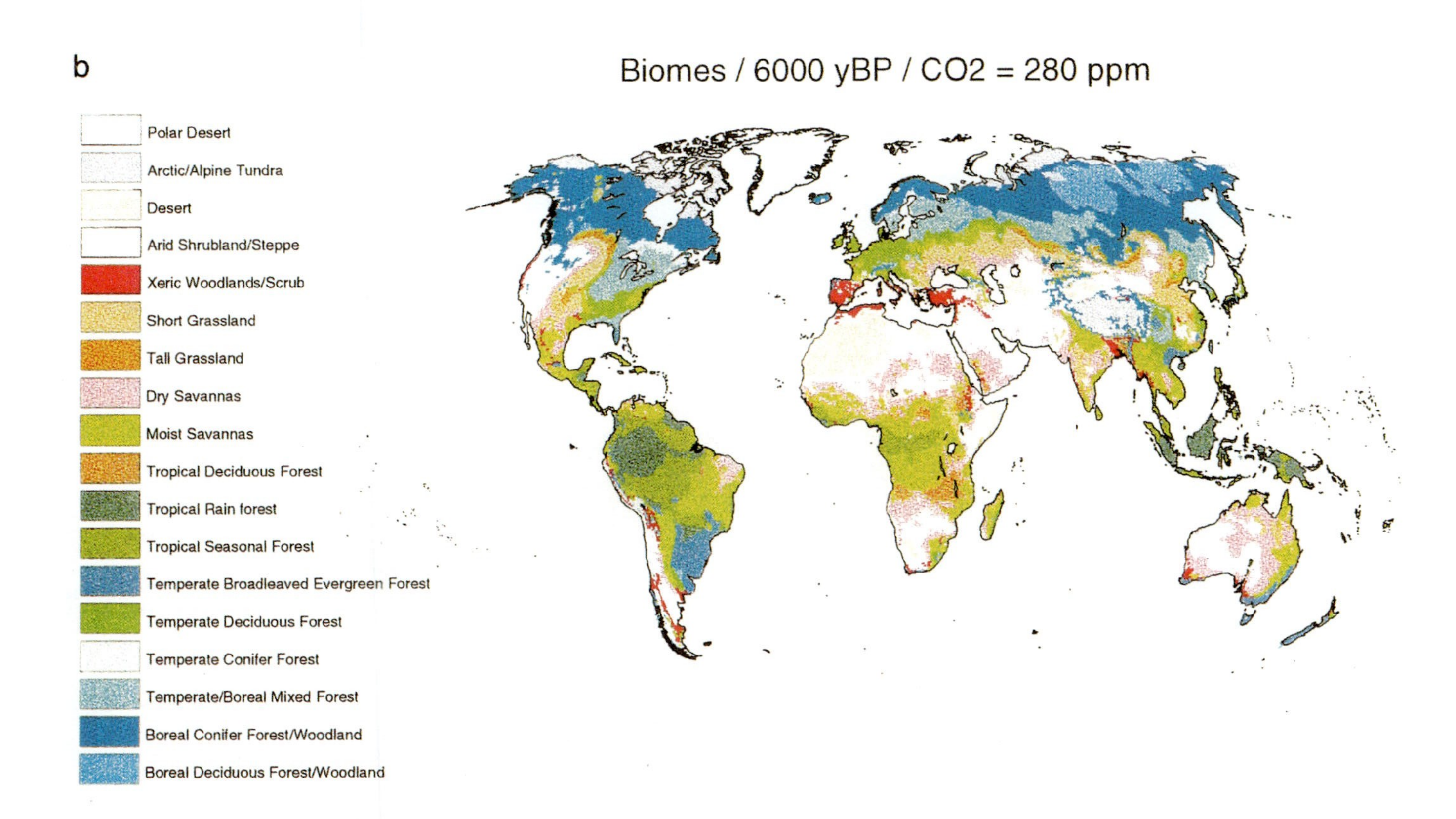

Figure 20-5 Simulated vegetation distribution (biomes) using BIOME3 (a) driven by modern climatology assuming a CO_2 concentration of 340 ppm and (b) for 6000 yr BP driven by climate simulation from the CCM1 model (Kutzbach *et al.*, 2000) assuming a CO_2 concentration of 280 ppm.

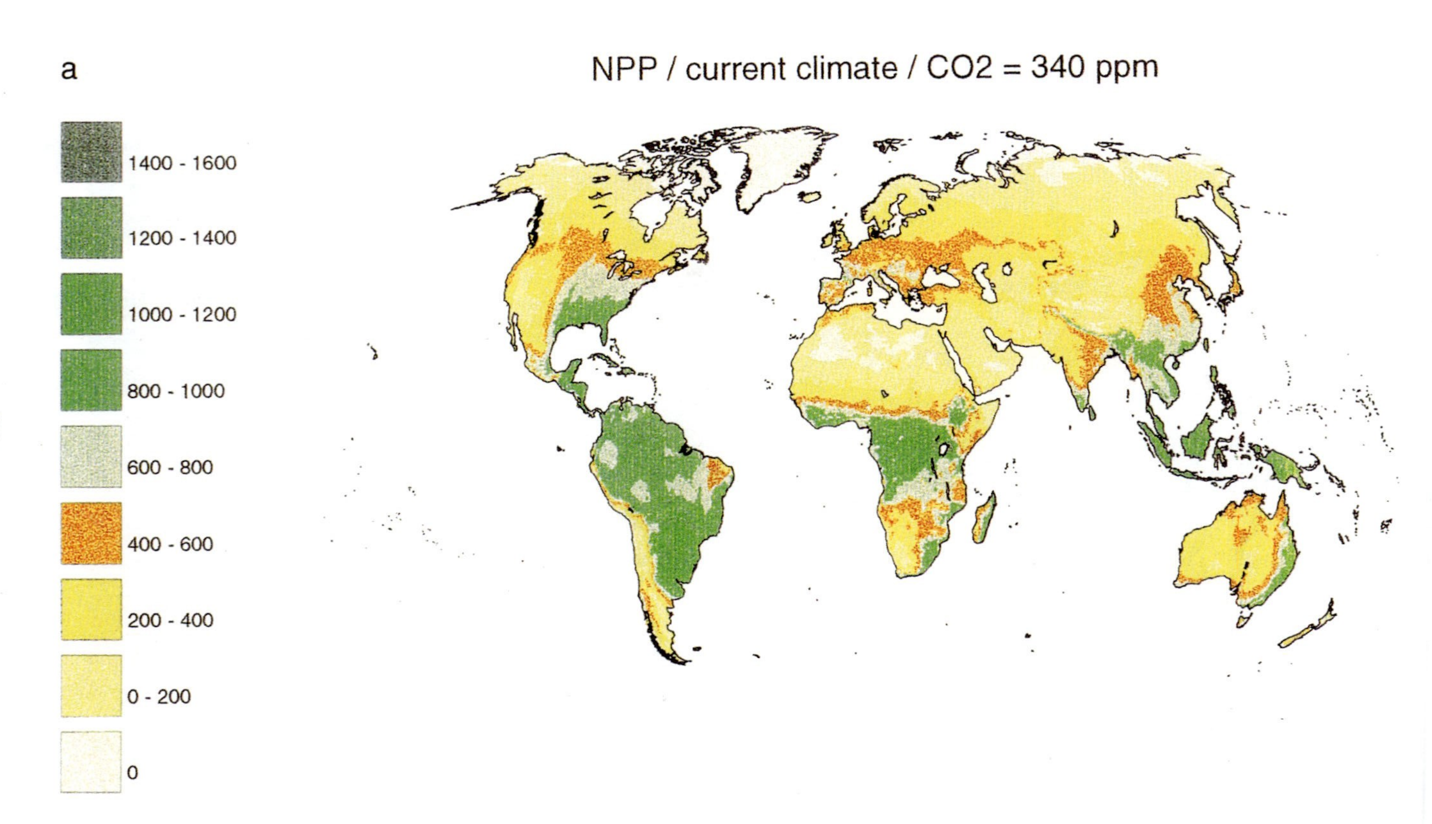
a
NPP / current climate / CO2 = 340 ppm
1400 - 1600
1200 - 1400
1000 - 1200
800 - 1000
600 - 800
400 - 600
200 - 400
0 - 200
0

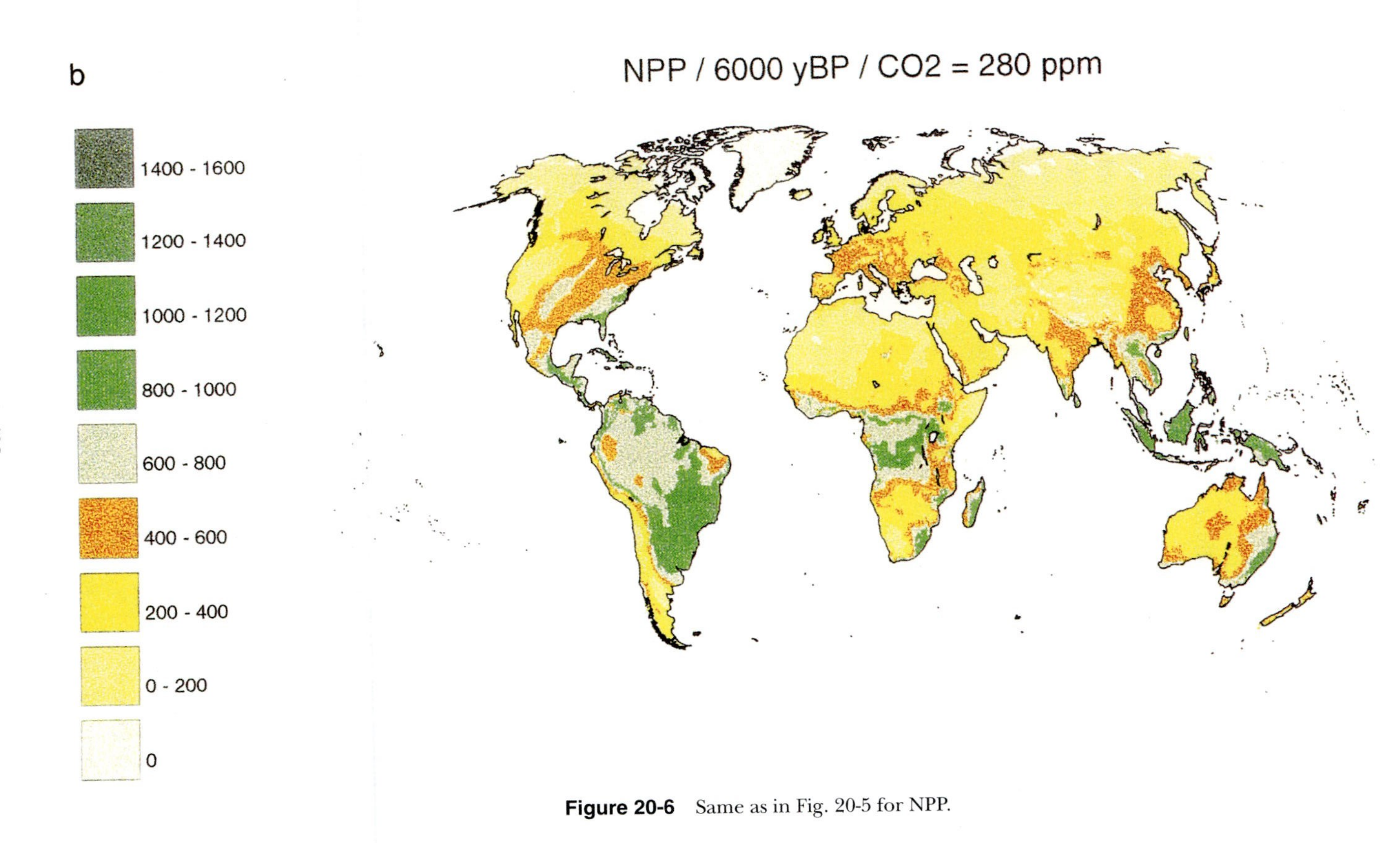

Figure 20-6 Same as in Fig. 20-5 for NPP.

The northern warming is confirmed by the data in Fig. 20-3, in which where we observe a clear shift of forest toward the north and a reduction of the tundra. In the Mediterranean area, however, the model conflicts with the data (Prentice *et al.*, 1998). In reality, the Mediterranean vegetation area was reduced and the deciduous forest increased, implying a more humid climate. In North Africa (Jolly *et al.*, 2000), the data confirm the monsoon increase not only in east Africa but also show that it was substantial in the west. Thus, although the simulated biomes show some realistic features, there are also specific biases. Similar biases are apparent in comparable simulations using other AGCMs (Harrison *et al.*, 1998). Much current research in paleoclimatology focuses on identifying the causes of such systematic biases in model results, which may include the neglect of feedbacks involving both biospheric changes and ocean surface changes (Claussen and Gayler, 1997; Texier *et al.*, 1997)

The simulated NPP values (Fig. 20-6) shift along with the simulated biome distribution. Extension of forest in northern Europe goes along with an NPP increase, and the apparent extension of xeric vegetation in southern Europe goes along with a decreased NPP. But this latter feature is presumably unreliable and contradicts the data-based reconstruction in Fig. 20-3.

IV. Conclusions

Evidently, there is as yet no reliable way to reconstruct past global patterns of NPP. Data-based reconstruction is possible in regions where the data network is exceptionally dense (e.g., Europe at 6000 yr BP). Biogeochemical models can be used to convert empirical regional climate and biome reconstructions to NPP and carbon storage, taking into account external factors including atmospheric CO_2 content. Global analysis, however, requires a complete "forward modeling" approach, for example, based on climate model simulations coupled to an equilibrium biogeography/biogeochemical model such as BIOME3 or to a fully dynamic biosphere model such as IBIS (Foley *et al.*, 1996). To date, however, climate models do not adequately reproduce past biome patterns at a regional scale, as evidenced by the paleoecological data, probably due to the omission of significant feedbacks involving biospheric and ocean circulation changes. Recent results obtained by including additional feedbacks (e.g., Claussen and Gayler, 1997; Z. Liu and J. E. Kutzbach, personal communication) offer the promise of more realistic simulation of past climates and ecosystems, and the potential for comprehensive simulation of long-term coupled changes in the carbon cycles and climate.

The study of the past cannot be directly extrapolated to the future, for example, by looking for past analogs of what is expected for the next century, because it is difficult to find a period where all the climate forcings and boundary conditions are similar to the future ones. It can be understood from the 6000-yr BP period, which was warmer than today, but with a higher summer insolation and a much higher CO_2 concentration. Nevertheless, the past data can be used to test the climate and vegetation models by experimentally assigning to them conditions drastically different from the present one. If the models are able reasonably to simulate these paleoenvironments, they prove to be robust and we can have some confidence in them for predicting the future.

References

Adams, J. M, Faure, H., Faure-Denard, L., McGlade, J. M., and Woodward, F. I. (1990). Increases in terrestrial carbon storage from the Last Glacial Maximum to the Present. *Nature (London)* **348,** 711–714.

Barnola, J. M., Raynaud, D., Korotkevitch, Y. S., and Lorius, C. (1987). Vostok Ice Core provides 160000 years record of atmospheric CO_2. *Nature (London)* **329,** 408–414.

Beaulieu, J. L. de, Pons, A., and Reille, M. (1988). Histoire de la flore et de la végétation du massif Central (France) depuis la fin de la dernière glaciation. *Cahiers Micropaléontol.* **3,** 5–36.

Branchu, P., Faure, H., Ambrosi, J. P., Bakker, E. M. V., and Fauredenard, L. (1993). Africa as source and sink for atmospheric carbon dioxide. *Global Planet. Change* **7,** 41–49.

Cheddadi, R., Yu, G., Guiot, J., Harrison, S. P., and Prentice, I. C. (1996). The climate of Europe 6000 years ago. *Climate Dynam.* **13,** 1–9.

Claussen, M. (1997). Modeling bio-geophysical feedback in the African and Indian monsoon region. *Climate Dynam.* **13,** 247–258.

Claussen, M., and Gayler, V. (1997). The greening of Sahara during the mid-Holocene: Results of an interactive atmosphere–biosphere model. *Global Ecol. Biogeogr. Lett.* **6,** 369–377.

COHMAP members (1988). Climatic changes of the last 18,000 years: Observations and model simulations. *Science* **241,** 1043–1052.

Crowley, T. J. (1995). Ice age terrestrial carbon changes revisited. *Global Biogeochem. Cycles* **9,** 377–389.

Delcourt, P. A., and Delcourt, H. R. (1984). Late-Quaternary palaeoclimates and biotic responses across eastern North America and the northwestern Atlantic Ocean. *Palaogeogr. Palaeoclimatol. Palaeoecol.* **48,** 263–284.

Delcourt, P. A, Delcourt, H. R, and Davidson, J. L., 1983. Mapping and calibration of modern pollen–vegetation relationships in the southeastern United States. *Rev. Palaeobot. Palynol.* **39,** 1–45.

Delcourt, P. A, Delcourt, H. R., and Webb III, T. (1984). Atlas of mapped distribution of dominance and modern pollen percentages for important tree taxa of eastern North America. *Am. Assoc. Strat. Palynol. Contrib. Ser.* **14,** 1–131.

Duplessy, J. C., Shackleton, N. J., Fairbanks, R. G., Labeyrie, L., Oppo, D., Kallel, N. (1988). Deepwater source variations during the last climatic cycle and their impact on the global deepwater cicrculation. *Palaeoceanography* **3,** 343–360.

Esser, G. (1991). Osnabrück Biosphere Model: Structure, construction, results. *In* "Modern Ecology: Basic and Applied Aspects" (G. Esser and D. Overdick, eds.), pp. 679–709. Elsevier, Amsterdam.

Esser, G., and Lautenschlager, M. (1993). Estimating the change of carbon in the terrestrial biosphere from 18,000 BP to present using a carbon cycle model. *Environ. Pollut.* **83,** 45–53.

Ewel, K. C., Cropper, W. P., Jr., and Gholz, H. L. (1987). Soil CO_2 evolution in Florida slash pine plantations; change through time. *Can. J. For. Res.* **17,** 325–329.

Foley, J. A., Prentice, I. C., Ramankutty, N., Levis, S., Pollard, D., Sitch, S., and Haxeltine, A. (1996). An integrated biosphere model of land surface processes, terrestrial carbon balance, and vegetation dynamics. *Global Biogeochem. Cycles* **10,** 603–628.

Friedlingstein, C., Delire, C., Muller, J. F., and Gérard, J. C. (1992). The climate induced variation of the continental biosphere: A model simulation of the last glacial maximum. *Geophy. Res. Lett.* **19,** 897–900.

Friedlingstein, P., Prentice, K. C., Fung, I. Y., John, J. G., and Brasseur, G. P. (1995). Carbon-biosphere-climate interactions in the last glacial maximum climate. *Journal of Geophysical Research - Atmos.* **100,** 7203–7221.

Gordon, A. M., Schlentner, R. E., and Van Cleve, K. (1987). Seasonal patterns of soil respiration and CO_2 evolution following harvesting in the white spruce forests of interior Alaska. *Can. J. For. Res.* **17,** 304–310.

Harrison, S. P., Jolly, D., Laarif, F., Abe-Ouchi, A., Dong, B., Herterich, K., Hewitt, C., Joussaume, S., Kutzbach, J. E., Mitchell, J. F. B., de Noblet, N., and Valdes, P. (1998). Intercomparison of simulated global vegetation distributions in response to 6 kyr BP orbital forcing. *J. Climate* **11,** 2721–2742.

Haxeltine, A., and Prentice, I. C. (1996). BIOME3: An equilibrium terrestrial biosphere model based on ecophysiological constraints, resource availability, and competition among plant functional types. *Global Biogeochem. Cycles* **10,** 693–709.

Haxeltine, A., and Prentice, I. C. (1997). A general model for the light use efficiency of primary production. *Funct. Ecol.* **10,** 551–561.

Huntley, B. (1990). European vegetation history: Palaeovegetation maps from pollen data—13000 yr BP to present. *J. Q. Sci.* **5,** 103–122.

Huntley, B., and Prentice, I. C. (1988). July temperatures in Europe from pollen data, 6000 years before present. *Science* **241,** 687–690.

Huntley, B., and Prentice, I. C. (1994). Holocene vegetation and climates of Europe. *In* "Global Climates since the Last Glacial Maximum" (H. E. Wright, J. E. Kutzbach, F. A. Street-Perrott, W. F. Ruddiman, and T. Webb III, eds.), pp. 550–575. Univ. Minnesota Press, Minneapolis.

Jolly, D., Prentice, I. C., Bonnefille, R., Ballouche, A., Bengo, M., Brenac, P., Buchet, G., Burney, D., Cazet, J. P., Cheddadi, R., Edorh, T., Elenga, H., Elmoutaki, S., Guiot, J., Laarif, F., Lamb, H., Lezine, A.-M., Maley, J., Mbenza, M., Peyron, O., Reille, M., Reynaud-Farrera, I., Riollet, G., Ritchie, J.-C., Roche, E., Scott, L., Ssemmanda, I., Straka, H., Umer, M., Van Campo, E., Vilimumbalo, S., Vincens, A., and Waller, M. (2000). Biome reconstruction from pollen and plant macrofossil data for Africa and the Arabian peninsula at 0 and 6 ka. *J. Biogeogr.* (in press).

Joussame, S., and Taylor, K. (1995). Status of the PMIP. *In* "Proceedings of the First International AMIP Scientific Conference," pp. 425–430. WMO, Geneva.

Kutzbach, J. E., and Guetter, P. J. (1980). On the design of paleoenvironmental data networks for estimating large-scale patterns of climate. *Q. Res.* **14,** 169–187.

Kutzbach, J. E., Gallimore, R., Harrison, S. P., Behling P. J., Selin, R., and Laarif, F. (1999). Climate and biome simulations for the past 21,000 years. *Q. Sci. Rev.* **18,** 3777–3791.

Lieth, H. (1975). Modeling the primary production of the world. *In* "Primary Productivity of the Biosphere" (H. Lieth and R. H. Whittaker, eds.), pp. 237–263. Springer-Verlag, New York.

Lieth, H., and Box E. (1972). Evapotranspiration and primary productivity, C. W. Thornthwaite Memorial Model. *Publ. Climatol.* **25,** 37–46.

Maslin, M. A., Adams, J. M., Thomas, E., Faure, H., and Hainesyoung, R. (1995). Estimating the carbon transfer between the ocean, atmosphere and the terrestrial biosphere since the last glacial maximum. *Terra Nova* **7,** 358–366.

Meentemeyer, V., Gardner, J., and Box, E. (1985). World patterns and amounts of detritus soil carbon. *Earth Surf. Processes Landforms* **10,** 557–567.

Olson, J. S., Watts, J. A., and Allison, L. J. (1985). Major world ecosystem complexes ranked by carbon in live vegetation, a database. NPD-017. Carbon Dioxide Information Center, Oak Ridge National Laboratory, Oak Ridge, Tennessee.

Overpeck, J. T., Prentice, I. C., and Webb III, T. (1985). Quantitative interpretation of fossil pollen spectra: Dissimilarity coefficients and the method of modern analogs. *Q. Res.* **23,** 87–108.

Overpeck, J. T., Webb, R. S., and Webb, T. (1992). Mapping eastern North-American vegetation change of the past 18 KA—No analogs and the future. *Geology* **20,** 1071–1074.

Parsons, R. W., Gordon, A. D., and Prentice, I. C. (1984). Statistical uncertainty in forest composition estimates obtained from fossil pollen spectra via the R-value model. *Rev. Palaeobot. Palynol.* **40,** 177–189.

Peng, C. H., and Apps, M. J. (1997). Contribution of China to the global carbon cycle since Last Glacial Maximum: Reconstruction from palaeodata and empirical biosphere model. *Tellus* **49B,** 393–408.

Peng, C. H., Guiot, J., Van Campo, E., and Cheddadi, R. (1994). The vegetation carbon storage variation since 6000 yr B.P.: Reconstruction from pollen. *J. Biogeogr.* **21,** 19–31.

Peng, C. H., Guiot, J., Van Campo, E., and Cheddadi, R. (1995a). The variation of terrestrial carbon storage at 6000 yr BP in Europe: Reconstruction from pollen data using two empirical biosphere models. *J. Biogeogr.* **22,** 863–873.

Peng, C. H., Guiot, J., Van Campo, E., and Cheddadi, R. (1995b). Temporal and spatial variations of terrestrial biomes and carbon storage since 13 000 yr BP in Europe: reconstruction from pollen data and statistical models. *Water Air Soil Pollut.* **82,** 375–391.

Peng, C. H., Guiot, J., and Van Campo, E. (1995c). Reconstruction of the past terrestrial carbon storage of the Northern Hemisphere from the Osnabrück Biosphere Model and paleodata. *Climate Res.* **5,** 107–118.

Peng, C. H., Guiot, J., and Van Campo, E. (1998). Past and future carbon balance of European ecosystems from pollen data and climatic models simulations. *Global Planet. Change* **18,** 189–200.

Post, W. M., Emmanuel, W. R., Zinke, P. J., and Sangenberger, A. G. (1982). Soil carbon pool and world life zones. *Nature (London)* **298,** 156–159.

Prentice, I. C (1982). Calibration of pollen spectra in terms of species abundance. *In* "Palaeohydrological Changes in the Temperate Zone in the Last 15,000 Years—Vol. III: Specific Methods" (B. E. Berglund, ed.), pp. 25–51. Lund University, Lund, Sweden.

Prentice, I. C (1986). Forest-composition calibration of pollen data. *In* "Handbook of Holocene Palaeoecology and Palaeohydrology" (B. E. Berglund, ed.), pp. 25–51. John Wiley & Sons, New York.

Prentice, K. C., and Fung, I. Y. (1990). The sensitivity of terrestrial carbon storage to climate change. *Nature (London)* **346,** 48–51.

Prentice, I. C., and Webb III, T. (1998). Biome 6000: Reconstructing global mid-Holocene vegitation patterns from palaeoecological records. *J. Biogeogr.* **25,** 997–1005.

Prentice, I. C., Bartlein, P. J., and Webb III, T. III (1991). Vegetation and climate change in Eastern North America since the last glacial maximum. *Ecology* **72,** 2038–2056.

Prentice, I. C., Cramer, W., Harrison, S. P., Leemans, R., Monserud, R. A., and Solomon, A. M. (1992). A global biome model based on plant physiology and dominance, soil properties and climate. *J. Biogeogr.* **19,** 117–134.

Prentice, I. C., Sykes, M. T., Lautenschlager, M., Harrison, S. P., Denissenko, O., and Bartlein, P. J. (1993). Modelling the increase in terrestrial carbon storage after the last glacial maximum. *Global Ecol. Biogeogr. Lett.* **3,** 67–76.

Prentice, I. C., Guiot, J., Huntley, B., Jolly, D., and Cheddadi, R. (1996). Reconstructing biomes from palaeoecological data: A general method and its application to European pollen data at 0 and 6 ka. *Climate Dynam.* **12,** 185–194.

Prentice, I. C., Harrison, S. P., Jolly, D. and Guiot, J. (1998). The climate and biomes of Europe at 6000 yr BP: Comparison of model simulations and pollen-based reconstructions. *Q. Sci. Rev.* **17,** 659–668.

Raich, J. W., and Schlesinger, W. H. (1992). The global carbon dioxide flux in soil respiration and its relationship to vegetation and climate. *Tellus* **44B,** 81–99.

Schlesinger, W. H. (1977). Carbon balance in terrestrial detritus. *Annu. Rev. Ecol. Syst.* **8,** 51–81.

Texier, D., de Noblet, N., Harrison, S. P., Haxeltine, A., Jolly, D., Joussaume, S., Laarif, F., Prentice, I. C., and Tarasov, P. (1997). Quantifying the role of biosphere–atmosphere feedbacks in climate change: Coupled model simulations for 6000 years BP and comparison with palaeodata for northern Eurasia and northern Africa. *Climate Dynam.* **13,** 865–882.

Van Campo, E., Guiot, J., and Peng, C. H. (1993). A data-based re-appraisal of the terrestrial carbon budget at the Last Glacial Maximum. *Global Planetary Change* **8,** 189–201.

Webb III, T., and Bryson, R. A. (1972). Late- and Postglacial climatic change in the northern Midwest, USA: Quantitative estimates derived from fossil pollen spectra by multivariate statistical analysis. *Q. Res.* **2,** 70–115.

Webb III, T., Howe, S., Bradshaw, R. H. W., and Heide, K. (1981). Estimating plant abundances from pollen percentages: The use of regression analysis. *Rev. Palaeobot. Palynol.,* **34,** 269–300.

Wright, H. E., Jr., Kutzbach, J. E. , Webb III, T., Ruddiman, W. F., Street-Perrott, F. A., and Bartlein, P. J. (1994). "Global Climates since the Last Glacial Maximum," Univ. Minnesota Press, Mineapolis.

21

Global Terrestrial Productivity and Carbon Balance

Richard A. Houghton

I. Introduction

Changes in land use over the past 150 years reduced the area of forests worldwide and released about 115 Pg C to the atmosphere. Other changes in terrestrial ecosystems, independent of land use change, have led to an accumulation of about 90 Pg C on land. The factors responsible for this accumulation are not known, but most attention has focused on three factors: CO_2 fertilization, the mobilization of nitrogen through human activities, and changes in climate. All of these factors, including land use change, are expected to change dramatically in the future, and the net effect on global productivity is difficult to predict. The approach taken here to suggest the importance of different factors is to explore temporal and spatial variations in carbon flux as revealed by analyses of the global carbon cycle. For example, a northern midlatitude accumulation of carbon is consistent with N deposition, whereas an uptake largely in tropical regions suggests the importance of CO_2 fertilization. Unfortunately, analyses of atmospheric data are divided with respect to the geographic distribution of a terrestrial carbon sink. Data from forest inventories are not consistent with a large carbon sink in northern midlatitudes. The emphasis of this analysis is on net changes in carbon storage. The changes are not the same as net ecosystem productivity (NEP) because they include the fate of carbon harvested from forests. In addition to the 115 Pg C lost from terrestrial ecosystems to the atmosphere, about 15 Pg C are calculated to have accumulated in wood products. Changes in carbon storage and in NEP are better known globally than are changes in net primary productivity (NPP), especially for much of the world's managed lands. Monitoring NPP is critical for determining the sustainability of current and future uses of land.

The challenge of this chapter is to predict the future productivity of the globe. There seems to be general agreement about the factors important in affecting or limiting terrestrial productivity (temperature, moisture, nutrients, atmospheric CO_2, and human management, including both deliberate and inadvertent distribution of fertilizers and toxins) but less agreement about the relative importance of these factors. Indeed, their importance varies in time and space. The further uncertainty of how these factors may change in the future makes predictions of future productivity extremely risky. The factors important in determining productivity, however, are likely to be those responsible for the current accumulation of carbon on land, and thus an exploration of the temporal and spatial variability of this accumulation balance may help suggest the productivity of a future world. Such an exploration is the subject here.

This chapter is divided into three sections: the past century (1850 to 1990), recent years (1980 to 1995), and the future. In each section a comparison is made between the flux of carbon resulting from changes in land use and the flux resulting from other changes, presumably changes resulting from environmental changes. The comparison allows one to see the importance of land use change to global net ecosystem production and suggests the mechanisms responsible for the current, and by implication the future, accumulation of carbon on land. A terrestrial sink implies that terrestrial productivity is changing. The question is why is it changing. What are the mechanisms responsible? The answer may be suggested by variations in the rate of accumulation over time or by geographic differences in the source/sink of terrestrial ecosystems.

The emphasis here is on changes in carbon storage, which are not equivalent to net ecosystem productivity (NEP). Changes in carbon storage are equal to the sum of NEP plus changes in wood products. Harvests increase the amount of carbon held in wood products; decay decreases the amount. If NEP is defined as the difference between net primary production and heterotrophic respiration, it ignores harvests altogether (Fig. 21-1a). In this figure the value of NEP integrated over 100 years is 100 g C m^{-2}. On the other hand, if NEP is defined by changes in the stocks of biomass and soil carbon, NEP must account for import or export or organic matter. Harvests are an export from the system and, when included, reduce NEP. The decay of these products, on the other hand, is never considered as a part of NEP. The effect of this harvested material on NEP is shown in Fig. 21-1b, where a forest is disturbed (as though harvested) and all of the biomass is left on site. The subsequent decay of this material makes NEP initially negative (for 20 years in Fig. 21-1b) and reduces it overall. In Fig. 21-1b NEP is zero when integrated over years 101 to 200 (decay of the disturbed biomass equals re-

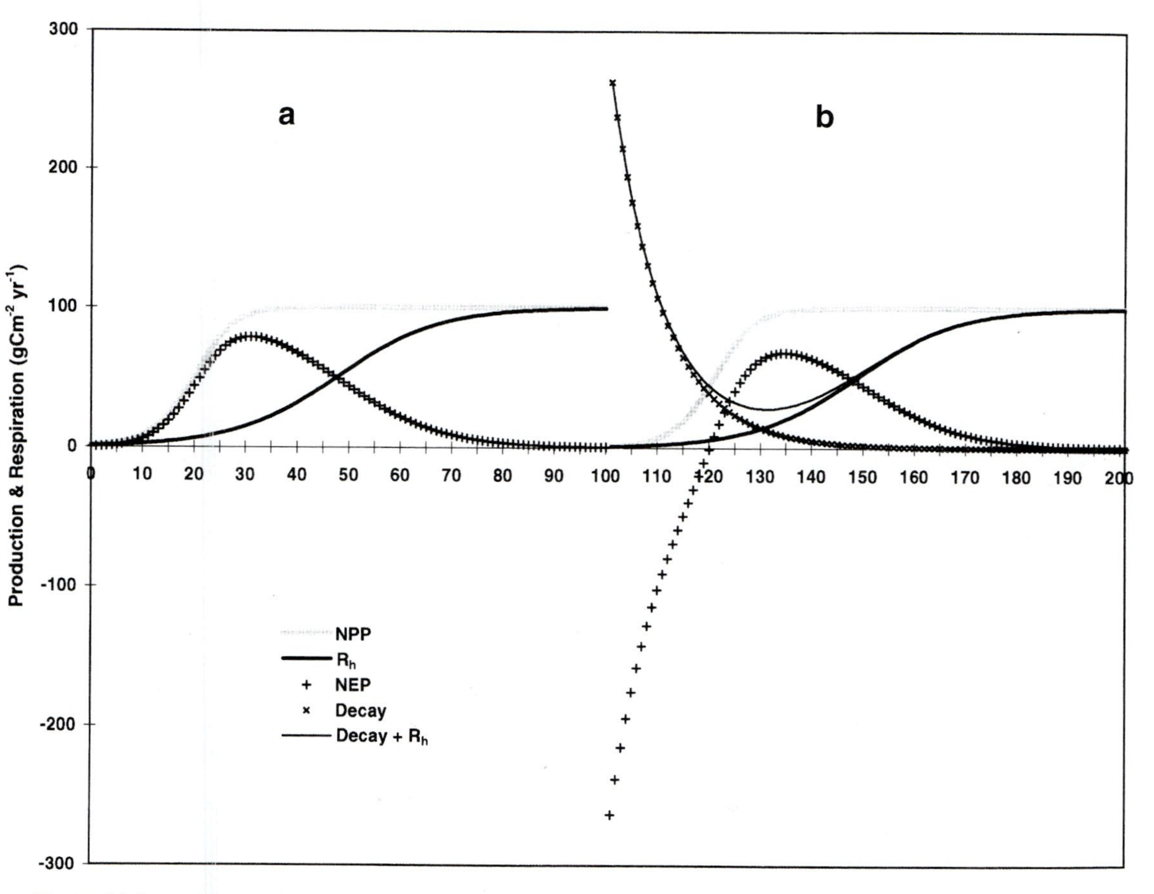

Figure 21-1 NPP, R_h, and NEP during forest succession. (a) The decay of organic matter accumulated before the year of disturbance (year 1) is not included. (b) The decay of organic matter accumulated before the disturbance in year 101 decays in the years following that disturbance.

growth of the new biomass). NPP and R_h are identical in a and b in Fig. 21-1. Figure 21-1b also includes decay of the 100 g C m^{-2} present at the time of disturbance. This decay is part of the respiration (Decay + R_h) that defines the new NEP. The emphasis in this paper on changes in carbon storage treats decay of harvested material as though it were still a part of the forest from which it was removed.

II. Carbon Fluxes, 1850 to 1990

Estimates of the net terrestrial flux of carbon can be derived from inversion calculations with ocean models and historic variations of atmospheric CO_2 concentrations (Siegenthaler and Oeschger, 1987; Keeling *et al.*, 1989; Sarmiento *et al.*, 1992). Forward calculations with an ocean model calculate the atmospheric concentrations of CO_2 that result from annual emissions of carbon. Inverse calculations determine the net annual additions or removals of carbon from the atmosphere required by observed concentrations. After subtracting the annual fossil fuel emissions from the calculated exchange, the residual source or sink is assumed to represent a change in the carbon content of terrestrial ecosystems. By convention, sources are represented as positive values because they add carbon to the atmosphere (although they represent a loss of carbon from land). The inversion by Sarmiento *et al.* (1992) shows that between 1850 and 1990 terrestrial ecosystems were nearly neutral with respect to carbon storage. They lost carbon to the atmosphere early in the period and accumulated it more recently (Fig. 21-2). The net flux over the entire 140-yr period was a loss of 25 Pg C. In contrast to the release calculated by inversion, the estimate calculated from a reconstruction of land use changes suggests that approximately 112 Pg C were released to the atmosphere (Houghton and Hackler, 1995). The annual loss of carbon increased from approximately 0.4 Pg C yr^{-1} in the middle of last century to 1.7 Pg C yr^{-1} in 1990 (Fig. 21-2).

It has been argued previously (Houghton, 1995) that the difference between the two estimates of change (between the inversion techniques and land use change, respectively) may represent real changes in terrestrial carbon storage that are unrelated to land use change. The inversion analysis defines a total net terrestrial flux of carbon, whereas the flux calculated on the basis of land use includes only that portion of the flux attributable to human management of the land. Thus the difference may be interpreted to indicate a flux attributable to changes in NPP or Rse (ecosystem respiration = autotrophic + heterotrophic respiration) that are not a direct result of human activity. The factors responsible include changes in the rates of natural disturbances, CO_2 fertilization of NPP (thereby potentially increasing

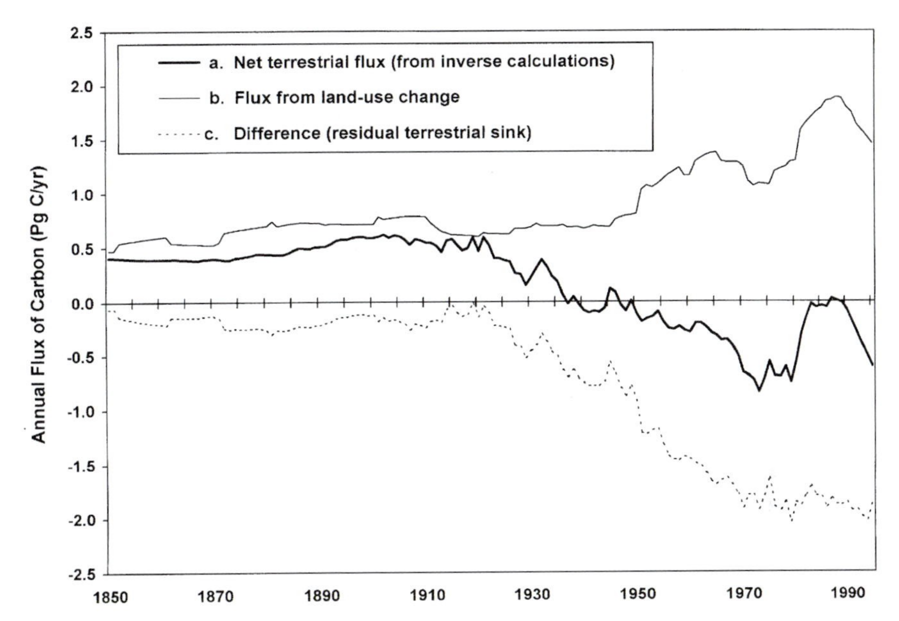

Figure 21-2 The annual loss of carbon from ecosystems affected by land use (from Houghton and Hackler, 1995), the annual net loss or accumulation of carbon in all terrestrial ecosystems [calculated by inversion (Sarmiento *et al.*, 1992)], and the difference between the two estimates of flux (or the residual terrestrial flux) (from Houghton, 1995). Positive values indicate a loss of carbon from land; negative values indicate an accumulation of carbon on land.

NEP), fertilization of NPP with nitrogen or other nutrients mobilized through human activity, variations in climate, or a combination of these factors.

If this interpretation of the difference is correct, Fig. 21-2 shows that prior to about 1920 the annual flux of carbon from agricultural and forestry practices seemed to account for the total flux of carbon. Other changes were either not important or offsetting in their effects on carbon storage. Only after about 1920 did the difference between the two estimates become significant. Since 1920 the annual difference has generally increased, interpreted here to indicate an accumulation of carbon on land in response to factors unrelated to land use.

In sum, changes in land use are estimated to have caused a loss of 112 Pg C from land, whereas other factors appear to have led to an accumulation of about 87 Pg C. The net effect was a loss of about 25 Pg C (Sarmiento *et al.*, 1992). The effects of land use change on the loss of carbon from land and the effects of other factors offsetting this loss have been as large as other terms in the global carbon equation (net exchange of carbon between 1850 and 1990, in Pg):

$$\text{Emissions from fossil fuels} + \text{Emissions from land-use change} = \text{Atmospheric increase} + \text{Oceanic uptake} - \text{Residual terrestrial uptake}$$

$$220 \quad + \quad 112 \quad = \quad 145 \quad + \quad 100 \quad - \quad 87. \qquad (1)$$

The direct effects of land use change released more than four times as much carbon as the net loss in terrestrial carbon storage over this period. Land use change is an important determinant of NEP. The factors responsible for the nonanthropogenic accumulation of 87 Pg C over this period are uncertain. However, the temporal pattern of this source/sink suggests a correlation with several environmental variables. For example, over the 140-yr period 75% of the annual variation in flux was explained with a linear regression against CO_2 concentrations (Houghton, 1995). Because the increase in CO_2 concentration is related to industrial activity, the regression does not allow the possible effects of rising CO_2 to be separated from the possible effects of increasing nitrogen mobilization. For the more recent decades (1940–1990) temperature was more important than CO_2 in explaining the annual variation in flux (Houghton, 1995). Warmer years appeared to be followed up to 7 yr later by reduced accumulations of carbon on land. The correlation implies that a warmer earth will reduce the accumulation of carbon, at least during the transition to a new climate.

III. Carbon Fluxes, 1980 to 1995

A. Analyses Based on Atmospheric Data

An equation similar to Eq. (1), but expressed on an annual basis, can be constructed for the 1980s (from Schimel *et al.*, 1995):

$$\text{Emissions from fossil fuels} + \text{Emissions from land-use change} = \text{Atmospheric increase} + \text{Oceanic uptake} - \text{Residual terrestrial uptake}$$

$$5.5\ (\pm 0.5) + 1.6\ (\pm 0.7) = 3.3\ (\pm 0.2) + 2.0\ (\pm 0.8) - 1.8\ (\pm 1.2). \quad (2)$$

Notice that the net terrestrial flux averaged approximately zero in the 1980s, a release of 1.6 Pg C yr^{-1} from land use change and a residual uptake of 1.8 Pg C yr^{-1}.

The previous section addressed temporal variations in the global carbon imbalance, but a number of independent analyses based on atmospheric data and models provide a geographic, rather than temporal, resolution of the current imbalance. For example, the analysis by Ciais *et al.* (1995) found a net global terrestrial sink of about 2 Pg C for the years 1992/1993. This

Table 21-1 Terrestrial Fluxes of Carbon Calculated from Inverse Modeling of Atmospheric Data[a]

Approach: net flux from atmospheric data and models	Year	Latitude				
		Global	30°–60°N	30°N–30°S	0°–30°N	0°–30°S
CO_2 and $^{13}CO_2$ (Ciais *et al.*, 1995)	1992/1993	−1.8	−3.5 (±0.9)	1.7 (±1.6) [0.1][b]	2.0 (±1.3)	−0.3 (±1.0)
O^2/N^2 (Keeling *et al.*, 1996)	1991–1994	−2	−2	0 [−1.6]	—	—
CO_2, $^{13}CO_2$, and N_2/O_2 (Rayner *et al.*, 1999)	1980–1995	−0.7	−0.7	0 [−1.6]	−0.2	0.2
Flux from changes in land use (Houghton and Hackler, 1995)	1990	1.7	0.0	1.6	—	—
Flux calculated with data from forest inventories (Houghton, 1998)	1980s	—	−0.6	—	—	—

[a]Calculated from changes in land use, and from forest inventories (units are Pg C yr^{-1}). Negative values indicate an accumulation of carbon on land.
[b]Brackets indicate the residual flux, that is, net terrestrial flux minus flux from changes in land use. Summed values may not equal totals because of rounding.

global sink was partitioned between a northern midlatitude sink of 3.5 Pg C and a tropical source of 1.7 Pg C (Table 21-1). Perhaps fortuitously, the net flux attributable to land use change in 1990 (Houghton, 1996) was identical: essentially all of the 1.7-Pg C release was from the tropics. However, Ciais *et al.* (1995) found that the northern tropics were a source of carbon and the southern tropics were a sink. The flux of carbon from deforestation, on the other hand, is estimated to have been evenly distributed about the equator (Houghton, 1996). Part of the difference between the two analyses may result from the different years (1992/1993 vs. 1990). The difference also suggests the importance of nonanthropogenic factors in affecting metabolic processes.

The large sink of 3.5 Pg C in northern midlatitudes is also difficult to explain except as a short-term anomaly resulting from the unusually cold (and wet) year (Keeling *et al.*, 1995) following the eruption of Mount Pinatubo. The year 1992 was atypical for the decade of the 1980s in that the growth rate of atmospheric CO_2 was low: 1.2 Pg C yr^{-1} rather than the average of 3.2 Pg C yr^{-1} for the 1980s (Conway *et al.*, 1994; Lambert *et al.*, 1995). The difference suggests a global sink in 1992 that was 2 Pg C greater than observed during the 1980s.

Other analyses based on atmospheric data and modeling have suggested that the major accumulation of carbon on land occurs in the tropics rather than the northern midlatitudes. One such analysis, based on latitudinal variations in O_2/N_2 ratios over the period 1991–1994, found that northern lands were accumulating about 2 Pg C yr^{-1} and that tropical lands were approximately in balance (Keeling *et al.*, 1996) (Table 21-1). The net flux of 0 for the tropics requires a terrestrial sink of about 1.6 Pg C yr^{-1} to balance the source from deforestation. Thus the terrestrial sink was evenly distributed over northern and tropical lands. The inversion analyses by Enting *et al.* (1995), based on spatial and seasonal variation in CO_2 concentrations, and by Rayner *et al.* (1997), based on CO_2, $^{13}CO_2$, and O_2/N_2 ratios, also suggest that the net accumulation of carbon on land was distributed evenly. If the estimates of emissions from tropical deforestation are correct, a major sink for carbon must be tropical lands.

Although the results of these atmospheric analyses are not entirely consistent, they all indicate a net terrestrial sink of 1 to 2 Pg C yr^{-1} (Table 21-1). The carbon balance that many of them show for tropical lands suggests that processes unrelated to land use change are offsetting the release of carbon from deforestation, and the net uptake in northern midlatitude regions suggests that the uptake there is also unrelated to the recovery of forests from harvests in earlier years (Houghton, 1996). The processes responsible for the accumulation of carbon in terrestrial ecosystems, whether in the tropics or in temperate zone and boreal lands, are thought to involve some combination of a recovery of forests from past disturbances (Kurz *et al.*, 1995), CO_2 fertilization, and the deposition of nitrogen mobilized through human activities.

The spatial distribution of the residual terrestrial flux might help resolve the relative importance of CO_2, as opposed to mobilized nitrogen, in explaining the accumulation of carbon on land. A northern distribution of the carbon sink would be consistent with the geographic distribution of nitrogen deposition (Holland *et al.*, 1997). The distribution of N fixed by internal combustion and fertilizers is largely confined to eastern North America and Europe rather than to the tropics. The history of N deposition is also consistent with the increasing residual terrestrial sink since 1920 (Fig. 21-2), but the correlation is not helpful in distinguishing between the effects of N and CO_2. If a major portion of the terrestrial sink were in the tropics, on the other hand, CO_2 fertilization would be implicated (Lloyd, 1997), in part because its effect is greater at higher temperatures (Long, 1991). At present the geographic distribution of carbon accumulation is in question; the calibration of ^{13}C among different laboratories has discrepancies yet to be resolved (R. J. Francey, personal communication, 1997).

B. Forest Inventories

Until recently, one might have argued that the accumulation of carbon on land was not directly measurable because the changes are small relative to the background stocks of carbon in trees and soil. However, foresters in most countries of the northern temperate zone have been conducting forest inventories for years. Rates of forest growth have increased in Europe (Spiecker *et al.*, 1996), for example, and the question is to what extent the increase is a result of management, as opposed to environmental changes? Forest inventories exist for most of the countries in northern midlatitudes, and carbon budgets have been constructed for the forests of Canada (Apps and Kurz, 1994), the United States (Turner *et al.*, 1995), Europe (Kauppi *et al.*, 1992), and the former Soviet Union (Kolchugina and Vinson, 1993, 1995; Shvidenko and Nilsson, 1994, 1997, 1998). The measurements seem to indicate an accumulation of carbon that is greater than expected from the regrowth associated with previous logging. However, the magnitude of the sink in northern temperate zone forests is small relative to the 2 to 3.5 Pg C yr^{-1} sink calculated by some of the analyses based on atmospheric data and models (Ciais *et al.*, 1995; Keeling *et al.*, 1996). Forest inventories provide data for a more direct estimate of changes in carbon storage than estimates based on rates of harvest and regrowth, and a comparison of the two approaches is revealing.

Figure 21-3 shows estimates of carbon exchange for the forests of North America, Europe, and the former Soviet Union, combined. The estimates are from two types of analyses, one calculated from changes in land use (largely wood harvest) and the other based on data from forest inventories. For each of the analyses carbon is divided among components of the forest ecosystem: living biomass, slash (logging debris), wood products removed from the forest, and soils. Values above the line represent an annual loss of carbon to the atmosphere; values below the line indicate the average annual accumulation of carbon in these forests. A comparison of the two approaches shows, first, that measured changes in carbon stocks (from forest inventories) indicate a net sink (580 Tg C yr^{-1}), whereas changes calculated on the basis of harvests and regrowth indicate a small source (40 Tg C yr^{-1}) not different from zero. The absolute difference of about 600 Tg C yr^{-1} suggests that these forests are accumulating carbon more rapidly than would be expected from past rates of harvest. The difference is smaller than the estimate derived recently by Houghton (1996) (800 Tg C yr^{-1}) because of revisions in the Russian inventories (Kolchugina and Vinson, 1993, 1995; Shvidenko and Nilsson, 1994, 1997). Even the sink of 600 Tg C yr^{-1} may be high because it is based, in part, on an average increase in the growing stocks of Russian forests between 1961 and 1983, whereas the stocks of eastern Siberia have declined since 1983 (Shvidenko and Nilsson, 1994). Neverthe-

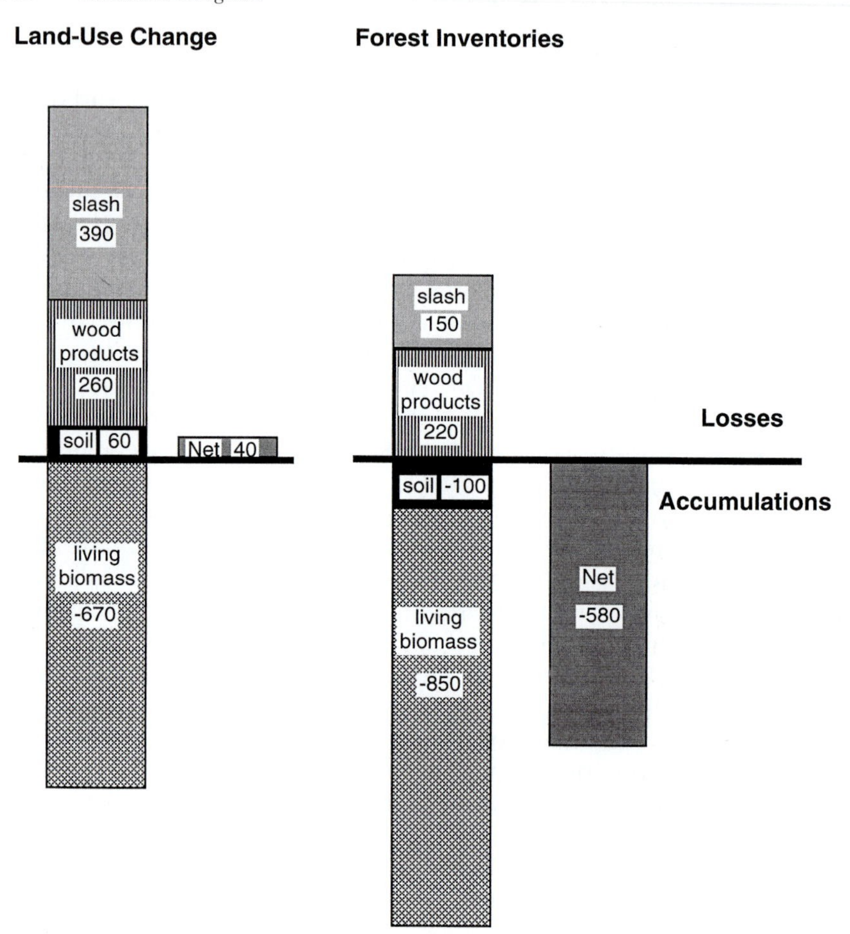

Figure 21-3 Average annual losses and accumulations of carbon in components of northern midlatitude forests according to two types of analyses (from Houghton, 1998). Histograms are scaled to units: Tg C yr^{-1}. Positive values indicate a release of carbon to the atmosphere from a component; negative values indicate an accumulation of carbon.

less, the results agree reasonably well with independent analyses. For example, the forest inventories summarized here indicate an accumulation of 300 Tg C yr^{-1} in living biomass (not in Fig. 21-3), whereas the value computed by Kohlmaier *et al.* (1995) with data from the ECE/FAO (1992) ranged between 210 and 330 Tg C yr^{-1}. Furthermore, the annual release of carbon from oxidation of wood products is approximately similar (260 and 220 Tg C yr^{-1} for the approaches based on harvests and inventories, respectively) (Fig. 21-3). The similarity gives confidence that the two approaches are consistent when they address the same process, namely, harvest.

The difference between the two approaches is greatest for the losses of carbon attributed to decay of slash (logging debris) (390 and 150 Tg C yr^{-1}, respectively) and for the uptake of carbon in forest growth (670 and 850 Tg C yr^{-1}, respectively). Estimates of the release of carbon from slash are consistently higher in analyses based on land use change than in analyses based on forest inventories. To resolve the difference, further work is needed to determine the amount of woody debris left behind at harvest, historically as well as currently. The uptake of carbon in living biomass refers to the annual growth of forests prior to harvest. If both analyses are accurate, about 80% (670/850) of the carbon taken up by forest growth can be attributed to the regrowth expected as a result of previous harvests. Only 20%, or 180 Tg C yr^{-1}, is "unexplained." The unexplained uptake may result from changes in environmental factors or from management, including fire suppression. It is important to note that silvicultural and other management techniques, including suppression of fire, were not included in the analyses based on land use change. To the extent that such changes in management have contributed to an increased accumulation of carbon in regrowing forests, there is little need to invoke either CO_2 fertilization or nitrogen deposition to explain the observed accumulation of carbon. Even without including management in the analyses based on land use change, the difference between the uptake of carbon unexplained by regrowth and the uptake calculated by analyses of atmospheric data leads one to conclude that either forest inventories have underestimated the accumulation of carbon, or the missing carbon sink must be elsewhere: in soils, in ecosystems other than forests, or outside the northern midlatitudes altogether.

In the tropics systematic forest inventories are rare, and so there exist no independent measurements of carbon storage besides those determined on the basis of land use change. There are, however, a small number of local inventories (Brown *et al.*, 1994; Phillips *et al.*, 1998) and a few sites where the flux of CO_2 has been measured directly during both wet and dry seasons (Grace *et al.*, 1995). Both of these measurements suggest that carbon is accumulating in tropical forests. The findings are intriguing but preliminary. Further studies are needed to determine whether a net uptake in growing vegetation is the result of recovery from an earlier disturbance, or whether growth is enhanced by a change in the environment (CO_2 or N fertilization).

In many parts of the tropics selective harvests appear to be reducing biomass. Although the transformation of forests to agricultural lands is the most visible change and releases the greatest amount of carbon to the atmosphere, harvests of wood are, nevertheless, significant in increasing R_h relative to NPP and, thus, in losing carbon. The evidence for this reduction in carbon stocks comes from field studies (Uhl and Vieira, 1989; Woods, 1989; Gajaseni and Jordan, 1990), from recent forest inventories in southeast Asia

(Brown *et al.*, 1994) and Africa (Brown and Gaston, 1995), and from historical analyses of tropical Asia (Flint and Richards, 1994; Houghton and Hackler, 1994). Selective logging is also increasing in the forests of Amazonia (Nepstad *et al.*, 1997). The logging is apparently reducing biomass more rapidly than growth restores it. The net effect accounts for a loss of carbon that is about 50% as large as the loss from the conversion of forests to agriculture (Brown *et al.*, 1994; Flint and Richards, 1994; Houghton and Hackler, 1994; Brown and Gaston, 1995).

IV. Future Changes in Carbon Storage and NPP Likely as a Result of Human Activity

The world of the future is not difficult to imagine, qualitatively. It is likely to have more people, perhaps twice as many as exist at present. This increase in human population is likely to increase the area of agricultural land and the rate at which wood is harvested. More carbon is likely to be released to the atmosphere as a result. The world is also likely to be warmer with higher levels of atmospheric CO_2. The changes in temperature and moisture that result are difficult to predict, especially for individual regions, but the changes are likely to affect the productivity of ecosystems, especially those not used or managed by the increasing number of human inhabitants. At the same time, concern about continued climatic change is likely to put additional pressures on terrestrial productivity as renewable forms of energy are substituted for fossil fuels. Forests may be set aside as carbon sinks, and timber and fuel plantations may be expanded to provide substitutes for energy-intensive products and fossil fuels. The competition for land will be intense. There will be fewer undisturbed ecosystems. Some of the changes may increase NPP, but the intensification of land use, whether for agriculture or for wood products, will require more energy and will probably lead to a reduction in the amount of carbon held on land. At the same time, the cycling or turnover of carbon (NPP and R_h) may be higher.

A. Conversion of Unmanaged Ecosystems to Agricultural Ecosystems

The largest changes in carbon storage from human activity in the past and, most likely in the future, are losses of carbon as a result of the transformation of forests to agricultural lands. Over the period 1850 to 1990 the transformation accounted for about 75% of the carbon lost through land use change; harvest of wood, with subsequent regrowth of harvested forests, accounted for the remaining 25% of the long-term loss (Houghton and Hackler, 1995) (Fig. 21-4).

The fraction of the land surface covered with agriculture is relatively small (Fig. 21-5), and the changes in these human-dominated systems, even over

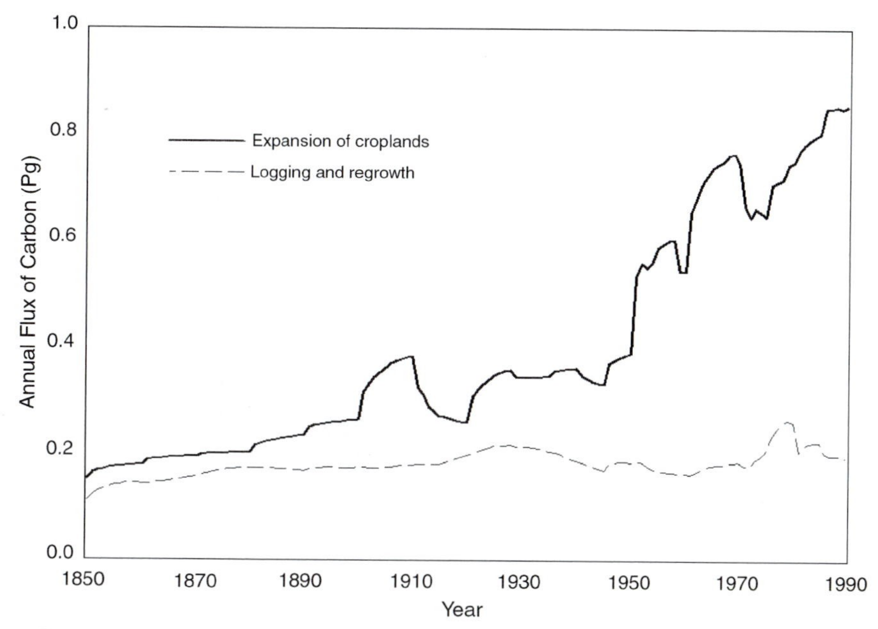

Figure 21-4 The annual, global flux of carbon from ecosystems converted to agriculture and from harvested forests (including decay of products) (from Houghton and Hackler, 1995). Positive values of flux indicate a loss of carbon from land.

140 yr, appear small relative to the large areas of natural systems. These natural systems include desert, rock, tundra, and ice, however, as well as boreal forests and other areas unlikely to be suitable for agriculture. The inset in Fig. 21-5 gives a different scale for the change in forest area, and shows the long-term change more dramatically.

To project future changes in croplands, these cropland areas, together with population data (McEvedy and Jones, 1978; FAO, 1995), were used to express changes in cropland as a function of population (Fig. 21-6). Population projections from the World Bank (1991) to the year 2100 were used to define a range of future cropland areas. Future areas of pasture were ignored. It was assumed that pastures would be converted to croplands if more food production were needed. The extrapolations of future cropland area given below may thus be underestimates of total agricultural area. One set of projections was based on the assumption that world per capita cropland area would average 0.1 ha/person, the current average for China (Fig. 21-6). Under this assumption, world populations of 7.6, 11.3, and 17.6 billion people would require 1.5, 1.6, and 1.8 $\times$ 10^9 ha of croplands. These areas represent increases of 1, 15, and 28%, respectively, over the 1990 cropland area (approximately 1.4 $\times$ 10^9 ha).

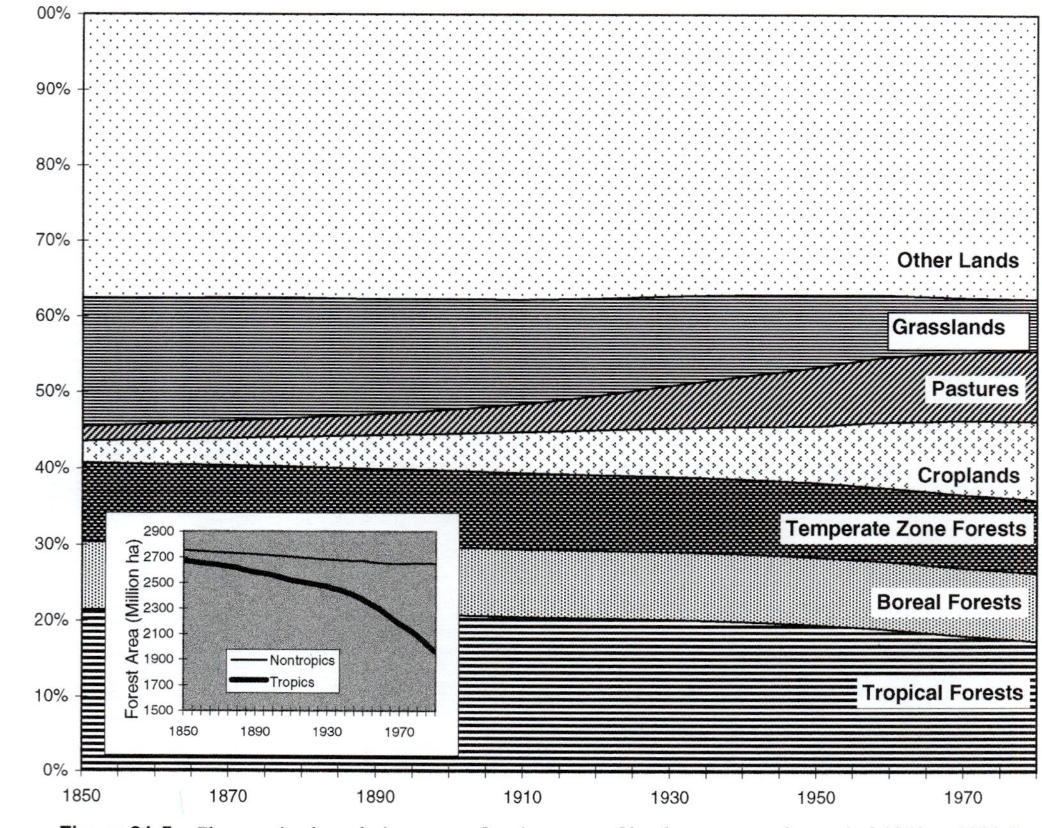

Figure 21-5 Changes in the relative areas of major types of land cover over the period 1850 to 1990. Inset: Changes in the area of forests.

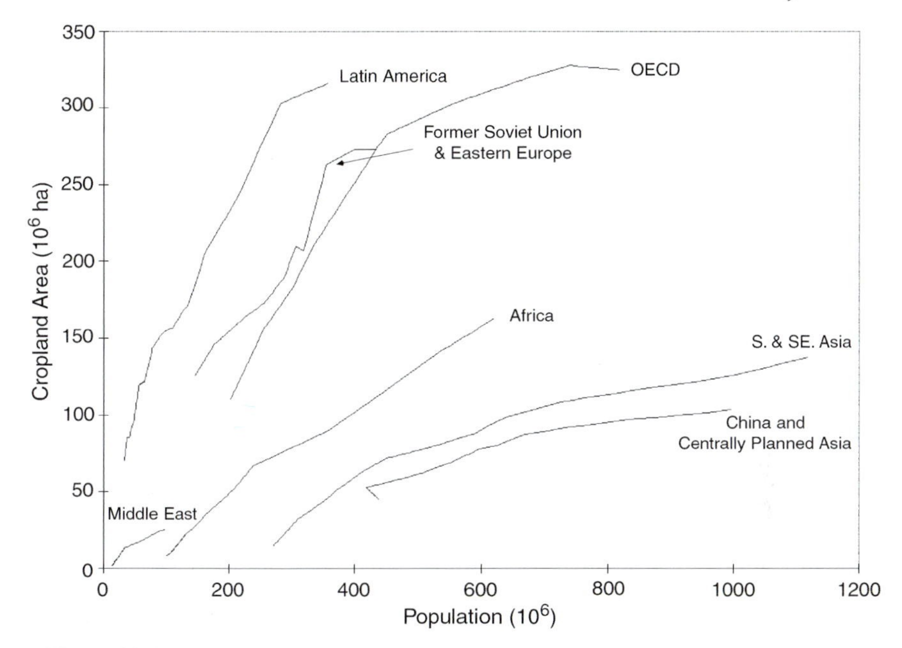

Figure 21-6 Croplands expressed as a function of population size for economic regions of the world over the period 1850 to 1990.

Another extrapolation was based on the assumption that the current per capita area of croplands would remain constant into the future. This extrapolation gave 1.7, 2.1, and 2.8 × 10^9 ha of croplands for the low, medium, and high projections of population increase (World Bank, 1991). The area of croplands required in this case for the year 2100 increased by 21, 50, and 100%. The total range of increase in croplands projected by these extrapolations varies between essentially no change and a doubling of croplands for increases in world population that vary between 35 and 215% above current numbers.

The projections are conservative. They assume that the current use of croplands is sustainable. On the contrary, current practices in many parts of the world seem to be leading to a loss of soil or soil fertility that essentially precludes that land from further use in agriculture. Thus, although per capita agricultural area may not be expanding, new croplands are, nevertheless, being added each year to replace worn-out lands. If the abandoned lands do not recover their productivity, then the transformation of agricultural lands to degraded lands will contribute importantly to reducing the world's productivity. This process seemed to be indicated by data (FAO, 1991) showing that the net loss of forest area in the tropics was twice as large as the net expansion of croplands and pastures combined (Houghton, 1994). The difference was made up by an expansion of the FAO category "other land." More recent data in the FAO's Production Yearbooks do not confirm this trend. The area of other land seems now to be decreasing as croplands increase. To the extent that degradation is important at present and becomes even more so in the future, the extrapolations underestimate future changes in land cover and future declines in NPP.

The argument can also be made that the extrapolations above are pessimistic. The development of new technologies and the greater application of existing technologies to regions where they are lacking may lower per capita cropland worldwide. A more populated world might require even less croplands than are in existence currently.

B. Harvest of Wood from Forests

Rates of harvest may either increase or decrease NPP in the future. Over the period 1850–1990 the harvest of wood from boreal, temperate zone, and tropical forests (increasing from about 5×10^6 ha yr^{-1} in 1850 to almost 20×10^6 ha yr^{-1} in 1980) was responsible for affecting about 1200 10^6 ha over the 140-yr interval. The area of forests harvested was larger than the area transformed to croplands.

About 25 Pg C were released to the atmosphere between 1850 and 1990 as a result of harvest of wood. The net release includes the recovery of logged forests and the storage and oxidation of wood products removed from the forest. The nearly constant net release of about 0.2 Pg C yr^{-1} from harvests over the period 1850 to 1990 (Fig. 21-4) is surprising because the world's population increased about fivefold over this period. Global rates of harvest also increased by a factor of five (Houghton and Hackler, 1995). Annual emissions of carbon remained fairly constant because the accumulations of carbon in forests recovering from logging generally lag the emissions. Thus, with increasing rates of harvest, emissions are always ahead of (and higher than) accumulations, and the net difference has been relatively constant over time.

The nearly constant rate of carbon loss from global harvest and regrowth obscures regional variations. In the decade of the 1980s, for example, the global net release of about 200 Tg C yr^{-1} from harvest of wood was composed of a tropical source of 240 Tg C yr^{-1} and a north temperate zone sink of 40 Tg C yr^{-1}. The uptake by northern forests illustrates that harvests may result in an accumulation of carbon on land (positive NEP), especially as rates of harvest decline. On the other hand, if rates of harvest are high relative to rates of recovery, as they are throughout much of the tropics, carbon will be lost and NEP, defined as a change in stocks, will also decrease.

Global productivity in the future is likely to be affected by increased rates of harvest. Increased rates will lower the storage of carbon in forests, and the average age of forests will also decrease. In regions where harvests are sustainable, NPP may be increased. In areas where it is not, NPP can be expected to decrease.

C. Productivity in Unmanaged or Natural Ecosystems

Several chapters in this volume discuss how NPP and NEP are affected by environmental variables. Here the approach has been to explore regional and global variations in terrestrial carbon accumulation. Much of the emphasis has been on changes in land use, but the difference between total carbon flux (as deduced from analyses of atmospheric data or forest inventories) and that flux attributable to land use suggests that the flux of carbon to unmanaged lands has also changed (Fig. 21-2). The patterns are consistent with the arguments that N deposition (in northern forests) and CO_2 fertilization (in the tropics) have increased NEP. CO_2 fertilization may be expected to continue to increase NEP as concentrations continue to rise. Nitrogen deposition may also increase NEP in tropical regions as fertilizer use and industrialization increase there; it may reduce NPP and NEP in northern regions if they become saturated with N (Aber *et al.*, 1989, 1995). It is important to note, however, that the terrestrial carbon sinks currently important in reducing the overall emissions of carbon from land, whether they operate through CO_2 or N fertilization, depend largely on forests. And forests, in the tropics, are being converted to agricultural and degraded ecosystems at about 1% yr^{-1} (FAO, 1993). Not only is carbon released, but the function of these forests as carbon sinks also decreases. The accumulation of carbon on land that has been increasing annually since 1920 may diminish in the future as forests continue to be cleared. It may also diminish, or reverse from a sink to a source, if the temperature of the earth increases rapidly and stimulates R_h relative to NPP (Woodwell *et al.*, 1998).

V. Conclusions

The types of land use change that affect NPP and NEP may be divided into three categories:

1. Conversion of natural systems to long-term ("permanently") managed agricultural lands, and the reverse (reforestation, afforestation, and the abandonment of agricultural lands)
2. Alternate clearing and abandonment of forests and woodlands (temporary clearing), including shifting cultivation and harvest of wood
3. Degradation of agricultural lands and lands unsustainably harvested

The three categories actually form a continuum without sharp distinctions. Nevertheless, they are different with respect to their effects on productivity and the extent to which they are known.

With regard to NEP, all three categories of change are followed by a negative NEP, if NEP is defined as the net change in stocks. The obvious exception, of course, is the conversion of open or cleared lands to forested lands. Although growth of forests following harvest increases NEP, the net effect of harvests and regrowth often results in a reduced storage of terrestrial carbon when averaged over an entire rotation, because only a small fraction of the initial biomass remains in long-lasting products.

With regard to NPP, it is not clear how the first two categories affect global NPP—that is, whether the net effect has been positive or negative. Locally eroded or salinized soils that support almost no vegetation in areas once highly productive indicate a reduction in NPP, whereas irrigated and fertilized croplands in areas formerly occupied by semiarid grasslands indicate increased productivity locally. The third category has clearly reduced NPP globally.

Ironically, the first two categories of change (those for which the effect on NPP may be either positive or negative, and for which the net global effect is probably small) are, or have the potential to be, relatively well determined either with land use statistics or with satellite data. The third category of change, on the other hand, for which the effect on NPP is most clear, is poorly determined, being neither well documented in land use statistics nor readily revealed with analysis of satellite data. There are, or course, exceptions. There have been attempts to estimate the global distribution and severity of soil erosion (UNEP/ISRIC, 1990) and the effect of erosion on reducing agricultural productivity (Kovda, 1983; Lal and Stewart, 1990). There are also local studies of degradation. For example, Sharma and Bhargava (1988) successfully used Landsat MSS data to measure the areas of saline and waterlogged soils in Uttar Pradesh, India, and Dwivedi *et al.* (1997) recently monitored erosion in northeastern India with a combination of satellite data. Despite some noteworthy exceptions, however, the degradation of land globally and the concomitant reduction in global NPP and NEP are poorly known. The process of degradation is relatively simple to recognize, but the changes in NPP along its path are difficult to quantify over large areas. The process seems to be widespread in subhumid tropical and semiarid areas (Syers *et al.*, 1996). It may be largely irreversible in some instances. It may include temporary recoveries in NPP through fallows, or it may spiral down rapidly. Estimates of the amount of agricultural land lost through degradation vary from 3×10^6 (Buringh, 1981), to 10×10^6 (Pimentel *et al.*, 1995), to 12×10^6 ha yr^{-1} (Lal and Stewart, 1990). An indirect estimate for the tropics alone suggested a rate of loss equivalent to about half the rate of tropical deforestation (Houghton, 1994), or $7–8 \times 10^6$ ha yr^{-1} (FAO, 1993), but the estimate is extremely sensitive to uncertainties in definition and measurement. Given the importance of NPP to the human enterprise, quantification of changes in NPP, globally, is one of the most fundamental and useful measures of change that can be made. Such changes are not limited to managed systems, of course, but changes in managed systems are likely to have the most immediate effects on the growing numbers of people dependent on them. Measurement of changes in NPP is both critically important and a major challenge.

Acknowledgments

Eric Davidson raised questions and made suggestions that improved the precision and clarity of this analysis. Joe Hackler produced the graphics. The research was supported through the Joint Program on Terrestrial Ecology and Global Change (TECO), Grant Number NAGW-4748, from the Terrestrial Ecology Program in NASA's Office of Mission to Planet Earth.

References

Aber, J. D., Nadelhoffer, K. J., Steudler, P., and Melillo, J. M. (1989). Nitrogen saturation in northern forest ecosystems. *BioScience* **39,** 378–386.

Aber, J. D., Magill, A., McNulty, S. G., Boone, R. D., Nadelhoffer, K. J., Downs, M., and Hallett, R. (1995). Forest biogeochemistry and primary production altered by nitrogen saturation. *Water Air Soil Pollut.* **85,** 1665–1670.

Apps, M. J., and Kurz, W. A. (1994). The role of Canadian forests in the global carbon budget. *In* "Carbon Balance of World's Forested Ecosystems: Towards a Global Assessment" (M. Kanninen, ed.), pp. 14–39. Publications of the Academy of Finland 3/93, Helsinki.

Brown, S., and Gaston, G. (1995). Use of forest inventories and geographic information systems to estimate biomass density of tropical forests: Application to tropical Africa. *Environ. Monitoring Assess.* **38,** 157–168.

Brown, S., Iverson, L. R., and Lugo, A. E. (1994). Land use and biomass changes of forests in Peninsular Malaysia from 1972 to 1982: A GIS Approach. *In* "Effects of Land Use Change on Atmospheric CO_2 Concentrations: South and Southeast Asia as a Case Study" (V. H. Dale, ed.), pp. 117–143. Springer-Verlag, New York.

Buringh, P. (1981). "An Assessment of Losses and Degradation of Productive Agricultural Land in the World." FAO Working Group on Soils Policy. FAO, Rome, Italy.

Ciais, P., Tans, P. P., White, J. W. C., Trolier, M., Francey, R. J., Berry, J. A., Randall, D. R., Sellers, P. J., Collatz, J. G., and Schimel, D. S. (1995). Partitioning of ocean and land uptake of CO_2 as inferred by $\delta^{13}C$ measurements from the NOAA Climate Monitoring and Diagnostics Laboratory global air sampling network. *J. Geophys. Res.* **100,** 5051–5070.

Conway, T. J., Tans, P. P., Waterman, L. S. Thoning, K. W., Kitzis, D. R., Masarie, K. A., and Zhang, N. (1994). Evidence for interannual variability of the carbon cycle from the National Oceanic and Atmospheric Administration/Climate Monitoring and Diagnostics Laboratory Global Air Sampling Network. *J. Geophys. Res.* **99,** 22,831–22,855.

Dwivedi, R. S., Sankar, T. R., Venkataratnam, L., Karale, R. L., Gawande, S. P., Rao, K. V. S., Senchaudhary, S., Bhaumik, K. R., and Mukharjee, K. K. (1997). The inventory and monitoring of eroded lands using remote sensing data. *Int. J. Remote Sensing* **18,** 107–119.

ECE/FAO (1992). "The Forest Resources of the Temperate Zones. Main Findings of the UN-ECE/FAO 1990 Forest Resource Assessment." ECE/TIM/60, United Nations, New York.

Enting, I. G., Trudinger, C. M., and Francey, R. J. (1995). A synthesis inversion of the concentration and $\delta^{13}C$ of atmospheric CO_2. *Tellus* **47B,** 35–52.

FAO (1991). "1990 FAO Production Yearbook." FAO, Rome.

FAO (1993). "Forest Resources Assessment 1990. Tropical Countries." FAO Forestry Paper No. 112. FAO, Rome.

FAO (1995). "1994 FAO Production Yearbook." FAO, Rome.

Flint, E. P., and Richards, J. F. (1994). Trends in carbon content of vegetation in south and southeast Asia associated with changes in land use. *In* "Effects of Land Use Change on Atmospheric CO_2 Concentrations: South and Southeast Asia as a Case Study" (V. H. Dale, ed.), pp. 201–299. Springer-Verlag, New York.

Gajaseni, J., and Jordan, C. F. (1990). Decline of teak yield in northern Thailand: Effects of selective logging on forest structure. *Biotropica* **22,** 114–118.

Grace, J., Lloyd, J., McIntyre, J., Miranda, A. C., Meir, P., Miranda, H. S., Nobre, C., Moncrieff, J., Massheder, J., Malhi, Y., Wright, I., and Gash, J. (1995). Carbon dioxide uptake by an undisturbed tropical rain forest in southwest Amazonia, 1992 to 1993. *Science* **270,** 778–780.

Holland, E. A., Braswell, B. H., Lamarque, J.-F., Townsend, A., Sulzman, J., Muller, J.-F., Dentener, F., Brasseur, G., Levy, H., Penner, J. E., and Roelofs, G.-J. (1997). Variations in the predicted spatial distribution of atmospheric nitrogen deposition and their impact on carbon uptake by terrestrial ecosystems. *J. Geophys. Res.* **102(D13),** 15,849–15,866.

Houghton, R. A. (1994). The worldwide extent of land-use change. *BioScience* **44,** 305–313.

Houghton, R. A. (1995). Effects of land-use change, surface temperature, and CO_2 concentration on terrestrial stores of carbon. *In* "Biotic Feedbacks in the Global Climatic System: Will the Warming Feed the Warming?" (G. M. Woodwell and F. T. Mackenzie, eds.), pp. 333–350. Oxford Univ. Press, New York.

Houghton, R. A. (1996). Terrestrial sources and sinks of carbon inferred from terrestrial data. *Tellus* **48,** 420–432.

Houghton, R. A. (2000). Historic role of forests in the global carbon cycle. *In* "Carbon Mitigation Potentials of Forestry and Wood Industry" (G. H. Kohlmaier, M. Weber, and R. A. Houghton, eds.). Springer-Verlag, New York.

Houghton, R. A., and Hackler, J. L. (1994). The net flux of carbon from deforestation and degradation in south and southeast Asia. *In* "Effects of Land Use Change on Atmospheric CO_2 Concentrations: South and Southeast Asia as a Case Study" (V. H. Dale, ed.), pp. 301–327. Springer-Verlag, New York.

Houghton, R. A., and Hackler, J. L. (1995). "Continental Scale Estimates of the Biotic Carbon Flux from Land Cover Change: 1850–1980." ORNL/CDIAC-79, NDP-050. Oak Ridge National Laboratory, Oak Ridge, Tennessee.

Kauppi, P. E., Mielikainen, K., and Kuusela, K. (1992). Biomass and carbon budget of European forests, 1971–1990. *Science* **256,** 70–74.

Keeling, C. D., Bacastow, R. B., Carter, A. F., Piper, S. C., Whorf, T. P., Heimann, M., Mook, W. G., and Roeloffzen, H. (1989). A three-dimensional model of atmospheric CO_2 transport based on observed winds: 1. Analysis of observational data. *In* "Aspects of Climate Variability in the Pacific and the Western Americas" (D. H. Peterson, ed.), pp. 165–236. Geophysical Monograph 55. American Geophysical Union, Washington, D.C.

Keeling, C. D., Whorf, T. P., Wahlen, M., and van der Pilcht, J. (1995). Interannual extremes in the rate of rise of atmospheric carbon dioxide since 1980. *Nature* **375,** 666–670.

Keeling, R. F., Piper, S. C., and Heimann, M. (1996). Global and hemispheric CO_2 sinks deduced from changes in atmospheric O_2 concentration. *Nature (London)* **381,** 218–221.

Kohlmaier, G. H., Hager, C., Wurth, G., Ludeke, M. K. B., Ramge, P., Badeck, F.-W., Kindermann, J., and Lang, T. (1995). Effects of the age class distributions of the temperate and boreal forests on the global CO_2 source-sink function. *Tellus* **47B,** 212–231.

Kolchugina, T. P., and Vinson, T. S. (1993). Carbon sources and sinks in forest biomes of the former Soviet Union. *Global Biogeochem. Cycles* **7,** 291–304.

Kolchugina, T. P., and Vinson, T. S. (1995). Role of Russian forests in the global carbon balance. *Ambio* **24,** 258–264.

Kovda, V. A. (1983). Loss of productive land due to salinization. *Ambio* **12,** 91–93.

Kurz, W. A., Apps, M. J., Beukema, S. J., and Lekstrum, T. (1995). 20th century carbon budget of Canadian forests. *Tellus* **47B,** 170–177.

Lal, R., and Stewart, B. A. (1990). "Soil Degradation." Springer-Verlag, New York.

Lambert, G., Monfray, P., Ardouin, B., Bonsang, G., Gaudry, A., Kazan, V., and Polian, G. (1995). Year-to-year changes in atmospheric CO_2. *Tellus* **47B,** 53–55.

Lloyd, J. (1997). Physiological constraints on the nature and location of the terrestrial sink for anthropogenically released CO_2. *In* "Fifth International Carbon Dioxide Conference," p. 161. CSIRO, Aspendale, Australia.

Long, S. P. (1991). Modification of the response of photosynthetic productivity to rising temperature by atmospheric CO_2 concentrations—Has its importance been underestimated? *Plant Cell Environ.* **14,** 729–739

McEvedy, C., and Jones, R. (1978). "Atlas of World Population History." Penguin Books, Middlesex, England.

Nepstad, D. C., Klink, C. A., Uhl, C., Vieira, I. C., Lefebvre, P., Pedlowski, M., Matricardi, E., Negreiros, G., Brown, I. F., Amaral, E., Homma, A., and Walker, R. (1997). Land-use in Amazonia and the cerrado of Brazil. *Ciencia Cultura* **49,** 73–86.

Phillips, O. L., Malhi, Y., Higuchi, N., Laurance, W. F., Nunez, P. V., Vasquez, R. M., Laurance, S. G., Ferreira, S. V., Stern, M., Brown, S., and Grace, J. (1998). Changes in the carbon balance of tropical forests: Evidence from long-term plots. *Science* **282,** 439–442.

Pimentel, D., Harvey, C., Resosudarmo, P., Sinclair, K., Kurz, D., McNair, M., Crist, S., Shpritz, L., Fitton, L., Safouri, R., and Blair, R. (1995). Environmental and economic costs of soil erosion and conservation benefits. *Science* **267,** 1117–1123.

Rayner, P. J., Enting, I. G., Francey, R. J., and Langenfelds, R. (1999). Reconstructing the recent carbon cycle from atmospheric CO_2, $\delta^{13}C$ and O_2/N_2 observations. *Tellus* **51B,** 213–232.

Sarmiento, J. L, Orr, J. C., and Siegenthaler, U. (1992). A perturbation simulation of CO_2 uptake in an ocean general circulation model. *J. Geophys. Res.* **97,** 3621–3645.

Schimel, D. S., Enting, I. G., Heimann, M., Wigley, T. M. L., Raynaud, D., Alves, D., and Siegenthaler, U. (1995). CO_2 and the carbon cycle. *In* "Climate Change 1994" (J. T. Houghton, L. G. Meira Filho, J. Bruce, Hoesung Lee, B. A. Callander, E. Haites, N. Harris, and K. Maskell, eds.), pp. 35–71. Cambridge Univ. Press, Cambridge.

Sharma, R. C., and Bhargava, G. P. (1988). Landsat imagery for mapping saline soils and wet lands in north-west India. *Int. J. Remote Sensing* **9,** 39–44.

Shvidenko, A., and Nilsson, S. (1994). What do we know about the Siberian forests? *Ambio* **23,** 396–404.

Shvidenko, A., and Nilsson, S. (1997). Are the Russian forests disappearing? *Unasylva* **188,** 57–64.

Shvidenko, A., and Nilsson, S. (1998). Dynamics of forest resources of the former Soviet Union with respect to the carbon budget. *In* "Carbon Dioxide Mitigation in Forestry and Wood Industry" (G. H. Kohlmaier, M. Weber, and R. A. Houghton, eds.). Springer-Verlag, New York.

Siegenthaler, U., and Oeschger, H. (1987). Biospheric CO_2 emissions during the past 200 years reconstructed by deconvolution of ice core data. *Tellus* **39B,** 140–154.

Spiecker, H. Mielikainen, K., Kohl, M., and Skovsgaard, J. (1996). "Growth Trends in European Forests—Studies from 12 Countries." Springer-Verlag, Berlin.

Syers, J. K., Lingard, J., Pieri, C., Ezcurra, E., and Faure, G. (1996). Sustainable land management for the semiarid and sub-humid tropics. *Ambio* **25,** 484–491.

Turner, D. P., Koerper, G. J., Harmon, M. E., and Lee, J. J. (1995). A carbon budget for forests of the conterminous United States. *Ecol. Appl.* **5,** 421–436.

Uhl, C., and Vieira, I. C. G. (1989). Ecological impacts of selective logging in the Brazilian Amazon: A case study from the Paragominas region of the State of Para. *Biotropica* **21,** 98–106.

UNEP/ISRIC (United Nations Environment Programme and International Soil Reference and Information Center) (1990). "World Map on Status of Human-Induced Soil Degradation." UNEP/ISRIC, Nairobi, Kenya.

Woods, P. (1989). Effects of logging, drought, and fire on structure and composition of tropical forests in Sabah, Malaysia. *Biotropica* **21,** 290–298.

Woodwell, G. M., Mackenzie, F. T., Houghton, R. A., Apps, M., Gorham, E., and Davidson, E. A. (1998). Biotic feedbacks in the warming of the earth. *Climat. Change* **40,** 495–518.

World Bank (1991). "World Development Report 1991." Oxford Univ. Press, New York.

22

Predicting the Future Productivity and Distribution of Global Terrestrial Vegetation

F. Ian Woodward, Mark R. Lomas, and Susan E. Lee

I. Introduction

The climatological community of scientists has provided predictions of how the global climate will change in a future of increasing atmospheric concentrations of greenhouse gases (Houghton *et al.*, 1996). These predictions emerge from general circulation models (GCMs), of which there are a number. Early applications of the outputs from these models, in the ecological sphere, were predictions of changes in the distribution of terrestrial vegetation in response to climates at equilibrium with particular changes in atmospheric CO_2 (or CO_2 equivalent) concentrations (Emanuel *et al.*, 1985). These static or equilibrium predictions were the obvious first steps in these ecological explorations.

GCMs are now capable of providing predictions of the more realistic transient changes in climate that are expected as the concentrations of greenhouse gases steadily increase in the atmosphere (Mitchell *et al.*, 1995). The ecological models that are necessary to interact with these transient climatic changes are those that can predict natural or dynamic changes in the distribution of vegetation. These dynamic global vegetation models (DGVMs) (Steffen *et al.*, 1992) must therefore be able to predict processes such as vegetation disturbance and succession, in addition to processes that are now generally included in ecosystem models, such as biogeochemical cycling and CO_2 and water fluxes.

This chapter provides general detail and predictions from a DGVM, the Sheffield dynamic global vegetation model (SDGVM), of terrestrial vegetation responses to transient changes in climate over the next 100 yr.

Table 22-1 Components of Dynamic Global Vegetation Model Leaf- and Plant-Level Processes

Component	Comment
Nitrogen uptake	Dependent on climate, soil C and N, and feedback from the CENTURY model
Photosynthesis	Coefficients for the Farquhar *et al.* C_3 model determined from N uptake and temperature; modified Collatz *et al.* for C_4 photosynthesis; individual leaf photosynthesis; controlled by within-canopy PAR and stomatal conductance
Leaf respiration	Coefficients for model determined from N uptake and temperature
Stomatal conductance	Determined by climate, soil water status, and photosynthesis

II. Methods

The SDGVM is an expansion of a previously published biogeochemical and phytogeographic model (Woodward *et al.*, 1995). The original model predicted vegetation leaf area index (LAI) and net primary productivity using only climatic and edaphic data as inputs. The SDGVM now also includes a range of vegetation processes (Tables 22-1–22-4). The new developments include the incorporation (Table 22-1) of the CENTURY model (Parton *et al.*, 1993) of carbon and nitrogen cycling, including litterfall and decomposition. This inclusion then closes the nitrogen and carbon cycling of the original model.

The SDGVM scales up from leaf and plant processes (Table 22-1), through canopy-level processes (Table 22-2), to vegetation processes (Table 22-3). New inclusions at the vegetation scale include predictions of leaf phenology, photosynthate allocation, and net ecosystem productivity (NEP).

Table 22-2 Components of Dynamic Global Vegetation Model Canopy-Level Process

Component	Comment
Canopy conductance and evapotranspiration	Conductance based on canopy height, LAI, stomatal conductance profile through canopy; precipitation interception based on LAI; transpiration and evaporation using Penman–Monteith model
Canopy CO_2 exchange	Integral of leaf photosynthetic profile through canopy of fixed/variable extinction coefficient
Soil water uptake	Soil water content reduced by transpiration and soil surface evaporation; precipitation throughfall influenced by LAI

Table 22-3 Components of Dynamic Global Vegetation Model Vegetation Processes

Component	Comment
Vegetation phenology	For cold-deciduous vegetation leaf phenology determined primarily by temperature; for cold-evergreen vegetation leaf phenology determined primarily by temperature; for drought-deciduous vegetation leaf phenology determined by a time integral of soil water
Leaf area index	Vegetation leaf area index the minimum as determined by annual hydrological balance, or net primary productivity of lowest leaf layer
Net primary productivity	NPP calculated after accounting for the annual maintenance and synthetic costs of roots, stems, and leaves
Photosynthate allocation	Root allocation determined by root mass necessary to support maximum transpiration rates; leaf allocation determined from LAI and shoot allocation by difference; shoot NPP used, with maximum transpiration rate, wood density, and xylem conductivity to determine maximum vegetation height
Litter production and decomposition	Root, leaf, and shoot litter production determined from allocation, phenology, and NPP patterns
Net ecosystem productivity	NPP minus heterotrophic respiration, calculated using the CENTURY model

Predictions of the likely nature of future vegetation require that the species composition of the vegetation should be known. This is clearly an impossibility, because little is known about the ~250,000 flowering plants of the world, either in the context of their responses to changes in CO_2 and climate or even in terms of their responses to current climate. Therefore this taxonomic complexity is markedly reduced to consider the distribution of a few functional types of species (Woodward and Cramer, 1996). The SDGVM predicts the growth and responses of seven functional types—evergreen broadleaf trees, deciduous broadleaf trees, evergreen needleleaf trees, deciduous needleleaf trees, shrubs of arid climates, grasses with the C_3 pathway of photosynthesis, and grasses with the C_4 pathway of photosynthesis (Table 22-4).

Disturbance of the vegetation is a major feature of natural vegetation change, and in the SDGVM disturbance is generated by fire, by drought-dependent mortality, and by a small random component, which crudely simulates other forms of disturbance, such as windthrow, which are not explicitly modeled (Table 22-4). A major problem with DGVMs is testing their capabilities. A number of approaches have been used to test submodules of the SDGVM, but a major output of the model is the prediction of the geo-

Table 22-4 Components of Dynamic Global Vegetation Model Vegetation/Biome Dynamics

Component	Comment
Functional type occurrence	Determined from climatic extremes (absolute minimum temperature); grasses and also C_3 and C_4 pathways determined from annual NPP and allocation to leaves
Disturbance generator	Major disturbance is fire, with probability increasing with soil surface dryness and litter amount; drought-dependent mortality; other disturbances considered to equal ≤2% of pixel per annum
Functional type growth	Time to maturity and annual growth rate determined from canopy height and plant density, proportion of different functional types determined from height dominance
Classifying biome/vegetation types	Using IGBP-DIS vegetation classification

graphical distribution of different vegetation types and so the output of the SDGVM, in terms of the occurrence of different functional types, has been classified into different biomes, using the same classification as the International Geosphere–Biosphere Program Data and Information System (IGBP-DIS) satellite classification of the present-day vegetation (Table 22-5).

III. Model Projections

Model predictions of plant and canopy processes, such as photosynthesis, respiration, stomatal conductance, soil water content, runoff, and soil carbon, have been successfully completed against data obtained from a whole catchment experiment with CO_2 enrichment and warming (Beerling *et al.*, 1997). Therefore that component of the model operates effectively under current climate and under a scenario of future warming and CO_2 enrichment, at least for a high-latitude ecosystem.

The SDGVM models the rates of growth and times to maturity of different functional types (Table 22-4) on the basis of a prediction of aboveground vegetation biomass at maturity, divided by the aboveground net primary productivity from the biogeochemical and phytogeography model (Woodward *et al.*, 1995). Because the SDGVM is a new type of vegetation model, none of which is yet published in detail, then it is important to investigate the accuracy and capacity of the SDGVM in order that the reader can glean some idea of its capabilities.

Table 22-5 The IGBP-DIS Land Cover Classification

Classification	Comment
Evergreen needleleaf forests	Lands dominated by trees with a percent canopy cover >60% and height exceeding 2 m. Almost all trees remain green all year. Canopy is never without green foliage
Evergreen broadleaf forests	Lands dominated by trees with a percent canopy cover >60% and height exceeding 2m. Almost all trees remain green all year. Canopy is never without green foliage
Deciduous needleleaf forest	Lands dominated by trees with a percent canopy cover >60% and height exceeding 2m. Consists of seasonal needleleaf tree communities with an annual cycle of leaf-on and leaf-off periods
Deciduous broadleaf forests	Lands dominated by trees with a percent canopy cover >60% and height exceeding 2m. Consists of seasonal broadleaf tree communities with an annual cycle of leaf-on and leaf-off periods
Mixed forests	Lands dominated by trees with a percent canopy cover >60% and height exceeding 2m. Consists of tree communities with interspersed mixtures or mosaics of the other four forest cover types. None of the forest types exceeds 60% of landscape
Closed shrublands	Lands with woody vegetation less than 2 m tall and with shrub canopy cover is >60%. The shrub foliage can be either evergreen or deciduous
Open shrublands	Lands with woody vegetation less than 2 m tall and with shrub canopy cover between 10 and 60%. The shrub foliage can be either evergreen or deciduous
Woody savannas	Lands with herbaceous and other understorey systems, and with forest canopy cover between 30 and 60%. The forest cover height exceeds 2 m
Savannas	Lands with herbaceous and other understorey systems, and with forest canopy cover between 10 and 30%. The forest cover height exceeds 2 m
Grassland	Lands with herbaceous types of cover. Tree and shrub cover is less than 10%

Testing Predictions of Canopy Height and Plant Density

Aboveground vegetation biomass at maturity is calculated as the product of maximum canopy height, aboveground biomass associated with this height and plant density. Maximum canopy height, h, is calculated as follows (from Beerling *et al.*, 1996; Margolis *et al.*, 1995):

$$h = \frac{\sqrt{\Delta\Psi}}{\sqrt{E_m}} \frac{\sqrt{x}}{\sqrt{w}} N_s, \tag{1}$$

where $\Delta\Psi$ is the maximum drop in water potential across the tree trunk/stem, E_m is the maximum predicted transpiration rate for the canopy, x is xylem water conductivity, w is stem/wood density, and N_s is shoot net primary productivity.

The aboveground biomass, b_a, for a maximum canopy height h is calculated as:

$$b_a = \left[\pi(d/2)^2 h\right] w, \tag{2}$$

where d is stem/trunk diameter, which may be calculated from one of the methods described by Niklas (1992).

The plant density, p (numbers per unit area), is calculated in two stages. The first stage calculates the minimum biomass, b_m, required to replace annually the xylem necessary to carry water to the top of the canopy, at its maximum height h:

$$b_m = \left[\pi\left(\frac{d + r_m}{2}\right)^2 hw\right] - b_a, \tag{3}$$

where r_m is the minimum annual radial increment of xylem necessary to supply water to the canopy of maximum height h and at a maximum transpiration rate E_m. Plant density is then calculated as shoot net primary productivity divided by b_m, and total aboveground biomass is the product of plant density and b_a.

The predictions of maximum canopy height for tree and forest canopies have been compared with observations (Fig. 22-1) presented in Cannell (1982), using appropriate climatic data from the Leemans and Cramer (1991) global climate database. There is good agreement between observation and prediction, although the predictions increasingly underestimate canopy heights above about 30 m. Some of this disagreement lies in the fact that individual tree height, rather than canopy height, is recorded in the field and this will almost certainly overestimate the mean canopy height.

The predictions of tree density are compared with observations (Fig. 22-2) using data compiled in Cannell (1982). A few observations on shrub plant density are also found in Cannell (1982). These are of high values, at about 150,000 for a *Cupressus* shrubland in California. The model projection is 130,000 for the same region. This value has not been included in Fig. 22-2 because the high values strongly influence the patterns for tree density. In

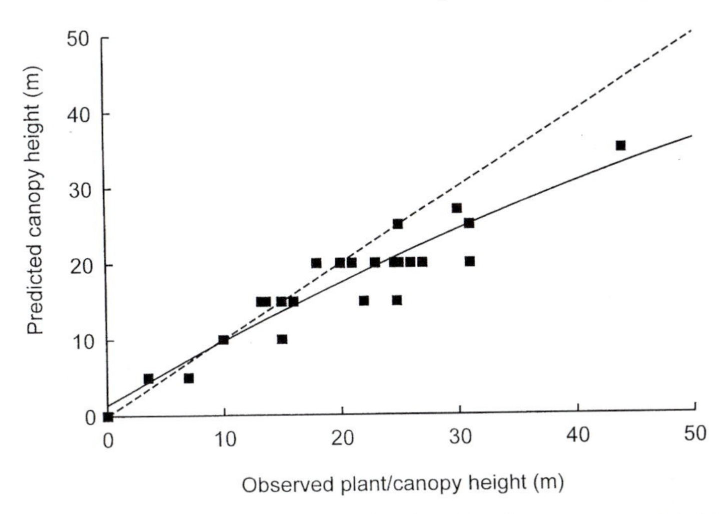

Figure 22-1 Relationship beween observed and predicted canopy or tree heights. Observations from Cannell (1982). Correlation coefficient 0.93; dashed line shows the 1:1 relationship.

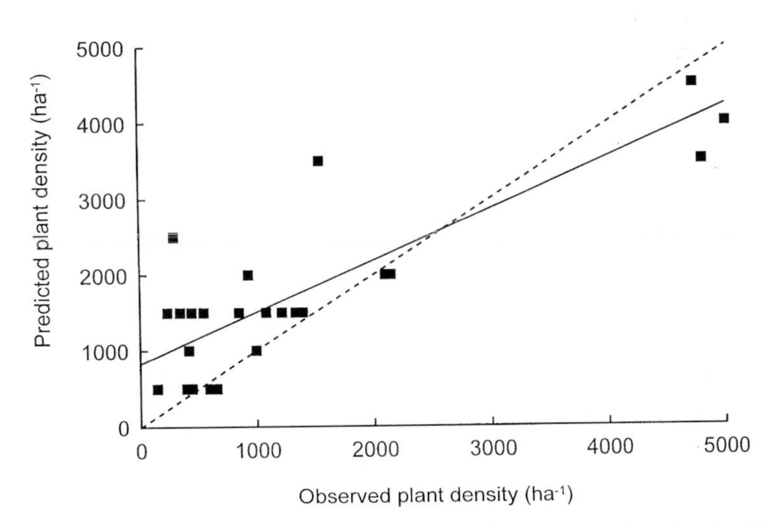

Figure 22-2 Relationship beween observed and predicted canopy plant density. Observations from Cannell (1982). Correlation coefficient 0.88; dashed line shows the 1:1 relationship.

general there is quite good agreement between observations and predictions, although the model tends to underpredict high densities.

1. Testing the Fire Model Fire is a major global-scale disturbance of vegetation. Modeling the occurrence of fire is complex, especially when considering its spread, its degree of stand replacement, and its spatial and temporal occurrence (Clark, 1990; Johnson, 1992; Johnson and Gutsell, 1994). A key feature of fires is the threshold dryness of, for example, the soil surface litter and also the total volume of litter that can be burned. A very simple approach to modeling fires has been taken in this model in which the water content of the soil litter layer is predicted, and this value and the availability of litter are taken together to provide an estimate of a fire return interval. A random-number generator in the model is then used to select whether an area is burned or not. The relationship between the litter water content and the fire interval was made by calibrating model predictions of the litter water content, for the period of 1970 to 1980, using the transient climate predictions of the United Kingdom Meteorological Office (UKMO) (Mitchell *et al.*, 1995), against literature publications of the fire interval (Archibold, 1995). Global-scale predictions have then been made and compared with a separate source of fire return data (Olson, 1981). The observations and predictions match quite well (Fig. 22-3), although it is really rather difficult to explain any differences, because observations of fire return intervals are not without significant errors.

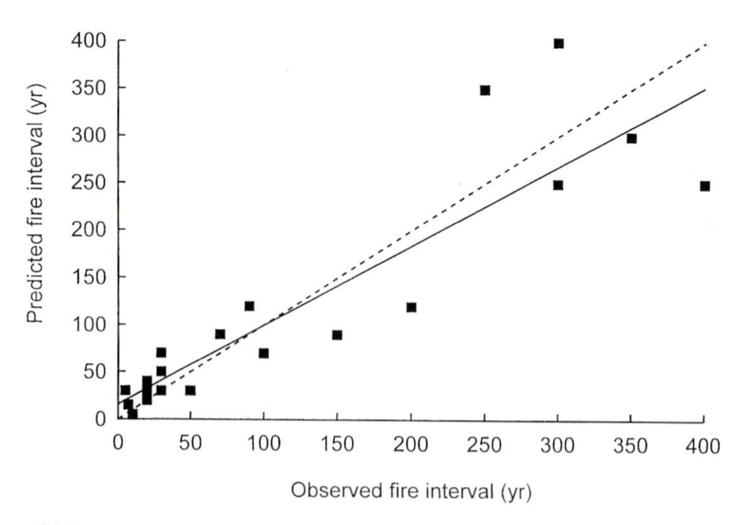

Figure 22-3 Relationship beween observed and predicted fire interval. Observations from Olson (1981). Correlation coefficient 0.89; dashed line shows the 1:1 relationship.

2. Testing Vegetation Projections The patch/landscape-scale projections indicated above scale to larger, pixel scales (0.5° to 2.5° scale), through processes of carbon accumulation (Table 22-3). At this scale of projection, model outputs include leaf area index, net primary productivity, and net ecosystem productivity. The leaf area index and net primary production predictions have already been tested (Woodward *et al.*, 1995). The net ecosystem productivity has been tested for the years 1987 and 1988, in which 1987 was an El Niño year (Diaz and Markgraf, 1992) and, according to Keeling *et al.* (1995), the terrestrial biosphere was a net carbon source of about 2.0 Gt C. The following year, the terrestrial biosphere was a much smaller carbon source of about 0.8 Gt C. Model projections of NEP were 2.1 Gt C in 1987 and 0.4 Gt C in 1988, indicating good agreement with the interpretations of the net carbon fluxes by Keeling *et al.* (1995).

The highest scale of the vegetation model predicts the occurrences of different functional types of species (Woodward, 1987; Woodward and Cramer, 1996) and also predicts the type of vegetation, or biome, that occurs in a particular pixel (Table 22-4). A major problem in testing this aspect of the model is the current lack of a precise vegetation map of the world, although a number of activities are leading to a new product. One such new product is just becoming available from the World Wide Fund for Nature and the World Conservation Monitoring Centre. These data are available as maps on the World Wide Web (for example, maps of America can be found at http://forests.lic.wisc.edu/forests/america.html). Not all areas of the world are yet mapped, however, a useful test-bed for the current vegetation model is provided by the map of Russian vegetation, from ~40° to 160°E. A transect has been selected (Fig. 22-4) at a latitude of 60°N and the species have been characterized to indicate four functional types of vegetation: (1) grass/shrub tundra, (2) deciduous needleleaf forest, (3) evergreen needleleaf forest, and (4) deciduous broadleaf forest. Mixed vegetation has been indicated by intermediate values between the integer scale.

The vegetation model has been used to predict the distribution of the dominant functional types and vegetation, using the UKMO transient climate run (Mitchell *et al.*, 1995) and extracting the data, along the transect, for the 1990s. There is good agreement between model and observation, although problems emerge at the boundaries between vegetation types.

IV. Model Projections for the 1860s to the 2090s

A. Global Totals

The UKMO transient GCM simulates a global warming of 1.5°C by the year 2050, in response to increases in greenhouse gases, of which CO_2 (Fig. 22-5) is the primary component, which reaches a doubling of the preindustri-

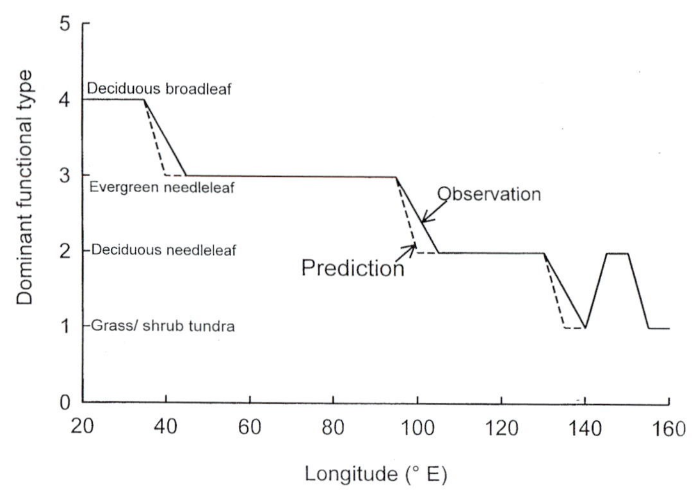

Figure 22-4 Transect of modeled and observed vegetation types for a latitude of 60°N.

al concentration by about 2050. Global precipitation (Fig. 22-5) has no significant trend, with a marked interannual variability.

Global net primary productivity (NPP) is predicted to increase with the same trend as atmospheric CO_2 concentration (Fig. 22-6), in keeping with other estimates (Melillo *et al.*, 1993). Global net ecosystem productivity slowly increases throughout the period of the simulations; in these simulations

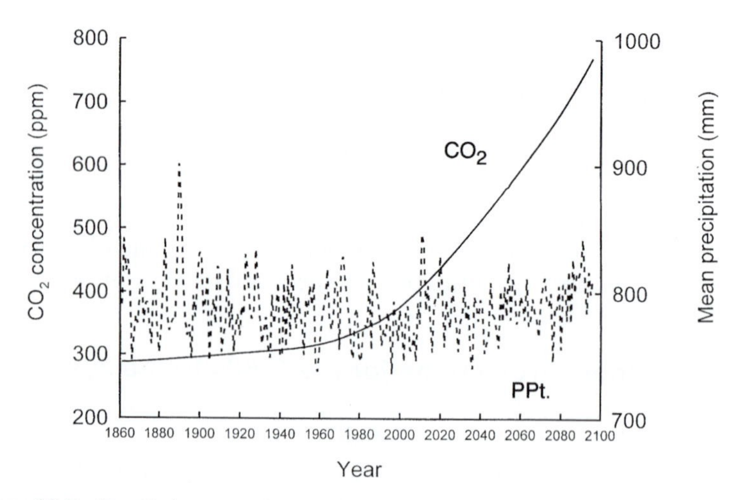

Figure 22-5 Trends in atmospheric CO_2 concentration and global mean precipitation (PPt.) through the UKMO transient climate projections.

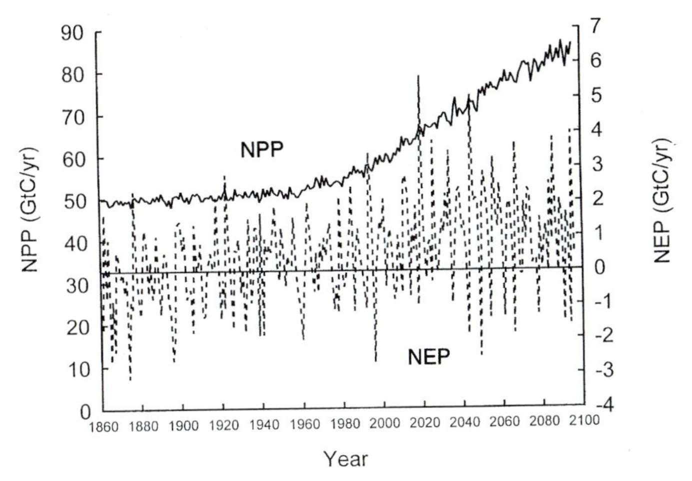

Figure 22-6 Trends in predicted annual terrestrial net primary productivity (NPP) and net ecosystem productivity (NEP).

NEP is positive when it is a sink and negative when it is a source of CO_2, with the terrestrial surface becoming an overall net sink of carbon. The predicted rate of increase in NEP is 8 Mt C yr^{-1}. It should be pointed out that the model assumes natural/potential vegetation at all sites, with no provision for conversion to agriculture, or managed landscapes.

Biomass in vegetation is also predicted to increase in a fashion similar to the increase in atmospheric CO_2 concentration (Fig. 22-7), with a present-

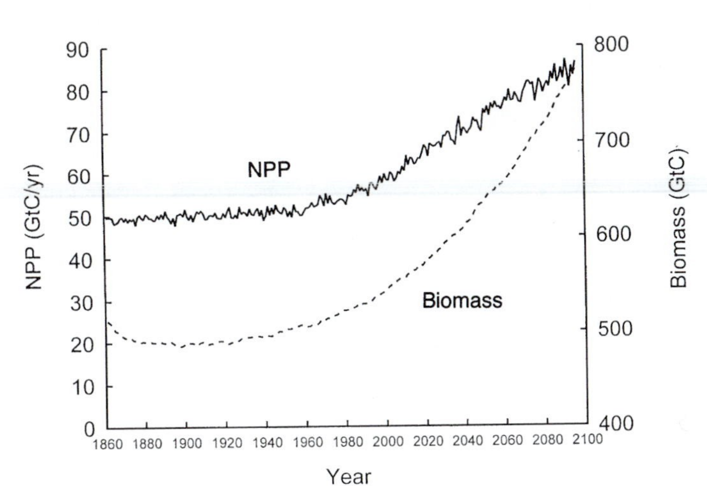

Figure 22-7 Predicted trends in terrestrial net primary productivity (NPP) and biomass.

day estimate of about 540 Gt C, which is broadly in keeping with present-day estimates of global biomass (Siegenthaler and Sarmiento, 1993).

B. Global Distributions

1. Productivity Global-scale distribution maps of NPP and NEP are presented as averages for the 1990s and 2090s and are used to indicate possible regional-scale changes in these estimates of productivity. It should be kept in mind that differences between these maps and others, produced either by modeling or by composites of observational data, will occur through a mixture of sources, particularly errors in the global distribution of climate and errors in the process-bases of the vegetation model.

There are marked increases in NPP from the 1990s to 2090s (Figs. 22-8 22-9), with particular areas to note between the tropics, the southeast Unit-

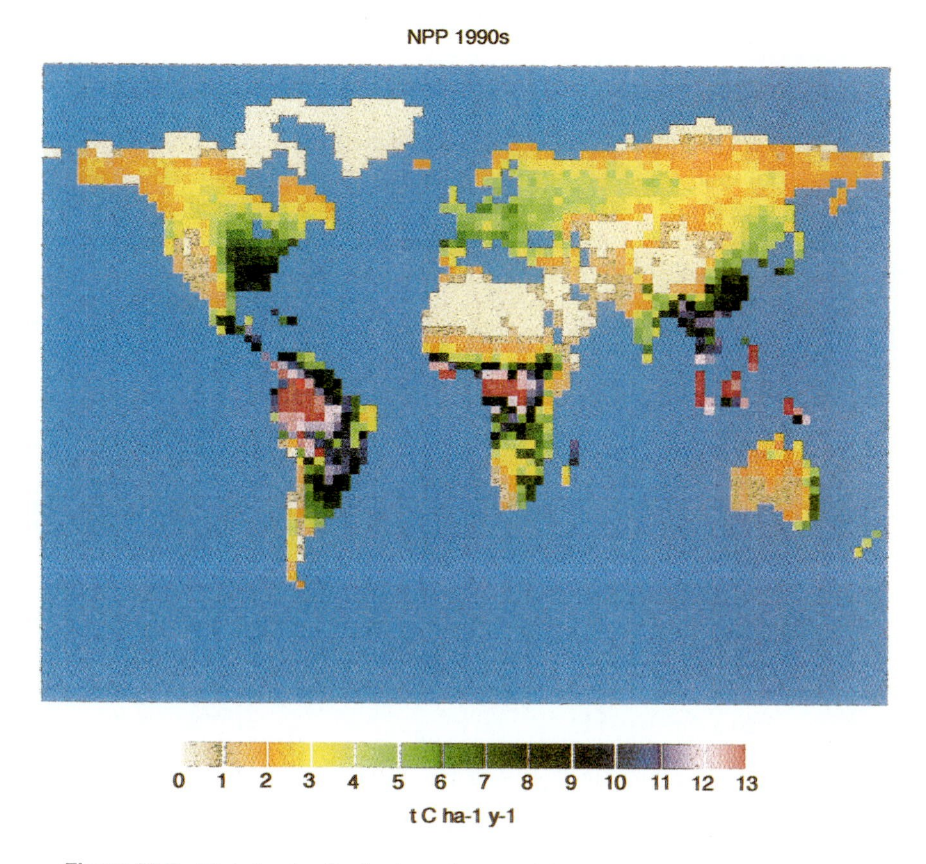

Figure 22-8 Global-scale distribution of terrestrial net primary productivity (t C ha^{-1} yr^{-1}) averaged for the 1990s.

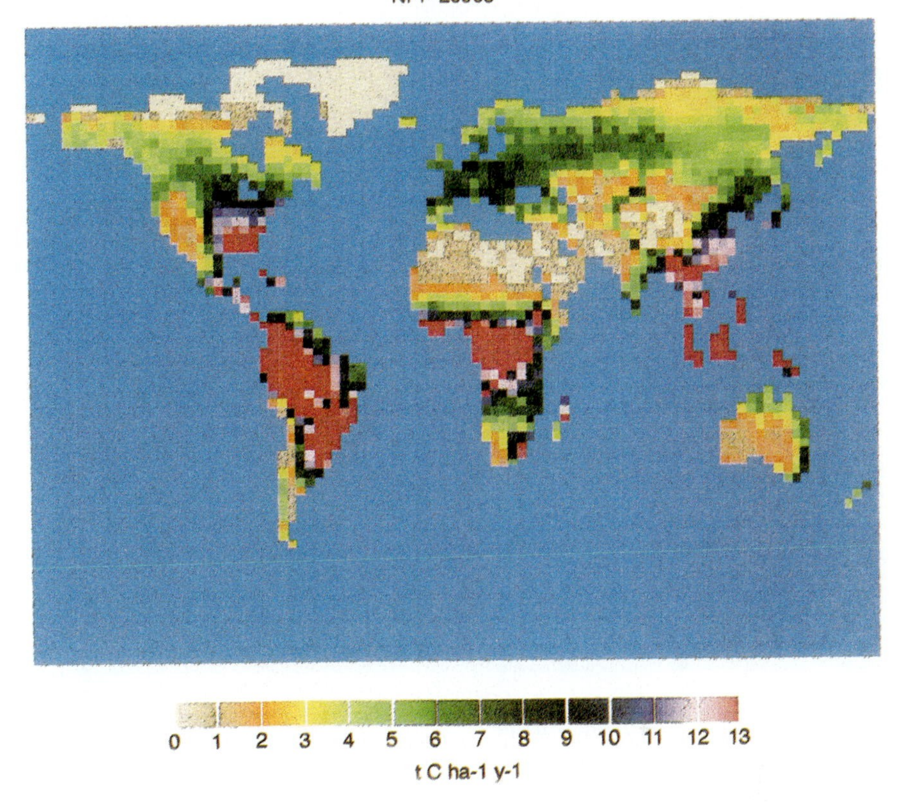

Figure 22-9 Global-scale distribution of terrestrial net primary productivity (t C ha^{-1} yr^{-1}) averaged for the 2090s.

ed States, and from the high latitudes between 40° and 70°N. In contrast, there are smaller relative increases for equatorial zones and west and south Australia. The general increases are caused by the direct impacts of increasing atmospheric CO_2 concentrations on photosynthesis plus increased rates of nitrogen mineralization in high-latitude sites, which have substantial reserves of organic carbon and nitrogen in the soil. The increased rates of nitrogen mineralization are driven by increasing temperatures, which increase nitrogen mineralization rates (Parton *et al.*, 1993).

The global patterns of NEP (Figs. 22-10 and 22-11) are more complex than NPP and reflect the fact that the NEP fluxes of carbon are the difference between two large fluxes, NPP and heterotrophic respiration. Over the period from the 1990s to the 2090s some areas are projected to become increasing sources of carbon, in particular Venezuela, Brazil, Europe, India,

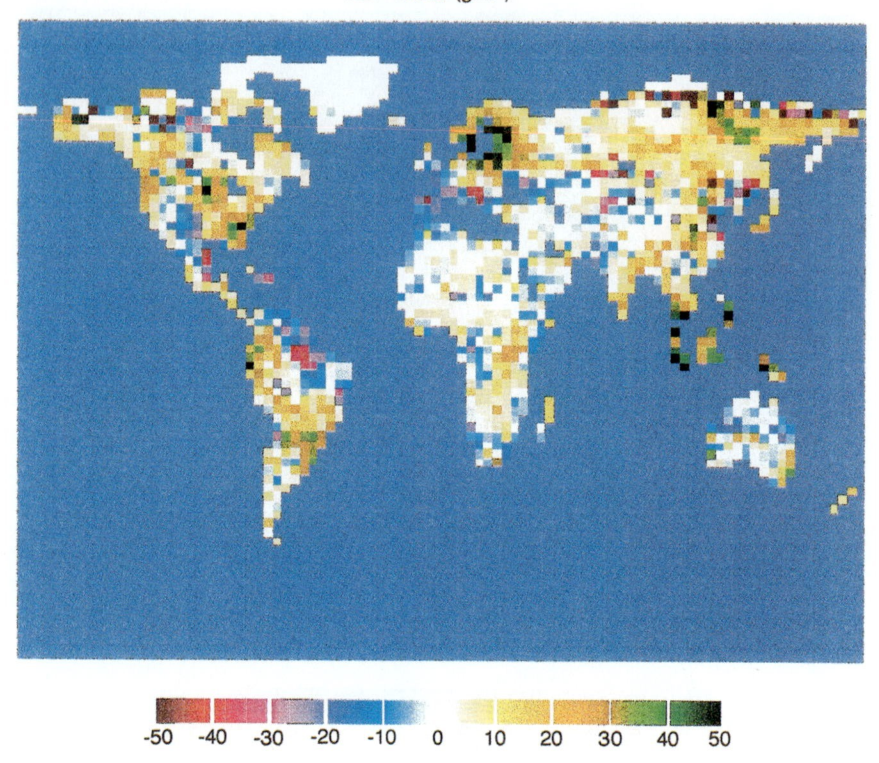

Figure 22-10 Global-scale distribution of annual terrestrial net ecosystem productivity (g C m^{-2}) averaged for the 1990s. Positive figures are carbon sinks, negative figures are carbon sources.

north China, east Australia, and north Canada. In contrast, some areas are projected to become increasing sinks, such as central Canada, the southeast United States, central Russia, and southeast Asia. The areas of increasing sink capacity occur through significant increases in NPP relative to heterotrophic respiration. The source areas occur where NPP responses to environmental change are small, usually through reductions in precipitation and increases in temperature, which increase rates of heterotrophic respiration. Such changes may be relatively short lived where soil carbon reserves are small.

*2. **Vegetation Distribution*** The patch model (Table 22-4) predicts the relative occurrence of seven functional types of plants: evergreen broadleaf trees, deciduous (cold and dry) broadleaf trees, evergreen needleleaf trees, deciduous needleleaf trees, shrubs (excluding tundra), grasses with the C_3 pathway of photosynthesis, and grasses with the C_4 pathway of photosyn-

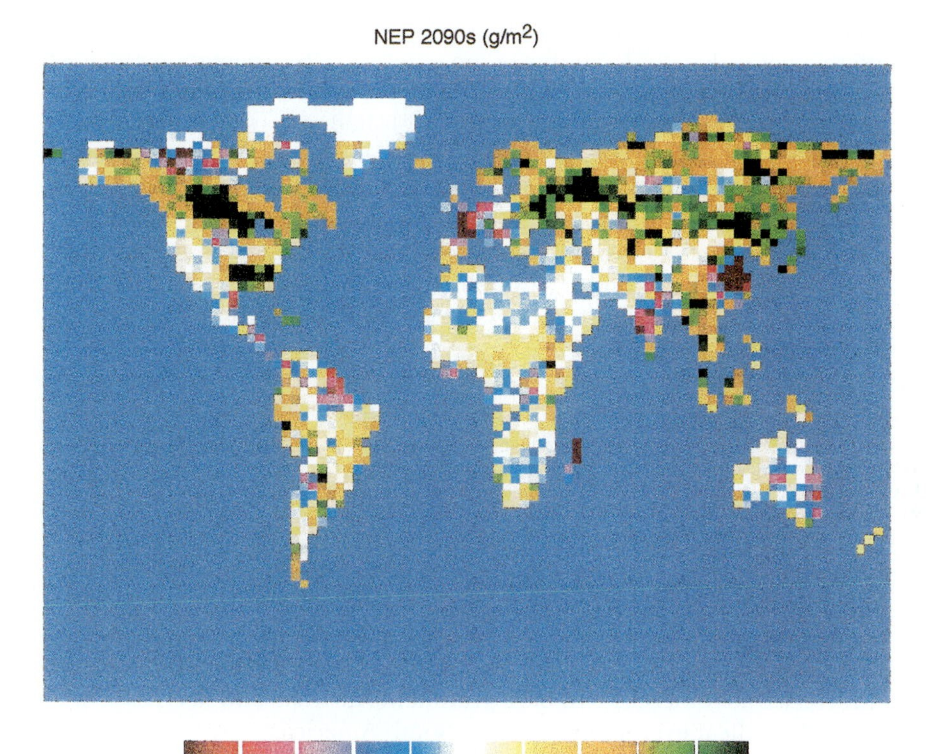

Figure 22-11 Global-scale distribution of annual terrestrial net ecosystem productivity (g C m^{-2}) averaged for the 2090s. Positive figures are carbon sinks, negative figures are carbon sources.

thesis. The relative occurrences of these functional types at any pixel need to be integrated to describe the occurrence of a particular vegetation type or biome. This is achieved using the IGBP-DIS scheme of vegetation description (Table 22-5; also available on the World Wide Web at the following site: http://edcwww.cr.usgs.gov/landdaac/glcc/glcc—na.html). This particular classification has been used so that the vegetation maps predicted by the SDGVM can, at least for the present time, be compared with a global-scale land cover map for the present day, as produced from satellite images. One extra vegetation category has been included for the present simulations, and this is mixed woodland, which is essentially the same as savanna but for cooler climates. The production of the IGBP-DIS global land cover map is close to completion and this can then be used to test aspects of the SDGVM predictions.

The predicted distributions of vegetation types for the 1990s (Fig. 22-12) and the 2090s (Fig. 22-13) indicate some potential for vegetation change in response to changes in climate and atmospheric CO_2 concentrations. Visual comparisons of the 1990s maps with published maps such as those of Olson (1981) or Emanuel *et al.* (1985) indicate reasonable agreement, although the large pixel size of the present predictions, determined by the GCM pixel size, does mask significant regional-scale detail. The major thrust of the present work is to determine the degree of vegetation change after a century of warming and with a doubling of the present-day CO_2 concentration, and so it is the relative change between the two periods that is of particular interest.

Significant changes in vegetation types are predicted to occur in a number of regions. In the 2090s, broadleaf deciduous vegetation is predicted to dominate the former mixed forest with evergreen needleleaf species in Russia and Canada. Needleleaf deciduous forest is predicted to move northward in Siberia, either replacing tundra or forming a mixed woodland with a tundra understorey. The UKMO GCM clearly predicts increased precipitation in a number of the arid areas of the world and consequently there are predicted increases in the expansion of shrublands in the west United States and central Asia and increases in the area of forests in the Sahel of Africa. In contrast, Australia and north South America are predicted to be drier, with consequent increases in bare ground in Australia and decreases in the area of broadleaf evergreen forests in South America.

Overall, 39% of the pixels show a change in vegetation type, according to the IGBP-DIS classification (Table 22-5). Of this 39%, about 30% indicates an increase in vegetation biomass or stature, or an increase in the proportion of thermophilic vegetation. A smaller percentage, 9% of the 39%, shows a decrease in biomass or vegetation stature. In fact, this quite marked degree of vegetation change is significantly due to the method of vegetation classification. When maps of the dominant functional types are compared, then the change is only about 18%, just less than half of that based on the vegetation classification. This indicates that the vegetation dominants change rather little, but that mixed vegetation types in particular can change their composition rather readily. It also appears that some rates of vegetation change can be in the order of 1 to 5 km yr^{-1}, which are high rates for vegetation (Huntley and Birks, 1983), indicating that the model must now consider including rates of species migration as a major limitation on the rate of vegetation change.

3. Dynamic and Equilibrium Vegetation Models By definition, DGVMs include those dynamics that are important in affecting vegetation and that measure dynamic vegetation effects. A critical component to be included is disturbance, typically fire, but also windthrow and also some measurement

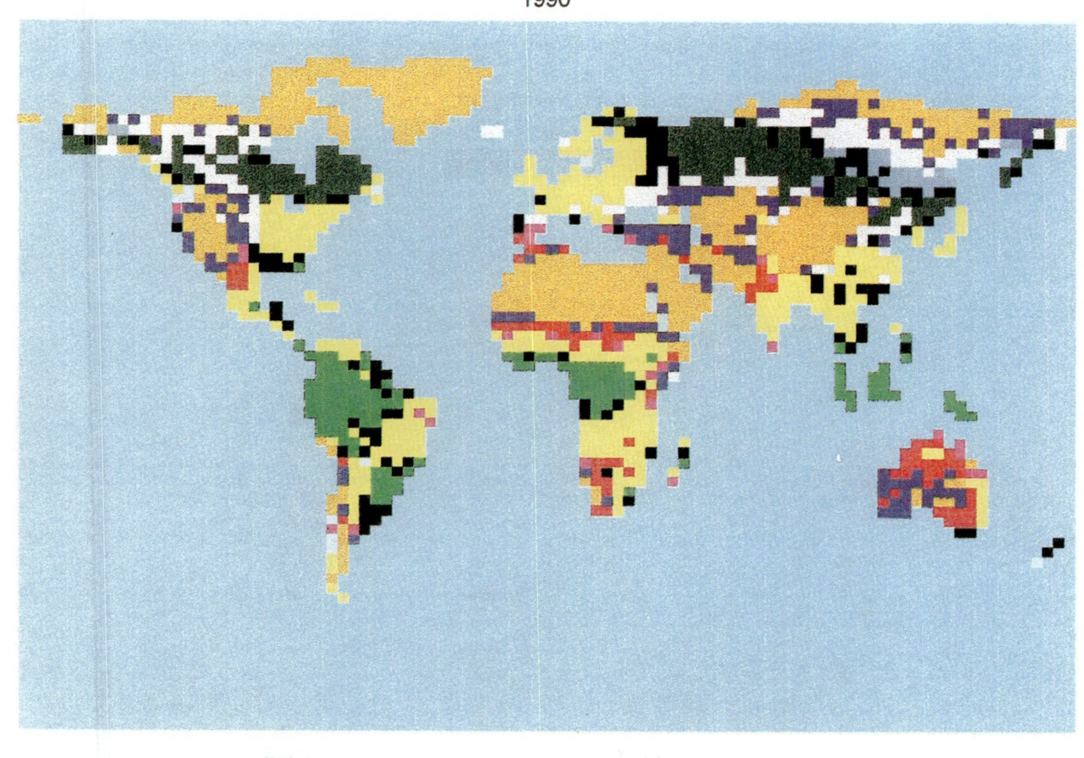

Figure 22-12 Global distribution of terrestrial vegetation types for the 1990s. EN, Evergreen needleleaf forest; EB, evergreen broadleaf forest; DN, deciduous needleleaf forest; DB, deciduous broadleaf forest; MF, mixed forests; SB, shrublands (not tundra); SV, savanna; GS, grassland; BG bare ground; MW, mixed woodland.

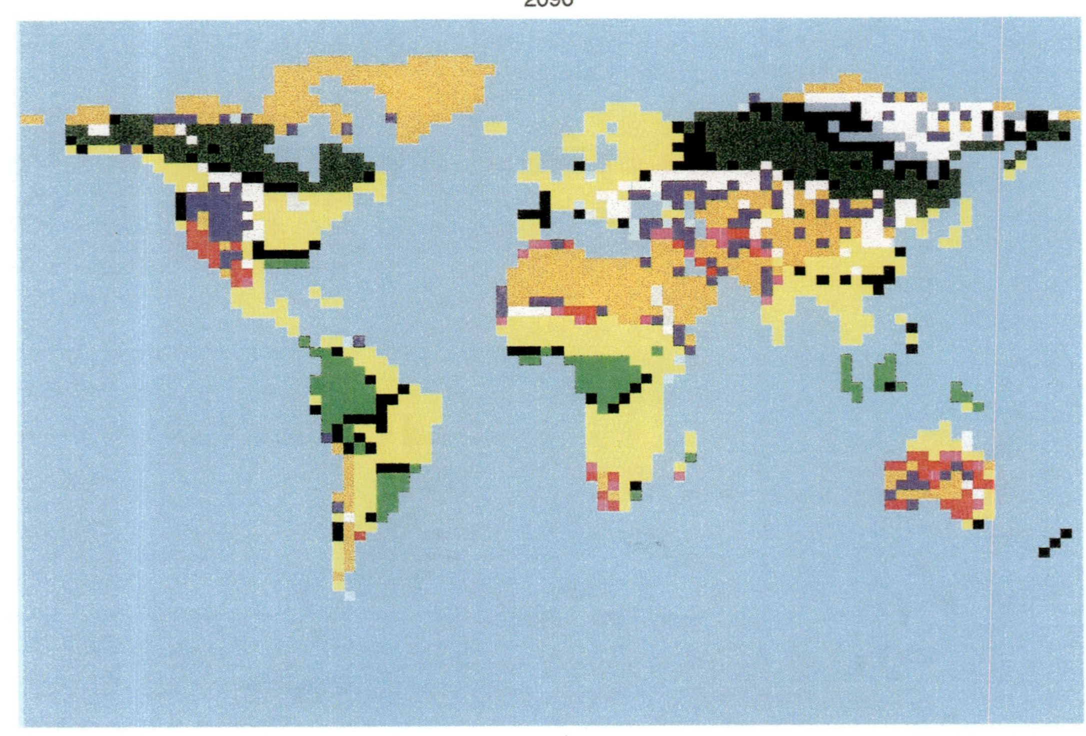

Figure 22-13 Global distribution of terrestrial vegetation types for the 2090s. EN, Evergreen needleleaf forest; EB, evergreen broadleaf forest; DN, deciduous needleleaf forest; DB, deciduous broadleaf forest; MF, mixed forests; SB, shrublands (not tundra); SV, savanna; GS, grassland; BG bare ground; MW, mixed woodland.

of individual plant mortality. When DGVMs are used to predict the dynamic responses of vegetation to transient changes in climate and atmospheric CO_2 concentration, there is therefore a whole suite of predicted responses that differ markedly in their response times and impacts of productivity. NPP is influenced following simulated fires, both through changes in nutrient uptake and through the possibilities of changes in functional type composition. It is to be expected, therefore, that DGVMs and equilibrium vegetation models, such as the terrestrial ecosystem model (TEM) (McGuire *et al.*, 1995), will provide differing responses to CO_2 doubling and associated warming. This will be an instantaneous change for the equilibrium model, but the DGVMs must be run in a transient mode, in order for the natural ecosystem dynamics to occur. For a doubling of the CO_2 concentration Melillo *et al.* (1996) predict a 360 Gt C increase in terrestrial carbon storage. In the version of the SDGVM reported here the increase in terrestrial carbon storage is 505 Gt C, for the same change in CO_2 concentration. In both cases global warming, in addition to CO_2 doubling, reduced this carbon storage. However, a quantitative comparison cannot be made because different GCMs have been used to derived the global climate fields. The greater carbon storage with CO_2 doubling predicted by the SDGVM occurs because of a greater stimulation of NPP than shown by TEM, particularly so in tropical and semiarid areas. In the semiarid areas, with both cool and warm climates, the SDGVM also predicts increases in woody biomass, indicating the expansion of trees or shrubs in these climates.

Other outputs from the SDGVM, such as changes in plant functional type distributions (Figs. 22-12 and 22-13) can be compared with equilibrium vegetation models, such as MAPSS (Neilson, 1993). Not surprisingly, the equilibrium models show considerably greater changes in the distribution of vegetation types than is the case for the DGVM. This indicates the critical importance of including the mechanisms of vegetation change in vegetation models.

*4. **Future Research Needs*** A number of DGVMs are presently being constructed by different research groups. The models are complex and attempt to predict a wide range of processes, some of which are poorly understood. Methods of predicting disturbances are still crude, but are essential and require more development. Models tend to use only a small number of functional types of plants (usually less than 10) and so the relevance of DGVMs for biodiversity studies is currently low, but should be a major growth area for DGVMs. A final and continuing problem is that of testing the models and establishing their accuracies. This will entail more creative approaches to testing, with boundary conditions established from time series of atmospheric CO_2 concentrations as critical features. DGVMs are also being developed for integration within GCMs so that vegetation feedbacks can be

23

Estimations of Global Terrestrial Productivity: Converging toward a Single Number?

Bernard Saugier, Jacques Roy, and Harold A. Mooney

I. Introduction

Terrestrial ecosystems significantly contribute to the global carbon cycle. They constitute large exchangeable stocks of carbon (living organisms, litter, and soil organic matter contain about 2200 Pg C, three times more than in the atmosphere) and provide large fluxes with the atmosphere (CO_2 uptake by land photosynthesis is 20 times larger than CO_2 release from the burning of fossil fuel, and is roughly compensated by respiration fluxes) (Schimel *et al.*, 1995). As discussed in several chapters of this book, these stocks and fluxes are increasingly altered by human activities, through changes in land use, in atmospheric composition, and in climate. This alteration is presently removing about 2 Pg C yr^{-1} (the terrestrial carbon sink) from the atmosphere, significantly decreasing the growth in atmospheric CO_2 (Schimel *et al.*, 1995). This carbon sink results from an increase in global terrestrial net primary productivity (NPP). Thus it is essential to know what is the present value of global NPP and if it will continue to increase to sustain an important terrestrial carbon sink. Here we will use the more recent information provided in each of the biome chapters to derive updated estimates of global values for NPP and phytomass. This NPP value will be compared to previous estimates, and to those derived from models based on vegetation structure or on satellite data. Finally, the current and expected changes in global NPP and net ecosystem productivity are discussed.

II. Synthesis of Biome Data

A. Average Biomass and NPP per Biome

Table 23-1 presents the estimates of biomass and productivity given in the biome chapters (Chapters 9–17) of this book. The novelty is the emphasis on the root compartment, which was often ignored in previous estimates. Measuring root biomass is difficult, and measuring root productivity is even more difficult. The root figures thus contain a high relative uncertainty, perhaps 20% for biomass and 40% for productivity. In addition, for biomes with a high spatial heterogeneity such as tundra, desert, or savanna, carbon allocation to roots is quite variable, depending on local conditions of water and nutrient supply and on biological types (herbaceous versus woody species). In spite of these shortcomings, root figures will be found quite useful to studies on the carbon cycle. The ratio of root biomass to total biomass varies from 0.13 in crops to 0.62 in arctic tundra and 0.67 in temperate grasslands, with intermediate figures (0.21–0.27) in forests. The highest values are found in biomes experiencing severe constraints, such as drought (deserts, mediterranean shrublands, temperate grasslands, semiarid savannas), or nutrient shortage (tundra). The relatively small values found in humid savannas likely come from the fast turnover of grass roots, which prevents biomass accumulation.

With respect to productivity, the root fraction varies from 0.13 to 0.67. If we remove crops, all figures range from 0.39 to 0.67. It is striking that figures are similar for deserts and forests (0.39 to 0.44), a bit higher for savannas and mediterranean shrublands (about 0.5), and highest for tundra (0.57) and temperate grasslands (0.67).

These biomass and NPP figures given are compared (Table 23-1) to those given by Ajtay *et al.* (1979). Biomass figures are similar for tropical and temperate forests, but our figure for boreal forests is much smaller than that of Ajtay *et al.,* and likely more realistic, being based on many recent measurements. Our figure for mediterranean shrublands is higher than proposed by Ajtay *et al.* (1979) for chaparral essentially because roots were taken into account.

With regard to total NPP, we find about 20% higher values than Ajtay *et al.* for tropical forests because roots were taken into account, and 50% less than Ajtay *et al.* for boreal forest. Savannas have a lower NPP in our estimate and we believe it is justified by many measurements reported in Chapter 16 (this volume). Other discrepancies are of minor importance; even our higher estimate of desert NPP has a relatively small impact at the global scale because desert NPP is small in any case.

B. Biophysical and Ecophysiological Parameters

Tables 23-2 and 23-3 give some parameters that are useful for people involved in models of vegetation functioning. It is not possible to give a single

Table 23-1 Average Values of Living Phytomass and Net Primary Productivity per Biome[a]

Biome	Phytomass [kg (DM) m^{-2}]					Net primary productivity [kg (DM) m^{-2} yr^{-1}]				
	Shoot	Root	Root/ total	Total	Total Ajtay	ANPP	BNPP	Root/ total	Total NPP	Total Ajtay
Tropical forests	30.4	8.4	0.22	38.8	36.70	1.40	1.10	0.44	2.50	2.08
Temperate forests	21	5.7	0.21	26.7	29.00	0.95	0.60	0.39	1.55	1.4
Boreal forests	6.1	2.2	0.27	8.3	22.80	0.23	0.15	0.39	0.38	0.79
Arctic tundra	0.25	0.4	0.62	0.65	1.37	0.08	0.10	0.57	0.18	0.22
Mediterranean shrublands	6	6	0.50	12	7.00	0.50	0.50	0.50	1.00	0.8
Crops	0.53	0.08	0.13	0.61	0.42	0.53	0.08	0.13	0.61	0.94
Tropical savannas and grasslands	4.0	1.7	0.30	5.8	6.5	0.54	0.54	0.50	1.08	1.75
Temperate grasslands	0.25	0.50	0.67	0.75	1.62	0.25	0.50	0.67	0.75	0.78
Deserts	0.35	0.35	0.50	0.70	0.58	0.15	0.10	0.40	0.25	0.1

[a]Derived from the biome chapters in this volume and comparison with values given by Ajtay *et al.* (1979).

Table 23-2 Biophysical Parameters for Each Biome

Biome	Albedo	Height (m)	z_0 (m)[a]	Max LAI[b]	Rooting depth (m)
Tropical forests	0.12–0.14	30–50	2–2.2	4–7.5	1.0–8.0
Temperate forests	0.1–0.18	15 (5–50)	1.0–3.0	7 (3–15)	0.5–3.0
Boreal forests	0.1–0.3	10 (2–20)	1 to 3	2 (1–6)	0.5 to 1
Arctic tundra	0.2–0.8	<0.5 (0–3)	<0.05	1 (0–3)	0.4–0.8
Mediterranean shrublands	0.12–0.2	0.3–10	0.03–0.5	3 (1–6)	1.0–6.0
Crops	0.1–0.2	Variable	Variable	4	0.2–1.5
Tropical savannas and grasslands	0.07–0.4	0.3–9	Variable	0.5–4	0.5–2 (grass)
Temperate grasslands	0.15–0.25	0.2–1	0.02–0.1	1.0–3.0	0.5–1.5
Deserts	0.2–0.4	<0.5	<0.05	1	0.2–15
Ice	0.5–0.8				

[a] z_0, roughness length.
[b] LAI, Leaf area index.

value per biome, because of time and space variations. Thus the albedo of green vegetation is often near 0.2. However, Table 23-2 shows important departures from this average value. In the absence of snow, forests have a low albedo, ranging from 0.1 to 0.18, because the space between tree crowns acts as a radiation trap, especially effective in coniferous forests. Snow-covered boreal forests have an albedo near 0.3, a relatively small value compared to that of tundra in winter (0.5–0.8). This is mainly caused by the shadows cast on the ground by tree crowns. As a result, the expected northward extension of boreal forests will lead to a decrease in albedo and thus to an increase in the absorption of solar radiation, acting as a positive feedback to global warming.

In water-limited ecosystems, albedo varies with the fractions of ground covered by bare soil, by green leaves, and by dead leaves. The albedo of bare soil may vary from 0.07 for a savanna soil covered with dark ashes after fire, to 0.4 for a bright sand. The albedo of green leaves is between 0.15 and 0.2, and that of dead leaves is close to 0.25. These differences explain figures given for mediterranean shrubland, crops, savanna, temperate grassland, and deserts.

Vegetation height mainly depends on the presence of trees. Herbaceous vegetation, in contrast, does not exceed 1 m in the temperate zone and 2 m in the tropics. Roughness length, which controls the intensity of turbulent transport for a given wind velocity, increases with vegetation height, and also depends on tree density in open canopies such as boreal forests, savannas, and mediterranean shrublands. Maximum leaf area index (LAI) is controlled by resource availability, being the highest in tropical and temperate forests and in crops.

Rooting depth is especially difficult to estimate, because, in most cases, deep roots exist in natural ecosystems, even if only a small fraction is present below 1 or 2 m. Thus what is useful to modelers is a sort of equivalent root depth, which may not be deduced easily from profiles of root biomass. Values given in Table 23-2 are quite variable, especially in trees or shrubs growing in dry habitats.

Table 23-3 is not complete because there were missing values for some biomes. Maximum surface conductance and maximum CO_2 flux during the daytime are related. It is somewhat surprising to find similar values for biomes as different as tropical forests, temperate forests, and savannas, with maximum CO_2 flux near 25 $\mu mol\ m^{-2}\ s^{-1}$. The relative constancy of radiation use efficiency among biomes has been noted by Ruimy *et al.* (1994, 1996). Strong constraints act to decrease CO_2 flux, which is smaller in boreal forests and even smaller in arctic tundra and deserts. CO_2 flux, as does productivity, has two major determinants: leaf area index and leaf photosynthetic capacities. In most cases a shortage in nutrients or in water will first affect leaf area, and leaf photosynthesis only when the shortage becomes severe. CO_2 flux is thus strongly linked to LAI, and only secondarily to leaf photosynthesis.

C. Biome Areas

The concept of biome was introduced by Clements (1916) as a large vegetation unit retaining uniformity in its physiognomy. In this book, we have used the classical biome names, but with the difficulty that there is no good agreement on their location and area. Most atlases contain a vegetation map giving such information in relation to climate variables, but the biome areas are not quantified, and what is generally supplied is a potential vegetation map, i.e., a map of what the vegetation would look like if humans had not changed the land use. Also, the biome definition is dependent on the authors of a report. To suit the needs of global climate models and later of global vegetation models, several biome maps were produced during the 1980s (i.e., Matthews, 1983; Olson *et al.*, 1983). A biome map has now been produced by the International Geosphere–Biosphere Program (http://www.igbp.kva.se/progelem.html) from satellite data (1-km resolution from NOAA–AVHRR sensors). The information it uses is the seasonal variation of the vegetation index, which allows a biome separation based on leaf phenology. It is very useful, but this biome definition has yet to be reconciled with the conventional definitions used in this book, and for that reason it has not been used here. Future progress will come from comparing the conventional biome distribution to the digital vegetation maps produced from satellite observations.

Table 23-3 Ecophysiological Parameters for Each Biome[a]

Biome	Max g_s (mol m^{-2} s^{-1})	g_a (mol m^{-2} s^{-1})	Max CO_2 flux, day (μmol m^{-2} s^{-1})	Max CO_2 flux, night (μmol m^{-2} s^{-1})	RUE [g (DM)/MJ (PAR)]
Tropical forests	0.5–1	0–4	25	5.0–8.0	0.9 (0.45–1.5)
Temperate forests	0.5	1.0–4.0	25	1.0–6.0	1
Boreal forests	0.2	10	12	0–4.0	0.3–0.5
Arctic tundra			0.5–2	1.0–2.0	
Mediterranean shrublands	0.5–1		12–15	6–7	
Crops	1.2	1.0–3.0	40	2.0–8.0	1–1.5 (pulses 0.8)
Tropical savannas and grasslands	0.2–1	0.1–4	4–25	2.0–5.0	0.4–1.8
Temperate grasslands	0.4–1	0.2–1.5	13–20	0.5–4	
Deserts					

[a]Max g_s, Maximum value of surface conductance for water vapor; g_a, canopy boundary layer conductance; RUE, radiation use efficiency (total net primary productivity per unit of intercepted radiation). CO_2 fluxes are vertical fluxes measured by the eddy-covariance technique; they are normally downward during the day and upward during the night. Absolute values are given.

Olson (1975) reviewed the extent of forest ecosystems and found global figures between 40 and 66 million km^2 (M km^2). The former figure came from "various United Nations sources," whereas the latter was an estimate by Basilevich *et al.* (1971), based on a careful study of a physical–geographical atlas, indicating 17 M km^2 for boreal, subalpine, and "semiboreal" zones, over 18 M km^2 for temperate wooded zones, 17 M km^2 for tropical forests, and 14 M km^2 for tropical woody savanna and scrubland. Ajtay *et al.* (1979) reviewed the knowledge on terrestrial primary productivity and phytomass for a SCOPE book on the global carbon cycle. They compared various estimates of biome areas. For total forest area the estimates had ranged from 36 to 64 million km^2. They revised these figures downward to 31 M km^2, the lowest figure ever noted. The Food and Agriculture Organisation (FAO) regularly issues reports on world forest resources based on national inventories. Detailed inventories are made every 10 yr, with updates every 5 yr, the most recent in 1995. The FAO forest includes forest and forest plantations, and is defined as land with tree crown cover of more than 20% of the area, and with trees growing taller than 7 m in height. It does not include open woodland (tree crown cover between 5 and 20%), nor scrub, shrub, and bushland, where woody elements cover more than 20% of the area, but does include heights between 0.5 and 7 m. According to the FAO, world forests covered 34.54 M km^2 in 1995, of which 17.34 M km^2 were tropical, and 17.20 M km^2 were temperate and boreal. Spatial information is not available from the FAO, but the World Conservation Monitoring Center (WCMC) compiled in 1996 a world forest map showing forest extent and protected areas. They found a total forest area of 40 M km^2, 16% higher than the FAO estimate. The WCMC figure is close to that given by Dixon *et al.* (1994) (41.65 M km^2), which will be used from now on, until there is better agreement between the various estimates. Dixon *et al.* (1994) gave an area of 17.55 M km^2 for tropical forests, in close agreement with the FAO figure of 17.34 M km^2. The agreement is not fortuitous because they use, at least partly, the FAO data. Their figure of 13.7 M km^2 for boreal forests is based on Russian (Dixon and Krankina, 1993) and Canadian (Apps and Kurz, 1991) estimates. Their figure of 10.4 M km^2 for temperate forests comes from a compilation of data for the continental United States (Birdsey *et al.*, 1993), Europe (Kauppi *et al.*, 1992), and China and Australia (WCMC).

Our estimates of biome areas are compared to those of Ajtay *et al.* (1979) in Table 23-4. Clearly the forest areas by Dixon *et al.* (1994) are larger than those given by Ajtay *et al.* (1979) for all forest types, especially for boreal forests. By contrast, tundra and deserts are more limited in our estimate, but we find higher areas for savanna and temperate grassland. Very dry grasslands or savannas (including tropical grasslands) may be ranged in deserts, which explains differences with data of Ajtay *et al.* (1979). Crop area (from FAO data) is more limited in our values than in work of Ajtay *et al.*, and even

Table 23-4 Area, Carbon in Living Phytomass, and Carbon Net Primary Productivity per Biome[a]

	Area (million km^2)		Total C pool [Pg (C)]		Total C NPP [Pg (C) yr^{-1}]	
Biome	RSM	Ajtay	RSM	Ajtay	RSM	Ajtay
Tropical forests	17.5	14.8	340	244	21.9	13.7
Temperate forests	10.4	9.5	139	108	8.1	6.3
Boreal forests	13.7	9.0	57	92	2.6	3.2
Arctic tundra	5.6	9.5	2	6	0.5	1.0
Mediterranean shrublands	2.8	2.5	17	8	1.4	0.9
Crops	13.5	16.0	4	3	4.1	6.8
Tropical savanna and grasslandss	27.6	22.5	79	66	14.9	17.7
Temperate grasslands	15	12.5	6	9	5.6	4.4
Deserts	27.7	30.0	10	8	3.5	1.4
Ice	15.5	15.5				
Other		7.5		16		4.5
Total	**149.3**	**149.3**	**652**	**560**	**62.6**	**59.9**

[a]Comparison with the values given by Ajtay *et al.* (1979); RSM, this volume. Values for C phytomass and C NPP are the total DM values given in Table 23-1 multiplied by the area of each biome and converted to C using a conversion factor of 0.5.

more so considering that out of the 13.5 M km^2 of crops, 2 M km^2 are fallow (see Chapter 13, this volume). In contrast to Ajtay *et al.* (1979), we did not specifically consider areas such as urban areas, lakes, streams, etc. that are classified as "other" in Table 23-4.

D. Global Phytomass and NPP: Residence Time of Carbon and Carbon Sink

Table 23-4 gives the area, living phytomass, and NPP for each biome. The global figures for all terrestrial ecosystems derived from the information collected in this book are 652 Pg of carbon for the living biomass and 62.6 Pg yr^{-1} for carbon NPP. These figures are not very different from those given by Ajtay *et al.* in 1979 (560 and 59.9 Pg yr^{-1}, respectively). The larger total biomass in our estimate comes from larger areas of all forest types and from a higher estimate of biomass in tropical forests, not compensated for by a smaller biomass of boreal forest (−63%). In our estimate, forests represent 28% of continental area, 82% of total biomass, and 52% of total NPP. Corresponding figures in Ajtay *et al.* were noticeably smaller, with 22% of the area, 79% of total biomass, and 39% of total NPP. From Table 23-4 the average biomass on continents is found to be 8.7 kg dry matter (DM) m^{-2} and the average NPP is 0.84 kg DM m^{-2} yr^{-1}.

Dividing biomass by NPP in Table 23-4 gives the average residence time of carbon in biomass. Averaged over all biomes, this value is 10.4 yr. It is longer in forests, reaching a maximum of 22 yr in boreal forests. Residence time is an important parameter for calculating the carbon sink of terrestrial ecosystems as NPP increases. Then it is the residence time of carbon in the terrestrial biosphere (including soil) that matters. Dividing the global carbon stock given in Chapter 1 (2200 Pg) by the NPP (62.6 Pg) gives 35 yr.

To understand the role of this residence time, let us assume that NPP increases linearly with time. Thus

$$\mathrm{NPP}(t) = \mathrm{NPP}_0(1 + at), \tag{1}$$

where NPP_0 is the NPP at time zero and a is a constant.

We will assume that heterotrophic respiration is equal to the NPP at time $t - t_r$, where t_r is the residence time of carbon in the ecosystem (i.e., defined as carbon in vegetation and soil divided by NPP). We might thus write:

$$R_h\,(t) = \mathrm{NPP}(t - t_r). \tag{2}$$

The net ecosystem productivity is equal to the difference between NPP and R_h, thus

$$\mathrm{NEP}(t) = \mathrm{NPP}(t) - \mathrm{NPP}(t - t_r) = \mathrm{NPP}_0 \cdot a t_r. \tag{3}$$

For NPP_0 and t_r we can use our figures, 62.6 Pg yr^{-1} and 35 yr^{-1}, respectively. To compute the value of a we may assume that the increase in NPP results entirely from an increase in atmospheric CO_2. We know the increase in CO_2 is about 0.4% yr^{-1} and for forests at least the beta factor (the fractional increase in NPP due to a doubling in CO_2) is about 0.3. This leads to a value for a of 0.3 (0.004) = 0.0012 yr^{-1}. When these values are used in Eq. (3) they lead to a NEP of 2.63 Pg yr^{-1}. This shows that a relatively large carbon sink may be explained by a small but sustained increase in NPP. Such an argument was already made by Gifford (1994).

III. Comparison of NPP and Phytomass Estimated by Different Methods

Interestingly enough, in the mid-nineteenth century, von Liebig (1862), from measurements made at the back of his garden, derived an estimate of global NPP quite close to the latest estimates (Lieth, 1975) (Fig. 23-1). NPP of herbaceous vegetation is indeed close to NPP averaged over all biomes (Table 23-1). During the twentieth century, using field measurements on an increasing number of ecosystem types, estimates varied widely, from 13 to 77 Pg C yr^{-1}. However, the last (most recent) two estimates from field data

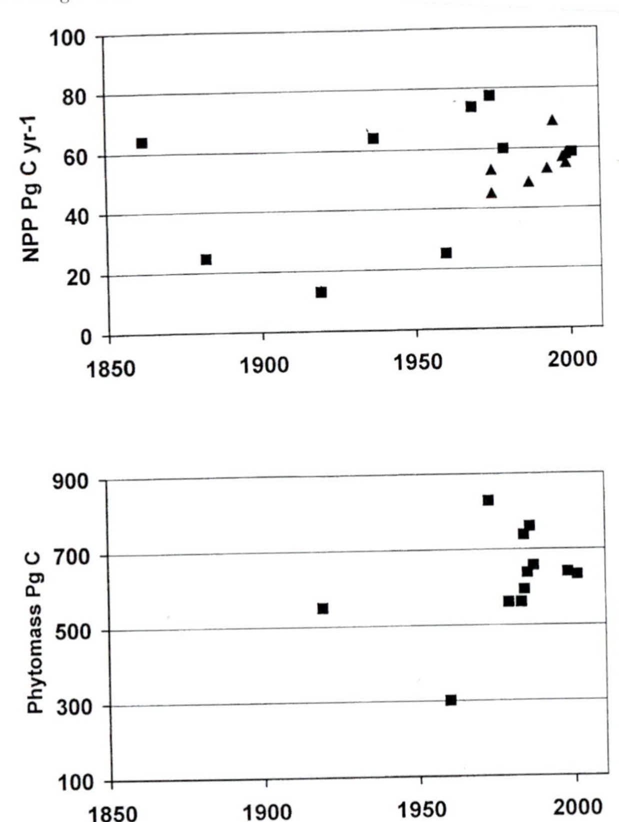

Figure 23-1 Historic estimates of global terrestrial NPP (top) and living phytomass (bottom). Estimates are based on field data (■) or models (▲). Data from von Liebig (1862), Ebermayer (1882), Schroeder (1919), Noddack (1937), Müller (1960), Whittaker and Likens [1969; in Lieth (1975)], Whittaker and Likens (1973), Rodin *et al.* (1975), Lieth (1975), Whittaker and Likens (1975), Atjay *et al.* (1979), Olson *et al.* (1983), Matthews (1984), Goudriaan and Ketner (1984); Aselmann [in Esser (1987)], Esser (1986), Esser (1987), Melillo *et al.* (1993), Prince and Goward (1995), Cao and Woodward (1998), Potter *et al.* (1999), and Cramer *et al.* (1999); also, this book (last square on the right). Literature before 1975 is quoted from Lieth (1975) or from Esser (1987).

(Ajtay *et al.*, 1979; this book), despite major differences in the estimation of individual components as detailed above, yield similar global values.

Starting in the mid-1970s, estimations of NPP were made using models. Some are empirical or physiological models simulating NPP from environmental variables (e.g., Lieth, 1975), others are remote-sensing-based models interpreting the spectrum of reflected radiation by the land surfaces (e.g., Prince and Goward, 1995). The latest estimates using models are close to our estimate from field data (Fig. 23-1). Cramer *et al.* (1999) compared

17 global models of terrestrial biogeochemistry. The estimate of global NPP averaged over all models was 54.9 Pg C yr^{-1}.

This convergence of the estimates of global NPP may, however, be partly fortuitous. There are still a lot of uncertainties in both methods and the convergence of the means should not mask the large range of outputs. After removal of two outliers, the NPP values obtained by Cramer *et al.* (1999) ranged from 44.3 to 66.3 Pg C yr^{-1}. Models, whether satellite based or ecosystem-process based, all provided high and low values of NPP, but estimates were dependent on whether nutrient constrains were considered and on the way the effect of water stress was modeled.

Estimates of living phytomass also fluctuated during the twentieth century, although to a lesser extent than estimates of NPP (Fig. 23-1). The latest three phytomass estimates obtained from models (Esser, 1987; Cao and Woodward, 1998) or field data (this volume) converge toward a value in the range of 611 to 657 Pg C.

IV. Global Change-Induced Variations in NPP and NEP

The convergent values of NPP or of phytomass mentioned above need to be put in the context of their dynamics imposed by global changes. Increase in atmospheric CO_2 concentration, increase in rates of atmospheric nitrogen deposition, and climate variability (temperature and precipitation) are the main drivers of NPP variation. Some reconstructed or anticipated changes in NPP are shown in Fig. 23-2. The short-term reconstruction of Potter *et al.*

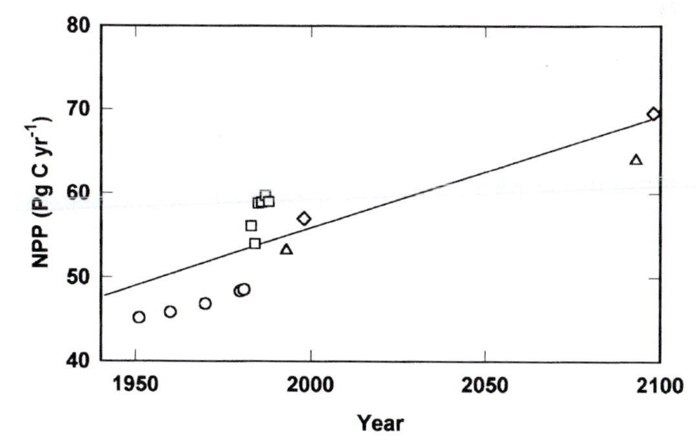

Figure 23-2 Reconstructed or predicted values of global terrestrial NPP. Reconstructed values take into account land use changes [Esser (1987), ○] and actual climatic data [Potter *et al.* (1999), □]. Predicted values [Melillo *et al.* (1993), △; Cao and Woodward (1998), ◇] correspond to the impact of a doubling of CO_2 and associated climatic changes. This doubling of CO_2 has been proposed to occur over the next century.

(1999) indicates that interannual variability, driven in the period analyzed mainly by temperature, can be substantial, on the order of 10%. The long-term simulations, although based on different models, have consistent slopes, of about 0.14 Pg C, which would lead to a 25% increase over the coming century. In the simulations of Cao and Woodward (1998), the effect of CO_2 alone would be a 25.8% increase in NPP. Adding the associated climatic changes brings this increase down to 22.1% due to less favorable hydric conditions in the tropics.

Inversion models reconstruct net ecosystem carbon fluxes (NEP) from atmospheric measurements of CO_2, O_2, N_2, and isotopic signatures. Using such a model, Rayner *et al.* (1999) calculated an average net sink of C for the period 1980–1995 of 0.7 Pg C yr^{-1} (0.3 in 1980–1987 and 1.1 in 1988–1995). During the same period, oceans were a net sink of 2.1 Pg C. Locations for the sinks on land are thought to be in the tropical regions, in relation to atmospheric CO_2, and in temperate regions in response to atmospheric N deposition (Rayner *et al.*, 1999; Lloyd, 1999). However, there is absolutely no certainty on this point, because global warming has been shown to increase the duration of the vegetative period, and thus NPP of temperate and boreal ecosystems (Keeling *et al.*, 1996; Myneni *et al.*, 1997). A fine discussion of the carbon balance of forests is found in Malhi *et al.* (1999). In a literature review, Lloyd (1999) estimated the terrestrial carbon sink in response to CO_2 fertilization of the order of 1.2 Pg C yr^{-1}. Changes in carbon storage in vegetation and soil for a doubling of CO_2 and associated climatic changes were simulated to be between 250 Pg C (Melillo, 1996) and 293 Pg C (Cao and Woodward, 1998), which correspond to an average NEP sink of between 2.5 and 2.9 Pg C yr^{-1}, respectively, if the doubling of atmospheric CO_2 is assumed to take place over the next 100 yr. The response to N deposition is also estimated to be a net sink of 1.2 Pg C yr^{-1} (Townsend *et al.*, 1996; Holland *et al.*, 1997), but this prediction is based on very crude models and insufficient knowledge (Lloyd, 1999). Nadelhoffer *et al.* (1999) suggested a smaller C sink (not exceeding 0.5 Pg C yr^{-1}) on the basis of the fate of ^{15}N, used as a tracer in forest soils. Their estimate is also open to criticism because it does not take into account a possible redistribution of nitrogen after 1 yr.

V. Conclusions

There were numerous attempts to quantify biomass and NPP in various biomes during the early days of the International Biosphere Programme (1968–1974), but there was no real attempt at the time to provide biome averages and global averages. To understand the global carbon cycle such information is necessary, and the work of Ajtay *et al.* (1979), done under the SCOPE, provided such information. In this book as in Atjay *et al.*, global terrestrial NPP has been computed from data on biome NPP and on biome area. Figures given for biome NPP come from the most recent information available. The main source of uncertainty on global NPP lies in biome area and especially in forest area. Significant progress on biome maps is expected in the near future from the use of satellite data, once the classification done using such data can be given biome names that can receive a large acceptance. There is also some uncertainty on the current and future rates of change in NPP. Research on this aspect, at all scales, is developing fast. We urgently need to know and predict more precisely the impact of human activities on NPP and on the carbon cycle in general. This information is crucial to design and implement sustainable development of human societies.

References

Ajtay, G. L., Ketner, P., and Duvigneaud, P. (1979). Terrestrial primary production and phytomass. *In* "The Global Carbon Cycle, SCOPE 13" (B. Bolin, E. T. Degens, and P. Ketner, eds.), pp. 129–182. John Wiley & Sons, New York.

Apps, M. J., and Kurz, W. A. (1991). Assessing the role of Canadian forests and forest sector activities in the global carbon balance. *World Resour. Rev.* **3,** 333–344.

Basilevich, N. I., Rodin, L. Y., and Rozov, M. M. (1971). Geographical aspects of biological productivity. *Sov. Geogr. Rev. (Transl.)* **12,** 293–317.

Birdsey, R. A., Plantiga, A. J., and Heath, L. S. (1993). Past and prospective carbon storage in United States forests. *For. Ecol. Manag.* **58,** 33–40.

Cao, M., and Woodward, F. I. (1998). New primary and ecosystem production and carbon stockes of terrestrial ecosystems and their responses to climate change. *Global Change Biol.* **4,** 185–198.

Clements, F. E. (1916). Plant succession, an analysis of the development of vegetation. *Carnegie Inst. Wash. Publ.* **242,** 1–512.

Cramer, W., Kicklighter, D. W., Bondeau, A., Moore III, B., Churkina, G., Nemry, B., Ruimy, A., Schloss, A. L., and the participants of the Postdam NPP Model Intercomparison (1999). Comparing global models of terrestrial net primary productivity (NPP): overview and key results. *Global Change Biol.* **5** (Suppl. 1), 1–15.

Dixon, R. K., and Krankina, O. N. (1993). Forest fires in Russia: Carbon dioxide emissions to the atmosphere. *Can. J. For. Res.* **23,** 700–705.

Dixon, R. K., Brown, S., Houghton, R. A., Solomon, A. M., Trexler, M. C., and Wisniewski, J. (1994). Carbon pools and flux of global forest ecosystems. *Science* **263,** 185–190.

Ebermayer, E. (1882). "Naturgesetzliche Grundlagen des Wald-und Ackerbaues, Vol. 1: Die Bestandteile des Pflanzen, Part I Physiologische Chemie der Pflanzen." Springer-Verlag, Berlin–Heidelberg.

Esser, G. (1986). The carbon budget of the biosphere—Structure and preliminary results of the Osnabrück Biosphere Model (in German with an extended English summary). Veröff. Naturf. Ges. zu Emden von 1814, New Series, Vol. 7.

Esser, G. (1987). Sensitivity of global carbon pools and fluxes to human and potential climatic impacts. *Tellus* **39B,** 245–260.

FAO: http://WWW.FAO.ORG/waicent/faoinfo/forestry/forestry.htm.

Gifford, R. M. (1994). The global carbon cycle—A viewpoint on the missing sink. *Austr. J. Plant Physiol.* **21,** 1–15.

Goudriaan, J., and Ketner, P. (1984). Are land biota a source or sink for CO_2? Interactions between climate and biosphere. *Prog. Biometeorol.* **3,** 247–252.

Holland, E. A., Braswell, B. H., Lamarque, J.-F., Townsend, A., Sulzman, J., Müler, J.-M., Dentener, F., Brasseur, G., Levy II, H., Penner, J. E., and Roelofs, G.-J. (1997). Variation in the predicted spatial distribution of atmospheric nitrogen deposition and their impact on carbon uptake by terrestrial ecosystems. *J. Geophys. Res.* **102,** 15849–15866.

Kauppi, P. E., Mielikänen, K., and Kuusela, K. (1992). Biomass and carbon budget of European forests, 1971–1990. *Science* **256,** 70–74.

Keeling, C. D., Chin, J. F. S., and Whorf, T. P. (1996). Increased activity of northern vegetation inferred from atmospheric CO_2 measurements. *Nature (London)* **382,** 146–149.

Lieth, H. (1975). Historical survey of primary productivity research. *In* "Primary Productivity of the Biosphere" (H. Lieth and R. H. Whittaker eds.), pp. 7–16. Springer-Verlag, New York.

Lloyd, J. (1999). Current perspectives on the terrestrial carbon cycle. *Tellus* **51B,** 336–342.

Malhi, Y., Baldocchi, D. D., and Jarvis, P. G. (1999). The carbon balance of tropical, temperate and boreal forests. *Plant Cell Environ.* **22,** 715–740.

Matthews, E. (1983). Global vegetation and land use: New high resolution data bases for climate studies. *J. Climate Appl. Meteorol.* **22,** 474–487.

Matthews, E. (1984). Global inventory of the pre-agricultural and present biomass. Interactions between climate and biosphere. *Prog. Biometeorol.* **3,** 237–247.

Melillo, J. M. (1996). Carbon and nitrogen interactions in the terrestrial biosphere: Anthropogenic effects. *In* "Global Change and Terrestrial Ecosystems" (B. Walker and W. Steffen, eds.), pp. 431–450. Cambridge Univ. Press, Cambridge.

Melillo, J. M., McGuire, A. D., Kicklighter, D. W., Moore III, B., Vorosmarty, C. J., and Schloss, A. L. (1993). Global climate change and terrestrial net primary production. *Nature (London)* **363,** 234–240.

Müller, D. (1960). Kreislauf des Kohlenstoffs. *Handb. Pflanzenphysiol.* **12,** 934–948.

Myneni, R. B., Keeling, C. D., Tucker, C. J., Asrar, G., and Nemani, R. R. (1997). Increased plant growth in the northern high latitudes from 1981 to 1991. *Nature (London)* **386,** 698–702.

Nadelhoffer, K. J., Emmett, B. A., Gundersen, P., Kjonaas, O. J., Koopmans, C. J., Schleppi, P., Tietema, A., and Wright, R. F. (1999). Nitrogen deposition makes a minor contribution to carbon sequestration in temperate forests. *Nature (London)* **398,** 145–148.

Noddack, W. (1937). Der Kohlenstoff im Haushalt der Natur. *Angew. Chem.* **50,** 505–510.

Olson, J. S. (1975). Productivity of forest ecosystems. *In* "Productivity of World Ecosystems" (D. E. Reichle, J. F. Franklin, and D. W. Goodall, eds.), pp. 33–43. National Academy of Sciences, Washington, D.C.

Olson, J. S., Watts, J. A., and Allison, L. J. (1983). Carbon in live vegetation of major world ecosystems. Environmental Science Division Publ. No. 1997. Oak Ridge National Laboratory, Oak Ridge, Tennessee.

Potter, C. S., Klooster, S., and Brooks, V. (1999). Interannual variability in terrestrial net primary production: Exploration of trends and controls on regional to global scales. *Ecosystems* **2,** 36–48.

Prince, S. D., and Goward, S. N. (1995). Global primary production: A remote sensing approach. *J. Biogeogr.* **22,** 815–835.

Rayner, P. J., Enting, I. G., Francey, R. J., and Lagenfelds, R. (1999). Reconstructing the recent carbon cycle from atmospheric CO_2, $\delta^{13}C$ and O_2/N_2 observations. *Tellus* **51B,** 213–232.

Rodin, L. E., Bazilevich, N. I., and Rozov, M. M. (1975). Productivity of the world's main ecosystems. *In* "Productivity of the World Ecosystems: Proceedings of the Seattle Symposium 1974" (D. E. Reichle, J. F. Franklin, and D. W. Goodall, eds.), pp. 13–26. National Academy of Sciences, Washington, D.C.

Ruimy, A., Dedieu, G., and Saugier, B. (1994). Methodology for the estimation of terrestrial net primary production from remotely sensed data. *J. Geophys. Res. (Atmospheres)* **99, D3,** 5263–5283.

Ruimy, A., Dedieu, G., and Saugier, B. (1996). TURC: A diagnostic model of continental gross primary productivity and new primary productivity. *Global Biogeochem. Cycles* **10,** 269–285.

Schimel, D., Enting, I. G., Heimann, M., Wigley, T. M. L., Raynaud, D., Alves, D., and Siegenthaler, U. (1995). CO_2 and the carbon cycle. *In* "Climate Change 1994. Radiative Forcing of Climate Change and an Evaluation of the IPCC IS92 Emission Scenarios" (J. T. Houghton, L. G. Meira Filho, J. Bruce, J. Lee, B. A. Callander, E. Haties, N. Harris, and K. Maskell, eds.), pp. 39–71. Cambridge Univ. Press, Cambridge.

Schroeder, H. (1919). Die jährliche Gesamtproduktion der grünen Pflanzendecke der Erde. *Naturwissenschaften* **7,** 8–12.

Townsend, A. R., Braswell, B. H., Holland, E. A., and Penner, J. E. (1996). Spatial and temporal patterns in terrestrial carbon storage due to deposition of fossil fuel nitrogen. *Ecological Appl.* **6,** 806–814.

von Liebig, J. (1862). "Die Naturgesetze des Feldbaues". p. 467. Vieweg, Braunschweig.

WCMC (World Conservation Monitoring Center): http://www.wcmc.org.uk/forest/data/.

Whittaker, R. H., and Likens, G. E. (1969). World productivity estimate, Brussels Symp. 1969. *Cited in:* Whittaker 1970, "Communities and ecosystems", Macmillan, New York.

Whittaker, R. H., and Likens, G. E. (1973). Carbon in the biota. *In* "Carbon and the Biosphere" (G. M. Woodwell and E. V. Pecan, eds.), pp. 281–302. U.S. National Techn. Inf. Center, Springfield, U.S.A.

Whittaker, R. H., and Likens, G. E. (1975). The biosphere and man. *In* "Primary Productivity of the Biosphere," (H. Lieth and R. H. Whittaker, eds.), pp. 305–328. Ecology Studies, **14,** Springer-Verlag, Berlin, Heidelberg, New York.

Subject Index

Physiological Ecology

A Series of Monographs, Texts, and Treatises

T. T. KOZLOWSKI. Growth and Development of Trees, Volumes I and II, 1971
D. HILLEL. Soil and Water: Physical Principles and Processes, 1971
V. B. YOUNGER and C. M. McKELL (Eds.). The Biology and Utilization of Grasses, 1972
J. B. MUDD and T. T. KOZLOWSKI (Eds.). Responses of Plants to Air Pollution, 1975
R. DAUBENMIRE. Plant Geography, 1978
J. LEVITT. Responses of Plants to Environmental Stresses, Second Edition
Volume I: Chilling, Freezing, and High Temperature Stresses, 1980
Volume II: Water, Radiation, Salt, and Other Stresses, 1980
J. A. LARSEN (Ed.). The Boreal Ecosystem, 1980
S. A. GAUTHREAUX, JR. (Ed.). Animal Migration, Orientation, and Navigation, 1981
F. J. VERNBERG and W. B. VERNBERG (Eds.). Functional Adaptations of Marine Organisms, 1981
R. D. DURBIN (Ed.). Toxins in Plant Disease, 1981
C. P. LYMAN, J. S. WILLIS, A. MALAN, and L. C. H. WANG. Hibernation and Torpor in Mammals and Birds, 1982
T. T. KOZLOWSKI (Ed.). Flooding and Plant Growth, 1984
E. L. RICE. Allelopathy, Second Edition, 1984
M. L. CODY (Ed.). Habitat Selection in Birds, 1985
R. J. HAYNES, K. C. CAMERON, K. M. GOH, and R. R. SHERLOCK (Eds.). Mineral Nitrogen in the Plant–Soil System, 1986
T. T. KOZLOWSKI, P. J. KRAMER, and S. G. PALLARDY. The Physiological Ecology of Woody Plants, 1991
H. A. MOONEY, W. E. WINNER, and E. J. PELL (Eds.). Response of Plants to Multiple Stresses, 1991
F. S. CHAPIN III, R. L. JEFFERIES, J. F. REYNOLDS, G. R. SHAVER, and J. SVOBODA (Eds.). Arctic Ecosystems in a Changing Climate: An Ecophysiological Perspective, 1991

T. D. SHARKEY, E. A. HOLLAND, and H. A. MOONEY (Eds.). Trace Gas Emissions by Plants, 1991

U. SEELIGER (Ed.). Coastal Plant Communities of Latin America, 1992

JAMES R. EHLERINGER and CHRISTOPHER B. FIELD (Eds.). Scaling Physiological Processes: Leaf to Globe, 1993

JAMES R. EHLERINGER, ANTHONY E. HALL, and GRAHAM D. FARQUHAR (Eds.). Stable Isotopes and Plant Carbon–Water Relations, 1993

E.-D. SCHULZE (Ed.). Flux Control in Biological Systems, 1993

MARTYN M. CALDWELL and ROBERT W. PEARCY (Eds.). Exploitation of Environmental Heterogeneity by Plants: Ecophysiological Processes Above- and Belowground, 1994

WILLIAM K. SMITH and THOMAS M. HINCKLEY (Eds.). Resource Physiology of Conifers: Acquisition, Allocation, and Utilization, 1995

WILLIAM K. SMITH and THOMAS M. HINCKLEY (Eds.). Ecophysiology of Coniferous Forests, 1995

MARGARET D. LOWMAN and NALINI M. NADKHARNI (Eds.). Forest Canopies, 1995

BARBARA L. GARTNER (Ed.). Plant Stems: Physiology and Functional Morphology, 1995

GEORGE W. KOCH and HAROLD A. MOONEY (Eds.). Carbon Dioxide and Terretrial Ecosystems, 1996

CHRISTIAN KÖRNER and FAKHRI A. BAZZAZ (Eds.). Carbon Dioxide, Populations, and Communities, 1996

THEODORE T. KOZLOWSKI and STEPHEN G. PALLARDY. Growth Control in Woody Plants, 1997

J. J. LANDSBERG and S. T. GOWER. Application of Physiological Ecology to Forest Management, 1997

FAKHRI A. BAZZAZ and JOHN GRACE (Eds.). Plant Resource Allocation, 1997

LOUISE E. JACKSON (Eds.). Ecology in Agriculture, 1997

ROWAN F. SAGE and RUSSELL K. MONSON (Eds.). C_4 Plant Biology, 1999

JACQUES ROY, BERNARD SAUGIER, and HAROLD A. MOONEY (Eds.). Terrestrial Global Productivity, 2001